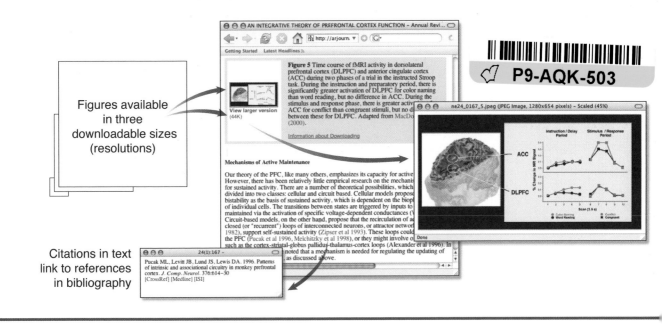

Figures available in three downloadable sizes (resolutions)

Citations in text link to references in bibliography

P9-AQK-503

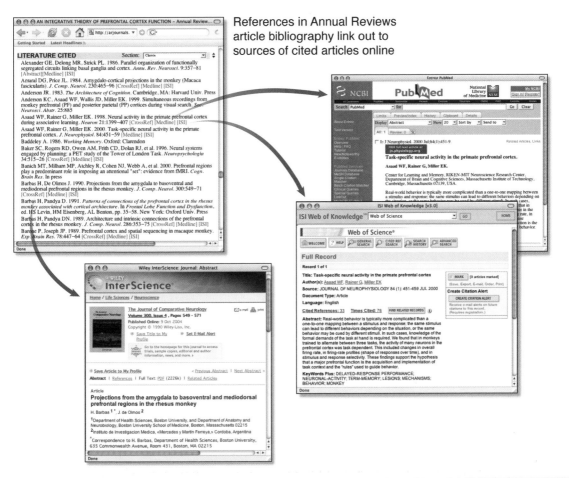

References in Annual Reviews article bibliography link out to sources of cited articles online

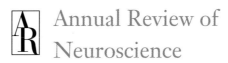

Annual Review of
Neuroscience

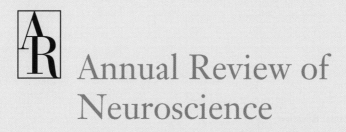

Annual Review of Neuroscience

Volume 31, 2008

Steven E. Hyman, *Editor*
Harvard University

Thomas M. Jessell, *Associate Editor*
Columbia University

Carla J. Shatz, *Associate Editor*
Stanford University

Charles F. Stevens, *Associate Editor*
Salk Institute for Biological Studies

www.annualreviews.org • science@annualreviews.org • 650-493-4400

Annual Reviews
4139 El Camino Way • P.O. Box 10139 • Palo Alto, California 94303-0139

Annual Reviews
Palo Alto, California, USA

International Standard Serial Number: 0147-006X
International Standard Book Number: 978-0-8243-2431-5

TYPESET BY APTARA
PRINTED AND BOUND BY SHERIDAN BOOKS, INC., CHELSEA, MICHIGAN

Annual Review of
Neuroscience

Volume 31, 2008

Contents

Indexes

Errata

An online log of corrections to *Annual Review of Neuroscience* articles may be found at
http://neuro.annualreviews.org/

Related Articles

Cerebellum-Like Structures and Their Implications for Cerebellar Function

Curtis C. Bell,[1] Victor Han,[2]
and Nathaniel B. Sawtell[1]

[1]Neurological Sciences Institute, Oregon Health and Science University,
Beaverton, Oregon 97006; email: bellc@ohsu.edu, sawtelln@ohsu.edu

[2]Oregon Regional Primate Center, Oregon Health and Science University,
Beaverton, Oregon 97006; email: hanv@ohsu.edu

Annu. Rev. Neurosci. 2008. 31:1–24

First published online as a Review in Advance on
February 14, 2008

The *Annual Review of Neuroscience* is online at
neuro.annualreviews.org

This article's doi:
10.1146/annurev.neuro.30.051606.094225

Key Words

forward model, synaptic plasticity, electric fish, cerebellum

Abstract

The nervous systems of most vertebrates include both the cerebellum
and structures that are architecturally similar to the cerebellum.
The cerebellum-like structures are sensory structures that receive
input from the periphery in their deep layers and parallel fiber input
in their molecular layers. This review describes these cerebellum-
like structures and compares them with the cerebellum itself. The
cerebellum-like structures in three groups of fish act as adaptive
sensory processors in which the signals conveyed by parallel fibers in
the molecular layer predict the patterns of sensory input to the deep
layers through a process of associative synaptic plasticity. Similarities
between the cerebellum-like structures and the cerebellum suggest
that the cerebellum may also generate predictions about expected
sensory inputs or states of the system, as suggested also by clinical,
experimental, and theoretical studies of the cerebellum. Understanding
the process of predicting sensory patterns in cerebellum-like structures
may therefore be a source of insight into cerebellar function.

Contents

LOCAL CIRCUITRY, GENE EXPRESSION, AND EVOLUTION OF CEREBELLUM-LIKE STRUCTURES

General Features

A distinctive molecular layer is a key identifying feature of all cerebellum-like structures (**Figure 1**). The molecular layer is composed of parallel fibers together with the dendrites and cell bodies on which the fibers terminate. The parallel fibers are numerous and closely packed. The granule cells that give rise to the parallel fibers in cerebellum-like structures are morphologically similar to cerebellar granular cells (Mugnaini et al. 1980a,b) but are usually located in an external granule cell mass rather than in a granule cell layer beneath the molecular layer as in the cerebellum. Unipolar brush cells and Golgi cells similar to those present in the granular layer of the cerebellum are also present in some cerebellum-like structures (Campbell et al. 2007, Mugnaini et al. 1997).

Functionally, the parallel fibers convey a rich variety of information from other central structures, which includes corollary discharge information associated with motor commands, information from higher levels of the same sensory modality represented in the deep layers, and information from other sensory modalities. In general, the types of signals conveyed by parallel fibers are signals that are likely to be associated with changes in the sensory input to the deep layers and that can therefore serve to predict such sensory input ("predictive inputs" in **Figure 1**).

The parallel fibers terminate on the dendritic spines of principal cells and on the smooth dendrites of inhibitory stellate cells in a manner very similar to the termination of parallel fibers on Purkinje cells and molecular layer interneurons of the cerebellum. We use the term principal cells to refer to large cells with spine-covered dendrites that extend throughout the molecular layer. Some of these principal cells are excitatory efferent cells that project to higher levels of the sensory system, whereas others are inhibitory neurons that terminate locally on each other and on the efferent cells. The latter are sometimes referred to as "Purkinje-like." The cell bodies of principal cells are usually located in a separate layer below the molecular layer, like the Purkinje cell layer of the cerebellum.

Afferent input from the periphery terminates in the deep layers of cerebellum-like structures, on basilar dendrites of principal

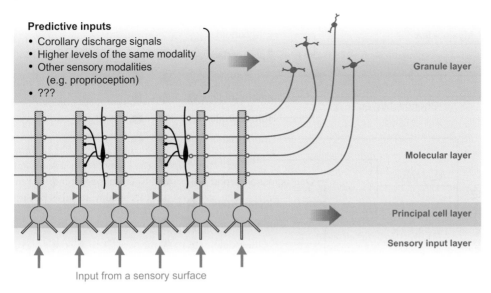

Predictive inputs
- Corollary discharge signals
- Higher levels of the same modality
- Other sensory modalities
  (e.g. proprioception)
- ???

Granule layer

Molecular layer

Principal cell layer

Sensory input layer

Input from a sensory surface

Figure 1

Schematic drawing showing major features of cerebellum-like sensory structures. Inhibitory stellate cells of the molecular layer are shown in black. Blue upward arrows indicate afferent input from the periphery terminating in the sensory input layer. In some cerebellum-like structures the afferent input also terminates on the smooth proximal portion of the apical dendrites as indicated by the small blue arrowheads.

cells, on proximal apical dendrites of principal cells, or on interneurons that relay the information from the periphery to the principal cells. Some of the interneurons of the deep layers are inhibitory, allowing for a change of sign, whereby excitation in the periphery is converted into inhibition of some principal cells. The peripheral input to the deep layers forms a map of a sensory surface, such as the skin surface, the retina, or the cochlea.

Local Circuitry of Different Cerebellum-Like Structures

The brains of all major groups of craniates except reptiles and birds have cerebellum-like structures (**Figures 2** and **3**). The similarities among the different cerebellum-like structures are clear, but so are the differences. Different structures may have different types of cells in addition to the principal cells, stellate cells, and granule cells that are present in all cerebellum-like structures. Moreover, some structures have additional inputs besides the inputs from the periphery and the parallel fibers.

This review describes major features of the different cerebellum-like structures of craniates but is not exhaustive. Recent reviews (Bell 2002, Bell & Maler 2005, Montgomery et al. 1995) and the original papers on individual structures, as provided below, should be consulted for more complete descriptions. Some of the structures are also much better known than others, which is reflected in the level of detail in the following descriptions.

Medial octavolateral nucleus. The medial octavolateral nucleus (MON) processes primary afferent input from the mechanical lateral line system and, in some fish, from eighth nerve end organs (Bell 1981b, McCormick 1999). It is present in all basal aquatic craniates with mechanical lateral line sensory systems (**Figures 2, 3a–d, 4a**). Myxinoids (atlantic hagfish; C.B. Braun, personal communication) and aquatic amniotes (reptiles, birds, and mammals; Montgomery et al. 1995) do not have lateral line systems and do not have an MON.

The efferent cells of the MON extend their spiny apical dendrites up into a molecular

MON: medial octavolateral nucleus

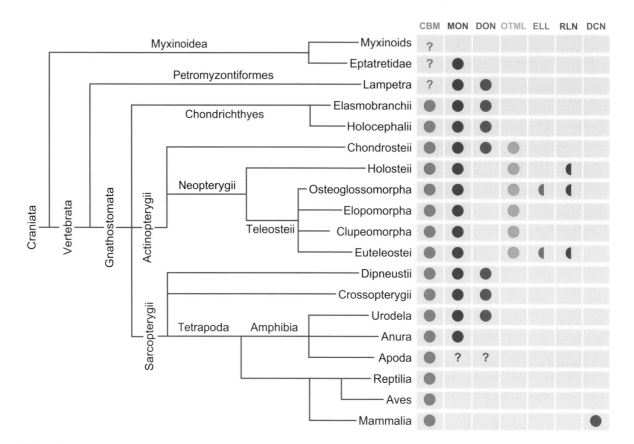

	CBM	MON	DON	OTML	ELL	RLN	DCN
Myxinoids	?						
Eptatretidae	?	●					
Lampetra	?	●	●				
Elasmobranchii	●	●	●				
Holocephalii	●	●	●				
Chondrosteii	●	●	●	●			
Holosteii	●	●		●		◖	
Osteoglossomorpha	●	●		●	◖	◖	
Elopomorpha	●	●		●			
Clupeomorpha	●	●		●			
Euteleostei	●	●		●	◖	◖	
Dipneustii	●	●	●				
Crossopterygii	●	●	●				
Urodela	●	●	●				
Anura	●	●					
Apoda	●	?	?				
Reptilia	●						
Aves	●						
Mammalia	●						●

Figure 2

Distribution of cerebellum-like structures and the cerebellum in different craniate groups. A filled circle means the structure is present in all or almost all the members of that group. A filled half circle means the structure is present only sporadically in that group. A question mark means that presence of the structure in that group is controversial. CBM, cerebellum; DCN, dorsal cochlear nucleus; DON, dorsal octavolateral nucleus; ELL, electrosensory lobe; MON, medial octavolateral nucleus; OTML, marginal layer of the optic tectum; RLN, rostrolateral nucleus of thalamus.

layer known as the cerebellar crest (**Figure 3a–d**). The parallel fibers of the cerebellar crest descend from an anterior granule cell mass known as the lateral granular mass in elasmobranchs and the eminentia granularis in other fish. The inputs to these granule cells include lateral line primary afferents (Bodznick & Northcutt 1980), eighth nerve primary afferents (Puzdrowski & Leonard 1993), input from the spinal cord (Schmidt & Bodznick 1987), and descending input from higher-order lateral line and acoustic centers (Bell 1981c, McCormick 1997, Tong & Finger 1983). The basilar dendrites of MON efferent cells are affected by primary afferent input.

DON: dorsal octavolateral nucleus

Dorsal octavolateral nucleus (DON). The dorsal octavolateral nucleus (DON) processes primary afferent input from electroreceptors and is present in many basal vertebrates with an electrosense (**Figures 2, 3a**) Electroreception is a vertebrate sense that may have originated as early as the lateral line or vestibular senses (Bullock et al. 1983). The Myxinoidea do not have electroreceptors and do not have a DON (Ronan 1986). Electroreception was lost during the evolution of neopterygian bony fish, and these fish do not have a DON. Electroreception reappeared independently at least twice during the evolution of the teleost radiation: once during the evolution of the two

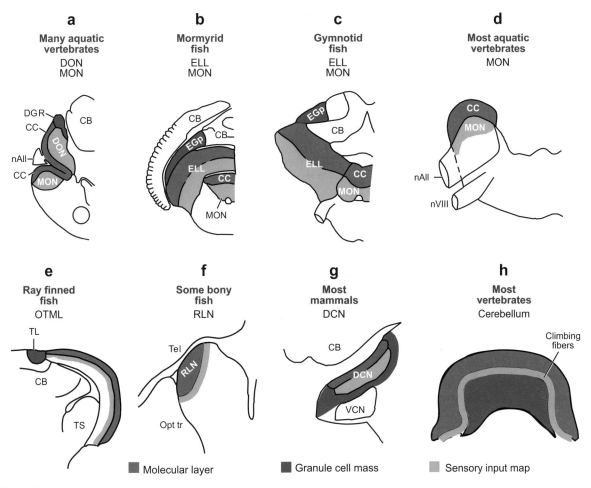

a
Many aquatic vertebrates
DON
MON

b
Mormyrid fish
ELL
MON

c
Gymnotid fish
ELL
MON

d
Most aquatic vertebrates
MON

e
Ray finned fish
OTML

f
Some bony fish
RLN

g
Most mammals
DCN

h
Most vertebrates
Cerebellum

■ Molecular layer   ■ Granule cell mass   ■ Sensory input map

Figure 3

Cerebellum-like structures in different vertebrate groups. The molecular layer, granule cell mass, and sensory input map are shown in different colors, as indicated at the bottom of the figure. The climbing fiber input to the cerebellum is shown here as a sensory input (*see text*). CB, cerebellum; CC, cerebellar crest; DCN, dorsal cochlear nucleus; DGR, dorsal granular ridge; DON, dorsal octavolateral nucleus; EGp, eminentia granularis posterior; ELL, electrosensory lobe; gran, granular layer; MON, medial octavolateral nucleus; mol, molecular layer; nAll, anterior lateral line nerve; nVIII, eighth nerve; Opt tr, optic tract; RLn, rostrolateral nucleus; Tel, telencephalon; TL, torus longitudinalis; TS, torus semicircularis; VCN, ventral cochlear nucleus.

related groups, Mormyriformes and Xeno-mystinae, and a second time during the evolution of the other two related groups, Gymnotiformes and Siluriformes (Bullock et al. 1983). However, the more recently derived electroreceptors and associated electrosensory central structures of teleosts are quite different from those of other aquatic vertebrates (see electrosensory lobe below).

The DON is located just dorsal to the MON and is similar to the MON in its structure

and connections. Primary afferent input from electroreceptors terminates on the basilar dendrites of efferent cells and inhibitory neurons of the deep layers, as in the MON (Bodznick & Northcutt 1980, Puzdrowski & Leonard 1993). The spine-covered apical dendrites of efferent cells extend up into the overlying cerebellar crest.

Parallel fibers of the DON cerebellar crest arise from the dorsal granular ridge, which receives proprioceptive input, recurrent

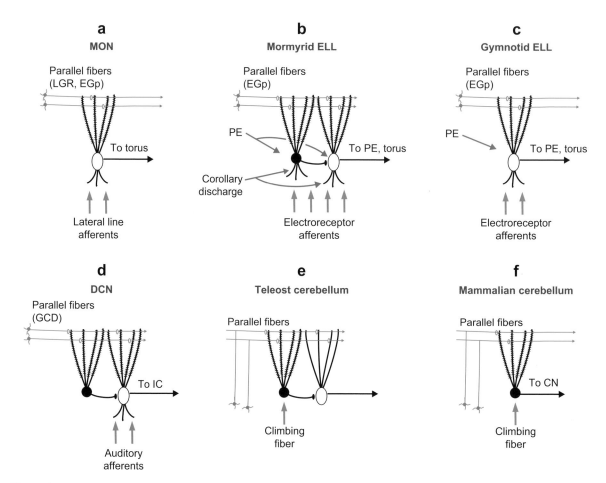

Figure 4

Local circuits of some cerebellum-like structures, the teleost cerebellum, and the mammalian cerebellum. Granule cells and parallel fibers are in red, afferent input from the periphery is in blue, and the additional inputs to the mormyrid and gymnotid ELLs are in green. The Purkinje-like cells of the mormyrid ELL and mammalian DCN as well as the Purkinje cells of the teleost and mammalian cerebellums are black. Excitatory efferent cells are white. IC, inferior colliculus; CN, cerebellar nucleus.

electrosensory input, and corollary discharge input associated with motor commands (Bodznick & Boord 1986, Conley & Bodznick 1994, Hjelmstad et al. 1996). All three types of input are active in relation to the fish's respiratory cycle. Electroreceptors in elasmobranchs are strongly affected by the fish's own respiration (Montgomery & Bodznick 1993). The activity in parallel fibers can therefore be used to predict the effect of these cyclic changes on electroreceptive input to the deep layers of DON (see Adaptive Processing in Cerebellum-Like Structures, below).

Marginal layer of the optic tectum. The optic tectum of actinopterygian (ray-finned) fishes is distinctive in that its outer layers are cerebellum like (**Figures 2, 3e**) (Meek 1983, Vanegas et al. 1979). The external layer of the optic tectum in these fish is a molecular layer known as the optic tectum marginal layer (OTML). The cell bodies of principal cells, the type I neurons of Meek (1983), are located below the marginal layer. The type I neurons extend their spine-covered apical dendrites up into the marginal layer and input from the retina maps onto their basilar dendrites and

OTML: marginal layer of the optic tectum

the smooth proximal portions of their apical dendrites.

The parallel fibers of the marginal layer arise from a medially located granule cell mass known as the torus longitudinalis. Granule cells of the torus longitudinalis respond to corollary discharge signals associated with the motor commands that evoke eye movements and respond to visual stimuli as well (Northmore et al. 1983). Parallel fiber activity driven by corollary discharge signals associated with eye movements could predict changes in retinal input to the deep layers, a possible interaction between the two types of input similar to that described above for the DON.

Electrosensory lobe (ELL). Electroreception is present in four groups of teleosts: Mormyriformes, an order of electric fish from Africa; Gymnotiformes, a superorder of electric fish from South America; Siluriformes, the order of catfish; and Xenomystinae, an African subfamily of the family Notopteridae (Bullock & Heiligenberg 1986). All these fish have a cerebellum-like electrosensory lobe (ELL) that receives primary afferent input from electroreceptors (**Figures 2**, **3b,c**) (Bell & Russell 1978, Braford 1982, Finger & Tong 1984, Maler et al. 1981).

The Mormyriformes and Gymnotiformes are electric fish with electric organs as well as electroreceptors. The order Mormyriformes includes the family Mormyridae, all of which have electric organ discharges (EODs) that are brief and pulse like, and the single-species family Gymnarchidae, which has a continuous wave-like EOD. The order Gymnotiformes includes some families with wave-like EODs and other families with pulsatile EODs. The ELLs of pulsatile mormyrids and wave gymnotids have been studied most extensively, although some work has been done on the ELLs of wave mormyriforms (Kawasaki & Guo 1998, Matsushita & Kawasaki 2005) and pulse gymnotiforms (Caputi et al. 2002, Schlegel 1973).

The spine-covered apical dendrites of ELL principal cells extend up into the overlying molecular layer. Primary afferent fibers from

electroreceptors in the skin map onto the deep layers, terminating on the basilar dendrites of principal cells or on interneurons (Bell & Maler 2005). The ELL efferent cells of mormyrid (Bell et al. 1997b), gymnotid (Saunders & Bastian 1984), and silurid fish (McCreery 1977) are of two main types: E-cells, which are excited by an increase in peripheral stimulus strength in the center of their receptive fields, and I-cells, which are inhibited by such an increase. These two functionally distinct cell types are also morphologically distinct; the E-cells have more extensive basilar dendrites.

Parallel fibers of ELLs arise from granule cells of the eminentia granularis posterior (EGp), which in mormyrids, at least, also contains Golgi cells and unipolar brush cells similar to the same cell types in the mammalian cerebellum (Campbell et al. 2007). The inputs to EGp in mormyrid and gymnotid fish include proprioceptive signals associated with bending of the body or the fins, recurrent electrosensory input from a higher levels of the system, and in mormyrids only, a corollary discharge signal associated with the motor command that elicits the electric organ (corollary) discharge (EOCD) (Bastian & Bratton 1990; Bell et al. 1992; Carr & Maler 1986; Szabo et al. 1979, 1990). These different inputs to EGp are relayed to ELL as parallel fiber inputs, where they can predict changes in electroreceptor input to the deep layers associated with tail movements, some other electrosensory input, or the EOD (see Adaptive Processing in Cerebellum-Like Structures).

The mormyrid (Bell et al. 1981), gymnotid (Carr & Maler 1986), and silurid (Tong 1982) ELLs receive additional input aside from the peripheral and parallel fiber inputs. They receive direct recurrent input from a higher-order electrosensory nucleus just rostral to ELL, the nucleus preeminentialis dorsalis (PE) (**Figures 4b,c**). The deep layers of the mormyrid ELL also receive EOCD input directly from an EOD motor command–related nucleus (Bell & von der Emde 1995). This input is in addition to the EOCD input conveyed via parallel fibers.

ELL: electrosensory lobe

EOD: electric organ discharge

EOCD: electric organ corollary discharge

RLN: rostrolateral
nucleus of the
thalamus

DCN: dorsal cochlear
nucleus

The ELLs of mormyrid and gymnotid fish have differences as well as similarities. Most important, the mormyrid ELL includes a principal cell that is not present in the gymnotid ELL (**Figures 4b,c**), the medium ganglion cell (Meek et al. 1996). These cells are referred to as Purkinje-like because they are GABAergic with extensive spine-covered dendrites in the overlying molecular layer. However, they differ from Purkinje cells because they have basilar dendrites and do not receive climbing fiber input. The medium ganglion cells are interneurons that inhibit both nearby efferent cells and each other (**Figure 4b**). They are more numerous than the efferent cells and have many more dendrites and spines in the molecular layer (Meek et al. 1996). They must therefore have a central role in the integration of peripheral and parallel fiber inputs in the mormyrid ELL. These and other differences between the mormyrid and gymnotid ELLs are consistent with their independent evolutionary origins.

Rostrolateral nucleus of the thalamus. The rostrolateral nucleus (RLN) (**Figures 2, 3f**) of the thalamus is a small, cerebellum-like structure found in the thalamus of a few widely scattered neopterygian fish (**Figure 2**) (Butler & Saidel 1992). The principal cells of RLN receive topographically organized direct input from the retina on the smooth proximal parts of their apical dendrites. The more distal apical dendrites are covered with spines and receive parallel fiber input from the torus longitudinalis.

Dorsal cochlear nucleus. All mammals possess a dorsal cochlear nucleus (DCN) (**Figures 2, 3g, 4d**). The DCN is laminated and cerebellum-like in marsupials and eutherian mammals but not in monotremes (Cant 1992, Nieuwenhuys et al. 1997). Fusiform cells are the major efferent cell type of the DCN. Their basilar dendrites are contacted by primary afferent fibers from the cochlea, which form a topographic map of the cochlea in the deeper layers below the molecular layer. The fusiform cells extend their spine-covered apical dendrites up into the molecular layer where they are contacted by parallel fibers. The parallel fibers arise from granule cells located around the margins of the nucleus. The parallel fibers course at right angles to the isofrequency bands in the deeper layers. Thus, parallel fibers cross through different frequency-specific regions of DCN.

The cartwheel cell is a second type of principal cell in the DCN (Cant 1992, Nieuwenhuys et al. 1997). These cells are Purkinje-like because they are GABAergic, have extensive spine-covered dendrites in the molecular layer, and inhibit the efferent fusiform cells. The cell bodies of cartwheel cells are in the molecular layer, and their dendrites are restricted to the molecular layer.

The local circuits of the DCN and the mormyrid ELL are very similar to the local circuit of the cerebellar cortex in actinopteryrian fish where most Purkinje cells are interneurons that terminate locally on efferent cells (**Figure 4e**) (Finger 1978, Meek 1998). The parallel fibers of the DCN, the mormyrid ELL, and the actinopterygian cerebellum pass through and excite the dendrites of both efferent cells and Purkinje or Purkinje-like cells. In all three cases, the Purkinje cells or Purkinje-like cells inhibit nearby efferent cells (**Figures 4b,d,e**). The efferent neurons of the actinopterygian cerebellum are equivalent to the cerebellar nucleus neurons of mammals (**Figure 4f**).

The granule cells of the DCN receive various types of input: recurrent auditory input from the inferior colliculus (Caicedo & Herbert 1993) and auditory cortex (Weedman & Ryugo 1996); primary vestibular afferent input (Burian & Gstoettner 1988); input from the pontine nuclei (Ohlrogge et al. 2001); somatosensory input from the dorsal column nuclei (Weinberg & Rustioni 1987), the trigeminal nuclei (Zhou & Shore 2004), and the somatosensory cortex (Wolff & Kunzle 1997); and direct input from the cochlea via fine unmyelinated Type II afferents (Brown et al. 1988). DCN granule cells also receive input from brainstem nuclei associated with vocalization and respiration that may convey corollary discharge signals (Shore & Zhou 2006). Proprioceptive input from the pinna has particularly strong effects on DCN granule cells in the cat (Kanold & Young 2001). Movements of the animal's pinna, head, or body have predictable effects on how the cochlea responds to an external sound source, and an animal's own vocalization and respiration will have predictable consequences on auditory input. Thus the signals conveyed by the parallel fibers in the DCN molecular layer could generate predictions about changes in afferent activity from the cochlea that arrive at the deep layers, as in other cerebellum-like structures.

Comparison of the Local Circuitries of Cerebellum-Like Structures and the Cerebellum

Many similarities in cell types and local circuitry between the cerebellum and cerebellum-like structures have been described in the preceding section. The similar cellular elements include the granule cells, the Golgi cells, the unipolar brush cells, the parallel fibers, the stellate cells, and the spine-covered molecular layer dendrites of principal cells.

The most crucial similarity is that between the two inputs to cerebellum-like structures and the two inputs to cerebellar Purkinje cells. Cerebellum-like structures receive parallel fiber and peripheral input, whereas Purkinje cells of the cerebellum receive parallel fiber input and climbing fiber input. In both cases, one input, the parallel fibers, conveys a rich variety of information to an entire set of principal cells or Purkinje cells. In both cases, a second input—peripheral input for cerebellum-like structures and climbing fiber input for the cerebellum—conveys specific information that subdivides the set of Purkinje cells that share the same parallel fiber input.

Olivary input to Purkinje cells is more specific than the peripheral input to the deep layers of cerebellum-like structures insofar as it is conveyed by just a single climbing fiber. Efferent cells and Purkinje-like cells in cerebellum-like structures do not have such single fiber inputs. The cerebellums of different vertebrates can vary markedly, but all the cerebellums that have been closely examined have a specific input from the inferior olive that terminates as climbing fibers. We suggest that the presence of a climbing fiber is the defining characteristic of the cerebellum that distinguishes it from cerebellum-like structures.

Climbing fibers and the peripheral sensory input to cerebellum-like structures are similar in many respects. Climbing fibers signal rather specific sensory events in most of the cases where the information they convey has been identified. Such sensory signals include retinal slip in a particular direction (Maekawa & Simpson 1972), somatosensory stimulation within a small region of skin (Ekerot & Jorntell 2001, Robertson 1985), and vestibular stimulation with tilt in a particular direction (Barmack & Shojaku 1992). Moreover, the climbing fibers of vertebrates other than mammals do not terminate throughout the molecular layer as in mammals. They terminate instead on smooth, proximal dendrites at the base of the molecular layer (Nieuwenhuys et al. 1997) in a manner similar to that of retinal input onto the smooth, proximal dendrites of principal cells in the OTML and RLN. This is not to say that the inferior olive is a simple sensory relay. It is not. But clearly sensory stimuli have a strong influence on the inferior olive and on climbing fibers, a result consistent with the origin of the inferior olive from the embryo's alar or sensory plate. Devor (2002) has in fact suggested that the inferior olive has been interposed between peripheral sensory structures and the cerebellum to gate sensory signals by motor commands and by the inferior olive's own intrinsic rhythmicity.

As noted in the previous section, the parallel fibers of cerebellum-like structures convey information that is associated with sensory input changes to the deep layers and that can therefore predict such changes. The parallel fibers of the cerebellum similarly convey information that can predict the occurrence of climbing fiber input. Climbing fibers in the flocculonodular lobe of the mammalian cerebellum, for example, signal retinal slip (Maekawa

LTD: long-term depression

& Simpson 1972), and the parallel fibers in this region convey vestibular information about head movement (Lisberger & Fuchs 1974), corollary discharge information about eye movement (Noda & Warabi 1982), and proprioceptive information from the neck (Matsushita & Tanami 1987), all of which could be used to predict movement of an image on the retina.

The presence of a climbing fiber is perhaps the critical difference between the cerebellum and cerebellum-like structures. Other differences include the presence of basilar dendrites on most principal cells of cerebellum-like structures but not on Purkinje cells; the presence of planar dendritic trees in most Purkinje cells but not in most principal cells; the presence of cell types in cerebellum-like structures not present in the cerebellum; and the presence of other inputs besides parallel fibers and climbing fibers in cerebellum-like structures not present in the cerebellum, such as the preeminential input in electroreceptive teleosts (**Figures 4*b*,*c***).

Patterns of Gene Expression in Cerebellum-Like Structures and the Cerebellum

Similarities and differences between the different cerebellum-like structures and the cerebellum itself are also revealed in gene expression patterns. Some genes are expressed in many different cerebellar and cerebellum-like structures, whereas others are expressed in only a few of these structures (Bell 2002). Common patterns of gene expression between cerebellar Purkinje cells and cartwheel cells of the DCN are particularly prominent, and many mutations affect both cell types (Berrebi et al. 1990).

One gene, the *GluRdelta2* gene, may be expressed in most if not all cerebellum-like structures and also in the cerebellum, but not in other structures. This gene is structurally related to the ionotropic glutamate receptors but does not form ion channels (Yuzaki 2003). The gene is necessary for long-term depression (LTD) at the parallel fiber to Purkinje cell synapse (Yawata et al. 2006). In mammals, the *GluR-delta2* gene is expressed in Purkinje cells (Yuzaki

2003) and in the principal cells of the DCN (Petralia et al. 1996). In zebrafish, the *GluR-delta2* gene is expressed in the molecular layers of the cerebellum, the MON, and the OTML, but not elsewhere in the brain as shown for both the gene and the protein (Mikami et al. 2004). Similarly, in the mormyrid brain, the GluRdelta2 protein is present in the molecular layers of the cerebellum, the ELL, the MON, and the OTML, but not elsewhere in the brain (J. Zhang & C. Bell, unpublished observations). Expression of the *GluRdelta2* gene in still other cerebellum-like structures remains to be established.

Some genes are expressed in some of the cerebellum-like structures or the cerebellum in the adult but are expressed only in other such structures during development. The *zebrin II* gene, for example, is expressed only in Purkinje cells in adult mammals, birds, and fish (Hawkes & Herrup 1995, Lannoo et al. 1991) but is expressed transiently during development in the MON and in part of the ELL of gymnotid fish (Lannoo et al. 1992). Similarly, functional N-methyl-D-aspartate (NMDA) receptors are present on principal cells of the adult mormyrid and gymnotid ELLs (Grant et al. 1998, Berman et al. 2001), as well as principal cells of the adult DCN (Manis & Molitor 1996), but are present on cerebellar Purkinje only during development (Dupont et al. 1987).

The common features in the local circuitry and in the gene expression patterns suggest the presence of a shared genetic-developmental program in all craniates, a program that once activated can generate a cerebellum or cerebellum-like structure. Some findings from experimental embryology support this idea. Thus, ectopic cerebellum-like structures develop in the forebrain or midbrain of a chick embryo if beads are coated with fibroblast growth factor 8 and placed at those sites in the embryo (Martinez et al. 1999). Similarly, cerebellar tissue will develop ectopically in the midbrain and forebrain of a mouse embryo with a genome that is Otx1+/- and Otx2+/- (*Drosophila* orthodenticle protein, a transcription factor) (Acampora et al. 1997).

Evolution of Cerebellum-Like Structures and the Cerebellum

The similarities between all the different cerebellum-like and cerebellar structures cannot be explained solely by homology in the sense of historical or phylogenetic homology (Butler & Saidel 2000). In this usage of the term, a feature is considered homologous across different taxa if the taxa have inherited the feature from a common ancestor that also had the feature. However, some of the individual structures described here are homologous. Thus the most parsimonious explanation for the presence of a cerebellum in all vertebrates is that it was present in a common ancestor. A common ancestor is also the most parsimonious explanation for the presence of an MON, a DON, or a DCN in some groups of craniates. However, we find no evidence for an ancestral cerebellum-like structure from which the cerebellum, MON, DON, marginal layer of the tectum, ELL, RLN, and DCN all evolved. (See Bell 2002 for a more complete analysis of the evolution of cerebellum-like structures.)

How then can we explain the clear similarities among the different cerebellums and cerebellum-like structures? The best explanation may be the presence of a developmental-genetic program that can generate a cerebellum or cerebellum-like structure, as described previously, together with evolutionary pressure for the type of information processing that these structures can perform.

Cerebellum-like structures may have evolved before the cerebellum itself. An MON is clearly present in some myxinoids, and both an MON and a DON are clearly present in lampreys, but the presence of a cerebellum is not well established in either of these groups. Some comparative anatomists affirm the presence of a cerebellum in myxinoids (Larsell 1967), whereas others deny it (Nieuwenhuys et al. 1997), and arguments have also been made both for (Larsell 1967, Nieuwenhuys et al. 1997) and against (Crosby 1969) the presence of a cerebellum in lampreys. As suggested previously, the identification of climbing fibers on putative Purkinje cells could indicate the presence of a cerebellum, but no efforts to identify climbing fibers have been made in myxinoids and lampreys. Purkinje cell–specific markers that do not stain cerebellum-like structures could also help determine the presence of a cerebellum. Thus the finding that the Zebrin II antibody does not stain cells in what some consider to be the lamprey cerebellum is of interest (Lannoo & Hawkes 1997) but is not conclusive because the Zebrin II antibody does not stain all Purkinje cells.

PREDICTIONS AND PLASTICITY IN CEREBELLUM-LIKE STRUCTURES AND THE CEREBELLUM

Predictions and Plasticity in Cerebellum-Like Structures

Cerebellum-like structures process information from peripheral sensory receptors in combination with an array of central signals conveyed by parallel fibers. If a common function exists among all cerebellum-like structures, it must involve the interaction between these two types of inputs. Progress toward understanding these interactions has been made in cerebellum-like structures concerned with the processing of electrosensory information in three distinct groups of fish: elasmobranchs, gymnotiform teleosts, and mormyrid teleosts. The cerebellum-like structures of these fish act as adaptive filters, removing predictable features of the sensory input (for reviews, see Bastian & Zakon 2005, Bell 2001, Bell et al. 1997a).

In these systems, the animals' own behavior strongly affects electroreceptors and could interfere with sensing weak electrosensory signals from the environment. In the passive electrosensory system of elasmobranch fish, for example, ventilatory movements modulate the fish's standing bioelectric field and can drive electroreceptor afferents through their entire dynamic range (Montgomery & Bodznick

1999). In the active electrosensory systems of mormyrid and gymnotid fish, movements of the electric organ (located in the tail) relative to sensory surface cause large changes in EOD-evoked electroreceptor input that could over-whelm the small changes resulting from nearby objects.

Parallel fiber inputs to cerebellum-like structures involved in electrolocation convey proprioceptive, corollary discharge, and elec-trosensory signals that could be used to pre-dict the electrosensory consequences of the an-imals' own behavior. Direct evidence for the generation of such predictions has been ob-tained from in vivo recordings from princi-pal cells in the mormyrid and gymnotid ELL and elasmobranch DON (Bastian 1996a, Bell 1981a, Bell et al. 1997b, Bodznick et al. 1999). In each case, pairing artificial electrosensory stimuli with central predictive signals—a corol-lary discharge signal at a particular delay after the EOD motor command in the case of the mormyrid ELL (**Figure 5a**), a proprioceptive signal at a particular tail angle in the case of the gymnotid ELL (**Figure 5b**), and a propriocep-tive or corollary discharge signal at a particular phase of the ventilatory cycle in the case of the elasmobranch DON (**Figure 5c**)—results in a change in the response to the predictive sig-nals alone that resembles a negative image of the response to the previously paired (and now predicted) stimulus. The negative images de-velop rapidly over the course of a few minutes of pairing and are specific to the sign as well as to the spatial and temporal patterns of activ-ity evoked by the stimulus. On the basis of these results investigators suggested that cerebellum-like circuitry could operate as an adaptive filter by continually generating and updating sensory predictions on the basis of associations between central signals and current sensory inputs and subtracting these predictions from the neural response. Adaptive filtering could thus allow external electrosensory signals to be detected more easily.

Several lines of evidence confirm that for-mation of negative images is due, at least in large part, to plastic changes occurring within

the cerebellum-like structures themselves (Bell 2001). Pairing predictive signals with intracel-lular current injections in vivo results in the formation of negative images in principal cells in all three groups of fish, indicating that the inputs to the recorded cell are plastic (Bastian 1996b, Bell et al. 1993, Bodznick et al. 1999). Given the types of predictive signals involved in negative image formation, synapses between parallel fibers and principal cells are the most natural candidates for the site of plastic changes. Negative image formation requires that the plasticity be anti-Hebbian in character, i.e., cor-relations between pre- and postsynaptic ac-tivity should decrease synaptic strength, and researchers have obtained evidence for anti-Hebbian plasticity at parallel fiber synapses with principal cells in all three classes of fish. Anti-Hebbian plasticity at parallel fiber synapses has also been shown recently in the DCN of mam-mals (Fujino & Oertel 2003, Tzounopoulos et al. 2004) but has not yet been connected to systems-level adaptive filtering.

Modeling studies have helped to link the properties of negative image formation with mechanisms of synaptic plasticity (Nelson & Paulin 1995, Roberts 1999, Roberts & Bell 2000). Temporal specificity is a key feature of negative image formation. In the mormyrid ELL, parallel fibers convey corollary discharge signals related to the motor command that drives the EOD. Pairing with electrosensory stimuli at various delays relative to the motor command results in negative images that are specific to the paired delay (Bell 1982). Re-sults of modeling studies suggest that tempo-rally specific negative images could be gener-ated using an anti-Hebbian learning rule similar to that observed experimentally (see below) to-gether with an array of parallel fiber inputs active at different delays following the mo-tor command (Roberts 1999, Roberts & Bell 2000). The mechanisms for generating tem-porally specific negative images in this model are quite similar to those proposed for some forms of cerebellar learning, such as the learn-ing of adaptively timed responses in classical eye-blink conditioning (Medina et al. 2000) or

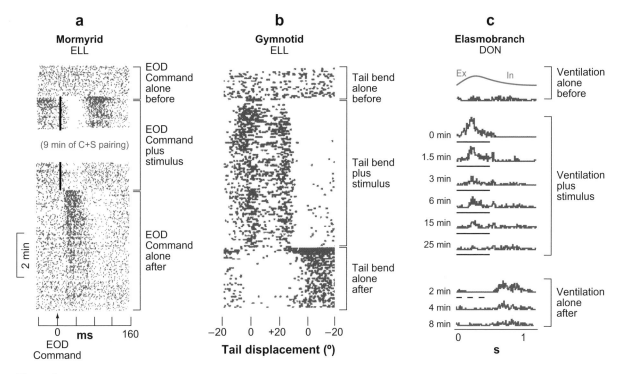

a
Mormyrid
ELL

EOD
Command
alone
before

(9 min of C+S pairing)

EOD
Command
plus
stimulus

EOD
Command
alone
after

2 min

0   **ms**   160
EOD
Command

b
Gymnotid
ELL

Tail bend
alone
before

Tail bend
plus
stimulus

Tail bend
alone
after

−20   0   +20   0   −20
Tail displacement (°)

c
Elasmobranch
DON

Ex   In   Ventilation
alone
before

0 min
1.5 min
3 min   Ventilation
6 min   plus
15 min   stimulus
25 min

2 min   Ventilation
4 min   alone
8 min   after

0   1
s

Figure 5

Formation of negative images of predicted sensory responses in three different cerebellum-like structures. (*a*) Raster display of the responses of a cell in the ampullary region of the mormyrid ELL. Each dot represents an action potential. The EOD motor command occurs at time 0. The command alone initially has no effect on the cell. An electrosensory stimulus (*vertical black line*) that evokes a pause-burst response is then paired with the command. After several minutes of pairing, the stimulus is turned off and a response to command alone is revealed, which was not present before the pairing and which is a negative image of the previously paired sensory response. From Bell 1986. (*b*) Raster display of responses of cell in the gymnotid ELL. The tail is moved back and forth passively. Each row of dots shows response to one movement cycle. Initially the tail bend has no effect on the cell. An electrosensory stimulus that evokes a burst-pause is then delivered in phase with the movement. The electrosensory stimulus is turned off after several minutes of pairing, which reveals a response to tail bending alone that was not present before the pairing and which is opposite to the previously paired sensory response. From Bastian 1995. (*c*) Histogram display of responses of a cell in the elasmobranch DON. Initially the cell does not respond to the exhalation (Ex)–inhalation (In) ventilatory cycle of the fish (*top histogram*). An electrosensory stimulus that evokes a burst-pause is then delivered in phase with the ventilatory cycle. The response to ventilation plus the electrosensory stimulus decreases during 25 min of pairing. Turning off the electrosensory stimulus after pairing reveals the presence of a response to ventilation alone, which was not present before and which is a negative image of the previously paired sensory response. From Bodznick 1993.

of appropriate phase relations in the vestibular ocular reflex (Raymond & Lisberger 1998).

The cellular properties of anti-Hebbian synaptic plasticity have been studied in some detail at synapses between parallel fibers and Purkinje-like medium ganglion cells in an in vitro preparation of the mormyrid ELL (Bell et al. 1997c, Han et al. 2000). Synaptic depression requires a postsynaptic dendritic spike and depends on the precise timing of the spike rela-

tive to the parallel fiber evoked excitatory postsynaptic potential (EPSP) onset. Depression develops when a postsynaptic dendritic spike occurs within 50 ms of EPSP onset, whereas other timing relations yield potentiation or no effect. Potentiation as measured in vitro is nonassociative and likely depends on simple repetition of the parallel fiber stimuli at a sufficiently high rate, although in vivo experiments suggest a spike timing–dependent component to the potentiation (Bell et al. 1997b, Sawtell et al. 2007). The depression requires activation of NMDA receptors and changes in postsynaptic calcium. The potentiation can reverse the depression and vice versa, with both potentiation and depression having a presynaptic locus of expression. Plasticity at parallel fiber synapses onto Purkinje-like cartwheel cells of the DCN is also anti-Hebbian, spike timing–dependent, NMDA dependent, and presynaptically expressed (Tzounopoulos et al. 2004, 2007).

Investigators have observed both similarities and differences between plasticity in cerebellum-like structures and plasticity in the cerebellum itself. The depression of responses to signals conveyed by parallel fibers following the pairing of these signals with postsynaptic excitation in cerebellum-like structures is similar to the depression of responses to parallel fiber stimulation in the mammalian Purkinje cells following pairing with climbing fiber input or with postsynaptic depolarization (Ito 2001). Such depression has been linked to the formation of negative images of predicted sensory input in cerebellum-like structures and to motor learning in the mammalian cerebellum (Ito 1984). It is of interest in this regard that the timing of stimulus-driven parallel fiber–evoked simple spike activity is consistently close to the inverse of climbing fiber responses in almost all the systems where this relation has been examined (Barmack & Shojaku 1992, Ebner et al. 2002, Graf et al. 1988, Kobayashi et al. 1998, Stone & Lisberger 1990). Thus in many systems, simple spike activity is a kind of negative image of predicted climbing fiber activity. Plasticity at parallel fiber synapses may play a role in generating the antiphase relation, but it is only part of the explanation because the antiphase relation is still present when parallel fiber LTD is blocked (Goossens et al. 2004).

The timing constraints on parallel fiber plasticity may be more restrictive in cerebellum-like structures than in the cerebellum. LTD in the cerebellum-like structures where timing relations have been tested occurred only when the postsynaptic spike followed the presynaptic spike by 50 ms or less (Bell et al. 1997c, Tzounopoulos et al. 2004). In the cerebellum, however, depression of the parallel fiber synapse is present after pairings with climbing fiber input in which delays varied between occurrence of the climbing fiber 50 ms before the parallel fiber stimulus and occurrence of the climbing fiber 200 ms after the parallel fiber stimulus (Safo & Regehr 2007, Wang et al. 2000).

The mechanisms of synaptic plasticity are clearly not the same in the cerebellum and in the cerebellum-like structures where it has been studied. Plasticity at parallel fiber synapses onto efferent or Purkinje-like cells in the mormyrid ELL (Han et al. 2000) and the mammalian DCN (Tzounopoulos et al. 2004) depends on activation of NMDA receptors, but synaptic plasticity at parallel fiber synapses onto Purkinje cells does not (Ito 2001). However, some aspects of the plasticity mechanisms may be shared as indicated by the presence of the *GluRdelta2* gene in the cerebellum and in cerebellum-like structures, and by the involvement of this gene in plasticity at Purkinje cell synapses (Hirano et al. 1995).

Adaptive processes in the cerebellum appear similar to those in cerebellum-like structures. In cerebellum-like structures, the pairing of parallel fiber signals with excitatory input from the periphery results in such signals eliciting a predictive reduction in principal cell activity. In the cerebellum, the pairing of parallel fiber signals with climbing fiber input likely leads to such signals eliciting a reduction in the firing of Purkinje cells (but see Steuber et al. 2007 for a contrary view). If the climbing fibers convey some type of sensory signal, gated through the inferior olive, then the parallel fiber signals that are paired with the climbing fibers, and which predict their occurrence, will reduce Purkinje cell activity, as shown by Jirenhed et al. (2007) during eye-blink conditioning.

This review focuses on sensory predictions through mechanisms of associative synaptic plasticity and with those features of cerebellum-like structures that are particularly relevant to cerebellar function. Cerebellum-like structures are also excellent sites for addressing other important issues in neuroscience, which cannot be discussed here because of space constraints. These include the roles of recurrent feedback from higher to lower levels of the same sensory system (Chacron et al. 2003, 2005; Doiron et al. 2003), the effects of motor commands on sensory processing (Bell & Grant 1992), the preservation and analysis of temporal information (Kawasaki 2005), and the neural processing of spectral cues for sound localization in the DCN (Young & Davis 2002).

Predictions in the Cerebellum

The many similarities between cerebellum-like structures and the cerebellum suggest that the cerebellum too may be involved in generating predictions concerning expected sensory input or states of the system (Bell et al. 1997a, Devor 2000), and a variety of experimental, clinical, and theoretical studies of the cerebellum support this hypothesis (Diedrichsen et al. 2007, Nixon 2001, Paulin 2005, Wolpert et al. 1998).

The probable involvement of the cerebellum in predictive or feedforward control through learning is well recognized (Bastian 2006, Ito 1984, Miall et al. 1993, Ohyama et al. 2003, Wolpert et al. 1998). Predictive control allows for prior knowledge to shape an action, as in knowing if a cup is full or empty before picking it up. Several studies indicate that predictive feedforward control is deficient in cerebellar patients (Morton & Bastian 2006, Smith & Shadmehr 2005). Such patients do not adapt their responses to predictable perturbations, although they respond quite well to sudden unpredictable perturbations of a movement, indicating that feedback control from the periphery is functional.

Theoreticians have proposed that the cerebellum may act in an adaptive and predictive manner through the generation of two types of models: forward models and inverse models (Wolpert et al. 1998). In a forward model, copies of a motor command are conveyed to the cerebellum together with information about the current state of the system such as positions and velocities of the limbs. The cerebellum then generates a prediction about the sensory consequences of the commanded motor act in the current context. In an inverse model, the desired goal of an action together with information about the current state are conveyed to the cerebellum, which then generates the precise motor commands that will yield the desired goal. Both types of models must be capable of plastic change or learning to adapt to changes in the task or in the system, such as changes in load or initial limb position.

Forward models are particularly important in generating fast, coordinated movement sequences. Feedback from peripheral sensory receptors is slow. An appropriate command for one phase of a movement must often be issued before peripheral feedback can arrive about the consequences of a motor command that evoked a previous phase of the movement. A forward model that predicts the sensory consequences of a motor command, accounting for all that is known about the current state of the system, allows the next motor command in a sequence to be issued appropriately and in accord with the expected consequences of previous commands. Such a process allows for the chunking of separate components of a motor sequence and their automatization, as described by Nixon (2001). Moreover, classic symptoms of cerebellar damage such as decomposition of movement, slowness, and tremor can all be understood as due to the absence of predictive forward models and reliance on peripheral feedback (Bastian 2006, Nixon 2001).

What is required in such automatization of a sequence of movements is the predicted effect of the motor command: the sensory consequences or state that results from the action, not simply the motor command itself. Recent experiments by Pasalar et al. (2006) suggest that the Purkinje cell output from large regions of the cerebellar hemispheres is indeed more tightly coupled with predictions about consequences of the movement than with the motor commands themselves (but see Yamamoto et al. 2007). Pasalar et al. (2006) recorded from Purkinje cells over a wide area of the hemisphere in monkeys that had been trained to

Forward model: predicts the future state of the system on the basis of the current state and the motor command

Inverse model: generates an appropriate motor command that will cause a desired change in the state of the system

control a cursor on a screen with a manipulandum and to make the cursor track a circularly moving stimulus. They then altered the forces required to move the manipulandum. The electromyograms in the arm muscles varied systematically with the changes in required forces, but Purkinje cell simple spike activity was unaffected by the changes in force. Purkinje cell simple spikes depended only on the position, direction, and velocity of the movement. Purkinje cell activity was phase advanced, that is, predictive of the movement parameters or state of the arm (T. Ebner, personal communication).

Pasalar et al. (2006) took their results as an argument against an inverse model in the cerebellum because Purkinje cell activity had little relation to the motor commands to the muscles. Although one could argue that the activity reflects a high-level motor command, in movement rather than muscle coordinates, the simpler explanation is that the Purkinje cell activity reflects a forward model of expected consequences, as required for the automatization of movement sequences. Their experiments suggest that not all sensory consequences are predicted; only those critical for accomplishing the task are predicted. Thus presumed changes in touch or muscle receptors associated with force changes were not predicted by Purkinje cell activity; only velocity and position of the limb were predicted.

Examples of what are, in effect, forward models in the cerebellum-like structures of mormyrid and elasmobranch fish are described in the previous section, showing that forward models can indeed be generated within structures such as the cerebellum. In these systems, corollary discharge signals come to elicit a prediction about the sensory input pattern that is expected to follow the motor command. The possibility of such corollary discharge effects in the OTML and DCN was also mentioned.

Cerebellum-like structures can generate predictions on the basis of other sensory inputs (Bastian 1996a, Bodznick et al. 1999), not just on the basis of motor commands, and the cerebellum may do so also. For example, in eye-blink conditioning, which is thought to involve the cerebellum, the timing of one sensory signal, an air puff to the cornea (signaled by the climbing fiber), is predicted from another sensory signal, a tone (signaled by mossy fibers) (Kim & Thompson 1997). Similarly, cerebellar modulation of the vestibular ocular reflex involves the prediction of one sensory stimulus, retinal slip (signaled by climbing fibers), by the occurrence of another sensory stimulus, vestibular input (signaled by mossy fibers). More broadly, Paulin (1993, 2005) has suggested that the cerebellum estimates future states of the organism or environment using a combination of sensory, motor, and possibly other types of information.

In simpler systems, such as the vestibular ocular reflex, in which Purkinje cell output is coupled quite directly with motor pathways, the adaptive alteration in Purkinje cell activity after pairing with the climbing fiber can be viewed as either a prediction about a sensory input or as a motor command. In more complex systems, where Purkinje cell output is less tightly coupled with motor pathways, as in the tracking task studied by Pasalar et al. (2006), the hypothesis of Purkinje cell activity as a predictor of consequences may provide a more useful perspective.

DIRECTIONS FOR FUTURE RESEARCH

Our understanding of adaptive processing in cerebellum-like structures is far from complete, and future work will be useful both for understanding the neural mechanisms of sensory processing and for understanding the cerebellum. Promising lines of research are outlined briefly below.

Activity Patterns in Granule Cells

How the different types of predictive inputs are combined and represented in the granule cells that are associated with cerebellum-like structures remains unclear, as is also the case for cerebellar granule cells.

Adaptive Filtering in Electrosensory Systems

Several aspects of adaptive filtering in cerebellum-like structures require further investigation, including (*a*) the behavioral consequences of adaptive filtering; (*b*) the effects of adaptive filtering on encoding naturalistic stimuli in the presence of self-generated interference; (*c*) the mechanisms of plasticity and the presence of plasticity at other sites, such as inhibitory synapses; and (*d*) the possible generation of more complex expectations such as those based on memories of entire scenes or sequences. The possibility of more complex expectations is suggested by the massive descending inputs that cerebellum-like structures receive from higher levels of the same sensory systems.

Adaptive Filtering in the DCN and Less-Studied Cerebellum-Like Structures

Recent studies have found synaptic plasticity at parallel fiber synapses onto Purkinje-like cartwheel cells and fusiform cells in the DCN in vitro. Yet very little is known at the systems level regarding the role of such plastic parallel fiber inputs in auditory processing. Similarly, very little is known about adaptive filtering in the MON or OTML.

Purkinje-Like Cells

The functional roles of Purkinje-like cells remain unclear. Recent work has shown that dendritic spikes that drive anti-Hebbian plasticity in Purkinje-like MG cells of the mormyrid ELL are strongly regulated by central signals, suggesting a parallel to supervised learning mediated by climbing fiber inputs to the cerebellum (Sawtell et al. 2007). In addition, the mormyrid ELL, the DCN, and the teleost cerebellum all provide excellent opportunities for examining interactions between Purkinje or Purkinje-like cells and neighboring efferent cells (analogous to deep cerebellar nuclear cells in the mammalian cerebellum).

Primitive Cerebellums

As discussed previously, the earliest craniates possess cerebellum-like structures, but it is not clear if they possess a cerebellum. Identification of a structure similar to the inferior olive in hagfish or lampreys would help to resolve this issue.

DISCLOSURE STATEMENT

The authors are not aware of any biases that might be perceived as affecting the objectivity of this review.

ACKNOWLEDGMENTS

We thank Drs. Neal Barmack, Timothy Ebner, and Johannes Meek for their critical reviews of the manuscript. The work was supported by grants from the National Institutes of Health (MH 49792 to C.C.B. and NS44961 to V.H.) and the National Science Foundation (IOB 0618212 to N.B.S.) and by a National Research Service Award (NS049728 to N.B.S.).

LITERATURE CITED

Acampora D, Avantaggiato V, Tuorto F, Simeone A. 1997. Genetic control of brain morphogenesis through Otx gene dosage requirement. *Development* 124(18):3639–50

Barmack NH, Shojaku H. 1992. Vestibularly induced slow oscillations in climbing fiber responses of Purkinje cells in the cerebellar nodulus of the rabbit. *Neuroscience* 50:1–5

Bastian AJ. 2006. Learning to predict the future: the cerebellum adapts feedforward movement control. *Curr. Opin. Neurobiol.* 16(6):645–49

Bastian J. 1995. Pyramidal-cell plasticity in weakly electric fish: a mechanism for attenuating responses to reafferent electrosensory inputs. *J. Comp. Physiol.* 176:63–78

Bastian J. 1996a. Plasticity in an electrosensory system. I. General features of dynamic sensory filter. *J. Neurophysiol.* 76:2483–96

Bastian J. 1996b. Plasticity in an electrosensory system. II. Postsynaptic events associated with a dynamic sensory filter. *J. Neurophysiol.* 76:2497–507

Bastian J, Bratton B. 1990. Descending control of electroreception. I. Properties of nucleus praeeminentialis neurons projecting indirectly to the electrosensory lateral line lobe. *J. Neurosci.* 10:1226–40

Bastian J, Zakon H. 2005. Plasticity of sense organs and brain. See Bullock et al. 2005, pp. 195–228

Bell CC. 1981a. An efference copy modified by reafferent input. *Science* 214:450–53

Bell CC. 1981b. Central distribution of octavolateral afferents and efferents in a teleost (Mormyridae). *J. Comp. Neurol.* 195:391–414

Bell CC. 1981c. Some central connections of medullary octavolateral centers in a mormyrid fish. In *Hearing and Sound Communication in Fishes*, ed. RR Fay, AN Popper, WN Tavolga, pp. 383–92. Berlin: Heidelberg, Springer-Verlag

Bell CC. 1982. Properties of a modifiable efference copy in electric fish. *J. Neurophysiol.* 47:1043–56

Bell CC. 1986. Duration of plastic change in a modifiable efference copy. *Brain Res.* 369:29–36

Bell CC. 2001. Memory-based expectations in electrosensory systems. *Curr. Opin. Neurobiol.* 11:481–87

Bell CC. 2002. Evolution of cerebellum-like structures. *Brain Behav. Evol.* 59:312–26

Bell CC, Bodznick D, Montgomery J, Bastian J. 1997a. The generation and subtraction of sensory expectations within cerebellum-like structures. *Brain Behav. Evol.* 50:17–31

Bell CC, Caputi A, Grant K. 1997b. Physiology and plasticity of morphologically identified cells in the mormyrid electrosensory lobe. *J. Neurosci.* 17:6409–22

Bell CC, Caputi A, Grant K, Serrier J. 1993. Storage of a sensory pattern by anti-Hebbian synaptic plasticity in an electric fish. *Proc. Natl. Acad. Sci. USA* 90:4650–54

Bell CC, Finger TE, Russell CJ. 1981. Central connections of the posterior lateral line lobe in mormyrid fish. *Exp. Brain Res.* 42:9–22

Bell CC, Grant K. 1992. Corollary discharge effects and sensory processing in the mormyromast regions of the mormyrid electrosensory lobe: II. Cell types and corollary discharge plasticity. *J. Neurophysiol.* 68:859–75

Bell CC, Grant K, Serrier J. 1992. Corollary discharge effects and sensory processing in the mormyrid electrosensory lobe: I. Field potentials and cellular activity in associated structures. *J. Neurophysiol.* 68:843–58

Bell CC, Han VZ, Sugawara S, Grant K. 1997c. Synaptic plasticity in a cerebellum-like structure depends on temporal order. *Nature* 387:278–81

Bell CC, Maler L. 2005. Central neuroanatomy of electrosensory systems in fish. See Bullock et al. 2005, pp. 68–111

Bell CC, Russell CJ. 1978. Termination of electroreceptor and mechanical lateral line afferents in the mormyrid acousticolateral area. *J. Comp. Neurol.* 182:367–82

Bell CC, von der Emde G. 1995. Electric organ corollary discharge pathways in mormyrid fish: II. The medial juxtalobar nucleus. *J. Comp. Physiol. A.* 177:463–79

Berman N, Dunn RJ, Maler L. 2001. Function of NMDA receptors in a feedback pathway of the electrosensory system. *J. Neurophysiol.* 86:1612–21

Berrebi AS, Morgan JI, Mugnaini E. 1990. The Purkinje cell class may extend beyond the cerebellum. *J. Neurocytol.* 19(5):643–54

Bodznick D. 1993. The specificity of an adaptive filter that suppresses unwanted reafference in electrosensory neurons of the skate medulla. *Biol. Bull.* 185:312–14

Bodznick D, Boord RL. 1986. Electroreception in Chondrichthyes: central anatomy and physiology. See Bullock & Heiligenberg 1986, pp. 225–56

Bodznick D, Montgomery JC, Carey M. 1999. Adaptive mechanisms in the elasmobranch hindbrain. *J. Exp. Biol.* 202:1357–64

Bodznick D, Northcutt RG. 1980. Segregation of electro- and mechanoreceptive inputs to the elasmobranch medulla. *Brain Res.* 195:313–21

Braford MR. 1982. African, but not Asian, notopterid fishes are electroreceptive: evidence from brain characters. *Neurosci. Lett.* 32:35–39

Brown MC, Berglund AM, Kiang NY, Ryugo DK. 1988. Central trajectories of type II spiral ganglion neurons. *J. Comp. Neurol.* 278(4):581–90

Bullock TH, Bodznick DA, Northcutt RG. 1983. The phylogenetic distribution of electroreception: evidence for convergent evolution of a primitive vertebrate sense modality. *Brain Res. Rev.* 6:25–46

Bullock TH, Heiligenberg W. 1986. *Electroreception.* New York: Wiley

Bullock TH, Hopkins CD, Popper AN, Fay RR, eds. 2005. *Electroreception.* New York: Springer

Burian M, Gstoettner W. 1988. Projection of primary vestibular afferent fibres to the cochlear nucleus in the guinea pig. *Neurosci. Lett.* 84(1):13–17

Butler AB, Saidel WM. 1992. Tectal projection to an unusual nucleus in the diencephalon of a teleost fish, *Pantodon buchholzi. Neurosci. Lett.* 145:193–96

Butler AB, Saidel WM. 2000. Defining sameness: historical, biological, and generative homology. *BioEssays* 22(9):846–53

Caicedo A, Herbert H. 1993. Topography of descending projections from the inferior colliculus to auditory brainstem nuclei in the rat. *J. Comp. Neurol.* 328(3):377–92

Campbell HR, Meek J, Zhang J, Bell CC. 2007. Anatomy of the posterior caudal lobe of the cerebellum and the eminentia granularis posterior in a mormyrid fish. *J. Comp. Neurol.* 502(5):714–35

Cant NB. 1992. The cochlear nucleus: neuronal types and their synaptic organization. In *The Mammalian Auditory Pathway: Neuroanatomy*, ed. DB Webster, AN Popper, RR Fay, pp. 66–116. New York: Springer

Caputi AA, Castello ME, Aguilera P, Trujillo-Cenoz O. 2002. Electrolocation and electrocommunication in pulse gymnotids: signal carriers, prereceptor mechanisms and the electrosensory mosaic. *J. Physiol. Paris* 96(5–6):493–505

Carr CE, Maler L. 1986. Electroreception in gymnatiform fish: central anatomy and physiology. See Bullock & Heiligenberg 1986, pp. 319–74

Chacron MJ, Doiron B, Maler L, Longtin A, Bastian J. 2003. Non-classical receptive field mediates switch in a sensory neuron's frequency tuning. *Nature* 423(6935):77–81

Chacron MJ, Maler L, Bastian J. 2005. Feedback and feedforward control of frequency tuning to naturalistic stimuli. *J. Neurosci.* 25(23):5521–32

Conley RA, Bodznick D. 1994. The cerebellar dorsal granular ridge in an elasmobranch has proprioceptive and electroreceptive representations and projects homotopically to the medullary electrosensory nucleus. *J. Comp. Physiol. A* 174:707–21

Crosby EC. 1969. Comparative aspects of cerebellar morphology. In *Neurobiology of Cerebellar Evolution and Development*, ed. R Llinas, pp. 19–41. Chicago: Am. Med. Assoc.

Devor A. 2000. Is the cerebellum like cerebellar-like structures? *Brain Res. Rev.* 34(3):149–56

Devor A. 2002. The great gate: control of sensory information flow to the cerebellum. *Cerebellum* 1(1):27–34

Diedrichsen J, Criscimagna-Hemminger SE, Shadmehr R. 2007. Dissociating timing and coordination as functions of the cerebellum. *J. Neurosci.* 27(23):6291–301

Doiron B, Chacron MJ, Maler L, Longtin A, Bastian J. 2003. Inhibitory feedback required for network oscillatory responses to communication but not prey stimuli. *Nature* 421(6922):539–43

Dupont JL, Gardette R, Crepel F. 1987. Postnatal development of the chemosensitivity of rat cerebellar Purkinje cells to excitatory amino acids. An in vitro study. *Brain Res.* 431(1):59–68

Ebner TJ, Johnson MT, Roitman A, Fu Q. 2002. What do complex spikes signal about limb movements? *Ann. N.Y. Acad. Sci.* 978:205–18

Ekerot CF, Jorntell H. 2001. Parallel fibre receptive fields of Purkinje cells and interneurons are climbing fibre-specific. *Eur. J. Neurosci.* 13:1303–10

Finger TE. 1978. Efferent neurons of the teleost cerebellum. *Brain Res.* 153:608–14

Finger TE, Tong SL. 1984. Central organization of eighth nerve and mechanosensory lateral line systems in the brainstem of ictalurid catfish. *J. Comp. Neurol.* 229:129–51

Fujino K, Oertel D. 2003. Bidirectional synaptic plasticity in the cerebellum-like mammalian dorsal cochlear nucleus. *Proc. Natl. Acad. Sci. USA* 100(1):265–70

Goossens HH, Hoebeek FE, van Alphen AM, Van Der SJ, Stahl JS, et al. 2004. Simple spike and complex spike activity of floccular Purkinje cells during the optokinetic reflex in mice lacking cerebellar long-term depression. *Eur. J. Neurosci.* 19(3):687–97

Graf W, Simpson JI, Leonard CS. 1988. Spatial organization of visual messages of the rabbit's cerebellar flocculus. II. Complex and simple spike responses of Purkinje cells. *J. Neurophysiol.* 60(6):2091–121

Grant K, Sugawara S, Gomez L, Han VZ, Bell CC. 1998. The Mormyrid electrosensory lobe in vitro: physiology and pharmacology of cells and circuits. *J. Neurosci.* 18:6009–25

Han VZ, Grant G, Bell CC. 2000. Reversible associative depression and nonassociative potentiation at a parallel fiber synapse. *Neuron* 27:611–22

Hawkes R, Herrup K. 1995. Aldolase C/zebrin II and the regionalization of the cerebellum. *J. Mol. Neurosci.* 6(3):147–58

Hirano T, Kasono K, Araki K, Mishina M. 1995. Suppression of LTD in cultured Purkinje cells deficient in the glutamate receptor d2 subunit. *NeuroReport* 6:524–26

Hjelmstad GO, Parks G, Bodznick D. 1996. Motor corollary discharge activity and sensory responses related to ventilation in the skate vestibulolateral cerebellum: implications for electrosensory processing. *J. Exp. Biol.* 199:673–81

Ito M. 1984. *The Cerebellum and Neural Control.* New York: Raven

Ito M. 2001. Cerebellar long-term depression: characterization, signal transduction, and functional roles. *Physiol. Rev.* 81(3):1143–95

Jirenhed DA, Bengtsson F, Hesslow G. 2007. Acquisition, extinction, and reacquisition of a cerebellar cortical memory trace. *J. Neurosci.* 27(10):2493–502

Kanold PO, Young ED. 2001. Proprioceptive information from the pinna provides somatosensory input to cat dorsal cochlear nucleus. *J. Neurosci.* 21(19):7848–58

Kawasaki M. 2005. Physiology of tuberous electrosensory systems. See Bullock et al. 2005, pp. 154–194

Kawasaki M, Guo YX. 1998. Parallel projection of amplitude and phase information from the hindbrain to the midbrain of the African electric fish *Gymnarchus niloticus. J. Neurosci.* 18(18):7599–611

Kim JJ, Thompson RF. 1997. Cerebellar circuits and synaptic mechanisms involved in classical eyeblink conditioning. *Trends Neurosci.* 20:177–81

Kobayashi Y, Kawano K, Takemura A, Inoue Y, Kitama T, et al. 1998. Temporal firing patterns of Purkinje cells in the cerebellar ventral paraflocculus during ocular following responses in monkeys II. Complex spikes. *J. Neurophysiol.* 80(2):832–48

Lannoo M, Hawkes R. 1997. A search for primitive Purkinje cells: zebrin II expression in sea lampreys (*Petromyzon marinus*). *Neurosci. Lett.* 237:53–55

Lannoo MJ, Maler L, Hawkes R. 1992. Zebrin II distinguishes the ampullary organ receptive map from the tuberous organ receptive maps during development in the teleost electrosensory lateral line lobe. *Brain Res.* 586:176–80

Lannoo MJ, Ross L, Maler L, Hawkes R. 1991. Development of the cerebellum and its extracellular Purkinje cell projection in teleost fishes as determined by zebrin II immunocytochemistry. *Prog. Neurobiol.* 37:329–63

Larsell O. 1967. *The Comparative Anatomy and Histology of the Cerebellum from Myxinoids through Birds*. Minneapolis: Univ. Minn. Press

Lisberger SG, Fuchs AF. 1974. Response of flocculus Purkinje cells to adequate vestibular stimulation in the alert monkey: fixation vs compensatory eye movements. *Brain Res.* 69:347–53

Maekawa K, Simpson JI. 1972. Climbing fiber activation of Purkinje cells in the flocculus by impulses transferred through the visual pathway. *Brain Res.* 39:245–51

Maler L, Sas EKB, Rogers J. 1981. The cytology of the posterior lateral line lobe of high frequency weakly electric fish (Gymnotidae): dendritic differentiation and synaptic specificity in a simple cortex. *J. Comp. Neurol.* 195:87–140

Manis PB, Molitor SC. 1996. N-methyl-D-aspartate receptors at parallel fiber synapses in the dorsal cochlear nucleus. *J. Neurophysiol.* 76:1639–55

Martinez S, Crossley PH, Cobos I, Rubenstein JL, Martin GR. 1999. FGF8 induces formation of an ectopic isthmic organizer and isthmocerebellar development via a repressive effect on Otx2 expression. *Development* 126(6):1189–200

Matsushita A, Kawasaki M. 2005. Neuronal sensitivity to microsecond time disparities in the electrosensory system of Gymnarchus niloticus. *J. Neurosci.* 25(49):11424–32

Matsushita M, Tanami T. 1987. Spinocerebellar projections from the central cervical nucleus in the cat, as studied by anterograde transport of wheat germ agglutinin-horseradish peroxidase. *J. Comp. Neurol.* 266(3):376–97

McCormick CA. 1997. Organization and connections of octaval and lateral line centers in the medulla of a clupeid, *Dorosoma cepedianum. Hear Res.* 110(1–2):39–60

McCormick CA. 1999. Anatomy of the central auditory pathways of fish and amphibians. In *Comparative Hearing: Fish and Amphibians*, ed. RR Fay, AN Popper, pp. 155–217. New York: Springer Verlag

McCreery DB. 1977. Two types of electroreceptive lateral lemniscal neurons of the lateral line lobe of the catfish *Ictalurus nebulosus*; connections from the lateral line nerve and steady-state frequency response characteristics. *J. Comp. Physiol.* 113:317–39

Medina JF, Garcia KS, Nores WL, Taylor NM, Mauk MD. 2000. Timing mechanisms in the cerebellum: testing predictions of a large-scale computer simulation. *J. Neurosci.* 20(14):5516–25

Meek J. 1983. Functional anatomy of the tectum mesencephali of the goldfish. An explorative analysis of the functional implication of the laminar structural organization of the tectum. *Brain Res. Rev.* 6:247–97

Meek J. 1998. Holosteans and teleosts. In *The Central Nervous System of Vertebrates*, ed. R Nieuwenhuys, HJ Ten Donkelaar, C Nicholson, Vol. 15, pp. 759–937. Berlin: Springer

Meek J, Grant K, Sugawara S, Hafmans TGM, Veron M, Denizot JP. 1996. Interneurons of the ganglionic layer in the mormyrid electrosensory lateral line lobe: morphology, immunocyto-chemistry, and synaptology. *J. Comp. Neurol.* 375:43–65

Miall RC, Weir DJ, Wolpert DM, Stein JF. 1993. Is the Cerebellum a smith predictor. *J. Motor Behav.* 25(3):203–16

Mikami Y, Yoshida T, Matsuda N, Mishina M. 2004. Expression of zebrafish glutamate receptor delta2 in neurons with cerebellum-like wiring. *Biochem. Biophys. Res. Commun.* 322(1):168–76

Montgomery JC, Bodznick D. 1993. Hindbrain circuitry mediating common mode suppression of ventilatory reafference in the electrosensory system of the little skate *Raja erinacea.* *J. Exp. Biol.* 183:203–15

Montgomery JC, Bodznick D. 1999. Signals and noise in the elasmobranch electrosensory system. *J. Exp. Biol.* 202(Pt. 10):1349–55

Montgomery JC, Coombs S, Conley RA, Bodznick D. 1995. Hindbrain sensory processing in lateral line, electrosensory, and auditory systems: a comparative overview of anatomical and functional similarities. *Auditory Neurosc.* 1:207–31

Morton SM, Bastian AJ. 2006. Cerebellar contributions to locomotor adaptations during splitbelt treadmill walking. *J. Neurosci.* 26(36):9107–16

Mugnaini E, Dino MR, Jaarsma D. 1997. The unipolar brush cells of the mammalian cerebellum and cochlear nucleus: cytology and microcircuitry. *Prog. Brain Res.* 114:131–50

Mugnaini E, Osen KK, Dahl AL, Friedrich VL Jr, Korte G. 1980a. Fine structure of granule cells and related interneurons (termed Golgi cells) in the cochlear nuclear complex of cat, rat and mouse. *J. Neurocytol.* 9(4):537–70

Mugnaini E, Warr WB, Osen KK. 1980b. Distribution and light microscopic features of granule cells in the cochlear nuclei of cat, rat, and mouse. *J. Comp. Neurol.* 191(4):581–606

Nelson ME, Paulin MG. 1995. Neural simulations of adaptive reafference suppression in the elasmobranch electrosensory system. *J. Comp. Physiol. A* 177:723–36

Nieuwenhuys R, Ten Donkelaar HJ, Nicholson C. 1997. *The Central Nervous System of Vertebrates.* Heidelberg: Springer

Nixon PD, Passingham RE. 2001. Predicting sensory events: the role of the cerebellum in motor learning. *Exp. Brain. Res.* 138:251–57

Noda H, Warabi T. 1982. Eye position signals in the flocculus of the monkey during smooth-pursuit eye movements. *J. Physiol.* 324:187–202

Northmore DPM, Williams B, Vanegas H. 1983. The teleostean torus longitudinalis: responses related to eye movements, visuotopic mapping, and functional relations with the optic tectum. *J. Comp. Physiol. A* 150:39–50

Ohlrogge M, Doucet JR, Ryugo DK. 2001. Projections of the pontine nuclei to the cochlear nucleus in rats. *J. Comp. Neurol.* 436(3):290–303

Ohyama T, Nores WL, Murphy M, Mauk MD. 2003. What the cerebellum computes. *Trends Neurosci.* 26(4):222–27

Pasalar S, Roitman AV, Durfee WK, Ebner TJ. 2006. Force field effects on cerbellar Purkinje cell discharge with implications for internal models. *Nat. Neurosci.* 9(11):1404–11

Paulin MG. 1993. The role of the cerebellum in motor control and perception. *Brain Behav. Evol.* 41:39–50

Paulin MG. 2005. Evolution of the cerebellum as a neuronal machine for Bayesian state estimation. *J. Neural. Eng.* 2(3):S219–34

Petralia RS, Wang YX, Zhao HM, Wenthold RJ. 1996. Ionotropic and metabotropic glutamate receptors show unique postsynaptic, presynaptic, and glial localizations in the dorsal cochlear nucleus. *J. Comp. Neurol.* 372(3):356–83

Puzdrowski RL, Leonard RB. 1993. The octavolateral systems in the stingray, *Dasyatis sabina*. I. Primary projections of the octaval and lateral line nerves. *J. Comp. Neurol.* 332(1):21–37

Raymond JL, Lisberger SG. 1998. Neural learning rules for the vestibulo-ocular reflex. *J. Neurosci.* 18(21):9112–29 (Abstr.)

Roberts PD. 1999. Computational consequences of temporally asymmetric learning rules: I. differential Hebbian learning. *J. Comp. Neurosci.* 7:235–46

Roberts PD, Bell CC. 2000. Computational consequences of temporally asymmetric learning rules: II. sensory image cancellation. *J. Comput. Neurosci.* 9:67–83

Robertson LT. 1985. Somatosensory representation of the climbing fiber system in the rostral intermediate cerebellum. *Exp. Brain Res.* 61(1):73–86

Ronan M 1986. Electroreception in cyclostomes. See Bullock & Heiligenberg 1986, pp. 209–24

Safo P, Regehr WG. 2008. Timing dependence of the induction of cerebellar LTD. *Neuropharmacology*. In press

Saunders J, Bastian J. 1984. The physiology and morphology of two types of electrosensory neurons in the weakly electric fish *Apteronotus leptorhynchus*. *J. Comp. Physiol. A* 154:199–209

Sawtell NB, Williams A, Bell CC. 2007. Central control of dendritic spikes shapes the responses of Purkinje-like cells through spike timing-dependent synaptic plasticity. *J. Neurosci.* 27(7):1552–65

Schlegel P. 1973. Perception of objects in weakly electric fish *Gymnotus carapo* as studied in recordings from rhombencephalic neurons. *Exp. Brain Res.* 18:340–54

Schmidt AW, Bodznick D. 1987. Afferent and efferent connections of the vestibulolateral cerebellum of the little skate, *Raja erinacea*. *Brain Behav. Evol.* 30:282–302

Shore SE, Zhou J. 2006. Somatosensory influence on the cochlear nucleus and beyond. *Hear. Res.* 216–17:90–99

Smith MA, Shadmehr R. 2005. Intact ability to learn internal models of arm dynamics in Huntington's disease but not cerebellar degeneration. *J. Neurophysiol.* 93(5):2809–21

Steuber V, Mittman W, Hoebeek FE, Silver RA, De Zeeuw CI, et al. 2007. Cerebellar LTD and pattern recognition by Purkinje cells. *Neuron* 54(1):121–36

Stone LS, Lisberger SG. 1990. Visual responses of Purkinje cells in the cerebellar flocculus during smooth-pursuit eye movements in monkeys. II. Complex spikes. *J. Neurophysiol.* 63(5):1262–75

Szabo T, Libouban S, Haugede-Carre F. 1979. Convergence of common and specific sensory afferents to the cerebellar auricle (auricula cerebelli) in the teleost fish *Gnathonemus* demonstrated by HRP method. *Brain Res.* 168:619–22

Szabo T, Libouban S, Denizot JP. 1990. A well defined spinocerebellar system in the weakly electric teleost fish *Gnathonemus petersii*. *Arch. Ital. Biol.* 128:229–47

Tong S. 1982. The nucleus praeeminentialis: an electro- and mechanoreceptive center in the brainstem of the catfish. *J. Comp. Physiol. A* 145:299–309

Tong S, Finger TE. 1983. Central organization of the electrosensory lateral line system in bullhead catfish *Ictalurus nebulosus*. *J. Comp. Neurol.* 217:1–16

Tzounopoulos T, Kim Y, Oertel D, Trussell LO. 2004. Cell-specific, spike timing-dependent plasticities in the dorsal cochlear nucleus. *Nat. Neurosci.* 7(7):719–25

Tzounopoulos T, Rubio ME, Keen JE, Trussell LO. 2007. Coactivation of pre- and postsynaptic signaling mechanisms determines cell-specific spike-timing-dependent plasticity. *Neuron* 54(2):291–301

Vanegas H, Williams B, Freeman JA. 1979. Responses to stimulation of marginal fibers in the teleostean optic tectum. *Exp. Brain Res.* 34(2):335–49

Wang SS, Denk W, Hausser M. 2000. Coincidence detection in single dendritic spines mediated by calcium release. *Nat. Neurosci.* 3:1266–73

Weedman DL, Ryugo DK. 1996. Projections from auditory cortex to the cochlear nucleus in rats: synapses on granule cell dendrites. *J. Comp. Neurol.* 371:311–24

Weinberg RJ, Rustioni A. 1987. A cuneocochlear pathway in the rat. *Neuroscience* 20(1):209–19

Wolff A, Kunzle H. 1997. Cortical and medullary somatosensory projections to the cochlear nuclear complex in the hedgehog tenrec. *Neurosci. Lett.* 221(2–3):125–28

Wolpert DM, Miall C, Kawato M. 1998. Internal models in the cerebellum. *Trends Cogn. Sci.* 2:338–47

Yamamoto K, Kawato M, Kotosaka S, Kitazawa S. 2007. Encoding coding of movements dynamics by Purkinje cell simple spike activity during fast arm movements under resistive and assistive force fields. *J. Neurophysiol.* 97:1588–99

Yawata S, Tsuchida H, Kengaku M, Hirano T. 2006. Membrane-proximal region of glutamate receptor delta2 subunit is critical for long-term depression and interaction with protein interacting with C kinase 1 in a cerebellar Purkinje neuron. *J. Neurosci.* 26(14):3626–33

Young ED, Davis KA. 2002. Circuitry and function of the dorsal cochlear nucleus. In *Integrative Functions in the Mammalian Auditory Pathway*, ed. D Oertel, AN Popper, RR Fay, pp. 160–206. New York: Springer-Verlag

Yuzaki M. 2003. The delta2 glutamate receptor: 10 years later. *Neurosci. Res.* 46(1):11–22

Zhou J, Shore S. 2004. Projections from the trigeminal nuclear complex to the cochlear nuclei: a retrograde and anterograde tracing study in the guinea pig. *J. Neurosci. Res.* 78(6):901–7

Spike Timing–Dependent Plasticity: A Hebbian Learning Rule

Natalia Caporale and Yang Dan

Division of Neurobiology, Department of Molecular and Cell Biology, and Helen Wills Neuroscience Institute, University of California, Berkeley, California 94720; email: caporale@socrates.berkeley.edu, ydan@berkeley.edu

Annu. Rev. Neurosci. 2008. 31:25–46

First published online as a Review in Advance on February 14, 2008

The *Annual Review of Neuroscience* is online at neuro.annualreviews.org

This article's doi: 10.1146/annurev.neuro.31.060407.125639

0147-006X/08/0721-0025$20.00

Key Words

long-term potentiation, long-term depression, synapse, memory, backpropagating action potential

Abstract

Spike timing–dependent plasticity (STDP) as a Hebbian synaptic learning rule has been demonstrated in various neural circuits over a wide spectrum of species, from insects to humans. The dependence of synaptic modification on the order of pre- and postsynaptic spiking within a critical window of tens of milliseconds has profound functional implications. Over the past decade, significant progress has been made in understanding the cellular mechanisms of STDP at both excitatory and inhibitory synapses and of the associated changes in neuronal excitability and synaptic integration. Beyond the basic asymmetric window, recent studies have also revealed several layers of complexity in STDP, including its dependence on dendritic location, the nonlinear integration of synaptic modification induced by complex spike trains, and the modulation of STDP by inhibitory and neuromodulatory inputs. Finally, the functional consequences of STDP have been examined directly in an increasing number of neural circuits in vivo.

Contents

INTRODUCTION

Electrical activity plays crucial roles in the structural and functional refinement of neural circuits throughout an organism's lifetime (Buonomano & Merzenich 1998, Gilbert 1998, Karmarkar & Dan 2006, Katz & Shatz 1996). Manipulations of sensory experience that disrupt normal activity patterns can lead to large-scale network remodeling and marked changes in neural response properties. Learning and memory are also likely to be mediated by activity-dependent circuit modifications. Understanding the cellular mechanisms underlying such functional plasticity has been a long-standing challenge in neuroscience (Martin et al. 2000).

In his influential postulate on the cellular basis for learning, Hebb stated that "when an axon of cell A is near enough to excite a cell B and repeatedly or persistently takes part in firing it, some growth process or metabolic change takes place in one or both cells such that A's efficiency, as one of the cells firing B, is increased" (Hebb 1949). This postulate gained strong experimental support with the finding of long-term potentiation (LTP) of synaptic transmission, initially discovered in the hippocampus (Bliss & Gardner-Medwin 1973, Bliss & Lomo 1973) and subsequently reported in a large number of neural circuits, including various neocortical areas (Artola & Singer 1987, Iriki et al. 1989, Hirsch et al. 1992), the amygdala (Chapman et al. 1990, Clugnet & LeDoux 1990), and the midbrain reward circuit (Liu et al. 2005, Pu et al. 2006). Traditionally, LTP is induced by high-frequency stimulation (HFS) of the presynaptic afferents or by pairing low-frequency stimulation (LFS) with large postsynaptic depolarization (>30 mV). In contrast, long-term depression (LTD) is induced by LFS, either alone or paired with a small postsynaptic depolarization (Artola et al. 1990, Dudek & Bear 1993, Kirkwood & Bear 1994, Linden & Connor 1995, Mulkey & Malenka 1992, Stanton & Sejnowski 1989). Together, LTP and LTD allow activity-dependent bidirectional modification of synaptic strength, thus serving as promising candidates for the synaptic basis of learning and memory (Bliss & Collingridge 1993; Ito 2005; Siegelbaum & Kandel 1991).

To characterize the temporal requirements for the induction of LTP and LTD, Levi & Steward (1983) varied the relative timing of a strong and a weak input from the entorhinal cortex to the dental gyrus and found that synaptic modification depended on the temporal order of the two inputs. Potentiation was produced when the weak input preceded the strong input by less than 20 ms, and reversing the order led to depression. Subsequent studies further demonstrated the importance of the temporal order of pre- and postsynaptic spiking in synaptic modification and delineated the critical window on the order of tens of milliseconds (Bi & Poo 1998, Debanne et al. 1998, Magee & Johnston 1997, Markram et al. 1997,

LTP: long-term potentiation

HFS: high-frequency stimulation

LFS: low-frequency stimulation

LTD: long-term depression

Zhang et al. 1998) (**Figure 1***a*, I). Such spike-timing-dependent plasticity (STDP) (Abbott & Nelson 2000) has now been observed at excitatory synapses in a wide variety of neural circuits (Boettiger & Doupe 2001, Cassenaer & Laurent 2007, Egger et al. 1999, Feldman 2000, Froemke & Dan 2002, Sjostrom et al. 2001, Tzounopoulos et al. 2004). Compared with the correlational forms of synaptic plasticity, STDP captures the importance of causality in determining the direction of synaptic modification, which is implied in Hebb's original postulate.

Recent studies have further characterized the mechanism and function of STDP in both in vitro and in vivo preparations, addressing the following questions: Which cellular mechanisms determine the STDP window, and how similar are they to the mechanisms underlying LTP and LTD induced by HFS and LFS, respectively? Does the window depend on the dendritic location of the input, and can it be regulated by neuromodulatory inputs? Does a similar learning rule apply to the inhibitory circuits? Can we observe the consequences of the asymmetric window in vivo, and can it account for the synaptic modifications induced by complex, naturalistic spike trains? In this review we summarize recent progress in these areas.

CELLULAR MECHANISMS

For many glutamatergic synapses, the inductions of LTP by HFS and LTD by LFS both require the activation of NMDA (N-methyl-d-aspartate) receptors and a rise in postsynaptic Ca^{2+} level (Malenka & Bear 2004). The NMDA receptor is thought to serve as the coincidence detector: The presynaptic activation provides glutamate and the postsynaptic depolarization causes removal of the Mg^{2+} block (Mayer et al. 1984, Nowak et al. 1984), which together allow Ca^{2+} influx though the NMDA receptors. The level and time course of postsynaptic Ca^{2+} rise depend on the induction protocol: HFS leads to fast, large Ca^{2+} influx, whereas LFS leads to prolonged, modest Ca^{2+} rise (Malenka & Bear 2004, Yang et al. 1999). In the Ca^{2+} hypothesis (Artola & Singer 1993, Lisman 1989, Yang et al.

a   Excitatory to excitatory

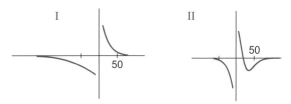

b   Excitatory to inhibitory

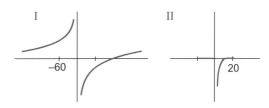

c   Inhibitory to excitatory

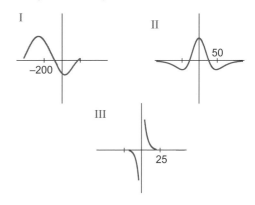

Figure 1

Diversity of temporal windows for STDP induction. *a:* Windows for excitatory to excitatory connections. *b:* Windows for excitatory to inhibitory connections. *c:* Windows for inhibitory to excitatory connections. Temporal axis is in milliseconds.

1999), these two types of Ca^{2+} signals cause the activation of separate molecular pathways. Activation of Ca^{2+}/calmodulin-dependent protein kinase II (CaMKII) by large Ca^{2+} rise is required for LTP, whereas recruitment of phosphatases such as protein phosphatase 1 (PP1) and calcineurin by modest Ca^{2+} increase is necessary for LTD (Malenka & Bear 2004). Spike timing–dependent LTP (tLTP) and LTD (tLTD) also depend on NMDA

STDP: spike timing–dependent plasticity

N-methyl-d-aspartate (NMDA) receptor: subtype of glutamate receptors

receptor activation and the rise in postsynaptic Ca^{2+} level (Bi & Poo 1998, Debanne et al. 1998, Feldman 2000, Magee & Johnston 1997, Markram et al. 1997, Sjostrom et al. 2001, Zhang et al. 1998). Can this simple model for conventional LTP and LTD account for STDP, in particular for its temporal window on the time scale of tens of milliseconds?

tLTP Window

Induction of tLTP requires activation of the presynaptic input milliseconds before the backpropagating action potential (BAP) in the postsynaptic dendrite (pre → post, positive intervals). The BAP can facilitate Mg^{2+} unblocking of NMDA receptors and thus allow Ca^{2+} influx, leading to tLTP induction. However, the width of the tLTP window cannot be explained solely by the time course of NMDA receptor activation. The dissociation of glutamate from the NMDA receptors occurs on the order of hundreds of milliseconds (Lester et al. 1990), much longer than the observed tLTP windows (**Figure 1a**). The short duration of the window may be due to the kinetics of Mg^{2+} unblocking NMDA receptors (Kampa et al. 2004), such that the BAPs arriving soon after the onset of the excitatory postsynaptic potential (EPSP) are better able to open the NMDA receptors.

In addition to the Mg^{2+} unblock of NMDA receptors, the tLTP window could also be shaped by other types of interactions between the EPSP and the BAP. For example, the EPSP can cause changes in the dendritic conductances that affect the action potential (AP) backpropagation into the dendrites. In the hippocampus, the distal dendrites of CA1 pyramidal neurons express a high density of A-type K^+ channels, which regulate the BAP amplitude (Hoffman et al. 1997). An EPSP that depolarizes the dendrite and inactivates these channels can boost the BAPs arriving within tens of milliseconds (Magee & Johnston 1997, Watanabe et al. 2002). This boosting of the BAPs can in turn increase the Ca^{2+} influx through voltage-dependent Ca^{2+} channels (VDCCs), which can modulate the

magnitude of tLTP (Bi & Poo 1998, Froemke et al. 2006, Magee & Johnston 1997). In the neocortex, a similar boosting of the BAP by the preceding EPSP is achieved by voltage-gated Na^{2+} channel activation in the distal dendrites (Stuart & Hausser 2001). Such nonlinear interactions between the EPSP and BAP at short positive intervals could explain the supralinear summation of Ca^{2+} influx to the active synapse in both hippocampal (Magee & Johnston 1997) and neocortical (Koester & Sakmann 1998) (Nevian & Sakmann 2004) neurons.

tLTD Window

Models based on the Ca^{2+} hypothesis have also been used to explain the tLTD window (post → pre, negative intervals) (Karmarkar & Buonomano 2002, Shouval et al. 2002). Assuming that the BAP contains an afterdepolarization lasting for tens of milliseconds and that all relevant Ca^{2+} enters the postsynaptic cell through NMDA receptors, the tLTD window can be explained by the interaction between the EPSP and the BAP. Unlike pairing of the BAP and the EPSP at positive intervals, which causes large Ca^{2+} influx through the NMDA receptors, the EPSP coinciding with the afterdepolarization leads to a moderate Ca^{2+} influx, resulting in tLTD. It should be noted that this model predicts an additional tLTD window at positive intervals outside the tLTP window (**Figure 1a**, II), where the rise in postsynaptic Ca^{2+} falls within the range for LTD induction. This additional tLTD window has indeed been observed in hippocampal CA1 neurons (Nishiyama et al. 2000, Wittenberg & Wang 2006) but not at other synapses. This suggests a distinct form of STDP at hippocampal synapses, or it could reflect insufficient sampling of long positive intervals in the experimental studies of STDP in other circuits.

In another model for tLTD based on the Ca^{2+} hypothesis (Froemke et al. 2005), a BAP preceding an EPSP induces Ca^{2+} influx through VDCCs, which inactivates the NMDA receptors (Rosenmund et al. 1995, Tong et al. 1995). The reduced Ca^{2+} influx through NMDA receptors in turn leads to tLTD. This model is supported by the observations that tLTD induction requires activation of VDCCs (Bender et al. 2006, Bi & Poo 1998, Froemke et al. 2005, Nevian & Sakmann 2006) and that pairing EPSPs and BAPs at negative intervals leads to sublinear summation of Ca^{2+} influx (Koester & Sakmann 1998, Nevian & Sakmann 2004). Furthermore, in L2/3 pyramidal neurons in visual cortical slices, BAP-induced Ca^{2+}-dependent NMDA receptor inactivation varied with dendritic location, mirroring the location dependence of the tLTD window at these synapses (Froemke et al. 2005).

In some other synapses, tLTD induction does not depend on activation of postsynaptic NMDA receptors (Bender et al. 2006, Egger et al. 1999, Nevian & Sakmann 2006, Sjostrom et al. 2003). These studies suggest a model involving two coincidence detectors, with the NMDA receptor for tLTP and an additional co-incidence detector for tLTD. In a two-detector model proposed by Karmarkar & Buonomano (2002), tLTD induction requires activation of postsynaptic mGluRs (metabotropic glutamate receptors) and Ca^{2+} influx through VDCCs, a premise supported by experimental findings in the barrel cortex (Bender et al. 2006, Egger et al. 1999, Nevian & Sakmann 2006). Signaling through mGluRs can lead to phospholipase C (PLC) activation, and Ca^{2+} influx through VDCCs can facilitate mGluR-dependent-PLC activation (Hashimotodani et al. 2005, Maejima et al. 2005). Thus, PLC can serve as a potential coincidence detector for tLTD.

Downstream of coincidence detection, PLC may generate inositol 1,4,5-triphosphate (IP_3), which in turn triggers release of Ca^{2+} from internal stores through IP_3 receptors (IP_3Rs) (Bender et al. 2006). Both PLC activation and Ca^{2+} level elevation (due to influx through VDCCs and/or NMDA receptors, or release from internal stores) can promote endocannabinoid synthesis and release (Hashimotodani et al. 2007). Endocannabinoids play important roles in both short- and long-term depression of many synapses (Chevaleyre et al. 2006). Signaling

through presynaptic CB1 endocannabinoid receptors is also required for tLTD for several excitatory–excitatory (Bender et al. 2006; Nevian & Sakmann 2006; Sjostrom et al. 2003) and excitatory–inhibitory connections (Tzounopoulos et al. 2007), presumably by inhibiting presynaptic transmitter release. In **Figure 2**, we have outlined the major signaling pathways implicated in STDP.

mGluR: metabotropic glutamate receptor

STDP OF INHIBITION

Balanced excitation and inhibition are crucial for normal brain functions (Shu et al. 2003) and

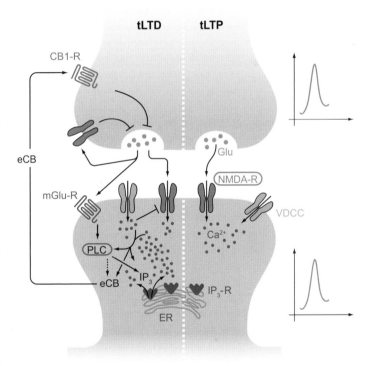

Figure 2

Schematic representation of signaling pathways involved in STDP induction. In tLTP induction (*right*), the NMDA receptors act as coincidence detectors for pre- and postsynaptic spiking. In tLTD induction (*left*) the coincidence detector may vary across synapses. The diagram includes several pathways that have been suggested to play a role in tLTD. Red oval indicates possible coincidence detectors. Arrow indicates activation/potentiation. Blunt-ended line indicates inhibition/suppression. Abbreviations: eCB, endocannabinoids; ER, endoplasmic reticulum; Glu, glutamate; IP_3, inositol 1,4,5-triphosphate; PLC, phospholipase C; VDCCs, voltage-dependent Ca^{2+} channels.

for regulating experience-dependent developmental plasticity (Hensch 2005). Although the strength of excitatory synapses can be modified through STDP, an important question is whether and how correlated pre- and postsynaptic activity affects inhibitory circuits. Inhibition in a network depends on both the excitatory synapses onto inhibitory neurons and the inhibitory synapses themselves. Spike timing–dependent plasticity has been studied at both of these synapses.

STDP of Excitatory Synapses onto Inhibitory Neurons

In a cerebellum-like structure in the electric fish, Bell and colleagues (1997) measured the excitatory inputs to Purkinje-like GABAergic neurons to study the dependence of synaptic modification on the temporal order of pre- and postsynaptic spiking. Pre $\to$ post pairing within a 60-ms window induces LTD, whereas post $\to$ pre pairing leads to LTP (**Figure 1b**, I). This asymmetrical window is thus opposite in polarity to the STDP window for the synapses between excitatory neurons (**Figure 1a**, I). However, given the difference in the postsynaptic neurons, the functional consequences of the two learning rules may be similar, and they could act cooperatively in activity-dependent network modifications. Mechanistically, LTD induced by pre $\to$ post pairings required NMDA receptor activation and postsynaptic Ca^{2+} elevation (Han et al. 2000), similarly to tLTD for excitatory-excitatory connections. However, LTP of these synapses can be induced by EPSPs alone without postsynaptic spiking, indicating a nonassociative component of the synaptic plasticity.

Another study of excitatory inputs to inhibitory neurons was conducted in mouse brain stem slices by pairing parallel fiber stimulation with cartwheel neuron spiking (Tzounopoulos et al. 2004, 2007). Pre $\to$ post pairing within a narrow window (<10 ms) induces LTD, whereas post $\to$ pre pairing causes no change in synaptic strength

(**Figure 1b**, II). For these synapses, LTD depends on postsynaptic NMDA receptor activation, Ca^{2+} influx, and endocannabinoid signaling, similar to the findings at excitatory synapses onto pyramidal neurons (see previous section). Interestingly, synapses from the same presynaptic fibers onto excitatory postsynaptic neurons (fusiform principal neurons) exhibit a STDP window similar to that of other excitatory-excitatory connections (**Figure 1a**, I) (Tzounopoulos et al. 2004, 2007). This target specificity of the learning rule can be attributed to the selective distribution of presynaptic endocannabinoid CB1 receptors in different axonal terminals.

STDP of GABAergic Synapses

Compared with the glutamatergic synapses, the learning rules for GABAergic synapses appear more variable. In a study of inhibitory inputs to neocortical L2/3 pyramidal neurons, synaptic modification was induced by pairing single presynaptic spikes with high-frequency postsynaptic bursts. Overlapping pre- and postsynaptic spiking induced LTD, and nonoverlapping post $\to$ pre spiking within hundreds of milliseconds induced LTP (Holmgren & Zilberter 2001) (**Figure 1c**, I). In the hippocampus, GABAergic synapses onto CA1 pyramidal neurons exhibit a symmetrical window, with pairing of single pre- and postsynaptic spikes at short intervals (within ±20 ms) leading to LTP, and pairing at long intervals leading to LTD (Woodin et al. 2003) (**Figure 1c**, II). In contrast, in the entorhinal cortex GABAergic inputs to layer II excitatory stellate cells exhibit an asymmetric window similar to the STDP window for excitatory-excitatory connections: LTP was found at positive intervals and LTD at negative intervals (Haas et al. 2006) (**Figure 1c**, III). Despite the differences between these temporal windows for GABAergic synapses, both the induction mechanism and the loci of expression have similarities. In both hippocampal CA1 (Woodin et al. 2003) and the entorhinal cortex (Haas et al. 2006), the induction of synaptic modification depends on postsynaptic Ca^{2+} influx through the L-type Ca^{2+} channels, and presynaptic expression was excluded because no change was observed in the paired pulse ratio. In the hippocampus (Woodin et al. 2003), the changes in inhibitory postsynaptic current (IPSC) amplitude are due to changes in the Cl^- reversal potential mediated by modification of the KCC2 K^+-Cl^- cotransporter, further indicating that the expression is postsynaptic.

STDP WITH COMPLEX SPIKE PATTERNS

To study synaptic plasticity, the induction paradigms are often selected for their effectiveness rather than for their physiological relevance, thus providing limited information on how circuits are modified by natural patterns of activity. Although most induction protocols for STDP consisted of repetitive pairing of pre- and postsynaptic spikes at regular intervals, neuronal activity in vivo is far from regular (Softky & Koch 1993), with periods of almost no activity intermingled with short bouts of high-frequency spike bursts. During each presynaptic burst, transmitter release is likely to be affected by short-term plasticity (Zucker & Regehr 2002), and in each postsynaptic burst the efficacy of individual spike propagation may depend on the spike pattern (Spruston et al. 1995; Williams & Stuart 2000). How well does the STDP learning rule measured with simple spike patterns account for the synaptic changes induced by naturalistic spike trains? When multiple spike pairs fall within the STDP window, how are the contributions of individual spikes integrated?

One simple strategy to study the interaction among multiple spikes is to add one spike at a time to the existing pairing protocol. In L2/3 of visual cortical slices (Froemke & Dan 2002) and in hippocampal cultures (Wang et al. 2005), spike "triplets" (pre → post → pre or post → pre → post) and "quadruplets" (pre → post → post → pre or post → pre → pre → post) were used to induce synaptic modifications. In both studies, the interaction

between multiple spikes was nonlinear, but the specific forms of nonlinearity were different. In cortical L2/3, the nonlinear interactions could be accounted for by a suppression model, in which the efficacy of later spikes in each train for synaptic modification is reduced by the preceding spikes (Froemke & Dan 2002). This model accurately predicted the synaptic changes induced by natural spike trains recorded in vivo in response to visual stimulation. In cultured hippocampal neurons, the "pre → post → pre" triplets induce no synaptic change, which suggests that LTP and LTD cancel each other, but the "post → pre → post" triplets induce LTP, which suggests that LTP "wins over" LTD under this condition. A third study using spike triplets showed that in hippocampal slices, different learning rules are revealed with different numbers of spike pairings (Wittenberg & Wang 2006). With 20–30 pairings at 5 Hz, LTP was induced regardless of the temporal order of the spikes. With 70–100 repeats, however, LTP was observed at short positive intervals (<30 ms), and LTD was found at both negative intervals and at long positive intervals (>30 ms) (**Figure 1a**, II). These results suggest that the integration across multiple spike pairs depends on the activity patterns over several minutes.

The effects of pre- and/or postsynaptic spike bursts on synaptic modification have also been examined. Paired recordings from L5 pyramidal neurons in visual cortical slices showed that the synaptic change depends on both the spike frequency within each burst and the interval between the pre- and postsynaptic spikes (Sjostrom et al. 2001). At high frequencies (≥50 Hz), LTP is induced regardless of the pre/post interval, whereas at intermediate frequencies (10–40 Hz), the pre/post interval determines the sign and magnitude of synaptic modification as described by the STDP window (**Figure 1a**, I). Pairing at low frequencies (<1 Hz) notably fails to induce LTP. This is likely caused by the small EPSPs evoked by activating a single presynaptic neuron in paired recordings because LTP can be rescued by adding extracellular stimulation that provides additional depolarization. The combined dependence of synaptic modification on burst timing and frequency can be accounted for by a model in which LTP wins over LTD, and only the interactions between neighboring spikes contribute to synaptic modification (Sjostrom et al. 2001). In another study in L2/3 neurons in rat visual cortical slices (Froemke et al. 2006), pairing of pre- and postsynaptic bursts at high frequencies also favored LTP regardless of the pre/post spike timing. However, systematic examination of the dependence of synaptic modification on both the number and the timing of pre- and postsynaptic spikes led to a modified suppression model (Froemke et al. 2006), which incorporates short-term depression of the presynaptic input (Zucker & Regehr 2002) and frequency-dependent attenuation of postsynaptic spikes (Spruston et al. 1995). Note that in both models described above, burst-induced synaptic modification is accounted for by integrating the contributions of individual spike pairs. However, in some synapses the learning rule for bursts seems to be completely different from that for individual spikes (Birtoli & Ulrich 2004, Kampa et al. 2006, Pike et al. 1999).

Although the above studies focused on synaptic modifications induced by short bursts lasting for tens of milliseconds, in some circuits bursts can last for hundreds of milliseconds to several seconds. In the developing retinogeniculate synapse, bursts of retinal ganglion cells lasting seconds are believed to be critical for circuit refinement (Butts & Rokhsar 2001). Temporally overlapping pre- and postsynaptic bursts (interval within a window of ~1 s) result in synaptic potentiation, whereas nonoverlapping bursts cause a slight depression (Butts et al. 2007). The degree of potentiation can be predicted by a model in which LTP depends on the interval but not the order between the pre- and postsynaptic bursts, and it increases linearly with the number of spikes in the burst. This is reminiscent of the classic correlation-based learning rule for synaptic plasticity (Stent 1973). A strikingly similar window for burst timing was found in the hippocampal CA3 re-

gion for correlated activation of the associational/commissural (A/C) fibers and the mossy fibers (Kobayashi & Poo 2004), although no depression was observed. In both studies, the width of the temporal window seems to scale with the duration of the spike bursts used in the induction protocol, and the changes in synaptic strength depend on the interburst interval rather than the precise timing of individual spikes. Such burst timing–dependent plasticity rules may be functionally advantageous for the circuits in which the information relevant for synaptic refinement is contained in the timing of the bursts rather than that of individual spikes (Butts & Rokhsar 2001).

Together, the studies described above indicate that the integration across multiple spike pairs for the induction of synaptic modification is highly nonlinear. The nature of the nonlinear interaction is likely to depend on short-term plasticity of the presynaptic neurons, on the biophysical properties of the postsynaptic dendrites, and on the downstream signaling pathways present in different cell types. Further characterization of the diversity of integration mechanisms for STDP will allow better understanding of circuit remodeling induced by natural patterns of neuronal activity.

DEPENDENCE ON DENDRITIC LOCATION

In the central nervous system, each neuron may receive thousands of synaptic inputs distributed throughout its dendritic tree. The processing of each input depends on the dendritic location (Hausser & Mel 2003) owing to both the passive cable properties (Rall 1967) and the nonuniform distribution of active conductances (Migliore & Shepherd 2002). Such location-dependent processing and integration of synaptic inputs are believed to be essential aspects of neuronal computation. Since a hallmark of STDP is its dependence on the BAPs, which are strongly attenuated along the dendrite (Stuart & Sakmann 1994, Stuart et al. 1997b, Waters et al. 2005), synaptic modification is likely to vary with

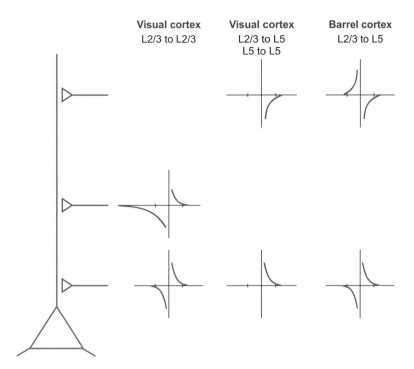

Visual cortex
L2/3 to L2/3

Visual cortex
L2/3 to L5
L5 to L5

Barrel cortex
L2/3 to L5

Figure 3

Dependence of STDP on dendritic location.

dendritic location (Rao & Sejnowski 2001b). Recent studies have examined the location dependence of both tLTP and tLTD. In L2/3 of rat visual cortex, the magnitude of tLTP induced by pre → post pairing of single spikes was smaller at intermediate-distal (100–150 μm) than at proximal (<50 μm) segments of the apical dendrite (Froemke et al. 2005) (**Figure 3**, left column). This reduction of tLTP amplitude is likely due to distance-dependent attenuation of the BAP. In experiments with paired recordings from a L5 and a L2/3 pyramidal neuron or from two L5 neurons (Sjostrom & Hausser 2006), burst pairing at positive intervals led to LTP at the proximal synapses but LTD at the distal synapses (**Figure 3**, middle column). Similar location dependence was also found among L2/3 to L5 connections by pairing a single EPSP with a postsynaptic burst at positive intervals (Letzkus et al. 2006) (**Figure 3**, right column). BAP boosting by subthreshold local dendritic depolarization or extracellular stimulation recovered tLTP at distal synapses (Letzkus et al. 2006, Sjostrom & Hausser 2006), which sug-

gests that distal tLTP requires cooperativity among inputs.

Two distinct effects have been reported for post → pre pairing. In L2/3 pyramidal neurons, the width of the tLTD window measured with single spike pairing is broader for intermediate-distal than for proximal inputs (Froemke et al. 2005) (**Figure 3**, left column). This difference in width is correlated with the window for AP-induced suppression of NMDA receptor activation, which suggests that the suppression plays an important role in setting the tLTD window. In L2/3–L5 synapses in rat barrel cortex, pairing single presynaptic spikes with postsynaptic bursts at negative intervals leads to LTD at proximal locations but LTP of distal inputs (Letzkus et al. 2006) (**Figure 3**, right column). This distal LTP could be explained by the induction of dendritic Ca^{2+} spikes by the later BAPs in the burst (Larkum et al. 1999a, Stuart et al. 1997a), such that the EPSP coincides with the peak postsynaptic depolarization. Local dendritic spikes can also play a prominent role in coincidence detection in the neocortex

(Larkum et al. 1999b) and in LTP induction in hippocampal CA1 (Golding et al. 2002) and the amygdala (Humeau & Luthi 2007).

Comparison across these studies suggests that the degree of spatial variation of the learning rule depends on the dendritic morphology, with quantitative changes over short distances (e.g., dendrite of L2/3 neurons) and qualitative differences along long dendrites (e.g., apical dendrites of L5 pyramids). Although the dendritic variations of STDP summarized above can be explained largely by differences in the local active conductances, the backpropagation of APs, or the local generation of Ca^{2+} spikes, differential distribution of other pre- and postsynaptic molecular machineries could also contribute to the observed heterogeneity. Functionally, the spatial variation of the STDP rule may lead to differential input selection at distal and proximal dendrites. For example, the relative paucity of LTP at distal dendrites after pre → post pairing predicts that proximal inputs should be stronger than distal inputs (Sjostrom & Hausser 2006). The involvement of locally generated Ca^{2+} spikes in LTP induction (Golding et al. 2002, Kampa et al. 2006) likely rewards cooperativity among distal inputs because their synchronous activation is known to evoke dendritic spikes. Furthermore, the broader LTD window for intermediate distal inputs to L2/3 neurons suggests that the distal dendrites strongly favor transient over prolonged inputs (Froemke et al. 2005).

MODULATION OF STDP BY OTHER INPUTS

In addition to the spiking of the pre- and postsynaptic neurons, STDP is also regulated by other inputs. In particular, neuromodulators and inhibitory activity in the network can affect both the magnitude and the temporal window of STDP.

Neuromodulators such as norepinephrine and acetylcholine (ACh) play important roles in experience-dependent neural plasticity (Bear & Singer 1986, Kilgard & Merzenich 1998).

At the cellular level, neuromodulators can influence AP backpropagation by modulating the activation and inactivation of various active conductances (Johnston et al. 1999). For example, agonists to muscarinic ACh receptors can reduce spike attenuation during high-frequency bursts, probably through reduction of Na^+ channel inactivation (Johnston et al. 1999, Tsubokawa & Ross 1997). Both β-adrenergic and muscarinic ACh receptor agonists can boost AP backpropagation by downregulating transient K^+ channels through protein kinase A (PKA) and protein kinase C (PKC) activation, respectively (Hoffman & Johnston 1998, 1999). Dopamine also has a similar effect on the BAP (Hoffman & Johnston 1999).

Such modulations of the BAPs are likely to have profound effects on STDP, particularly at distal dendritic locations. In the Schaffer collateral pathway to hippocampal CA1, pairing a weak and a strong input (which evokes postsynaptic spiking) at positive intervals can induce NMDA receptor–dependent tLTP within a narrow window of 3–10 ms. Bath application of isoproterenol, a β-adrenergic receptor agonist, broadens the window to 15 ms without changing the magnitude of tLTP (Lin et al. 2003), an effect that depends on PKA and mitogen-activated protein kinase (MAPK) signaling. In the amygdala, dopamine can gate the induction of tLTP by suppressing feedforward inhibitory inputs to the postsynaptic cell (Bissiere et al. 2003). In L5 pyramidal neurons of the prefrontal cortex, nicotine application converted tLTP to tLTD by reducing dendritic Ca^{2+} signals during spike pairing (Couey et al. 2007), and this reduction is mediated by an enhancement of GABAergic synaptic transmission. In L2/3 pyramidal neurons, activation of M1 muscarinic receptors promotes tLTD induction through a PLC-dependent pathway, whereas β-adrenergic receptor activation promotes tLTP through the adenylate cyclase cascade (Seol et al. 2007). Thus, neuromodulators can regulate both the magnitude and the polarity of synaptic modifications.

The timing and location of inhibitory inputs can also affect STDP. Somatic inhibition can prevent AP propagation through hyperpolarization and shunting (Miles et al. 1996, Tsubokawa & Ross 1996), which may preclude STDP induction. In contrast, inhibitory inputs to the dendrites have a variety of effects, from reducing dendritic depolarization through shunting to facilitating depolarization and even spike generation (Gulledge & Stuart 2003). An additional layer of complexity is added by the fact that the strength and distribution of inhibition are developmentally regulated (Hensch 2005), predicting that the learning rule can vary considerably across developmental stages. In hippocampal CA1 pyramidal neurons, pairing single pre- and postsynaptic spikes at positive intervals leads to tLTP in juvenile (p9–p14) but not in young (p22–p28) rats (Meredith et al. 2003). However, in young rats tLTP can be rescued by replacing the single postsynaptic spike with a burst or by adding GABA$_A$ antagonists, suggesting that the change in tLTP threshold might be due to a developmental enhancement of inhibition in this circuit.

PLASTICITY OF NEURONAL EXCITABILITY AND SYNAPTIC INTEGRATION

Information processing by neuronal networks depends not only on the connectivity between neurons, but also on the intrinsic conductances in each neuron that determine its excitability and synaptic integration. Changes in neuronal excitability have been reported in a variety of invertebrate and vertebrate neural circuits during associative learning (Daoudal & Debanne 2003, Zhang & Linden 2003). At the cellular level, LTP induction by tetanic stimulation also leads to increases in intrinsic excitability in both the hippocampus and the cerebellum (Aizenman & Linden 2000, Armano et al. 2000, Bliss & Gardner-Medwin 1973, Bliss & Lomo 1973). These activity-dependent changes in intrinsic neuronal properties may interact synergistically with synaptic plasticity to mediate learning and memory.

Changes in neuronal excitability have also been examined in the context of STDP. In hippocampal cell cultures (Ganguly et al. 2000) and neocortical slices (Li et al. 2004), repeated pre → post pairing of single spikes leads to LTP and to an enhancement of excitability and spike time reliability of the presynaptic neurons. Pairings at negative intervals result in LTD and a reduction in presynaptic excitability (Li et al. 2004). Mechanistically, these presynaptic changes require NMDA receptor activation and Ca^{2+} influx to the postsynaptic neuron, suggesting the involvement of retrograde signaling. On the presynaptic side, PKC is necessary for the increase in excitability (Ganguly et al. 2000), and both PKC and PKA are required for the decrease (Li et al. 2004). Interestingly, the changes in excitability can be dissociated from the changes in synaptic strength because presynaptic blockage of PKC and/or PKA abolished the excitability changes with little effect on the synaptic modifications.

Activity-dependent changes in intrinsic membrane properties can also affect synaptic integration (Magee & Johnston 2005). A recent study examined the changes in spatial summation between two input pathways in hippocampal CA1 neurons following STDP induction (Wang et al. 2003). Induction of tLTP in one pathway resulted in an increase in the linearity of spatial summation of the two pathways, whereas induction of tLTD produced the opposite effect. The observed changes depend on NMDA receptor activation and may be mediated by modifications of the I$_h$ channels. In another study in hippocampal CA1, LTP induction by paired theta bursts causes an increase in the linearity of temporal summation between the potentiated input and a neighboring input (Xu et al. 2006); the temporal specificity of this effect varied with dendritic location. For distal inputs, the increase in linearity is limited to EPSPs arriving within 5 ms of each other, favoring summation of coincident inputs. In contrast, for proximal inputs the increase can be observed for EPSPs arriving within a broader window of 20 ms. Such location-dependent modulation of synaptic integration may interact with the location dependence of the STDP learning rule (see above) to further enrich dendritic processing.

STDP IN VIVO

Whereas most of the early experiments on STDP were conducted in slices and cell cultures, an increasing number of studies have begun to address the functional consequences of STDP in intact nervous systems. Neural circuits in vivo exhibit both spontaneous activity and sensory-evoked responses, modulated by the behavioral states of the animal. Backpropagation of the APs may also be more variable in vivo, as the neurons receive barrages of excitatory and inhibitory inputs (Destexhe et al. 2003). These factors could significantly complicate the rules for synaptic plasticity. How well does the STDP learning rule described in vitro apply to activity-dependent synaptic modification in vivo?

Electrical Stimulation

The first demonstration of STDP in vivo came from a study at the retinotectal projection in the developing *Xenopus* (Zhang et al. 1998). Repetitive electrical stimulation of the retinal ganglion cells within 20 ms before tectal neuron spiking leads to LTP, whereas pairings at negative intervals lead to LTD. Both LTP and LTD are NMDA receptor dependent, and the temporal window is similar to the STDP windows measured in vitro (e.g., Bi & Poo 1998, Froemke & Dan 2002, Tzounopoulos et al. 2004). In addition to the strength of the retinotectal connection, the amplitude of the tectal visual response can also be modified by pairing visual stimulation with postsynaptic spiking (Mu & Poo 2006, Vislay-Meltzer et al. 2006).

Plasticity with similar asymmetric windows has also been demonstrated in the mammalian visual cortex. Optical imaging in the kitten visual cortex showed that pairing visual stimula-tion at a given orientation with cortical electrical stimulation leads to changes in the orientation map (Schuett et al. 2001). Electrical activation after the arrival of the visual input causes expansion of the cortical representation of the paired orientation, whereas the reverse order causes a reduction. Whole-cell recordings in juvenile rat visual cortex showed that pairing visual stimulation with single neuron spiking leads to potentiation or depression of the visual response, depending on the order between the visual inputs and the postsynaptic spiking (Meliza & Dan 2006).

STDP has also been described in other sensory modalities in vivo. In the somatosensory cortex of anesthetized rats, pairing subthreshold whisker deflections with postsynaptic spiking at negative intervals leads to LTD of the paired whisker (Jacob et al. 2007). In an olfactory circuit of the locusts (β-lobe in the mushroom body), pairing odor-induced synaptic activity with postsynaptic spiking results in robust synaptic modifications, with a temporal window similar to those for vertebrate excitatory synapses (**Figure 1a**, I) (Cassenaer & Laurent 2007).

In the motor system, STDP has been demonstrated in human subjects. Pairing electrical stimulation of a somatosensory afferent nerve with transcranial magnetic stimulation (TMS) of the motor cortex leads to long-lasting changes in the motor-evoked potentials (MEPs) elicited by TMS (Wolters et al. 2003). The direction and magnitude of the change depend on the relative timing between the afferent stimulation and the TMS within a window of tens of milliseconds, comparable to the STDP windows measured in vitro. The potentiation induced by pairing at positive intervals can be blocked by NMDA receptor antagonists (Stefan et al. 2002), and the depression at negative intervals is blocked by both NMDA receptor and VDCC antagonists (Wolters et al. 2003), consistent with the pharmacological properties of STDP found in several studies (Bi & Poo 2001). Wolters et al. (2005) also used a similar experimental protocol to demonstrate STDP in human somatosensory cortex.

Paired Sensory Stimulation

Although electrical stimulation affords excellent control of spike timing in the study of STDP, an important question is whether the temporal requirements of this learning rule can be satisfied under natural conditions, as spiking responses to sensory stimuli are known to be highly variable (Shadlen & Newsome 1994). Several studies on the functional role of STDP in vivo have been performed with pure sensory stimulation. In anesthetized adult cats, repetitive presentation of gratings at a pair of orientations induced shifts in orientation tuning of individual V1 neurons; the direction of the shift depended on the temporal order of the two orientations (Yao et al. 2004, Yao & Dan 2001). In a parallel set of experiments in the space domain, repeated visual stimulation in two adjacent retinal regions induced shifts in V1 receptive fields (Fu et al. 2002), with a similar dependence on the stimulus order. In both the orientation and space domain, significant changes in cortical response properties were observed at intervals within ± 40 ms, similar to the STDP windows observed in vitro. For the shift in orientation tuning, the effect showed complete interocular transfer, indicating that the underlying neuronal modifications occur largely in the cortex, after the inputs from the two eyes converge (Yao et al. 2004). Psychophysical experiments in human subjects using analogous induction protocols showed perceptual changes consistent with the electrophysiological effects (Fu et al. 2002, Yao et al. 2004, Yao & Dan 2001), which suggests that the neuronal changes have direct consequences in visual perception.

Motion Stimuli

Compared with the repetitively flashed stimuli used in the above studies, moving stimuli are much more common in nature. Motion stimuli are intrinsically sequential (e.g., an object moving across the visual field should sequentially enter the neuronal receptive fields distributed along its trajectory) and are thus ideally suited for interacting with the STDP learning rule. In the *Xenopus* tadpole, repeated presentation of a moving bar in a given direction selectively potentiated the response to the conditioned direction, resulting in the emergence of direction sensitivity in the tectal neurons (Engert et al. 2002). Induction of direction selectivity through STDP has indeed been predicted in a theoretical study (Rao & Sejnowski 2001a). A follow-up experiment using both sequentially flashed bars and moving bars provided further support for the role of STDP in the induction of direction selectivity (Mu & Poo 2006). The selective enhancement at the conditioned direction manifests as a potentiation of the early phase and a reduction of the late phase of the visual response, consistent with the prediction from STDP. Blocking the cellular signaling pathways underlying STDP abolished the effect of unidirectional motion stimuli in inducing direction selectivity.

In the visual cortex, the interaction between motion stimuli and STDP has been used to predict two receptive field properties and to explain two motion-position illusions. Model simulations predicted that the prevalence of motion stimuli in various directions during visual cortical development would lead to a spatial asymmetry in the direction-selective inputs to each cortical neuron (e.g., inputs preferring rightward motion are biased toward the left side of the receptive field) (Fu et al. 2004). This asymmetry in the mature cortex in turn predicts that (*a*) receptive field position depends on the local motion signals within the test stimuli, and (*b*) motion adaptation causes the receptive field position to shift. Both effects were confirmed experimentally in anesthetized cat V1. Psychophysical measurement using matching stimulus parameters showed that these physiological effects could each explain a known visual illusion involving the interaction between motion and perceived object position (De Valois & De Valois 1991, Nishida & Johnston 1999, Ramachandran & Anstis 1990, Snowden 1998, Whitaker et al. 1999).

In addition to the motion signals in sensory inputs, locomotion of the animal may also induce circuit modification through STDP. The place fields of hippocampal neurons are known to be dynamically modified as the animal navigates in a novel environment. During repeated running of a linear track, the place fields of both CA1 and CA3 cells are initially symmetrical, but they experience a gradual asymmetric expansion against the direction of locomotion (Lee et al. 2004; Mehta et al. 1997, 2000). Simulation with a simple feedforward network model showed that this effect can be explained by STDP (Blum & Abbott 1996, Mehta et al. 2000). In the orientation domain, Yu et al. (2006) recently reported a similar shift in head-direction tuning curves in thalamic head-direction cells as the animal runs in a circular track.

Sensory Deprivation

STDP may also play a role in other forms of experience-dependent plasticity, even if the sensory inputs do not explicitly involve timing on the order of tens of milliseconds. In an experiment measuring the neural activity during sensory deprivation, rats were chronically implanted with electrode arrays to monitor the spiking activity in L4 and L2/3 of the barrel cortex during free-moving behaviors (Celikel et al. 2004). Stimulus deprivation induced by trimming a single whisker, a manipulation known to induce whisker map reorganization, caused an immediate reversal of the firing order and decreased correlation between L4 and L2/3 neurons. Both of these changes are known to drive tLTD in barrel cortical slices (Feldman 2000), thus providing a plausible explanation for deprivation-induced LTD of L4 to L2/3 connections (Allen et al. 2003). In addition to the somatosensory system, sensory deprivation induces circuit reorganization in the visual and auditory systems (Buonomano & Merzenich 1998, Gilbert 1998). It would be interesting to test whether deprivation in these modalities (e.g., monocular deprivation of visual input) also induces changes in the relative spike timing among neurons that could cause the observed circuit modifications through STDP.

FINAL REMARKS

Over the past decade, the STDP learning rule has been demonstrated in a range of species from insects to humans, and our understanding of its cellular mechanisms and functional implications has progressed significantly. However, many questions remain unresolved.

Regarding the mechanism, it remains unclear whether a single model can explain STDP at different synapses or whether different neurons employ distinct molecular machineries to achieve similar outcomes. Studies are only beginning to examine whether and how STDP depends on several signaling events that have been strongly implicated in conventional LTP and LTD, including secretion of brain-derived neurotrophic factor (BDNF) and nitric oxide (Mu & Poo 2006), activation of CaMKII (Tzounopoulos et al. 2007) and phosphatases (Froemke et al. 2005), and modification and insertion/removal of AMPA receptors. It would also be interesting to investigate whether the type of NMDA receptor subunits (NR2A/NR2B) and their synaptic location play a role in STDP (Sjostrom et al. 2003), as has been suggested for LTP/LTD induced by HFS/LFS (Cull-Candy & Leszkiewicz 2004, Liu et al. 2004). In addition, whereas several molecules have been proposed as coincidence detectors at excitatory synapses (**Figure 2**), there is so far no candidate for inhibitory synapses. Furthermore, although postsynaptic Ca^{2+} signals are required for STDP in most cell types, recent imaging experiments showed that volume-averaged Ca^{2+} transients in the dendritic spines are poorly correlated with the direction of synaptic modification (Nevian & Sakmann 2006). Perhaps new techniques that allow measurement of Ca^{2+} signals at a more microscopic scale (e.g., microdomains) will shed new light on the cellular mechanisms of STDP.

To understand the functional consequences of STDP, an important factor to consider is the high level of ongoing activity in vivo. Spontaneous activity can significantly affect membrane potential, conductance, and intracellular

Ca^{2+} levels, and in some cases it can boost AP backpropagation in vivo (Waters & Helmchen 2004). These effects will likely modulate the rules for synaptic modification. Furthermore, spontaneous postsynaptic spiking reduces the persistence of synaptic potentiation and depression (Zhou et al. 2003). An important question is how experience-dependent synaptic modifications can persist in vivo in the face of the ongoing network activity. Recent studies have suggested that sensory-evoked activity patterns can reverberate in subsequent spontaneous activity in early sensory circuits (Galan et al. 2006, Yao et al. 2007) or be replayed in the hippocampus during sleep (Ji & Wilson 2007, Louie & Wilson 2001, Nadasdy et al. 1999, Ribeiro et al. 2004, Wilson & McNaughton 1994). These reactivated patterns may serve to consolidate the transient effects of sensory stimulation into long-lasting circuit modifications. Characterization of neuronal plasticity at the network level during natural behaviors is a crucial step in understanding the neural basis for learning and memory.

DISCLOSURE STATEMENT

The authors are not aware of any biases that might be perceived as affecting the objectivity of this review.

LITERATURE CITED

Abbott LF, Nelson SB. 2000. Synaptic plasticity: taming the beast. *Nat. Neurosci.* 3(Suppl):1178–83

Aizenman CD, Linden DJ. 2000. Rapid, synaptically driven increases in the intrinsic excitability of cerebellar deep nuclear neurons. *Nat. Neurosci.* 3:109–11

Allen CB, Celikel T, Feldman DE. 2003. Long-term depression induced by sensory deprivation during cortical map plasticity in vivo. *Nat. Neurosci.* 6:291–99

Armano S, Rossi P, Taglietti V, D'Angelo E. 2000. Long-term potentiation of intrinsic excitability at the mossy fiber-granule cell synapse of rat cerebellum. *J. Neurosci.* 20:5208–16

Artola A, Brocher S, Singer W. 1990. Different voltage-dependent thresholds for inducing long-term depression and long-term potentiation in slices of rat visual cortex. *Nature* 347:69–72

Artola A, Singer W. 1987. Long-term potentiation and NMDA receptors in rat visual cortex. *Nature* 330:649–52

Artola A, Singer W. 1993. Long-term depression of excitatory synaptic transmission and its relationship to long-term potentiation. *Trends Neurosci.* 16:480–87

Bear MF, Singer W. 1986. Modulation of visual cortical plasticity by acetylcholine and noradrenaline. *Nature* 320:172–76

Bell CC, Han VZ, Sugawara Y, Grant K. 1997. Synaptic plasticity in a cerebellum-like structure depends on temporal order. *Nature* 387:278–81

Bender VA, Bender KJ, Brasier DJ, Feldman DE. 2006. Two coincidence detectors for spike timing-dependent plasticity in somatosensory cortex. *J. Neurosci.* 26:4166–77

Bi G, Poo M. 2001. Synaptic modification by correlated activity: Hebb's postulate revisited. *Annu. Rev. Neurosci.* 24:139–66

Bi GQ, Poo MM. 1998. Synaptic modifications in cultured hippocampal neurons: dependence on spike timing, synaptic strength, and postsynaptic cell type. *J. Neurosci.* 18:10464–72

Birtoli B, Ulrich D. 2004. Firing mode-dependent synaptic plasticity in rat neocortical pyramidal neurons. *J. Neurosci.* 24:4935–40

Bissiere S, Humeau Y, Luthi A. 2003. Dopamine gates LTP induction in lateral amygdala by suppressing feedforward inhibition. *Nat. Neurosci.* 6:587–92

Bliss TV, Collingridge GL. 1993. A synaptic model of memory: long-term potentiation in the hippocampus. *Nature* 361:31–39

Bliss TV, Gardner-Medwin AR. 1973. Long-lasting potentiation of synaptic transmission in the dentate area of the unanaesthetized rabbit following stimulation of the perforant path. *J. Physiol.* 232:357–74

Bliss TV, Lomo T. 1973. Long-lasting potentiation of synaptic transmission in the dentate area of the anaesthetized rabbit following stimulation of the perforant path. *J. Physiol.* 232:331–56

Blum KI, Abbott LF. 1996. A model of spatial map formation in the hippocampus of the rat. *Neural. Comput.* 8:85–93

Boettiger CA, Doupe AJ. 2001. Developmentally restricted synaptic plasticity in a songbird nucleus required for song learning. *Neuron* 31:809–18

Buonomano DV, Merzenich MM. 1998. Cortical plasticity: from synapses to maps. *Annu. Rev. Neurosci.* 21:149–86

Butts DA, Kanold PO, Shatz CJ. 2007. A burst-based "Hebbian" learning rule at retinogeniculate synapses links retinal waves to activity-dependent refinement. *PLoS Biol.* 5:e61

Butts DA, Rokhsar DS. 2001. The information content of spontaneous retinal waves. *J. Neurosci.* 21:961–73

Cassenaer S, Laurent G. 2007. Hebbian STDP in mushroom bodies facilitates the synchronous flow of olfactory information in locusts. *Nature* 448:709–13

Celikel T, Szostak VA, Feldman DE. 2004. Modulation of spike timing by sensory deprivation during induction of cortical map plasticity. *Nat. Neurosci.* 7:534–41

Chapman PF, Kairiss EW, Keenan CL, Brown TH. 1990. Long-term synaptic potentiation in the amygdala. *Synapse* 6:271–78

Chevaleyre V, Takahashi KA, Castillo PE. 2006. Endocannabinoid-mediated synaptic plasticity in the CNS. *Annu. Rev. Neurosci.* 29:37–76

Clugnet MC, LeDoux JE. 1990. Synaptic plasticity in fear conditioning circuits: induction of LTP in the lateral nucleus of the amygdala by stimulation of the medial geniculate body. *J. Neurosci.* 10:2818–24

Couey JJ, Meredith RM, Spijker S, Poorthuis RB, Smit AB, et al. 2007. Distributed network actions by nicotine increase the threshold for spike-timing-dependent plasticity in prefrontal cortex. *Neuron.* 54:73–87

Cull-Candy SG, Leszkiewicz DN. 2004. Role of distinct NMDA receptor subtypes at central synapses. *Sci. STKE* 2004:re16

Daoudal G, Debanne D. 2003. Long-term plasticity of intrinsic excitability: learning rules and mechanisms. *Learn. Mem.* 10:456–65

Debanne D, Gahwiler BH, Thompson SM. 1998. Long-term synaptic plasticity between pairs of individual CA3 pyramidal cells in rat hippocampal slice cultures. *J. Physiol.* 507(Pt. 1):237–47

Destexhe A, Rudolph M, Pare D. 2003. The high-conductance state of neocortical neurons in vivo. *Nat. Rev. Neurosci.* 4:739–51

De Valois RL, De Valois KK. 1991. Vernier acuity with stationary moving Gabors. *Vision Res.* 31:1619–26

Dudek SM, Bear MF. 1993. Bidirectional long-term modification of synaptic effectiveness in the adult and immature hippocampus. *J. Neurosci.* 13:2910–18

Egger V, Feldmeyer D, Sakmann B. 1999. Coincidence detection and changes of synaptic efficacy in spiny stellate neurons in rat barrel cortex. *Nat. Neurosci.* 2:1098–105

Engert F, Tao HW, Zhang LI, Poo MM. 2002. Moving visual stimuli rapidly induce direction sensitivity of developing tectal neurons. *Nature* 419:470–75

Feldman DE. 2000. Timing-based LTP and LTD at vertical inputs to layer II/III pyramidal cells in rat barrel cortex. *Neuron* 27:45–56

Froemke RC, Dan Y. 2002. Spike-timing-dependent synaptic modification induced by natural spike trains. *Nature* 416:433–38

Froemke RC, Poo MM, Dan Y. 2005. Spike-timing-dependent synaptic plasticity depends on dendritic location. *Nature* 434:221–25

Froemke RC, Tsay IA, Raad M, Long JD, Dan Y. 2006. Contribution of individual spikes in burst-induced long-term synaptic modification. *J. Neurophysiol.* 95:1620–29

Fu YX, Djupsund K, Gao H, Hayden B, Shen K, Dan Y. 2002. Temporal specificity in the cortical plasticity of visual space representation. *Science* 296:1999–2003

Fu YX, Shen Y, Gao H, Dan Y. 2004. Asymmetry in visual cortical circuits underlying motion-induced perceptual mislocalization. *J. Neurosci.* 24:2165–71

Galan RF, Weidert M, Menzel R, Herz AV, Galizia CG. 2006. Sensory memory for odors is encoded in spontaneous correlated activity between olfactory glomeruli. *Neural. Comput.* 18:10–25

Ganguly K, Kiss L, Poo M. 2000. Enhancement of presynaptic neuronal excitability by correlated presynaptic and postsynaptic spiking. *Nat. Neurosci.* 3:1018–26

Gilbert CD. 1998. Adult cortical dynamics. *Physiol. Rev.* 78:467–85

Golding NL, Staff NP, Spruston N. 2002. Dendritic spikes as a mechanism for cooperative long-term potentiation. *Nature* 418:326–31

Gulledge AT, Stuart GJ. 2003. Excitatory actions of GABA in the cortex. *Neuron* 37:299–309

Haas JS, Nowotny T, Abarbanel HD. 2006. Spike-timing-dependent plasticity of inhibitory synapses in the entorhinal cortex. *J. Neurophysiol.* 96:3305–13

Han VZ, Grant K, Bell CC. 2000. Reversible associative depression and nonassociative potentiation at a parallel fiber synapse. *Neuron* 27:611–22

Hashimotodani Y, Ohno-Shosaku T, Tsubokawa H, Ogata H, Emoto K, et al. 2005. Phospholipase Cbeta serves as a coincidence detector through its Ca^{2+} dependency for triggering retrograde endocannabinoid signal. *Neuron* 45:257–68

Hashimotodani Y, Ohno-Shosaku T, Watanabe M, Kano M. 2007. Roles of phospholipase C{beta} and NMDA receptor in activity-dependent endocannabinoid release. *J. Physiol.* 584:373–80

Hausser M, Mel B. 2003. Dendrites: bug or feature? *Curr. Opin. Neurobiol.* 13:372–83

Hebb DO. 1949. *The Organization of Behavior; A Neuropsychological Theory*. New York: Wiley. xix, 335 pp.

Hensch TK. 2005. Critical period plasticity in local cortical circuits. *Nat. Rev. Neurosci.* 6:877–88

Hirsch JC, Barrionuevo G, Crepel F. 1992. Homo- and heterosynaptic changes in efficacy are expressed in prefrontal neurons: an in vitro study in the rat. *Synapse* 12:82–85

Hoffman DA, Johnston D. 1998. Downregulation of transient K^+ channels in dendrites of hippocampal CA1 pyramidal neurons by activation of PKA and PKC. *J. Neurosci.* 18:3521–28

Hoffman DA, Johnston D. 1999. Neuromodulation of dendritic action potentials. *J. Neurophysiol.* 81:408–11

Hoffman DA, Magee JC, Colbert CM, Johnston D. 1997. K^+ channel regulation of signal propagation in dendrites of hippocampal pyramidal neurons. *Nature* 387:869–75

Holmgren CD, Zilberter Y. 2001. Coincident spiking activity induces long-term changes in inhibition of neocortical pyramidal cells. *J. Neurosci.* 21:8270–77

Humeau Y, Luthi A. 2007. Dendritic calcium spikes induce bi-directional synaptic plasticity in the lateral amygdala. *Neuropharmacology* 52:234–43

Iriki A, Pavlides C, Keller A, Asanuma H. 1989. Long-term potentiation in the motor cortex. *Science* 245:1385–87

Ito M. 2005. Bases and implications of learning in the cerebellum—adaptive control and internal model mechanism. *Prog. Brain Res.* 148:95–109

Jacob V, Brasier DJ, Erchova I, Feldman D, Shulz DE. 2007. Spike timing-dependent synaptic depression in the in vivo barrel cortex of the rat. *J. Neurosci.* 27:1271–84

Ji D, Wilson MA. 2007. Coordinated memory replay in the visual cortex and hippocampus during sleep. *Nat. Neurosci.* 10:100–7

Johnston D, Hoffman DA, Colbert CM, Magee JC. 1999. Regulation of back-propagating action potentials in hippocampal neurons. *Curr. Opin. Neurobiol.* 9:288–92

Kampa BM, Clements J, Jonas P, Stuart GJ. 2004. Kinetics of Mg^{2+} unblock of NMDA receptors: implications for spike-timing dependent synaptic plasticity. *J. Physiol.* 556:337–45

Kampa BM, Letzkus JJ, Stuart GJ. 2006. Requirement of dendritic calcium spikes for induction of spike-timing-dependent synaptic plasticity. *J. Physiol.* 574:283–90

Karmarkar UR, Buonomano DV. 2002. A model of spike-timing dependent plasticity: one or two coincidence detectors? *J. Neurophysiol.* 88:507–13

Karmarkar UR, Dan Y. 2006. Experience-dependent plasticity in adult visual cortex. *Neuron* 52:577–85

Katz LC, Shatz CJ. 1996. Synaptic activity and the construction of cortical circuits. *Science* 274:1133–38

Kilgard MP, Merzenich MM. 1998. Cortical map reorganization enabled by nucleus basalis activity. *Science* 279:1714–18

Kirkwood A, Bear MF. 1994. Hebbian synapses in visual cortex. *J. Neurosci.* 14:1634–45

Kobayashi K, Poo MM. 2004. Spike train timing-dependent associative modification of hippocampal CA3 recurrent synapses by mossy fibers. *Neuron* 41:445–54

Koester HJ, Sakmann B. 1998. Calcium dynamics in single spines during coincident pre- and postsynaptic activity depend on relative timing of back-propagating action potentials and subthreshold excitatory postsynaptic potentials. *Proc. Natl. Acad. Sci. USA* 95:9596–601

Larkum ME, Kaiser KM, Sakmann B. 1999a. Calcium electrogenesis in distal apical dendrites of layer 5 pyramidal cells at a critical frequency of back-propagating action potentials. *Proc. Natl. Acad. Sci. USA* 96:14600–4

Larkum ME, Zhu JJ, Sakmann B. 1999b. A new cellular mechanism for coupling inputs arriving at different cortical layers. *Nature* 398:338–41

Lee I, Rao G, Knierim JJ. 2004. A double dissociation between hippocampal subfields: differential time course of CA3 and CA1 place cells for processing changed environments. *Neuron* 42:803–15

Lester RA, Clements JD, Westbrook GL, Jahr CE. 1990. Channel kinetics determine the time course of NMDA receptor-mediated synaptic currents. *Nature* 346:565–67

Letzkus JJ, Kampa BM, Stuart GJ. 2006. Learning rules for spike timing-dependent plasticity depend on dendritic synapse location. *J. Neurosci.* 26:10420–29

Levy WB, Steward O. 1983. Temporal contiguity requirements for long-term associative potentiation/depression in the hippocampus. *Neuroscience* 8:791–97

Li CY, Lu JT, Wu CP, Duan SM, Poo MM. 2004. Bidirectional modification of presynaptic neuronal excitability accompanying spike timing-dependent synaptic plasticity. *Neuron* 41:257–68

Lin YW, Min MY, Chiu TH, Yang HW. 2003. Enhancement of associative long-term potentiation by activation of beta-adrenergic receptors at CA1 synapses in rat hippocampal slices. *J. Neurosci.* 23:4173–81

Linden DJ, Connor JA. 1995. Long-term synaptic depression. *Annu. Rev. Neurosci.* 18:319–57

Lisman J. 1989. A mechanism for the Hebb and the anti-Hebb processes underlying learning and memory. *Proc. Natl. Acad. Sci. USA* 86:9574–78

Liu L, Wong TP, Pozza MF, Lingenhoehl K, Wang Y, et al. 2004. Role of NMDA receptor subtypes in governing the direction of hippocampal synaptic plasticity. *Science* 304:1021–24

Liu QS, Pu L, Poo MM. 2005. Repeated cocaine exposure in vivo facilitates LTP induction in midbrain dopamine neurons. *Nature* 437:1027–31

Louie K, Wilson MA. 2001. Temporally structured replay of awake hippocampal ensemble activity during rapid eye movement sleep. *Neuron* 29:145–56

Maejima T, Oka S, Hashimotodani Y, Ohno-Shosaku T, Aiba A, et al. 2005. Synaptically driven endocannabinoid release requires Ca^{2+}-assisted metabotropic glutamate receptor subtype 1 to phospholipase Cbeta4 signaling cascade in the cerebellum. *J. Neurosci.* 25:6826–35

Magee JC, Johnston D. 1997. A synaptically controlled, associative signal for Hebbian plasticity in hippocampal neurons. *Science* 275:209–13

Magee JC, Johnston D. 2005. Plasticity of dendritic function. *Curr. Opin. Neurobiol.* 15:334–42

Malenka RC, Bear MF. 2004. LTP and LTD: an embarrassment of riches. *Neuron* 44:5–21

Markram H, Lubke J, Frotscher M, Sakmann B. 1997. Regulation of synaptic efficacy by coincidence of postsynaptic APs and EPSPs. *Science* 275:213–15

Martin SJ, Grimwood PD, Morris RG. 2000. Synaptic plasticity and memory: an evaluation of the hypothesis. *Annu. Rev. Neurosci.* 23:649–711

Mayer ML, Westbrook GL, Guthrie PB. 1984. Voltage-dependent block by Mg^{2+} of NMDA responses in spinal cord neurones. *Nature* 309:261–63

Mehta MR, Barnes CA, McNaughton BL. 1997. Experience-dependent, asymmetric expansion of hippocampal place fields. *Proc. Natl. Acad. Sci. USA* 94:8918–21

Mehta MR, Quirk MC, Wilson MA. 2000. Experience-dependent asymmetric shape of hippocampal receptive fields. *Neuron* 25:707–15

Meliza CD, Dan Y. 2006. Receptive-field modification in rat visual cortex induced by paired visual stimulation and single-cell spiking. *Neuron* 49:183–89

Meredith RM, Floyer-Lea AM, Paulsen O. 2003. Maturation of long-term potentiation induction rules in rodent hippocampus: role of GABAergic inhibition. *J. Neurosci.* 23:11142–46

Migliore M, Shepherd GM. 2002. Emerging rules for the distributions of active dendritic conductances. *Nat. Rev. Neurosci.* 3:362–70

Miles R, Toth K, Gulyas AI, Hajos N, Freund TF. 1996. Differences between somatic and dendritic inhibition in the hippocampus. *Neuron* 16:815–23

Mu Y, Poo MM. 2006. Spike timing-dependent LTP/LTD mediates visual experience-dependent plasticity in a developing retinotectal system. *Neuron* 50:115–25

Mulkey RM, Malenka RC. 1992. Mechanisms underlying induction of homosynaptic long-term depression in area CA1 of the hippocampus. *Neuron* 9:967–75

Nadasdy Z, Hirase H, Czurko A, Csicsvari J, Buzsaki G. 1999. Replay and time compression of recurring spike sequences in the hippocampus. *J. Neurosci.* 19:9497–507

Nevian T, Sakmann B. 2004. Single spine Ca^{2+} signals evoked by coincident EPSPs and back-propagating action potentials in spiny stellate cells of layer 4 in the juvenile rat somatosensory barrel cortex. *J. Neurosci.* 24:1689–99

Nevian T, Sakmann B. 2006. Spine Ca^{2+} signaling in spike-timing-dependent plasticity. *J. Neurosci.* 26:11001–13

Nishida S, Johnston A. 1999. Influence of motion signals on the perceived position of spatial pattern. *Nature* 397:610–12

Nishiyama M, Hong K, Mikoshiba K, Poo MM, Kato K. 2000. Calcium stores regulate the polarity and input specificity of synaptic modification. *Nature* 408:584–88

Nowak L, Bregestovski P, Ascher P, Herbet A, Prochiantz A. 1984. Magnesium gates glutamate-activated channels in mouse central neurones. *Nature* 307:462–65

Pike FG, Meredith RM, Olding AW, Paulsen O. 1999. Rapid report: postsynaptic bursting is essential for "Hebbian" induction of associative long-term potentiation at excitatory synapses in rat hippocampus. *J. Physiol.* 518(Pt. 2):571–76

Pu L, Liu QS, Poo MM. 2006. BDNF-dependent synaptic sensitization in midbrain dopamine neurons after cocaine withdrawal. *Nat. Neurosci.* 9:605–7

Rall W. 1967. Distinguishing theoretical synaptic potentials computed for different soma-dendritic distributions of synaptic input. *J. Neurophysiol.* 30:1138–68

Ramachandran VS, Anstis SM. 1990. Illusory displacement of equiluminous kinetic edges. *Perception* 19:611–16

Rao RP, Sejnowski TJ. 2001a. Predictive learning of temporal sequences in recurrent neocortical circuits. *Novartis Found Symp.* 239:208–29; discussion 29–40

Rao RP, Sejnowski TJ. 2001b. Spike-timing-dependent Hebbian plasticity as temporal difference learning. *Neural Comput.* 13:2221–37

Ribeiro S, Gervasoni D, Soares ES, Zhou Y, Lin SC, et al. 2004. Long-lasting novelty-induced neuronal reverberation during slow-wave sleep in multiple forebrain areas. *PLoS Biol.* 2:E24

Rosenmund C, Feltz A, Westbrook GL. 1995. Calcium-dependent inactivation of synaptic NMDA receptors in hippocampal neurons. *J. Neurophysiol.* 73:427–30

Schuett S, Bonhoeffer T, Hubener M. 2001. Pairing-induced changes of orientation maps in cat visual cortex. *Neuron* 32:325–37

Seol GH, Ziburkus J, Huang S, Song L, Kim IT, et al. 2007. Neuromodulators control the polarity of spike-timing-dependent synaptic plasticity. *Neuron* 55:919–29

Shadlen MN, Newsome WT. 1994. Noise, neural codes and cortical organization. *Curr. Opin. Neurobiol.* 4:569–79

Shouval HZ, Bear MF, Cooper LN. 2002. A unified model of NMDA receptor-dependent bidirectional synaptic plasticity. *Proc. Natl. Acad. Sci. USA* 99:10831–36

Shu Y, Hasenstaub A, Badoual M, Bal T, McCormick DA. 2003. Barrages of synaptic activity control the gain and sensitivity of cortical neurons. *J. Neurosci.* 23:10388–401

Siegelbaum SA, Kandel ER. 1991. Learning-related synaptic plasticity: LTP and LTD. *Curr. Opin. Neurobiol.* 1:113–20

Sjostrom PJ, Hausser M. 2006. A cooperative switch determines the sign of synaptic plasticity in distal dendrites of neocortical pyramidal neurons. *Neuron* 51:227–38

Sjostrom PJ, Turrigiano GG, Nelson SB. 2001. Rate, timing, and cooperativity jointly determine cortical synaptic plasticity. *Neuron* 32:1149–64

Sjostrom PJ, Turrigiano GG, Nelson SB. 2003. Neocortical LTD via coincident activation of presynaptic NMDA and cannabinoid receptors. *Neuron* 39:641–54

Snowden RJ. 1998. Shifts in perceived position following adaptation to visual motion. *Curr. Biol.* 8:1343–45

Softky WR, Koch C. 1993. The highly irregular firing of cortical cells is inconsistent with temporal integration of random EPSPs. *J. Neurosci.* 13:334–50

Spruston N, Schiller Y, Stuart G, Sakmann B. 1995. Activity-dependent action potential invasion and calcium influx into hippocampal CA1 dendrites. *Science* 268:297–300

Stanton PK, Sejnowski TJ. 1989. Associative long-term depression in the hippocampus induced by Hebbian covariance. *Nature* 339:215–18

Stefan K, Kunesch E, Benecke R, Cohen LG, Classen J. 2002. Mechanisms of enhancement of human motor cortex excitability induced by interventional paired associative stimulation. *J. Physiol.* 543:699–708

Stent GS. 1973. A physiological mechanism for Hebb's postulate of learning. *Proc. Natl. Acad. Sci. USA* 70:997–1001

Stuart G, Schiller J, Sakmann B. 1997a. Action potential initiation and propagation in rat neocortical pyramidal neurons. *J. Physiol.* 505(Pt. 3):617–32

Stuart G, Spruston N, Sakmann B, Hausser M. 1997b. Action potential initiation and backpropagation in neurons of the mammalian CNS. *Trends Neurosci* 20:125–31

Stuart GJ, Hausser M. 2001. Dendritic coincidence detection of EPSPs and action potentials. *Nat. Neurosci.* 4:63–71

Stuart GJ, Sakmann B. 1994. Active propagation of somatic action potentials into neocortical pyramidal cell dendrites. *Nature* 367:69–72

Tong G, Shepherd D, Jahr CE. 1995. Synaptic desensitization of NMDA receptors by calcineurin. *Science* 267:1510–12

Tsubokawa H, Ross WN. 1996. IPSPs modulate spike backpropagation and associated $[Ca^{2+}]i$ changes in the dendrites of hippocampal CA1 pyramidal neurons. *J. Neurophysiol.* 76:2896–906

Tsubokawa H, Ross WN. 1997. Muscarinic modulation of spike backpropagation in the apical dendrites of hippocampal CA1 pyramidal neurons. *J. Neurosci.* 17:5782–91

Tzounopoulos T, Kim Y, Oertel D, Trussell LO. 2004. Cell-specific, spike timing-dependent plasticities in the dorsal cochlear nucleus. *Nat. Neurosci.* 7:719–25

Tzounopoulos T, Rubio ME, Keen JE, Trussell LO. 2007. Coactivation of pre- and postsynaptic signaling mechanisms determines cell-specific spike-timing-dependent plasticity. *Neuron* 54:291–301

Vislay-Meltzer RL, Kampff AR, Engert F. 2006. Spatiotemporal specificity of neuronal activity directs the modification of receptive fields in the developing retinotectal system. *Neuron* 50:101–14

Wang HX, Gerkin RC, Nauen DW, Bi GQ. 2005. Coactivation and timing-dependent integration of synaptic potentiation and depression. *Nat. Neurosci.* 8:187–93

Wang Z, Xu NL, Wu CP, Duan S, Poo MM. 2003. Bidirectional changes in spatial dendritic integration accompanying long-term synaptic modifications. *Neuron* 37:463–72

Watanabe S, Hoffman DA, Migliore M, Johnston D. 2002. Dendritic K^+ channels contribute to spike-timing dependent long-term potentiation in hippocampal pyramidal neurons. *Proc. Natl. Acad. Sci. USA* 99:8366–71

Waters J, Helmchen F. 2004. Boosting of action potential backpropagation by neocortical network activity in vivo. *J. Neurosci.* 24:11127–36

Waters J, Schaefer A, Sakmann B. 2005. Backpropagating action potentials in neurones: measurement, mechanisms and potential functions. *Prog. Biophys. Mol. Biol.* 87:145–70

Whitaker D, McGraw PV, Pearson S. 1999. Non-veridical size perception of expanding and contracting objects. *Vision Res.* 39:2999–3009

Williams SR, Stuart GJ. 2000. Backpropagation of physiological spike trains in neocortical pyramidal neurons: implications for temporal coding in dendrites. *J. Neurosci.* 20:8238–46

Wilson MA, McNaughton BL. 1994. Reactivation of hippocampal ensemble memories during sleep. *Science* 265:676–79

Wittenberg GM, Wang SS. 2006. Malleability of spike-timing-dependent plasticity at the CA3-CA1 synapse. *J. Neurosci.* 26:6610–17

Wolters A, Sandbrink F, Schlottmann A, Kunesch E, Stefan K, et al. 2003. A temporally asymmetric Hebbian rule governing plasticity in the human motor cortex. *J. Neurophysiol.* 89:2339–45

Wolters A, Schmidt A, Schramm A, Zeller D, Naumann M, et al. 2005. Timing-dependent plasticity in human primary somatosensory cortex. *J. Physiol.* 565:1039–52

Woodin MA, Ganguly K, Poo MM. 2003. Coincident pre- and postsynaptic activity modifies GABAergic synapses by postsynaptic changes in Cl- transporter activity. *Neuron* 39:807–20

Xu NL, Ye CQ, Poo MM, Zhang XH. 2006. Coincidence detection of synaptic inputs is facilitated at the distal dendrites after long-term potentiation induction. *J. Neurosci.* 26:3002–9

Yang SN, Tang YG, Zucker RS. 1999. Selective induction of LTP and LTD by postsynaptic $[Ca^{2+}]i$ elevation. *J. Neurophysiol.* 81:781–87

Yao H, Dan Y. 2001. Stimulus timing-dependent plasticity in cortical processing of orientation. *Neuron* 32:315–23

Yao H, Shen Y, Dan Y. 2004. Intracortical mechanism of stimulus-timing-dependent plasticity in visual cortical orientation tuning. *Proc. Natl. Acad. Sci. USA* 101:5081–86

Yao H, Shi L, Han F, Gao H, Dan Y. 2007. Rapid learning in cortical coding of visual scenes. *Nat. Neurosci.* 10:772–78

Yu X, Yoganarasimha D, Knierim JJ. 2006. Backward shift of head direction tuning curves of the anterior thalamus: comparison with CA1 place fields. *Neuron* 52:717–29

Zhang LI, Tao HW, Holt CE, Harris WA, Poo M. 1998. A critical window for cooperation and competition among developing retinotectal synapses. *Nature* 395:37–44

Zhang W, Linden DJ. 2003. The other side of the engram: experience-driven changes in neuronal intrinsic excitability. *Nat. Rev. Neurosci.* 4:885–900

Zhou Q, Tao HW, Poo MM. 2003. Reversal and stabilization of synaptic modifications in a developing visual system. *Science* 300:1953–57

Zucker RS, Regehr WG. 2002. Short-term synaptic plasticity. *Annu. Rev. Physiol.* 64:355–405

Balancing Structure and Function at Hippocampal Dendritic Spines

Jennifer N. Bourne and Kristen M. Harris

Center for Learning and Memory, Department of Neurobiology, University of Texas, Austin, Texas 78712-0805; email: jbourne@mail.clm.utexas.edu, kharris@mail.clm.utexas.edu

Annu. Rev. Neurosci. 2008. 31:47–67

First published online as a Review in Advance on February 19, 2008

The *Annual Review of Neuroscience* is online at neuro.annualreviews.org

This article's doi: 10.1146/annurev.neuro.31.060407.125646

0147-006X/08/0721-0047$20.00

Key Words

serial section transmission electron microscopy, long-term potentiation, long-term depression, development, morphological plasticity

Abstract

Dendritic spines are the primary recipients of excitatory input in the central nervous system. They provide biochemical compartments that locally control the signaling mechanisms at individual synapses. Hippocampal spines show structural plasticity as the basis for the physiological changes in synaptic efficacy that underlie learning and memory. Spine structure is regulated by molecular mechanisms that are fine-tuned and adjusted according to developmental age, level and direction of synaptic activity, specific brain region, and exact behavioral or experimental conditions. Reciprocal changes between the structure and function of spines impact both local and global integration of signals within dendrites. Advances in imaging and computing technologies may provide the resources needed to reconstruct entire neural circuits. Key to this endeavor is having sufficient resolution to determine the extrinsic factors (such as perisynaptic astroglia) and the intrinsic factors (such as core subcellular organelles) that are required to build and maintain synapses.

Contents

INTRODUCTION

Since Golgi and Cajal first revealed the intricate structure of dendrites more than 100 years ago, scientists have pondered several questions: Why are dendritic spines distributed nonuniformly along dendrites? Why do dendrites become grossly distorted among individuals with severe neuropathology and mental retardation? Is the number of spines limited by size? Does the number reach saturation? Do more or less spiny dendrites have a greater capacity for plasticity? Which intrinsic and extrinsic features control dendritic plasticity or allow for homeostatic regulation? As protrusions with diverse lengths and shapes, spines allow more connections to form in a compact neuropil. A constricted neck compartmentalizes molecular signals in the spine head and imparts synapse specificity, promotes plasticity, and protects the parent dendrite from excitotoxicity. Spine shape can reflect different inputs in some brain regions such as the lateral nucleus of the amygdala, where cortical inputs synapse on thin spines and thalamic inputs synapse on mushroom spines (Humeau et al. 2005). Conversely, both thin and mushroom spines can synapse with the same CA3 inputs in the hippocampus (Harris & Stevens 1989). Furthermore, cerebellar Purkinje cell spines appear club-shaped even without synaptic input (Cesa & Strata 2005). Live imaging with two-photon microscopy has revealed rapid, activity-dependent spine turnover common during development, but as an animal matures more spines stabilize (Alvarez & Sabatini 2007). This form of imaging also reveals dynamic changes in the shapes of individual spines but is not of sufficient resolution to measure dimensions, count numbers, determine local subcellular or molecular composition, or identify exactly where synapses occur. Electron microscopy is needed to reveal these features (Harris et al. 2006, Rostaing et al. 2006, Masugi-Tokita & Shigemoto 2007). New approaches to combine light and electron microscopy are promising (Zito et al. 1999, Knott et al. 2006, Nagerl et al. 2007), although refinement is needed because the reaction products used to track the dendrites often obscure synapses and subcellular organelles.

This review concentrates on hippocampal dendritic spines. Spatial training (Moser et al. 1997) and exposure to enriched environments (Kozorovitskiy et al. 2005) alter hippocampal spine numbers. Long-term potentiation (LTP) alters spine number, shape, and subcellular composition in both the immature (Maletic-Savatic et al. 1999, Engert & Bonhoeffer 1999, Ostroff et al. 2002, Lang et al. 2004, Matsuzaki et al. 2004, Kopec et al. 2006, Nagerl et al. 2007) and the mature hippocampus (Van Harreveld & Fifkova 1975, Trommald et al. 1996, Popov et al. 2004, Stewart et al. 2005, Bourne et al. 2007b). Conversely, long-term depression (LTD) decreases spine number and size (Chen et al. 2004, Nagerl et al. 2004, Zhou et al. 2004). Structural spine plasticity in the hippocampus involves a change in the size and composition of the postsynaptic density (PSD); assembly and disassembly of actin filaments; exocytosis and endocytosis of glutamate

---

Thin spines: spines that have constricted necks and small heads

Mushroom spines: spines with constricted necks and heads exceeding 0.6 microns in diameter

LTP: long-term potentiation

LTD: long-term depression

PSD: postsynaptic density

receptors and ion channels; regulation of local protein synthesis by redistribution of polyribosomes and proteasomes; dynamic repositioning of smooth endoplasmic reticulum (SER) and mitochondria; and metabolic and structural interactions between spines and perisynaptic astroglia. The extent and type of structural change depend partly on experimental methods, developmental age, and regional differences in synaptic organization. This review discusses factors that regulate spine structure and function during hippocampal synaptogenesis and plasticity (**Table 1**).

STRUCTURE AND COMPOSITION OF DENDRITIC SPINES

In the hippocampus, spines vary greatly in size and shape even along short dendritic segments (**Figure 1**). Most spines have constricted necks and are either mushroom shaped with heads exceeding 0.6 microns in diameter or thin shaped with smaller heads (Harris et al. 1992). Other spines are stubby protrusions with head widths equal to neck lengths, branched protrusions with two or more heads, or single protrusions with multiple synapses along the head and neck. These features provide measurably distinct shape categories (**Figure 1a**) that might reflect functional histories of the spines. Mushroom spines have larger, more complex PSDs (Harris et al. 1992) with a higher density of glutamate receptors (Matsuzaki et al. 2001, Nicholson et al. 2006). Larger spines are more likely to have SER (Spacek & Harris 1997), polyribosomes (Ostroff et al. 2002, Bourne et al. 2007b), endosomal compartments (Cooney et al. 2002, Park et al. 2006), and perisynaptic astroglia (Witcher et al. 2007). These features suggest that larger spines are functionally stronger in their response to glutamate, local regulation of intracellular calcium, endosomal recycling, protein translation and degradation, and interaction with astroglia. Smaller spines may be more flexible, rapidly enlarging or shrinking in response to subsequent activation (Bourne & Harris 2007).

Postsynaptic Density

Spine heads provide a local biochemical compartment where ions and signaling molecules become concentrated following synaptic activation. The PSD is an electron-dense thickening on spine heads that is apposed to the presynaptic active zone. The PSD contains hundreds of proteins including NMDA (N-methyl-d-aspartate), AMPA (α-amino-3-hydroxyl-5-methyl-4-isoxazole-propionate), and metabotropic glutamate receptors; scaffolding proteins such as PSD-95; and signaling proteins such as calcium/calmodulin-dependent kinase II (CamKII) (Okabe 2007). The PSD surfaces vary from small discs to large irregular shapes that can be perforated by electron lucent regions. Differences in PSD dimensions can reflect distance-dependent differences in dendritic function (Magee & Johnston 2005). Relatively more of the distal synapses on CA1 pyramidal cells have perforated synapses; however, perforated synapses associated with the distal input of entorhinal cortex host a lower density of AMPA receptors than do perforated synapses at proximal CA3 input of the same CA1 cells (Nicholson et al. 2006). PSDs appear larger and are more likely to have perforations shortly after the induction of LTP (Geinisman et al. 1991, Toni et al. 1999, Mezey et al. 2004, Popov et al. 2004, Dhanrajan et al. 2004, Stewart et al. 2005), consistent with the idea that perforations are transient structural perturbations responding to activation (Lisman & Harris 1994, Sorra et al. 1998, Fiala et al. 2002, Spacek & Harris 2004). Larger spines with more AMPA and NMDA receptors in the PSD are more sensitive to glutamate (Takumi et al. 1999a,b; Matsuzaki et al. 2001). Small "silent" spine synapses contain only NMDA receptors, and LTP activates them with exocytic insertion of AMPA receptors (Isaac et al. 1995, Liao et al. 1995, Liao et al. 1999, Petralia et al. 1999, Lu et al. 2001, Park et al. 2004, Kopec et al. 2006). AMPA receptors must be constitutively exchanged to sustain the newly active spines; fortunately, lateral diffusion of AMPA receptors out of a spine is limited by the constricted

SER: smooth endoplasmic reticulum

Stubby spines: spines that have head widths equal to the neck length

NMDA: N-methyl-d-aspartate, glutamate receptor

AMPA: α-amino-3-hydroxyl-5-methyl-4-isoxazole-propionate, glutamate receptor

CamKII: calcium/calmodulin-dependent kinase II

Perforated synapse: PSD surface is irregularly shaped with electron lucent region(s) dividing it

Table 1 Molecular mediators of spine morphology

Protein	Function	References
PSD-95	Stabilizes nascent spines and anchors receptors and scaffolding proteins at the synapse.	Ehrlich et al. 2007, Marrs et al. 2001, Okabe et al. 2001
CamKII	Increases the thickness of the PSD and phosphorylates signaling molecules involved in plasticity.	Aakalu et al. 2001; Havik et al. 2003; Kennedy et al. 1983, 1990; Liao et al. 1995; Lledo et al. 1995; Martone et al. 1996; McGlade-McCulloh et al. 1993; Ouyang et al. 1997, 1999; Pettit et al. 1994
Actin	Regulates the extension of filopodia and mediates the expansion of spine heads with LTP and the shrinkage of spine heads with LTD.	Chen et al. 2004, Fukazawa et al. 2003, Matus 2000, Kim & Lisman 1999, Krucker et al. 2000, Lin et al. 2005, Nagerl et al. 2004, Ouyang et al. 2005, Star et al. 2002, Zhou et al. 2004
Profilin	Promotes activity-dependent actin polymerization and stabilizes actin.	Ackermann & Matus 2003, Ethell & Pasquale 2005, Tada & Sheng 2006
Cofilin	Depolymerizes actin filaments, but LTP or learning-induced phosphorylation decreases its affinity for actin, promoting polymerization and spine enlargement.	Chen et al. 2007, Fedulov et al. 2007
Rap1/AF-6	Elongates spines and removes AMPA receptors with activation, whereas inactivation enlarges spines and recruits AMPA receptors.	Xie et al. 2005, Zhu et al. 2002
Myosin IIb	Stabilizes mushroom spines.	Ryu et al. 2006
Myosin VI	Regulates clathrin-mediated endocytosis of AMPA receptors.	Osterweil et al. 2005
Synaptopodin	Binds to the spine apparatus and may mediate interactions between the actin cytoskeleton and calcium signaling. Synaptopodin-deficient mice have normal spine morphology and density, but all spines lack a spine apparatus.	Deller et al. 2007
Telencephalin	Slows the development of dendritic spines with overexpression, whereas deletion accelerates the spine development, suggesting a role in maintaining filopodia during development.	Matsuno et al. 2006
SynGAP	Maintains filopodia during development and localizes to the synapse to negatively regulate Ras signaling pathways, which promote spine formation and growth.	Chen et al. 1998, Kim et al. 1998, Krapivinsky et al. 2004, Oh et al. 2004, Vazquez et al. 2004
miR-134	Negatively regulates spine development by inhibiting translation of Limk1. Overexpression of miR-134 results in a decrease of spine volume.	Schratt et al. 2006
N-cadherin	Stabilizes mature synapses and regulates spine morphology and synaptic efficacy.	Abe et al. 2004, Bozdagi et al. 2000, Kosik et al. 2005, Nuriya & Huganir 2006, Tai et al. 2007, Togashi et al. 2002
EphB/EphrinB	Clusters receptors and mediates spine morphology by recruiting molecules involved in actin polymerization.	Contractor et al. 2002, Dalva et al. 2000, Irie & Yamaguchi 2004, Grunwald et al. 2004, Penzes et al. 2003
EphA/EphrinA	Regulates neuro-glial signaling and induces the retraction of spines. Expression decreases during development and is inactive in mature brains, suggesting a potential role in synaptic pruning.	Allen & Barres 2005, Grunwald et al. 2004, Murai et al. 2003

spine neck (Adesnik et al. 2005, Ashby et al. 2006). AMPA receptors can also be actively removed via endocytosis during LTD (Beattie et al. 2000, Man et al. 2000, Snyder et al. 2001, Xiao et al. 2001, Lee et al. 2002, Brown et al. 2005). Both exo- and endocytic processes alter spine shape. Because the PSD's size is well correlated with spine head volume and the number of presynaptic vesicles (Harris & Stevens 1989, Harris et al. 1992), there is likely a trans-synaptic mechanism to coordinate them during plasticity (Lisman & Harris 1993, Spacek & Harris 2004).

Actin Cytoskeleton

Spine formation and morphology are regulated by actin filaments (Matus 2000, Zito et al. 2004). Filamentous actin (F-actin) forms organized bundles in spine necks, and altered polymerization-depolymerization states accompany changes in head shapes (Star et al. 2002). Induction of LTP briefly depolymerizes actin filaments (Ouyang et al. 2005), whereas maintenance of LTP and sustained spine enlargement require polymerization of F-actin (Kim & Lisman 1999, Krucker et al. 2000, Fukazawa et al. 2003, Lin et al. 2005). In contrast, LTD results in the depolymerization of actin and spine elongation or shrinkage of spine heads (Chen et al. 2004, Nagerl et al. 2004, Zhou et al. 2004). The actin cytoskeleton is regulated by actin-binding proteins (Ethell & Pasquale 2005, Tada & Sheng 2006). Profilin is a promoter of actin polymerization that could facilitate LTP-induced actin assembly and spine enlargement (Ackermann & Matus 2003). Cofilin is an actin-binding protein that causes actin depolymerization; induction of LTP or exposure to enriched environments causes phosphorylation-mediated inhibition of cofilin and promotes spine enlargement (Chen et al. 2007, Fedulov et al. 2007). Rap1 is an actin-binding protein that localizes AF-6 to the synaptic membrane, where it induces rearrangement of actin filaments and promotes removal of AMPA receptors (Xie et al. 2005) and spine elongation, a morphological correlate of

LTD (Zhu et al. 2002). Conversely, inactivation of Rap1 releases AF-6 from the synaptic membrane to regulate a different pool of actin filaments that promote recruitment of AMPA receptors to the synapse and spine enlargement with LTP (Xie et al. 2005). Myosins IIb and VI are motor proteins enriched in the PSD that translocate along, and regulate contractility of, actin filaments and spine shape (Osterweil et al. 2005, Ryu et al. 2006). Myosin VI–deficient spines have disrupted clathrin-mediated endocytosis of AMPA receptors, suggesting a role in LTD (Osterweil et al. 2005).

Recycling Endosomes

LTP requires exocytosis-mediated insertion of AMPA receptors (Lu et al. 2001, Park et al. 2004, Kopec et al. 2006) and is accompanied by endocytosis of Kv4.2 subunits of voltage-gated A-type K^+ channels, which enhances local dendritic excitability (Kim et al. 2007). Patches of preassembled clathrin provide hot spots of endocytosis along spine and dendritic membranes (Blanpied et al. 2002, Racz et al. 2004). Spine shape is regulated by recycling endosomes, and blocking this pathway results in significant spine loss (Park et al. 2006). Following the induction of LTP, live imaging and serial section transmission electron microscopy (ssTEM) revealed translocation into spines of endosomes having sufficient surface area to provide an abundant resource for spine growth. Two membrane pools were identified: recycling endosomes with tubules, vesicles, and clathrin-coated pits or buds and large amorphous vesicular clumps (AVC). Quantification suggested that AVCs provided membrane for new or enlarged spines, and recycling endosomes maintained them. LTD results in AMPA receptor internalization and reduced spine and synapse size (Man et al. 2000, Chen et al. 2004, Nagerl et al. 2004, Zhou et al. 2004, Brown et al. 2005). Interference with this AMPA receptor internalization leads to excitotoxicity via increased sensitivity to glutamate and eventual spine loss (Halpain et al. 1998, Hasbani et al. 2001). Thus, exo- and endocytosis must maintain an

ssTEM: serial section transmission electron microscopy

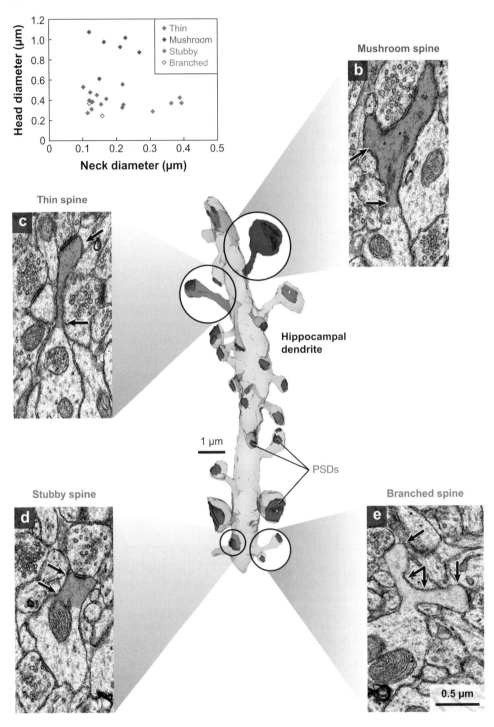

a

Head diameter (μm) vs Neck diameter (μm)

Legend:
- ◆ Thin
- ◆ Mushroom
- ◆ Stubby
- ◇ Branched

Mushroom spine

b

Thin spine

c

Hippocampal dendrite

1 μm

PSDs

Stubby spine

d

Branched spine

e

0.5 μm

activity-dependent balance to fine-tune the physiological and structural responses of spines to synaptic plasticity.

Polyribosomes and Proteasomes

Dendritic spine response to synaptic plasticity relies on spines' ability to regulate protein synthesis and degradation. Treatment with anisomycin prevents spine enlargement during LTP (Fifkova et al. 1982, Kelleher et al. 2004). Other findings show that polyribosomes, the machinery necessary to translate proteins, occur at the base of dendritic spines (Steward & Levy 1982) and preferentially redistribute into dendritic spines with enlarged heads and synapses during LTP (Ostroff et al. 2002, Bourne et al. 2007b). Which plasticity-related proteins could be translated by these local polyribosomes to increase the PSD size? One candidate is CamKII, a cytoplasmic protein highly enriched in the PSD (Kennedy et al. 1983, 1990; Otmakhov et al. 2004). CamKII becomes autophosphorylated (Miller & Kennedy 1986) following activation and can regulate glutamate receptors both directly and indirectly long after calcium levels have returned to baseline during LTP (McGlade-McCulloh et al. 1993, Pettit et al. 1994, Liao et al. 1995, Lledo et al. 1995). Furthermore, the mRNA transcripts for CamKII are present in dendrites (Martone et al. 1996, Havik et al. 2003), and translation of CamKII is upregulated (Ouyang et al. 1997, Ouyang et al. 1999, Aakalu et al.

2001) and more CamKII is present in the PSD after LTP (Otmakhov et al. 2004). Induction of LTD through activation of metabotropic glutamate receptors (mGluRs) is dependent on protein synthesis in adolescent but not neonatal rats (Huber et al. 2001, Nosyreva & Huber 2005). Stimulation of mGluRs in synaptoneurosomes triggers the aggregation of polyribosomes and the translation of proteins, including the fragile X mental retardation protein (Weiler et al. 1997), although the dendritic distribution of polyribosomes following induction of LTD has not yet been examined.

Rough endoplasmic reticulum (RER) and Golgi have been identified in dendrites, where they could locally synthesize and regulate integral membrane proteins (Steward & Reeves 1988, Gardiol et al. 1999, Cooney et al. 2002, Horton & Ehlers 2004, Grigston et al. 2005). One intriguing possibility is that the enigmatic spine apparatus, which occurs in ~10%–15% of mature hippocampal spines (Spacek & Harris 1997), may also be an extension of the Golgi apparatus (Pierce et al. 2000). Localized synthesis of the GluR1 and GluR2 subunits for AMPA glutamate receptors has been demonstrated in hippocampal dendrites (Kacharmina et al. 2000, Ju et al. 2004, Grooms et al. 2006), and the mRNAs for other integral membrane and secretory proteins are found throughout the dendritic arbor (Steward & Schuman 2003).

Maintenance of LTP also relies on proteasomes to degrade proteins (Fonseca et al. 2006, Karpova et al. 2006). Lysosomes and

mGluR:
metabotropic
glutamate receptor

Figure 1

Variability in spine shape and size. A three-dimensional reconstruction of a hippocampal dendrite (*gray*) illustrating different spine shapes including mushroom (*blue*), thin (*red*), stubby (*green*), and branched (*yellow*). PSDs (*red*) also vary in size and shape. (*a*) A graph plotting the ratio of head diameters to neck diameters for the spines on the reconstructed dendrite. Spine heads were measured at their widest point parallel to the PSD, and spine necks were measured just above the base of the spine to give a uniform location of measurement across all spines. Mushroom spines (*blue diamonds*), stubby spines (*green diamonds*), and thin spines (*red diamonds*) segregated into distinct groups. Both branches of the branched spine were of a thin shape and were situated among the thin spine dimensions (*yellow diamonds*). (*b*) An example of a mushroom spine (*blue*) with a head diameter exceeding 0.6 microns and a narrow neck. (*c*) An example of a thin spine (*red*) with a small head and narrow neck. (*d*) An example of a stubby spine (*green*) with an equal head and neck diameter and an overall length that equals its width. (*e*) An example of a branched spine (*yellow*) where both branches are thin spines. Scale bar = 0.5 μm, and arrows indicate where the head and neck diameters were measured for each spine in *b–e*.

multivesicular bodies also occur in dendritic spines (Spacek & Harris 1997, Cooney et al. 2002). It will be interesting to learn whether the balance of protein synthesis and degradation is shifted depending on whether a synapse is potentiated or depressed.

SER

Many dendritic spines contain SER, which likely regulates calcium. SER is present in all dendritic spines of cerebellar Purkinje neurons (Harris & Stevens 1988) but in less than half of cortical or hippocampal spines (Spacek 1985a, Spacek & Harris 1997). Calcium influx can trigger release from SER, thereby extending its elevation in stimulated spine heads (Sabatini et al. 2001). The elevated calcium facilitates remodeling of the actin cytoskeleton (Oertner & Matus 2005). Laminae of SER and dense-staining material form a spine apparatus in ~10%–20% of hippocampal and cortical spines. Synaptopodin is an actin-associated protein that occurs in the spine apparatus, and mice lacking synaptopodin also lack a spine apparatus and display deficits in synaptic plasticity (Deller et al. 2007). SER can shift throughout the dendrite (Toresson & Grant 2005), and it will be interesting to learn whether these dynamics are influenced by synaptic plasticity.

Mitochondria

Mitochondria are abundant in dendritic shafts, and the ATP they produce likely diffuses into spines to provide energy for signal transduction. In contrast, mitochondria are rarely found in dendritic spines and are usually restricted to very large and complex spines, such as the branched spines or "thorny excrescences" located on proximal dendrites of CA3 pyramidal cells (Chicurel & Harris 1992). In cultured neurons from area CA1, mitochondria occasionally migrate into some dendritic spines during periods of intense synaptic remodeling (Li et al. 2004). The enzymes involved in the glycolic generation and regulation of ATP have been localized to isolated PSDs, suggesting a mechanism for direct synthesis of ATP at individual synapses even in the absence of mitochondria in spines (Rogalski-Wilk & Cohen 1997, Wu et al. 1997). Synaptic ATP could provide an energy source for signaling via protein kinases found at the PSD, such as protein kinase A, protein kinase C, and CamKII, and for local protein synthesis by polyribosomes. Although enzymes localized to the PSD are a potentially important source of ATP, it would be interesting to know whether the distances between dendritic mitochondria and spines are altered in response to input-specific plasticity, such as LTP and LTD.

THE FORMATION AND STABILIZATION OF NEW SPINES

New spines are formed in the hippocampus during development and some forms of adult plasticity. Filopodia are nonsynaptic or multisynaptic, actin-rich protrusions with pointy tips (Fiala et al. 1998) that tend to be transient and last ~10 min during development (Ziv & Smith 1996). With maturation, the density of the neuropil increases and additional mechanisms may be required for new spines to find, compete for, and maintain presynaptic partners.

Development

During the first few weeks of postnatal life, hippocampal dendrites have numerous filopodia (Papa & Segal 1996, Ziv & Smith 1996, Fiala et al. 1998). Some filopodia become spines with synapses (Marrs et al. 2001), whereas others withdraw into the dendrite to form synapses on the dendritic shaft (Fiala et al. 1998, Marrs et al. 2001). These shaft synapses either reemerge as spines or are preferentially eliminated later in life (Harris 1999, Bourne & Harris 2007).

Stabilization of hippocampal spines requires assembly of pre- and postsynaptic elements, although the timing of these events may vary (Harris et al. 2003, Ostroff & Harris 2004, Risher et al. 2006, Nagerl et al. 2007). Dense

Filopodia: dynamic protrusions from dendrites that may become spines

core vesicles containing piccolo and bassoon appear in axonal processes within 2 days and cluster along dendritic profiles by 4 days in vitro in cultured hippocampal neurons, which suggests that presynaptic active zones are prepackaged (Zhai et al. 2001, Shapira et al. 2003). PSD-95 is necessary to stabilize the spine, as evidenced by RNAi knockdowns that cause spine loss (Ehrlich et al. 2007). Assembly of PSD-95 is spatially and temporally correlated with spine morphogenesis (Marrs et al. 2001) and the clustering of presynaptic vesicle proteins (Okabe et al. 2001). Stabilization of dendritic spines also relies on the insertion and activation of glutamate receptors; AMPA receptor activation in particular decreases spine motility and stabilizes spine shape (Fischer et al. 2000). Blocking NMDA receptor signaling does not affect the emergence or density of spines during development (Rao & Craig 1997, Kirov et al. 2004a, Alvarez et al. 2007), but knocking down NMDA receptors through RNA interference (RNAi) results in increased spine motility and eventual elimination (Alvarez et al. 2007).

Synaptogenesis requires that filopodia be maintained long enough to find appropriate presynaptic partners. Telencephalin is an adhesion molecule of the Ig superfamily and SynGAP is a Ras-GTPase activating protein; both of these proteins maintain filopodia in a dynamic state during synaptogenesis, and mice deficient in either protein show accelerated spine development and larger spine heads (Vazquez et al. 2004, Matsuno et al. 2006). Once filopodia become spines, telencephalin relocates to the dendritic shaft and is replaced with adhesion molecules, N-cadherin and α-catenin, which stabilize the new spine (Bozdagi et al. 2000, Togashi et al. 2002, Abe et al. 2004). SynGAP remains at the synapse and is bound to PSD-95 through its PDZ (PSD-95/Discs large/zona occludens-1) domain (Chen et al. 1998, Kim et al. 1998). Activation of NMDA receptors alters the phosphorylation state of different SynGAP isoforms, linking NMDA receptor activation and Ras signaling pathways

(Chen et al. 1998, Krapivinsky et al. 2004, Oh et al. 2004).

Spinogenesis is also regulated by microRNAs, small noncoding RNAs that control the translation of messenger RNAs. miR-134 is a brain-specific microRNA localized to dendritic spines that negatively regulates spine size by inhibiting protein kinase Limk1 translation (Schratt et al. 2006). Treatment with brain-derived neurotrophic factor (BDNF) relieves miR-134-mediated inhibition of Limk1 translation, which suggests that synaptic stimuli and extracellular signals can regulate spine development through local translation mechanisms.

Spinogenesis in the Mature Hippocampus

Filopodia are rarely observed in the mature hippocampus; however, blocking synaptic transmission in mature hippocampal slices triggers filopodia and new spines in an apparent attempt to compensate for the loss of synaptic input (Kirov & Harris 1999). Chilling hippocampal slices during preparation results in an immediate disappearance of spines, but upon rewarming new spines proliferate beyond levels found in vivo (Kirov et al. 1999, Kirov et al. 2004b). Instead, if slices are prepared rapidly at room temperature, then spine density matches that found in perfusion-fixed hippocampus even several hours later (Bourne et al. 2007a). Hibernating ground squirrels also show substantial spine loss at near-freezing temperatures, but rapid spinogenesis occurs within minutes of awakening and return to warmer body temperatures (Popov et al. 1992, Popov & Bocharova 1992). Telencephalin levels remain high in adulthood, suggesting an ongoing involvement in transforming filopodia to new spines in the mature brain (Matsuno et al. 2006).

Adhesion and Trans-Synaptic Signaling

Cell-adhesion molecules, such as N-cadherins, catenins, neurexins, and neuroligins, and Ephs

RNAi: ribonucleic acid interference

and ephrins begin to cluster on the pre- and postsynaptic sides and help stabilize the nascent spines and their synapses (Calabrese et al. 2006). N-cadherin is an adhesive molecule that links pre- and postsynaptic elements through calcium-dependent homophilic interactions. N-cadherin and β-catenin form a calcium-regulated complex with AMPA receptors, and overexpression of N-cadherin increases the surface expression of the AMPA receptor subunit GluR1 (Nuriya & Huganir 2006, Tai et al. 2007). NMDA receptor activation increases the concentration of unphosphorylated β-catenin and inhibits endocytosis of N-cadherin (Tai et al. 2007). N-cadherin also regulates spine morphology via its binding proteins, α- and β-catenin, which interact with the actin cytoskeleton (Kosik et al. 2005). Thus synaptic activity stabilizes synapse structure via N-cadherin, which in turn recruits AMPA receptors and maintains synaptic efficacy. Prolonged stability of N-cadherin abolishes NMDA receptor–induced LTD, perhaps because N-cadherin prevents the internalization of AMPA receptors associated with synaptic depression (Tai et al. 2007).

Eph receptor–ephrin binding results in multimeric clusters that bridge juxtaposed cell surfaces and mediate cell-cell adhesion and bidirectional signaling. Trans-endocytosis of the eph-ephrin complex loosens the adhesion between the pre- and postsynaptic elements, which may permit structural synaptic plasticity. EphB receptors directly associate with NMDA receptors at synapses, and ephrinB-induced activation of EphB receptors causes NMDA receptor clustering (Dalva et al. 2000). At the mossy fiber synapse in CA3, postsynaptic EphB2 receptors interact with a PDZ-domain protein, glutamate receptor interacting protein (GRIP), to mediate AMPA receptor-dependent LTP (Contractor et al. 2002). EphB2 also associates with the GTP exchange factors intersectin and kalirin (Penzes et al. 2003, Irie & Yamaguchi 2004). The intersectin-Cdc42-Wasp-actin and kalirin-Rac-Pak-actin pathways may regulate the EphB receptor–mediated morphogenesis and maturation of dendritic

spines in cultured hippocampal and cortical neurons. Perhaps the interaction of presynaptic ephrins with postsynaptic Eph receptors coordinates the establishment of the well-known correlation between presynaptic vesicle number and postsynaptic size during structural synaptic plasticity.

Trans-synaptic signaling may also be mediated by the formation of spinules. Spinules are double-membrane structures that emerge primarily from dendritic spines into presynaptic or neighboring axons or astroglial processes (Spacek & Harris 2004). Spinules are likely involved in active trans-endocytosis, as evidenced by the presence of clathrin-like coats along the cytoplasmic surface of the engulfing structure, such as the presynaptic axons, across from the spinule tip. In particular, this trans-endocytosis could be the morphological correlate of retrograde signaling via cell surface molecules such as Ephs and ephrins, which must remain in the plasma membrane while signaling. Spinules may also be involved in remodeling the postsynaptic membrane, as suggested by their transient increase shortly after LTP induction (Applegate & Landfield 1988, Schuster et al. 1990, Geinisman et al. 1993, Toni et al. 1999).

Perisynaptic Astroglia

The development and stabilization of synapses also require astroglia (Allen & Barres 2005). Astroglia form nonoverlapping domains in the hippocampus and cortex, and a single astrocyte contacts hundreds of dendrites and thousands of synapses, which suggests that it coordinates multiple neuronal networks (Bushong et al. 2002, Halassa et al. 2007). Transient interactions between the ephrin-A3 ligand and the EphA4 receptor regulate the structure of excitatory synaptic connections through neuroglial cross talk (Murai et al. 2003, Grunwald et al. 2004). Activation of EphA4 by ephrin-A3 induces spine retraction, whereas inhibiting ephrin/EphA4 interactions distorts spine shape and organization (Murai et al. 2003). Expression of EphA4 decreases during maturation, suggesting its role in synaptic elimination and connection refinement. Astrocytes also secrete soluble factors such as thrombospondins and cholesterol, which influence spine formation and synapse maturation (Ullian et al. 2004, Christopherson et al. 2005). In the mature neocortex and hippocampus, fewer than half the synapses have perisynaptic astroglial processes (Spacek 1985b, Ventura & Harris 1999); however, synapses with astroglial processes at their perimeter are larger and presumably more effective than those without (Witcher et al. 2007). Synapse size is associated with the presence of an astroglial process juxtaposed to the postsynaptic spine and/or the synaptic cleft, not with the degree to which the astroglial process surrounds the synapse. Even the largest hippocampal or neocortical synapses might have only a small fraction of their perimeters surrounded by an astroglial process, which suggests that cross talk via spillover of neurotransmitters between synapses might be functionally significant. Thus, interactions between cell surface molecules and the release of various soluble factors by astroglia may be crucially important to the turnover and enlargement of spines observed with synaptic plasticity.

CONCLUSIONS

Modern molecular biology, electrophysiology, and imaging studies have provided many insights into the mechanisms of the morphological alterations undergone by dendritic spines during development and synaptic plasticity. Nevertheless, fundamental structural questions remain. Presently, only three-dimensional (3D) reconstruction from ssTEM provides sufficient resolution to determine how intrinsic and extrinsic factors might interact to control the structure and function of spines and synapses. Advances in imaging and computing technologies may soon provide resources to reconstruct entire neural circuits (e.g., projectomes or connectomes; Kasthuri & Lichtman 2007). It is not sufficient, however, to have just the wiring diagram because we also need to know what controls the switches. Determining the extrinsic factors that regulate connectivity along dendrites and axons and the intrinsic factors that regulate the availability of core subcellular structures required to build and maintain synapses is necessary to formulate a comprehensive understanding of neural circuits that underlie perception, memory, and cognition.

SUMMARY POINTS

1. Dendritic spines are complex biochemical compartments that integrate individual synaptic inputs into complex neural networks.

2. Dendritic spines in the hippocampus undergo genesis, elimination, and structural modification in response to a variety of stimuli.

3. Spines coordinate the activation of glutamate receptors with calcium regulation, cytoskeletal remodeling, membrane trafficking, protein synthesis and degradation, and trans-synaptic signaling.

4. The dynamic balance of the molecular machinery within spines is manifested by morphological changes in spine shape and density and by the translocation of necessary organelles into and out of spines.

5. Although light level microscopy can provide information on real-time dynamics of spines and proteins, ssTEM is required to detect small but crucial changes in spine dimensions and interspine spacing and the presence and distribution of subcellular organelles and perisynaptic astroglia.

DISCLOSURE STATEMENT

The authors are not aware of any biases that might be perceived as affecting the objectivity of this review.

ACKNOWLEDGMENTS

This work was supported by NIH grants NS21184, NS33574, and EB002170 to K.M.H.

LITERATURE CITED

Aakalu G, Smith WB, Nguyen N, Jiang C, Schuman EM. 2001. Dynamic visualization of local protein synthesis in hippocampal neurons. *Neuron* 30(2):489–502

Abe K, Chisaka O, Van Roy F, Takeichi M. 2004. Stability of dendritic spines and synaptic contacts is controlled by alpha N-catenin. *Nat. Neurosci.* 7(4):357–63

Ackermann M, Matus A. 2003. Activity-induced targeting of profilin and stabilization of dendritic spine morphology. *Nat. Neurosci.* 6(11):1194–200

Adesnik H, Nicoll RA, England PM. 2005. Photoinactivation of native AMPA receptors reveals their real-time trafficking. *Neuron* 48(6):977–85

Allen NJ, Barres BA. 2005. Signaling between glia and neurons: focus on synaptic plasticity. *Curr. Opin. Neurobiol.* 15(5):542–48

Alvarez VA, Ridenour DA, Sabatini BL. 2007. Distinct structural and ionotropic roles of NMDA receptors in controlling spine and synapse stability. *J. Neurosci.* 27(28):7365–76

Alvarez VA, Sabatini BL. 2007. Anatomical and physiological plasticity of dendritic spines. *Annu. Rev. Neurosci.* 30:79–97

Applegate MD, Landfield PW. 1988. Synaptic vesicle redistribution during hippocampal frequency potentiation and depression in young and aged rats. *J. Neurosci.* 8(4):1096–111

Ashby MC, Maier SR, Nishimune A, Henley JM. 2006. Lateral diffusion drives constitutive exchange of AMPA receptors at dendritic spines and is regulated by spine morphology. *J. Neurosci.* 26(26):7046–55

Beattie EC, Carroll RC, Yu X, Morishita W, Yasuda H, et al. 2000. Regulation of AMPA receptor endocytosis by a signaling mechanism shared with LTD. *Nat. Neurosci.* 3(12):1291–300

Blanpied TA, Scott DB, Ehlers MD. 2002. Dynamics and regulation of clathrin coats at specialized endocytic zones of dendrites and spines. *Neuron* 36(3):435–49

Bourne J, Harris KM. 2007. Do thin spines learn to be mushroom spines that remember? *Curr. Opin. Neurobiol.* 17(3):381–86

Bourne JN, Kirov SA, Sorra KE, Harris KM. 2007a. Warmer preparation of hippocampal slices prevents synapse proliferation that might obscure LTP-related structural plasticity. *Neuropharmacology.* 52(1):55–59

Bourne JN, Sorra KE, Hurlburt J, Harris KM. 2007b. Polyribosomes are increased in spines of CA1 dendrites 2 h after the induction of LTP in mature rat hippocampal slices. *Hippocampus.* 17(1):1–4

Bozdagi O, Shan W, Tanaka H, Benson DL, Huntley GW. 2000. Increasing numbers of synaptic puncta during late-phase LTP: N-cadherin is synthesized, recruited to synaptic sites, and required for potentiation. *Neuron.* 28(1):245–59

Brown TC, Tran IC, Backos DS, Esteban JA. 2005. NMDA receptor-dependent activation of the small GTPase Rab5 drives the removal of synaptic AMPA receptors during hippocampal LTD. *Neuron* 45(1):81–94

Bushong EA, Martone ME, Jones YZ, Ellisman MH. 2002. Protoplasmic astrocytes in CA1 stratum radiatum occupy separate anatomical domains. *J. Neurosci.* 22(1):183–92

Calabrese B, Wilson MS, Halpain S. 2006. Development and regulation of dendritic spine synapses. *Physiology* 21:38–47

Cesa R, Strata P. 2005. Axonal and synaptic remodeling in the mature cerebellar cortex. *Prog. Brain Res.* 148:45–56

Chen HJ, Rojas-Soto M, Oguni A, Kennedy MB. 1998. A synaptic Ras-GTPase activating protein (p135 SynGAP) inhibited by CaM kinase II. *Neuron* 20(5):895–904

Chen LY, Rex CS, Casale MS, Gall CM, Lynch G. 2007. Changes in synaptic morphology accompany actin signaling during LTP. *J. Neurosci.* 27(20):5363–72

Chen YC, Bourne J, Pieribone VA, Fitzsimonds RM. 2004. The role of actin in the regulation of dendritic spine morphology and bidirectional synaptic plasticity. *NeuroReport* 15(5):829–32

Chicurel ME, Harris KM. 1992. Three-dimensional analysis of the structure and composition of CA3 branched dendritic spines and their synaptic relationships with mossy fiber boutons in the rat hippocampus. *J. Comp. Neurol.* 325:169–82

Christopherson KS, Ullian EM, Stokes CC, Mullowney CE, Hell JW, et al. 2005. Thrombospondins are astrocyte-secreted proteins that promote CNS synaptogenesis. *Cell* 120(3):421–33

Contractor A, Rogers C, Maron C, Henkemeyer M, Swanson GT, Heinemann SF. 2002. Trans-synaptic Eph receptor-ephrin signaling in hippocampal mossy fiber LTP. *Science* 296(5574):1864–69

Cooney JR, Hurlburt JL, Selig DK, Harris KM, Fiala JC. 2002. Endosomal compartments serve multiple hippocampal dendritic spines from a widespread rather than a local store of recycling membrane. *J. Neurosci.* 22(6):2215–24

Dalva MB, Takasu MA, Lin MZ, Shamah SM, Hu L, et al. 2000. EphB receptors interact with NMDA receptors and regulate excitatory synapse formation. *Cell* 103(6):945–56

Deller T, Bas Orth C, Del Turco D, Vlachos A, Burbach GJ, et al. 2007. A role for synaptopodin and the spine apparatus in hippocampal synaptic plasticity. *Ann. Anat.* 189(1):5–16

Dhanrajan TM, Lynch MA, Kelly A, Popov VI, Rusakov DA, Stewart MG. 2004. Expression of long-term potentiation in aged rats involves perforated synapses but dendritic spine branching results from high-frequency stimulation alone. *Hippocampus* 14:255–64

Ehrlich I, Klein M, Rumpel S, Malinow R. 2007. PSD-95 is required for activity-driven synapse stabilization. *Proc. Natl. Acad. Sci. USA* 104(10):4176–81

Engert F, Bonhoeffer T. 1999. Dendritic spine changes associated with hippocampal long-term synaptic plasticity. *Nature* 399(6731):66–70

Ethell IM, Pasquale EB. 2005. Molecular mechanisms of dendritic spine development and remodeling. *Prog. Neurobiol.* 75:161–205

Fedulov V, Rex CS, Simmons DA, Palmer L, Gall CM, Lynch G. 2007. Evidence that long-term potentiation occurs within individual hippocampal synapses. *J. Neurosci.* 27(30):8031–39

Fiala JC, Allwardt B, Harris KM. 2002. Dendritic spines do not split during hippocampal LTP or maturation. *Nat. Neurosci.* 5(4):297–98

Fiala JC, Feinberg M, Popov V, Harris KM. 1998. Synaptogenesis via dendritic filopodia in developing hippocampal area CA1. *J. Neurosci.* 18(21):8900–11

Fifkova E, Anderson CL, Young SJ, Van Harreveld A. 1982. Effect of anisomycin on stimulation-induced changes in dendritic spines of the dentate granule cells. *J. Neurocytol.* 11(2):183–210

Fischer M, Kaech S, Wagner U, Brinkhaus H, Matus A. 2000. Glutamate receptors regulate actin-based plasticity in dendritic spines. *Nat. Neurosci.* 3(9):887–94

Fonseca R, Vabulas RM, Hartl FU, Bonhoeffer T, Nagerl UV. 2006. A balance of protein synthesis and proteasome-dependent degradation determines the maintenance of LTP. *Neuron* 52(2):239–45

Fukazawa Y, Saitoh Y, Ozawa F, Ohta Y, Mizuno K, Inokuchi K. 2003. Hippocampal LTP is accompanied by enhanced F-actin content within the dendritic spine that is essential for late LTP maintenance in vivo. *Neuron* 38(3):447–60

Gardiol A, Racca C, Triller A. 1999. Dendritic and postsynaptic protein synthetic machinery. *J. Neurosci.* 19(1):168–79

Geinisman Y, de Toledo-Morrell L, Morrell F. 1991. Induction of long-term potentiation is associated with an increase in the number of axospinous synapses with segmented postsynaptic densities. *Brain Res.* 566(1–2):77–88

Geinisman Y, de Toledo-Morrell L, Morrell F, Heller RE, Rossi M, Parshall RF. 1993. Structural synaptic correlate of long-term potentiation: formation of axospinous synapses with multiple, completely partitioned transmission zones. *Hippocampus* 3(4):435–45

Grigston JC, VanDongen HM, McNamara JO II, VanDongen AM. 2005. Translation of an integral membrane protein in distal dendrites of hippocampal neurons. *Eur. J. Neurosci.* 21(6):1457–68

Grooms SY, Noh KM, Regis R, Bassell GJ, Bryan MK, et al. 2006. Activity bidirectionally regulates AMPA receptor mRNA abundance in dendrites of hippocampal neurons. *J. Neurosci.* 26(32):8339–51

Grunwald IC, Korte M, Adelmann G, Plueck A, Kullander K, et al. 2004. Hippocampal plasticity requires postsynaptic ephrinBs. *Nat. Neurosci.* 7(1):33–40

Halassa MM, Fellin T, Takano H, Dong J-H, Haydon PG. 2007. Synaptic islands defined by the territory of a single astrocyte. *J. Neurosci.* 27(24):6473–77

Halpain S, Hipolito A, Saffer L. 1998. Regulation of F-actin stability in dendritic spines by glutamate receptors and calcineurin. *J. Neurosci.* 18(23):9835–44

Harris KM. 1999. Structure, development, and plasticity of dendritic spines. *Curr. Opin. Neurobiol.* 9:343–48

Harris KM, Fiala JC, Ostroff L. 2003. Structural changes at dendritic spine synapses during long-term potentiation. *Philos. Trans. R. Soc. London B Biol. Sci* 358(1432):745–48

Harris KM, Jensen FE, Tsao B. 1992. Three-dimensional structure of dendritic spines and synapses in rat hippocampus (CA1) at postnatal day 15 and adult ages: implications for the maturation of synaptic physiology and long-term potentiation. *J. Neurosci.* 12(7):2685–705

Harris KM, Perry E, Bourne J, Feinberg M, Ostroff L, Hurlburt J. 2006. Uniform serial sectioning for transmission electron microscopy. *J. Neurosci.* 26(47):12101–3

Harris KM, Stevens JK. 1988. Dendritic spines of rat cerebellar Purkinje cells: serial electron microscopy with reference to their biophysical characteristics. *J. Neurosci.* 8(12):4455–69

Harris KM, Stevens JK. 1989. Dendritic spines of CA1 pyramidal cells in the rat hippocampus: serial electron microscopy with reference to their biophysical characteristics. *J. Neurosci.* 9(8):2982–97

Hasbani MJ, Schlief ML, Fisher DA, Goldberg MP. 2001. Dendritic spines lost during glutamate receptor activation reemerge at original sites of synaptic contact. *J. Neurosci.* 21(7):2393–403

Havik B, Rokke H, Bardsen K, Davanger S, Bramham CR. 2003. Bursts of high-frequency stimulation trigger rapid delivery of pre-existing alpha-CaMKII mRNA to synapses: a mechanism in dendritic protein synthesis during long-term potentiation in adult awake rats. *Eur. J. Neurosci.* 17(12):2679–89

Horton AC, Ehlers MD. 2004. Secretory trafficking in neuronal dendrites. *Nat. Cell Biol.* 6(7):585–91

Huber KM, Roder JC, Bear MF. 2001. Chemical induction of mGluR5- and protein synthesis—dependent long-term depression in hippocampal area CA1. *J. Neurophysiol.* 86(1):321–25

Humeau Y, Herry C, Kemp N, Shaban H, Fourcaudot E, et al. 2005. Dendritic spine heterogeneity determines afferent-specific Hebbian plasticity in the amygdala. *Neuron* 45(1):119–31

Irie F, Yamaguchi Y. 2004. EPHB receptor signaling in dendritic spine development. *Front Biosci.* 9:1365–73

Isaac JT, Nicoll RA, Malenka RC. 1995. Evidence for silent synapses: implications for the expression of LTP. *Neuron* 15(2):427–34

Ju W, Morishita W, Tsui J, Gaietta G, Deerinck TJ, et al. 2004. Activity-dependent regulation of dendritic synthesis and trafficking of AMPA receptors. *Nat. Neurosci.* 7(3):244–53

Kacharmina JE, Job C, Crino P, Eberwine J. 2000. Stimulation of glutamate receptor protein synthesis and membrane insertion within isolated neuronal dendrites. *Proc. Natl. Acad. Sci. USA* 97(21):11545–50

Karpova A, Mikhaylova M, Thomas U, Knopfel T, Behnisch T. 2006. Involvement of protein synthesis and degradation in long-term potentiation of Schaffer collateral CA1 synapses. *J. Neurosci.* 26(18):4949–55

Kasthuri N, Lichtman JW. 2007. The rise of the 'projectome.' *Nat. Methods* 4(4):307–8

Kelleher RJ III, Govindarajan A, Jung HY, Kang H, Tonegawa S. 2004. Translational control by MAPK signaling in long-term synaptic plasticity and memory. *Cell* 116(3):467–79

Kennedy MB, Bennett MK, Bulliet RF, Erondu NE, Jennings VR, et al. 1990. Structure and regulation of type II calcium/calmodulin-dependent protein kinase in central nervous system neurons. *Cold Spring Harbor Symp. Quant. Biol.* 55:101–10

Kennedy MB, Bennett MK, Erondu NE. 1983. Biochemical and immunochemical evidence that the "major postsynaptic density protein" is a subunit of a calmodulin-dependent protein kinase. *Proc. Natl. Acad. Sci. USA* 80:7357–61

Kim CH, Lisman JE. 1999. A role of actin filament in synaptic transmission and long-term potentiation. *J. Neurosci.* 19(11):4314–24

Kim J, Jung SC, Clemens AM, Petralia RS, Hoffman DA. 2007. Regulation of dendritic excitability by activity-dependent trafficking of the A-type K^+ channel subunit Kv4.2 in hippocampal neurons. *Neuron* 54(6):933–47

Kim JH, Liao D, Lau LF, Huganir RL. 1998. SynGAP: a synaptic RasGAP that associates with the PSD-95/SAP90 protein family. *Neuron* 20(4):683–91

Kirov SA, Goddard CA, Harris KM. 2004a. Age-dependence in the homeostatic upregulation of hippocampal dendritic spine number during blocked synaptic transmission. *Neuropharmacology* 47(5):640–48

Kirov SA, Harris KM. 1999. Dendrites are more spiny on mature hippocampal neurons when synapses are inactivated. *Nat. Neurosci.* 2(10):878–83

Kirov SA, Petrak LJ, Fiala JC, Harris KM. 2004b. Dendritic spines disappear with chilling but proliferate excessively upon rewarming of mature hippocampus. *Neurosci.* 127(1):69–80

Kirov SA, Sorra KE, Harris KM. 1999. Slices have more synapses than perfusion-fixed hippocampus from both young and mature rats. *J. Neurosci.* 19(8):2876–86

Knott GW, Holtmaat A, Wilbrecht L, Welker E, Svoboda K. 2006. Spine growth precedes synapse formation in the adult neocortex in vivo. *Nat. Neurosci.* 9(9):1117–24

Kopec CD, Li B, Wei W, Boehm J, Malinow R. 2006. Glutamate receptor exocytosis and spine enlargement during chemically induced long-term potentiation. *J. Neurosci.* 26:2000–9

Kosik KS, Donahue CP, Israely I, Liu X, Ochiishi T. 2005. Delta-catenin at the synaptic-adherens junction. *Trends Cell Biol.* 15(3):172–78

Kozorovitskiy Y, Gross CG, Kopil C, Battaglia L, McBreen M, et al. 2005. Experience induces structural and biochemical changes in the adult primate brain. *Proc. Natl. Acad. Sci. USA* 102:17478–82

Krapivinsky G, Medina I, Krapivinsky L, Gapon S, Clapham DE. 2004. SynGAP-MUPP1-CaMKII synaptic complexes regulate p38 MAP kinase activity and NMDA receptor-dependent synaptic AMPA receptor potentiation. *Neuron* 43(4):563–74

Krucker T, Siggins GR, Halpain S. 2000. Dynamic actin filaments are required for stable long-term potentiation (LTP) in area CA1 of the hippocampus. *Proc. Natl. Acad. Sci. USA* 97(12):6856–61

Lang C, Barco A, Zablow L, Kandel ER, Siegelbaum SA, Zakharenko SS. 2004. Transient expansion of synaptically connected dendritic spines upon induction of hippocampal long-term potentiation. *Proc. Natl. Acad. Sci. USA* 101(47):16665–70

Lee SH, Liu L, Wang YT, Sheng M. 2002. Clathrin adaptor AP2 and NSF interact with overlapping sites of GluR2 and play distinct roles in AMPA receptor trafficking and hippocampal LTD. *Neuron* 36(4):661–74

Li Z, Okamoto K, Hayashi Y, Sheng M. 2004. The importance of dendritic mitochondria in the morphogenesis and plasticity of spines and synapses. *Cell* 119(6):873–87

Liao D, Hessler NA, Malinow R. 1995. Activation of postsynaptically silent synapses during pairing-induced LTP in CA1 region of hippocampal slice. *Nature* 375(6530):400–4

Liao D, Zhang X, O'Brien R, Ehlers MD, Huganir RL. 1999. Regulation of morphological postsynaptic silent synapses in developing hippocampal neurons. *Nat. Neurosci.* 2(1):37–43

Lin B, Kramar EA, Bi X, Brucher FA, Gall CM, Lynch G. 2005. Theta stimulation polymerizes actin in dendritic spines of hippocampus. *J. Neurosci.* 25(8):2062–69

Lisman J, Harris KM. 1993. Quantal analysis and synaptic anatomy—integrating two views of hippocampal plasticity. *Trends Neurosci.* 16:141–47

Lisman J, Harris KM. 1994. Who's been nibbling on my PSD: Is it LTD? *J. Physiol. (Paris)* 88:193–95

Lledo PM, Hjelmstad GO, Mukherji S, Soderling TR, Malenka RC, Nicoll RA. 1995. Calcium calmodulin-dependent kinase-II and long term potentiation enhance synaptic transmission by the same mechanism. *Proc. Natl. Acad. Sci. USA* 92:11175–79

Lu W, Man H, Ju W, Trimble WS, MacDonald JF, Wang YT. 2001. Activation of synaptic NMDA receptors induces membrane insertion of new AMPA receptors and LTP in cultured hippocampal neurons. *Neuron* 29(1):243–54

Magee JC, Johnston D. 2005. Plasticity of dendritic function. *Curr. Opin. Neurobiol.* 15(3):334–42

Maletic-Savatic M, Malinow R, Svoboda K. 1999. Rapid dendritic morphogenesis in CA1 hippocampal dendrites induced by synaptic activity. *Science* 283(5409):1923–27

Man HY, Lin JW, Ju WH, Ahmadian G, Liu L, et al. 2000. Regulation of AMPA receptor-mediated synaptic transmission by clathrin-dependent receptor internalization. *Neuron* 25(3):649–62

Marrs GS, Green SH, Dailey ME. 2001. Rapid formation and remodeling of postsynaptic densities in developing dendrites. *Nat. Neurosci.* 4(10):1006–13

Martone ME, Pollock JA, Jones YZ, Ellisman MH. 1996. Ultrastructural localization of dendritic messenger RNA in adult rat hippocampus. *J. Neurosci.* 16(23):7437–46

Masugi-Tokita M, Shigemoto R. 2007. High-resolution quantitative visualization of glutamate and GABA receptors at central synapses. *Curr. Opin. Neurobiol.* 17(3):387–93

Matsuno H, Okabe S, Mishina M, Yanagida T, Mori K, Yoshihara Y. 2006. Telencephalin slows spine maturation. *J. Neurosci.* 26(6):1776–86

Matsuzaki M, Ellis-Davies GC, Nemoto T, Miyashita Y, Iino M, Kasai H. 2001. Dendritic spine geometry is critical for AMPA receptor expression in hippocampal CA1 pyramidal neurons. *Nat. Neurosci.* 4(11):1086–92

Matsuzaki M, Honkura N, Ellis-Davies GC, Kasai H. 2004. Structural basis of long-term potentiation in single dendritic spines. *Nature* 429(6993):761–66

Matus A. 2000. Actin-based plasticity in dendritic spines. *Science* 290(5492):754–58

McGlade-McCulloh E, Yamamoto H, Tan SE, Brickey DA, Soderling TR. 1993. Phosphorylation and regulation of glutamate receptors by calcium/calmodulin-dependent protein kinase II. *Nature* 362(6421):640–42

Mezey S, Doyere V, De Souza I, Harrison E, Cambon K, et al. 2004. Long-term synaptic morphometry changes after induction of long-term potentiation and long-term depression in the dentate gyrus of awake rats are not simply mirror phenomena. *Eur. J. Neurosci.* 19(8):2310–18

Miller SG, Kennedy MB. 1986. Regulation of brain type II Ca^{2+}/calmodulin-dependent protein kinase by autophosphorylation: a Ca^{2+}-triggered molecular switch. *Cell* 44(6):861–70

Moser MB, Trommald M, Egeland T, Andersen P. 1997. Spatial training in a complex environment and isolation alter the spine distribution differently in rat CA1 pyramidal cells. *J. Comp. Neurol.* 380(3):373–81

Murai KK, Nguyen LN, Irie F, Yamaguchi Y, Pasquale EB. 2003a. Control of hippocampal dendritic spine morphology through ephrin-A3/EphA4 signaling. *Nat. Neurosci.* 6(2):153–60

Nagerl UV, Eberhorn N, Cambridge SB, Bonhoeffer T. 2004. Bidirectional activity-dependent morphological plasticity in hippocampal neurons. *Neuron* 44(5):759–67

Nagerl UV, Kostinger G, Anderson JC, Martin KA, Bonhoeffer T. 2007. Protracted synaptogenesis after activity-dependent spinogenesis in hippocampal neurons. *J. Neurosci.* 27(30):8149–56

Nicholson DA, Trana R, Katz Y, Kath WL, Spruston N, Geinisman Y. 2006. Distance-dependent differences in synapse number and AMPA receptor expression in hippocampal CA1 pyramidal neurons. *Neuron* 50(3):431–42

Nosyreva ED, Huber KM. 2005. Developmental switch in synaptic mechanisms of hippocampal metabotropic glutamate receptor-dependent long-term depression. *J. Neurosci.* 25(11):2992–3001

Nuriya M, Huganir RL. 2006. Regulation of AMPA receptor trafficking by N-cadherin. *J. Neurochem.* 97(3):652–61

Oertner TG, Matus A. 2005. Calcium regulation of actin dynamics in dendritic spines. *Cell Calcium* 37(5):477–82

Oh JS, Manzerra P, Kennedy MB. 2004. Regulation of the neuron-specific Ras GTPase-activating protein, synGAP, by Ca^{2+}/calmodulin-dependent protein kinase II. *J. Biol. Chem.* 279(17):17980–88

Okabe S. 2007. Molecular anatomy of the postsynaptic density. *Mol. Cell Neurosci.* 34(4):503–18

Okabe S, Miwa A, Okado H. 2001. Spine formation and correlated assembly of presynaptic and postsynaptic molecules. *J. Neurosci.* 21(16):6105–14

Osterweil E, Wells DG, Mooseker MS. 2005. A role for myosin VI in postsynaptic structure and glutamate receptor endocytosis. *J. Cell Biol.* 168(2):329–38

Ostroff LE, Fiala JC, Allwardt B, Harris KM. 2002. Polyribosomes redistribute from dendritic shafts into spines with enlarged synapses during LTP in developing rat hippocampal slices. *Neuron* 35(3):535–45

Ostroff LE, Harris KM. 2004. Dynamic dendrites: More and bigger spines and synapses and proliferation of polyribosomes during early LTP. *Soc. Neurosci.* (Abstr.) 637.8

Otmakhov N, Tao-Cheng JH, Carpenter S, Asrican B, Dosemeci A, et al. 2004. Persistent accumulation of calcium/calmodulin-dependent protein kinase II in dendritic spines after induction of NMDA receptor-dependent chemical long-term potentiation. *J. Neurosci.* 24(42):9324–31

Ouyang Y, Kantor D, Harris KM, Schuman EM, Kennedy MB. 1997. Visualization of the distribution of autophosphorylated calcium/calmodulin-dependent protein kinase II after tetanic stimulation in the CA1 area of the hippocampus. *J. Neurosci.* 17(14):5416–27

Ouyang Y, Rosenstein A, Kreiman G, Schuman EM, Kennedy MB. 1999. Tetanic stimulation leads to increased accumulation of $Ca^{(2+)}$/calmodulin-dependent protein kinase II via dendritic protein synthesis in hippocampal neurons. *J. Neurosci.* 19(18):7823–33

Ouyang Y, Wong M, Capani F, Rensing N, Lee CS, et al. 2005. Transient decrease in F-actin may be necessary for translocation of proteins into dendritic spines. *Eur. J. Neurosci.* 22(12):2995–3005

Papa M, Segal M. 1996. Morphological plasticity in dendritic spines of cultured hippocampal neurons. *Neurosci.* 71(4):1005–11

Park M, Penick EC, Edwards JG, Kauer JA, Ehlers MD. 2004. Recycling endosomes supply AMPA receptors for LTP. *Science* 305(5692):1972–75

Park M, Salgado JM, Ostroff L, Helton TD, Robinson CG, et al. 2006. Plasticity-induced growth of dendritic spines by exocytic trafficking from recycling endosomes. *Neuron* 52(5):817–30

Penzes P, Beeser A, Chernoff J, Schiller MR, Eipper BA, et al. 2003. Rapid induction of dendritic spine morphogenesis by trans-synaptic ephrinB-EphB receptor activation of the Rho-GEF kalirin. *Neuron* 37(2):263–74

Petralia RS, Esteban JA, Wang Y-X, Partridge JG, Zhao H-M, et al. 1999. Selective acquisition of AMPA receptors over postnatal developement suggests a molecular basis for silent synapses. *Nat. Neurosci.* 2(1):31–36

Pettit DL, Perlman S, Malinow R. 1994. Potentiated transmission and prevention of further LTP by increased CaMKII activity in postsynaptic hippocampal slice neurons. *Science* 266(5192):1881–85

Pierce JP, van Leyen K, McCarthy JB. 2000. Translocation machinery for synthesis of integral membrane and secretory proteins in dendritic spines. *Nat. Neurosci.* 3(4):311–13

Popov VI, Bocharova LS. 1992. Hibernation-induced structural changes in synaptic contacts between mossy fibres and hippocampal pyramidal neurons. *Neuroscience* 48:53–62

Popov VI, Bocharova LS, Bragin AG. 1992. Repeated changes of dendritic morphology in the hippocampus of ground squirrels in the course of hibernation. *Neuroscience* 48:45–51

Popov VI, Davies HA, Rogachevsky VV, Patrushev IV, Errington ML, et al. 2004. Remodelling of synaptic morphology but unchanged synaptic density during late phase long-term potentiation (LTP): a serial section electron micrograph study in the dentate gyrus in the anaesthetised rat. *Neuroscience* 128(2):251–62

Racz B, Blanpied TA, Ehlers MD, Weinberg RJ. 2004. Lateral organization of endocytic machinery in dendritic spines. *Nat. Neurosci.* 7(9):917–18

Rao A, Craig AM. 1997. Activity regulates the synaptic localization of the NMDA receptor in hippocampal neurons. *Neuron* 19(4):801–12

Risher WC, Ostroff LE, Harris KM. 2006. What dendritic filopodia induced by LTP encounter along their path through the neuropil of PN15 rat hippocampus. *Soc. Neurosci. Abs.* (Abstr.) 135.4

Rogalski-Wilk AA, Cohen RS. 1997. Glyceraldehyde-3-phosphate dehydrogenase activity and F-actin associations in synaptosomes and postsynaptic densities of porcine cerebral cortex. *Cell Mol. Neurobiol.* 17(1):51–70

Rostaing P, Real E, Siksou L, Lechaire JP, Boudier T, et al. 2006. Analysis of synaptic ultrastructure without fixative using high-pressure freezing and tomography. *Eur. J. Neurosci.* 24(12):3463–74

Ryu J, Liu L, Wong TP, Wu DC, Burette A, et al. 2006. A critical role for myosin IIb in dendritic spine morphology and synaptic function. *Neuron* 49(2):175–82

Sabatini BL, Maravall M, Svoboda K. 2001. $Ca^{(2+)}$ signaling in dendritic spines. *Curr. Opin. Neurobiol.* 11(3):349–56

Schratt GM, Tuebing F, Nigh EA, Kane CG, Sabatini ME, et al. 2006. A brain-specific microRNA regulates dendritic spine development. *Nature* 439(7074):283–89

Schuster T, Krug M, Wenzel J. 1990. Spinules in axospinous synapses of the rat dentate gyrus: changes in density following long-term potentiation. *Brain Res.* 523(1):171–74

Shapira M, Zhai RG, Dresbach T, Bresler T, Torres VI, et al. 2003. Unitary assembly of presynaptic active zones from Piccolo-Bassoon transport vesicles. *Neuron* 38(2):237–52

Snyder EM, Philpot BD, Huber KM, Dong X, Fallon JR, Bear MF. 2001. Internalization of ionotropic glutamate receptors in response to mGluR activation. *Nat. Neurosci.* 4(11):1079–85

Sorra KE, Fiala JC, Harris KM. 1998. Critical assessment of the involvement of perforations, spinules, and spine branching in hippocampal synapse formation. *J. Comp. Neurol.* 398(2):225–40

Spacek J. 1985a. Three-dimensional analysis of dendritic spines. II. Spine apparatus and other cytoplasmic components. *Anat. Embryol.* 171:235–43

Spacek J. 1985b. Three-dimensional analysis of dendritic spines. III. Glial sheath. *Anat. Embryol. (Berl.)* 171(2):245–52

Spacek J, Harris KM. 1997. Three-dimensional organization of smooth endoplasmic reticulum in hippocampal CA1 dendrites and dendritic spines of the immature and mature rat. *J. Neurosci.* 17(1):190–203

Spacek J, Harris KM. 2004. Trans-endocytosis via spinules in adult rat hippocampus. *J. Neurosci.* 24(17):4233–41

Star EN, Kwiatkowski DJ, Murthy VN. 2002. Rapid turnover of actin in dendritic spines and its regulation by activity. *Nat. Neurosci.* 5(3):239–46

Steward O, Levy WB. 1982. Preferential localization of polyribosomes under the base of dendritic spines in granule cells of the dentate gyrus. *J. Neurosci.* 2(3):284–91

Steward O, Reeves TM. 1988. Protein-synthetic machinery beneath postsynaptic sites on CNS neurons: association between polyribosomes and other organelles at the synaptic site. *J. Neurosci.* 8(1):176–84

Steward O, Schuman EM. 2003. Compartmentalized synthesis and degradation of proteins in neurons. *Neuron* 40(2):347–59

Stewart MG, Medvedev NI, Popov VI, Schoepfer R, Davies HA, et al. 2005. Chemically induced long-term potentiation increases the number of perforated and complex postsynaptic densities but does not alter dendritic spine volume in CA1 of adult mouse hippocampal slices. *Eur. J. Neurosci.* 21(12):3368–78

Tada T, Sheng M. 2006. Molecular mechanisms of dendritic spine morphogenesis. *Curr. Opin. Neurobiol.* 16(1):95–101

Tai CY, Mysore SP, Chiu C, Schuman EM. 2007. Activity-regulated N-cadherin endocytosis. *Neuron* 54(5):771–85

Takumi Y, Matsubara A, Rinvik E, Ottersen OP. 1999a. The arrangement of glutamate receptors in excitatory synapses. *Ann. N. Y. Acad. Sci.* 868:474–82

Takumi Y, Ramirez-Leon V, Laake P, Rinvik E, Ottersen OP. 1999b. Different modes of expression of AMPA and NMDA receptors in hippocampal synapses. *Nat. Neurosci.* 2(7):618–24

Togashi H, Abe K, Mizoguchi A, Takaoka K, Chisaka O, Takeichi M. 2002. Cadherin regulates dendritic spine morphogenesis. *Neuron* 35(1):77–89

Toni N, Buchs PA, Nikonenko I, Bron CR, Muller D. 1999. LTP promotes formation of multiple spine synapses between a single axon terminal and a dendrite. *Nature* 402(6760):421–25

Toresson H, Grant SG. 2005. Dynamic distribution of endoplasmic reticulum in hippocampal neuron dendritic spines. *Eur. J. Neurosci.* 22(7):1793–98

Trommald M, Hulleberg G, Andersen P. 1996. Long-term potentiation is associated with new excitatory spine synapses on rat dentate granule cells. *Learn. Mem.* 3(2–3):218–28

Ullian EM, Christopherson KS, Barres BA. 2004. Role for glia in synaptogenesis. *Glia* 47(3):209–16

Van Harreveld A, Fifkova E. 1975. Swelling of dendritic spines in the fascia dentata after stimulation of the perforant fibers as a mechanism of post-tetanic potentiation. *Exp. Neurol.* 49(3):736–49

Vazquez LE, Chen HJ, Sokolova I, Knuesel I, Kennedy MB. 2004. SynGAP regulates spine formation. *J. Neurosci.* 24(40):8862–72

Ventura R, Harris KM. 1999. Three-dimensional relationships between hippocampal synapses and astrocytes. *J. Neurosci.* 19(16):6897–906

Weiler IJ, Irwin SA, Klintsova AY, Spencer CM, Brazelton AD, et al. 1997. Fragile X mental retardation protein is translated near synapses in response to neurotransmitter activation. *Proc. Natl. Acad. Sci. USA* 94(10):5395–400

Witcher MR, Kirov SA, Harris KM. 2007. Plasticity of perisynaptic astroglia during synaptogenesis in the mature rat hippocampus. *Glia* 55(1):13–23

Wu K, Aoki C, Elste A, Rogalski-Wilk AA, Siekevitz P. 1997. The synthesis of ATP by glycolytic enzymes in the postsynaptic density and the effect of endogenously generated nitric oxide. *Proc. Natl. Acad. Sci. USA* 94(24):13273–78

Xiao MY, Zhou Q, Nicoll RA. 2001. Metabotropic glutamate receptor activation causes a rapid redistribution of AMPA receptors. *Neuropharmacology* 41(6):664–71

Xie Z, Huganir RL, Penzes P. 2005. Activity-dependent dendritic spine structural plasticity is regulated by small GTPase Rap1 and its target AF-6. *Neuron* 48(4):605–18

Zhai RG, Vardinon-Friedman H, Cases-Langhoff C, Becker B, Gundelfinger ED, et al. 2001. Assembling the presynaptic active zone: a characterization of an active one precursor vesicle. *Neuron* 29(1):131–43

Zhou Q, Homma KJ, Poo MM. 2004. Shrinkage of dendritic spines associated with long-term depression of hippocampal synapses. *Neuron* 44(5):749–57

Zhu JJ, Qin Y, Zhao M, Van Aelst L, Malinow R. 2002. Ras and Rap control AMPA receptor trafficking during synaptic plasticity. *Cell* 110(4):443–55

Zito K, Knott G, Shepherd GM, Shenolikar S, Svoboda K. 2004. Induction of spine growth and synapse formation by regulation of the spine actin cytoskeleton. *Neuron* 44(2):321–34

Zito K, Parnas D, Fetter RD, Isacoff EY, Goodman CS. 1999. Watching a synapse grow: noninvasive confocal imaging of synaptic growth in *Drosophila*. *Neuron* 22(4):719–29

Ziv NE, Smith SJ. 1996. Evidence for a role of dendritic filopodia in synaptogenesis and spine formation. *Neuron* 17(1):91–102

Place Cells, Grid Cells, and the Brain's Spatial Representation System

Edvard I. Moser,[1] Emilio Kropff,[1,2] and May-Britt Moser[1]

[1]Kavli Institute for Systems Neuroscience and Centre for the Biology of Memory, Norwegian University of Science and Technology, 7489 Trondheim, Norway

[2]Cognitive Neuroscience Sector, International School for Advanced Studies, Trieste, Italy; email: edvard.moser@cbm.ntnu.no

Annu. Rev. Neurosci. 2008. 31:69–89

First published online as a Review in Advance on February 19, 2008

The *Annual Review of Neuroscience* is online at neuro.annualreviews.org

This article's doi: 10.1146/annurev.neuro.31.061307.090723

Key Words

hippocampus, entorhinal cortex, path integration, attractor, memory, phase precession

Abstract

More than three decades of research have demonstrated a role for hippocampal place cells in representation of the spatial environment in the brain. New studies have shown that place cells are part of a broader circuit for dynamic representation of self-location. A key component of this network is the entorhinal grid cells, which, by virtue of their tessellating firing fields, may provide the elements of a path integration–based neural map. Here we review how place cells and grid cells may form the basis for quantitative spatiotemporal representation of places, routes, and associated experiences during behavior and in memory. Because these cell types have some of the most conspicuous behavioral correlates among neurons in nonsensory cortical systems, and because their spatial firing structure reflects computations internally in the system, studies of entorhinal-hippocampal representations may offer considerable insight into general principles of cortical network dynamics.

Contents

INTRODUCTION

Questions about how we perceive space and our place in that space have engaged epistemologists for centuries. Although the British empiricists of the seventeenth and eighteenth centuries thought that all knowledge about the world was ultimately derived from sensory impressions, Kant argued that some ideas exist as a priori intuitions, independent of specific experience. One of these ideas is the concept of space, which he considered an innate organizing principle of the mind, through which the world is, and must be, perceived. With the birth of experimental psychology and neuroscience a century later, the organization and development of spatial behavior and cognition could be analyzed experimentally. We review evidence from the past three decades that indicates the presence of a preconfigured or semipreconfigured brain system for representation and storage of self-location relative to the external environment. In agreement with the general ideas of Kant, place cells and grid cells in the hippocampal and entorhinal cortices may determine how we perceive and remember our position in the environment as well as the events we experience in that environment.

PLACE CELLS AND THE HIPPOCAMPAL MAP

The experimental study of spatial representations in the brain began with the discovery of place cells. More than 35 years ago, O'Keefe & Dostrovsky (1971) reported spatial receptive fields in complex-spiking neurons in the rat hippocampus, which are likely to be pyramidal cells (Henze et al. 2000). These place cells fired whenever the rat was in a certain place in the local environment (the place field of the cell; **Figure 1a**). Neighboring place cells fired at different locations such that, throughout the hippocampus, the entire environment was represented in the activity of the local cell population (O'Keefe 1976, Wilson & McNaughton 1993). The same place cells participated in representations for different environments, but the relationship of the firing fields differed from one setting to the next (O'Keefe & Conway 1978). Inspired by Tolman (1948), who suggested that local navigation is guided by internal "cognitive maps" that flexibly represent the overall spatial relationships between landmarks in the environment, O'Keefe & Nadel (1978) proposed that place cells are the basic elements of a distributed noncentered map-like representation. Place cells were suggested to provide the animal with a dynamic, continuously updated representation of allocentric space and the animal's own position in that space. We now have abundant evidence from a number of mammalian species demonstrating that the hippocampus plays a key role in spatial representation and spatial memory (Nadel 1991, Rolls 1999, Ekstrom et al. 2003, Ulanovsky & Moss 2007), although new evidence suggests that position is only one of several facets of experience stored in the hippocampal network (Eichenbaum et al. 1999, Leutgeb et al. 2005b).

GRID CELLS AND THE ENTORHINAL MAP

All subfields of the hippocampal region contain place-modulated neurons, but the most distinct firing fields are found in the CA areas (Barnes et al. 1990). On the basis of the apparent amplification of spatial signals from the entorhinal cortex to the CA fields (Quirk et al. 1992), many investigators thought, until recently, that place signals depended primarily on computations within the hippocampal network. This view was challenged by the observation that spatial firing persisted in CA1 neurons after removal of intrahippocampal inputs from the dentate gyrus (McNaughton et al. 1989) and CA3 (Brun et al. 2002). This raised the possibility that spatial signals were conveyed to CA1 by the direct perforant-path projections from layer III of the entorhinal cortex. Projection neurons in layers II and III of the medial entorhinal cortex (MEC) were subsequently shown to exhibit sharply tuned spatial firing, much like place cells in the hippocampus, except that each cell had multiple firing fields (Fyhn et al. 2004). The many fields of each neuron formed a periodic triangular array, or grid, that tiled the entire environment explored by the animal (Hafting et al. 2005) (**Figure 1***b*). Such grid cells collectively signaled the rat's changing position with a precision similar to that of place cells in the hippocampus (Fyhn et al. 2004). The graphics paper–like shape of the grid immediately indicated grid cells as possible elements of a metric system for spatial navigation (Hafting et al. 2005), with properties similar to that of the allocentric map proposed for the hippocampus more than 25 years earlier (O'Keefe & Nadel 1978).

How do grid representations map onto the surface of the entorhinal cortex? Each grid is characterized by spacing (distance between fields), orientation (tilt relative to an exter-

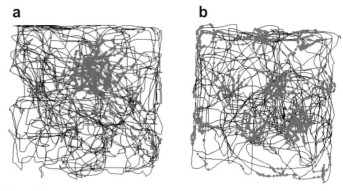

a **b**

Figure 1

Place cell in the hippocampus (*a*) and grid cell in the medial entorhinal cortex (MEC) (*b*). Spike locations (*red*) are superimposed on the animal's trajectory in the recording enclosure (*black*). Whereas most place cells have a single firing location, the firing fields of a grid cell form a periodic triangular matrix tiling the entire environment available to the animal.

nal reference axis), and phase (*xy* displacement relative to an external reference point). Although cells in the same part of the MEC have similar grid spacing and grid orientation, the phase of the grid is nontopographic, i.e., the firing vertices of colocalized grid cells appear to be shifted randomly, just like the fields of neighboring place cells in the hippocampus. The spacing increases monotonically from dorsomedial to ventrolateral locations in MEC (Hafting et al. 2005, Solstad et al. 2007), mirroring the increase in size of place fields along the dorsoventral axis of the hippocampus (Jung et al. 1994, Maurer et al. 2005, Kjelstrup et al. 2007). Cells in different parts of the MEC may also have different grid orientations (Hafting et al. 2005), but the underlying topography, if there is one, has not been established. Thus, we do not know whether the entorhinal map has discrete subdivisions. The entorhinal cortex has several architectonic features suggestive of a modular arrangement, such as periodic bundling of pyramidal cell dendrites and axons and cyclic variations in the density of immunocytochemical markers (Witter & Moser 2006), but whether the anatomical cell clusters correspond to functionally segregated grid maps, each with their own spacing and orientation, remains to be determined.

MEC: medial entorhinal cortex

SENSORY CUES AND PATH INTEGRATION

Which factors control the spatial discharge pattern of place cells and grid cells? Early on, it became apparent that place fields are strongly influenced by distal sensory cues. When rats walked in a circle, rotation of the circumferential cues caused rotation of the place fields, whereas rotation of the proximal environment itself, in the presence of fixed distal cues, failed to change the firing locations (O'Keefe & Conway 1978, O'Keefe & Speakman 1987, Muller & Kubie 1987). Extending the sides of a rectangular recording box stretched or split the fields in the extended direction (O'Keefe & Burgess 1996). These observations indicate a primary role for extrinsic cues, and especially geometric boundaries, in defining the firing location of a place cell, although individual proximal landmarks do exert some influence under certain conditions (Muller & Kubie 1987, Gothard et al. 1996b, Cressant et al. 1997, Shapiro et al. 1997, Zinyuk et al. 2000, Fyhn et al. 2002). A similar dependence on distal landmarks and boundaries has since been demonstrated for entorhinal grid cells (Hafting et al. 2005, Barry et al. 2007). However, place cells and grid cells do not merely mirror sensory stimuli. When salient landmarks are removed while the animal is running in a familiar environment, both cell types continue to fire in the original location (Muller & Kubie 1987, Hafting et al. 2005). Moreover, representations in the hippocampus are often maintained when the recording box is transformed smoothly into another familiar box (J.K. Leutgeb et al. 2005), suggesting that the history of testing may sometimes exert stronger control on place cell activity than would the actual sensory stimuli.

Discrete representations of individual places would not be sufficient to support navigation from one place to another. The brain needs algorithms for linking the places in metric terms. When animals move away from a start position, they can keep track of their changing positions by integrating linear and angular self-motion (Barlow 1964, Mittelstaedt & Mittelstaedt 1980, Etienne & Jeffery 2004).

This process, referred to as path integration, is a primary determinant of firing in place cells. When rats are released from a movable start box on a linear track with a fixed goal at the end, the firing is initially determined by the distance from the start box (Gothard et al. 1996a, Redish et al. 2000). Although the activity is soon corrected against the external landmarks, this initial firing pattern suggests that place-selective firing can be driven by self-motion information alone.

Where is the path integrator? The hippocampus itself is not a good candidate. Without invoking a separate hippocampal circuit for this process, it would be hard to imagine how the algorithm could be adapted for each of the many overlapping spatial maps stored in the very same population of place cells. Place cells may instead receive inputs from a general metric navigational system outside the hippocampus (O'Keefe 1976). Grid cells may be part of this system. The persistence of grid fields after removal or replacement of major landmarks points to self-motion information as the primary source for maintaining and updating grid representations (Hafting et al. 2005, Fyhn et al. 2007). The system has access to direction and speed information required to transform the representation during movement (Sargolini et al. 2006b). However, whereas path integration may determine the basic structure of the firing matrix, the grid map is anchored to geometric boundaries and landmarks unique to each environment. These associations may, under some circumstances, override concurrent path integration–driven processes, such as when the sides of a familiar rectangular recording environment are extended moderately (Barry et al. 2007). The origin of the self-motion signals and the mechanisms for integration of self-motion signals with extrinsic sensory inputs have not been determined.

THEORETICAL MODELS OF GRID FORMATION

Most current models of grid field formation suggest that MEC neurons path-integrate speed and direction signals provided by specialized cells, whereas sensory information related to the environment is used for setting the initial parameters of the grid or adjusting it to correct the cumulative error intrinsically associated with the integration of velocity.

One class of models suggests that grid formation is a result of local network activity. In these models, a single position is represented by an attractor, a stable firing state sustained by recurrent connections with robust performance in the presence of noise (Hopfield 1982, Amit 1989, Rolls & Treves 1998). A network can store several attractor states associated with different locations and retrieve any of them in response to sensory or path integration cues. When a large number of very close positions are represented, a continuous attractor emerges, which permits a smooth variation of the representation in accordance with the rat's trajectory (Tsodyks & Sejnowski 1995, Samsonovich & McNaughton 1997, Battaglia & Treves 1998). We review two of the models in this class.

Fuhs & Touretzky (2006) proposed that MEC is roughly topographically organized; neighboring grid cells display similar activity, and the representation of a single place forms a grid pattern on the neuron layer. Such a pattern emerges naturally at a population level in a network with Mexican hat connectivity, where every neuron is excited by its neighbors, inhibited by neurons at an intermediate distance, and unaffected by those far away. The authors include several alternating excitation and inhibition ranges and an overall decay of the synaptic strength with distance. With this connectivity rule, a grid of activity appears spontaneously in the MEC layer, except at the borders of the layer, where the lack of balance overexcites neurons and additional attenuation is required. To transform the grid pattern in the MEC layer into a grid firing field in each neuron, the representation of a single place must be rigidly displaced along the topographically organized network following the movements of the rat (**Figure 2a**). This happens if any given neuron receives an input proportional to the running speed only when the rat runs in a preferred direction, which is different for each neuron. Increased speed produces increased excitation and a faster displacement of the grid pattern across the neural layer. The proposed mechanism may fail to displace the initial representation accurately for realistic trajectories of the rat, resulting in neurons that do not express grid fields (Burak & Fiete 2006; see authors' response at **http://www.jneurosci.org/cgi/data/26/37/9352/DC1/1**).

The above model is challenged by the apparent lack of topography in grid phases of neighboring MEC neurons (Hafting et al. 2005). Motivated by this experimental observation, McNaughton et al. (2006) proposed, in an alternative model, that a topographically arranged network is present in the cortex during early postnatal development and serves as a tutor to train MEC cell modules with randomly distributed Hebbian connections and no topographical organization (**Figure 2b**). Attractor representations of space are formed in each of these modules during training. Because the inputs from the tutor are scrambled, neurons in a similar phase are not necessarily neighbors, but they are associated through synaptic plasticity. If after the training period neurons in a module were rearranged according to their firing phase or connection strength, a single bump of activity would be observed at any time. Because the tutor has the periodicity of a grid, the rearranged network has no borders and resembles the surface of a torus (**Figure 2b**). To displace representations along the abstract space of the continuous attractor, the model introduces an additional layer of cells whose firing is modulated by place, head direction, and speed. Sargolini et al. (2006b) have identified neurons with such properties. Neurons in this hidden layer may receive input from currently firing grid cells and project back selectively to grid cells that fire next along the trajectory; the activation of target cells depends on the current head direction and velocity of the rat (Samsonovich & McNaughton 1997, McNaughton et al. 2006; **Figure 2b**).

In a second class of models, path integration occurs at the single cell level and is intimately

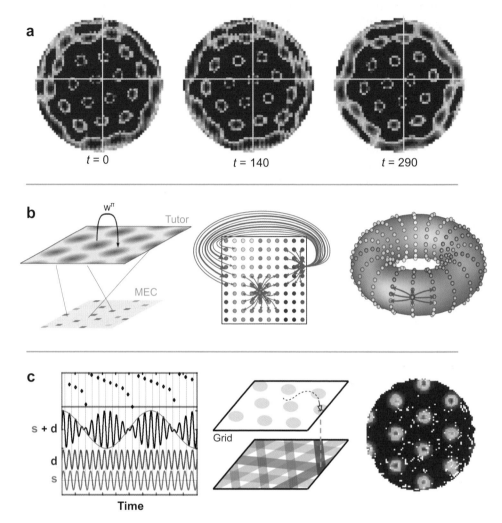

Figure 2

Different models of grid field formation. (*a*) Three consecutive snapshots of the activity in an MEC network adapted from Fuhs & Touretzky (2006). A stable grid pattern of activity emerges spontaneously, and following rightward movement of the rat, the activity is rigidly displaced in the corresponding direction across the network. In the origin of the white axes, a neuron is not firing at time $t = 0$, fires in the maximum of a grid node at $t = 140$, and is at rest again at $t = 290$. (*b*) Geometry of the connectivity inside a module of grid cells, adapted from McNaughton et al. (2006). A topographically arranged network similar to the one in *a* serves as a tutor to train an MEC module with no topographical arrangement (*left*). Owing to the periodicity of the tutor, if after training the module was rearranged so as to make strongly connected neurons neighbors (*center*), the effective geometry would be that of a toroidal surface (*right*). (*c*) Sum of three or more linear interference maps, adapted from Burgess et al. (2007). While the animal is running on a linear track, the sum of a somatic (s) and a dendritic (d) oscillation with slightly different frequencies results in an interference pattern exhibiting phase precession and slow periodic spatial modulation (*left*). Combining three or more linear interference maps, responding to different projections of the velocity, results in a grid map (*center*). Simulated grid map after 10 min of a rat's actual trajectory (*right*). All panels reprinted with permission (Fuhs & Touretzky 2006, McNaughton et al. 2006, Burgess et al. 2007).

related to phase precession, a progressive advance of spike times relative to the theta phase observed in hippocampal place cells (O'Keefe & Recce 1993) and entorhinal grid cells (Hafting et al. 2006) when rats run through a localized firing field. O'Keefe & Reece (1993) modeled phase precession on a linear track as the sum between two oscillatory signals with frequencies around the theta rhythm but slightly differing by an amount proportional to the rat's running speed. The resulting interference pattern can be decomposed into an oscillation at the mean of the two frequencies, which advances with respect to the slower (theta) rhythm, and a slow periodical modulation with a phase that integrates the rat's speed and thus reflects its position along the track (O'Keefe & Recce 1993, O'Keefe & Burgess 2005, Lengyel et al. 2003). Burgess et al. (2007) extended the interference model to two dimensions by considering the interaction of one somatic intrinsic oscillator of frequency w_s (~ theta rhythm) with several dendritic oscillators, each with a frequency equal to w_s plus a term proportional to the projection of the rat velocity in some characteristic preferred direction. The interference of the somatic signal with each of these dendritic oscillators has a slow modulation that integrates the preferred component of the velocity into a linear spatial interference pattern (a plane wave). When combining several of these linear patterns, a triangular grid map is obtained, provided that their directions differ in multiples of $60°$ and the phases are set in such a way that all maxima coincide, a choice of parameters that could result from a self-organization process maximizing the neuron's overall activity. In a variant of this idea, Blair et al. (2007) proposed that the interference sources could be two theta-grid cells (grid cells with a high spatial frequency associated with the theta rhythm), differing in either their relative size or their orientation, although evidence for such cells has not yet been reported.

The above models make different predictions about the organization of the entorhinal grid map. The network models explicitly or implicitly rely on a discontinuous module arrangement with different grid spacing and grid orientation. Fuhs & Touretzky (2006) propose large clusters of MEC cells each with fixed grid spacing and orientation but with continuous topographic variation of phase among neighbors. They estimate ~17 such clusters in layer II, restricting the variability in grid parameter values in a given environment. Possibly smaller and larger in number, the attractor networks proposed by McNaughton and colleagues inherit in principle the orientation and spacing from their tutor, whereas the phase varies randomly inside each cluster. However, if two networks trained by the same tutor were fed with speed signals of different gain, the displacement of the representations in response to a rat movement would differ, resulting in grids with different spacing, as observed along the dorsoventral axis. Burgess et al. associate such a modulation in spacing with a gradient in the frequency of subthreshold membrane potential oscillations along this axis, a prediction that was recently verified (Giocomo et al. 2007), whereas their model is agnostic to neighboring cells' orientation and phase.

PLACE FIELDS MAY BE EXTRACTED FROM GRID FIELDS

Inspired by Fourier analysis, researchers have proposed that grid fields of different spacing, playing the role of periodic basis functions, combine linearly to generate place fields in the hippocampus (O'Keefe & Burgess 2005, Fuhs & Touretzky 2006, McNaughton et al. 2006). The resulting hippocampal representation would be periodic but because the period would be equal to the least common multiple of the grid spacings, and because it would be further enhanced by differences in grid orientation, only single fields would be observed in standard experimental settings. The peak of the representation would be at the location where most of the contributing grids are in phase. In a computational model, Solstad et al. (2006) showed that in small two-dimensional environments, single place fields can be formed by summing the activity of a modest number (10–50) of grid cells with relatively similar grid phases, random orientations, and a biologically plausible range of spacings corresponding with convergence of inputs from ~25% of the dorsoventral axis of MEC (Dolorfo & Amaral 1998). Rolls et al. (2006) showed that the choice of grid cells contributing to a given place field need not be hardwired but can result from a competitive Hebbian learning process starting from random connectivity, provided that enough variability in orientation, phase, and spacing is available in the afferent population of grid cells. They also showed that if variability in the in-peak frequency of grid cells is considered, more place cells have a single field, whereas trace learning (a variant of Hebbian learning that uses short time averages of, for example, the postsynaptic firing rate) produces broader fields, more similar to the ones observed in hippocampus.

The idea that place fields emerge through LTP-like competitive learning mechanisms receives only partial support from studies of place field formation in the presence of NMDA receptor blockers. When animals explore new environments, it takes several minutes for the firing fields to reach a stable state (Hill 1978, Wilson & McNaughton 1993, Frank et al. 2004). During this period, place fields may fade in or out, or their fields may expand toward earlier parts of the trajectory (Mehta et al. 1997, 2000). The stabilization is slower in CA3 than in CA1 (Leutgeb et al. 2004). Although synaptic plasticity is necessary for experience-dependent field expansions (Ekstrom et al. 2001), synaptic modifications may not be required for manifestation of place-specific firing as such. In the absence of functional NMDA receptors, CA1 place cells continue to express spatially confined firing fields, although some selectivity and stability may be lost (McHugh et al. 1996, Kentros et al. 1998). Whether hardwired connections are sufficient for place cell formation in all parts of the circuit remains to be determined, however. Preliminary data suggest that during exploration of new environments, systemic blockade of NMDA receptors disrupts spatial selectivity in dentate granule cells while CA3 cells continue to exhibit localized activity (Leutgeb et al. 2007b), raising the possibility

that on dentate granule cells, unlike hippocampal pyramidal cells, spatial selectivity may be established by competitive selection of active inputs.

PLACE CELLS AND HIPPOCAMPAL MEMORY

Following the discovery of place cells, several studies indicated a broader role for the hippocampus in representation and storage of experience, consistent with a long tradition of work on humans implying hippocampal involvement in declarative and episodic memory (Scoville & Milner 1957, Squire et al. 2004). Not only are hippocampal neurons triggered by location cues, but also they respond to salient events in a temporal sequence (Hampson et al. 1993) and nonspatial stimuli such as texture or odors (Young et al. 1994, Wood et al. 1999). However, nonspatial variables are not represented primarily by a dedicated subset of neurons or a nonspatial variant of the place cells. Fenton and colleagues observed that when a rat passed through a neuron's place field, the rate variation across traversals substantially exceeded that of a random model with Poission variance (Fenton & Muller 1998; Olypher et al. 2002). This excess variance, or overdispersion, raised the possibility that nonspatial signals are represented in place cells on top of the location signal by continuous rate modulation within the field. More direct support for this idea comes from the observation that, in hippocampal cell assemblies, spatial and nonspatial variables (place and color) are represented independently by variation in firing location and firing rate, respectively (Leutgeb et al. 2005a). Together these studies indicate a conjunctive spatial-nonspatial code for representation of experience in the hippocampus.

How could memories be stored in the place cell system? On the basis of the extensive intrinsic connectivity and modifiability of the CA3 network, theoretical work has indicated attractor dynamics (Hopfield 1982, Amit 1989) as a potential mechanism for low-interference storage of arbitrary input patterns to the hippocampus (McNaughton & Morris 1987, Treves & Rolls 1992, Hasselmo et al. 1995, McClelland & Goddard 1996, Rolls & Treves 1998). In networks with discrete attractor states (a Hopfield network), associative connections would allow stored memories to be recalled from degraded versions of the original input (pattern completion) without mixing up the memory with other events stored in the network (pattern separation).

Many observations suggest that place cells perform both pattern completion and pattern separation. Pattern completion is apparent from the fact that place cells maintain their location specificity after removing many of the landmarks that originally defined the environment (O'Keefe & Conway 1978, Muller & Kubie 1987, O'Keefe & Speakman 1987, Quirk et al. 1990, Nakazawa et al. 2002). Representations are regenerated with greater strength in CA3 than in CA1 (Lee et al. 2004, Vazdarjanova & Guzowski 2004, J.K. Leutgeb et al. 2005), possibly because the relative lack of recurrent collaterals in CA1 makes firing patterns more sensitive to changes in external inputs. As predicted from the theoretical models, pattern completion is disrupted by blockade of NMDA receptor–dependent synaptic plasticity in CA3 (Nakazawa et al. 2002). Pattern separation can be inferred from the ability of place cells to undergo substantial remapping after only minor changes in the sensory input, such as a change in color or shape of the recording enclosure or a change in the overall motivational context (Muller & Kubie 1987, Bostock et al. 1991, Markus et al. 1995, Wood et al. 2000). Two forms of remapping have been reported (Leutgeb et al. 2005b): The cell population may undergo complete orthogonalization of both firing locations and firing rates (global remapping), or the rate distribution may be changed selectively in the presence of stable firing locations (rate remapping). In each instance, remapping tends to be instantaneous (Leutgeb et al. 2006, Fyhn et al. 2007), although delayed transitions occur under some training conditions (Lever et al. 2002). The disambiguation of the firing patterns is stronger in CA3 than in CA1

(Leutgeb et al. 2004, Vazdarjanova & Guzowski 2004). Pattern separation in the CA areas may be facilitated by prior orthogonalization of hippocampal input patterns in the dentate gyrus (Leutgeb et al. 2007a, McHugh et al. 2007, Leutgeb & Moser 2007). The sparse firing of the granule cells (Jung & McNaughton 1993, Chawla et al. 2005, Leutgeb et al. 2007a) and the formation of one-to-one detonator synapses between granule cells and pyramidal cells in CA3 (Claiborne et al. 1986, Treves & Rolls 1992) may jointly contribute to decorrelation of incoming cortical signals in the dentate gyrus (McNaughton & Morris 1987, Treves & Rolls 1992).

Although place cells exhibit both pattern completion and pattern separation, we cannot discount that firing is maintained by unidentified stimuli that are present both during encoding with the full set of landmarks and during retrieval with a smaller subset of landmarks. A more direct way to test the attractor properties of the network is to measure the response of the place cell population to continuous or step-wise transformations of the recording environment. Wills et al. (2005) trained rats, on alternating trials, in a square and a circular version of a morph box with flexible walls. Different place cell maps were formed for the two environments. On the test day, rats were exposed to multiple intermediate shapes. A sharp transition from square-like representations to circle-like representations was observed near the middle in the geometric sequence, as predicted if the network had discrete attractor states corresponding to each of the familiar square and circular shapes. Whether the implied attractor dynamics occurs in the hippocampus itself or in upstream areas such as the MEC, or both, remains to be determined. Because the changes in firing patterns indicate global remapping and global remapping is invariably accompanied by realignment and rotation of the entorhinal grid map (Fyhn et al. 2007), it may well be that some of the underlying attractor dynamics lies in the MEC (S. Leutgeb et al. 2007).

Hippocampal representations cannot always be discontinuous as in a Hopfield network.

Hippocampal memories are characterized by events that are tied together in sequences (Tulving & Markowitsch 1998, Shapiro et al. 2006), just like positions are tied together in two dimensions as spatial maps. A continuous attractor network (Tsodyks & Sejnowski 1995) may be needed to preserve the continuity of both types of representations. Recent observations support this possibility. When two recording environments are morphed in the presence of salient distal landmarks, the firing locations remain constant across the sequence of intermediate shapes, but the rate distribution changes smoothly between the preestablished states (J.K. Leutgeb et al. 2005). Stable states can thus be attained along the entire continuum between two preexisting representations (Blumenfeld et al. 2006). This ability to represent continua may provide hippocampal networks with a capacity for encoding and retrieving consecutive inputs as uninterrupted, distinguishable episodes.

SEQUENCE CODING IN PLACE CELLS AND GRID CELLS

The idea that neuronal representations have a temporal dimension can be traced back to Hebb (1949), who suggested that cell assemblies are activated in sequences, and that such phase sequences may provide the neural basis of thought. More recent theoretical studies have proposed a number of mechanisms by which temporal sequences could be formed and stored as distinct entities. Such mechanisms include potentiation of asymmetric connections between serially activated neurons (Blum & Abbott 1996) and orthogonalization of the individual sequence elements (McNaughton & Morris 1987). However, although these and related ideas have nourished important experiments, the mechanisms of sequence coding are still not well understood.

The most intriguing example of a temporal code in the rodent hippocampus is probably the expression of theta phase precession in place cells when animals follow a fixed path in a linear environment (O'Keefe & Recce

1993, Skaggs et al. 1996). The reliable tendency of place cells to fire at progressively earlier phases of the theta rhythm during traversal of the place field increases the information about the animal's location in the environment, both in one-dimensional (Jensen & Lisman 2000, Harris et al. 2003, Huxter et al. 2003) and two-dimensional (Huxter et al. 2007) environments. Moreover, when the rat runs through a sequence of overlapping place fields from different cells, the firing sequence of the cells will be partially replicated in compressed form within individual theta periods (Jensen & Lisman 1996, Skaggs et al. 1996, Tsodyks et al. 1996, Dragoi & Buzsaki 2006). The repeated compression of discharges within windows of some tens of milliseconds provides a mechanism for associating temporally extended path segments on the basis of the rules of spike-dependent plasticity (Dan & Poo 2004). Such associations may be necessary for storage of route and event representations in the network.

We do not know the neuronal mechanisms of phase precession or their locations in the brain. O'Keefe & Recce (1993) suggested that phase precession is caused by interference between intrinsic and extrinsic neuronal membrane potential oscillations with slightly different frequencies (**Figure 2c**). Spike times are determined, in this view, by the high-frequency wave of the interference pattern, which advances progressively with reference to the field theta activity. An alternative set of models suggests that phase precession occurs when theta-modulated inhibition interacts with progressively increasing excitation of the place cell over the extent of the firing field (Harris et al. 2002, Mehta et al. 2002). Because of the ramp-up of the excitation, cells would discharge at successively earlier points in the theta cycle. A third type of models puts the mechanism at the network level (Jensen & Lisman 1996, Tsodyks et al. 1996, Wallenstein & Hasselmo 1998). By intrinsic connections between place cells, cells that are strongly activated by external excitation may initiate, at each location, a wave of activity that spreads toward the cells with place fields that are further along the animal's path.

Experimental data do not rule out any of these models. To understand the underlying mechanisms, it may be necessary to identify the neural circuits that support phase precession and then determine which of those are able to generate phase precession on their own. Recent observations imply that phase precession is not exclusively hippocampal. Phase precession in CA1 is not frozen by brief inhibition of hippocampal activity (Zugaro et al. 2005), which suggests that the precession may, at least under some circumstances, be imposed on hippocampal place cells by cells from other areas. One such area could be the MEC. Grid cells in this region exhibit phase precession (Hafting et al. 2006). Of particular interest is the stellate cell in layer II of MEC. Because stellate cells exhibit voltage-dependent intrinsic oscillations that may be faster than the field theta rhythm (Alonso & Llinas 1989, Klink & Alonso 1993, Giocomo et al. 2007), these cells may express interference patterns of the type proposed by O'Keefe and colleagues (O'Keefe & Recce 1993, Burgess et al. 2007). Additional assumptions must be invoked, however, to explain the stronger correlation of phase with position, and the presence of intrinsic oscillations does not rule out other potential mechanisms including rampant excitation or neural network properties.

REPLAY AND PREPLAY IN PLACE CELL ENSEMBLES

After a memory trace is encoded during an experience, the memory is thought to undergo further consolidation off-line when the subject is sleeping or is engaged in consummatory activities. Although the cascade of events leading to consolidation of hippocampal memory is not well understood, ensemble recordings in sleeping rats have provided some clues. Cells that are coactivated in the hippocampus during awake behavior continue to exhibit correlated activity during sleep episodes subsequent to behavioral testing (Wilson & McNaughton 1994). The order of firing is generally preserved but the rate may be faster (Skaggs & McNaughton

1996, Lee & Wilson 2002). Such replay or reactivation is associated with hippocampal sharp waves, which are bursts of synchronous pyramidal-cell activity during slow-wave sleep and awake rest (Buzsáki et al. 1983). Sharp-wave bursts can induce plasticity in downstream areas and may therefore be involved in information transfer from the hippocampus to the neocortex during the consolidation time window (Buzsáki 1989). Direct evidence for this hypothesis is still lacking. The correlation of sharp waves in CA1 and upstates in the neocortex suggests interaction between these brain areas during sleep, but the slight delay of membrane potential changes in CA1 compared with neocortex suggests that signals are transferred from neocortex to CA1 and not vice versa (Isomura et al. 2006, Hahn et al. 2007, Ji & Wilson 2007). Much work is still needed to establish the potential significance of sleep-associated reactivation in memory consolidation.

Reactivation of place cell discharges is observed not just during sleep. Recent studies have reported reactivation during sharp waves that are "interleaved" in waking activities (Foster & Wilson 2006, O'Neill et al. 2006), which indicates possible mechanisms for maintaining recent memories on shorter time scales when rest is not possible. When the animal stops at the turning points of a linear track, the hippocampus enters sharp-wave mode, and the preceding sequence of place-cell activity on the track is replayed, but now in a time-reversed order (Foster & Wilson 2006). Reverse replay may facilitate the storage of goal-directed behavioral sequences by allowing reinforcement signals that coincide with reward induction (e.g., dopamine release) to strengthen primarily the later parts of the behavioral sequence. The point at which sharp wave-associated reactivation is forward or backward has not been determined, and further work is needed to establish how the two forms of reactivation contribute to memory, if at all.

Not all nonlocal activity is retrospective. During theta-associated behaviors, place-cell discharges may sometimes correlate with future locations. Training rats to choose the correct goal location on discrete trials in a plus maze, Ferbinteanu & Shapiro (2003) found that firing on the start arm was in some cells determined by the subsequent choice of goal arm. Johnson & Redish (2007) showed that when rats reach the choice point of a modified T maze, representations sweep ahead of the animal in the direction of the reward location. The forward-looking activity was experience dependent. Together, these observations suggest that before animals choose between alternative trajectories, future locations are preplayed in the hippocampal place-cell ensembles, probably by retrieving stored representations. This interpretation implies a direct involvement of hippocampal networks in active problem solving and evaluation of possible futures, consistent with the recently reported failure to imagine new experiences in patients with hippocampal amnesia (Hassabis et al. 2007).

SPATIAL MAPS INCLUDE MORE THAN HIPPOCAMPUS AND ENTORHINAL CORTEX

Spatial representation engages a wide brain circuit. A key component is the network of head direction–modulated cells in presubiculum (Ranck 1985, Taube et al. 1990) and upstream areas such as the anterior thalamus (see Taube 1998 for review). Axons from the presubiculum terminate in layers III and V of MEC (Witter & Amaral 2004), where grid cells are modulated by head direction (Sargolini et al. 2006b), possibly as a consequence of the presubicular input. Head-direction cells may also control grid field orientation. The MEC has strong connections with the parasubiculum and the retrosplenial cortex (Witter & Amaral 2004), which contain cells that are tuned to position or head direction (Chen et al. 1994, Taube 1995, Sargolini et al. 2006a). Lesion studies suggest that the retrosplenial cortex is necessary for path integration–based navigation and topographic memory (Sutherland et al. 1988, Takahashi et al. 1997, Cooper & Mizumori 1999), although little is known about the specific role of this area in these processes. The MEC also interacts closely with the lateral entorhinal cortex, where cells apparently do not show spatial modulation (Hargreaves et al. 2005). Although the lateral entorhinal cortex provides a major component of the cortical input to the hippocampus, its function is not known.

The parietal cortex may be an important element of the spatial representation and navigation system. Similar to rats with lesions in the entorhinal cortex, rats with parietal cortex lesions fail to navigate back to a refuge under conditions where the return pathway can be computed only on the basis of the animal's own movement (Save et al. 2001, Parron & Save 2004). Rats and humans with parietal cortex lesions also fail to acquire spatial tasks and remember positional relationships (Kolb et al. 1983, DiMattia & Kesner 1988, Takahashi et al. 1997). The parietal cortex of the rat contains neurons that map navigational epochs when the animal follows a fixed route (Nitz 2006). These cells fire in a reliable order at specific stages of the route, but the firing is not determined by landmarks or movement direction. Much additional work is required to determine whether these discharge patterns are path integration based and contribute to a representation of self-location, and whether they are dependent on grid cells (Hafting et al. 2005) and path-associated firing (Frank et al. 2000) in the MEC. Finally, it is possible that navigation can be aided also by action-based neural computations in the striatum, where key positions along the trajectory are reflected in the local activity (Jog et al. 1999; see also Packard & McGaugh 1996, Hartley et al. 2003).

DEVELOPMENT OF THE SPATIAL REPRESENTATION SYSTEM

Is space an organizing principle of the mind, imposed on experience according to brain preconfiguration, as Kant suggested? Some properties of the spatial representation system certainly indicate a preconfigured network. Grid fields appear from the very first moment of exploration in a new environment and persist following major landmark removal (Hafting et al.

2005), and the spatial phase relationship of different grid cells remains constant across environments (Fyhn et al. 2007). This suggests a rigidly structured map, but whether animals are born with it remains to be determined. Grid structure may appear from genetically specified properties of the entorhinal circuit, but specific maturational programs and experiences may also be necessary for the development of an adult map-like organization (Hubel et al. 1977, McNaughton et al. 2006). Unfortunately, we do not know much about the ontogeny of the spatial representation system. Apparently, only one study has systematically explored the development of spatial representations. Martin & Berthoz (2002) found that sharp and confined place fields were not expressed in CA1 until ~P50 in the rat. The lack of functional studies is matched by a similarly fragmented understanding of how intrinsic and extrinsic connections of the entorhinal and hippocampal cortices develop relative to each other. The few existing studies suggest that, in the rat, some connections of the system, such as the cholinergic innervation of the entorhinal cortex, appear only at ~35 days of age (Ritter et al. 1972, Matthews et al. 1974). Although most other connections are apparently present a few days after birth, functional entorhinal-hippocampal circuits may not emerge until all connections are in place. The slow development of the entorhinal-hippocampal system leaves considerable possibilities for postnatal shaping of the spatial map. A major challenge, if we want to address the Kantian question about the a priori nature of space perception, will be to identify the factors that control map formation in young animals.

CONCLUSION

The past few years have witnessed radical advances in our understanding of the brain's spatial representation system. We are beginning to see the contours of a modularly organized network with grid cells, place cells, and head-direction cells as key computational units. Interactions between grid cells and place cells may underlie the unique ability of the hippocampus to store large amounts of orthogonalized information. The mechanisms of this interaction, their significance for memory storage, and their interactions with representations in other cortical regions remain to be determined. More work is also required to establish the computational principles by which grid maps are formed and by which self-location is mapped dynamically as animals move through spatial environments. Perhaps the largest knowledge gap is concerned with how grid structure emerges during ontogenesis of the nervous system. With the emerging arsenal of genetic tools for time-limited selective activation and inactivation of specific neuronal cell types and circuits and with new possibilities for in utero application of these techniques (Callaway 2005, Tervo & Karpova 2007, Zhang et al. 2007), we should be able to address these issues in the near future. If so, spatial navigation may become one of the first nonsensory cognitive functions to be understood in reasonable mechanistic detail at the microcircuit level.

SUMMARY POINTS

1. The hippocampal-entorhinal spatial representation system contains place cells, grid cells, and head-direction cells.

2. Grid cells have periodic firing fields that form a regular triangular grid across the environment. Grid fields are likely generated by path integration to serve as part of a neural map of self-location.

3. The integration of speed and direction signals may take place at the population level by virtue of recurrent connectivity or at the single neuron level as a consequence of interference among temporal oscillators with frequencies that depend on speed.

4. Hippocampal place fields may be formed by summing convergent input from grid cells with a range of different spacings, similar to a Fourier transform.

5. Unlike ensembles of grid cells, place cells participate in a number of highly orthogonalized environment-specific representations. Attractor dynamics may play a role in storing and reactivating representations during memory retrieval.

6. Theta-phase precession provides a possible mechanism for sequence representation in place cells.

FUTURE ISSUES

1. How is the grid pattern generated during nervous system development? Are specific maturational events required, and does grid formation depend on specific experience? Which elements of the map, if any, remain plastic in the adult brain?

2. What are the cellular and neural network mechanisms of path integration, and how is the grid representation updated in accordance with the rat's own movement? Which mechanism corrects cumulative error, and on what information does it rely?

3. How is the entorhinal map organized? Is the map modular or continuous? How are modularity and continuity generated during development, and how do modules interact in the mature nervous system?

4. What is the mechanism of phase precession, where does precession originate, and what is its relationship with spatial periodicity?

5. What is the function of the lateral entorhinal cortex, and how does it contribute to representation in the hippocampus?

6. How do hippocampal memories influence neocortical memory formation, and what is the function of grid cells or other entorhinal neurons in this process?

7. How does the entorhinal-hippocampal spatial map interact with other cortical systems required for spatial navigation, such as the parietal cortex or the striatum?

DISCLOSURE STATEMENT

The authors are not aware of any biases that might be perceived as affecting the objectivity of this review.

ACKNOWLEDGMENTS

We thank Neil Burgess, Mark Fuhs, Dave Touretzky, and Alessandro Treves for discussion. The authors were supported by The Kavli Foundation, the Norwegian Research Council, and the Fondation Bettencourt Schueller.

LITERATURE CITED

Alonso A, Llinas RR. 1989. Subthreshold Na$^+$-dependent theta-like rhythmicity in stellate cells of entorhinal cortex layer II. *Nature* 342:175–77

Amit DJ. 1989. *Modelling Brain Function: The World of Attractor Networks*. New York: Cambridge Univ. Press

Barlow JS. 1964. Inertial navigation as a basis for animal navigation. *J. Theor. Biol.* 6:76–117

Barnes CA, McNaughton BL, Mizumori SJ, Leonard BW, Lin LH. 1990. Comparison of spatial and temporal characteristics of neuronal activity in sequential stages of hippocampal processing. *Prog. Brain Res.* 83:287–300

Barry C, Hayman R, Burgess N, Jeffery KJ. 2007. Experience-dependent rescaling of entorhinal grids. *Nat. Neurosci.* 10:682–84

Battaglia FP, Treves A. 1998. Attractor neural networks storing multiple space representations: a model for hippocampal place fields. *Phys. Rev. E* 58:7738–53

Blair HT, Welday AC, Zhang K. 2007. Scale-invariant memory representations emerge from moire interference between grid fields that produce theta oscillations: a computational model. *J. Neurosci.* 27:3211–29

Blum KI, Abbott LF. 1996. A model of spatial map formation in the hippocampus of the rat. *Neural Comp.* 8:85–93

Blumenfeld B, Preminger S, Sagi D, Tsodyks M. 2006. Dynamics of memory representations in networks with novelty-facilitated synaptic plasticity. *Neuron* 52:383–94

Bostock E, Muller RU, Kubie JL. 1991. Experience-dependent modifications of hippocampal place cell firing. *Hippocampus* 1:193–205

Brun VH, Otnass MK, Molden S, Steffenach HA, Witter MP, et al. 2002. Place cells and place recognition maintained by direct entorhinal-hippocampal circuitry. *Science* 296:2243–46

Burak Y, Fiete I. 2006. Do we understand the emergent dynamics of grid cell activity? *J. Neurosci.* 26:9352–54

Burgess N, Barry C, O'Keefe J. 2007. An oscillatory interference model of grid cell firing. *Hippocampus* 17:801–12

Buzsáki G. 1989. Two-stage model of memory trace formation: a role for "noisy" brain states. *Neuroscience* 31:551–70

Buzsáki G, Leung LW, Vanderwolf CH. 1983. Cellular bases of hippocampal EEG in the behaving rat. *Brain Res.* 287:139–71

Callaway EM. 2005. A molecular and genetic arsenal for systems neuroscience. *Trends Neurosci.* 28:196–201

Chawla MK, Guzowski JF, Ramirez-Amaya V, Lipa P, Hoffman KL, et al. 2005. Sparse, environmentally selective expression of Arc RNA in the upper blade of the rodent fascia dentata by brief spatial experience. *Hippocampus* 15:579–86

Chen LL, Lin LH, Green EJ, Barnes CA, McNaughton BL. 1994. Head-direction cells in the rat posterior cortex. I. Anatomical distribution and behavioral modulation. *Exp. Brain Res.* 101:8–23

Claiborne BJ, Amaral DG, Cowan WM. 1986. A light and electron microscopic analysis of the mossy fibers of the rat dentate gyrus. *J. Comp. Neurol.* 246:435–58

Cooper BG, Mizumori SJ. 1999. Retrosplenial cortex inactivation selectively impairs navigation in darkness. *Neuroreport* 10:625–30

Cressant A, Muller RU, Poucet B. 1997. Failure of centrally placed objects to control the firing fields of hippocampal place cells. *J. Neurosci.* 17:2531–42

Dan Y, Poo MM. 2004. Spike timing-dependent plasticity of neural circuits. *Neuron* 44:23–30

DiMattia BV, Kesner RP. 1988. Role of the posterior parietal association cortex in the processing of spatial event information. *Behav. Neurosci.* 102:397–403

Dolorfo CL, Amaral DG. 1998. Entorhinal cortex of the rat: topographic organization of the cells of origin of the perforant path projection to the dentate gyrus. *J. Comp. Neurol.* 398:25–48

Dragoi G, Buzsáki G. 2006. Temporal encoding of place sequences by hippocampal cell assemblies. *Neuron* 50:145–57

Eichenbaum H, Dudchenko P, Wood E, Shapiro M, Tanila H. 1999. The hippocampus, memory, and place cells: Is it spatial memory or a memory space? *Neuron* 23:209–26

Ekstrom AD, Kahana MJ, Caplan JB, Fields TA, Isham EA, et al. 2003. Cellular networks underlying human spatial navigation. *Nature* 425:184–88

Ekstrom AD, Meltzer J, McNaughton BL, Barnes CA. 2001. NMDA receptor antagonism blocks experience-dependent expansion of hippocampal "place fields." *Neuron* 31:631–38

Etienne AS, Jeffery KJ. 2004. Path integration in mammals. *Hippocampus* 14:180–92

Fenton AA, Muller RU. 1998. Place cell discharge is extremely variable during individual passes of the rat through the firing field. *Proc. Natl. Acad. Sci. USA* 95:3182–87

Ferbinteanu J, Shapiro ML. 2003. Prospective and retrospective memory coding in the hippocampus. *Neuron* 40:1227–39

Foster DJ, Wilson MA. 2006. Reverse replay of behavioural sequences in hippocampal place cells during the awake state. *Nature* 440:680–83

Frank LM, Brown EN, Wilson M. 2000. Trajectory encoding in the hippocampus and entorhinal cortex. *Neuron* 27:169–78

Frank LM, Stanley GB, Brown EN. 2004. Hippocampal plasticity across multiple days of exposure to novel environments. *J. Neurosci.* 24:7681–89

Fuhs MC, Touretzky DS. 2006. A spin glass model of path integration in rat medial entorhinal cortex. *J. Neurosci.* 26:4266–76

Fyhn M, Hafting T, Treves A, Moser M-B, Moser EI. 2007. Hippocampal remapping and grid realignment in entorhinal cortex. *Nature* 446:190–94

Fyhn M, Molden S, Hollup SA, Moser M-B, Moser EI. 2002. Hippocampal neurons responding to first-time dislocation of a target object. *Neuron* 35:555–66

Fyhn M, Molden S, Witter MP, Moser EI, Moser M-B. 2004. Spatial representation in the entorhinal cortex. *Science* 305:1258–64

Giocomo LM, Zilli EA, Fransen E, Hasselmo ME. 2007. Temporal frequency of subthreshold oscillations scales with entorhinal grid cell field spacing. *Science* 315:1719–22

Gothard KM, Skaggs WE, McNaughton BL. 1996a. Dynamics of mismatch correction in the hippocampal ensemble code for space: interaction between path integration and environmental cues. *J. Neurosci.* 16:8027–40

Gothard KM, Skaggs WE, Moore KM, McNaughton BL. 1996b. Binding of hippocampal CA1 neural activity to multiple reference frames in a landmark-based navigation task. *J. Neurosci.* 16:823–35

Hafting T, Fyhn M, Molden S, Moser M-B, Moser EI. 2005. Microstructure of a spatial map in the entorhinal cortex. *Nature* 436:801–6

Hafting T, Fyhn M, Moser M-B, Moser EI. 2006. Phase precession and phase locking in entorhinal grid cells. *Soc. Neurosci. Abstr.* 32:68.8

Hahn TT, Sakmann B, Mehta MR. 2007. Differential responses of hippocampal subfields to cortical up-down states. *Proc. Natl. Acad. Sci. USA* 104:5169–74

Hampson RE, Heyser CJ, Deadwyler SA. 1993. Hippocampal cell firing correlates of delayed-match-to-sample performance in the rat. *Behav. Neurosci.* 107:715–39

Hargreaves EL, Rao G, Lee I, Knierim JJ. 2005. Major dissociation between medial and lateral entorhinal input to the dorsal hippocampus. *Science* 308:1792–94

Harris KD, Csicsvari J, Hirase H, Dragoi G, Buzsáki G. 2003. Organization of cell assemblies in the hippocampus. *Nature* 424:552–56

Harris KD, Henze DA, Hirase H, Leinekugel X, Dragoi G, et al. 2002. Spike train dynamics predicts theta-related phase precession in hippocampal pyramidal cells. *Nature* 417:738–41

Hartley T, Maguire EA, Spiers HJ, Burgess N. 2003. The well-worn route and the path less traveled: distinct neural bases of route following and wayfinding in humans. *Neuron* 37:877–88

Hassabis D, Kumaran D, Vann SD, Maguire EA. 2007. Patients with hippocampal amnesia cannot imagine new experiences. *Proc. Natl. Acad. Sci. USA* 104:1726–31

Hasselmo ME, Schnell E, Barkai E. 1995. Dynamics of learning and recall at excitatory recurrent synapses and cholinergic modulation in rat hippocampal region CA3. *J. Neurosci.* 15:5249–62

Hebb DO. 1949. *The Organization of Behavior*. New York: Wiley

Henze DA, Borhegyi Z, Csicsvari J, Mamiya A, Harris KD, Buzsáki G. 2000. Intracellular features predicted by extracellular recordings in the hippocampus in vivo. *J. Neurophysiol.* 84:390–400

Hill AJ. 1978. First occurrence of hippocampal spatial firing in a new environment. *Exp. Neurol.* 62:282–97

Hopfield JJ. 1982. Neural networks and physical systems with emergent collective computational abilities. *Proc. Natl. Acad. Sci. USA* 79:2554–58

Hubel DH, Wiesel TN, LeVay S. 1977. Plasticity of ocular dominance columns in monkey striate cortex. *Philos. Trans. R. Soc. London B Biol. Sci.* 278:377–409

Huxter J, Burgess N, O'Keefe J. 2003. Independent rate and temporal coding in hippocampal pyramidal cells. *Nature* 425:828–32

Huxter J, Senior T, Allen K, Csicsvari J. Trajectory and heading in theta-organized spike timing. *Soc. Neurosci. Abstr.* 33:640.13

Isomura Y, Sirota A, Ozen S, Montgomery S, Mizuseki K, et al. 2006. Integration and segregation of activity in entorhinal-hippocampal subregions by neocortical slow oscillations. *Neuron* 52:871–82

Jensen O, Lisman JE. 1996. Hippocampal CA3 region predicts memory sequences: accounting for the phase precession of place cells. *Learn. Mem.* 3:279–87

Jensen O, Lisman JE. 2000. Position reconstruction from an ensemble of hippocampal place cells: contribution of theta phase coding. *J. Neurophysiol.* 83:2602–9

Ji D, Wilson MA. 2007. Coordinated memory replay in the visual cortex and hippocampus during sleep. *Nat. Neurosci.* 10:100–7

Jog MS, Kubota Y, Connolly CI, Hillegaart V, Graybiel AM. 1999. Building neural representations of habits. *Science* 286:1745–49

Johnson A, Redish AD. 2007. Neural ensembles in CA3 transiently encode paths forward of the animal at a decision point. *J. Neurosci.* 27:12176–89

Jung MW, McNaughton BL. 1993. Spatial selectivity of unit activity in the hippocampal granular layer. *Hippocampus* 3:165–82

Jung MW, Wiener SI, McNaughton BL. 1994. Comparison of spatial firing characteristics of units in dorsal and ventral hippocampus of the rat. *J. Neurosci.* 14:7347–56

Kentros C, Hargreaves E, Hawkins RD, Kandel ER, Shapiro M, Muller RU. 1998. Abolition of long-term stability of new hippocampal place cell maps by NMDA receptor blockade. *Science* 280:2121–26

Kjelstrup KB, Solstad T, Brun VH, Fyhn M, Hafting T, et al. 2007. Very large place fields at the ventral pole of the hippocampal CA3 area. *Soc. Neurosci. Abstr.* 33:93.1

Klink R, Alonso A. 1993. Ionic mechanisms for the subthreshold oscillations and differential electroresponsiveness of medial entorhinal cortex layer II neurons. *J. Neurophysiol.* 70:144–57

Kolb B, Sutherland RJ, Whishaw IQ. 1983. A comparison of the contributions of the frontal and parietal association cortex to spatial localization in rats. *Behav. Neurosci.* 97:13–27

Lee AK, Wilson MA. 2002. Memory of sequential experience in the hippocampus during slow wave sleep. *Neuron* 36:1183–94

Lee I, Yoganarasimha D, Rao G, Knierim JJ. 2004. Comparison of population coherence of place cells in hippocampal subfields CA1 and CA3. *Nature* 430:456–59

Lengyel M, Szatmary Z, Erdi P. 2003. Dynamically detuned oscillations account for the coupled rate and temporal code of place cell firing. *Hippocampus* 13:700–14

Leutgeb JK, Leutgeb S, Moser M-B, Moser EI. 2007a. Pattern separation in dentate gyrus and CA3 of the hippocampus. *Science* 315:961–66

Leutgeb JK, Leutgeb S, Tashiro A, Moser EI, Moser M-B. 2007b. The encoding of novelty in the dentate gyrus and CA3 network. *Soc. Neurosci. Abstr.* 33:93.9

Leutgeb JK, Leutgeb S, Treves A, Meyer R, Barnes CA, et al. 2005. Progressive transformation of hippocampal neuronal representations in "morphed" environments. *Neuron* 48:345–58

Leutgeb JK, Moser EI. 2007. Pattern separation and the function of the dentate gyrus. *Neuron* 55:176–78

Leutgeb S, Colgin LL, Jezek K, Leutgeb JK, Fyhn M, et al. 2007. Path integration-based attractor dynamics in the entorhinal cortex. *Soc. Neurosci. Abstr.* 33:93.8

Leutgeb S, Leutgeb JK, Barnes CA, Moser EI, McNaughton BL, Moser M-B. 2005a. Independent codes for spatial and episodic memory in hippocampal neuronal ensembles. *Science* 309:619–23

Leutgeb S, Leutgeb JK, Moser M-B, Moser EI. 2005b. Place cells, spatial maps and the population code for memory. *Curr. Opin. Neurobiol.* 15:738–46

Leutgeb S, Leutgeb JK, Moser EI, Moser M-B. 2006. Fast rate coding in hippocampal CA3 cell ensembles. *Hippocampus* 16:765–74

Leutgeb S, Leutgeb JK, Treves A, Moser M-B, Moser EI. 2004. Distinct ensemble codes in hippocampal areas CA3 and CA1. *Science* 305:1295–98

Lever C, Wills T, Cacucci F, Burgess N, O'Keefe J. 2002. Long-term plasticity in hippocampal place-cell representation of environmental geometry. *Nature* 416:90–94

Markus EJ, Qin YL, Leonard B, Skaggs WE, McNaughton BL, Barnes CA. 1995. Interactions between location and task affect the spatial and directional firing of hippocampal neurons. *J. Neurosci.* 15:7079–94

Martin PD, Berthoz A. 2002. Development of spatial firing in the hippocampus of young rats. *Hippocampus* 12:465–80

Matthews DA, Nadler JV, Lynch GS, Cotman CW. 1974. Development of cholinergic innervation in the hippocampal formation of the rat. I. Histochemical demonstration of acetylcholinesterase activity. *Dev. Biol.* 36:130–41

Maurer AP, VanRhoads SR, Sutherland GR, Lipa P, McNaughton BL. 2005. Self-motion and the origin of differential spatial scaling along the septo-temporal axis of the hippocampus. *Hippocampus* 15:841–52

McClelland JL, Goddard NH. 1996. Considerations arising from a complementary learning systems perspective on hippocampus and neocortex. *Hippocampus* 6:654–65

McHugh TJ, Blum KI, Tsien JZ, Tonegawa S, Wilson MA. 1996. Impaired hippocampal representation of space in CA1-specific NMDAR1 knockout mice. *Cell* 87:1339–49

McHugh TJ, Jones MW, Quinn JJ, Balthasar N, Coppari R, et al. 2007. Dentate gyrus NMDA receptors mediate rapid pattern separation in the hippocampal network. *Science* 317:94–99

McNaughton BL, Barnes CA, Meltzer J, Sutherland RJ. 1989. Hippocampal granule cells are necessary for normal spatial learning but not for spatially-selective pyramidal cell discharge. *Exp. Brain Res.* 76:485–96

McNaughton BL, Battaglia FP, Jensen O, Moser EI, Moser M-B. 2006. Path integration and the neural basis of the "cognitive map." *Nat. Rev. Neurosci.* 7:663–78

McNaughton BL, Morris RGM. 1987. Hippocampal synaptic enhancement and information storage within a distributed memory system. *Trends Neurosci.* 10:408–15

Mehta MR, Barnes CA, McNaughton BL. 1997. Experience-dependent, asymmetric expansion of hippocampal place fields. *Proc. Natl. Acad. Sci. USA* 94:8918–21

Mehta MR, Lee AK, Wilson MA. 2002. Role of experience and oscillations in transforming a rate code into a temporal code. *Nature* 417:741–46

Mehta MR, Quirk MC, Wilson MA. 2000. Experience-dependent asymmetric shape of hippocampal receptive fields. *Neuron* 25:707–15

Mittelstaedt ML, Mittelstaedt H. 1980. Homing by path integration in a mammal. *Naturwissenschaften* 67:566–67

Muller RU, Kubie JL. 1987. The effects of changes in the environment on the spatial firing of hippocampal complex-spike cells. *J. Neurosci.* 7:1951–68

Nadel L. 1991. The hippocampus and space revisited. *Hippocampus* 1:221–29

Nakazawa K, Quirk MC, Chitwood RA, Watanabe M, Yeckel MF, et al. 2002. Requirement for hippocampal CA3 NMDA receptors in associative memory recall. *Science* 297:211–18

Nitz DA. 2006. Tracking route progression in the posterior parietal cortex. *Neuron* 49:747–56

O'Keefe J. 1976. Place units in the hippocampus of the freely moving rat. *Exp. Neurol.* 51:78–109

O'Keefe J, Burgess N. 1996. Geometric determinants of the place fields of hippocampal neurons. *Nature* 381:425–28

O'Keefe J, Burgess N. 2005. Dual phase and rate coding in hippocampal place cells: theoretical significance and relationship to entorhinal grid cells. *Hippocampus* 15:853–66

O'Keefe J, Conway DH. 1978. Hippocampal place units in the freely moving rat: why they fire where they fire. *Exp. Brain Res.* 31:573–90

O'Keefe J, Dostrovsky J. 1971. The hippocampus as a spatial map. Preliminary evidence from unit activity in the freely-moving rat. *Brain Res.* 34:171–75

O'Keefe J, Nadel L. 1978. *The Hippocampus as a Cognitive Map.* Oxford: Clarendon

O'Keefe J, Recce ML. 1993. Phase relationship between hippocampal place units and the EEG theta rhythm. *Hippocampus* 3:317–30

O'Keefe J, Speakman A. 1987. Single unit activity in the rat hippocampus during a spatial memory task. *Exp. Brain Res.* 68:1–27

Olypher AV, Lansky P, Fenton AA. 2002. Properties of the extrapositional signal in hippocampal place cell discharge derived from the overdispersion in location-specific firing. *Neuroscience* 111:553–66

O'Neill J, Senior T, Csicsvari J. 2006. Place-selective firing of CA1 pyramidal cells during sharp wave/ripple network patterns in exploratory behavior. *Neuron* 49:143–55

Packard MG, McGaugh JL. 1996. Inactivation of hippocampus or caudate nucleus with lidocaine differentially affects expression of place and response learning. *Neurobiol. Learn. Mem.* 65:65–72

Parron C, Save E. 2004. Evidence for entorhinal and parietal cortices involvement in path integration in the rat. *Exp. Brain Res.* 159:349–59

Quirk GJ, Muller RU, Kubie JL. 1990. The firing of hippocampal place cells in the dark depends on the rat's recent experience. *J. Neurosci.* 10:2008–17

Quirk GJ, Muller RU, Kubie JL, Ranck JB Jr. 1992. The positional firing properties of medial entorhinal neurons: description and comparison with hippocampal place cells. *J. Neurosci.* 12:1945–63

Ranck JB. 1985. Head direction cells in the deep cell layer of dorsal presubiculum in freely moving rats. In *Electrical Activity of the Archicortex*, ed. G Buzsáki, CH Vanderwolf, pp. 217–20. Budapest: Akademiai Kiado

Redish AD, Rosenzweig ES, Bohanick JD, McNaughton BL, Barnes CA. 2000. Dynamics of hippocampal ensemble activity realignment: time versus space. *J. Neurosci.* 20:9298–309

Ritter J, Meyer U, Wenk H. 1972. Chemodifferentiation of the hippocampus formation in the postnatal development of albino rats. II. Transmitter enzymes. *J. Hirnforsch.* 13:254–78

Rolls ET. 1999. Spatial view cells and the representation of place in the primate hippocampus. *Hippocampus* 9:467–80

Rolls ET, Stringer SM, Elliot T. 2006. Entorhinal cortex grid cells can map to hippocampal place cells by competitive learning. *Network* 17:447–65

Rolls ET, Treves A. 1998. *Neural Networks and Brain Function.* Oxford, UK: Oxford Univ. Press

Samsonovich A, McNaughton BL. 1997. Path integration and cognitive mapping in a continuous attractor neural network model. *J. Neurosci.* 17:272–75

Sargolini F, Boccara C, Witter MP, Moser M-B, Moser EI. 2006a. Grid cells outside the medial entorhinal cortex. *Soc. Neurosci. Abstr.* 32:68.11

Sargolini F, Fyhn M, Hafting T, McNaughton BL, Witter MP, et al. 2006b. Conjunctive representation of position, direction and velocity in entorhinal cortex. *Science* 312:754–58

Save E, Guazzelli A, Poucet B. 2001. Dissociation of the effects of bilateral lesions of the dorsal hippocampus and parietal cortex on path integration in the rat. *Behav. Neurosci.* 115:1212–23

Scoville WB, Milner B. 1957. Loss of recent memory after bilateral hippocampal lesions. *J. Neurol. Neurosurg. Psychiatry* 20:11–21

Shapiro ML, Kennedy PJ, Ferbinteanu J. 2006. Representing episodes in the mammalian brain. *Curr. Opin. Neurobiol.* 16:701–9

Shapiro ML, Tanila H, Eichenbaum H. 1997. Cues that hippocampal place cells encode: dynamic and hierarchical representation of local and distal stimuli. *Hippocampus* 7:624–42

Skaggs WE, McNaughton BL. 1996. Replay of neuronal firing sequences in rat hippocampus during sleep following spatial experience. *Science* 271:1870–73

Skaggs WE, McNaughton BL, Wilson MA, Barnes CA. 1996. Theta phase precession in hippocampal neuronal populations and the compression of temporal sequences. *Hippocampus* 6:149–72

Solstad T, Brun VH, Kjelstrup KB, Fyhn M, Witter MP, et al. 2007. Grid expansion along the dorso-ventral axis of the medial entorhinal cortex. *Soc. Neurosci. Abstr.* 33:93.2

Solstad T, Moser EI, Einevoll GT. 2006. From grid cells to place cells: a mathematical model. *Hippocampus* 16:1026–31

Squire LR, Stark CE, Clark RE. 2004. The medial temporal lobe. *Annu. Rev. Neurosci.* 27:279–306

Sutherland RJ, Whishaw IQ, Kolb B. 1988. Contributions of cingulate cortex to two forms of spatial learning and memory. *J. Neurosci.* 8:1863–72

Takahashi N, Kawamura M, Shiota J, Kasahata N, Hirayama K. 1997. Pure topographic disorientation due to right retrosplenial lesion. *Neurology* 49:464–69

Taube JS. 1995. Place cells recorded in the parasubiculum of freely moving rats. *Hippocampus* 5:569–83

Taube JS. 1998. Head direction cells and the neurophysiological basis for a sense of direction. *Prog. Neurobiol.* 55:225–56

Taube JS, Muller RU, Ranck JB Jr. 1990. Head-direction cells recorded from the postsubiculum in freely moving rats. I. Description and quantitative analysis. *J. Neurosci.* 10:420–35

Tervo DGR, Karpova AY. 2007. Rapidly inducible, genetically targeted inactivation of neural and synaptic activity in vivo. *Curr. Opin. Neurobiol.* 17:In press

Tolman EC. 1948. Cognitive maps in rats and men. *Psychol. Rev.* 55:189–208

Treves A, Rolls ET. 1992. Computational constraints suggest the need for two distinct input systems to the hippocampal CA3 network. *Hippocampus* 2:189–99

Tsodyks M, Sejnowski T. 1995. Associative memory and hippocampal place cells. *Int. J. Neural Syst.* 6(Suppl.):81–86

Tsodyks MV, Skaggs WE, Sejnowski TJ, McNaughton BL. 1996. Population dynamics and theta rhythm phase precession of hippocampal place cell firing: a spiking neuron model. *Hippocampus* 6:271–80

Tulving E, Markowitsch HJ. 1998. Episodic and declarative memory: role of the hippocampus. *Hippocampus* 8:198–204

Ulanovsky N, Moss CF. 2007. Hippocampal cellular and network activity in freely moving echolocating bats. *Nat. Neurosci.* 10:224–33

Vazdarjanova A, Guzowski JF. 2004. Differences in hippocampal neuronal population responses to modifications of an environmental context: evidence for distinct, yet complementary, functions of CA3 and CA1 ensembles. *J. Neurosci.* 24:6489–96

Wallenstein GV, Hasselmo ME. 1998. GABAergic modulation of hippocampal population activity: sequence learning, place field development, and the phase precession effect. *J. Neurophysiol.* 78:393–408

Wills TJ, Lever C, Cacucci F, Burgess N, O'Keefe J. 2005. Attractor dynamics in the hippocampal representation of the local environment. *Science* 308:873–76

Wilson MA, McNaughton BL. 1993. Dynamics of the hippocampal ensemble code for space. *Science* 261:1055–58

Wilson MA, McNaughton BL. 1994. Reactivation of hippocampal ensemble memories during sleep. *Science* 265:676–79

Witter MP, Amaral DG. 2004. Hippocampal formation. In *The Rat Nervous System*, ed. G. Paxinos, pp. 637–703. San Diego: Academic. 3rd ed.

Witter MP, Moser EI. 2006. Spatial representation and the architecture of the entorhinal cortex. *Trends Neurosci.* 29:671–78

Wood ER, Dudchenko PA, Eichenbaum H. 1999. The global record of memory in hippocampal neuronal activity. *Nature* 397:613–16

Wood ER, Dudchenko PA, Robitsek RJ, Eichenbaum H. 2000. Hippocampal neurons encode information about different types of memory episodes occurring in the same location. *Neuron* 27:623–33

Young BJ, Fox GD, Eichenbaum H. 1994. Correlates of hippocampal complex-spike cell activity in rats performing a nonspatial radial maze task. *J. Neurosci.* 14:6553–63

Zhang F, Wang LP, Brauner M, Liewald JF, Kay K, et al. 2007. Multimodal fast optical interrogation of neural circuitry. *Nature* 446:633–39

Zinyuk L, Kubik S, Kaminsky Y, Fenton AA, Bures J. 2000. Understanding hippocampal activity by using purposeful behavior: place navigation induces place cell discharge in both task-relevant and task-irrelevant spatial reference frames. *Proc. Natl. Acad. Sci. USA* 97:3771–76

Zugaro MB, Monconduit L, Buzsáki G. 2005. Spike phase precession persists after transient intrahippocampal perturbation. *Nat. Neurosci.* 8:67–71

Mitochondrial Disorders in the Nervous System

Salvatore DiMauro[1] and Eric A. Schon[1,2]

Departments of Neurology[1] and Genetics and Development,[2] Columbia University Medical Center, New York, NY 10032; email: sd12@columbia.edu, eas3@columbia.edu

Annu. Rev. Neurosci. 2008. 31:91–123

First published online as a Review in Advance on March 10, 2008

The *Annual Review of Neuroscience* is online at neuro.annualreviews.org

This article's doi:
10.1146/annurev.neuro.30.051606.094302

0147-006X/08/0721-0091$20.00

Key Words

mitochondrial DNA, maternal inheritance, oxidative stress, apoptosis, oxidative phosphorylation, aging

Abstract

Mitochondrial diseases (encephalomyopathies) have traditionally been ascribed to defects of the respiratory chain, which has helped researchers explain their genetic and clinical complexity. However, other mitochondrial functions are greatly important for the nervous system, including protein importation, organellar dynamics, and programmed cell death. Defects in genes controlling these functions are attracting increasing attention as causes not only of neurological (and psychiatric) diseases but also of age-related neurodegenerative disorders. After discussing some pathogenic conundrums regarding the neurological manifestations of the respiratory chain defects, we review altered mitochondrial dynamics in the etiology of specific neurological diseases and in the physiopathology of more common neurodegenerative disorders.

Contents

INTRODUCTION

Mitochondrial dysfunction plays a crucial role in neurology. This notion became apparent three decades ago when pediatric neurologists coined the term mitochondrial encephalomyopathies to call attention to the frequent occurrence of brain disease in children with mitochondrial alterations in their muscle biopsies (Shapira et al. 1977). The selective vulnerability of skeletal muscle and of the nervous system was confirmed in 1988, when the first pathogenic mutations in the mitochondrion's own DNA (mtDNA) were discovered (Holt et al. 1988, Wallace et al. 1988). These discoveries heralded the era of mitochondrial genetics and led to the recognition of a multitude of mtDNA-related disorders, mostly maternally inherited and mostly manifesting as encephalomyopathies (DiMauro & Davidzon 2005, DiMauro & Schon 2003). Because mtDNA encodes only 13 proteins, all of them subunits of the mitochondrial respiratory chain—the business end in terms of ATP production—another notion became widely accepted: The term mitochondrial encephalomyopathies was reserved for defects of the respiratory chain.

Even within these boundaries, the classification of the mitochondrial encephalomyopathies soon became quite cumbersome, including two flavors of primary mtDNA mutations (i.e., the impairment of global mitochondrial protein synthesis and of the translation of specific respiratory chain subunits) and a much larger menu of Mendelian disorders (**Table 1**). Also, genetic errors in other fundamental mitochondrial functions that do not affect the respiratory chain directly have major deleterious effects on the nervous system, including impaired importation of mitochondrial proteins and defects of mitochondrial dynamics, such as motility, fission, fusion, and distribution.

Another topic of current interest is the role of progressive mitochondrial dysfunction in normal aging and in the pathogenesis of late-onset neurodegenerative disorders.

In this review, we discuss first the nervous system disorders caused by mitochondrial respiratory chain defects, emphasizing how their pathogenesis is still largely terra incognita. We then consider the burgeoning new group of disorders attributed to defects of mitochondrial

Table 1 Mitochondrial respiratory chain disease targets

Mutations in mtDNA	Mutations in nDNA
R.C. subunits	R.C. subunits
	Complex I, II, III
Protein synthesis genes	Ancillary proteins
Rearrangements	Complex I, III, IV, V; CoQ
tRNAs	Intergenomic communication
rRNAs	Multiple mtDNA deletions
	Depletion of mtDNA
	Translation of mt-mRNAs
	Mitochondrial lipids

dynamics. Last, we review the neurodegenerative disorders in which mitochondrial dysfunction is either primary or seems to be at least involved in pathogenesis. We do not discuss mitochondrial metabolic pathway defects other than the respiratory chain, such as pyruvate dehydrogenase complex (PDHC) deficiency or ß-oxidation defects, although the nervous system is frequently affected in those disorders, too.

DISEASES OF THE MITOCHONDRIAL RESPIRATORY CHAIN

These diseases can be caused by mutations in mtDNA (sporadic or maternally inherited traits) or by mutations in nuclear DNA (nDNA; Mendelian diseases).

Disorders Caused by Mutations in mtDNA

Human mtDNA (**Figure 1**) is a 16.6-kb circular, double-stranded molecule, which contains 37 genes: 2 rRNA genes, 22 tRNA genes, and 13 structural genes encoding subunits of the mitochondrial respiratory chain (Anderson et al. 1981). Reducing equivalents produced in the Krebs cycle and in the ß-oxidation "spiral" are passed along a series of protein complexes embedded in the inner mitochondrial membrane (the electron transport chain), which consists of four multimeric complexes (I, II, III, and IV) plus two small electron carriers, coenzyme Q (or ubiquinone) and cytochrome c (**Figure 2**). The energy generated by the reactions of the electron transport chain is used to pump protons from the mitochondrial matrix into the intermembrane space (IMS) located between the inner and outer mitochondrial membranes. This process creates an electrochemical proton gradient, which is utilized by complex V (or ATP synthase), a tiny rotary machine that generates ATP as protons flow back into the matrix through its membrane-embedded F_0 portion, the rotor of the turbine. The motor's stator (called the F_1 portion) protrudes into the matrix and converts ADP and

inorganic phosphate (Pi) to ATP in a tripartite series of catalytic reactions [three sets of α/β dimeric subunits alter their conformations via a rotating "cam" that connects F_0 to F_1 so as to bind ADP +Pi first, then to convert ADP and Pi to form ATP, and finally to release the ATP into the matrix, where it is exported from the organelle into the cytoplasm via the adenine nucleotide translocator (ANT)].

At this point, a brief reminder of the rules of mitochondrial genetics is "de rigueur."

1. Heteroplasmy and threshold effect. Each cell contains hundreds or thousands of mtDNA copies, which, at cell division, distribute randomly among daughter cells. In normal tissues, all mtDNA molecules are identical (homoplasmy). Deleterious mutations of mtDNA usually affect some but not all mtDNAs (heteroplasmy), and the clinical expression of a pathogenic mtDNA mutation is determined largely by the relative proportions of normal and mutant genomes in different tissues. A minimum critical mutation load (typically above 80%–90%) is required to cause mitochondrial dysfunction in a particular organ or tissue and mitochondrial disease in an individual: This is the threshold effect.

2. Mitotic segregation. At cell division, the proportion of mutant mtDNAs in daughter cells may shift and the phenotype may change accordingly. This phenomenon, called mitotic segregation, explains how the clinical phenotype in patients with mtDNA-related disorders may change as patients grow older.

3. Maternal inheritance. At fertilization, all mtDNA derives from the oocyte. Therefore, the mode of transmission of mtDNA and of mtDNA point mutations (single deletions of mtDNA are usually sporadic events) differs from Mendelian inheritance. A mother carrying a mtDNA point mutation will pass it on to all her children (males and females), but only the daughters will transmit it to their progeny. Thus, a disease expressed in both sexes

mtDNA: mitochondrial DNA

PDHC: pyruvate dehydrogenase complex

nDNA: nuclear DNA

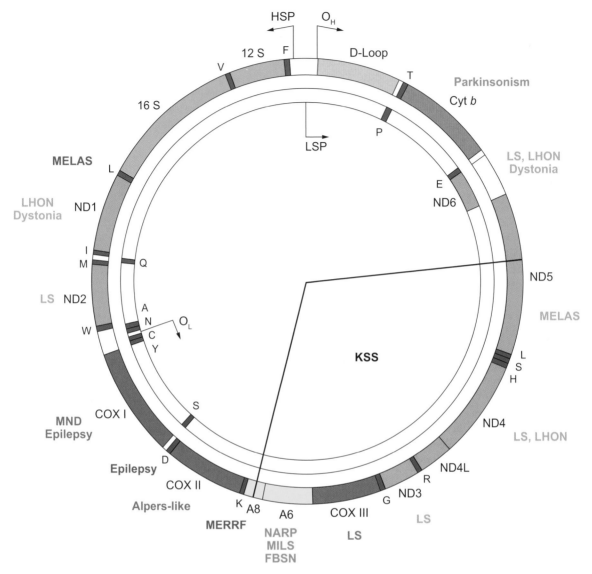

Figure 1

The human mitochondrial genome. The mtDNA-encoded gene products for the 12S and 16S ribosomal RNAs, the subunits of NADH-coenzyme Q oxidoreductase (ND), cytochrome *c* oxidase (COX), cytochrome *b* (Cyt b), and ATP synthase (A), and 22 tRNAs (1-letter amino acid nomenclature) are shown, as are the origins of heavy- and light-strand replication (O_H and O_L) and the promoters of heavy- and light-strand transcription (HSP and LSP). Some pathogenic mutations (for expanded versions of all the key terms in this article, see **Supplemental Term List**; follow the **Supplemental Material link** from the Annual Reviews home page at **http://www.annualreviews.org**) that affect the nervous system in particular are indicated (*colors correspond to those of the affected genes*).

but with no evidence of paternal transmission is strongly suggestive of an mtDNA point mutation.

About 200 mtDNA point mutations and innumerable single large-scale (kilobase-sized) partial deletions have been associated with human diseases, most of which affect the central and peripheral nervous system, especially if myopathies are considered—as they should—the domain of peripheral neurology. This concept

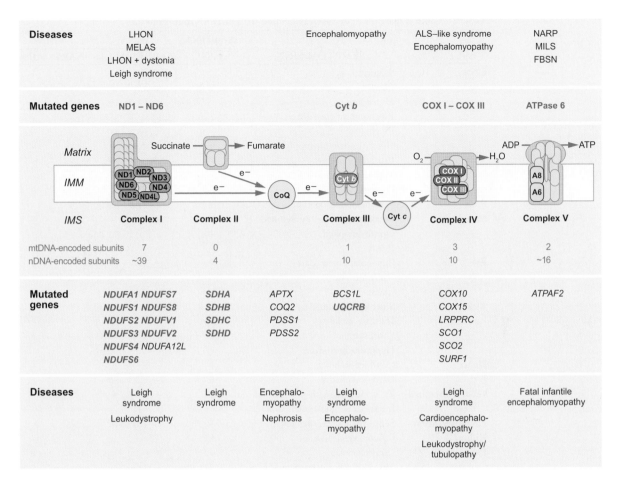

Diseases

Complex I	Complex III	Complex IV	Complex V
LHON MELAS LHON + dystonia Leigh syndrome	Encephalomyopathy	ALS–like syndrome Encephalomyopathy	NARP MILS FBSN

Mutated genes

Complex I	Complex III	Complex IV	Complex V
ND1 – ND6	Cyt b	COX I – COX III	ATPase 6

Matrix — Succinate → Fumarate · ADP → ATP

IMM · ND1 ND2 ND3 ND4 ND5 ND4L ND6 · CoQ · Cyt b · O_2 H_2O · COX I COX II COX III · A8 A6 · e^-

IMS · Complex I · Complex II · Complex III · Cyt c · Complex IV · Complex V

	Complex I	Complex II	Complex III	Complex IV	Complex V
mtDNA-encoded subunits	7	0	1	3	2
nDNA-encoded subunits	~39	4	10	10	~16

Mutated genes

Complex I	Complex II	Complex III	Complex III	Complex IV	Complex V
NDUFA1 NDUFS7 NDUFS1 NDUFS8 NDUFS2 NDUFV1 NDUFS3 NDUFV2 NDUFS4 NDUFA12L NDUFS6	SDHA SDHB SDHC SDHD	APTX COQ2 PDSS1 PDSS2	BCS1L UQCRB	COX10 COX15 LRPPRC SCO1 SCO2 SURF1	ATPAF2

Diseases

Complex I	Complex II	Complex III	Complex III	Complex IV	Complex V
Leigh syndrome Leukodystrophy	Leigh syndrome	Encephalo-myopathy Nephrosis	Leigh syndrome Encephalo-myopathy	Leigh syndrome Cardioencephalo-myopathy Leukodystrophy/tubulopathy	Fatal infantile encephalomyopathy

Figure 2

The mitochondrial respiratory chain (RC), showing nDNA-encoded subunits (*blue*) and mtDNA-encoded subunits (*colors corresponding to the genes in the map in* **Figure 1**). Protons are pumped from the matrix to the intermembrane space through complexes I, III, and IV, and are pumped back to the matrix through complex V to produce ATP. Coenzyme Q and cytochrome c are electron (e^-) transfer carriers. Diseases (see **Supplemental Term List**) caused by mutations in mtDNA (above the RC) and in nDNA (below the RC) are listed according to the correspondingly affected RC complex. Genes in bold encode RC subunits; those in plain text encode ancillary or assembly proteins.

is illustrated in **Figure 1** and, in more detail, in **Table 2**, which highlights the typical clinical features of the five most common mtDNA-related syndromes of neurological interest. These are not described here in any more detail because the features can be found in textbook reviews (Hays et al. 2006, Hirano et al. 2006a).

The human mitochondrial genome is saturated with mutations. Does this mean that we are scraping the bottom of the barrel as far

as our understanding of mtDNA-related diseases is concerned? Not by a long stretch. Although, understandably, the pace at which new pathogenic mutations are discovered has slackened in recent years, novel mutations are still being reported, and several questions still await answers in the field of mitochondrial genetics. For example, whereas most pathogenic mtDNA mutations are heteroplasmic and clinical severity is usually related to mutation load, some mutations are homoplasmic, and yet the severity

Table 2 Clinical features in diseases associated with mtDNA mutations

TISSUE	SYMPTOM/SIGN	Δ-mtDNA KSS	tRNA MERRF	tRNA MELAS	ATPase6 NARP	ATPase6 MILS
CNS	Seizures	−	+	+	−	+
	Ataxia	+	+	+	+	±
	Myoclonus	−	+	±	−	−
	Psychomotor retardation	−	−	−	−	+
	Psychomotor regression	+	±	+	−	−
	Hemiparesis/hemianopia	−	−	+	−	−
	Cortical blindness	−	−	+	−	−
	Migraine-like headaches	−	−	+	−	−
	Dystonia	−	−	+	−	+
PNS	Peripheral neuropathy	±	±	±	+	−
Muscle	Weakness/exercise intolerance	+	+	+	+	+
	Ophthalmoplegia	+	−	−	−	−
	Ptosis	+	−	−	−	−
Eye	Pigmentary retinopathy	+	−	−	+	±
	Optic atrophy	−	−	−	±	±
	Cataracts	−	−	−	−	−
Blood	Sideroblastic anemia	±	−	−	−	−
Endocrine	Diabetes mellitus	±	−	±	−	−
	Short stature	+	+	+	−	−
	Hypoparathyroidism	±	−	−	−	−
Heart	Conduction block	+	−	±	−	−
	Cardiomyopathy	±	−	±	−	±
Gastro/intestinal	Exocrine pancreatic dysfunction	±	−	−	−	−
	Intestinal pseudo-obstruction	−	−	−	−	−
Ear/nose/throat	Sensorineural hearing loss	−	+	+	±	−
Kidney	Fanconi syndrome	±	−	±	−	−
Laboratory	Lactic acidosis	+	+	+	−	±
	Muscle biopsy: Ragged-red fibers	+	+	+	−	−
Inheritance	Maternal	−	+	+	+	+
	Sporadic	+	−	−	−	−

of the syndromes they cause differs in different families or even in members of the same family. A related question concerns the functional significance of mtDNA haplotypes. In the migration out of Africa, human beings have accumulated distinctive variations on the mtDNA of the ancestral "mitochondrial Eve," resulting in several haplotypes characteristic of different ethnic groups (Wallace et al. 1999). Different mtDNA haplotypes may modulate oxidative phosphorylation, thus influencing the overall physiology of individuals and predisposing them to—or protecting them from—certain diseases (Carelli et al. 2006). Clearly, much work remains to be done to define better both the pathogenic role of homoplasmic mutations and the modulatory role of haplotypes in health and disease.

A major problem in mtDNA-related neurological diseases is our woeful ignorance about genotype-phenotype correlations. In fact, it is surprising that mtDNA mutations should cause different syndromes in the first place. If, as conventional wisdom dictates, both large-scale mtDNA rearrangements and point mutations in rRNA or tRNA genes impair mitochondrial protein synthesis and ATP production, it would be logical to expect a clinical swamp of ill-defined and overlapping symptoms and signs, as originally predicted by the "lumpers" (Rowland 1994). Although clinical overlap does occur in mtDNA-related diseases, it is fair to say that the "splitters" won the day, in that most mutations result in well-defined syndromes, including mutations associated with KSS/PEO, DAD, MERRF, MELAS, NARP/MILS, LHON, and SNHL.

To explain the distinctive brain symptoms in patients with KSS, MERRF, and MELAS, the different mutations have been "mapped" indirectly through immunohistochemical techniques. Consistent with clinical symptoms and laboratory data, immunohistochemical evidence suggests that the mtDNA deletion (Δ-mtDNA) of KSS abounds in the choroid plexus (Tanji et al. 2000) (**Figure 3a–d**), the 3243-MELAS mutation is abundant in the walls of cerebral arterioles (Betts et al. 2006) (**Figure 3e–f**), and the 8344-MERRF mutation is abundant in the olivary nucleus of the cerebellum (Tanji et al. 2001) (**Figure 3g–j**). Direct evidence of the accumulation of Δ-mtDNA in the choroid plexus of KSS patients was provided by Tanji et al. (2000) using in situ hybridization (**Figure 4**). However, these data fail to explain what directs each mutation to a particular area of the brain, how the mutation correlates with the clinical syndrome, or why the syndromes differ from each other.

That mutations in different tRNA genes may have different mechanisms of action is suggested by the apparently selective tissue vulnerability associated with mutations in some tRNAs: For example, cardiomyopathy is often associated with mutations in tRNA^{Ile}, diabetes is a frequent manifestation of the T14709C mutation in tRNA^{Glu}, and multiple lipomas have been reported only in patients with mutations in tRNA^{Lys}. However these are mere associations, not explanations. It is fair to conclude that the pathogenesis of mtDNA-related disorders is still largely unexplained.

Disorders Caused by Mutations in nDNA

During the many millennia of symbiotic relation with the nDNA, the mtDNA has lost more than 99% of its original genes and most

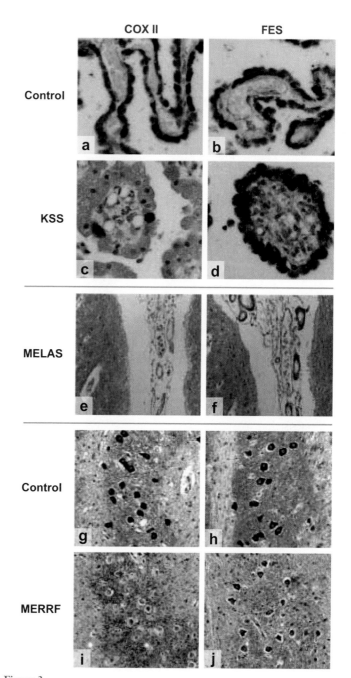

Figure 3

Immunohistochemistry to detect the mtDNA-encoded COX II subunit of complex IV (*left panels*) and the nDNA-encoded FeS subunit of complex III (*right panels*) in brain structures in three mtDNA-related diseases. Choroid plexus from a control (A, B) and a KSS patient (C, D). Sub-pial arterioles from a MELAS patient (E, F). Olivary nucleus from a control (G, H) and a MERRF patient (I, J). Courtesy of Drs. Eduardo Bonilla and Kurenai Tanji, Columbia University Medical Center.

a

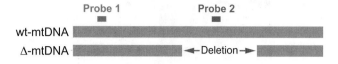

Probe 1   Probe 2

wt-mtDNA

Δ-mtDNA   ◄─Deletion─►

b

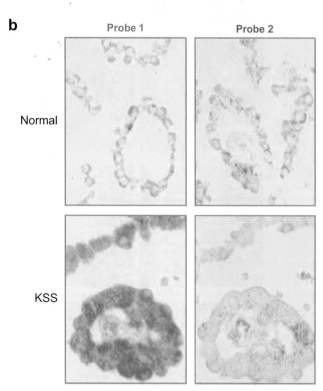

Probe 1   Probe 2

Normal

KSS

Figure 4

In-situ hybridization to detect mtDNAs in the choroid plexus from a KSS patient. (*a*) Map of wt- and Δ-mtDNAs from the patient showing the two probes: Probe 1 (*red*) detects both wt- and Δ-mtDNAs; probe 2 (*blue*) detects only wt-mtDNA. (*b*) As opposed to the uniform signal with both probes in the control (*upper panels*), there is a much stronger signal in the patient with probe 1 than with probe 2 (*lower panels*), indicating a massive accumulation of Δ-mtDNAs. Courtesy of Drs. Eduardo Bonilla and Kurenai Tanji, Columbia University Medical Center.

can be divided into at least four subgroups (**Table 1**).

Mutations in genes encoding respiratory chain subunits. These mutations (direct hits) have been found predominantly in the first two complexes of the respiratory chain, suggesting that deleterious mutations in the terminal complexes are either rare or incompatible with life. One explanation suggests that complexes I and II are in parallel, allowing for some residual electron transport even when one complex is out of commission, whereas complexes III, IV, and V are in series (**Figure 2**). Although direct hits do occur in the mtDNA-encoded subunits of complexes III (cytochrome *b*), IV (COX I, II, or III), and V (ATPase 6), the heteroplasmic nature of these mutations may permit some residual activity. However, the series/parallel hypothesis has been undercut by the finding of a homozygous frameshift mutation in the ubiquinone-binding subunit of complex III UQCRB (Haut et al. 2003), which is located at the C-terminus of the protein and still allows for some residual complex III activity. It is more difficult to explain why severe COX deficiency with recessive mutations in assembly proteins (for example SCO2) is still compatible with life, albeit a very abbreviated life.

Most mutations in nDNA-encoded complex I or in complex II subunits cause Leigh syndrome (LS) (**Table 3**). The hallmark neuropathological lesions of this devastating neurodegenerative disorder of infancy or early

Table 3 Causes of Leigh syndrome[a]

Defect	Transmission	Frequency
Complex I	AR, M	+++
Complex II	AR	+
Complex IV	AR	+++
Complex V	M	++
tRNA$^{Leu(UUR)}$	M	+
tRNA^{Lys}	M	+
CoQ$_{10}$	AR	+
PDHC	XR, AR	+++

[a]Abbreviations: AR, autosomal recessive; M, maternal; X, X-linked.

of its autonomy, and it now depends on nuclear factors for all its basic functions, including replication, translation, synthesis of most respiratory chain subunits, and assembly of respiratory chain complexes, and for the synthesis of the phospholipids that constitute the inner mitochondrial membrane (IMM). This is why the Mendelian defects of the respiratory chain

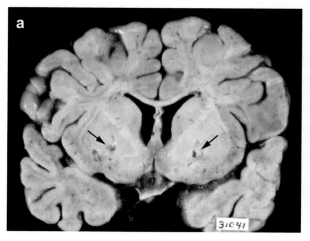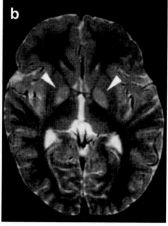

Figure 5

Typical brain lesions in
Leigh syndrome.
(*a*) Coronal section
showing bilateral
symmetrical cavitating
lesions in the basal
ganglia (*arrows*).
(*b*) T2-weighted MRI
showing abnormal
bilateral symmetrical
hyperintense signals in
the lenticular nuclei
(*arrowheads*).

childhood [bilaterally symmetrical foci of cys-
tic cavitation **(Figure 5)**, vascular proliferation,
neuronal loss, and demyelination in the basal
ganglia, brainstem, and posterior columns of
the spinal cord] probably reflect the stereo-
typical ravages caused by defective oxidative
metabolism on the developing nervous system.
This concept is supported by the observation
that LS is also caused by mtDNA mutations
when they are sufficiently abundant (MILS;
Table 2) or severe enough to impair oxidative
phosphorylation early in life (Kirby et al. 2003,
Sarzi et al. 2007, Tatuch et al. 1992).

Although some mutations in mtDNA complex I genes cause LS, most do not, but
rather cause Leber hereditary optic neuropathy
(LHON), a maternally inherited optic atrophy
that causes blindness in young adults with challenging contradictions. First, all pathogenic
LHON mutations are in complex I genes, and
yet the complex I deficiency is not particularly severe. Second, whereas some LHON
mutations are heteroplasmic (as in most mitochondrial diseases), most are homoplasmic,
and yet the pathology is confined, on the
whole, to the retinal ganglion cells (Carelli
et al. 2007). Third, even though the mutation is often homoplasmic, the blindness
usually does not occur until the patient is
older than age 20, and then each eye is affected sequentially within months. Fourth, although LHON is maternally inherited, men
are affected far more frequently, and more
severely, than are women, implying an X-linked
modifier effect (Hudson et al. 2005). Also,
rarely, the blindness is partially reversible.

**Mutations in genes encoding ancillary pro-
teins.** This group of disorders is caused by indirect hits, that is, mutations in proteins that
are not part of any complex but are needed
to synthesize and direct the proper assembly
of the various nDNA- and mtDNA-encoded
subunits, together with their prosthetic groups.
Important clues to the molecular etiology of
these disorders, and especially COX deficiency,
came from yeast genetics because most genes
needed for COX assembly in yeast have human homologues. Another shortcut to finding mutant genes without sequencing multiple
candidate COX-assembly genes was the search
for complementation in COX-deficient cultured cells from patients via monochromosomal
hybrid fusion or microcell-mediated chromosome transfer, which led to the identification of
the most common gene responsible for COX-
deficient LS, *SURF1* (Tiranti et al. 1998, Zhu
et al. 1998). Integrative genomics, on the basis
of information derived from DNA, mRNA, and
proteomics studies, led to the identification of
LRPPRC, the gene responsible for LS-French-
Canadian type (LSFC), another COX-deficient
form of LS associated with liver diseases and

KSS: Kearns-Sayre
syndrome

PEO: progressive
external
ophthalmoplegia

MERRF: Myoclonus
epilepsy ragged-red
fibers

MELAS:
mitochondrial
encephalomyopathy,
lactic acidosis, and
strokelike episodes

NARP: Neuropathy,
ataxia, retinitis
pigmentosa

LS: Leigh syndrome

MILS: Maternally
inherited Leigh
syndrome

LHON: Leber
hereditary optic
neuropathy

prevalent in the Saguenay-Lac Saint-Jean region of Quebec (Mootha et al. 2003, Morin et al. 1993). A bioinformatics approach was also used to identify the first mutant assembly gene responsible for complex I deficiency, *NDUFA12L* (formerly called *B17.2L*), in a child with severe cavitating leukoencephalopathy (Ogilvie et al. 2005). Knowledge of the molecular defects in these fatal infantile neurological disorders offers young parents who have lost one child the option of prenatal diagnosis.

Primary coenzyme Q_{10} (CoQ_{10}) deficiency encompasses disorders caused by blocks in the biosynthetic pathway of this small ubiquinone carrier. CoQ_{10} transfers electrons from complexes I and II to complex III and receives electrons from the β-oxidation pathway via the electron transfer flavoprotein dehydrogenase (ETF-DH) **(Figure 6)**. Mutations in two CoQ_{10} biosynthetic enzymes (in the PDSS1 and PDSS2 subunits of of COQ1, and in COQ2) have been identified in infants or children with encephalomyopathy (one of them had LS) and nephrotic syndrome (López et al. 2006, Mollet et al. 2007, Quinzii et al. 2006). Because at least nine enzymes are needed to synthesize CoQ_{10}, mutations in the other seven enzymes will probably also be associated with encephalomyopathic syndromes (DiMauro et al. 2007). Several syndromes have also been associated with a presumed secondary CoQ_{10} deficiency. These include autosomal recessive

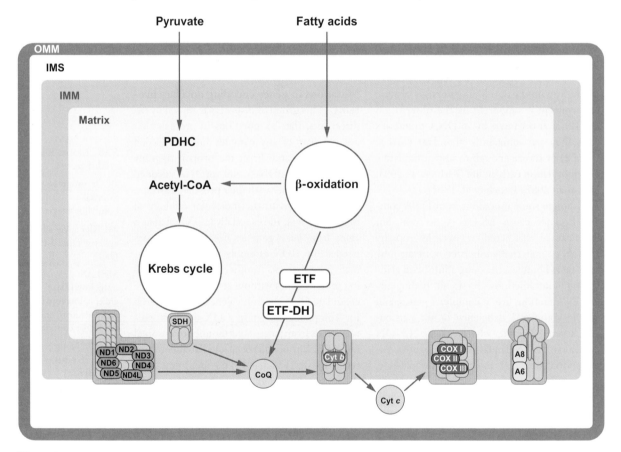

Figure 6

Schematic of mitochondrial intermediate metabolism showing the relationships between pyruvate and fatty acid metabolism and ATP synthesis. Note that the electron-transfer flavoprotein (ETF) delivers electrons from the β-oxidation pathway to CoQ_{10} via the ETF-dehydrogenase (ETF-DH).

cerebellar ataxia of unknown etiology in children, the syndrome of ataxia and oculomotor apraxia (AOA1) caused by mutations in the aprataxin gene (*APTX*) (Quinzii et al. 2005), and a predominantly myopathic form of glutaric aciduria type II (GAII) caused by mutations in the electron transfer flavoprotein dehydrogenase gene (*ETFDH*) (Gempel et al. 2007) (**Figure 6**). Aside from its scientific importance, knowledge of CoQ$_{10}$ deficiency syndromes is important for physicians because most patients improve with CoQ$_{10}$ supplementation.

The other respiratory complexes obviously also require assembly, and mutations in assembly factor BCS1L for complex III (Visapaa et al. 2002) and ATPAF2 for complex V (De Meirleir et al. 2004) have also been found. Clearly, the pool of available candidate genes has yet to be exhausted (DiMauro & Hirano 2005).

Defects of intergenomic communication. The alterations of mtDNA of some disorders are not caused by primary mutations of the mitochondrial genome, but rather are the result of garbled messages from the nuclear genome, which controls mtDNA replication, maintenance, and translation. The resulting Mendelian disorders are characterized by qualitative (multiple deletions) or quantitative (depletion) alterations of mtDNA, or by defective translation of mtDNA-encoded respiratory chain components. Of note, most of these disorders are caused by alterations in the pools of nucleotides required to synthesize mtDNA, or in enzymes associated with mtDNA replication itself (Spinazzola & Zeviani 2005) (**Figure 7**).

Multiple mtDNA deletions. From the clinical point of view, multiple mtDNA deletion syndromes share the cardinal features of ocular and limb myopathy (PEO, ptosis, proximal weakness), which are almost invariably associated with extramuscular system involvement, including peripheral nerves (sensorimotor neuropathy), the brain (ataxia, dementia, psychosis), the ear (sensorineural hearing loss), and the eye (cataracts). Mutations in several genes, all involved in the homeostasis of the mitochondrial nucleotide pools, have been associated with PEO and multiple mtDNA

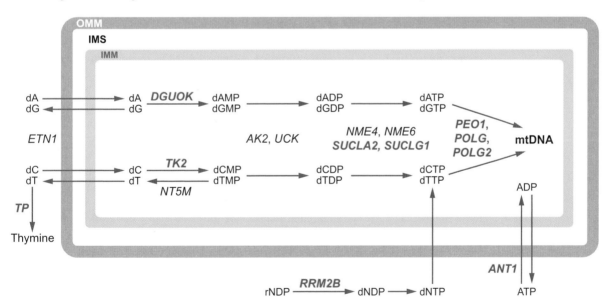

Figure 7

Schematic of nucleotide metabolism for mtDNA synthesis and replication. Genes in bold have been associated with diseases characterized by multiple mtDNA deletions and/or with mtDNA depletion.

deletions. These include *ANT1*, which encodes the adenosine nucleoside translocator; *PEO1*, which encodes a helicase called Twinkle; *ECGF1*, which encodes the cytosolic enzyme thymidine phosphorylase (TP); *POLG*, which encodes the mitochondrial polymerase γ catalytic subunit; and *POLG2*, which encodes the dimeric accessory subunit of POLG (Spinazzola & Zeviani 2005). Two of these disorders are of special interest to neurologists.

The first is MNGIE (mitochondrial neurogastrointestinal encephalomyopathy), an autosomal recessive multisystem disease of young adults caused by mutations in TP (Nishino et al. 1999) and characterized clinically by PEO, neuropathy, leukoencephalopathy, and intestinal dysmotility leading to cachexia and early death. The lack of TP activity damages mtDNA synthesis, causing not only multiple deletions, but also depletion and point mutations, which are evident in skeletal muscle, although muscle expresses little TP (Hirano et al. 2005). This muscle paradox suggests that TP deficiency acts through toxic intermediates. Two such toxic intermediates, thymidine and deoxyuridine, accumulate massively in the blood of MNGIE patients. Hemodialysis, an obvious therapeutic approach, has only transient effects, as do platelet infusions, but allogeneic bone marrow transplantation in one patient restored TP activity in buffy coat cells and normalized blood levels of thymidine and deoxyuridine. Although the patient has improved subjectively 18 months after the procedure, clinical efficacy remains to be firmly documented (Hirano et al. 2006b).

Disorders associated with mutations in *POLG* are inherited as either autosomal-recessive or autosomal-dominant traits. Both forms of inheritance are encountered in adults with PEO and multiple mtDNA deletions: Clinical manifestations include ataxia, peripheral neuropathy, parkinsonism, psychiatric disorders, myoclonus epilepsy, and gastrointestinal symptoms (DiMauro et al. 2006a). Autosomal recessive inheritance of mutations in *POLG* is the rule in children with Alpers syndrome, a severe hepatocerebral disease associated with mtDNA depletion and extreme vulnerability to valproate administration (Naviaux & Nguyen 2004). This clinical heterogeneity can, at least in part, be attributed to the site of the mutation in the catalytic subunit, which has a polymerase (i.e., replicating) domain and an exonuclease (i.e., proofreading) domain joined by a "linker" region: Most patients with Alpers syndrome have at least one mutation in the linker region and another in the polymerase domain, whereas adults with PEO tend to have mutations solely in the polymerase domain. To complicate matters further, mutations in the dimeric accessory subunit POLG2, which is responsible for processive DNA synthesis and tight binding of the POLG complex to DNA, can also cause autosomal dominant PEO (Longley et al. 2006).

Depletion of mtDNA. We have seen how some mutations in POLG predominantly cause mtDNA depletion and result in a severe infantile hepatocerebral disorder (Alpers syndrome). In fact, mutations in other proteins controlling the mitochondrial nucleotide pool also cause mtDNA depletion. For reasons that are not completely clear, the degree of depletion varies in different tissues, but two major syndromes have emerged: (*a*) hepatocerebral syndrome, caused by mutation either in *POLG* or in *DGUOK*, which encodes the enzyme deoxyguanosine kinase (dGK); and (*b*) a purely or predominantly myopathic syndrome associated with mutations in *TK2*, which encodes the mitochondrial form of the enzyme thymidine kinase, with mutations in *SUCLA2*, encoding the β subunit of the mitochondrial matrix enzyme succinyl-CoA synthetase (Elpeleg et al. 2005), and with mutations in *RRM2B*, encoding the cytosolic p53-inducible ribonucleotide reductase small subunit (p53R2) (Bourdon et al. 2007). However, not all cases of mtDNA depletion are explained by mutations in these four genes, and not all mutated genes are involved in nucleotide pool homeostasis. For example, some children with hepatocerebral syndrome harbored pathogenic mutations in a gene on chromosome 2, *MPV17*, which encodes an

IMM protein of unknown function (Spinazzola et al. 2006). The importance of this gene was bolstered by the finding that the same homozygous mutation encountered in a southern Italian family is the cause of a disease endemic in the Navajo population of the American southwest (Karadimas et al. 2006). The disease is called Navajo neurohepatopathy (NNH) to stress that neuropathy rather than encephalopathy accompanies the liver dysfunction in this condition, probably because of some as-yet-unknown genetic modifier. We can now provide sound genetic counseling to the Navajo population in the hopes of eradicating this dreadful disease.

Defects of mtDNA translation. Faithful translation of the 13 mtDNA-encoded subunits of the respiratory chain requires not only intact mtDNA, a trustworthy polymerase, and the availability of nucleotide building blocks, but also ribosomal proteins, RNA modification enzymes, and initiation, elongation, and termination factors, all encoded by nDNA. Defects in mtDNA translation result in severe combined respiratory chain complex defects, and it is important to think of this pathogenic mechanism in infants or children with hepatocerebral syndrome, encephalopathy, infantile cavitating leukoencephalopathy, or cardiomyopathy and otherwise unexplained multiple respiratory chain defects. Thus far, investigators have described mutations in four genes, but this number will certainly increase in the years to come. The first gene, *GFM1*, encodes one of four ribosomal elongation factors (Coenen et al. 2004, Valente et al. 2007); the second, *MRPS16*, encodes the mitochondrial ribosomal protein subunit 16 (Miller et al. 2004); the third, *TSFM*, encodes the mitochondrial elongation factor EFTs (Smeitink et al. 2006); and the fourth gene, *TUFM*, encodes the elongation factor Tu (Valente et al. 2007). A different syndrome is caused by defective pseudouridylation of mitochondrial tRNAs and is characterized by myopathy, lactic acidosis, and sideroblastic anemia (MLASA): Mutations in this gene, *PUS1*, which encodes the mitochondrial enzyme pseudouridine synthase 1, have

been identified in three families (Bykhovskaya et al. 2004, Fernandez-Vizarra et al. 2006).

Mutations affecting the lipid milieu of the respiratory chain. The complexes of the respiratory chain are embedded in the lipid milieu of the IMM, whose major component is cardiolipin, an acidic phospholipid. Cardiolipin does not have merely a scaffolding function, but also participates in the formation of supercomplexes (stoichiometric assemblies of individual respiratory chain complexes into functional units) (Zhang et al. 2005b) and interacts directly with COX (Sedlak et al. 2006); conversely, intact respiratory chain function is essential for cardiolipin biosynthesis (Gohil et al. 2004). Therefore, genetic abnormalities of cardiolipin could impair respiratory chain function in humans. The best candidate for this role is Barth syndrome, an X-linked recessive disorder characterized by mitochondrial myopathy, cardiomyopathy, and growth retardation, and caused by mutations in the gene encoding a phospholipid acyltransferase called tafazzin (*TAZ*) (Schlame & Ren 2006). Tafazzin promotes structural uniformity and molecular symmetry among cardiolipin molecular species, and mutations in *TAZ* alter the concentration and composition of cardiolipin, leading to altered mitochondrial architecture and function. Some *TAZ* mutations cause mislocalization of cardiolipin from the outer mitochondrial membrane (OMM) and IMM to the mitochondrial matrix (Claypool et al. 2006).

DISEASES CAUSED BY IMPAIRED MITOCHONDRIAL PROTEIN IMPORT

Of the 1300+ proteins found in mammalian mitochondria (Schon 2007), only 13 are encoded by mtDNA. All others are encoded by nDNA genes, synthesized in the cytoplasm, and imported into the organelle. Mitochondrial import is a complex process, with different pathways for protein targeting and sorting to each of the four mitochondrial compartments (OMM, IMM, IMS, and the matrix enveloped by the IMM). Among the components of the import machinery, composed of nearly 60 polypeptides, are members of the heat shock protein (HSP) family, chaperones needed for the unfolding and refolding of mitochondrially targeted proteins as they transit through the import receptors and are directed to the appropriate compartment. Most, but not all, mitochondrial proteins (especially those destined for the IMM and the matrix) have well-defined targeting signals, usually located at the N-terminus of the protein. Once inside the mitochondrion, the mitochondrial targeting signal (MTS, or leader peptide) is cleaved to release the mature protein. The import machinery consists of polymeric translocases in the outer membrane (TOM) or the inner membrane (TIM). In collaboration with a sorting and assembly machinery (SAM), a presequence translocation-associated motor (PAM), and a mitochondrial import and assembly (MIA) pathway specific for a subset of IMS proteins (Gabriel et al. 2006), TOM and TIM sort out incoming polypeptides to the proper compartments (Chacinska & Rehling 2004).

Although a few mutations in leader peptides have been associated with specific enzyme defects, such as methylmalonic acidemia (Ledley et al. 1990) and PDHC deficiency (Takakubo et al. 1995), remarkably few human diseases have been attributed to genetic defects of the general importation machinery. One of these is an X-linked recessive deafness-dystonia syndrome (Mohr-Tranebjaerg syndrome) caused by mutations in the gene (*TIMM8A*) encoding the deafness/dystonia protein (DDP), an MIA pathway protein located in the IMS (Roesch et al. 2002). Another is an autosomal dominant form of hereditary spastic paraplegia (HSP type 13; SPG13) caused by mutations in the import chaperonin HSP60 (Hansen et al. 2002).

Unless most disorders caused by disruption of the general importation machinery are incompatible with life, as suggested by Fenton (1995), we can expect more such disorders to be identified in the near future.

DISEASES CAUSED BY ABERRANT MITOCHONDRIAL DYNAMICS

This relatively new area of interest for clinical neuroscientists has already yielded instructive results and is sure to provide many more in the coming years. Remembering their bacterial origin, mitochondria move, fuse, and divide within cells, where they often form tubular networks that may favor the delivery of organelles to areas of high energy demand (Bossy-Wetzel et al. 2003). The need for mitochondrial motility is nowhere more evident than in motor neurons of the anterior horn cells, where mitochondria must travel a huge distance from the cell soma to the neuromuscular junction. Mitochondria travel on microtubular rails, propelled by motor proteins, usually GTPases, called kinesins (when mitochondria travel downstream) or dyneins (when they travel upstream). The first mitochondrial motility defect was identified in a family with autosomal dominant hereditary spastic paraplegia type 10 (SPG10) and mutations in a gene encoding one of the kinesins (*KIF5A*): The mutation affects a region of the protein involved in microtubule binding (Fichera et al. 2004) (**Figure 8**).

In yeast, at least four proteins are required for mitochondrial fission: Dnm1p (dynamin-related protein), Fis1p (fission-related protein), Mdv1p (mitochondrial division protein), and Caf4p (carbon catabolite repression-associated factor). Of the four, only Fis1p is an integral part of mitochondria, located in the outer membrane. Upon a signal to divide, Fis1p recruits Dnm1p to the organelle via the bridge proteins, Mdv1p and Caf4p; Dnm1p then forms an ever-tightening spiral collar around the organelle, which severs the mitochondrion by strangulation (Chan 2006). For the opposite process of mitochondrial fusion, two proteins are required in yeast: Fzo1p (the yeast homolog of the *Drosophila* fuzzy onion protein) and Ugo1p (ugo is Japanese for fusion). For fission to occur, the OMM and IMM must establish contact sites, apparently through the action of yet another protein called Mgm1p (mitochondrial genome maintenance protein 1).

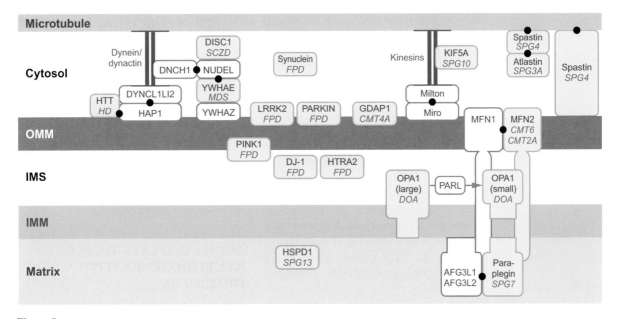

Figure 8

Selected genes associated with mitochondrial dynamics. Genes boxed in yellow have been associated with neurodegenerative or psychiatric diseases (*disease abbreviations in red*). Mitochondrially targeted gene products are in bold. Black dots denote reported interactions between proteins.

Mutations in the human orthologs of Mgm1p (OPA1) and Fzo1p (MFN2 or mitofusin 2) have been associated with human diseases. Mutations in OPA1 cause autosomal dominant optic atrophy (DOA), the Mendelian counterpart, as it were, of LHON and are characterized by maldistribution of mitochondria in affected cells (Alexander et al. 2000, Delettre et al. 2000). Notably, OPA1 interacts with mitofusin 1 (MFN1) to promote fusion (Cipolat et al. 2004). However, beyond its role in fusion, OPA1, an IMM protein, is also required for remodeling the cristae (Cipolat et al. 2006) together with PARL (presenilin-associated rhomboid like), an IMS-localized protein (Pellegrini & Scorrano 2007).

Mutations in the second mitofusin gene, *MFN2*, cause an autosomal dominant axonal variant of Charcot-Marie-Tooth disease (CMT type 2A) (Lawson et al. 2005, Zuchner et al. 2004). A recent review of 62 unrelated axonal CMT families revealed *MFN2* mutations in 26 patients from 15 families, which suggests that this is a major cause of axonal CMT2A (Chung

et al. 2006). In addition, mutations in *GDAP1*, the gene encoding ganglioside-induced differentiation protein 1, which is located in the OMM and which regulates the mitochondrial network (Niemann et al. 2005), cause CMT type 4A, an autosomal recessive, severe, early-onset form of either demyelinating or axonal neuropathy (Pedrola et al. 2005) (**Figure 8**).

A remarkable example of the underlying connections between mitochondrial movement and ostensibly disparate diseases is Charcot-Marie-Tooth disease type 6 (CMT6), which is characterized by the coexistence of peripheral neuropathy and optic atrophy. Moreover, optic atrophy onset is followed in many patients by slow vision recovery, as sometimes seen in LHON patients. Suchner et al. (2006a) found mutations in MFN2 in affected members of six unrelated families with CMT6; one of them had a missense mutation (R94W) identical to that in some patients with CMT2A. An underlying problem in mitochondrial movement presumably causes both peripheral and optic neuropathy, even though most patients with

CMT: Charcot-Marie Tooth

MFN2 mutations do not have optic atrophy, and most patients with OPA1 mutations do not have CMT. The patient's nuclear background may influence the penetrance of the mitochondrial trafficking defect.

Because mitochondria are not the only cargoes to be moved around the cell, it is not too surprising that mutations in genes controlling mitochondrial motility may also affect other organelles. Both mitochondria and peroxisomes had abnormal size, shape, and distribution in fibroblasts from an infant with a syndrome of encephalopathy, optic atrophy, lactic acidosis, and a heterozygous dominant mutation in the human ortholog of yeast Dnm1p called dynamin-like or dynamin-related protein 1 (DLP1/DRP1; gene *DNM1L*) (Waterham et al. 2007).

A mitochondrial import defect may be related to impaired neuronal migration (**Figure 8**). Two neuronal migration disorders, isolated lissencephaly sequence (ILS) and the Miller-Dieker syndrome (MDS), are associated with deletions on chromosome 17p13.3. Mutations in LIS1 (gene *PAFAH1B1*) cause ILS, whereas a second gene at this locus, encoding the 14-3-3 protein ε isoform (*YWHAE*), is invariably deleted in patients with MDS lissencephaly (Toyo-oka et al. 2003). YWHAE is a cytoplasmic chaperone that targets precursor proteins to the mitochondria, which is why it is also called the mitochondrial import stimulating factor subunit L (MSFL) (Alam et al. 1994). YWHAE interacts with three other proteins that not only are required for neuronal migration but also are known to associate with mitochondria: NUDEL (nuclear distribution protein nudE-like 1) (Brandon et al. 2005, Ikuta et al. 2007), FEZ1 (fasciculation and elongation protein zeta-1) (Ikuta et al. 2007), and DISC1 (deleted in schizophrenia 1) (Millar et al. 2005). NUDEL targets dynein to microtubule ends through LIS1 (Li et al. 2005), whereas altered expression of both FEZ1 and DISC1 caused mitochondrial morphology and mobility defects (Ikuta et al. 2007, Millar et al. 2005). In yeast, the homolog of human YWHAE (Bmh1p; 14-3-3εαβγ) interacts with

the homolog of human YWHAZ (Bmh2p; 14-3-3ζ) (Chaudhri et al. 2003); YWHAZ is, in fact, present in mitochondria (Schindler et al. 2006, Taylor et al. 2003). Thus, loss of YWHAE may well affect neuronal migration either by disrupting the trafficking of these latter proteins to mitochondria or by interdicting the binding of mitochondria to dynein.

These diseases are only the proverbial tip of what will be found to be an iceberg of human neurodegenerative disorders directly or indirectly linked to abnormal mitochondrial motility, fusion, or fission (**Table 4**).

AGING AND LATE-ONSET NEURODEGENERATIVE DISORDERS

In the title of a 1992 *News & Views* article for *Nature Genetics*, the late Anita Harding posited the role of mitochondria in normal aging with her usual wit: "Growing Old: The Most Common Mitochondrial Disease of All?" (Harding 1992). Her comments concerned an article documenting the age-related accumulation of the "common" 4977-bp mtDNA deletion (Schon et al. 1989) in human brain, but especially in the caudate, putamen, and substantia nigra (Soong et al. 1992). Last year, using more sophisticated techniques (laser microdissection, single-molecule PCR, long-range PCR), two groups confirmed the age-related accumulation of somatic and clonal mtDNA deletions in substantia nigra and showed that neurons with high mutation loads were COX-negative (Bender et al. 2006, Kraytsberg et al. 2006). These findings are consistent with the almost 40-year-old "mitochondrial theory of aging" (Harman 1972), which postulates a vicious cycle whereby somatic mtDNA mutations [predominantly deletions (Pallotti et al. 1996)] generate excessive reactive oxygen species (ROS), and these, in turn, further damage mtDNA. The main objection to this hypothesis came from clinical experience because the mutation loads recorded in most postmitotic tissues during normal aging are at least one order of magnitude lower than those found

Table 4 Diseases associated with defects in mitochondrial dynamics[a]

Disease	Gene	Protein
Dominant optic atrophy (DOA)	**OPA1**	Dynamin-related GTPase
CMT type 2A	**MFN2**	Mitofusin 2
CMT type 4A	**GDAP1**	Ganglioside-induced differentiation protein 1
CMT type 6	**MFN2**	Mitofusin 2
AD-HDP type 3A	SPG3A	Atlastin (associated with spastin)
AD-HSP type 4	SPAST	Spastin (microtubule severing protein)
AR-HSP type 7	**SPG7**	Paraplegin (AAA protease)
AD-HSP type 10	KIF5A	Kinesin heavy chain
AR-HSP type 20	**SPG20**	Spartin (microtubule-interacting protein?)
AD-HSP type 31	**REEP1**	Receptor expression-enhancing protein
Infantile microcephaly	**DNM1L**	Dynamin-related protein DLP1
Huntington disease	HD	Huntingtin (binds HAP1)
Lissencephaly (Miller-Dieker)	YWHAE	14-3-3ε protein

[a]Genes encoding mitochondrially targeted proteins are in bold.

in patients with primary pathogenic mtDNA deletions (e.g., KSS; **Table 2**). However, the proportion of Δ-mtDNA measured in single neurons of the substantia nigra from aged normal individuals approaches or surpasses the estimated pathogenic threshold (Bender et al. 2006, Kraytsberg et al. 2006), although neurons from patients with Parkinson disease do not contain significantly more Δ-mtDNAs than did age-matched controls (Bender et al. 2006). The observation that many of these neurons are functionally impaired (COX-negative) makes conceivable the second step in the vicious cycle: excessive ROS generation. Although the mitochondrial theory of aging in and by itself does not explain either natural aging or late-onset neurodegenerative diseases, it almost certainly plays a role in both conditions, together with nuclear genetic factors. A dramatic example of the importance (but not necessarily the functional significance) of nuclear factors is the precocious, in fact precipitous, aging of transgenic mice that express a proofreading-deficient POLG (Khrapko et al. 2006, Kujoth et al. 2005, Trifunovic et al. 2004).

The role of nuclear-encoded mitochondrial factors in neurodegenerative disorders can be approached by considering first the general re-

YOU CAN PAY ME NOW OR YOU CAN PAY ME LATER

The classic mitochondrial diseases known as the mitochondrial encephalomyopathies are caused by mutations in the mitochondrial or nuclear genome that affect the respiratory chain directly. Overall, these disorders cause acute (e.g., seizures, strokes) or subacute (e.g., ataxia, neuropathy) clinical problems that manifest early in life, in children or in young adults. However, as a general rule, genetic defects in mitochondrial functions that do not directly impact the respiratory chain—such as protein import, organellar dynamics, and programmed cell death—cause chronic clinical problems of much later onset, highly reminiscent of the three more common age-related and apparently sporadic neurodegenerative disorders, Parkinson disease, Alzheimer disease, and amyotrophic lateral sclerosis. In fact, now that we have begun to appreciate that the familial forms of the Big Three involve mitochondrial function in the guise of altered organellar dynamics, it is no stretch of the imagination to envision the same kinds of mitochondrial involvement even in the far-more-common sporadic presentations of these devastating disorders.

lationship of mitochondrial biology to neurodegeneration and then the specific diseases attributed to mutations in nuclear-encoded proteins, most of them targeted to the mitochondria (Tieu & Przedborski 2006).

Mitochondria and Neurodegeneration

FRDA: Friedreich ataxia

Cell death in neurodegenerative diseases usually occurs by apoptosis, more commonly by the intrinsic mitochondrial pathway than by the extrinsic cell-signaling pathway. The intrinsic pathway controls activation of caspase 9, through the adaptor molecule Apaf-1, by regulating the release of cytochrome c from the IMS to the cytosol. Proapoptotic and antiapoptotic members of the Bcl-2 family, and also stress and survival signals, regulate the release of cytochrome c from the ISS into the cytoplasm. Proapoptotic signals can also release proteins such as Smac/DIABLO and Omi/HTRA2, which block IAP (inhibitor of apoptosis) proteins to activate cell death caspases. However, in the intrinsic pathway of apoptosis, mitochondria are not merely passive containers capable of leaking cytochrome c: Rather, their life-supporting functions are clearly linked to their death-promoting activity. These modulating factors include the respiratory chain activity, with the unavoidably associated generation of ROS; mitochondrial fusion and fission; calcium homeostasis; the lipid composition of the mitochondrial membranes; and the mitochondrial permeability transition.

As an obvious example of the respiratory chain influence, cytochrome c is a vital water-soluble electron carrier, not just an executioner in apoptosis.

Also, ROS are normal byproducts of the respiratory chain activity, and their concentration is controlled by mitochondrial antioxidant enzymes, such as manganese superoxide dismutase (SOD2) and glutathione peroxidase. Excessive ROS production (oxidative stress) is considered a central feature in the pathogenesis of all neurodegenerative disorders (Beal 2005), which explains the popularity of ROS-scavenging compounds, such as CoQ_{10} or analogous molecules, in therapeutic trials (DiMauro et al. 2006c, Shults & Schapira 2001). A pathogenic role for ROS in age-related neurodegeneration is also suggested by the correlation between rates of formation of mitochondrial reactive oxygen and nitrogen species (RONS), rates of neurodegeneration in brain and retina, and maximum lifespan potentials in five different mammalian species (Wright et al. 2004).

The observation that during apoptosis the normally tubular mitochondrial network becomes fragmented, and that the proapoptotic molecule Bax colocalizes with the fusion-related proteins DRP1 and MFN2 (Newmeyer & Ferguson-Miller 2003), suggests a regulatory role for mitochondrial fission and fusion. As mentioned above, cardiolipin has many functions beyond being a scaffold for the respiratory chain: One such function may be to favor apoptosis through Bax-mediated permeabilization of the OMM. Although cardiolipin is predominantly a component of the IMM, it may be present in the OMM at sites of contact with the IMM, where Bid and Bcl-2 also cluster (Newmeyer & Ferguson-Miller 2003).

The permeability transition (PT) refers to a still largely hypothetical pore composed of cyclophilin D and the ANT1 protein in the IMM and of the voltage-dependent anion channel (VDAC) and the peripheral benzodiazepine receptor in the OMM. Sustained opening of the PT pore is considered an obligatory step in apoptosis.

Neurodegenerative Diseases Caused by Mutations in Nuclear-Encoded Proteins Targeted to Mitochondria

Friedreich ataxia (FRDA) is an autosomal recessive disorder characterized clinically by early onset (before 25 years of age), progressive limb and gait ataxia, peripheral neuropathy with areflexia, pyramidal signs, hypertrophic cardiomyopathy, and increased incidence of diabetes. The hallmark neuropathology of FRDA is degeneration of the spinocerebellar tracts and large sensory neurons. The mutated mitochondrial protein, frataxin, is encoded by a gene (*FXN*) on chromosome 9q13, and most patients are homozygous for a GAA trinucleotide repeat expansion in the first intron of *FXN*. These are loss-of-function mutations, and residual frataxin expression level correlates with the severity of the clinical phenotype.

FRDA pathogenesis is controversial because frataxin is involved in the formation of non-heme iron-sulfur clusters (ISCs), heme biosynthesis, and the detoxification of iron. Loss of frataxin causes impaired mitochondrial iron storage and metabolism and defects in mitochondrial enzymes containing ISCs, including aconitase and complexes I, II, and III. Iron accumulation increases ROS generation by the Fenton reaction, causing oxidative damage and further mitochondrial enzyme inactivation. To worsen the situation, antioxidant defenses are decreased in cultured cells from FRDA patients (Chantrel-Groussard et al. 2001). Although the pathogenic role of oxidative stress in FRDA seemed bolstered by the beneficial effects of the antioxidant idebenone, at least on the cardiopathy (Schulz et al. 2000), paradoxically, a conditional neuronal frataxin knockout mouse showed neither evidence of oxidative stress nor improvement with antioxidants (Seznec et al. 2005).

Hereditary spastic paraplegia (HSP) is the term for a group of clinically similar disorders rather than a specific clinical entity. We have already discussed two different mitochondrial causes of autosomal dominant HSP: one a defect in mitochondrial protein importation caused by mutations in *SPG13*, encoding the chaperonin HSP60 (HSP here stands for heat shock protein), and the other a defect of mitochondrial behavior caused by mutations in the kinesin KIF5A. More controversial is the pathogenesis of an autosomal recessive form of HSP caused by mutations in a gene (*SPG7*) encoding paraplegin, a protein highly homologous to the AAA family of mitochondrial proteases (Casari et al. 1998). Because AAA proteases have a quality control function ensuring that unassembled respiratory chain subunits are degraded, a mutated paraplegin may result in an accumulation of defective subunits "choking" the importation machinery (similar to mutations in HSP60) and, ultimately, the respiratory chain (Claypool et al. 2006). However, another function of paraplegin seems to involve process-

ing MRPL32, a component of the large ribosomal subunit tightly bound to the IMM (Claypool et al. 2006, Nolden et al. 2005). Thus, a mutated paraplegin may impair mtDNA translation, in which case this form of HSP would belong with the subgroup of intergenomic communication disorders discussed above.

Autosomal dominant HSP type 4 (SPG4) is caused by mutations in spastin (gene *SPAST*), a microtubule-severing protein located in the cytoplasm. Because mitochondria must be attached to microtubules for them to travel down axons, disruption of this connection should affect mitochondrial mobility, and indeed, cells of SPG4 patients showed an abnormal perinuclear clustering of mitochondria, presumably a consequence of an inability of mutated spastin to sever microtubules (McDermott et al. 2003). Spastin's binding partner is known as atlastin (Sanderson et al. 2006), and mutations in the gene encoding this protein (*SPG3A*) also cause HSP (autosomal dominant HSP type 3A), again implicating cargo traffic on microtubules (and almost certainly mitochondria) in the pathogenesis of the disorder.

Autosomal recessive HSP type 20 (also called Troyer syndrome) is due to mutations in spartin (gene *SPG20*), an OMM protein (Lu et al. 2006). Spartin has a microtubule interacting and trafficking (MIT) domain at its N-terminus (interestingly, its mitochondrial targeting signal is located at the C-terminus), implying yet again the role of mitochondrial trafficking in the pathogenesis of this syndromic group (Lu et al. 2006).

Finally, autosomal dominant HSP type 31 (SPG31) is caused by mutations in receptor expression-enhancing protein 1 (*REEP1*), a mitochondrial protein of unknown function (Zuchner et al. 2006b).

Parkinson disease (PD) is a predominantly sporadic late-onset disorder, and the mitochondrial theory of aging, with its nonfamilial, age-related accumulation of somatic mtDNA deletions in the substantia nigra (coupled with biochemical evidence of complex I deficiency), provided an attractive pathogenic explanation. Although, as discussed above, this mechanism

PD: Parkinson disease

per se is not sufficient to explain sporadic PD, PD (or parkinsonism, if the diagnostic criteria of the London Brain Bank are applied strictly) is familial more often than thought until five years ago (Hardy et al. 2006). To date, six nuclear genes have been implicated: *PARK2* (disease locus PARK2), encoding parkin; *PINK1* (locus PARK6), encoding PTEN-induced putative kinase 1, or PINK1; *PARK7* (locus PARK7), encoding DJ-1 (Hardy et al. 2006); *SNCA* (locus PARK1/4), encoding α-synuclein; *LRRK2* (locus PARK8), encoding dardarin; and *HTRA2* (locus PARK13) encoding Omi/HTRA2. All these proteins interact directly or indirectly with mitochondria and seem to affect apoptosis.

Mutations in *PARK2* have been associated with autosomal recessive PD. Parkin is a ubiquitin E3 ligase associated with the OMM, where it has a protective role against mitochondrial swelling caused by ceramide-induced apoptosis (Darios et al. 2003). As further evidence of a mitochondrial role for parkin, patients with PD and parkin mutations have decreased complex I in leukocytes (Muftuoglu et al. 2003).

PINK1 is a mitochondrial kinase (Silvestri et al. 2005) whose precise function is unknown, but which, when mutated, causes early-onset recessive PD and, when overexpressed, protects against neuronal apoptosis (Petit et al. 2005).

Omi/HTRA2 is a serine protease localized to the mitochondrial IMS and released into the cytosol upon apoptosis induction. Strauss et al. (2005) found a mutation in *HTRA2* in four sporadic patients with PD, and a polymorphism in the same gene seems to predispose to PD development.

Although mutations in DJ-1 were thought to abolish the oxidation-induced localization of the protein to mitochondria (Canet-Aviles et al. 2004), good evidence demonstrates that both wild-type and mutant DJ-1 proteins are present in mitochondria (matrix and IMS) (Zhang et al. 2005a), where they likely have an antiapoptotic function.

α-synuclein is a cytosolic protein, but its functional relationship with mitochondria is revealed by several observations: (*a*) Overexpression of mutant α-synuclein in cell cultures im-
pairs the respiratory chain and induces oxidative damage; (*b*) transgenic mice overexpressing α-synuclein in neurons are overly sensitive to MPTP; (*c*) α-synuclein-deficient mice are more resistant to respiratory chain inhibitors; and (*d*) transgenic mice expressing mutant α-synuclein show neuronal degeneration, accumulation of intraneural inclusions, and complex IV deficiency in the spinal cord. In humans, mutations in the *SNCA* gene cause autosomal dominant PD (Polymeropoulos et al. 1997).

Autosomal recessive parkinsonism is not uncommon in patients with PEO and mutations in *POLG*, and it can be seen even in young patients without PEO (Davidzon et al. 2006).

Huntington disease (HD), an autosomal dominant disorder, penetrates fully by mid-adult life and is characterized by choreoathetotic movements, emotional problems, and dementia. Selective degeneration of striatal neurons and marked atrophy of caudate and putamen occur. HD is caused by abnormal expansion of a CAG repeat in the *HD* gene on chromosome 4, which encodes a protein called huntingtin (HTT). Although HTT is not a mitochondrial protein, four pathogenic scenarios all involve mitochondrial dysfunction.

The first scenario postulates an energy metabolism defect and is based on magnetic resonance spectroscopy (MRS) of the brain (showing lactate peaks in the occipital cortex and basal ganglia) and of muscle (showing decreased PCr/Pi ratios). Both direct and indirect biochemical evidence also show impaired energy production because the activities of respiratory chain complexes II and III were decreased in postmortem HD brains, and inhibition of complex II by malonate in experimental animals caused pathological lesions resembling those of human HD.

The second scenario is based on evidence that polyglutamine accumulation impairs calcium handling, causing calcium-induced permeability transition and cytochrome *c* release (Choo et al. 2004).

The third pathogenic mechanism in a sense includes the previous two and suggests that mutant HTT impairs mitochondrial function in a more general way, by repressing PGC-1α-regulated gene transcription of many nucleus-encoded mitochondrial genes; PGC-1α (peroxisome proliferators-activated receptor-γ coactivator 1α) is a master transcriptional coactivator that controls mitochondrial biogenesis and oxidative phosphorylation (Greenamyre 2007).

Fourth, and perhaps most provocative, a physical interaction that impacts organellar mobility in this disease may exist between HTT and mitochondria. HTT binds to huntingtin-interacting protein 1 (HAP1). HAP1 is a cytosolic protein that associates with microtubules and other membranous compartments of the cell, including mitochondria (Gutekunst et al. 1998). However, using immuno-electron microscopy, HAP1 was localized to small puncta in both the nucleus and the mitochondria (Gutekunst et al. 1998). In addition, HAP1 interacts with the p150Glued subunit of dynactin (*DYNC1LI2*) (Engelender et al. 1997, Li et al. 1998). The relationship between HTT, HAP1, and dynactin may explain the observation that microtubules are destabilized in HD (Trushina et al. 2003) and that mutant HTT impairs axonal trafficking in mammalian neurons (Trushina et al. 2004).

Amyotrophic lateral sclerosis (ALS) is a late-onset, sporadic disorder typically affecting both lower (anterior horn cells of the spinal cord) and upper (cortical) motor neurons, causing widespread paralysis and premature death. About 5%–10% of patients have a familial form of ALS (FALS), and ~20% of these harbor mutations in the Cu,Zn-superoxide dismutase 1 (*SOD1*) gene. SOD1 is present in both the cytosol and in the IMS (Sturtz et al. 2001). Transgenic mouse models overexpressing mutant SOD1 also develop motor neuron degeneration. Most pathogenic mutations do not impair SOD1 activity, and investigators assume that they cause a toxic gain of function.

Mitochondrial involvement in FALS is suggested by the early mitochondrial degeneration observed in motor neurons from patients and transgenic animals, by the presence of mutant SOD1 and of aggregates containing mutant SOD1 in the mitochondrial matrix and

IMS (Liu et al. 2004), and by the impaired mitochondrial functions (respiratory chain and calcium homeostasis) seen in transgenic mice.

Studies also report respiratory chain abnormalities in spinal cord of sporadic ALS patients (Borthwick et al. 1999, Wiedemann et al. 2002). Conversely, one patient with primary mitochondrial disease (a microdeletion in the COX I gene of mtDNA) had a typical, albeit early-onset, ALS phenotype (Comi et al. 1998).

Alzheimer disease (AD) is a neurodegenerative dementing disorder of late onset, with a relatively long course (Mattson 2004). Studies show progressive neuronal loss, especially in the cortex and the hippocampus. The two main histopathological hallmarks of AD are the accumulation of extracellular neuritic plaques, consisting mainly of β-amyloid (Aβ), and of neurofibrillary tangles, consisting mainly of hyperphosphorylated forms of the microtubule-associated protein tau (Goedert & Spillantini 2006, Roberson et al. 2007). Most AD cases are sporadic, but three genes have been identified in the familial form (FAD): amyloid precursor protein (*APP*), presenilin 1 (PS1; gene *PSEN1*), and presenilin 2 (PS2; gene *PSEN2*). Variants in two genes predispose people to SAD: apolipoprotein E isoform 4 (*APOE4*) (Corder et al. 1993) and *SORL1*, a neuronal sorting receptor (Rogaeva et al. 2007).

Abundant evidence indicates that mitochondria are affected in AD, including reduction in brain energy metabolism shown by positron emission tomography (Azari et al. 1993), mitochondrial metabolic enzyme deficiency (Mastrogiacomo et al. 1993, Sheu et al. 1985), and respiratory chain deficiency (Bonilla et al. 1999, Kish et al. 1992), etc. However, a direct role for mitochondria in AD pathogenesis has been controversial, hinging mainly on findings related to both APP and PS1.

The current view is that APP is located predominantly in the plasma membrane, where it is cleaved in a series of proteolytic events (e.g., by α-, β-, and γ-secretases) to release intra- and extracellular fragments of uncertain function. However, Avadhani's group showed by genetic dissection and expression of APP

ALS: amyotrophic lateral sclerosis

FALS: Familial ALS

AD: Alzheimer disease

FAD: familial AD

constructs in vitro that the APP protein contains a possible mitochondrial targeting signal at its N-terminus (Anandatheerthavarada et al. 2003). They then showed that nonglycosylated full-length and C-terminal truncated APP accumulates exclusively in the mitochondrial protein import channels in AD brains but not in age-matched controls (Devi et al. 2006). This result was consistent with (*a*) the identification of Aβ42 in mitochondria (Lustbader et al. 2004, Manczak et al. 2006); (*b*) the observation that Aβ binds to 17-β-hydroxysteroid dehydrogenase, type 10 (HADH2), a mitochondrial matrix protein involved in fatty acid metabolism (Lustbader et al. 2004); and (*c*) the finding that β-amyloid inhibits respiratory chain function in isolated rat brain mitochondria (Casley et al. 2002). However, it is unclear how Aβ could be derived from its precursor APP within mitochondria, given that the putative requisite initial processing proteases (e.g., α- and β-secretases) have not been found in mitochondria. Aβ could be imported into mitochondria, but at present there is no coherent explanation as to how this might be accomplished.

Researchers disagree with regard to PS1. Using immunohistochemical techniques, PS1 has been localized to numerous membranous compartments in cells. These include the endoplasmic reticulum (ER) (Walter et al. 1996), the Golgi apparatus (Annaert et al. 1999, Siman & Velji 2003), endosomes/lysosomes (Runz et al. 2002, Vetrivel et al. 2004), the nuclear envelope (Honda et al. 2000), and the plasma membrane (Schwarzman et al. 1999), where they are especially enriched at intercellular contacts known as adherens junctions (Marambaud et al. 2002). PS1 has not been found in mitochondria, except by one group that used Western blotting and immunoelectron microscopy, not immunohistochemistry, to localize PS1 to the rat mitochondrial inner membrane (Ankarcrona & Hultenby 2002, Hansson et al. 2004).

Besides PS1, the γ-secretase complex contains at least three other proteins: APH1, PEN2, and nicastrin (De Strooper 2003). Using immunoelectron microscopy and Western blotting, all three proteins have been localized to rat mitochondria (Hansson et al. 2004). However, localization of γ-secretase subunits, including PS1, to mitochondria has not been confirmed or demonstrated by other, more definitive, methods.

Given that mitochondria play a role in the pathogenesis of several neurodegenerative diseases (**Figure 8**), it would not be unreasonable to invoke similar mechanisms for AD, and especially a mechanism involving mitochondrial movement and localization (Kins et al. 2006). This possibility is supported by two observations. First, a PS1 mutation (M146V) in a mouse PS1 knock-in model impairs axonal transport and also increases tau phosphorylation (Pigino et al. 2003). Second, axonal defects, consisting of swellings that accumulate abnormal amounts of microtubule-associated and molecular motor proteins, organelles, and vesicles, have been found in both sporadic AD patients and in transgenic mouse models of FAD (Stokin et al. 2005).

MITOCHONDRIAL PSYCHIATRY

Given the brain's high dependence on oxidative metabolism, it is hardly surprising that primary mitochondrial disorders often cause cognitive deficits: dementia in adults and mental retardation or neuropsychological regression in children. Researchers have paid comparatively less attention to the relationship between psychiatric diseases and mitochondrial dysfunction. For the sake of order, let us consider separately the psychiatric manifestations of primary mitochondrial diseases and the evidence of mitochondrial dysfunction in patients with isolated primary psychiatric illnesses.

Although the literature is replete with anecdotal reports of psychiatric problems, mostly severe depression, in patients with mtDNA-related diseases (DiMauro et al. 2006b), there have been few systematic neuropsychiatric studies of large cohorts of patients with known mitochondrial diseases. We have reported preliminary data on a large group of MELAS and MERRF families (102 persons from 30 kindreds), including not only patients but also their oligosymptomatic or asymptomatic maternal relatives. In MELAS families (harboring the A3243G mutation), 42% of fully symptomatic carriers reported depressive symptoms, but 22% of asymptomatic and 29% of oligosymptomatic carriers also had depressive traits, compared with only 7% of the control group (Kaufmann et al. 2002). In MERRF families with the A8344G mutation, 80% of fully symptomatic patients, 20% of oligosymptomatic carriers, and none of the asymptomatic carriers reported depressive symptoms. The finding that even asymptomatic carriers of the MELAS mutation were prone to depression suggested that psychiatric problems might be an early clinical expression of the mutation, which is also supported by correlative neuropsychological and MRS studies, showing cerebral lactic acidosis in asymptomatic MELAS relatives and a correlation between neuropsychological scores and ventricular lactate levels (Kaufmann et al. 2004).

Among the Mendelian mitochondrial diseases, depression is frequent in patients with defects of intergenomic communication, PEO, and multiple mtDNA deletions. Psychiatric problems are especially common in patients harboring mutations in *ANT1*, *PEO1*, and *POLG* (DiMauro et al. 2006b).

Although maternal inheritance of bipolar disorder had been suggested by a higher-than-expected frequency of affected mothers and increased risk of illness in maternal relatives, mutations in mtDNA have been excluded as causes of the disease (Kirk et al. 1999). However, mitochondrial dysfunction may still be involved in bipolar disorder pathogenesis through several pathogenic mechanisms, many of which, not surprisingly, are similar to those proposed for neurodegenerative diseases. These include alterations of calcium homeostasis (Kato et al. 2002), downregulation of genes controlling mitochondrial energy metabolism (Konradi et al. 2004), and impaired mtDNA replication (Kakiuchi et al. 2005), all possibly related to a change (-116C > G) in the promoter region of *XBP1*, a pivotal gene in the endoplasmic reticulum stress response (Kakiuchi et al. 2003).

Another gene that has been highlighted in the pathogenesis of both bipolar disorder and schizophrenia is *Disrupted in schizophrenia 1* (*DISC1*), so called because a disruption of this gene by the chromosome 1 breakpoint of a balanced t(1;11) translocation did, in fact, cosegregate with schizophrenia and related mood disorders in a large Scottish family (St Clair et al. 1990). The association of DISC1 and schizophrenia was confirmed in Finnish, American, Japanese, and Taiwanese populations (Roberts 2007) and extended to bipolar disorder (Maeda et al. 2006). Although its precise role remains unclear, DISC1 bound predominantly to mitochondria (James et al. 2004) and interacted with several proteins, including FEZ1, LIS1, and NUDEL, which are also involved, at least indirectly, in neurodegenerative diseases (**Figure 8**). Overexpression of DISC1 in COS-7 cells disrupts mitochondrial organization and leads to the formation of ring-like structures, suggesting a role of this protein in controlling mitochondrial dynamics (Millar et al. 2005). The distribution of DISC1 in the developing and adult brain (frontal cortex, hippocampus, thalamus) and its involvement with neuronal migration, neurite outgrowth, and synaptic plasticity support the pathogenic role of genetic variants in psychiatric disorders.

CONCLUSIONS

Two main concepts emerge from this overview of mitochondrial disorders in the nervous system. The first is a sense of amazement that this small organelle, a foreign guest that took up permanent residence in all our cells, participates in such a wide array of neurological disorders, from LS in infancy to AD in old age. A second, and related, consideration is that there is much more to mitochondria than the standard textbook gloss that they are the powerhouses of the cell that only produce ATP. Besides ATP production, mitochondria perform varied functions that are important for cell life and death, including ROS generation, calcium homeostasis, and programmed cell death, and the pathogenesis of any mitochondrial disease is likely to involve, at least to some extent and at some point, all these functions. In fact, we have seen that even primary defects of the respiratory chain—the most likely causes of power failures—are so diverse as to defy a unitary mode of pathogenesis that invokes "merely" the loss of ATP capacity.

In the nervous system, mitochondrial dynamics are crucial to guarantee long distance delivery and balanced distribution of energy (in addition to all the other functions) to the farthest reaches of neurons (synapses and dendrites). Although the concept and the rules of mitochondrial dynamics have been latecomers to the field of mitochondrial diseases, alterations in the topology and topography of this most plastic of organelles may well be the unifying theme providing, more often than not, a common final pathogenic pathway for neurodegenerative and psychiatric diseases alike.

DISCLOSURE STATEMENT

The authors are not aware of any biases that might be perceived as affecting the objectivity of this review.

ACKNOWLEDGMENTS

This work has been supported by grants from the National Institutes of Health (NS 11766 and HD32062), from the Muscular Dystrophy Association, and from the Marriott Mitochondrial Disorder Clinical Research Fund (MMDCRF). The authors are grateful to Drs. Michio Hirano and Lewis P. Rowland for revising the manuscript and to Drs. Eduardo Bonilla and Kurenai Tanji for providing **Figures 3** and **4**.

LITERATURE CITED

Alam R, Hachiya N, Sakaguchi M, Kawabata S, Iwanaga S, et al. 1994. cDNA cloning and characterization of mitochondrial import stimulation factor (MSF) purified from rat liver cytosol. *J. Biochem.* 116:416–25

Alexander C, Votruba M, Pesch UEA, Thiselton DL, Mayer S, et al. 2000. OPA1, encoding a dynamin-related GTPase, is mutated in autosomal dominant optic atrophy linked to chromosome 3q28. *Nat. Genet.* 26:211–15

Anandatheerthavarada HK, Biswas G, Robin MA, Avadhani NG. 2003. Mitochondrial targeting and a novel transmembrane arrest of Alzheimer's amyloid precursor protein impairs mitochondrial function in neuronal cells. *J. Cell. Biol.* 161:41–54

Anderson S, Bankier AT, Barrel BG, DeBruijn M, Coulson AR, et al. 1981. Sequence and organization of the human mitochondrial genome. *Nature* 290:457–65

Ankarcrona M, Hultenby K. 2002. Presenilin-1 is located in rat mitochondria. *Biochem. Biophys. Res. Commun.* 295:766–70

Annaert WG, Levesque L, Craessaerts K, Dierinck I, Snellings G, et al. 1999. Presenilin 1 controls γ-secretase processing of amyloid precursor protein in pre-Golgi compartments of hippocampal neurons. *J. Cell Biol.* 147:277–94

Azari NP, Pettigrew KD, Schapiro MB, Haxby JV, Grady CL, et al. 1993. Early detection of Alzheimer's disease: a statistical approach using positron emission tomographic data. *J. Cereb. Blood Flow Metab.* 13:438–47

Beal MF. 2005. Mitochondria take center stage in aging and neurodegeneration. *Ann. Neurol.* 58:495–505

Bender A, Krishnan KJ, Morris CM, Taylor GA, Reeve AK, et al. 2006. High levels of mitochondrial DNA deletions in substantia nigra neurons in aging and Parkinson disease. *Nat. Genet.* 38:515–17

Betts J, Jaros E, Perry RH, Schaefer AM, Taylor RW, et al. 2006. Molecular neuropathology of MELAS: level of heteroplasmy in individual neurones and evidence of extensive vascular involvement. *Neuropath. Appl. Neurobiol.* 32:359–73

Bonilla E, Tanji K, Hirano M, Vu TH, DiMauro S, Schon EA. 1999. Mitochondrial involvement in Alzheimer's disease. *Biochim. Biophys. Acta* 1410:171–82

Borthwick GM, Johnson MA, Ince PG, Shaw PJ, Turnbull DM. 1999. Mitochondrial enzyme activity in amyotrophic lateral sclerosis: implications for the role of mitochondria in neuronal cell death. *Ann. Neurol.* 46:787–90

Bossy-Wetzel E, Barsoum MJ, Godzik A, Schwartzenbacher R, Lipton SA. 2003. Mitochondrial fission in apoptosis, neurodegeneration and aging. *Curr. Opin. Cell Biol.* 15:706–16

Bourdon A, Minai L, Serre V, Jais J-P, Sarzi E, et al. 2007. Mutation of RRM2B, encoding p53-controlled ribonucleotide reductase (p53R2), causes severe mitochondrial DNA depletion. *Nat. Genet.* 39:776–80

Brandon NJ, Schurov I, Camargo LM, Handford EJ, Duran-Jimeniz B, et al. 2005. Subcellular targeting of DISC1 is dependent on a domain independent from the Nudel binding site. *Mol. Cell. Neurosci.* 28:613–24

Bykhovskaya Y, Casas KA, Mengesha E, Inbal A, Fischel-Ghodsian N. 2004. Missense mutation in pseudouridine synthase 1 (*PUS1*) causes mitochondrial myopathy and sideroblastic anemia (MLASA). *Am. J. Hum. Genet.* 74:1303–8

Canet-Aviles RM, Wilson MA, Miller DW, Ahmad R, McLendon C, et al. 2004. The Parkinsons disease protein DJ-1 is neuroprotective due to cysteine-sulfinic acid-driven mitochondrial localization. *Proc. Natl. Acad. Sci. USA* 101:9103–8

Carelli V, Achilli A, Valentino ML, Rengo C, Semino O, et al. 2006. Haplogroup effects and recombination of mitochondrial DNA: novel clues from the analysis of Leber hereditary optic neuropathy pedigrees. *Am. J. Hum. Genet.* 78:564–74

Carelli V, La Morgia C, Iommarini L, Carroccia R, Mattiazzi M, et al. 2007. Mitochondrial optic neuropathies: how two genomes may kill the same cell type? *Biosci. Rep.* 27:173–84

Casari G, De Fusco M, Ciarmatori S, Zeviani M, Mora M, et al. 1998. Spastic paraplegia and OXPHOS impairment caused by mutations in paraplegin, a nuclear-encoded mitochondrial metalloprotease. *Cell* 93:973–83

Casley CS, Canevari L, Land JM, Clark JB, Sharpe MA. 2002. β-amyloid inhibits integrated mitochondrial respiration and key enzyme activities. *J. Neurochem.* 80:91–100

Chacinska A, Rehling P. 2004. Moving proteins from the cytosol into mitochondria. *Biochem. Soc. Trans.* 32:774–76

Chan DC. 2006. Mitochondria: dynamic organelles in disease, aging, and development. *Cell* 125:1241–52

Chantrel-Groussard K, Geromel V, Puccio H, Koenig M, Munnich A, et al. 2001. Disabled early recruitment of antioxidant defenses in Friedreich's ataxia. *Hum. Mol. Genet.* 10:2061–67

Chaudhri M, Scarabel M, Aitken A. 2003. Mammalian and yeast 14-3-3 isoforms form distinct patterns of dimers in vivo. *Biochem. Biophys. Res. Commun.* 300:679–85

Choo YS, Johnson GV, MacDonald M, Detloff PJ, Lesort M. 2004. Mutant huntingtin directly increases susceptibility of mitochondria to the calcium-induced permeability transition and cytochrome *c* release. *Hum. Mol. Genet.* 13:1407–20

Chung KW, Kim SB, Park KD, Choi KG, Lee JH, et al. 2006. Early onset severe and late-onset mild Charcot-Marie-Tooth disease with mitofusin 2 (MFN2) mutations. *Brain* 129:2103–18

Cipolat S, Martins de Brito O, Dal Zilio B, Scorrano L. 2004. OPA1 requires mitofusin 1 to promote mitochondrial fusion. *Proc. Natl. Acad. Sci. USA* 101:15927–32

Cipolat S, Rudka T, Hartmann D, Costa V, Serneels L, et al. 2006. Mitochondrial rhomboid PARL regulates cytochrome *c* release during apoptosis via OPA1-dependent cristae remodeling. *Cell* 126:163–75

Claypool SM, McCaffrey JM, Koehler CM. 2006. Mitochondrial mislocalization and altered assembly of cluster of Barth syndrome mutant tafazzins. *J. Cell Biol.* 174:379–90

Coenen MJH, Antonicka H, Ugalde C, Sasarman F, Rossi P, et al. 2004. Mutant mitochondrial elongation factor G1 and combined oxidative phosphorylation deficiency. *New Engl. J. Med.* 351:2080–86

Comi GP, Bordoni A, Salani S, Franceschina L, Sciacco M, et al. 1998. Cytochrome *c* oxidase subunit I microdeletion in a patient with motor neuron disease. *Ann. Neurol.* 43:110–16

Corder EH, Saunders AM, Strittmatter WJ, Schmechel DE, Gaskell PC, et al. 1993. Gene dose of apolipoprotein E type 4 allele and the risk of Alzheimer's disease in late onset families. *Science* 261:921–23

Darios F, Corti O, Lucking CB, Hampe C, Muriel M-P, et al. 2003. Parkin prevents mitochondrial swelling and cytochrome *c* release in mitochondria-dependent cell death. *Hum. Mol. Genet.* 12:517–26

Davidzon G, Greene P, Mancuso M, Klos KJ, Ahlskog JE, et al. 2006. Early-onset familial parkinsonism due to POLG mutations. *Ann. Neurol.* 59:859–62

Delettre C, Lenaers G, Griffoin J-M, Gigarel N, Lorenzo C, et al. 2000. Nuclear gene OPA1, encoding a mitochondrial dynamin-related protein, is mutated in dominant optic atrophy. *Nat. Genet.* 26:207–10

De Meirleir L, Seneca S, Lissens W, De Clercq I, Eyskens F, et al. 2004. Respiratory chain complex V deficiency due to a mutation in the assembly gene ATP12. *J. Med. Genet.* 41:120–24

De Strooper B. 2003. Aph-1, Pen-2, and nicastrin with presenilin generate an active γ-secretase complex. *Neuron* 38:9–12

Devi L, Prabhu BM, Galati DF, Avadhani NG, Anandatheerthavarada HK. 2006. Accumulation of amyloid precursor protein in the mitochondrial import channels of human Alzheimer's disease brain is associated with mitochondrial dysfunction. *J. Neurosci.* 26:9057–68

DiMauro S, Davidzon G. 2005. Mitochondrial DNA and disease. *Ann. Med.* 37:222–32

DiMauro S, Davidzon G, Hirano M. 2006a. A polymorphic polymerase. *Brain* 126:1637–39

DiMauro S, Hirano M. 2005. Mitochondrial encephalomyopathies: an update. *Neuromusc. Disord.* 15:276–86

DiMauro S, Hirano M, Kaufmann P, Mann JJ. 2006b. Mitochondrial psychiatry. In *Mitochondrial Medicine*, ed. S DiMauro, M Hirano, EA Schon, pp. 261–77. London: Informa Healthcare

DiMauro S, Hirano M, Schon EA. 2006c. Approaches to the treatment of mitochondrial diseases. *Muscle Nerve* 34:265–83

DiMauro S, Quinzii C, Hirano M. 2007. Mutations in coenzyme Q10 biosynthetic genes. *J. Clin. Invest.* 117:587–89

DiMauro S, Schon EA. 2003. Mitochondrial respiratory-chain diseases. *New Engl. J. Med.* 348:2656–68

Elpeleg O, Miller C, Hershkovitz E, Bitner-Glindzicz M, Bondi-Rubinstein G, et al. 2005. Deficiency of the ADP-forming succinyl-CoA synthase activity is associated with encephalomyopathy and mitochondrial DNA depletion. *Am. J. Hum. Genet.* 76:1081–86

Engelender S, Sharp AH, Colomer V, Tokito MK, Lanahan A, et al. 1997. Huntingtin-associated protein 1 (HAP1) interacts with the p150Glued subunit of dynactin. *Hum. Mol. Genet.* 6:2205–12

Fenton WA. 1995. Mitochondrial protein transport: a system in search of mutations. *Am. J. Hum. Genet.* 57:235–38

Fernandez-Vizarra E, Berardinelli A, Valente L, Tiranti V, Zeviani M. 2006. Nonsense mutation in pseudouridylate synthase 1 (PUS1) in two brothers affected by myopathy, lactic acidosis and sideroblastic anemia (MLASA). *J. Med. Genet.* 44:173–80

Fichera M, Lo Giudice M, Falco M, Sturnio M, Amata A, et al. 2004. Evidence of kinesin heavy chain (*KIF5A*) involvement in pure hereditary spastic paraplegia. *Neurology* 63:1108–10

Gabriel K, Milenkovic D, Chacinska A, Muller J, Guiard B, et al. 2006. Novel mitochondrial intermembrane space proteins as substrates of the MIA import pathway. *J. Mol. Biol.* 365:612–20

Gempel K, Topaloglu H, Talim B, Schneiderat P, Schoser BGH, et al. 2007. The myopathic form of coenzyme Q10 deficiency is caused by mutations in the electron-transferring-flavoprotein dehydrogenase (ETFDH) gene. *Brain* 130:2037–44

Goedert M, Spillantini MG. 2006. A century of Alzheimer's disease. *Science* 314:777–81

Gohil VM, Hayes P, Matsuyama S, Schagger H, Schlame M, Greenberg ML. 2004. Cardiolipin biosynthesis and mitochondrial respiratory chain function are interdependent. *J. Biol. Chem.* 279:42612–18

Greenamyre JT. 2007. Huntington's Disease—making connections. *New Engl. J. Med.* 356:518–20

Gutekunst CA, Li SH, Yi H, Ferrante RJ, Li XJ, Hersch SM. 1998. The cellular and subcellular localization of huntingtin-associated protein 1 (HAP1): comparison with huntingtin in rat and human. *J. Neurosci.* 18:7674–86

Hansen JJ, Durr A, Cournu-Rebeix I, Georgopoulos C, Ang D, et al. 2002. Hereditary spastic paraplegia SPG13 is associated with a mutation in the gene encoding the mitochondrial chaperonin Hsp60. *Am. J. Hum. Genet.* 70:1328–32

Hansson CA, Frykman S, Farmery MR, Tjernberg LO, Nilsberth C, et al. 2004. Nicastrin, presenilin, APH-1, and PEN-2 form active γ-secretase complexes in mitochondria. *J. Biol. Chem.* 279:51654–60

Harding AE. 1992. Growing old: the most common mitochondrial disease of all? *Nat. Genet.* 2:251–52

Hardy J, Cai H, Cookson MR, Gwinn-Hardy K, Singleton A. 2006. Genetics of Parkinson's disease and parkinsonism. *Ann. Neurol.* 60:389–98

Harman D. 1972. The biologic clock: the mitochondria? *J. Am. Geriat. Soc.* 20:145–47

Haut S, Brivet M, Touati G, Rustin P, Lebon S, et al. 2003. A deletion in the human QP-C gene causes a complex III deficiency resulting in hypoglycaemia and lactic acidosis. *Hum. Genet.* 113:118–22

Hays AP, Oskoui M, Tanji K, Kaufmann P, Bonilla E. 2006. Mitochondrial neurology II: myopathies and peripheral neuropathies. In *Mitochondrial Medicine*, ed. S DiMauro, M Hirano, EA Schon, pp. 45–74. London: Informa Healthcare

Hirano M, Kaufmann P, De Vivo DC, Tanji K. 2006a. Mitochondrial neurology I: Encephalopathies. In *Mitochondrial Medicine*, ed. S DiMauro, M Hirano, EA Schon, pp. 27–44. London: Informa Healthcare

Hirano M, Lagier-Tourenne C, Valentino ML, Marti R, Nishigaki Y. 2005. Thymidine phosphorylase mutations cause instability of mitochondrial DNA. *Gene* 354:152–56

Hirano M, Marti R, Casali C, Tadesse BS, Uldrick T, et al. 2006b. Allogeneic stem cell transplantation corrects biochemical derangements in MNGIE. *Neurology* 67:1458–60

Holt IJ, Harding AE, Morgan Hughes JA. 1988. Deletions of muscle mitochondrial DNA in patients with mitochondrial myopathies. *Nature* 331:717–19

Honda T, Nihonmatsu N, Yasutake K, Ohtake A, Sato K, et al. 2000. Familial Alzheimer's disease-associated mutations block translocation of full-length presenilin 1 to the nuclear envelope. *Neurosci. Res.* 37:101–11

Hudson G, Keers S, Yu Wai Man P, Griffiths P, Huoponen K, et al. 2005. Identification of an X-chromosomal locus and haplotype modulating the phenotype of a mitochondrial DNA disorder. *Am. J. Hum. Genet.* 77:1086–91

Ikuta J, Maturana A, Fujita T, Okajima T, Tatematsu K, et al. 2007. Fasciculation and elongation protein zeta-1 (FEZ1) participates in the polarization of hippocampal neuron by controlling the mitochondrial motility. *Biochem. Biophys. Res. Commun.* 353:127–32

James R, Adams RR, Christie S, Buchanan SR, Porteous DJ, Millar JK. 2004. Disrupted in schizophrenia 1 (DISC1) is a multicompartmentalized protein that predominantly localizes to mitochondria. *Mol. Cell. Neurosci.* 26:112–22

Kakiuchi C, Ishiwata M, Kametani M, Nelson C, Iwamoto K, Kato T. 2005. Quantitative analysis of mitochondrial DNA deletions in the brains of patients with bipolar disorder and schizophrenia. *Int. J. Neuropharmacol.* 8:1–8

Kakiuchi C, Iwamoto K, Ishiwata M, Bundo M, Kasahara T, et al. 2003. Impaired feedback regulation of XBP1 as a genetic risk factor for bipolar disorder. *Nat. Genet.* 35:171–75

Karadimas CL, Vu TH, Holve SA, Quinzii C, Tanji K, et al. 2006. Navajo neurohepatopathy is caused by a mutation in the *MPV17* gene. *Am. J. Hum. Genet.* 79:544–48

Kato T, Ishiwata M, Mori K, Washizuka S, Tajima O, et al. 2002. Mechanisms of altered calcium signalling in transformed lymphoblastoid cells from patients with bipolar disorder. *Int. J. Neuropharmacol.* 6:379–89

Kaufmann P, Sano MC, Jhung S, Engelstadt K, De Vivo DC. 2002. Psychiatric symptoms are common features of clinical syndromes associated with mitochondrial DNA point mutations. *Neurology* 58:A315 (Abstract)

Kaufmann P, Shungu D, Sano MC, Jhung S, Engelstad K, et al. 2004. Cerebral lactic acidosis correlates with neurological impairment in MELAS. *Neurology* 62:1297–302

Khrapko K, Kraytsberg Y, de Grey ADNJ, Vijg J, Schon EA. 2006. Does premature aging of the mtDNA mutator mouse prove that mtDNA mutations are involved in natural aging? *Aging Cell* 5:279–82

Kins S, Lauther N, Szodorai A, Beyreuther K. 2006. Subcellular trafficking of the amyloid precursor protein gene family and its pathogenic role in Alzheimer's disease. *Neurodegen. Dis.* 3:218–26

Kirby DM, Boneh A, Chow CW, Ohtake A, Ryan MT, et al. 2003. Low mutant load of mitochondrial DNA G13513A mutation can cause Leigh's disease. *Ann. Neurol.* 54:473–78

Kirk R, Furlong RA, Amos W, Cooper G, Rubinsztein JS, et al. 1999. Mitochondrial genetic analyses suggest selection against maternal lineages in bipolar affective disorder. *Am. J. Hum. Genet.* 65:508–18

Kish SJ, Bergeron C, Rajput A, Dozic S, Mastrogiacomo F, et al. 1992. Brain cytochrome oxidase in Alzheimer's disease. *J. Neurochem.* 59:776–79

Konradi C, Eaton M, MacDonald ML, Walsh J, Benes FM, Heckers S. 2004. Molecular evidence for mitochondrial dysfunction in bipolar disease. *Arch. Gen. Psychiat.* 61:300–8

Kraytsberg Y, Kudryavtseva E, McKee AC, Geula C, Kowall NW, Khrapko K. 2006. Mitochondrial DNA deletions are abundant and cause functional impairment in aged human substantia nigra neurons. *Nat. Genet.* 38:518–20

Kujoth GC, Hiona A, Pugh TD, Someya S, Panzer K, et al. 2005. Mitochondrial DNA mutations, oxidative stress, and apoptosis in mammalian aging. *Science* 309:481–84

Lawson VH, Graham BV, Flanigan KM. 2005. Clinical and electrophysiologic features of CMT2A with mutations in the mitofusin 2 gene. *Neurology* 65:197–204

Ledley FD, Jansen R, Nham SU, Fenton WA, Rosenberg LE. 1990. Mutation eliminating mitochondrial leader sequence of methylmalonyl-CoA mutase causes *mut°* methylmalonic acidemia. *Proc. Natl. Acad. Sci. USA* 87:3147–50

Li J, Lee WL, Cooper JA. 2005. NudEL targets dynein to microtubule ends through LIS1. *Nat. Cell Biol.* 7:686–90

Li SH, Gutekunst CA, Hersch SM, Li XJ. 1998. Interaction of huntingtin-associated protein with dynactin P150Glued. *J. Neurosci.* 18:1261–69

Liu JK, Lillo C, Jonsson PA, Velde CV, Ward CM, et al. 2004. Toxicity of familial ALS-linked SOD1 mutants from selective recruitment to spinal mitochondria. *Neuron* 43:5–17

Longley MJ, Clark S, Man CYW, Hudson G, Durham SE, et al. 2006. Mutant POLG2 disrupts DNA polymerase gamma subunits and causes progressive external ophthalmoplegia. *Am. J. Hum. Genet.* 78:1026–34

López LC, Quinzii C, Schuelke M, Kanki T, Naini A, et al. 2006. Leigh syndrome with nephropathy and CoQ10 deficiency due to decaprenyl diphosphate synthase subunit 2 (PDSS2) mutations. *Am. J. Hum. Genet.* 79:1125–29

Lu J, Rashid F, Byrne PC. 2006. The hereditary spastic paraplegia protein spartin localises to mitochondria. *J. Neurochem.* 98:1908–19

Lustbader JW, Cirilli M, Lin C, Xu HW, Takuma K, et al. 2004. ABAD directly links Aβ to mitochondrial toxicity in Alzheimer's disease. *Science* 304:448–52

Maeda K, Nwulia E, Chang J, Balkissoon R, Ishizuka K, et al. 2006. Differential expression of disrupted-in-schizophrenia (DISC1) in bipolar disorder. *Biol. Psych.* 60:929–35

Manczak M, Anekonda TS, Henson E, Park BS, Quinn J, Reddy PH. 2006. Mitochondria are a direct site of Aβ accumulation in Alzheimer's disease neurons: implications for free radical generation and oxidative damage in disease progression. *Hum. Mol. Genet.* 15:1437–49

Marambaud P, Shioi J, Serban G, Georgakopoulos A, Sarner S, et al. 2002. A presenilin-1/γ-secretase cleavage releases the E-cadherin intracellular domain and regulates disassembly of adherens junctions. *EMBO J.* 21:1948–56

Mastrogiacomo F, Bergeron C, Kish SJ. 1993. Brain α-ketoglutarate dehydrogenase complex activity in Alzheimer's disease. *J. Neurochem.* 61:2007–14

Mattson MP. 2004. Pathways towards and away from Alzheimer's disease. *Nature* 430:631–39

McDermott CJ, Grierson AJ, Wood JD, Bingley M, Wharton SB, et al. 2003. Hereditary spastic paraparesis: disrupted intracellular transport associated with spastin mutation. *Ann. Neurol.* 54:748–59

Millar JK, James R, Christie S, Porteous DJ. 2005. Disrupted in schizophrenia 1 (DISC1): subcellular targeting and induction of ring mitochondria. *Mol. Cell. Neurosci.* 30:477–84

Miller C, Saada A, Shaul N, Shabtai N, Ben-Shalom E, et al. 2004. Defective mitochondrial translation caused by a ribosomal protein (MRPS16) mutation. *Ann. Neurol.* 56:734–38

Mollet J, Giurgea I, Schlemmer D, Dallner G, Chretien D, et al. 2007. Prenyldiphosphate synthase (PDSS1) and OH-benzoate prenyltransferase (COQ2) mutations in ubiquinone deficiency and oxidative phosphorylation disorders. *J. Clin. Invest.* 117:765–72

Mootha VK, Lepage P, Miller K, Bunkenborg J, Reich M, et al. 2003. Identification of a gene causing human cytochrome *c* oxidase deficiency by integrative genomics. *Proc. Natl. Acad. Sci. USA* 100:605–10

Morin C, Mitchell G, Larochelle J, Lambert M, Ogier H, et al. 1993. Clinical, metabolic, and genetic aspects of cytochrome *c* oxidase deficiency in Saguenay-Lac-Saint-Jean. *Am. J. Hum. Genet.* 53:488–96

Muftuoglu M, Elibol B, Dalmizrak O, Ercan A, Kulaksiz G, et al. 2003. Mitochondrial complex I and IV activities in leukocytes from patients with parkin mutations. *Mov. Disord.* 19:544–48

Naviaux RK, Nguyen KV. 2004. *POLG* mutations associated with Alpers' syndrome and mitochondrial DNA depletion. *Ann. Neurol.* 55:706–12

Newmeyer DD, Ferguson-Miller S. 2003. Mitochondria: releasing power for life and unleashing the machinery of death. *Cell* 112:481–90

Niemann A, Ruegg M, La Padula V, Schenone A, Suter U. 2005. Ganglioside-induced differentiation associated protein 1 is a regulator of the mitochondrial network: new implications for Charcot-Marie-Tooth disease. *J. Cell Biol.* 170:1067–78

Nishino I, Spinazzola A, Hirano M. 1999. Thymidine phosphorylase gene mutations in MNGIE, a human mitochondrial disorder. *Science* 283:689–92

Nolden M, Ehses S, Koppen M, Bernacchia A, Rugarli EI, Langer T. 2005. The m-AAA protease defective in hereditary spastic paraplegia controls ribosome assembly in mitochondria. *Cell* 123:277–89

Ogilvie I, Kennaway NG, Shoubridge EA. 2005. A molecular chaperone for mitochondrial complex I assembly is mutated in a progressive encephalopathy. *J. Clin. Invest.* 115:2784–92

Pallotti F, Chen X, Bonilla E, Schon EA. 1996. Evidence that specific mtDNA point mutations may not accumulate in skeletal muscle during normal human aging. *Am. J. Hum. Genet.* 59:591–602

Pedrola L, Espert A, Wu X, Claramunt R, Shy ME, Palau F. 2005. GDAP1, the protein causing Charcot-Marie-Tooth disease type 4A, is expressed in neurons and is associated with mitochondria. *Hum. Mol. Genet.* 14:1087–94

Pellegrini L, Scorrano L. 2007. A cut short to death: Parl and Opa1 in the regulation of mitochondrial morphology and apoptosis. *Cell Death Differ.* 14:1275–84

Petit A, Kawarai T, Paitel E, Sanjo N, Maj M, et al. 2005. Wild-type PINK1 prevents basal and induced neuronal apoptosis, a protective effect abrogated by Parkinson disease-related mutations. *J. Biol. Chem.* 280:34025–32

Pigino G, Morfini G, Pelsman A, Mattson MP, Brady ST, Busciglio J. 2003. Alzheimer's presenilin 1 mutations impair kinesin-based axonal transport. *J. Neurosci.* 23:4499–508

Polymeropoulos MH, Lavedan C, Leroy E, Ide SE, Dehejia A, et al. 1997. Mutation in the α-synuclein gene identified in families with Parkinson's disease. *Science* 276:2045–47

Quinzii C, Kattah AG, Naini A, Akman HO, Mootha VK, et al. 2005. Coenzyme Q deficiency and cerebellar ataxia associated with an aprataxin mutation. *Neurology* 64:539–41

Quinzii C, Naini A, Salviati L, Trevisson E, Navas P, et al. 2006. A mutation in para-hydoxybenzoate-polyprenyl transferase (COQ2) causes primary coenzyme Q10 deficiency. *Am. J. Hum. Genet.* 78:345–49

Roberson ED, Scearce-Levie K, Palop JJ, Yan F, Cheng IH, et al. 2007. Reducing endogenous tau ameliorates amyloid β-induced deficits in an Alzheimer's disease mouse model. *Science* 316:750–54

Roberts RC. 2007. Schizophrenia in translation: disrupted in schizophrenia (DISC1): integrating clinical and basic findings. *Schizophr. Bull.* 33:11–15

Roesch K, Curran SP, Tranebjaerg L, Koehler CM. 2002. Human deafness dystonia syndrome is caused by a defect in assembly of the DDP1/TIMM8a-TIMM13 complex. *Hum. Mol. Genet.* 11:477–86

Rogaeva E, Meng Y, Lee JH, Gu Y, Kawarai T, et al. 2007. The neuronal sortilin-related receptor SORL1 is genetically associated with Alzheimer disease. *Nat. Genet.* 39:168–77

Rowland LP. 1994. Mitochondrial encephalomyopathies: lumping, splitting, and melding. In *Mitochondrial Disorders in Neurology*, ed. AHV Schapira, S DiMauro, pp. 116–29. Oxford: Butterworth-Heinemann

Runz H, Rietdorf J, Tomic I, de Bernard M, Beyreuther K, et al. 2002. Inhibition of intracellular cholesterol transport alters presenilin localization and amyloid precursor protein processing in neuronal cells. *J. Neurosci.* 22:1679–89

Sanderson CM, Connell JW, Edwards TL, Bright NA, Duley S, et al. 2006. Spastin and atlastin, two proteins mutated in autosomal-dominant hereditary spastic paraplegia, are binding partners. *Hum. Mol. Genet.* 15:307–18

Sarzi E, Brown MD, Lebon S, Chretien D, Munnich A, et al. 2007. A novel mitochondrial DNA mutation in ND3 gene is associated with isolated complex I deficiency causing Leigh syndrome and dystonia. *Am. J. Med. Genet.* 143A:33–41

Schindler CK, Heverin M, Henshall DC. 2006. Isoform- and subcellular fraction-specific differences in hippocampal 14-3-3 levels following experimentally evoked seizures and in human temporal lobe epilepsy. *J. Neurochem.* 99:561–69

Schlame M, Ren M. 2006. Barth syndrome, a human disorder of cardiolipin metabolism. *FEBS Lett.* 580:5450–55

Schon EA. 2007. Appendix 5. Gene products present in mitochondria of yeast and animal cells. *Meth. Cell Biol.* 80:835–76

Schon EA, Rizzuto R, Moraes CT, Nakase H, Zeviani M, DiMauro S. 1989. A direct repeat is a hotspot for large-scale deletions of human mitochondrial DNA. *Science* 244:346–49

Schulz JB, Dehmer T, Schols L, Hardt C, Vorgerd M, et al. 2000. Oxidative stress in patients with Friedreich ataxia. *Neurology* 55:1719–21

Schwarzman AL, Singh N, Tsiper M, Gregori L, Dranovsky A, et al. 1999. Endogenous presenilin 1 redistributes to the surface of lamellipodia upon adhesion of Jurkat cells to a collagen matrix. *Proc. Natl. Acad. Sci. USA* 96:7932–37

Sedlak E, Panda M, Dale MP, Weintraub ST, Robinson NC. 2006. Photolabeling of cardiolipin binding subunits within bovine heart cytochrome *c* oxidase. *Biochemistry* 45:746–54

Seznec H, Simon D, Bouton C, Reutenauer L, Hertzog A, et al. 2005. Friedreich ataxia: the oxidative stress paradox. *Hum. Mol. Genet.* 14:463–74

Shapira Y, Harel S, Russell A. 1977. Mitochondrial encephalomyopathies: a group of neuromuscular disorders with defects in oxidative metabolism. *Isr. J. Med. Sci.* 13:161–64

Sheu KF, Kim YT, Blass JP, Weksler ME. 1985. An immunochemical study of the pyruvate dehydrogenase deficit in Alzheimer's disease brain. *Ann. Neurol.* 17:444–49

Shults CW, Schapira AHV. 2001. A cue to queue for CoQ? *Neurology* 57:375–76

Silvestri L, Caputo V, Bellacchio E, Atorino L, Dallapiccola B, et al. 2005. Mitochondrial import and enzymatic activity of PINK1 mutants associated with recessive parkinsonism. *Hum. Mol. Genet.* 14:3477–92

Siman R, Velji J. 2003. Localization of presenilin-nicastrin complexes and γ-secretase activity to the trans-Golgi network. *J. Neurochem.* 84:1143–53

Smeitink JAM, Elpeleg O, Antonicka H, Diepstra H, Saada A, et al. 2006. Distinct clinical phenotypes associated with a mutation in the mitochondrial translation elongation factor EFTs. *Am. J. Hum. Genet.* 79:869–77

Soong NW, Hinton DR, Cortopassi G, Arnheim N. 1992. Mosaicism for a specific mitochondrial DNA mutation in adult human brain. *Nat. Genet.* 2:318–23

Spinazzola A, Viscomi C, Fernandez-Vizarra E, Carrara F, D'Adamo P, et al. 2006. MPV17 encodes an inner mitochondrial membrane protein and is mutated in infantile hepatic mitochondrial DNA depletion. *Nat. Genet.* 38:570–75

Spinazzola A, Zeviani M. 2005. Disorders of nuclear-mitochondrial intergenomic signaling. *Gene* 354:162–68

St Clair D, Blackwood D, Muir W, Carothers A, Walker M, et al. 1990. Association within a family of a balanced autosomal translocation with major mental illness. *Lancet* 336:13–16

Stokin GB, Lillo C, Falzone TL, Brusch RG, Rockenstein E, et al. 2005. Axonopathy and transport deficits early in the pathogenesis of Alzheimer's disease. *Science* 307:1282–88

Strauss KM, Martins LM, Plun-Favreau H, Marx FP, Kautzmann S, et al. 2005. Loss of function mutations in the gene encoding Omi/HtrA2 in Parkinson's disease. *Hum. Mol. Genet.* 14:2099–111

Sturtz LA, Diekert K, Jensen LT, Lill R, Culotta VC. 2001. A fraction of yeast Cu,Zn-superoxide dismutase and its metallochaperone, CCS, localize to the intermembrane space of mitochondria. A physiological role for SOD1 in guarding against mitochondrial oxidative damage. *J. Biol. Chem.* 276:38084–89

Takakubo F, Cartwright P, Hoogenraad N, Thorburn DR, Collins F, et al. 1995. An amino acid substitution in the pyruvate dehydrogenase E1α gene, affecting mitochondrial import of the precursor protein. *Am. J. Hum. Genet.* 57:772–80

Tanji K, Kunimatsu T, Vu TH, Bonilla E. 2001. Neuropathological features of mitochondrial disorders. *Cell Dev. Biol.* 12:429–39

Tanji K, Schon EA, DiMauro S, Bonilla E. 2000. Kearns-Sayre syndrome: oncocytic transformation of choroid plexus epithelium. *J. Neurol. Sci.* 178:29–36

Tatuch Y, Christodoulou J, Feigenbaum A, Clarke J, Wherret J, et al. 1992. Heteroplasmic mtDNA mutation (T > G) at 8993 can cause Leigh disease when the percentage of abnormal mtDNA is high. *Am. J. Hum. Genet.* 50:852–58

Taylor SW, Fahy E, Zhang B, Glenn GM, Warnock DE, et al. 2003. Characterization of the human heart mitochondrial proteome. *Nat. Biotech.* 21:281–86

Tieu K, Przedborski S. 2006. Mitochondrial dysfunction and neurodegenerative disorders. In *Mitochondrial Medicine*, ed. S DiMauro, M Hirano, EA Schon, pp. 279–307. London: Informa Healthcare

Tiranti V, Hoertnagel K, Carrozzo R, Galimberti C, Munaro M, et al. 1998. Mutations of SURF-1 in Leigh disease associated with cytochrome *c* oxidase deficiency. *Am. J. Hum. Genet.* 63:1609–21

Toyo-oka K, Shionoya A, Gambello MJ, Cardoso C, Leventer R, et al. 2003. 14-3-3ε is important for neuronal migration by binding to NUDEL: a molecular explanation for Miller-Dieker syndrome. *Nat. Genet.* 34:274–85

Trifunovic A, Wredenberg A, Falkenberg M, Spelbrink JN, Rovio AT, et al. 2004. Premature ageing in mice expressing defective mitochondrial DNA polymerase. *Nature* 429:417–23

Trushina E, Dyer RB, Badger JD 2nd, Ure D, Eide L, et al. 2004. Mutant huntingtin impairs axonal trafficking in mammalian neurons in vivo and in vitro. *Mol. Cell. Biol.* 24:8195–209

Trushina E, Heldebrant MP, Perez-Terzic CM, Bortolon R, Kovtun IV, et al. 2003. Microtubule destabilization and nuclear entry are sequential steps leading to toxicity in Huntington's disease. *Proc. Natl. Acad. Sci. USA* 100:12171–76

Valente L, Tiranti V, Marsano RM, Malfatti E, Fernandez-Vizarra E, et al. 2007. Infantile encephalopathy and defective mitochondrial DNA translation in patients with mutations of mitochondrial elongation factors EGF1 and EFTu. *Am. J. Hum. Genet.* 80:44–58

Vetrivel KS, Cheng H, Lin W, Sakurai T, Li T, et al. 2004. Association of γ-secretase with lipid rafts in post-Golgi and endosome membranes. *J. Biol. Chem.* 279:44945–54

Visapaa I, Fellman V, Vesa J, Dasvarma A, Hutton JL, et al. 2002. GRACILE syndrome, a lethal metabolic disorder with iron overload, is caused by a point mutation in BCS1L. *Am. J. Hum. Genet.* 71:863–76

Wallace DC, Singh G, Lott MT, Hodge JA, Shurr TG, et al. 1988. Mitochondrial DNA mutation associated with Leber's hereditary optic neuropathy. *Science* 242:1427–30

Wallace DC, Brown MD, Lott MT. 1999. Mitochondrial DNA variation in human evolution and disease. *Gene* 238:211–30

Walter J, Capell A, Grunberg J, Pesold B, Schindzielorz A, et al. 1996. The Alzheimer's disease-associated presenilins are differentially phosphorylated proteins located predominantly within the endoplasmic reticulum. *Mol. Med.* 2:673–91

Waterham HR, Koster J, van Roermund CWT, Mooyer PAW, Wanders RJA, Leonard JV. 2007. A lethal defect of mitochondrial and peroxisomal fission. *New Engl. J. Med.* 356:1736–41

Wiedemann FR, Manfredi G, Mawrin C, Beal MF, Schon EA. 2002. Mitochondrial DNA and respiratory chain function in spinal cords of ALS patients. *J. Neurochem.* 80:616–25

Wright AF, Jacobson SG, Cideciyan AV, Roman AJ, Shu X, et al. 2004. Lifespan and mitochondrial control of neurodegeneration. *Nat. Genet.* 36:1153–58

Zhang LS, Shimoji M, Thomas B, Moore D, Yu S-W, et al. 2005a. Mitochondrial localization of Parkinson's disease related protein DJ-1: implications for pathogenesis. *Hum. Mol. Genet.* 14:2063–73

Zhang M, Mileykovskaya E, Dowhan W. 2005b. Cardiolipin is essential for organization of complexes III and IV into a supercomplex in intact yeast mitochondria. *J. Biol. Chem.* 280:29403–8

Zhu Z, Yao J, Johns T, Fu K, De Bie I, et al. 1998. SURF1, encoding a factor involved in the biogenesis of cytochrome *c* oxidase, is mutated in Leigh syndrome. *Nat. Genet.* 20:337–43

Zuchner S, De Jonghe P, Jordanova A, Claeys KG, Guergueltcheva V, et al. 2006a. Axonal neuropathy with optic atrophy is caused by mutations in mitofusin 2. *Ann. Neurol.* 59:276–81

Zuchner S, Mersiyanova IV, Muglia M, Bissar-Tadmouri N, Rochelle J, et al. 2004. Mutations in the mitochondrial GTPase mitofusin 2 cause Charcot-Marie-Tooth neuropathy type 2A. *Nat. Genet.* 36:449–51

Zuchner S, Wang G, Tran-Viet KN, Nance MA, Gaskell PC, et al. 2006b. Mutations in the novel mitochondrial protein REEP1 cause hereditary spastic paraplegia type 31. *Am. J. Hum. Genet.* 79:365–69

Vestibular System: The Many Facets of a Multimodal Sense

Dora E. Angelaki[1] and Kathleen E. Cullen[2]

[1] Department of Anatomy and Neurobiology, Washington University School of Medicine, St. Louis, Missouri 63110; email: angelaki@wustl.edu

[2] Department of Physiology, McGill University, Montreal, Quebec H3G 1Y4, Canada email: kathleen.cullen@mcgill.ca

Annu. Rev. Neurosci. 2008. 31:125–150

First published online as a Review in Advance on March 13, 2008

The *Annual Review of Neuroscience* is online at neuro.annualreviews.org

This article's doi:
10.1146/annurev.neuro.31.060407.125555

Key Words

multisensory, reference frame, navigation, spatial orientation, computation, corollary discharge

Abstract

Elegant sensory structures in the inner ear have evolved to measure head motion. These vestibular receptors consist of highly conserved semicircular canals and otolith organs. Unlike other senses, vestibular information in the central nervous system becomes immediately multisensory and multimodal. There is no overt, readily recognizable conscious sensation from these organs, yet vestibular signals contribute to a surprising range of brain functions, from the most automatic reflexes to spatial perception and motor coordination. Critical to these diverse, multimodal functions are multiple computationally intriguing levels of processing. For example, the need for multisensory integration necessitates vestibular representations in multiple reference frames. Proprioceptive-vestibular interactions, coupled with corollary discharge of a motor plan, allow the brain to distinguish actively generated from passive head movements. Finally, nonlinear interactions between otolith and canal signals allow the vestibular system to function as an inertial sensor and contribute critically to both navigation and spatial orientation.

Contents

INTRODUCTION

Known as the balance organs of the inner ear, the vestibular system constitutes our sixth sense. Three roughly orthogonal semicircular canals sense rotational movements, and two otolith organs (the utricle and the saccule) sense linear accelerations. Vestibular afferents are continuously active even at rest and are strikingly sensitive for signaling motion accelerations as our head translates and rotates in space. Even when we remain motionless, the otolith organs sense the pull of gravity (a form of linear acceleration). The signals from the semicircular canals and the otolith organs are complementary; their combined activation is necessary to explore and comprehend the enormous range of physical motions experienced in everyday life.

The vestibular system differs from other senses in many respects. Most notably, central vestibular processing is highly convergent and strongly multimodal. For example, canal/otolith interactions take place in the brain stem and cerebellum immediately at the first synapse. Also, visual/vestibular and proprioceptive/vestibular interactions occur throughout the central vestibular pathways and are vital for gaze and postural control. Signals from muscles, joints, skin, and eyes are continuously integrated with vestibular inflow. Because of the strong and extensive multimodal convergence with other sensory and motor signals, vestibular stimulation does not give rise to a separate and distinct conscious sensation. Yet, the vestibular system plays an important role in everyday life because it contributes to a surprising range of functions, ranging from reflexes to the highest levels of perception and consciousness.

Experimental approaches in the vestibular system were traditionally framed by the perspective of the sensorimotor transformations required for reflex generation (for recent reviews see Angelaki 2004, Angelaki & Hess 2005, Boyle 2001, Cullen & Roy 2004, Raphan & Cohen 2002, Wilson & Schor 1999). Techniques based on control systems theory have been used to establish the sensorimotor transformations by which vestibular information is transformed into a motor output. This approach followed logically from the influential theory of a reflex chain made popular more than a century ago by Sherrington (1906). By recording from individual neurons at each successive stage in a reflex pathway (reviewed in Goldberg 2000), quantitative levels of sensorimotor processing were established. Using this approach, studies in reduced or in-vitro preparations have provided important insights into the functional circuitry, intrinsic electrophysiology, and signal

Semicircular canals: one of the two sets of vestibular end organs that measure angular acceleration of the head

processing of vestibularly driven reflexes. The relative simplicity of the neural circuits that mediate vestibular reflexes have also proven to be well suited for linking systems and cellular levels of analyses.

A unique feature of the vestibular system is that many second-order sensory neurons in the brain stem are also premotor neurons; the same neurons that receive afferent inputs send direct projections to motoneurons. An advantage of this streamlined circuitry is that vestibular sensorimotor responses have extraordinarily short latencies. For example, the latency of the vestibulo-ocular reflex (VOR) is as short as 5–6 ms. Simple pathways also mediate the vestibulo-spinal reflexes that are important for maintaining posture and balance. Recent studies, however, have emphasized the importance of extravestibular signals in shaping even these simple sensorimotor transformations. Moreover, multisensory and multimodal interactions play an essential role in higher-level functions such as self-motion perception and spatial orientation. Largely owing to their inherent complexity and strongly multimodal nature, these very intriguing vestibular-related functions have just begun to be explored.

The vestibular system represents a great forum in which to address several fundamental questions in neuroscience: multisensory integration, changes of coordinate systems, separation of active from passive head movements, and the role of corollary discharge. In this review we discuss some of these issues. We first summarize recent work addressing how semicircular canal and otolith signals interact to compute inertial motion (i.e., motion of the head relative to the world) and then explore how vestibular information converges with proprioceptive and other extravestibular signals to distinguish self-generated from passive head movements. Finally, we present a few examples showing that vestibular signals in the brain are expressed in multiple reference frames, a signature of a truly multimodal and multifunctional sense. To date, these processes have been characterized most extensively in the brain stem vestibular nuclei and vestibulo-cerebellum, the two areas that receive direct vestibular afferent signals, which they then process and distribute to oculomotor, skeletomotor, visceral, and thalamo-cortical systems.

COMPUTATION OF INERTIAL MOTION

The vestibular system constitutes an inertial sensor, i.e., it encodes the motion of the head relative to the outside world. However, this is not precisely true when considering signals separately from the semicircular canal and otolith organs. There exist two problems in interpreting information from the peripheral vestibular sensors. First is the rotation problem, which arises because vestibular sensors are physically fixed in the head. During rotation, semicircular canal afferents detect endolymph fluid motion relative to the skull-fixed bony ducts (Goldberg & Fernandez 1975), coding angular velocity (the integral of angular acceleration) in a head-centered reference frame but providing no information about how the head moves relative to the world. For example, a horizontal (yaw) rotation in an upright orientation activates canal afferents similarly as a yaw rotation in a supine orientation (**Figure 1***a*). Yet, these two movements differ in inertial (i.e., world-centered) space. The second problem, referred to as the linear acceleration problem, is due to a sensory ambiguity that arises because of physical laws, namely Einstein's equivalence principle: Otolith afferents detect net linear acceleration but cannot distinguish translational from gravitational components (**Figure 1***b*) (Fernandez & Goldberg 1976). That is, whether we actually walk forward or tilt our head backward is indistinguishable to primary otolith afferents.

Whereas each of the two sets of vestibular sensors alone is ambiguous, the brain can derive a reliable estimate of both attitude (i.e., orientation) and motion relative to the world by appropriately combining semicircular canal and otolith signals (Merfeld 1995; Glasauer & Merfeld 1997; Angelaki et al. 1999; Mergner & Glasauer 1999; Merfeld & Zupan 2002; Zupan et al. 2002; Green & Angelaki 2003, 2004;

Otolith organs: linear acceleration sensors with receptor hair cells having polarization vectors distributed over the utricular and saccular maculae

Vestibulo-ocular reflex (VOR): vestibular-related reflex by which a head movement is compensated by an eye rotation in order to keep retinal images stable

Vestibulo-cerebellum: areas of the cerebellar cortex that receive first- or second-order vestibular signals and contribute to vestibular signal processing

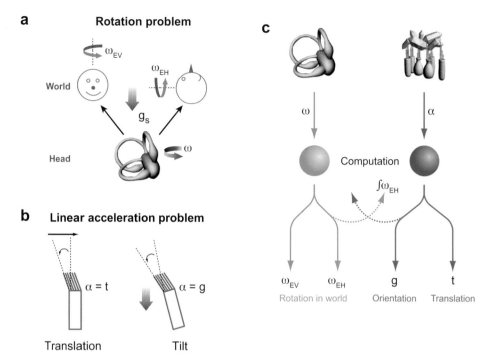

Figure 1

Schematic of the two computational problems in inertial motion detection. (*a*) The rotation problem involves calculation of head angular velocity, ω, relative to the world (as defined by gravity, g_s). (*b*) The linear acceleration problem involves the discrimination of net gravitoinertial acceleration, α, into translational, t, and gravitational acceleration, g. (*c*) Schematic of the computational solution. Angular velocity, ω, from the semicircular canals must be combined with gravitational information to be decomposed into two components, one parallel to gravity, ω_{EV}, and another perpendicular to gravity, ω_{EH}. In parallel, net linear acceleration, α, from the otolith organs must be combined with the temporal integral of ω_{EH} ($\int \omega_{EH}$), such that it is separated into translational, t, and gravitational acceleration, g. Replotted with permission from Yakusheva et al. (2007).

Green et al. 2005). The mathematical solution (for details see Green and Angelaki 2004, Green et al. 2005), schematized in **Figure 1c**, consists of two interdependent steps (Yakusheva et al. 2007). First, rotational signals from the semicircular canals (ω, coded relative to the head) must interact with a gravity signal (g) to construct an estimate of angular velocity relative to the world. Angular velocity can then be decomposed into two perpendicular components: an earth-vertical (i.e., parallel to gravity) component, ω_{EV}, and an earth-horizontal (perpendicular to gravity) component, ω_{EH} (**Figure 1c**, *left*). The former (ω_{EV}) signals only those rotations that do not change orientation relative to gravity (e.g., yaw from upright). The latter

(ω_{EH}) signals a rotation that changes head orientation relative to gravity (e.g., pitch/roll from upright). Temporal integration of ω_{EH} ($\int \omega_{EH}$) can yield an estimate of spatial attitude or tilt. In a second computational step this tilt signal, $\int \omega_{EH}$, can be combined with net linear acceleration from the otolith organs, α, to extract the linear acceleration component that is due to translation, t (**Figure 1c**, *right*). The logic behind these computations is simple: Using signals from the semicircular canals, the brain can generate an internal estimate of the linear accelerations that should be detected by the otolith system during head tilts relative to gravity. This signal can then be subtracted from the net activation of primary otolith afferents. Whatever is left is then interpreted as translational motion.

Evidence for Allocentric Coding of Angular Velocity

Strong evidence for a world-centered representation of rotational signals comes from reflexive eye movement studies and, in particular, a process known as the velocity storage mechanism, which dominates the rotational VOR at low frequencies (Raphan et al. 1979). In particular, eye velocity during low frequency rotation is driven by semicircular canal signals that have been spatially transformed to align with gravity (i.e., they represent an ω_{EV} signal; Merfeld et al. 1993, 1999; Angelaki & Hess 1994; Angelaki et al. 1995; Wearne et al. 1998; Zupan et al. 2000). More recently, Fitzpatrick et al. (2006) demonstrated that a world-referenced angular velocity signal is also available for perception and balance. Using galvanic (electrical) stimulation of vestibular receptors in the inner ear, Fitzpatrick, Day, and colleagues evoked a virtual rotation as subjects walked in the dark. Depending on head orientation, the authors could either steer walking or produce balance disturbances, concluding that the brain resolves the canal signal according to head posture into world-referenced orthogonal components. Each of these components could have a potentially different function: Rotations in vertical planes (i.e., an ω_{EH} signal) can be used to control balance, whereas rotations in the horizontal plane (i.e., an ω_{EV} signal) can be used primarily for navigation. In this particular experiment, such computation could be performed either entirely by vestibular signals or through contributions from both vestibular and nonvestibular estimates of head orientation (e.g., derived from somatosensory and motor information).

In line with such decomposition, whereby ω_{EH} contributes to orientation and balance and ω_{EV} contributes to navigation, is also a role of vestibular signals in the generation of head direction cell properties in the limbic system (for a recent review, see Taube 2007). Although details about the neural implementations are still missing, vestibular-driven angular velocity appears essential for generating the head direction signal (Stackman & Taube 1997, Muir et al. 2004). However, head direction cell firing is dependent only on the earth-vertical component of angular velocity (Stackman et al. 2000). These results (see also Calton & Taube 2005) suggest that the head direction signal is generated by temporal integration of an ω_{EV} (rather than ω) signal.

Evidence for Segregation of Head Attitude and Translational Motion

That the brain correctly interprets linear acceleration is obvious from everyday activities. As we swing in the play ground, for example, a motion that includes changes in both head attitude and translation, we properly perceive our motion. This is true even when our eyes are closed (thus excluding important visual cues). Quantitative evidence that a solution to the linear acceleration problem can exist using otolith/canal convergence comes from monkey and human studies. Angelaki and colleagues (Angelaki et al. 1999, Green & Angelaki 2003) showed that an extraotolith signal does contribute to the compensatory eye movements during mid/high frequency translation (>0.1 Hz). In line with the schematic of **Figure 1c**, these signals arise from temporal integration of angular velocity from the semicircular canals (Green & Angelaki 2003). Parallel studies by Merfeld and colleagues focused on low-frequency motion stimuli; they showed the contribution of canal cues in generating a neural estimate of translation by exploring erroneous behavioral responses (eye movements and perception) that are typically attributed to the velocity storage mechanism (Merfeld et al. 1999, 2001; Zupan et al. 2000). Tilt-translation ambiguities are not properly resolved at these lower frequencies because the semicircular canals do not provide a veridical estimate of angular velocity at low frequencies (<0.1 Hz) or when the head is statically tilted.

Similarly, the tilt-translation ambiguity is not always correctly resolved at the perceptual level; low-frequency linear accelerations

Velocity storage mechanism: the prolongation of the vestibular time constant during rotation compared with that in the vestibular eighth nerve

Navigation: the ability to move appropriately and purposefully through the environment

in the absence of other, extravestibular cues are incorrectly interpreted as tilt even when generated by translational motion. Thus the ability to discriminate between tilt and translation based solely on vestibular cues (e.g., during passive motion in darkness) deteriorates at low frequencies (Glasauer 1995; Seidman et al. 1998; Merfeld et al. 2005a,b; Kaptein & Van Gisbergen 2006). In fact, it is typically at these low frequencies that perceptual illusions occur (e.g., "somatogravic and oculogravic" illusions, often elicited during airplane landing and take-off; Graybiel 1952; Clark & Graybiel 1963, 1966). Under these circumstances extravestibular information (e.g., visual signals) is necessary to avoid illusions. For example, visual cues can significantly influence our percept of head orientation relative to gravity (Dichgans et al. 1972, Howard & Hu 2001). In addition, visual rotational cues contribute to estimation of inertial motion because they can substitute for canal-driven angular velocity information (Zupan & Merfeld 2003, McNeilage et al. 2007).

Neural Substrates for Inertial Motion Detection

To characterize whether and how neurons use canal and otolith information to separate ω_{EV} and ω_{EH}, and to distinguish translational from gravitational accelerations, otolith afferents and central neurons have been studied during combinations of tilt and translation stimuli, as shown in **Figure 2**. Stimulus conditions included translation only (e.g., left/right motion), tilt only (e.g., sinusoidal tilt toward right/left ear down without linear displacement), and combinations of the two (tilt − translation and tilt + translation, illustrated by cartoon

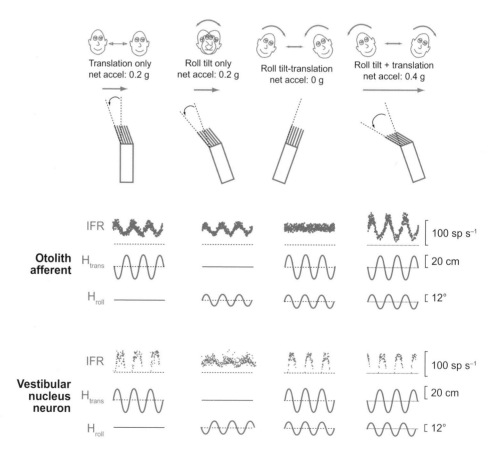

Figure 2

Instantaneous firing rate (IFR) from a primary otolith afferent (*top*) that encodes net linear acceleration, and a central vestibular nucleus neuron (*bottom*) that encodes translational motion during four movement protocols: Translation only, Tilt only, Tilt − Translation and Tilt + Translation (0.5 Hz). The stimulus (*bottom*) traces show sled position (H_{tran}) and roll tilt position (H_{roll}). Replotted with permission from Angelaki et al. (2004).

drawings, **Figure 2**, *top*). In the most important of these stimulus combinations (tilt – translation motion), roll tilt and translation stimuli were carefully matched to ensure that the gravitational and translational components of acceleration along the interaural axis canceled each other out. In this case, the body translated in space, but there was no net lateral linear acceleration stimulus to the otolith receptors. As expected (Fernandez & Goldberg 1976), primary otolith afferents encoded net linear acceleration, modulating similarly during translation and tilt (thereby emphasizing the linear acceleration ambiguity; **Figure 1*b***). Note that during the tilt – translation stimulus condition, primary otolith afferents transmit no information about the subject's translation (i.e., there is no sinusoidal modulation of firing rate) because net linear acceleration along the axis of motion is zero.

In contrast with primary otolith afferents, many central neurons selectively encode translational motion and remain relatively insensitive to changes in head orientation relative to gravity. For example, the vestibular nucleus neuron illustrated in **Figure 2** (*bottom*) modulated little during its firing rate the tilt-only condition, whereas combined motion protocols resulted in cell activation similar to that during pure translation. Neurons that selectively encode translation rather than net acceleration were found not only in the vestibular nuclei (VN) but also in the rostral fastigial (FN) cerebellar nuclei (Angelaki et al. 2004), as well as in the nodulus and uvula (NU) of cerebellar cortex (vermal lobules X and IX; Yakusheva et al. 2007). Results from all three areas are summarized and compared with primary otolith afferents in **Figure 3**, which illustrates partial correlation coefficients describing the degree to which responses to these stimuli corresponded to neural coding of translation (ordinate) or net acceleration (abscissa). Data points falling in the upper left quadrant represent neurons that were significantly more translation coding than afferent like ($p = 0.01$; *dashed lines*). In contrast, cells in the lower right quadrant were significantly more afferent like than trans-

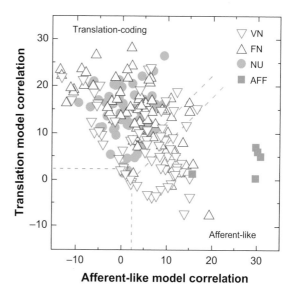

Figure 3

Summary of how well neurons in different subcortical areas discriminate translational from gravitational acceleration. Z-transformed partial correlation coefficients for fits of each cell's responses with a translation-coding model and an afferent-like model. Data from nodulus/uvula (NU) Purkinje cells (*orange circles*), rostral fastigial nucleus (FN, *purple up triangles*) and vestibular nucleus (VN, *green down triangles*) are compared with primary otolith afferents (AFF, *blue squares*). Dashed lines divide the plots into an upper-left area corresponding to cell responses that were a significantly better fit ($p < 0.01$) by the translation-coding model; a lower-right area including neurons that were a significantly better fit by the afferent-like model; and an intermediate area that indicates cells that were not a significantly better fit by either model. Modified and replotted with permission from Angelaki et al. (2004) and Yakusheva et al. (2007).

lation coding. VN and FN neurons tended to span the whole range, in contrast with NU Purkinje cells, all of which fell in the upper left quadrant (Angelaki et al. 2004, Yakusheva et al. 2007).

Inactivation of the canals completely eliminated the presence of translation-coding cells. All neurons in the VN/FN/NU became afferent like and encoded net linear acceleration after canal inactivation (Shaikh et al. 2005, Yakusheva et al. 2007). This occurs because, in addition to an otolith input, these neurons also receive a semicircular canal-driven signal. Yakusheva et al. (2007) showed that, in line with **Figure 1*c***, this canal-driven signal in the nodulus/uvula has been processed relative to canal afferents in two important aspects: (*a*) It represents an ω_{EH} (rather than ω) signal.

Accordingly, Purkinje cells modulate only during canal activation involving rotations that change orientation relative to gravity, e.g., during pitch and roll in upright orientation, but not during pitch/roll in ear-down/supine orientation (Yakusheva et al. 2007). (b) This canal-driven, spatially transformed signal has been temporally integrated, thus coding head position relative to gravity ($\int \omega_{EH}$, rather than rotational velocity). Such an earth-centered estimate of head attitude is then subtracted from net linear acceleration provided by the otoliths and used to estimate inertial motion during navigation. Next we show that neurons in these same areas (VN and FN) seem to be performing another important function: distinguishing between rotations that are self-generated and those that are externally applied.

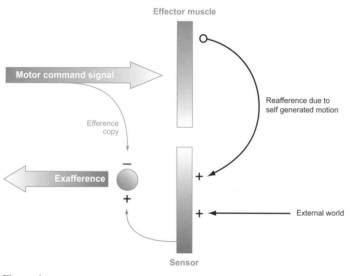

Figure 4

A simplified schematic of Von Holst & Mittelstaedt's reafference principle applied to the vestibular system. A motor command is sent to the effector muscle, and in turn, sensory activation, resulting from the effector's activation of the vestibular sensors, is returned. This reafference is then compared with an efference copy of the original motor command. Here, reafference is arbitrarily marked (+), and the efference copy is marked (−). When the reafference and efference copy signals are of equal magnitude, they cancel, and no sensory information is transmitted to the next processing levels. In contrast, a difference between reafference and efference copy indicates an externally generated event (i.e., exafference) that is considered behaviorally relevant and is thus further processed.

DISTINGUISHING PASSIVE FROM ACTIVE HEAD MOVEMENTS

The ability to navigate and orient through the environment requires knowledge not only of inertial motion, but also of which components of vestibular activation result from active (i.e., self-generated) and passive (i.e., externally applied) movements. How does the brain differentiate between sensory inputs that arise from changes in the world and those that result from our own voluntary actions? This question concerned many eminent scientists of the past century, including Helmholtz, Hering, Mach, and Sherrington. For example, Von Helmholtz (1925) made the salient and easily replicated observation that although targets rapidly jump across the retina as we move our eyes to make saccades, we never see the world move over our retina. Yet, tapping on the canthus of the eye to displace the retinal image (as during a saccadic eye movement) results in an illusory shift of the visual world.

More than 50 years ago, Von Holst & Mittelstaedt (1950) proposed the principle of reafference (**Figure 4**), in which a copy of the expected sensory results of a motor command (termed reafference) is subtracted from the actual sensory signal to create a perception of the outside world (termed exafference). Thus, the nervous system can distinguish sensory inputs that arise from external sources from those that result from self-generated movements. More recent behavioral investigations have generalized this original proposal by suggesting that an internal prediction of the sensory consequence of our actions, derived from motor efference copy, is compared with actual sensory input (Wolpert et al. 1995, Decety 1996, Farrer et al. 2003). In line with this proposal, work in several model systems, including the electrosensory systems of mormyrid fish (Bell 1981, Mohr et al. 2003) and elasmobranchii (i.e., sharks, skates, and rays; Hjelmstad et al. 1996), the mechanosensory system of the crayfish (Krasne & Bryan 1973, Edwards et al. 1999), and the auditory system of the cricket (Poulet & Hedwig

2003, 2006), demonstrated that sensory information arising from self-generated behaviors can be selectively suppressed at the level of afferent fibers or the central neurons to which they project.

Differential Processing of Active Versus Passive Head Movement

Until recently, the vestibular system had been exclusively studied in head-restrained animals by moving the head and body together (reviewed in Cullen & Roy 2004). Thus because neuronal responses were driven by an externally applied stimulus, our understanding of vestibular processing was limited to the neuronal encoding of vestibular exafference. More recently, investigators in the field have overcome the technical difficulties associated with recording single-unit responses during self-generated head movements. As shown in **Figure 5**, whereas vestibular afferents reliably encode active movements (Cullen & Minor 2002, Sadeghi et al. 2007), VN neuron responses can be dramatically attenuated (compare panels **5a** and **5b**) (McCrea

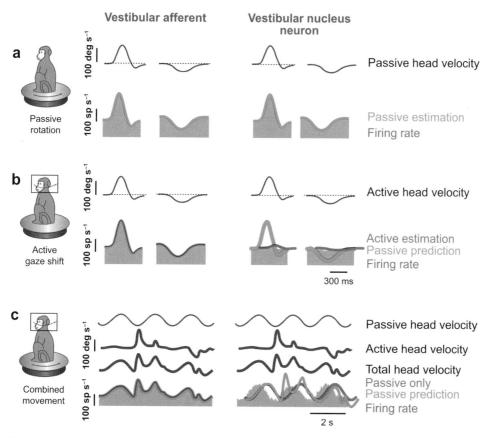

Figure 5

In the vestibular system, second-order neurons distinguish between sensory inputs that result from our own actions from those that arise externally. Representation of the activity of a horizontal canal afferent (*left panel*) and VN neuron (*right panel*) during (*a*) passive head movements, (*b*) active head movements, and (*c*) combined active and passive head movements. Afferents reliably encode head motion in all conditions. In contrast, VO neurons show significantly attenuated responses to the active component of head motion, but remain responsive to active head movements during combined stimulation.

et al. 1999, Roy & Cullen 2001). What is even more striking is that these same second-order vestibular neurons continue to respond selectively to passively applied head motion when a monkey generates active head-on-body movements (**Figure 5c**). Thus, consistent with Von Holst & Mittelstaedt's original proposal, vestibular information arising from self-generated movements is selectively suppressed early in sensory processing to create a neural representation of the outside world (i.e., vestibular exafference). This suppression of vestibular reafference is specific to a class of second-order neurons, which had been classically termed vestibular-only (VO) neurons on the basis of their lack of eye movement–related responses in head-restrained animals (e.g., Fuchs & Kimm 1975, Keller & Daniels 1975, Lisberger & Miles 1980, Chubb et al. 1984, Tomlinson & Robinson 1984, Scudder & Fuchs 1992, Cullen & McCrea 1993). However, given that they only reliably encode passively applied head velocity (i.e., vestibular exafference), this nomenclature is clearly deceptive. This is the same group of neurons that, as summarized earlier, is involved in the computation of inertial motion (Angelaki et al. 2004).

Neural Mechanisms Underlying the Differential Processing of Actively Generated Versus Passive Head Movement

How does the brain distinguish between active and passive head movements at the first stage of central processing in the vestibular system? Theoretically, the existence of extensive multimodal convergence of other sensory and motor signals with vestibular information in the VN provides several possible solutions. To frame this question better, it is important to note that neuronal responses were compared during self-generated head movements that were produced by activation of the neck musculature (i.e., voluntary head-on-body movements) and passive movements that were generated by whole body

rotations (i.e., the traditional stimulus for quantifying vestibular responses).

Consequently, recent studies in alert rhesus monkeys have focused on the implications of the difference between the extravestibular cues that were present in these two conditions. First, studies show a difference in the net sensory information that is available to the brain. Notably, during active head-on-body movements, neck proprioceptors as well as vestibular receptors are stimulated. Thus this additional information could alter neuronal responses during active head-on-body movements. Indeed, neck-related inputs are conveyed to the vestibular nuclei using a disynaptic pathway (Sato et al. 1997). In addition, activation of neck muscle spindle afferents has long been known to influence the VN neuron activity in decerebrate animals (Boyle and Pompeiano 1981, Anastasopoulos & Mergner 1982, Wilson et al. 1990). However, passive activation of neck proprioceptors alone does not significantly alter neuronal sensitivities to head rotation in alert rhesus monkeys (Roy & Cullen 2004). Second, during active head-on-body movements, the brain produces a command signal to activate the neck musculature. To quantify the influence of this additional cue, recordings were made in head-restrained monkeys who attempted to move their heads. The generation of neck torque, even reaching a level comparable to that issued to produce large active head movements, had no effect on neuronal responses (Roy & Cullen 2004).

Taken together, these results show that neither neck motor efference copy nor proprioception cues alone are sufficient to account for the elimination of neuronal sensitivity to active head rotation. However, a common feature of both these experiments was that neck motor efference copy and proprioceptive signals were not matched as they typically are during normal active head movements. By experimentally controlling the correspondence between intended and actual head movement, Roy & Cullen (2004) showed that a cancellation signal is generated only when the

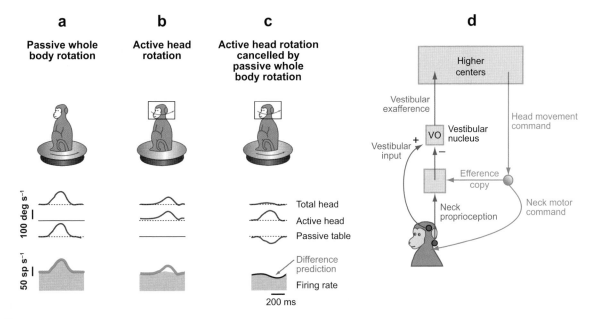

a
Passive whole
body rotation

b
Active head
rotation

c
Active head rotation
cancelled by
passive whole
body rotation

d

Higher centers

Vestibular exafference

Head movement command

VO Vestibular nucleus

Vestibular input

Efference copy

Neck proprioception

Neck motor command

—— Total head
—~·— Active head
·——· Passive table

Difference prediction

Firing rate

200 ms

Figure 6

An internal model of the sensory consequences of active head motion is used to suppress reafference selectively at the vestibular nuclei level. (*a*) Activity of an example VN neuron (*gray filled trace*) during passive whole body rotation. In this condition, only vestibular inputs are available to the central nervous system, and there is no motor efference copy signal because the monkey does not actively move its head. (*b*) Activity of the same neuron during active-head-on body movements. In this condition, the monkey commands an active head movement, so an efference copy signal is theoretically available. In addition, the head movement activates both vestibular and proprioceptive afferents. A prediction of the neuron's activity based on its response to passive head motion is superimposed (*blue trace*). (*c*) The neuron is then recorded as the monkey actively moves its head; however, the head velocity generated by the monkey (*red arrow in schema*) is experimentally cancelled by simultaneously rotating the monkey in the opposite direction (*blue arrow in schema*). Consequently, the head moves relative to the body but not to space. As a result, one finds an efference copy signal, and the neck proprioceptors are activated, but vestibular afferent input is greatly reduced. The neuron's response shows a marked inhibition, in excellent correspondence to that predicted from the difference in response during passive (*a*) vs. active (*b*) head movements (*black superimposed trace*; modified with permission from Roy & Cullen 2004). (*d*) Schematic to explain the selective elimination of vestibular sensitivity to active head-on-body rotations. Vestibular signals that arise from self-generated head movements are inhibited by a mechanism that compares the brain's internal prediction of the sensory consequences to the actual resultant sensory feedback. Accordingly, during active movements of the head on body, a cancellation signal is gated into the vestibular nuclei only in conditions where the activation of neck proprioceptors matches that expected on the basis of the neck motor command.

activation of neck proprioceptors matches the motor-generated expectation (**Figure 6***a–c*). This interaction among vestibular, proprioceptive, and motor efference copy signals occurs as early as the first-order vestibular neurons. In agreement with Von Holst's & Mittelstaedt's original hypothesis, an internal model of the sensory consequences of active head motion (**Figure 6***d*) is used to selectively suppress reafference at the vestibular nuclei level.

The finding that vestibular reafference is suppressed early in sensory processing has clear analogies with other sensory systems,

most notably the electrosensory system of the weakly electric fish (Bell 1981, Bastian 1999, Mohr et al. 2003). This is not unexpected because both systems have presumably evolved from the lateral line (Romer & Parsons 1977). Considerable evidence from work in electric fish demonstrates that cerebellum-like electrosensory lobes play a key role in the attenuation of sensory responses to self-generated stimulation (Bell et al. 1999, Mohr et al. 2003, Sawtell et al. 2007). Consistent with this idea, fMRI studies have suggested that the cerebellum plays a similar role in the suppression of tactile stimulation during self-produced tickle (Blakemore et al. 1998, 1999a,b). Identifying the neural representations of the cancellation signal for vestibular reafference promises to be an interesting area of investigation, and the cerebellum is a likely site. However, perhaps an even more interesting question is, how does the brain facilitate the temporal/spatial comparison between proprioceptive inputs and motor commands that is required to cancel reafference? This is a critical point, given that the ability to attenuate incoming vestibular afferent signals depends on this comparison. Yet, not only does peripheral feedback from the movement lag descending motor commands, but also it reflects the spatial complexity of the neck motor system.

The ability to distinguish actively generated and passive stimuli is not a general feature of all early vestibular processing. Position-vestibular-pause (PVP) neurons constitute the middle leg of the three neuron arc that generates the VOR and thus are both sensory and premotor neurons. Unlike VO cells, PVP neurons code head velocity in a manner that depends exclusively on the subject's current gaze strategy. Specifically, vestibular inputs arising from active and passive head movements are similarly encoded, as long as the goal is to stabilize gaze (Roy & Cullen 1998; Roy & Cullen 2002, 2003). In contrast, when the goal is to redirect gaze (e.g., during orienting gaze shifts), neuronal responses to active head movements are suppressed. This finding is logically consistent with the role of these neurons in stabilizing gaze; because during gaze shifts eye movements compensatory to head movement would be counterproductive, the VOR is significantly suppressed (see discussion by Cullen et al. 2004). Also consistent with the proposal that these neurons process vestibular inputs in a manner that depends on the current gaze strategy is the finding that their rotational head movement sensitivity depends on viewing distance (Chen-Huang & McCrea 1999). This is because larger rotational VOR gains are necessary to stabilize near vs. far targets (as a result of the differences in the translations of the target relative to the eye).

In summary, whereas the behaviorally dependent processing of vestibular inputs is a general feature of early vestibular areas, the ability to distinguish actively generated and passive head movements is specific to a distinct population of neurons. The functional significance of this ability to selectively suppress vestibular inputs that arise from self-generated movements is considered next.

Functional Implications: Consequences for Motor Control and Spatial Orientation

The differential processing of vestibular information during active vs. passive head movements is essential for ensuring accurate motor control. This point can be easily appreciated by considering that many of the same neurons that distinguish actively generated from passive head movements control the vestibulo-collic reflex (VCR) via their projections to the cervical segments of the spinal cord (Boyle 1993, Boyle et al. 1996, Gdowski & McCrea 1999). The function of the VCR is to assist stabilization of the head in space via activation of the neck musculature during head motion. In situations where it is helpful to stabilize head position in space, the compensatory head movements produced by this reflex are obviously beneficial. Yet, when the behavioral goal is to make an active head movement, the vestibular drive to the reflex pathway would command an inappropriate head movement to move the head in the direction opposite of the intended goal. Thus it is important that VN neurons that control the VCR are less responsive during active head movements. Furthermore, because neurons continue to reliably encode information about passive head-on-body rotations that occur during the execution of voluntary movements (McCrea et al. 1999, Roy & Cullen 2001), they can selectively respond to adjust postural tone in response to any head movements that the brain does not expect. This selectivity is fundamental because recovery from tripping over an obstacle while walking or running requires a selective but robust postural response to the unexpected vestibular stimulation.

The same VN neurons that distinguish actively generated from passive head movements are also reciprocally interconnected with the fastigial nucleus and nodulus/uvula of the cerebellum. As was detailed in the previous section, the same network (VN, FN, NU) also makes significant contributions to the computation of inertial motion (Angelaki & Hess 1995, Wearne et al. 1998). Results from a preliminary report in the rostral fastigial nucleus show that FN neurons also distinguish actively generated from passive head rotations (Brooks & Cullen 2007). This finding suggests that FN neurons do not compute an estimate of self-motion during active movements, but rather use multimodal information to compute an exafference signal (i.e., motions applied by the outside world). Because the rostral FN is generally thought to be involved in vestibulo-spinal control, this processing is most likely essential for the regulation of gait and posture.

During self-motion, the ability to distinguish between actively generated and passively applied head movements is not only important for shaping motor commands, but also critical for ensuring perceptual stability (reviewed in Cullen 2004). Notably, Roy & Cullen (2001) asked whether the head movement–related responses of VN neurons might be attenuated not only during active head movements, but also during a more cognitively demanding, less natural, self-motion task. Single-unit recordings were made in the VN of monkeys while they controlled a steering wheel to actively rotate their heads and bodies together through space (**Figure 7**). In contrast to what was observed during active head-on-body movements, all second-order vestibular neurons continued to respond robustly to angular head velocity during these self-generated rotations. Although this result further emphasizes the important role that movement commands and proprioceptive signals play in shaping the responses of secondary vestibular neurons (i.e., during

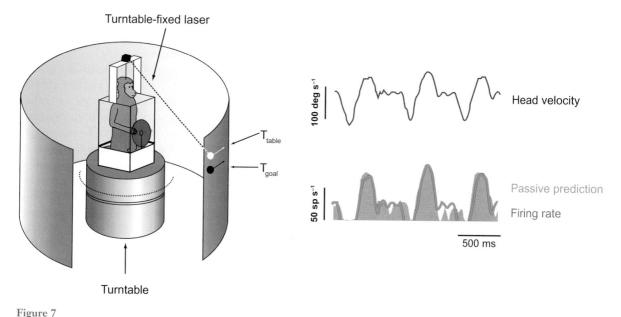

Turntable-fixed laser

T_{table}

T_{goal}

Turntable

100 deg s^{-1}

Head velocity

50 sp s^{-1}

Passive prediction

Firing rate

500 ms

Figure 7

Example of a VN neuron response to voluntary combined head-body motion. Head-restrained monkeys manually controlled a steering wheel to rotate the vestibular turntable relative to space. Their goal was to align a turntable-fixed laser target (T_{table}) with a computer-controlled target (T_{goal}). The example neuron is typical in that modulation was well predicted by its response to passive head movement. Modified with permission from Roy & Cullen (2001).

natural orienting movements), further training to control movement by steering might have ultimately resulted in the suppression of vestibular responses. For example, after extensive flight training with a particular aircraft, it is common for pilots to make comments such as "the aircraft began to feel like an extension of my limbs." Perhaps this sensation occurs once the brain has built an accurate internal model of the vehicle being driven and in turn is capable of canceling the sensory consequences of motion that result from the manual steering of the aircraft (or vestibular chair). Future studies of motor learning during self-motion tasks will be required to address this proposal.

These results might also be relevant to the generation of the properties of head direction cells. As summarized in the previous section, the spatial tuning of these cells is currently thought to be created through online integration of the animal's angular head velocity, generally assumed to arise from the vestibular nuclei (reviewed in Brown et al. 2002, Taube 2007). Results from VN studies comparing coding of active vs. passive head movements, however, remain to be incorporated into models of how heading direction is computed. Indeed, one apparent contradiction between these two lines of research is the finding that head direction neurons actually respond far more robustly to active than passive head rotations (Zugaro et al. 2002, Stackman et al. 2003, Bassett et al. 2005). Accordingly, the construction of an accurate internal representation of head direction for these neurons appears to require the integration of multimodal signals (proprioceptive, motor efference copy, and optic field flow) with vestibular inputs.

In summary, the multimodal interactions outlined so far served to mediate particular functions: (*a*) computation of inertial motion and (*b*) isolation of a vestibular exafference signal. However, multisensory interactions involving vestibular information are much more extensive and abundant throughout the brain. Although an explicit coverage of this topic is beyond the scope of this review, in the next section we touch on a fundamental concept that is relevant to these multisensory interactions: the concept of reference frames.

REFERENCE FRAMES FOR CODING VESTIBULAR SIGNALS

A reference frame can be defined as the particular perspective from which an observation of a spatial variable (e.g., position, velocity) is made. All sensorimotor systems that require the encoding or decoding of spatial information must face the issue of reference frames (for reviews, see Cohen & Andersen 2002, Pouget & Snyder 2000). As the otolith organs and semicircular canals are fixed inside the head, both linear acceleration and angular velocity are initially encoded in a head-centered reference frame by the primary receptors (similar to auditory information but unlike visual information, which is encoded in an eye-centered frame). This is fine for controlling eye movements through the VOR because the eyes are also locked in the head. However, because the head can adopt almost any position relative to the body or the world, there is a need to transform vestibular signals into reference frames relevant to the behavior being controlled.

Furthermore, vestibular signals in the brain become strongly multisensory. Investigators have traditionally thought that sensory information from disparate sources (e.g., visual and auditory or vestibular) needs to be brought into a common frame of reference (Cohen & Andersen 2002) before it can be combined in a useful way, although this assumption has recently been challenged (Deneve et al. 2001, Avillac et al. 2005). Next we discuss which reference frames are used to code vestibular signals, i.e., whether vestibular information remains invariant when expressed in eye-, head- or body-fixed reference frames. To date, this question has been mainly addressed in two ways: (*a*) Head- vs. body-centered reference frames have been examined in the brainstem and vestibulo-cerebellum; and (*b*) head- vs. eye-centered reference frames have been studied in extrastriate visual cortex.

Head- vs. Body-Centered Reference Frames: Vestibular/Neck Proprioceptive Interactions

Although the vestibular system alone may be sufficient to compute position and motion of the head, several daily functions including maintenance of posture and balance and perception of self-motion require knowledge of body position, orientation, and movement. By combining vestibular signals, which encode motion in a head-centered frame, with neck proprioceptive information that signals the static position of the head relative to the body, a coordinate transformation could take place to convert motion signals into a body-centered reference frame.

To test this, one must measure the spatial tuning of central neurons while the motion of the head and body are dissociated, e.g., during motion along different directions (defined relative to the body) while the head is fixed at different static positions relative to the trunk (**Figure 8a**). A body-centered reference frame assumes that spatial tuning should be independent of the change in head position and the three tuning curves should superimpose. In contrast, if a cell detects motion in a head-centered reference frame, its preferred movement direction should systematically shift to the left or to the right to reflect the shifted direction of motion in a head reference frame. **Figure 8b** and **c** show tuning curves from two representative neurons (Shaikh et al. 2004). For the cell in **Figure 8b**, the directions of maximum and minimum response gains (0° and 90° motion directions, respectively) were the same for all three head-on-body positions (*blue, orange and red lines superimpose*), indicating a body-fixed reference frame. For the other cell (**Figure 8c**),

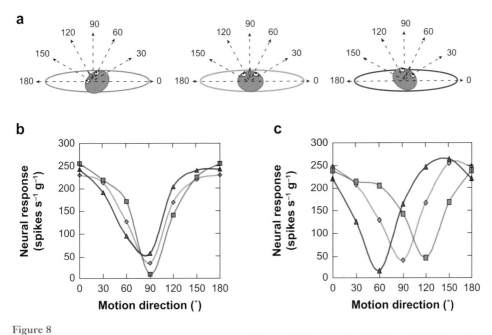

Figure 8

Head- vs. body-centered reference frames. (*a*) Schematic of the experimental manipulation. Head and body reference frames were dissociated by systematically varying both the direction of translation in the horizontal plane (0°, 30°, 60°, 90°, 120°, 150°, and 180°, defined relative to the body) and the static orientation of the head relative to the trunk (three different positions were used: straight-ahead and 30° rotated to the left/right). (*b, c*) Examples of cell response gain plotted as a function of motion direction. For a neuron coding motion in a body-centered reference frame, the spatial tuning curves for the three head-in-trunk positions superimpose (*b*). For a neuron coding motion in a head-centered frame, the three tuning curves are shifted accordingly (*c*). Blue, head 30°-left; orange, head straight ahead; red, head 30°-right. Data from Shaikh et al. (2004).

the directions of maximum and minimum response gain shifted for the three head-on-body positions, such that they remained fixed relative to the head.

Most neurons in the rostral VN were consistent with the spatial shift expected from a head-centered reference frame, but this was not the case for the rostral FN (Shaikh et al. 2004). In fact, many cells showed intermediate properties: Their tuning curves shifted through an angle that was in-between, suggesting intermediate or a mixture of reference frames. Kleine et al. (2004) reported similar findings regarding a mixture of head- and body-centered reference frames in the rostral FN using different head-on-trunk positions during rotation, as illustrated with an example cell that codes motion of the body in **Figure 9**. A body-centered reference frame in the FN might be beneficial because the rostral FN represents a main output of the anterior vermis (Voogd 1989) and nodulus/uvula (Voogd et al. 1996), both of which have been implicated in vestibular/proprioceptive interactions for limb and postural control.

To date, reference frame questions in the vestibular system have been studied using passive movements. Do the same coordinate transformations also characterize responses during active movements? Recent findings that FN cell responses are greatly attenuated during active head rotation (Brooks and Cullen 2007) suggest that the same computations may not be required during active movements. A similar logic might apply to higher levels of processing; active and passive information might be processed in ways appropriate to their functional roles. Supporting this idea, recent studies have shown that cortical neurons differentially process active and passive movements (e.g., Klam & Graf 2006, Fukushima et al. 2007).

Head- Versus Eye-Centered Reference Frames: Vestibular/ Visual Interactions

Another example of multisensory vestibular function that faces the problem of reference frames is that of visual/vestibular interactions. The dorsal subdivision of the medial superior temporal area (MSTd) is one of the likely candidates to mediate the integration of visual and vestibular signals for heading (i.e., translational motion) perception (Duffy 1998; Bremmer et al. 1999; Page & Duffy 2003; Gu et al. 2006a, 2007; Takahashi et al. 2007). It is commonly thought that multisensory neural populations should represent different sensory signals in a common reference frame (Stein & Meredith 1993, Cohen & Andersen 2002). Thus, it might be expected that both visual and vestibular signals in MSTd should code heading in a head-centered reference frame. This would enable neurons to encode a particular motion direction regardless of the sensory modality or eye position.

This hypothesis was recently tested and refuted (Fetsch et al. 2007). Head- vs. eye-centered reference frames were dissociated by manipulating static eye position while quantifying spatial tuning curves, constructed separately for translational inertial motion in the absence of visual motion (vestibular condition) and optic flow simulating translational motion (visual condition). As shown with an example cell in **Figure 10**, the reference frame for vestibular signals was close to head centered but at the population level shifted slightly toward eye centered (Fetsch et al. 2007). In contrast, visual signals continued to be represented in a retinal reference frame. These results contradict the conventional wisdom in two respects. First, reference frames for visual and vestibular heading signals in MSTd remain distinct, although evidence clearly shows that these neurons might mediate multisensory cue integration (Gu et al. 2006b). Thus, sensory signals might not be expressed in a common frame of reference for integration to occur. Second, rather than shifting the visual signals toward a head-centered representation, there was a modest shift of vestibular tuning toward an eye-centered representation. Similar to the results in the cerebellum, several MSTd neurons showed partial shifts and could thus be considered to represent motion direction in an

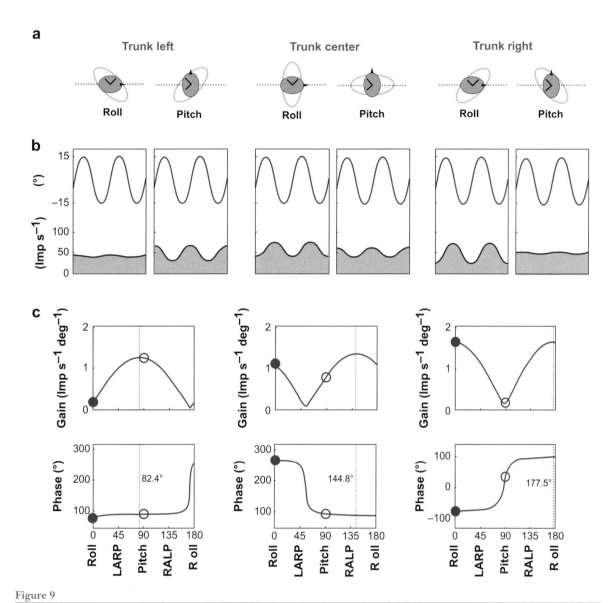

Figure 9

Head- vs. body-centered reference frames. (*a*) Schematic of the experimental protocol. (*b*) Rotation responses with sinusoidal fit.
(*c*) Tuning curves for neuronal gain and phase. Vertical dotted lines (and numbers) illustrate maximum response direction [which
changes from pitch (trunk left) to RALP (trunk center) to roll (trunk right)]. RALP: right anterior/left posterior canal axis orientation;
LARP: left anterior/right posterior canal axis orientation. Replotted with permission from Kleine et al. (2004).

intermediate frame of reference (Fetsch et al.
2007).

In summary, these results are not consistent
with the hypothesis that multisensory areas use
a common reference frame to encode visual
and vestibular signals. Similar conclusions
have also been reached in other cortical and
subcortical areas. For example, unlike visual
receptive fields, tactile receptive fields in the
ventral intraparietal (VIP) area are purely head
centered (Avillac et al. 2005). In addition, visual
and auditory receptive fields in VIP, as well as
the lateral and medial intraparietal areas (LIP
and MIP), exhibit a continuum of reference

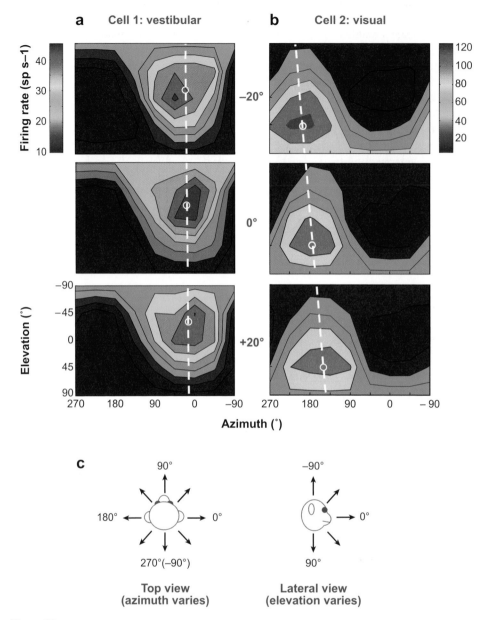

Figure 10

Three-dimensional heading tuning functions of two example MSTd neurons. (*a*, *b*) Cell 1 was tested in the vestibular condition and cell 2 in the visual condition. Tuning was measured at three different eye positions: −20° (*top*), 0° (*middle*), and +20° (*bottom*). Mean firing rate (*color contour plots*) is plotted as a function of the heading trajectory in spherical coordinates, with the azimuth and elevation of the heading vector represented on the abscissa and ordinate, respectively. For illustration purposes, small white circles are positioned at the preferred heading for each tuning function, computed as a vector sum of responses around the sphere. (*c*) Conventions for defining the real (vestibular) or simulated (visual) motion directions of three-dimensional heading stimuli. Replotted with permission from Fetsch et al. (2007).

frames from head centered to eye centered (Mullette-Gillman et al. 2005, Schlack et al. 2005). Investigators traditionally thought that intermediate frames may represent a middle stage in the process of transforming signals between different reference frames (Jay & Sparks 1987, Stricanne et al. 1996, Andersen et al. 1997). Alternatively, broadly distributed and/or intermediate reference frames may be computationally useful. According to this latter view, intermediate frames may arise naturally when a multimodal brain area makes recurrent connections with unimodal areas that encode space in their native reference frame (Pouget et al. 2002). Using a recurrent neural network architecture, Pouget and colleagues have shown that a multi-sensory layer expressing multiple reference frames, combined with an eye position signal, can optimally mediate multisensory integration in the presence of noise (Deneve et al. 2001, Deneve & Pouget 2004). This modeling framework predicts a robust relationship between the relative strength of visual and nonvisual signals and the respective reference frames in a particular brain area (Avillac et al. 2005); the stronger a sensory signal is, the more dominant its native reference frame is. Accordingly, in MSTd, where visual responses are stronger than vestibular responses, an eye-centered reference frame tends to dominate (Fetsch et al. 2007). Future studies of visual/vestibular interactions in other brain areas will be useful in further testing this framework.

SUMMARY POINTS

1. The vestibular system represents our sixth sense. Because of the need for these diverse, multimodal functions, computationally intriguing transformations of vestibular information occur as early as the first-order neurons in the brainstem vestibular nuclei and vestibulo-cerebellum.

2. Within a network consisting of the brainstem vestibular nuclei (VN), the most medial of the deep cerebellar (fastigial, FN) nuclei, and the most posterior lobulus (X and IX) of the cerebellar vermis, a critical computation of inertial motion (i.e., how our head moves in space) takes place. Nonlinear interactions between signals from the semicircular canals and otolith organs give the central vestibular system the ability to function as an inertial sensor and contribute critically to both navigation and spatial orientation.

3. Neurons at the first central stage of vestibular processing (VN and FN) can also distinguish between self-generated and passive movements. During active movements, a cancellation signal is generated when the activation of proprioceptors matches the motor-generated expectation. This mechanism eliminates self-generated movements from subsequent computation of orientation and postural control.

4. The ability to distinguish actively generated and passive stimuli is not a general feature of all early central vestibular processing; central vestibular neurons process vestibular information in a manner that is consistent with their functional role. For example, central neurons controlling gaze process vestibular information in a behaviorally dependent manner according to current gaze strategy.

5. The need for multisensory integration with both proprioceptive and visual signals necessitates that vestibular information is represented in widely different reference frames within the central nervous system. Here we summarize two such examples where vestibular information has been at least partially transformed from a head-fixed to a body-centered (cerebellar FN) and eye-centered (extrastriate area MSTd) reference frame.

FUTURE ISSUES

1. Most of the vestibular signal processing studies have concentrated on the brain stem vestibular nuclei and vestibulo-cerebellum. Vestibular information is also heavily present in the reticular formation, spinal cord, thalamus, and cortex. What are the properties and functions of vestibular information in these diverse brain areas?

2. What are the exact relationships between neurons that discriminate translation from tilt and those that have velocity storage properties?

3. How does an ω_{EV} signal generate head direction cell activity and contribute to navigation? Although ω_{EH} signals have been isolated in single-cell responses that selectively encode for translation, an ω_{EV} signal has yet to be identified in single-cell activity.

4. The distinction between passive and active head movements in neural activity has so far been tested only during rotation. Whether neurons respond differently during passive and active translational movements has yet to be explored.

5. Which information is encoded by cortical areas that contribute to the perception of self-motion? Can these areas distinguish actively generated from passive head movements? If so, which mechanisms underlie the computation, and what is the functional significance of the information that is ultimately encoded?

6. Prior studies describing the transformation of vestibular information from a head-centered to other reference frames (i.e., body-centered and eye-centered) considered only passive head movement stimuli. Is vestibular information encoded in the same reference frames during actively generated movements? Or alternatively, is the necessary transformation of vestibular information behavior dependent?

7. The only study of reference frames involving convergent visual and vestibular signals has been in extrastriate visual cortex (area MSTd). However, some form of visual/vestibular convergence (studied mainly by using optokinetic stimulation at low frequencies) already occurs as early as the vestibular nuclei and vestibulo-cerebellum. Which reference frames are used in these interactions? Have vestibular signals been transformed at least partially into an eye-centered reference frame (such as in MSTd)? Or, alternatively, are optokinetic signals coded in a head-centered reference frame?

DISCLOSURE STATEMENT

The authors are not aware of any biases that might be perceived as affecting the objectivity of this review.

ACKNOWLEDGMENTS

Work presented in this review was supported by NIH grants R01-DC04260 and R01-EY12814, NASA, the Canadian Institutes of Health Research (CIHR), Canadian Space Agency, and the McGill Dawson Chair Program.

LITERATURE CITED

Anastasopoulos D, Mergner T. 1982. Canal-neck interaction in vestibular nuclear neurons of the cat. *Exp. Brain Res.* 46:269–80

Andersen RA, Snyder LH, Bradley DC, Xing J. 1997. Multimodal representation of space in the posterior parietal cortex and its use in planning movements. *Annu. Rev. Neurosci.* 20:303–30

Angelaki DE. 2004. Eyes on target: what neurons must do for the vestibuloocular reflex during linear motion. *J. Neurophysiol.* 92:20–35

Angelaki DE, Hess BJ. 1994. Inertial representation of angular motion in the vestibular system of rhesus monkeys. I. Vestibuloocular reflex. *J. Neurophysiol.* 71:1222–49

Angelaki DE, Hess BJ. 1995. Inertial representation of angular motion in the vestibular system of rhesus monkeys. II. Otolith-controlled transformation that depends on an intact cerebellar nodulus. *J. Neurophysiol.* 73:1729–51

Angelaki DE, Hess BJ. 2005. Self-motion-induced eye movements: effects on visual acuity and navigation. *Nat. Rev. Neurosci.* 6:966–76

Angelaki DE, Hess BJ, Suzuki J. 1995. Differential processing of semicircular canal signals in the vestibulo-ocular reflex. *J. Neurosci.* 15:7201–16

Angelaki DE, McHenry MQ, Dickman JD, Newlands SD, Hess BJ. 1999. Computation of inertial motion: neural strategies to resolve ambiguous otolith information. *J. Neurosci.* 19:316–27

Angelaki DE, Shaikh AG, Green AM, Dickman JD. 2004. Neurons compute internal models of the physical laws of motion. *Nature* 430:560–64

Avillac M, Deneve S, Olivier E, Pouget A, Duhamel JR. 2005. Reference frames for representing visual and tactile locations in parietal cortex. *Nat. Neurosci.* 8:941–49

Bassett JP, Zugaro MB, Muir GM, Golob EJ, Muller RU, Taube JS. 2005. Passive movements of the head do not abolish anticipatory firing properties of head direction cells. *J. Neurophysiol.* 93:1304–16

Bastian J. 1999. Plasticity of feedback inputs in the apteronotid electrosensory system. *J. Exp. Biol.* 202:1327–37

Bell CC. 1981. An efference copy which is modified by reafferent input. *Science* 214:450–53

Bell CC, Han VZ, Sugawara Y, Grant K. 1999. Synaptic plasticity in the mormyrid electrosensory lobe. *J. Exp. Biol.* 202:1339–47

Blakemore SJ, Frith CD, Wolpert DM. 1999a. Spatio-temporal prediction modulates the perception of self-produced stimuli. *J. Cogn. Neurosci.* 11:551–59

Blakemore SJ, Wolpert DM, Frith CD. 1998. Central cancellation of self-produced tickle sensation. *Nat. Neurosci.* 1:635–40

Blakemore SJ, Wolpert DM, Frith CD. 1999b. The cerebellum contributes to somatosensory cortical activity during self-produced tactile stimulation. *Neuroimage* 10:448–59

Boyle R. 1993. Activity of medial vestibulospinal tract cells during rotation and ocular movement in the alert squirrel monkey. *J. Neurophysiol.* 70:2176–80

Boyle R. 2001. Vestibulospinal control of reflex and voluntary head movement. *Ann. N. Y. Acad. Sci.* 942:364–80

Boyle R, Belton T, McCrea RA. 1996. Responses of identified vestibulospinal neurons to voluntary eye and head movements in the squirrel monkey. *Ann. N. Y. Acad. Sci.* 781:244–63

Boyle R, Pompeiano O. 1981. Responses of vestibulospinal neurons to neck and macular vestibular inputs in the presence or absence of the paleocerebellum. *Ann. N. Y. Acad. Sci.* 374:373–94

Bremmer F, Kubischik M, Pekel M, Lappe M, Hoffmann KP. 1999. Linear vestibular self-motion signals in monkey medial superior temporal area. *Ann. N. Y. Acad. Sci.* 871:272–81

Brooks J, Cullen KE. 2007. Reference frames and reafference in the rostral fastigial nucleus. *Soc. Neurosci. Abstr.* 33:861.2

Brown JE, Yates BJ, Taube JS. 2002. Does the vestibular system contribute to head direction cell activity in the rat? *Physiol. Behav.* 77:743–48

Calton JL, Taube JS. 2005. Degradation of head direction cell activity during inverted locomotion. *J. Neurosci.* 25:2420–28

Chen-Huang C, McCrea RA. 1999. Effects of viewing distance on the responses of vestibular neurons to combined angular and linear vestibular stimulation. *J. Neurophysiol.* 81:2538–57

Chubb MC, Fuchs AF, Scudder CA. 1984. Neuron activity in monkey vestibular nuclei during vertical vestibular stimulation and eye movements. *J. Neurophysiol.* 52:724–42

Clark B, Graybiel A. 1963. Contributing factors in the perception of the oculogravic illusion. *Am. J. Psychol.* 76:18–27

Clark B, Graybiel A. 1966. Factors contributing to the delay in the perception of the oculogravic illusion. *Am. J. Psychol.* 79:377–88

Cohen YE, Andersen RA. 2002. A common reference frame for movement plans in the posterior parietal cortex. *Nat. Rev. Neurosci.* 3:553–62

Cullen KE. 2004. Sensory signals during active versus passive movement. *Curr. Opin. Neurobiol.* 14:698–706

Cullen KE, Huterer M, Braidwood DA, Sylvestre PA. 2004. Time course of vestibuloocular reflex suppression during gaze shifts. *J. Neurophysiol.* 92:3408–22

Cullen KE, McCrea RA. 1993. Firing behavior of brain stem neurons during voluntary cancellation of the horizontal vestibuloocular reflex. I. Secondary vestibular neurons. *J. Neurophysiol.* 70:828–43

Cullen KE, Minor LB. 2002. Semicircular canal afferents similarly encode active and passive head-on-body rotations: implications for the role of vestibular efference. *J. Neurosci.* 22:RC226

Cullen KE, Roy JE. 2004. Signal processing in the vestibular system during active versus passive head movements. *J. Neurophysiol.* 91:1919–33

Decety J. 1996. Neural representations for action. *Rev. Neurosci.* 7:285–97

Deneve S, Latham PE, Pouget A. 2001. Efficient computation and cue integration with noisy population codes. *Nat. Neurosci.* 4:826–31

Deneve S, Pouget A. 2004. Bayesian multisensory integration and cross-modal spatial links. *J. Physiol. Paris* 98:249–58

Dichgans J, Held R, Young LR, Brandt T. 1972. Moving visual scenes influence the apparent direction of gravity. *Science* 178:1217–19

Duffy CJ. 1998. MST neurons respond to optic flow and translational movement. *J. Neurophysiol.* 80:1816–27

Edwards DH, Heitler WJ, Krasne FB. 1999. Fifty years of a command neuron: the neurobiology of escape behavior in the crayfish. *Trends Neurosci.* 22:153–61

Farrer C, Franck N, Paillard J, Jeannerod M. 2003. The role of proprioception in action recognition. *Conscious Cogn.* 12:609–19

Fernandez C, Goldberg JM. 1976. Physiology of peripheral neurons innervating otolith organs of the squirrel monkey. I. Response to static tilts and to long-duration centrifugal force. *J. Neurophysiol.* 39:970–84

Fetsch CR, Wang S, Gu Y, DeAngelis GC, Angelaki DE. 2007. Spatial reference frames of visual, vestibular, and multimodal heading signals in the dorsal subdivision of the medial superior temporal area. *J. Neurosci.* 27:700–12

Fitzpatrick RC, Butler JE, Day BL. 2006. Resolving head rotation for human bipedalism. *Curr. Biol.* 16:1509–14

Fuchs AF, Kimm J. 1975. Unit activity in vestibular nucleus of the alert monkey during horizontal angular acceleration and eye movement. *J. Neurophysiol.* 38:1140–61

Fukushima K, Akao T, Saito H, Kurkin S, Fukushima J, Peterson BW. 2007. Neck proprioceptive signals in pursuit neurons in the frontal eye fields (FEF) of monkeys. *Soc. Neurosci. Abstr.* 33:398.5

Gdowski GT, McCrea RA. 1999. Integration of vestibular and head movement signals in the vestibular nuclei during whole-body rotation. *J. Neurophysiol.* 82:436–49

Glasauer S. 1995. Linear acceleration perception: frequency dependence of the hilltop illusion. *Acta Otolaryngol. Suppl.* 520(Pt. 1):37–40

Glasauer S, Merfeld DM. 1997. Modeling three-dimensional responses during complex motion stimulation. In *Three-Dimensional Kinematics of Eye, Head and Limb Movements*, ed. M Fetter, T Haslwanter, H Misslisch, D Tweed, pp. 387–98. Amsterdam: Harwood Acad.

Goldberg JM. 2000. Afferent diversity and the organization of central vestibular pathways. *Exp. Brain Res.* 130:277–97

Goldberg JM, Fernandez C. 1975. Vestibular mechanisms. *Annu. Rev. Physiol.* 37:129–62

Graybiel A. 1952. Oculogravic illusion. *AMA Arch. Ophthalmol.* 48:605–15

Green AM, Angelaki DE. 2003. Resolution of sensory ambiguities for gaze stabilization requires a second neural integrator. *J. Neurosci.* 23:9265–75

Green AM, Angelaki DE. 2004. An integrative neural network for detecting inertial motion and head orientation. *J. Neurophysiol.* 92:905–25

Green AM, Shaikh AG, Angelaki DE. 2005. Sensory vestibular contributions to constructing internal models of self-motion. *J. Neural. Eng.* 2:S164–79

Gu Y, Angelaki DE, DeAngelis GC. 2006a. Sensory integration for heading perception in area MSTd: I. Neuronal and psychophysical sensitivity to visual and vestibular heading cues. *Soc. Neurosci. Abstr.* 306.8

Gu Y, DeAngelis GC, Angelaki DE. 2007. A functional link between area MSTd and heading perception based on vestibular signals. *Nat. Neurosci.* 10(8):1038–47

Gu Y, Watkins PV, Angelaki DE, DeAngelis GC. 2006b. Visual and nonvisual contributions to three-dimensional heading selectivity in the medial superior temporal area. *J. Neurosci.* 26:73–85

Hjelmstad G, Parks G, Bodznick D. 1996. Motor corollary discharge activity and sensory responses related to ventilation in the skate vestibulolateral cerebellum: implications for electrosensory processing. *J. Exp. Biol.* 199:673–81

Howard IP, Hu G. 2001. Visually induced reorientation illusions. *Perception* 30:583–600

Jay MF, Sparks DL. 1987. Sensorimotor integration in the primate superior colliculus. II. Coordinates of auditory signals. *J. Neurophysiol.* 57:35–55

Kaptein RG, Van Gisbergen JA. 2006. Canal and otolith contributions to visual orientation constancy during sinusoidal roll rotation. *J. Neurophysiol.* 95:1936–48

Keller EL, Daniels PD. 1975. Oculomotor related interaction of vestibular and visual stimulation in vestibular nucleus cells in alert monkey. *Exp. Neurol.* 46:187–98

Klam F, Graf W. 2006. Discrimination between active and passive head movements by macaque ventral and medial intraparietal cortex neurons. *J. Physiol.* 574:367–86

Kleine JF, Guan Y, Kipiani E, Glonti L, Hoshi M, Buttner U. 2004. Trunk position influences vestibular responses of fastigial nucleus neurons in the alert monkey. *J. Neurophysiol.* 91:2090–100

Krasne FB, Bryan JS. 1973. Habituation: regulation through presynaptic inhibition. *Science* 182:590–92

Lisberger SG, Miles FA. 1980. Role of primate medial vestibular nucleus in long-term adaptive plasticity of vestibuloocular reflex. *J. Neurophysiol.* 43:1725–45

McCrea RA, Gdowski GT, Boyle R, Belton T. 1999. Firing behavior of vestibular neurons during active and passive head movements: vestibulo-spinal and other noneye-movement related neurons. *J. Neurophysiol.* 82:416–28

McNeilage PR, Banks MS, Berger DR, Bulthoff HH. 2007. A Bayesian model of the disambiguation of gravitoinertial force by visual cues. *Exp. Brain Res.* 179(2):263–90

Merfeld DM. 1995. Modeling the vestibulo-ocular reflex of the squirrel monkey during eccentric rotation and roll tilt. *Exp. Brain Res.* 106:123–34

Merfeld DM, Park S, Gianna-Poulin C, Black FO, Wood S. 2005a. Vestibular perception and action employ qualitatively different mechanisms. I. Frequency response of VOR and perceptual responses during Translation and Tilt. *J. Neurophysiol.* 94:186–98

Merfeld DM, Park S, Gianna-Poulin C, Black FO, Wood S. 2005b. Vestibular perception and action employ qualitatively different mechanisms. II. VOR and perceptual responses during combined Tilt&Translation. *J. Neurophysiol.* 94:199–205

Merfeld DM, Young LR, Paige GD, Tomko DL. 1993. Three dimensional eye movements of squirrel monkeys following postrotatory tilt. *J. Vestib. Res.* 3:123–39

Merfeld DM, Zupan L, Peterka RJ. 1999. Humans use internal models to estimate gravity and linear acceleration. *Nature* 398:615–18

Merfeld DM, Zupan LH. 2002. Neural processing of gravitoinertial cues in humans. III. Modeling tilt and translation responses. *J. Neurophysiol.* 87:819–33

Merfeld DM, Zupan LH, Gifford CA. 2001. Neural processing of gravito-inertial cues in humans. II. Influence of the semicircular canals during eccentric rotation. *J. Neurophysiol.* 85:1648–60

Mergner T, Glasauer S. 1999. A simple model of vestibular canal-otolith signal fusion. *Ann. N. Y. Acad. Sci.* 871:430–34

Mohr C, Roberts PD, Bell CC. 2003. The mormyromast region of the mormyrid electrosensory lobe. I. Responses to corollary discharge and electrosensory stimuli. *J. Neurophysiol.* 90:1193–210

Muir GM, Carey JP, Hirvonen TP, Minor LB, Taube JS. 2004. Head direction cell activity is unstable following plugging of the semicircular canals in the freely-moving chinchilla. *Soc. Neurosci. Abstr.* 868.11

Mullette-Gillman OA, Cohen YE, Groh JM. 2005. Eye-centered, head-centered, and complex coding of visual and auditory targets in the intraparietal sulcus. *J. Neurophysiol.* 94:2331–52

Page WK, Duffy CJ. 2003. Heading representation in MST: sensory interactions and population encoding. *J. Neurophysiol.* 89:1994–2013

Pouget A, Deneve S, Duhamel JR. 2002. A computational perspective on the neural basis of multisensory spatial representations. *Nat. Rev. Neurosci.* 3:741–47

Pouget A, Snyder LH. 2000. Computational approaches to sensorimotor transformations. *Nat. Neurosci.* 3(Suppl.):1192–98

Poulet JF, Hedwig B. 2003. Corollary discharge inhibition of ascending auditory neurons in the stridulating cricket. *J. Neurosci.* 23:4717–25

Poulet JF, Hedwig B. 2006. The cellular basis of a corollary discharge. *Science* 311:518–22

Raphan T, Cohen B. 2002. The vestibulo-ocular reflex in three dimensions. *Exp. Brain Res.* 145:1–27

Raphan T, Matsuo V, Cohen B. 1979. Velocity storage in the vestibulo-ocular reflex arc (VOR). *Exp. Brain Res.* 35:229–48

Romer A, Parsons TS. 1977. *The Vertebrate Body*. Philadelphia: Saunders

Roy JE, Cullen KE. 1998. A neural correlate for vestibulo-ocular reflex suppression during voluntary eye-head gaze shifts. *Nat. Neurosci.* 1:404–10

Roy JE, Cullen KE. 2001. Selective processing of vestibular reafference during self-generated head motion. *J. Neurosci.* 21:2131–42

Roy JE, Cullen KE. 2002. Vestibuloocular reflex signal modulation during voluntary and passive head movements. *J. Neurophysiol.* 87:2337–57

Roy JE, Cullen KE. 2003. Brain stem pursuit pathways: dissociating visual, vestibular, and proprioceptive inputs during combined eye-head gaze tracking. *J. Neurophysiol.* 90:271–90

Roy JE, Cullen KE. 2004. Dissociating self-generated from passively applied head motion: neural mechanisms in the vestibular nuclei. *J. Neurosci.* 24:2102–11

Sadeghi SG, Minor LB, Cullen KE. 2007. Response of vestibular-nerve afferents to active and passive rotations under normal conditions and after unilateral labyrinthectomy. *J. Neurophysiol.* 97(2):1503–14

Sato H, Ohkawa T, Uchino Y, Wilson VJ. 1997. Excitatory connections between neurons of the central cervical nucleus and vestibular neurons in the cat. *Exp. Brain Res.* 115:381–86

Sawtell NB, Williams A, Bell CC. 2007. Central control of dendritic spikes shapes the responses of Purkinje-like cells through spike timing-dependent synaptic plasticity. *J. Neurosci.* 27:1552–65

Schlack A, Sterbing-D'Angelo SJ, Hartung K, Hoffmann KP, Bremmer F. 2005. Multisensory space representations in the macaque ventral intraparietal area. *J. Neurosci.* 25:4616–25

Scudder CA, Fuchs AF. 1992. Physiological and behavioral identification of vestibular nucleus neurons mediating the horizontal vestibuloocular reflex in trained rhesus monkeys. *J. Neurophysiol.* 68:244–64

Seidman SH, Telford L, Paige GD. 1998. Tilt perception during dynamic linear acceleration. *Exp. Brain Res.* 119:307–14

Shaikh AG, Green AM, Ghasia FF, Newlands SD, Dickman JD, Angelaki DE. 2005. Sensory convergence solves a motion ambiguity problem. *Curr. Biol.* 15:1657–62

Shaikh AG, Meng H, Angelaki DE. 2004. Multiple reference frames for motion in the primate cerebellum. *J. Neurosci.* 24:4491–97

Sherrington CS. 1906. *Integrative Action of the Nervous System.* New York: Scribner

Stackman RW, Golob EJ, Bassett JP, Taube JS. 2003. Passive transport disrupts directional path integration by rat head direction cells. *J. Neurophysiol.* 90:2862–74

Stackman RW, Taube JS. 1997. Firing properties of head direction cells in the rat anterior thalamic nucleus: dependence on vestibular input. *J. Neurosci.* 17:4349–58

Stackman RW, Tullman ML, Taube JS. 2000. Maintenance of rat head direction cell firing during locomotion in the vertical plane. *J. Neurophysiol.* 83:393–405

Stein BE, Meredith MA. 1993. *The Merging of the Senses.* Cambridge, MA: Bradford

Stricanne B, Andersen RA, Mazzoni P. 1996. Eye-centered, head-centered, and intermediate coding of remembered sound locations in area LIP. *J. Neurophysiol.* 76:2071–76

Takahashi K, Gu Y, May PJ, Newlands SD, DeAngelis GC, Angelaki DE. 2007. Multi-modal coding of three-dimensional rotation in area MSTd: comparison of visual and vestibular selectivity. *J. Neurosci.* 27(36):9742–56

Taube JS. 2007. The head direction signal: origins and sensory-motor integration. *Annu. Rev. Neurosci.* 30:181–207

Tomlinson RD, Robinson DA. 1984. Signals in vestibular nucleus mediating vertical eye movements in the monkey. *J. Neurophysiol.* 51:1121–36

von Helmholtz H. 1925. *Handbuch der Physiologischen Optik [Treatise on Physiological Optics]*, ed. JPC Southall, pp. 44–51. Rochester: Opt. Soc. Am.

von Holst E, Mittelstaedt H. 1950. Das reafferenzprinzip. *Naturwissenschaften* 37:464–76

Voogd J. 1989. Parasagittal zones and compartments of the anterior vermis of the cat cerebellum. *Exp. Brain Res.* 17:3–19

Voogd J, Gerrits NM, Ruigrok TJ. 1996. Organization of the vestibulocerebellum. *Ann. N. Y. Acad. Sci.* 781:553–79

Wearne S, Raphan T, Cohen B. 1998. Control of spatial orientation of the angular vestibuloocular reflex by the nodulus and uvula. *J. Neurophysiol.* 79:2690–715

Wilson VJ, Schor RH. 1999. The neural substrate of the vestibulocollic reflex. What needs to be learned. *Exp. Brain Res.* 129:483–93

Wilson VJ, Yamagata Y, Yates BJ, Schor RH, Nonaka S. 1990. Response of vestibular neurons to head rotations in vertical planes. III. Response of vestibulocollic neurons to vestibular and neck stimulation. *J. Neurophysiol.* 64:1695–703

Wolpert DM, Ghahramani Z, Jordan MI. 1995. An internal model for sensorimotor integration. *Science* 269:1880–82

Yakusheva TA, Shaikh AG, Green AM, Blazquez PM, Dickman JD, Angelaki DE. 2007. Purkinje cells in posterior cerebellar vermis encode motion in an inertial reference frame. *Neuron* 54:973–85

Zugaro MB, Berthoz A, Wiener SI. 2002. Peak firing rates of rat anterodorsal thalamic head direction cells are higher during faster passive rotations. *Hippocampus* 12:481–86

Zupan LH, Merfeld DM. 2003. Neural processing of gravito-inertial cues in humans. IV. Influence of visual rotational cues during roll optokinetic stimuli. *J. Neurophysiol.* 89:390–400

Zupan LH, Merfeld DM, Darlot C. 2002. Using sensory weighting to model the influence of canal, otolith and visual cues on spatial orientation and eye movements. *Biol. Cybern.* 86:209–30

Zupan LH, Peterka RJ, Merfeld DM. 2000. Neural processing of gravito-inertial cues in humans. I. Influence of the semicircular canals following postrotatory tilt. *J. Neurophysiol.* 84:2001–15

Role of Axonal Transport in Neurodegenerative Diseases*

Kurt J. De Vos,[1] Andrew J. Grierson,[2] Steven Ackerley,[1] and Christopher C.J. Miller[1]

[1]MRC Center for Neurodegeneration Research, Institute of Psychiatry, King's College, London SE5 8AF, United Kingdom; email: Kurt.DeVos@iop.kcl.ac.uk, chris.miller@iop.kcl.ac.uk

[2]Academic Unit of Neurology, School of Medicine and Biomedical Sciences, University of Sheffield, Sheffield S10 2RX, United Kingdom; email: A.J.Grierson@sheffield.ac.uk

Annu. Rev. Neurosci. 2008. 31:151–173

First published online as a Review in Advance on April 2, 2008

The *Annual Review of Neuroscience* is online at neuro.annualreviews.org

This article's doi: 10.1146/annurev.neuro.31.061307.090711

0147-006X/08/0721-0151$20.00

*This review is dedicated to Steven Ackerley who tragically died on 14 January 2008.

Key Words

amyotrophic lateral sclerosis, motor neuron disease, Charcot-Marie-Tooth disease, Alzheimer's disease, Huntington's disease, Parkinson's disease

Abstract

Many major human neurodegenerative diseases, including Alzheimer's disease, Parkinson's disease, and amyotrophic lateral sclerosis (ALS), display axonal pathologies including abnormal accumulations of proteins and organelles. Such pathologies highlight damage to the axon as part of the pathogenic process and, in particular, damage to transport of cargoes through axons. Indeed, we now know that disruption of axonal transport is an early and perhaps causative event in many of these diseases. Here, we review the role of axonal transport in neurodegenerative disease.

Contents

AXONAL TRANSPORT

Intracellular transport of protein and organelle cargoes is an essential requirement for all mammalian cells, but this is the case especially for neurons. Neurons are polarized with axons and dendrites, and because most neuronal proteins are synthesized in cell bodies, mechanisms are required to direct axonal vs. dendritic transport. In addition, the distances over which cargoes have to be moved are longer than in other cell types (a human motor neuron axon can exceed 1 m). Finally, even within an individual axon, cargoes must be targeted to specific compartments, e.g., sodium channels are enriched at nodes of Ranvier, whereas synaptic proteins are targeted to the axon terminal. Thus the architecture of neurons makes them particularly dependent on intracellular transport processes.

Axonal transport: the movement of protein and organelle cargoes through axons

The main mechanism to deliver cellular components to their action site is long-range microtubule-based transport. The two major components of the transport machinery are the "engines," or molecular motors (**Figure 1**; see sidebar on Molecular Motors), and microtubules, the "rails" on which they run. Microtubules are polarized; the faster-growing end is referred to as the plus end and the slower growing end as the minus end. In axons, microtubule orientation is nearly uniform, with the plus ends pointing toward the synapse and the minus ends facing the cell body. As most molecular motors of the kinesin family unidirectionally move toward the microtubule plus end, they mostly mediate transport toward the synapse (anterograde). In contrast, the molecular motor cytoplasmic dynein moves toward the microtubule minus end and, accordingly, mediates transport of most cargoes toward the cell body (retrograde).

Classically, axonal transport is divided into fast and slow axonal transport on the basis of the bulk speeds of cargo movement; cargoes such as vesicles and mitochondria move by fast axonal transport at speeds of $\sim$1 μm/s, whereas cytoskeleton components move in slow axonal transport at speeds of $\sim$1 mm/day. However, it is now clear that both fast and slow axonal transport are mediated by the same "fast" molecular motors kinesin and cytoplasmic dynein and that the slower overall rate of slow axonal transport is due to prolonged pauses between movements (Roy et al. 2000, Wang et al. 2000).

AXONAL TRANSPORT AND NEURODEGENERATIVE DISEASE

Axonal and cell body accumulations of organelles and other proteins are hallmark pathologies for many human neurodegenerative diseases. Tau is present in the paired helical filaments (PHFs) of Alzheimer's and related diseases, α-synuclein is the principal component of Lewy bodies in Parkinson's disease, neurofilament accumulations are seen in amyotrophic lateral sclerosis (ALS), and more recently, TDP-43 accumulations have been observed in ALS and frontotemporal lobar degeneration (Ballatore et al. 2007, Neumann et al. 2006, Spillantini et al. 1997, Xiao et al. 2006). Furthermore, axonal swellings and spheroids have been described in a number of neurodegenerative diseases (Coleman 2005). Together, such pathologies suggest that defective functioning of the axon contributes to disease and, in particular, that damage to axonal transport may underlie the pathogenic accumulation of organelles. Indeed, we now have evidence that this is the case for many neurodegenerative diseases.

The mechanisms by which axonal transport is disrupted in disease are varied. Like a train journey, disruption to transport can occur via damage to the engines (kinesin and cytoplasmic dynein) (**Figure 1**), damage to the rails (microtubules) (**Figure 2**), damage to the cargoes (for example, to inhibit their attachment to motors) (**Figure 3**), and damage to the ATP fuel supply for the engines (mitochondria) (**Figure 4**). In fact, all these insults can contribute to neurodegeneration. Below we describe evidence linking defective axonal transport and disease.

MOTOR NEURON DISORDERS

Axonal transport defects are perhaps best characterized for motor neuron disorders. These disorders include ALS, distal hereditary motor neuropathy, spinal muscular atrophy (SMA), and hereditary spastic paraplegia (HSP).

Amyotrophic Lateral Sclerosis

Although most forms of ALS are sporadic, ~10% of cases are familial, and evidence now implicates mutations in 5 genes as causative for the disorder (Pasinelli & Brown 2006). Some of these disease mutants damage axonal transport.

Mutant SOD1. Mutations in the gene encoding Cu/Zn superoxide dismutase-1 (SOD1) cause ~20% of familial ALS cases. Expression of mutant SOD1 in transgenic mice induces motor neuron disease (MND), and anal-

MOLECULAR MOTORS

Molecular motors generate force from ATP hydrolysis to move cargoes along cytoskeleton tracks. The main microtubule-based motors are members of the kinesin superfamily (45 in humans) and cytoplasmic dynein (**Figure 1**).

Kinesin-1 (also known as KIF5) is the most studied and is a heterotetramer of two kinesin heavy chains (KHC) and two kinesin light chains (KLC). KHC is composed of the catalytic motor domain, a short neck linker region, the α-helical coiled-coil stalk that is interrupted by two hinge regions, and the tail. The motor domain binds microtubules, contains the ATPase activity, and, together with the neck, confers processivity and directionality; the stalk is involved in dimerization; the tail, together with KLC, binds cargo and regulates motor activity (Hirokawa & Takemura 2005).

Cytoplasmic dynein is a multisubunit complex that contains two heavy chains that are associated with intermediate chains, light intermediate chains, and light chains. The heavy chains harbor ATPase activity and bind microtubules, whereas the other chains are involved in cargo binding and binding to dynactin. Dynactin is a protein complex that contains p150[Glued], p62, dynamitin, actin-related protein 1, CapZα and CapZβ, p27, and p24. Dynactin is a processivity factor for dynein and is implicated in cargo binding (Pfister et al. 2006).

yses of such mice reveal that damage to axonal transport is an early pathogenic event (De Vos et al. 2007, Kieran et al. 2005, Williamson & Cleveland 1999, Zhang et al. 1997). The early nature of this damage argues that damage to axonal transport contributes to the disease process in a primary fashion and is not just an end-stage epiphenomenon. Mutant SOD1 damages both fast and slow axonal transport (De Vos et al. 2007, Williamson & Cleveland 1999, Zhang et al. 1997). However, mutant SOD1 differentially affects axonal transport of specific cargoes. Anterograde movement of the cytoskeleton including neurofilaments is slowed, fast transport of vesicles is inhibited in both anterograde and retrograde directions but inhibition of mitochondrial movement is anterograde specific (De Vos et al. 2007, Williamson & Cleveland 1999, Zhang et al. 1997).

SOD1: Cu/Zn superoxide dismutase-1

MND: motor neuron disease

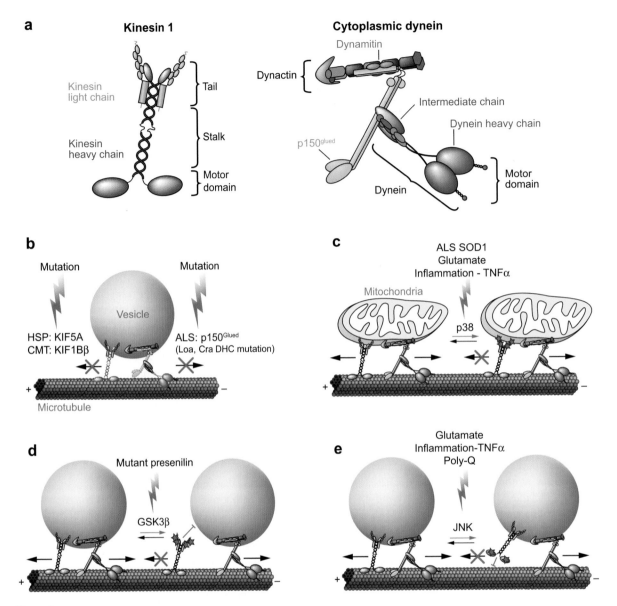

Figure 1

Mechanisms of axonal transport defects: damage to molecular motors. (*a*) Kinesin and cytoplasmic dynein are the main microtubule-based motors. (*b*) Mutations in kinesins or cytoplasmic dynein that inhibit their activity (or are predicted to do so) cause some familial forms of ALS (p150^Glued), hereditary spastic paralegia (HSP) (KIF5A), and Charcot-Marie-Tooth disease (CMT) (KIF1Bβ). Furthermore, Loa and Cra mice that carry mutations in dynein heavy chain (DHC) exhibit axonal transport defects and develop MND. Phosphorylation (*stars*) of kinesin 1 inhibits its activity at multiple levels. (*c*) Phosphorylation of KLC by a p38-dependent pathway inhibits mitochondria-bound kinesin 1 activity without affecting its binding to microtubules or mitochondria. (*d*) By contrast, mutant presenilin-induced phosphorylation of KLC by GSK3β inhibits kinesin 1–mediated transport of vesicles by disrupting attachment of kinesin 1 to vesicles. (*e*) Finally, phosphorylation of KHC by JNK inhibits kinesin 1–mediated vesicle transport by impeding the interactions of KHC with microtubules; JNK is activated by disease-associated signals including expanded polyglutamines (Poly-Q). Cytoplasmic dynein activity may also be regulated by phosphorylation. However, we do not know if this plays a role in neurodegeneration.

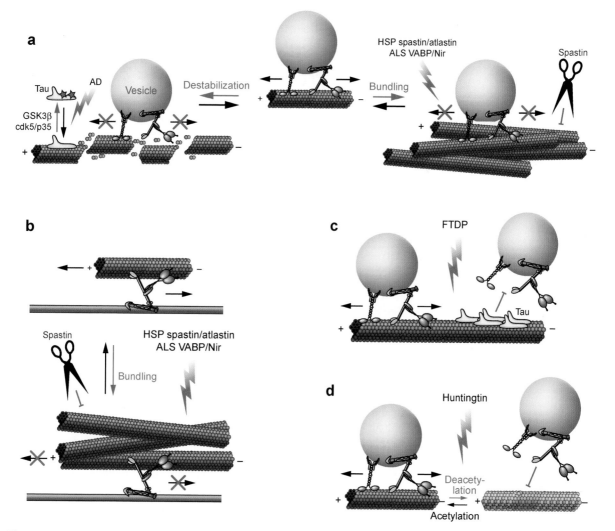

Figure 2

Mechanisms of axonal transport defects: damage to microtubules. Microtubules are highly dynamic structures that undergo rapid periods of growth and shrinkage, and their dynamic behavior is regulated by several mechanisms. Deregulation of these dynamic properties may lead to disruption of cargo transport. (*a*) Destabilization of microtubules by decreased binding of GSK3β and/or cdk5/p35-induced hyperphosphorylated tau may lead to a loss of microtubule "rails" for transport (*left*). Loss-of-function mutant spastin and mutant VAPB induce abnormal bundling of microtubules to misdirect transport (*right*).
(*b*) Because short microtubules are preferentially transported, mutant spastin/atlastin and VAPB may damage axonal transport of microtubules themselves. This could happen via inhibition of microtubule severing by mutant spastin or microtubule bundling by VAPB/Nir. (*c*) FTDP tau may damage transport by interfering with the interaction of kinesin 1 with microtubules. (*d*) Finally, mutant huntingtin may induce deacetylation of α-tubulin and subsequent release of motors from microtubules.

How mutant SOD1 perturbs axonal transport is not fully understood, but it is likely that it involves several different pathways. These mutant SOD1-induced pathways may include damage to mitochondria and reduced ATP supply to molecular motors, pathogenic signaling that alters phosphorylation of molecular motors, and altered phosphorylation of cargoes such as neurofilaments to disrupt their association with motors.

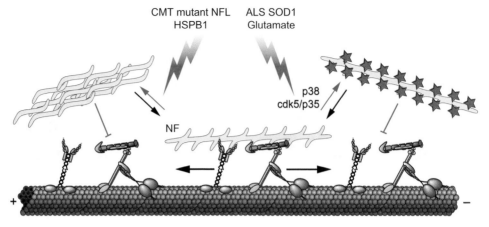

CMT mutant NFL
HSPB1

ALS SOD1
Glutamate

p38
cdk5/p35

NF

NF

+   −

Figure 3

Mechanisms of axonal transport defects: damage to cargoes. Regulation/disruption of motor-cargo interaction may disrupt axonal transport in a number of neurodegenerative diseases. Phosphorylation (*stars*) of neurofilaments (NF) by p38 and/or cdk5/p35 kinases slows their transport, possibly by preventing attachment of molecular motors (*right*). Alternatively, misassembly of NF by mutations in CMT mutant NFL or the NF chaperone HSP1B may disrupt their interaction with molecular motors and cause their defective axonal transport (*left*).

Damage to mitochondria. Many studies have demonstrated that mutant SOD1 selectively associates with and damages mitochondria (see Boillee et al. 2006). This damage is believed to severely impair the mitochondrial electron transfer chain and ATP synthesis (Mattiazzi et al. 2002). Damage to mitochondria has been linked to a reduction in their anterograde transport (Miller & Sheetz 2004), and recently, De Vos et al. (2007) presented formal evidence that mutant SOD1 perturbs mitochondrial anterograde movement. Inhibition of anterograde mitochondrial transport leads to a net increase in their retrograde transport, which results in depletion of mitochondria from axons. This could adversely affect axonal transport of other cargoes because of diminished ATP levels (**Figure 4**) (De Vos et al. 2007).

Inflammatory signals. Although studies of isolated motor neurons in culture have shown that mutant SOD1-induced damage to axonal transport can be neuron specific (De Vos et al. 2007, Kieran et al. 2005), elegant experiments involving chimeric mutant SOD1 mice and mice with a deletable mutant *SOD1* gene have demonstrated that other cell types contribute

to disease. Thus, although expression of mutant SOD1 in motor neurons is a primary determinant of disease onset, expression in microglia markedly influences disease progression (see Boillee et al. 2006). Inflammatory responses have been strongly linked to ALS, and mutant SOD1 itself can be secreted and activate microglia (Beers et al. 2006, Nguyen et al. 2004, Urushitani et al. 2006). Also, the anti-inflammatory agent minocycline is protective in mutant SOD1 transgenic mice (Kriz et al. 2002, Van Den Bosch et al. 2002, Zhu et al. 2002). Inflammatory signals from neighboring cells may therefore provide additional insults to axonal transport. Indeed, tumor necrosis factor-α (TNFα) signaling inhibits kinesin function via p38 stress-activated kinase-dependent phosphorylation of kinesin (**Figure 1**) (De Vos et al. 2000), and p38 is activated in mutant SOD1 transgenic mice (Ackerley et al. 2004, Raoul et al. 2002, Tortarolo et al. 2003).

Excitotoxic signaling. Excitotoxic insults likely contribute to ALS because alterations to proteins and metabolites involved in glutamate handling are seen in sporadic ALS cases (Van Den Bosch et al. 2006). Moreover, mutant

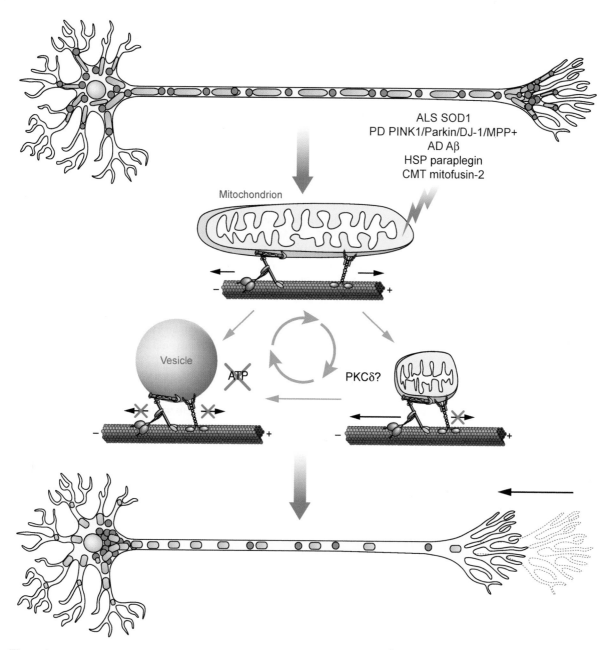

Figure 4

Mechanisms of axonal transport defects: damage to mitochondria. Damage to mitochondria is seen in many neurodegenerative diseases. Mitochondrial dysfunction likely affects axonal transport in at least two ways. First, inhibition of mitochondrial function reduces anterograde transport of both mitochondria and vesicles possibly by activating PKCδ. Second, the resulting diminution of mitochondria in axons will likely decrease ATP supply to molecular motors leading to decreased anterograde and retrograde movement of other axoplasmic cargoes. As mitochondrial dysfunction also reduces ATP production, the latter might be part of a vicious circle mechanism that ultimately leads to dying-back of axons.

JNKs: c-Jun-N-terminal kinases

SOD1 selectively damages the glial glutamate transporter EAAT2 (which removes synaptic glutamate), and EAAT2 levels are reduced in mutant SOD1 transgenic mice (Bruijn et al. 1997, Trotti et al. 1999). Excitotoxicity may therefore contribute to the pathogenic process in both sporadic and mutant SOD1 familial forms of ALS.

Two studies have now shown that excitoxic application of glutamate damages axonal transport (Ackerley et al. 2000, Hiruma et al. 2003). The mechanisms involved could damage both motors and cargoes. In particular, glutamate activates JNKs, p38, and cdk5/p35 (Ackerley et al. 2000, Brownlees et al. 2000, Kawasaki et al. 1997, Lee et al. 2000, Schwarzschild et al. 1997), and all are strongly linked to axonal transport. Thus JNKs and p38 phosphorylate kinesin heavy and light chains, respectively, to inhibit transport (**Figure 1**) (De Vos et al. 2000, Morfini et al. 2006), and JNKs, p38, and cdk5/p35 all phosphorylate neurofilament medium chain (NFM)/neurofilament heavy chain (NFH) sidearms, a process that is linked to slowing of neurofilament transport (**Figure 3**) (see below).

Damage to cargoes. Finally, mutant SOD1 may damage cargoes to inhibit their transport (possibly by promoting their release from motors) (**Figure 3**). One cargo that is intimately linked to ALS is neurofilaments. Neurofilament accumulations are a hallmark pathology of ALS, and overexpression of neurofilament light (NFL) and heavy (NFH) chain proteins, peripherin (a further neuronal intermediate filament protein), or a mutant NFL that disrupts neurofilament assembly induces MND in transgenic mice (Beaulieu et al. 1999, Cote et al. 1993, Lee et al. 1994, Xu et al. 1993). Moreover, axonal transport is defective in these neurofilament transgenics (Collard et al. 1995, Millecamps et al. 2006). Such overexpression of individual NF subunits likely alters neurofilament assembly properties to disrupt transport because proper assembly is required for transport (Millecamps et al. 2006, Yuan et al. 2003). Direct evidence to link mutant SOD1

toxicity with neurofilaments comes from experiments in which mutant SOD1 transgenic mice have been crossed with neurofilament transgenics. Modulating neurofilament expression in this way markedly alters disease onset (Couillard-Després et al. 1998, Williamson et al. 1998). Notably, deletion of NFM and NFH sidearm domains is strongly protective against mutant SOD1 disease (Lobsiger et al. 2005). The pathological neurofilament accumulations seen in ALS are aberrantly hyperphosphorylated on these NFM/NFH sidearm domains, and such phosphorylation slows neurofilament transport (Ackerley et al. 2003, Shea et al. 2004). This probably occurs by promoting neurofilament release from molecular motors (Jung et al. 2005, Wagner et al. 2004). p38 and cdk5/p35 phosphorylate NFM/NFH sidearms (Ackerley et al. 2004, Guidato et al. 1996, Sun et al. 1996), and both are activated in mutant SOD1 transgenic mice (Ackerley et al. 2004, Nguyen et al. 2001, Raoul et al. 2002, Tortarolo et al. 2003).

Mutant dynactin and disruption to cytoplasmic dynein function. The finding that mutations in the dynactin subunit p150^Glued cause disease demonstrates a direct role of molecular motor dysfunction in ALS (**Figure 1**) (Munch et al. 2005, Puls et al. 2003). One mutation lowers the affinity of p150^Glued for both microtubules and EB1 and may cause a loss of dynactin/dynein function and a gain of toxic function (Levy et al. 2006). The effects of additional mutations on cytoplasmic dynein function and axonal transport have not so far been described. Other evidence for a role of defective cytoplasmic dynein function in ALS involves the Loa and Cra mouse mutants and transgenic mice that express dynamitin; these animals develop MND (Hafezparast et al. 2003, LaMonte et al. 2002). The Loa and Cra mutations are in dynein heavy chain and are associated with defective retrograde transport (Hafezparast et al. 2003, Kieran et al. 2005). Expression of dynamitin perturbs dynein-dynactin interaction and, as such, dynein function (LaMonte et al. 2002).

Loa mice have been crossed with mutant SOD1 transgenics, and the phenotype is unexpected. Rather than accelerating disease progression, the presence of both mutant proteins is protective. Moreover, defective retrograde transport is partially corrected in the double mutant (Kieran et al. 2005). Although the mechanisms that underlie this effect are not clear, one possibility is that the Loa mutation counterbalances the increase in net retrograde transport of mitochondria induced by mutant SOD1 (De Vos et al. 2007). Thus, although mutant SOD1 may damage mitochondria, the mitochondria remain within axons to provide at least some ATP fuel for axonal function and transport.

ALS2/Alsin and VAPB. ALS2 and VAPB are mutated in some rare forms of ALS, and both are linked to vesicle trafficking within neurons. ALS2 is a guanine nucleotide exchange factor for both Rab5 and Rac (Jacquier et al. 2006, Otomo et al. 2003, Topp et al. 2004, Tudor et al. 2005); Rab5 functions in retrograde endosome transport (Deinhardt et al. 2006). Although ALS2 knockout mice (predicted to model ALS2 forms of ALS) have a mild phenotype, they display defects in endosomal trafficking consistent with a role in axonal transport (Devon et al. 2006, Hadano et al. 2005).

VAPB too is thought to have a role in membrane transport, particularly from endoplasmic reticulum (ER)/Golgi to the synapse (Skehel et al. 2000). VAPB also associates with microtubules, and evidence demonstrates that it has a role in microtubule organization via its interacting partner Nir (**Figure 2**) (Amarilio et al. 2005, Pennetta et al. 2002). In this context, it is noteworthy that a mutation in the tubulin chaperone Tbce, which alters microtubule organization, causes MND in the pmn mouse (Bommel et al. 2002, Martin et al. 2002).

Distal Hereditary Motor Neuropathy

Mutations in two small heat-shock proteins (HSPB1 and HSPB8) cause a form of MND that affects primarily lower motor neurons

(Evgrafov et al. 2004, Irobi et al. 2004). The small heat-shock proteins perform many functions including chaperoning activity but are particularly important in responding to cellular stresses. HSPB1 and HSPB8 interact with several intermediate filament proteins to facilitate formation of properly organized neurofilament networks, and mutant HSPB1 disrupts neurofilament assembly (Evgrafov et al. 2004). Direct analyses of the effect of mutant HSPB1 on axonal transport has shown that it disrupts movement of a number of cargoes including neurofilaments (**Figure 3**) (Ackerley et al. 2006).

Spinal Muscular Atrophy

SMA is the consequence of mutations within the survival of motor neuron 1 (*SMN1*) gene (Monani 2005). *SMN1* is ubiquitously expressed and has at least two functions. One is in the assembly and regeneration of spliceosomal small nuclear ribonucleoproteins (snRNPs), a task it presumably carries out in most cell types, but the other is linked to transport of specific mRNAs through axons (Rossoll et al. 2003).

Although most axonal proteins are synthesized within cell bodies and then transported through axons, a proportion are synthesized locally within the axon/axon terminal following mRNA transport (Kiebler & Bassell 2006). β-actin is one such mRNA and functions within the growth cone to promote axonal outgrowth. Loss of SMN leads to reduced axonal outgrowth that correlates with reductions in both β-actin mRNA and protein in the growth cone (Rossoll et al. 2003). Also, the SMN binding partner hnRNP-R binds to the 3′-UTR of β-actin mRNA, and so SMN may be complexed with β-actin mRNA (Rossoll et al. 2003). Thus, SMN may function in axonal transport of actin and other mRNAs, and its loss in SMA might induce disease as a consequence of disrupted mRNA transport. Defective axonal transport of RNA complexes has also been linked to another neurological disease, Fragile X syndrome. The Fragile X mental retardation protein is also present in RNA granules and has been linked to transport of several RNAs including that for β-actin (see Kiebler & Bassell 2006).

Hereditary Spastic Paraplegia

HSP is a group of heterogeneous hereditary neurological disorders that affect upper motor neurons. Thirty genetic loci have been linked to HSP and 15 responsible mutant genes so far identified (Fink 2006). Some of these mutants have been shown to, or are highly likely to, disrupt axonal transport, although the mechanisms involve a variety of insults.

Damage to kinesin. Mutations in kinesin 1 are the cause of a few rare forms of HSP (**Figure 1**). Four causative mutations have been described and, though not functionally tested, all are likely to influence motor function because they involve conserved residues in either the motor head domains or the coiled-coil stem region (Reid 2003).

Damage to microtubules. Mutations in the gene encoding spastin are responsible for 40% of autosomal dominant HSP cases. A range of mutations throughout the protein causes disease, and whereas some may represent loss-of-function mutants, others may act in a dominant-negative fashion. Spastin is part of a small family of AAA proteins that function to sever and bundle microtubules (Roll-Mecak & Vale 2005, Salinas et al. 2005). By severing microtubules, spastin likely plays a role in axonal transport; recent experimental studies support this hypothesis and the notion that HSP mutant spastin is defective in the process (McDermott et al. 2003, Tarrade et al. 2006). How mutant spastin might perturb axonal transport is not fully clear. One possibility is that the mutants alter the microtubule rails to disrupt movement of cargo-carrying motors (**Figure 2**). However, another possibility is that mutant spastin disrupts transport of microtubules themselves. The length of microtubules is related to their potential to be transported through axons; long microtubules are essentially stationary, whereas shorter ones move bidirectionally (Ahmad et al. 2006, Baas

et al. 2005, Wang & Brown 2002). Defective microtubule severing by mutant spastin may therefore block microtubule transport (**Figure 2**). Ultimately, this may damage axonal transport of other cargoes. Spastin also binds to membrane proteins of the endoplasmic reticulum and endosomes, so one possibility is that it functions in regulating subsets of microtubules responsible for axonal transport of endoplasmic reticulum/endosomes (Reid et al. 2005). Mutations in the gene encoding atlastin are responsible for further forms of HSP, and atlastin binds to spastin; thus mutant atlastin may also damage axonal transport (Evans et al. 2006, Sanderson et al. 2006).

Mutations in NIPA1 that cause a dominant pure form of HSP have also been linked to defective axonal transport. The *Drosophila* homologue is termed spichthyin, localizes to early endosomes, and regulates growth of the neuromuscular junction via inhibition of bone morphogenetic protein (BMP)-transforming growth factor-β signaling. In the fly, BMP signaling regulates microtubule dynamics and axonal transport, and spichthyin inhibits these functions (Wang et al. 2007). Thus, mammalian NIPA1 may also function in microtubule maintenance and axonal transport.

Damage to mitochondria. Loss-of-function mutations in the *SPG7* gene encoding the mitochondrial ATPase paraplegin disrupt axonal transport. Paraplegin-deficient mice have distal axonopathy of spinal and peripheral axons. Morphological abnormalities include abnormal hypertrophic mitochondria and, later, axonal swellings containing massive accumulation of organelles and neurofilaments that are consistent with a defect in axonal transport of these cargoes. Indeed, neurotracer studies reveal that the mice have a defect in retrograde axonal transport (Ferreirinha et al. 2004).

Damage to myelination. Mutations in the proteolipid protein 1 (*PLP1*) gene that encodes PLP and DM20, the major proteins of the myelin sheath, are responsible for X-linked HSP2. Mice in which PLP1 has been deleted are a model for some forms of HSP2. These animals exhibit axonal accumulations of mitochondria and other membranous organelles that are consistent with a disruption to axonal transport and show defective retrograde transport of labeled cholera toxin (Edgar et al. 2004). Many studies have demonstrated a link between myelination and axonal transport (e.g., de Waegh & Brady 1990, de Waegh et al. 1992, Sanchez et al. 2000). Although the mechanisms are not fully understood, myelin probably signals to induce changes to phosphorylation of a number of axonal proteins associated with transport. Neurofilaments are one such protein (de Waegh et al. 1992, Garcia et al. 2003). Defective myelination also causes some forms of Charcot-Marie-Tooth (CMT) disease (see below).

ALZHEIMER'S DISEASE AND RELATED DISORDERS

Two of the hallmark pathologies of Alzheimer's disease are neurofibrillary tangles containing paired helical filaments (PHFs) and amyloid plaques. PHFs are assembled from hyperphosphorylated tau, a microtubule-associated protein; amyloid plaques are areas of degenerating neurites surrounding a core of amyloid-β peptide (Aβ). Aβ is a 40-42 amino acid peptide that is derived by proteolytic cleavage from a larger precursor protein, APP. The Aβ and tau pathologies are central to the pathogenesis of Alzheimer disease and analyses of genetic forms of Alzheimer's disease, and some related fronto-temporal dementias support this notion. Thus, mutations and duplication of the *APP* gene and mutations in the presenilin genes (components of the γ-secretase complex involved in cleaving Aβ from APP) cause some familial forms of Alzheimer's disease. At least some of these genetic lesions increase production of Aβ or the relative proportion of Aβ(1-42) to Aβ(1-40); the longer Aβ(1-42) form is believed to be pathogenic. Additionally, mutations in the *tau* gene are responsible for some familial forms of fronto-temporal dementia with Parkinsonism (FTDP) (Hardy 2006).

Axonal transport defects have now been described as an early pathological feature in a variety of animal models of Alzheimer's disease and tauopathies. These include transgenic mouse models overexpressing APP or familial Alzheimer's disease mutant APP (Salehi et al. 2006, Stokin et al. 2005), mutant presenilin-1 (Lazarov et al. 2007), and both wild-type and FTDP mutant tau (Ishihara et al. 1999, Zhang et al. 2004). The mechanisms by which these different genetic insults disrupt axonal transport are not properly understood. Aβ itself disrupts transport of a variety of cargoes including mitochondria (Hiruma et al. 2003, Rui et al. 2006). Indeed, Aβ associates with and may damage mitochondria (**Figure 4**) (Manczak et al. 2006). As such, mutant APP and presenilin-1 may damage transport by altering neurotoxic Aβ production. However, evidence also shows that mutant presenilin-1 impairs transport via an effect on kinesin 1 motor function because it activates the KLC2 kinase glycogen synthase kinase-3β (GSK3β), which leads to a release of cargo from motors (**Figure 1**) (Pigino et al. 2003).

Both wild-type and FTDP tau also disrupt axonal transport of multiple cargoes (Stamer et al. 2002, Zhang et al. 2004). One possibility is that tau interferes with kinesin binding to microtubules to damage transport (**Figure 2**) (Seitz et al. 2002). However, tau is also involved in stabilizing microtubules, and hyperphosphorylated tau (such as in PHF) has a reduced affinity for microtubules and is less potent at stabilizing them (Wagner et al. 1996). Hyperphosphorylation of tau by GSK3β and/or cdk5/p35 in Alzheimer's disease (Lovestone et al. 1994, Patrick et al. 1999) may therefore lead to a loss of microtubule rails for axonal transport. As such, both loss and gain of tau function may disrupt axonal transport (**Figure 2**).

Disruption to anterograde axonal transport (including that of APP) via a depletion of KLC1 increases Aβ production, although the precise mechanisms that underlie this effect are unclear (Stokin et al. 2005). BACE1, the rate-limiting enzyme in APP processing to produce Aβ is

APP: amyloid precursor protein

γ-**secretase:** cleaves APP at the C-terminus of the Aβ sequence. It comprises at least four proteins (presenilin-1/2, Aph1, Pen2, and Nicastrin).

KLC: kinesin light chain

Anterograde axonal transport: movement of cargoes toward the synapse

Secretases: enzymes
that cleave APP

KHC: kinesin heavy
chain

most active in an acidic environment, and as
such, a significant proportion of Aβ is produced in endosomes and lysosomes (Wilquet
& De Strooper 2004). Thus, one possibility
is that defective anterograde transport alters
APP trafficking so that it moves to endosomes
and/or lysosomes. Whatever the precise mechanism, any primary damage to axonal transport
in Alzheimer's disease is likely to be amplified
by this related increase in neurotoxic Aβ.

Because disruption to anterograde axonal
transport of APP increases Aβ production
(Stokin et al. 2005), a proper understanding is
required of the mechanism by which APP and
its secretases are transported through axons.
Kamal et al. (2001) proposed that APP is a lig-
and for KLC that facilitates axonal transport of
a membranous compartment containing both
BACE1 and presenilin. However, others have
disputed these findings (Lazarov et al. 2005).
Also, direct analyses of YFP/CFP-tagged APP
and BACE1 movement in cotransfected neurons do not reveal coincident movement of
the two cargoes (Goldsbury et al. 2006). A
further possibility is that APP movement involves some of its binding partners. APP interacts with PTB-containing proteins, including
the adaptor proteins JNK-interacting proteins
(JIPs) and X11s, and these may be involved (see
Miller et al. 2006). JIPs are ligands for KLC
(Bowman et al. 2000, Verhey et al. 2001), and
some evidence shows that APP and JIP1 are
transported as a complex (Muresan & Muresan
2005). X11α interacts with the dendritic kinesin
2 family member KIF17, and both X11α and
X11β bind to a further ligand for KLC1, al-
cadein/calsyntenin (Araki et al. 2007, Konecna
et al. 2006, Setou et al. 2000). The X11s inhibit
Aβ production, which may be via an effect on
APP transport (Miller et al. 2006). Thus a variety of insults associated with Alzheimer's disease
are now known to damage axonal transport.
These insults include damage to motors (altered phosphorylation of KLC2 via mutant presenilin to release cargoes), blocking of kinesin
binding to microtubules (mutant tau), and damage to mitochondria via toxic Aβ (**Figures 1, 2,**
and **4**).

HUNTINGTON'S AND OTHER POLYGLUTAMINE EXPANSION DISEASES

Some familial neurodegenerative diseases are
caused by expansion of polyglutamine stretches.
Such diseases include Huntington's disease,
Kennedy's disease, and some spinocerebellar
ataxias. In Huntington's and Kennedy's diseases, the expansions occur within the huntingtin and androgen receptor proteins, respectively, and expression of mutant huntingtin
or androgen receptor proteins disrupts axonal transport in many models, including isolated squid axoplasm (Szebenyi et al. 2003),
Drosophila (Gunawardena et al. 2003, Lee et al.
2004), and mammalian neurons (Chang et al.
2006, Gauthier et al. 2004, Trushina et al. 2004).

The mechanisms by which they damage
transport are not fully understood, but the
polyglutamine-expanded androgen receptor
leads to phosphorylation of KHC by JNKs;
this phosphorylation reduces kinesin binding
to microtubules to inhibit transport (**Figure 1**)
(Morfini et al. 2006). However, at least for
Huntington's disease, evidence indicates other
mechanisms whereby the mutants may damage
transport, and some of these mechanisms
involve disruption to the normal function of
huntingtin. Huntingtin binds to HAP1
(huntingtin-associated protein), and HAP1 is
strongly implicated in transport processes; it
interacts with KLC1 (McGuire et al. 2005) and
the p150^Glued subunit of dynactin (Engelender
et al. 1997, Li et al. 2001). Also, huntingtin
enhances retrograde transport of brain-derived
neurotrophic factor (BDNF), which involves
HAP1 and dynactin, but mutant huntingtin is
defective in this process (Gauthier et al. 2004).

Clearance of aggregate-prone proteins such
as mutant huntingtin likely represents a key
neuronal defense mechanism. One route for
such clearance involves autophagosomes, and
proper functioning of cytoplasmic dynein is
likely essential for this process (Ravikumar
et al. 2005). One possibility is that cytoplasmic dynein retrogradely transports aggregate-
prone proteins to the cell body for clearance

by lysosomes. Thus, mutant huntingtin may damage retrograde transport not only of cargoes, such as BDNF, but also of toxic aggregates of itself. Finally, Dompierre et al. (2007) recently linked mutant huntingtin with axonal transport via an effect on α-tubulin acetylation. Tubulin acetylation promotes kinesin 1 binding to microtubules and stimulates axonal transport (Reed et al. 2006). Tubulin acetylation is reduced in Huntington disease brains, so mutant huntingtin may disrupt transport via an effect on microtubule acetylation (**Figure 2**) (Dompierre et al. 2007). Expanded polyglutamines may therefore disrupt axonal transport by pathological phosphorylation of KHC, and mutant huntingtin could damage transport by affecting both molecular motors and microtubules (**Figures 1** and **2**).

PARKINSON'S DISEASE

α-synuclein-containing Lewy bodies and Lewy neurites within dopaminergic neurons of the substantia nigra are principal pathologies of Parkinson's disease and related disorders (Tofaris & Spillantini 2007). Moreover, mutations or increased dosage of the α-synuclein gene are the cause of some familial forms of Parkinson's disease (Hardy et al. 2006). The mechanisms by which α-synuclein accumulates within Lewy bodies are not properly understood, but evidence now suggests that defective axonal transport of α-synuclein itself may contribute to the process. First, direct analyses of movement of wild-type and two familial Parkinson's disease-associated mutant α-synucleins through axons of cultured neurons have revealed reduced transport rates of the mutants (Saha et al. 2004). Second, serine-129 of α-synuclein is selectively hyperphosphorylated in Lewy bodies and other synucleinopathy lesions (Fujiwara et al. 2002), and mutation of this site to mimic permanent phosphorylation also reduces axonal transport of α-synuclein (Saha et al. 2004). Thus blocking axonal transport of α-synuclein may contribute to its accumulation within Lewy bodies in both sporadic and some familial cases of Parkinson's disease.

Mutant *parkin*, *PINK1*, *DJ-1*, and *LRRK2* are additional genes associated with familial forms of Parkinson's disease (Hardy et al. 2006). Although direct evidence to link these proteins with axonal transport is lacking, parkin, PINK1, and DJ-1 have all been associated with the maintenance of mitochondria and antioxidant defenses (Abou-Sleiman et al. 2006), and damage to mitochondria perturbs transport of mitochondria through axons (De Vos et al. 2007, Miller & Sheetz 2004). Moreover, inhibition of complex I of the electron transport chain with 1-methyl-4-phenylpyridinium (MPP+), the active metabolite of MPTP, which induces Parkinsonism, decreases anterograde and increases retrograde axonal transport of membranous vesicles in squid axoplasm. This effect involves caspase 3 and PKCδ activation (**Figure 4**) (Morfini et al. 2007).

CHARCOT-MARIE-TOOTH DISEASE

Charcot-Marie-Tooth disease (CMT) includes a heterogeneous group of hereditary motor and sensory neuropathies, the majority being classified as demyelinating (CMT1) or axonal (CMT2) on the basis of electrophysiology.

Axonal Forms of CMT (CMT2)

Dominantly inherited mutations in the gene encoding mitofusin 2 have been demonstrated as causative in CMT2A. Mitofusin 2 is present in the outer mitochondrial membrane and is required for mitochondrial fusion. Fibroblasts from patients and mitofusin 2 knockout mice show fragmented, dispersed mitochondria with severely reduced motility, and expression of CMT2A mutant mitofusin 2 impairs axonal transport of mitochondria in primary neurons (**Figure 4**) (Baloh et al. 2007, Chen et al. 2003). A loss-of-function mutation in the kinesin 3 family member KIF1bβ is also causative of CMT2A (Zhao et al. 2001). KIF1bβ transports some synaptic vesicles, and heterozygous KIF1bβ knockout mice develop a peripheral neuropathy (**Figure 1**) (Zhao et al. 2001).

Mutations in the small GTPase Rab7 cause CMT2B, which predominantly affects sensory neurons in patients. Rab7 regulates vesicle trafficking events from early to late endosomes in the endocytic pathway, and studies recently demonstrated its function in the retrograde axonal transport pathway for neurotrophins (Deinhardt et al. 2006). Thus CMT2B mutant Rab7 may disrupt retrograde transport. Finally, mutations in the neurofilament NFL are found in some CMT2E patients, and at least some of these disrupt neurofilament assembly and axonal transport of neurofilaments and mitochondria (**Figure 3**) (Brownlees et al. 2002, Perez-Olle et al. 2005).

Demyelinating Forms of CMT (CMT1)

Demyelinating forms of CMT are the most prevalent and often involve duplications of, or point mutations in, the peripheral myelin protein 22 gene (*PMP22*). In CMT1 patients, Schwann cells that ensheath motor and sensory axons in the peripheral nervous system show defective myelination. A wealth of data link Schwann cells with axonal transport, and indeed, PMP22 mutant Trembler mice display defective slow axonal transport of the cytoskeleton (de Waegh & Brady 1990). One suggestion is that signaling mechanisms from the Schwann cell impact the axon to alter phosphorylation of the cytoskeleton and axonal transport (Garcia et al. 2003).

MITOCHONDRIAL DAMAGE AND AXONAL TRANSPORT

Mitochondria use oxidative phosphorylation to produce ATP but as a by-product generate toxic reactive oxygen species (ROS). These ROS can damage many cellular components but also mitochondria and their DNA. Mitochondria accumulate mutations/damage over their lifetime, which results in reduced function with aging. As such, mitochondria represent a prime target for age-related toxic insults (Wallace 2005). As listed above, mitochondria are se-

lectively targeted for damage in many neurodegenerative diseases that involve defective axonal transport. Mutations in the mitochondrial proteins paraplegin, PINK1, and mitofusin 2 cause forms of HSP, Parkinson's disease, and CMT, respectively, and mutant forms of SOD1, Parkin, DJ-1, and huntingtin proteins all damage mitochondria. Finally Aβ, excitotoxic glutamate, and MPP+ all damage both mitochondria and axonal transport. Two recent studies have formally quantified the effect of mitochondrial damage on axonal transport. First, antimycin (an inhibitor of electron transport at complex III) was used to inhibit mitochondrial function; second, the effect of ALS mutant SOD1, which selectively associates with and damages mitochondria, was investigated. In both studies, damage to mitochondria resulted in a net increase in their retrograde movement (De Vos et al. 2007, Miller & Sheetz 2004). Such changes in mitochondrial dynamics predict that their numbers will be reduced in axons and that they will accumulate in cell bodies. For mutant SOD1, this prediction is upheld with the result that each remaining mitochondrion must serve an axonal segment of approximately twice the size of that in control neurons (De Vos et al. 2007). ALS-associated accumulation of mitochondria in cell bodies/axon hillocks are not restricted just to experimental systems but are also seen in human ALS cases (Sasaki & Iwata 2007). Thus not only are mitochondria damaged within mutant SOD1 axons, but also their numbers are dramatically reduced. Although mitochondrial transport and distribution have been quantified only in selected cases, some evidence shows that similar depletion of mitochondria numbers/function within axons occurs in other neurodegenerative diseases (Brownlees et al. 2002, Stamer et al. 2002). Also, mitochondria are a target for damage by Aβ in Alzheimer's disease, and Aβ disrupts mitochondrial transport (Manczak et al. 2006, Rui et al. 2006). Therefore, one possibility is that mitochondria represent one of the principal targets for age-related injury in at least some neurodegenerative diseases. The key role of mitochondria as suppliers of ATP fuel for molecular motors implies that any damage to these organelles and/or related defects in their distribution will likely have a profound effect on axonal transport and, as a consequence, axonal maintenance and function (**Figure 4**).

SUMMARY POINTS

1. Defects in axonal transport are early pathogenic events in a number of human neurode-generative diseases.

2. Disruption to axonal transport can occur via a number of routes. These disruptions include damage to (*a*) molecular motors, (*b*) microtubules, (*c*) cargoes (such as inhibiting their attachment to motors), and (*d*) mitochondria, which supply energy for molecular motors.

3. Age-related damage to mitochondria may amplify any primary defects to axonal transport. In this way, disruption to mitochondria may explain why many neurodegenerative diseases are diseases of old age.

FUTURE ISSUES

1. Which mechanisms regulate anterograde vs. retrograde transport of disease-associated cargoes?

2. Which mechanisms control attachment of disease-associated cargoes to kinesins and dynein? In particular, what role does phosphorylation of kinesins and dynein play in cargo binding, and how is this phosphorylation regulated?

3. Which mechanisms control mitochondrial transport? How does damage to mitochondria inhibit anterograde transport?

DISCLOSURE STATEMENT

The authors are not aware of any biases that might be perceived as affecting the objectivity of this review.

ACKNOWLEDGMENTS

Work in the authors' laboratories is supported by grants from the MRC, Wellcome Trust, MNDA, BBSRC, European Union 6th Framework NeuroNE Network of Excellence, ART, and Alzheimer's Association. We apologize that, owing to space restrictions, we were unable to directly cite many of the important contributions that have been made to this field.

LITERATURE CITED

Abou-Sleiman PM, Muqit MM, Wood NW. 2006. Expanding insights of mitochondrial dysfunction in Parkinson's disease. *Nat. Rev. Neurosci.* 7:207–19

Ackerley S, Grierson AJ, Banner S, Perkinton MS, Brownlees J, et al. 2004. p38a stress-activated protein kinase phosphorylates neurofilaments and is associated with neurofilament pathology in amyotrophic lateral sclerosis. *Mol. Cell Neurosci.* 26:354–64

Ackerley S, Grierson AJ, Brownlees J, Thornhill P, Anderton BH, et al. 2000. Glutamate slows axonal transport of neurofilaments in transfected neurons. *J. Cell Biol.* 150:165–75

Ackerley S, James PA, Kalli A, French S, Davies KE, Talbot K. 2006. A mutation in the small heat-shock protein HSPB1 leading to distal hereditary motor neuronopathy disrupts neurofilament assembly and the axonal transport of specific cellular cargoes. *Hum. Mol. Genet.* 15:347–54

Ackerley S, Thornhill P, Grierson AJ, Brownlees J, Anderton BH, et al. 2003. Neurofilament heavy chain side-arm phosphorylation regulates axonal transport of neurofilaments. *J. Cell Biol.* 161:489–95

Ahmad FJ, He Y, Myers KA, Hasaka TP, Francis F, et al. 2006. Effects of dynactin disruption and dynein depletion on axonal microtubules. *Traffic* 7:524–37

Amarilio R, Ramachandran S, Sabanay H, Lev S. 2005. Differential regulation of endoplasmic reticulum structure through VAP-Nir protein interaction. *J. Biol. Chem.* 280:5934–44

Araki Y, Kawano T, Taru H, Saito Y, Wada S, et al. 2007. The novel cargo Alcadein induces vesicle association of kinesin-1 motor components and activates axonal transport. *EMBO J.* 26:1475–86

Baas PW, Karabay A, Qiang L. 2005. Microtubules cut and run. *Trends Cell Biol.* 15:518–24

Ballatore C, Lee VM, Trojanowski JQ. 2007. Tau-mediated neurodegeneration in Alzheimer's disease and related disorders. *Nat. Rev. Neurosci.* 8:663–72

Baloh RH, Schmidt RE, Pestronk A, Milbrandt J. 2007. Altered axonal mitochondrial transport in the pathogenesis of Charcot-Marie-Tooth disease from mitofusin 2 mutations. *J. Neurosci.* 27:422–30

Beaulieu JM, Nguyen MD, Julien JP. 1999. Late onset death of motor neurons in mice overexpressing wild-type peripherin. *J. Cell Biol.* 147:531–44

Beers DR, Henkel JS, Xiao Q, Zhao W, Wang J, et al. 2006. Wild-type microglia extend survival in PU.1 knockout mice with familial amyotrophic lateral sclerosis. *Proc. Natl. Acad. Sci. USA* 103:16021–26

Boillee S, Vande Velde C, Cleveland DW. 2006. ALS: a disease of motor neurons and their nonneuronal neighbors. *Neuron* 52:39–59

Bommel H, Xie G, Rossoll W, Wiese S, Jablonka S, et al. 2002. Missense mutation in the tubulin-specific chaperone E (Tbce) gene in the mouse mutant progressive motor neuronopathy, a model of human motoneuron disease. *J. Cell Biol.* 159:563–69

Bowman AB, Kamal A, Ritchings BW, Philp AV, McGrail M, et al. 2000. Kinesin-dependent axonal transport is mediated by the sunday driver (SYD) protein. *Cell* 103:583–94

Brownlees J, Ackerley S, Grierson AJ, Jacobsen NJ, Shea K, et al. 2002. Charcot-Marie-Tooth disease neurofilament mutations disrupt neurofilament assembly and axonal transport. *Hum. Mol. Genet.* 11:2837–44

Brownlees J, Yates A, Bajaj NP, Davis D, Anderton BH, et al. 2000. Phosphorylation of neurofilament heavy chain side-arms by stress activated protein kinase-1b/Jun N-terminal kinase-3. *J. Cell Sci.* 113:401–7

Bruijn LI, Becher MW, Lee MK, Anderson KL, Jenkins NA, et al. 1997. ALS-linked SOD1 mutant G85R mediates damage to astrocytes and promotes rapidly progressive disease with SOD1-containing inclusions. *Neuron* 18:327–38

Chang DT, Rintoul GL, Pandipati S, Reynolds IJ. 2006. Mutant huntingtin aggregates impair mitochondrial movement and trafficking in cortical neurons. *Neurobiol. Dis.* 22:388–400

Chen H, Detmer SA, Ewald AJ, Griffin EE, Fraser SE, Chan DC. 2003. Mitofusins Mfn1 and Mfn2 coordinately regulate mitochondrial fusion and are essential for embryonic development. *J. Cell Biol.* 160:189–200

Coleman M. 2005. Axon degeneration mechanisms: commonality amid diversity. *Nat. Rev. Neurosci.* 6:889–98

Collard J-F, Cote F, Julien J-P. 1995. Defective axonal transport in a transgenic mouse model of amyotrophic lateral sclerosis. *Nature* 375:61–64

Cote F, Collard J-F, Julien J-P. 1993. Progressive neuronopathy in transgenic mice expressing the human neurofilament heavy gene: a mouse model of amyotrophic lateral sclerosis. *Cell* 73:35–46

Couillard-Després S, Zhu QZ, Wong PC, Price DL, Cleveland DW, Julien JP. 1998. Protective effect of neurofilament heavy gene overexpression in motor neuron disease induced by mutant superoxide dismutase. *Proc. Natl. Acad. Sci. USA* 95:9626–30

Deinhardt K, Salinas S, Verastegui C, Watson R, Worth D, et al. 2006. Rab5 and Rab7 control endocytic sorting along the axonal retrograde transport pathway. *Neuron* 52:293–305

Devon RS, Orban PC, Gerrow K, Barbieri MA, Schwab C, et al. 2006. Als2-deficient mice exhibit disturbances in endosome trafficking associated with motor behavioral abnormalities. *Proc. Natl. Acad. Sci. USA* 103:9595–600

De Vos K, Severin F, Van Herreweghe F, Vancompernolle K, Goossens V, et al. 2000. Tumor necrosis factor induces hyperphosphorylation of kinesin light chain and inhibits kinesin-mediated transport of mitochondria. *J. Cell Biol.* 149:1207–14

De Vos KJ, Chapman AL, Tennant ME, Manser C, Tudor EL, et al. 2007. Familial amyotrophic lateral sclerosis-linked SOD1 mutants perturb fast axonal transport to reduce axonal mitochondria content. *Hum. Mol. Genet.* 16:2720–28

de Waegh S, Brady ST. 1990. Altered slow axonal transport and regeneration in a myelin-deficient mutant mouse: the Trembler as an in vivo model for Schwann cell-axon interactions. *J. Neurosci.* 10:1855–65

de Waegh SM, Lee VM-Y, Brady ST. 1992. Local modulation of neurofilament phosphorylation, axonal caliber and slow axonal transport by myelinating Schwann cells. *Cell* 68:451–63

Dompierre JP, Godin JD, Charrin BC, Cordelieres FP, King SJ, et al. 2007. Histone deacetylase 6 inhibition compensates for the transport deficit in Huntington's disease by increasing tubulin acetylation. *J. Neurosci.* 27:3571–83

Edgar JM, McLaughlin M, Yool D, Zhang SC, Fowler JH, et al. 2004. Oligodendroglial modulation of fast axonal transport in a mouse model of hereditary spastic paraplegia. *J. Cell Biol.* 166:121–31

Engelender S, Sharp AH, Colomer V, Tokito MK, Lanahan A, et al. 1997. Huntingtin-associated protein 1 (HAP1) interacts with the p150[Glued] subunit of dynactin. *Hum. Mol. Genet.* 6:2205–12

Evans K, Keller C, Pavur K, Glasgow K, Conn B, Lauring B. 2006. Interaction of two hereditary spastic paraplegia gene products, spastin and atlastin, suggests a common pathway for axonal maintenance. *Proc. Natl. Acad. Sci. USA* 103:10666–71

Evgrafov OV, Mersiyanova I, Irobi J, Van Den Bosch L, Dierick I, et al. 2004. Mutant small heat-shock protein 27 causes axonal Charcot-Marie-Tooth disease and distal hereditary motor neuropathy. *Nat. Genet.* 36:602–6

Ferreirinha F, Quattrini A, Pirozzi M, Valsecchi V, Dina G, et al. 2004. Axonal degeneration in paraplegin-deficient mice is associated with abnormal mitochondria and impairment of axonal transport. *J. Clin. Invest.* 113:231–42

Fink JK. 2006. Hereditary spastic paraplegia. *Curr. Neurol. Neurosci. Rep.* 6:65–76

Fujiwara H, Hasegawa M, Dohmae N, Kawashima A, Masliah E, et al. 2002. alpha-synuclein is phosphorylated in synucleinopathy lesions. *Nat. Cell Biol.* 4:160–64

Garcia ML, Lobsiger CS, Shah SB, Deerinck TJ, Crum J, et al. 2003. NF-M is an essential target for the myelin-directed "outside-in" signaling cascade that mediates radial axonal growth. *J. Cell Biol.* 163:1011–20

Gauthier LR, Charrin BC, Borrell-Pages M, Dompierre JP, Rangone H, et al. 2004. Huntingtin controls neurotrophic support and survival of neurons by enhancing BDNF vesicular transport along microtubules. *Cell* 118:127–38

Goldsbury C, Mocanu MM, Thies E, Kaether C, Haass C, et al. 2006. Inhibition of APP trafficking by tau protein does not increase the generation of amyloid-beta peptides. *Traffic* 7:873–88

Guidato S, Tsai L-H, Woodgett J, Miller CCJ. 1996. Differential cellular phosphorylation of neurofilament heavy side-arms by glycogen synthase kinase-3 and cyclin-dependent kinase-5. *J. Neurochem.* 66:1698–706

Gunawardena S, Her LS, Brusch RG, Laymon RA, Niesman IR, et al. 2003. Disruption of axonal transport by loss of huntingtin or expression of pathogenic polyQ proteins in *Drosophila*. *Neuron* 40:25–40

Hadano S, Benn SC, Kakuta S, Otomo A, Sudo K, et al. 2005. Mice deficient in the Rab5 guanine nucleotide exchange factor ALS2/alsin exhibit age-dependent neurological deficits and altered endosome trafficking. *Hum. Mol. Genet.* 15:233–50

Hafezparast M, Klocke R, Ruhrberg C, Marquardt A, Ahmad-Annuar A, et al. 2003. Mutations in dynein link motor neuron degeneration to defects in retrograde transport. *Science* 300:808–12

Hardy J. 2006. A hundred years of Alzheimer's disease research. *Neuron* 52:3–13

Hardy J, Cai H, Cookson MR, Gwinn-Hardy K, Singleton A. 2006. Genetics of Parkinson's disease and parkinsonism. *Ann. Neurol.* 60:389–98

Hirokawa N, Takemura R. 2005. Molecular motors and mechanisms of directional transport in neurons. *Nat. Rev. Neurosci.* 6:201–14

Hiruma H, Katakura T, Takahashi S, Ichikawa T, Kawakami T. 2003. Glutamate and amyloid beta-protein rapidly inhibit fast axonal transport in cultured rat hippocampal neurons by different mechanisms. *J. Neurosci.* 23:8967–77

Irobi J, Van Impe K, Seeman P, Jordanova A, Dierick I, et al. 2004. Hot-spot residue in small heat-shock protein 22 causes distal motor neuropathy. *Nat. Genet.* 36:597–601

Ishihara T, Hong M, Zhang B, Nakagawa Y, Lee MK, et al. 1999. Age-dependent emergence and progression of a tauopathy in transgenic mice overexpressing the shortest human tau isoform. *Neuron* 24:751–62

Jacquier A, Buhler E, Schafer MK, Bohl D, Blanchard S, et al. 2006. Alsin/Rac1 signaling controls survival and growth of spinal motoneurons. *Ann. Neurol.* 60:105–17

Jung C, Lee S, Ortiz D, Zhu Q, Julien JP, Shea TB. 2005. The high and middle molecular weight neurofilament subunits regulate the association of neurofilaments with kinesin: inhibition by phosphorylation of the high molecular weight subunit. *Brain Res. Mol. Brain Res.* 141:151–55

Kamal A, Almenar-Queralt A, LeBlanc JF, Roberts EA, Goldstein LB. 2001. Kinesin-mediated axonal transport of a membrane compartment containing B-secretase and presenilin-1 requires APP. *Nature* 414:643–48

Kawasaki H, Morooka T, Shimohama S, Kimura J, Hirano T, et al. 1997. Activation and involvement of p38 mitogen-activated protein kinase in glutamate-induced apoptosis in rat cerebellar granule cells. *J. Biol. Chem.* 272:18518–21

Kiebler MA, Bassell GJ. 2006. Neuronal RNA granules: movers and makers. *Neuron* 51:685–90

Kieran D, Hafezparast M, Bohnert S, Dick JR, Martin J, et al. 2005. A mutation in dynein rescues axonal transport defects and extends the life span of ALS mice. *J. Cell Biol.* 169:561–67

Konecna A, Frischknecht R, Kinter J, Ludwig A, Steuble M, et al. 2006. Calsyntenin-1 docks vesicular cargo to kinesin-1. *Mol. Biol. Cell* 17:3651–63

Kriz J, Nguyen M, Julien J. 2002. Minocycline slows disease progression in a mouse model of amyotrophic lateral sclerosis. *Neurobiol. Dis.* 10:268–78

LaMonte BH, Wallace KE, Holloway BA, Shelly SS, Ascano J, et al. 2002. Disruption of dynein/dynactin inhibits axonal transport in motor neurons causing late-onset progressive degeneration. *Neuron* 34:715–27

Lazarov O, Morfini GA, Lee EB, Farah MH, Szodorai A, et al. 2005. Axonal transport, amyloid precursor protein, kinesin-1, and the processing apparatus: revisited. *J. Neurosci.* 25:2386–95

Lazarov O, Morfini GA, Pigino G, Gadadhar A, Chen X, et al. 2007. Impairments in fast axonal transport and motor neuron deficits in transgenic mice expressing familial Alzheimer's disease-linked mutant presenilin 1. *J. Neurosci.* 27:7011–20

Lee MK, Marszalek JR, Cleveland DW. 1994. A mutant neurofilament subunit causes massive, selective motor neuron death: implications for the pathogenesis of human motor neuron disease. *Neuron* 13:975–88

Lee MS, Kwon YT, Li M, Peng J, Friedlander RM, Tsai LH. 2000. Neurotoxicity induces cleavage of p35 to p25 by calpain. *Nature* 405:360–64

Lee WC, Yoshihara M, Littleton JT. 2004. Cytoplasmic aggregates trap polyglutamine-containing proteins and block axonal transport in a *Drosophila* model of Huntington's disease. *Proc. Natl. Acad. Sci. USA* 101:3224–29

Levy JR, Sumner CJ, Caviston JP, Tokito MK, Ranganathan S, et al. 2006. A motor neuron disease-associated mutation in p150[Glued] perturbs dynactin function and induces protein aggregation. *J. Cell Biol.* 172:733–45

Li BS, Sun MK, Zhang L, Takahashi S, Ma W, et al. 2001. Regulation of NMDA receptors by cyclin-dependent kinase-5. *Proc. Natl. Acad. Sci. USA* 98:12742–47

Lobsiger CS, Garcia ML, Ward CM, Cleveland DW. 2005. Altered axonal architecture by removal of the heavily phosphorylated neurofilament tail domains strongly slows superoxide dismutase 1 mutant-mediated ALS. *Proc. Natl. Acad. Sci. USA* 102:10351–56

Lovestone S, Reynolds CH, Latimer D, Davis DR, Anderton BH, et al. 1994. Alzheimer's disease-like phosphorylation of the microtubule-associated protein tau by glycogen synthase kinase-3 in transfected mammalian cells. *Curr. Biol.* 4:1077–86

Manczak M, Anekonda TS, Henson E, Park BS, Quinn J, Reddy PH. 2006. Mitochondria are a direct site of Abeta accumulation in Alzheimer's disease neurons: implications for free radical generation and oxidative damage in disease progression. *Hum. Mol. Genet.* 15:1437–49

Martin N, Jaubert J, Gounon P, Salido E, Haase G, et al. 2002. A missense mutation in Tbce causes progressive motor neuronopathy in mice. *Nat. Genet.* 32:443–47

Mattiazzi M, D'Aurelio M, Gajewski CD, Martushova K, Kiaei M, et al. 2002. Mutated human SOD1 causes dysfunction of oxidative phosphorylation in mitochondria of transgenic mice. *J. Biol. Chem.* 277:29626–33

McDermott CJ, Grierson AJ, Wood JD, Bingley M, Wharton SB, et al. 2003. Hereditary spastic paraparesis: disrupted intracellular transport associated with spastin mutation. *Ann. Neurol.* 54:748–59

McGuire JR, Rong J, Li SH, Li XJ. 2005. Interaction of huntingtin-associated protein-1 with kinesin light chain: implications in intracellular trafficking in neurons. *J. Biol. Chem.* 281:3552–59

Millecamps S, Robertson J, Lariviere R, Mallet J, Julien JP. 2006. Defective axonal transport of neurofilament proteins in neurons overexpressing peripherin. *J. Neurochem.* 98:926–38

Miller CC, McLoughlin DM, Lau KF, Tennant ME, Rogelj B. 2006. The X11 proteins, Abeta production and Alzheimer's disease. *Trends Neurosci.* 29:280–85

Miller KE, Sheetz MP. 2004. Axonal mitochondrial transport and potential are correlated. *J. Cell Sci.* 117:2791–804

Monani UR. 2005. Spinal muscular atrophy: a deficiency in a ubiquitous protein; a motor neuron-specific disease. *Neuron* 48:885–96

Morfini G, Pigino G, Opalach K, Serulle Y, Moreira JE, et al. 2007. 1-methyl-4-phenylpyridinium affects fast axonal transport by activation of caspase and protein kinase C. *Proc. Natl. Acad. Sci. USA* 104:2442–47

Morfini G, Pigino G, Szebenyi G, You Y, Pollema S, Brady ST. 2006. JNK mediates pathogenic effects of polyglutamine-expanded androgen receptor on fast axonal transport. *Nat. Neurosci.* 9:907–16

Munch C, Rosenbohm A, Sperfeld AD, Uttner I, Reske S, et al. 2005. Heterozygous R1101K mutation of the DCTN1 gene in a family with ALS and FTD. *Ann. Neurol.* 58:777–80

Muresan Z, Muresan V. 2005. Coordinated transport of phosphorylated amyloid-beta precursor protein and c-Jun NH2-terminal kinase-interacting protein-1. *J. Cell Biol.* 171:615–25

Neumann M, Sampathu DM, Kwong LK, Truax AC, Micsenyi MC, et al. 2006. Ubiquitinated TDP-43 in frontotemporal lobar degeneration and amyotrophic lateral sclerosis. *Science* 314:130–13

Nguyen MD, D'Aigle T, Gowing G, Julien JP, Rivest S. 2004. Exacerbation of motor neuron disease by chronic stimulation of innate immunity in a mouse model of amyotrophic lateral sclerosis. *J. Neurosci.* 24:1340–49

Nguyen MD, Lariviere RC, Julien JP. 2001. Deregulation of Cdk5 in a mouse model of ALS: toxicity alleviated by perikaryal neurofilament inclusions. *Neuron* 30:135–47

Otomo A, Hadano S, Okada T, Mizumura H, Kunita R, et al. 2003. ALS2, a novel guanine nucleotide exchange factor for the small GTPase Rab5, is implicated in endosomal dynamics. *Hum. Mol. Genet.* 12:1671–87

Pasinelli P, Brown RH. 2006. Molecular biology of amyotrophic lateral sclerosis: insights from genetics. *Nat. Rev. Neurosci.* 7:710–23

Patrick G, Zukerberg L, Nikolic M, de la Monte S, Dikkes P, Tsai L-H. 1999. Conversion of p35 to p25 deregulates Cdk5 activity and promotes neurodegeneration. *Nature* 402:615–22

Pennetta G, Hiesinger P, Fabian-Fine R, Meinertzhagen I, Bellen H. 2002. *Drosophila* VAP-33A directs bouton formation at neuromuscular junctions in a dosage-dependent manner. *Neuron* 35:291–306

Perez-Olle R, Lopez-Toledano MA, Goryunov D, Cabrera-Poch N, Stefanis L, et al. 2005. Mutations in the neurofilament light gene linked to Charcot-Marie-Tooth disease cause defects in transport. *J. Neurochem.* 93:861–74

Pfister KK, Shah PR, Hummerich H, Russ A, Cotton J, et al. 2006. Genetic analysis of the cytoplasmic dynein subunit families. *PLoS Genet.* 2:e1

Pigino G, Morfini G, Pelsman A, Mattson MP, Brady ST, Busciglio J. 2003. Alzheimer's presenilin 1 mutations impair kinesin-based axonal transport. *J. Neurosci.* 23:4499–508

Puls I, Jonnakuty C, LaMonte BH, Holzbaur EL, Tokito M, et al. 2003. Mutant dynactin in motor neuron disease. *Nat. Genet.* 33:455–56

Raoul C, Estevez AG, Nishimune H, Cleveland DW, deLapeyriere O, et al. 2002. Motoneuron death triggered by a specific pathway downstream of Fas: potentiation by ALS-linked SOD1 mutations. *Neuron* 35:1067–83

Ravikumar B, Acevedo-Arozena A, Imarisio S, Berger Z, Vacher C, et al. 2005. Dynein mutations impair autophagic clearance of aggregate-prone proteins. *Nat. Genet.* 37:771–76

Reed NA, Cai D, Blasius TL, Jih GT, Meyhofer E, et al. 2006. Microtubule acetylation promotes kinesin-1 binding and transport. *Curr. Biol.* 16:2166–72

Reid E. 2003. Science in motion: common molecular pathological themes emerge in the hereditary spastic paraplegias. *J. Med. Genet.* 40:81–86

Reid E, Connell J, Edwards TL, Duley S, Brown SE, Sanderson CM. 2005. The hereditary spastic paraplegia protein spastin interacts with the ESCRT-III complex-associated endosomal protein CHMP1B. *Hum. Mol. Genet.* 14:19–38

Roll-Mecak A, Vale RD. 2005. The *Drosophila* homologue of the hereditary spastic paraplegia protein, spastin, severs and disassembles microtubules. *Curr. Biol.* 15:650–55

Rossoll W, Jablonka S, Andreassi C, Kroning AK, Karle K, et al. 2003. Smn, the spinal muscular atrophy-determining gene product, modulates axon growth and localization of beta-actin mRNA in growth cones of motoneurons. *J. Cell Biol.* 163:801–12

Roy S, Coffee P, Smith G, Liem RKH, Brady ST, Black MM. 2000. Neurofilaments are transported rapidly but intermittently in axons: implications for slow axonal transport. *J. Neurosci.* 20:6849–61

Rui Y, Tiwari P, Xie Z, Zheng JQ. 2006. Acute impairment of mitochondrial trafficking by beta-amyloid peptides in hippocampal neurons. *J. Neurosci.* 26:10480–87

Saha AR, Hill J, Utton MA, Asuni AA, Ackerley S, et al. 2004. Parkinson's disease alpha-synuclein mutations exhibit defective axonal transport in cultured neurons. *J. Cell Sci.* 117:1017–24

Salehi A, Delcroix JD, Belichenko PV, Zhan K, Wu C, et al. 2006. Increased APP expression in a mouse model of Down's syndrome disrupts NGF transport and causes cholinergic neuron degeneration. *Neuron* 51:29–42

Salinas S, Carazo-Salas RE, Proukakis C, Cooper JM, Weston AE, et al. 2005. Human spastin has multiple microtubule-related functions. *J. Neurochem.* 95:1411–20

Sanchez I, Hassinger L, Sihag RK, Cleveland DW, Mohan P, Nixon RA. 2000. Local control of neurofilament accumulation during radial growth of myelinating axons in vivo. Selective role of site-specific phosphorylation. *J. Cell Biol.* 151:1013–24

Sanderson CM, Connell JW, Edwards TL, Bright NA, Duley S, et al. 2006. Spastin and atlastin, two proteins mutated in autosomal-dominant hereditary spastic paraplegia, are binding partners. *Hum. Mol. Genet.* 15:307–18

Sasaki S, Iwata M. 2007. Mitochondrial alterations in the spinal cord of patients with sporadic amyotrophic lateral sclerosis. *J. Neuropathol. Exp. Neurol.* 66:10–16

Schwarzschild MA, Cole RL, Hyman SE. 1997. Glutamate, but not dopamine, stimulates stress-activated protein kinase and AP-1-mediated transcription in striatal neurons. *J. Neurosci.* 17:3455–66

Seitz A, Kojima H, Oiwa K, Mandelkow EM, Song YH, Mandelkow E. 2002. Single-molecule investigation of the interference between kinesin, tau and MAP2c. *EMBO J.* 21:4896–905

Setou M, Nakagawa T, Seog D-H, Hirokawa N. 2000. Kinesin superfamiy motor protein KIF17 and mLin-10 in NMDA receptor-containing vesicle transport. *Science* 288:1796–802

Shea TB, Yabe JT, Ortiz D, Pimenta A, Loomis P, et al. 2004. Cdk5 regulates axonal transport and phosphorylation of neurofilaments in cultured neurons. *J. Cell Sci.* 117:933–41

Skehel PA, Fabian-Fine R, Kandel ER. 2000. Mouse VAP33 is associated with the endoplasmic reticulum and microtubules. *Proc. Natl. Acad. Sci. USA* 97:1101–6

Spillantini MG, Schmidt ML, Lee VM-Y, Trojanowski JQ, Jakes R, Goedert M. 1997. Alpha-synuclein in Lewy bodies. *Nature* 388:839–40

Stamer K, Vogel R, Thies E, Mandelkow E, Mandelkow E-M. 2002. Tau blocks traffic of organelles, neurofilaments, and APP vesicles in neurons and enhances oxidative stress. *J. Cell Biol.* 156:1051–63

Stokin GB, Lillo C, Falzone TL, Brusch RG, Rockenstein E, et al. 2005. Axonopathy and transport deficits early in the pathogenesis of Alzheimer's disease. *Science* 307:1282–88

Sun D, Leung CL, Liem RKH. 1996. Phosphorylation of the high molecular weight neurofilament protein (NF-H) by cdk-5 and p35. *J. Biol. Chem.* 271:14245–51

Szebenyi G, Morfini GA, Babcock A, Gould M, Selkoe K, et al. 2003. Neuropathogenic forms of huntingtin and androgen receptor inhibit fast axonal transport. *Neuron* 40:41–52

Tarrade A, Fassier C, Courageot S, Charvin D, Vitte J, et al. 2006. A mutation of spastin is responsible for swellings and impairment of transport in a region of axon characterized by changes in microtubule composition. *Hum. Mol. Genet.* 15:3544–58

Tofaris GK, Spillantini MG. 2007. Physiological and pathological properties of alpha-synuclein. *Cell Mol. Life Sci.* 64:2194–201

Topp JD, Gray NW, Gerard RD, Horazdovsky BF. 2004. Alsin is a Rab5 and Rac1 guanine nucleotide exchange factor. *J. Biol. Chem.* 279:24612–23

Tortarolo M, Veglianese P, Calvaresi N, Botturi A, Rossi C, et al. 2003. Persistent activation of p38 mitogen-activated protein kinase in a mouse model of familial amyotrophic lateral sclerosis correlates with disease progression. *Mol. Cell Neurosci.* 23:180–92

Trotti D, Rolfs A, Danbolt NC, Brown RHJ, Hediger MA. 1999. SOD1 mutants linked to amyotrophic lateral sclerosis selectively inactivate a glial glutamate transporter. *Nat. Neurosci.* 2:427–33

Trushina E, Dyer RB, Badger JD 2nd, Ure D, Eide L, et al. 2004. Mutant huntingtin impairs axonal trafficking in mammalian neurons in vivo and in vitro. *Mol. Cell Biol.* 24:8195–209

Tudor EL, Perkinton MS, Schmidt A, Ackerley S, Brownlees J, et al. 2005. ALS2/ALSIN regulates RAC-PAK signalling and neurite outgrowth. *J. Biol. Chem.* 280:34735–40

Urushitani M, Sik A, Sakurai T, Nukina N, Takahashi R, Julien JP. 2006. Chromogranin-mediated secretion of mutant superoxide dismutase proteins linked to amyotrophic lateral sclerosis. *Nat. Neurosci.* 9:108–18

Van Den Bosch L, Tilkin P, Lemmens G, Robberecht W. 2002. Minocycline delays disease onset and mortality in a transgenic model of ALS. *NeuroReport* 13:1067–70

Van Den Bosch L, Van Damme P, Bogaert E, Robberecht W. 2006. The role of excitotoxicity in the pathogenesis of amyotrophic lateral sclerosis. *Biochim. Biophys. Acta* 1762:1068–82

Verhey KJ, Meyer D, Deehan R, Blenis J, Schnapp BJ, et al. 2001. Cargo of kinesin identified as JIP scaffolding proteins and associated signaling molecules. *J. Cell Biol.* 152:959–70

Wagner OI, Ascano J, Tokito M, Leterrier JF, Janmey PA, Holzbaur EL. 2004. The interaction of neurofilaments with the microtubule motor cytoplasmic dynein. *Mol. Biol. Cell* 15:5092–100

Wagner U, Utton M, Gallo J-M, Miller CCJ. 1996. Cellular phosphorylation of tau by GSK-3beta influences tau binding to microtubules and microtubule organisation. *J. Cell Sci.* 109:1537–43

Wallace DC. 2005. A mitochondrial paradigm of metabolic and degenerative diseases, aging, and cancer: a dawn for evolutionary medicine. *Annu. Rev. Genet.* 39:359–407

Wang L, Brown A. 2002. Rapid movement of microtubules in axons. *Curr. Biol.* 12:1496–501

Wang L, Ho C-L, Sun D, Liem RKH, Brown A. 2000. Rapid movement of axonal neurofilaments interrupted by prolonged pauses. *Nat. Cell Biol.* 2:137–41

Wang X, Shaw WR, Tsang HT, Reid E, O'Kane CJ. 2007. *Drosophila* spichthyin inhibits BMP signaling and regulates synaptic growth and axonal microtubules. *Nat. Neurosci.* 10:177–85

Williamson TL, Bruijn LI, Zhu QZ, Anderson KL, Anderson SD, et al. 1998. Absence of neurofilaments reduces the selective vulnerability of motor neurons and slows disease caused by a familial amyotrophic lateral sclerosis-linked superoxide dismutase 1 mutant. *Proc. Natl. Acad. Sci. USA* 95:9631–36

Williamson TL, Cleveland DW. 1999. Slowing of axonal transport is a very early event in the toxicity of ALS-linked SOD1 mutants to motor neurons. *Nat. Neurosci.* 2:50–56

Wilquet V, De Strooper B. 2004. Amyloid-beta precursor protein processing in neurodegeneration. *Curr. Opin. Neurobiol.* 14:582–88

Xiao S, McLean J, Robertson J. 2006. Neuronal intermediate filaments and ALS: a new look at an old question. *Biochim. Biophys. Acta* 1762:1001–12

Xu Z, Cork LC, Griffin JW, Cleveland DW. 1993. Increased expression of neurofilament subunit NF-L produces morphological alterations that resemble the pathology of human motor neuron disease. *Cell* 73:23–33

Yuan A, Rao MV, Kumar A, Julien JP, Nixon RA. 2003. Neurofilament transport in vivo minimally requires hetero-oligomer formation. *J. Neurosci.* 23:9452–58

Zhang B, Higuchi M, Yoshiyama Y, Ishihara T, Forman MS, et al. 2004. Retarded axonal transport of R406W mutant tau in transgenic mice with a neurodegenerative tauopathy. *J. Neurosci.* 24:4657–67

Zhang B, Tu P, Abtahian F, Trojanowski JQ, Lee VM. 1997. Neurofilaments and orthograde transport are reduced in ventral root axons of transgenic mice that express human SOD1 with a G93A mutation. *J. Cell Biol.* 139:1307–15

Zhao C, Takita J, Tanaka Y, Setou M, Nakagawa T, et al. 2001. Charcot-Marie-Tooth disease type 2A caused by mutation in a microtubule motor KIF1Bbeta. *Cell* 105:587–97

Zhu S, Stavrovskaya IG, Drozda M, Kim BY, Ona V, et al. 2002. Minocycline inhibits cytochrome c release and delays progression of amyotrophic lateral sclerosis in mice. *Nature* 417:74–78

Active and Passive Immunotherapy for Neurodegenerative Disorders

David L. Brody and David M. Holtzman

Department of Neurology, Developmental Biology, Alzheimer's Disease Research Center, and Hope Center for Neurological Disorders, Washington University School of Medicine, St. Louis, Missouri 63110; email: brodyd@neuro.wustl.edu, Holtzman@neuro.wustl.edu

Annu. Rev. Neurosci. 2008. 31:175–93

First published online as a Review in Advance on March 19, 2008

The *Annual Review of Neuroscience* is online at neuro.annualreviews.org

This article's doi: 10.1146/annurev.neuro.31.060407.125529

Key Words

Alzheimer disease, vaccination, monoclonal antibody

Abstract

Immunotherapeutic strategies to combat neurodegenerative disorders have galvanized the scientific community since the first dramatic successes in mouse models recreating aspects of Alzheimer disease (AD) were reported. However, initial human trials of active amyloid-beta (Aβ) vaccination were halted early because of a serious safety issue: meningoencephalitis in 6% of subjects. Nonetheless, some encouraging preliminary data were obtained, and rapid progress has been made toward developing alternative, possibly safer active and passive immunotherapeutic approaches for several neurodegenerative conditions. Many of these are currently in human trials for AD. Despite these advances, our understanding of the essential mechanisms underlying the effects seen in preclinical models and human subjects is still incomplete. Antibody-induced phagocytosis of pathological protein deposits, direct antibody-mediated disruption of aggregates, neutralization of toxic soluble proteins, a shift in equilibrium toward efflux of specific proteins from the brain, cell-mediated immune responses, and other mechanisms may all play roles depending on the specific immunotherapeutic scenario.

Contents

INTRODUCTION

The remarkable power of therapeutics designed to enhance aspects of the immune response has been recognized at least since the time of Jenner (1798). Triumph after triumph has come from the realm of infectious diseases since then. More recently, the basic strategies of active vaccination, and by extension, the use of monoclonal antibodies for passive vaccination, have been applied to neoplastic, autoimmune, and atherosclerotic diseases. Considering that the number of individuals greater than age 65 is increasing markedly throughout the world, it is quite natural that the modern immunotherapeutic armamentarium should be brought to bear on the problems of Alzheimer disease (AD) and other neurodegenerative disorders. Hundreds of previous reviews have been published on the topic of immunotherapeutics in neurodegenerative diseases.[1] The pathogenesis of these diseases has also been reviewed extensively (e.g., Skovronsky et al. 2006). Here we address selected critical issues that bear directly on the development of effective active and passive immunotherapeutics for human patients with neurodegenerative disorders.

ACTIVE Aβ VACCINATION IN TRANSGENIC MICE MODELING ASPECTS OF AD PATHOLOGY: INITIAL FINDINGS

The first report of an immunotherapeutic approach to a neurodegenerative disease in vivo was published in 1999. The experiments involved active vaccination with aggregated amyloid-beta (Aβ) in PDAPP mice, transgenic animals that develop extracellular Aβ deposition in the form of plaques similar in many respects to those seen in human AD patients (Games et al. 1995, Johnson-Wood et al. 1997). The initial results were dramatic: Immunization starting either before or after plaque deposition had a clear beneficial effect on the

---

[1]A search of PubMed on June 2, 2007, for neurodegenerative disease AND immunotherapy yielded 222 review articles.

AD-like plaque pathology (**Figure 1a**). High anti-Aβ antibody titers in serum were detected, and phagocytic cells were found engulfing Aβ aggregates in vaccinated animals' brains (Schenk et al. 1999). Quickly thereafter, two groups simultaneously published independent reports showing improvements in behavioral abnormalities using similarly vaccinated amyloid precursor protein (APP) transgenic mice (**Figure 1b–c**; see **Supplemental Table 1**. Follow the Supplemental Material link from the Annual Reviews home page at **http://www.annualreviews.org**). Strikingly, behavioral performance was even more strongly improved than were brain Aβ levels or plaque pathology; some performance deficits in TgCRND8 (Chishti et al. 2001) and Tg2576 (Hsiao et al. 1996) mice were almost entirely normalized, whereas Aβ plaques were reduced but not eliminated (Janus et al. 2000, Morgan et al. 2000).

THE AN1792 TRIALS OF ACTIVE Aβ VACCINATION IN HUMAN AD PATIENTS

This approach was translated remarkably rapidly to the human clinical arena. A phase I (safety) study in patients with mild-moderate sporadic AD was performed in the year 2000. Eighty subjects, in their 70s on average, were randomly assigned to one of four combinations of AN1792 (the same aggregated Aβ$_{1-42}$ preparation used in the transgenic mice) plus the QS21 surface active saponin adjuvant or placebo. Four vaccinations over a six-month period were administered intramuscularly. Four deaths occurred in the active treatment group but none was considered related to the vaccination. Other adverse events possibly related to vaccination included confusion and hallucinations, hostility, and convulsions, but all of these have been reported to occur also in untreated AD patients (Bayer et al. 2005). Overall the treatment was well tolerated during this trial, and the strategy was advanced to a phase II trial.

After the phase II trial was well underway, one actively vaccinated patient from the phase

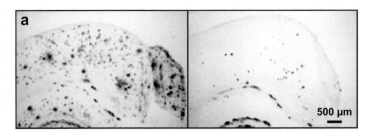

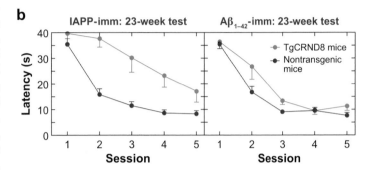

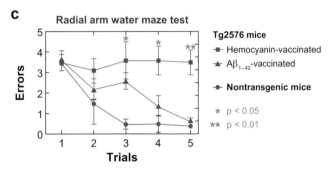

Figure 1

Effects of active vaccination in transgenic mice modeling aspects of Alzheimer disease. (*a*) Reduction in Aβ plaque deposition in the cortex of PDAPP mice following vaccination with aggregated Aβ$_{1-42}$ (*right*) compared with saline-injected mice (*left*) at the same age. Adapted from figure 4*b,c* of Schenk et al. (1999). (*b*) Improved behavioral performance in TgCRND8 mice following vaccination with aggregated Aβ$_{1-42}$ (*red circles, right panel*) compared with TgCRND8 mice vaccinated with the irrelevant islet-associated polypeptide (IAPP, *red circles, left panel*). Performance of similarly vaccinated nontransgenic mice (*gray circles*) was mostly unchanged. The figure shows latency to reach the hidden platform in the Morris water maze (Morris et al. 1984); each daily session consisted of four trials per mouse. Adapted from figure 2*e* of Janus et al. (2000). (*c*) Improved behavioral performance in Tg2576 mice following vaccination with aggregated Aβ$_{1-42}$ (*purple triangles*) compared with Tg2576 mice vaccinated with the irrelevant keyhole limpet hemocyanin (*blue squares*). Aβ-vaccinated Tg2576 mice performed nearly as well as nontransgenic mice (*green circles*). The figure shows number of errors in the radial arm water maze during working memory (trials 1–4) and retention (trial 5) testing on days 10 and 11 of training. Adapted from figure 1*b* of Morgan et al. (2000). Reprinted by permission of Macmillan Publishers Ltd.

I trial developed a subacute encephalopathy, did not recover to baseline, and died from non-CNS causes. At autopsy, the patient was found to have T-lymphocyte predominant meningoencephalitis (Nicoll et al. 2003). Meningoencephalitis is not an expected occurrence in AD and has not been reported in any unvaccinated AD patients to our knowledge, although investigators have reported perivascular inflammation in some cases of cerebral amyloid angiopathy with Aβ deposition in brain blood vessels (Eng et al. 2004). This event occurred after the vaccine formulation was changed to include the preservative polysorbate 80. At around the same time, another report showed similar encephalitis in 6% (18/300) of the actively immunized patients in the phase II study, which also used the polysorbate 80-containing formulation (Orgogozo et al. 2003). The phase II study was then terminated after only 1-3 of the planned 6 vaccinations had been given. Twelve of the patients recovered within weeks, whereas six patients suffered disabling cognitive or neurological sequelae. Analysis of the interrupted trial revealed that Aβ antibody titer was not correlated with the signs and symptoms of meningoencephalitis. These adverse events had not been predicted on the basis of the preclinical studies performed prior to initiation of the clinical trials.

In the phase I study, there was a trend toward slower cognitive decline in the actively vaccinated patients compared with controls. However, active vaccination did not bring the cognitive performance of these elderly, demented individuals back to near normal, as had been reported for some of the transgenic mice. One possible explanation was that in the phase I trial, serum anti-Aβ antibody titers were low, above 1:1000 in only ~60% of the elderly human AD patients, whereas most of the relatively young mice in the early preclinical vaccination trials had titers over 1:10,000 (Schenk et al., 1999). Of note, in a preclinical study of significantly older transgenic mice, Austin et al. (2003) found lower antibody titers following active vaccination, and they saw no effect on behavioral performance. Another study found that

immunization in older mice was less effective at clearing plaques than was immunization in younger mice, even though all ages produced high anti-Aβ antibody titers (Das et al. 2001).

Overall in the multicenter phase II study, there likewise was not a major effect on cognitive performance (Gilman et al. 2005). A single-center analysis of a subgroup of 30 patients appeared to indicate stabilization on some measures of cognitive status in 6 patients with robust antibody responses (Hock et al. 2003). However, the 9 patients who received active vaccination but did not mount a significant antibody response actually performed worse than would have been expected from the natural history of the disease; their mini-mental status exam scores declined by an average of 6 points out of 30 in 1 year (Hock et al. 2003), whereas the overall placebo group declined by an average of 1.5 points (Gilman et al. 2005). In the phase II study analysis, ~20% of the patients had significant serum antibody titers (>1:2200) after 1-3 vaccinations. Even among this subgroup of patients who were considered responders, the primary clinical outcome measures, including the mini-mental status examination, revealed no effect of active vaccination compared with placebo. A few of the tests in the neuropsychological battery relating to episodic memory showed differences favoring active vaccination, although the study found no dramatic cognitive improvement in any domain examined. Of note, analyses excluding the 22% of the responders who developed encephalitis did not change the reported results.

Importantly, active immunization resulted in generation of anti-Aβ antibodies that appeared predominantly directed against the N-terminus of Aβ and bound to both soluble and insoluble forms of Aβ (Lee et al. 2005). These antibodies were similar in many respects to those raised during active vaccination of transgenic mice (Bard et al. 2003).

Neuropathological analysis of the brains of a few deceased patients revealed several striking findings suggesting that the vaccination strategy was in fact having some effect on human AD pathology. The first patient to come to autopsy from the phase I study had a patchy distribution of cortical Aβ plaques, in contrast with the more uniform distribution of plaques in several cortical regions in typical AD (Nicoll et al. 2003). Some regions of cortex were nearly free of plaques. Plaque pathology was extensive in the basal ganglia and cerebellum, which are areas not especially prone to plaque pathology. Cerebral amyloid angiopathy and neurofibrillary tangles were persistent, even in areas relatively free of parenchymal Aβ plaques. Although this patient may have had an unusual form of the disease (Klunk et al. 2007), a more likely explanation is that the vaccination-induced immune response caused a clearance of Aβ deposits out of cortex, possibly resulting in a redistribution into other structures. As seen in vaccinated transgenic mice, the brain contained apparent phagocytic cells with closely associated punctate Aβ staining (Nicoll et al. 2003).

A second case report from a vaccination-related meningoencephalitis patient revealed similar findings, with the addition of multinucleated giant cells appearing to engulf Aβ deposits (Ferrer et al. 2004). Tris and guanidine-soluble Aβ levels in extracts from these brains were higher than in several unvaccinated patients (Patton et al. 2006). In the brain of a third patient who, importantly, did not develop clinical signs of meningoencephalitis (he died of unrelated causes 18 months after his first vaccination), there was still a marked reduction in Aβ plaque pathology in some regions of cortex as compared with typical AD patients. Cerebral amyloid angiopathy and tangle pathology were again persistent. This patient mounted a vigorous anti-Aβ antibody response, based on his serum anti-Aβ antibody titers. No evidence of multinucleated giant cells or subclinical T-cell meningoencephalitis was found (Masliah et al. 2005a).

An unresolved question exists about the nature of the effects of vaccination on AD pathology in the vast majority of patients who survived. Cerebrospinal fluid studies revealed decreases in tau levels in a subset of patients (Gilman et al. 2005). This decrease may suggest reduced neurodegeneration, but the interpretation of this finding is not clear. Amyloid imaging was not available at that time but could have been helpful. A volumetric MRI study over a one-year period somewhat surprisingly revealed that vaccinated patients had a greater decrease in brain volume than did placebo patients. When patients with encephalitis were excluded, the results remained significant. A moderate correlation between anti-Aβ antibody titer and volume loss was found; higher maximum IgG titers were associated with greater volume loss (Fox et al. 2005). What accounts for these MRI findings is again not clear. Acceleration of volume loss has been associated with progression to later clinical stages of AD, but there was no evidence of accelerated neurodegeneration in the vaccinated patients. More interestingly, if Aβ deposits take up a substantial portion of the substance of the cortex in AD, removal of these deposits could account for the reported volume loss.

PASSIVE Aβ VACCINATION IN TRANSGENIC MICE MODELING ASPECTS OF AD PATHOLOGY

Shortly after the publication of the initial preclinical active vaccination trial, two reports indicated that passive vaccination by systemic infusion of monoclonal antibodies directed at Aβ could similarly decrease Aβ plaque pathology in transgenic mice (Bard et al. 2000, DeMattos et al. 2001) (see also **Supplemental Table 2**). Passive vaccination strategies also successfully improved the behavioral performance of these transgenic animals. However, effects on plaque pathology were dissociated from effects on behavior. For example, six weekly injections of m266, the antibody to the central domain of Aβ used by DeMattos et al., produced marked normalization of certain behavioral impairments without any apparent effects on plaque pathology. Even a single injection of this antibody produced beneficial effects that were apparent one day after treatment (Dodart et al. 2002). Likewise, three injections of BAM-10, another monoclonal antibody, rapidly improved Morris

Morris Water Maze: rodent behavioral test that assesses spatial learning and memory by evaluating ability to swim to a small platform in a pool of opacified water

water maze performance in another transgenic mouse model without apparently affecting levels of any of the soluble species of Aβ measured by enzyme-linked immunosorbent assay (ELISA) in brain extracts. These mice had not yet developed plaques but already had worse behavioral performance than did their wild-type littermates (Kotilinek et al. 2002). Janus et al. (2000) had similarly demonstrated that active vaccination starting before the development of plaque pathology improved behavior without reduction in brain Aβ levels.

These reports and others that followed (Hartman et al. 2005, Lee et al. 2006, Oddo et al. 2006b, Chen et al. 2007) indicated that in the mouse models, effects on behavior did not necessarily correlate with effects on either plaque pathology or soluble brain Aβ levels as measured by ELISA. The first dissociation between effects on behavior and effects on plaques was initially considered likely to be due to neutralization of some form of toxic, soluble Aβ in the brain. The second dissociation between effects on behavior and effects on soluble Aβ levels as measured by ELISA suggested that this might not be the entire explanation either. A central role for rare but highly toxic soluble Aβ species (e.g., protofibrils, oligomers, ADDLs, Aβ* species) not easily measured by conventional ELISAs has been posited as an overarching framework to explain these findings (Walsh & Selkoe 2004). The potential importance of other aspects of AD pathology such as neurofibrillary tangles and the effects of tau abnormalities on axonal transport have also been revisited in this context (Lee & Trojanowski 2006, Oddo et al. 2006a). A fundamental issue that complicates interpretation of all these results is that we do not know whether behavioral abnormalities seen in APP transgenic mice are analogous to any of the cognitive deficits seen in humans with AD.

One potential concern associated with administration of certain anti-Aβ antibodies is intracerebral hemorrhage. Hemorrhages associated with cerebral amyloid angiopathy (CAA) were increased in APP transgenic mice treated with some antibodies, especially those that bind Aβ-containing plaques (Pfeifer et al. 2002, Wilcock et al. 2004b, Racke et al. 2005). This finding raises a new safety concern that may be relevant to humans because ~90% of human AD patients have concomitant CAA, and similar intracranial hemorrhages were seen in one actively vaccinated patient (Ferrer et al. 2004). Most of these hemorrhages were small, and the clinical significance of these microbleeds is not entirely certain; they can be found frequently in asymptomatic patients with CAA (Towfighi et al. 2005). Likewise, one preclinical passive vaccination study reported that although there were larger numbers of microhemorrhages and worsening CAA in antibody-treated mice, these did not negate the generally positive effects of treatment on behavioral performance (Wilcock et al. 2004b).

POTENTIAL MECHANISMS UNDERLYING EFFECTS OF IMMUNOTHERAPEUTICS

Investigators have proposed a wide variety of mechanisms to explain various aspects of the effects of passive and active vaccination strategies. Most of this work has been based on mouse models used for preclinical development of AD treatments. Early reports of active vaccination trials presented immunohistochemical evidence for phagocytosis of Aβ deposits; small, densely staining Aβ plaques were seen in association with major histocompatability complex (MHC) II-labeled cells resembling activated microglia and monocytes (Schenk et al. 1999). This and the high anti-Aβ titers found in these mice suggested that antibody-directed clearance of plaques by phagocytosis (**Figure 2a**) could be a central mechanism. When administered systemically, a small fraction of certain anti-Aβ monoclonal antibodies gained access to the brain and bound plaques. In vitro experiments using monoclonal antibodies and cultured microglia supported the possibility that these antibodies trigger phagocytosis of Aβ deposits (Bard et al. 2000). Apparent phagocytosis of Aβ plaques was also observed in the brains of two patients who were treated with AN1792

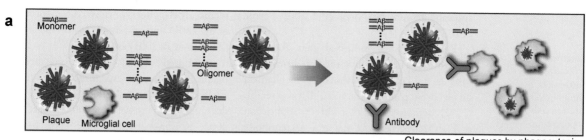

a Clearance of plaques by phagocytosis

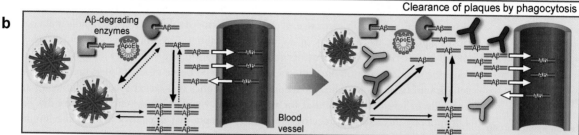

b Shifting equilibrium towards Aβ monomers, favoring clearance and degradation

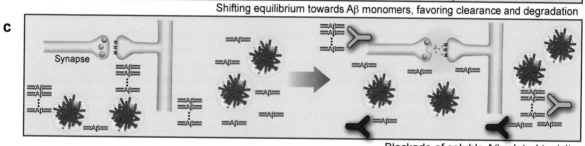

c Blockade of soluble Aβ-related toxicity

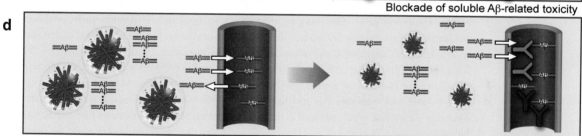

d Peripheral sink: prevention of reverse Aβ flux

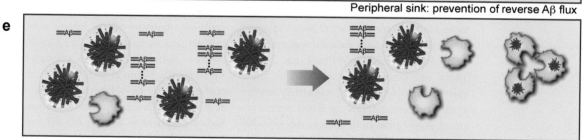

e Antibody-independent, cell mediated plaque clearance

Figure 2

Potential mechanisms underlying effects of immunotherapeutics in models of Alzheimer disease. These represent only a subset of the possibilities and are not meant to be considered mutually exclusive; more than one may be in operation at any given time or several may play important roles at different stages of therapy. (This figure was produced with the assistance of MedPIC at Washington University School of Medicine.) For further explanation, see Supplemental Material online.

and came to autopsy (Nicoll et al. 2003, Ferrer et al. 2004).

Further studies indicated that fragment crystalizable (Fc) receptor-mediated phagocytosis is unlikely to be the only mechanism involved in plaque clearance. Bacskai et al. (2002) demonstrated that direct application of F(ab')2 fragments of the anti-Aβ antibody 3D6 to the surface of the brains of plaque-bearing mice resulted in clearance of Aβ deposits equivalent to the effects of application of full-length 3D6. F(ab')$_2$ fragments generally have the same antigen-binding effects as full-length antibodies but lack the Fc portion of the antibody believed to be essential for triggering antibody-mediated phagocytosis via interactions with the Fc receptor on microglia and macrophages. This finding suggested that antibody binding could potentially directly disrupt Aβ aggregates, shifting the equilibrium between aggregates and monomeric Aβ toward the more readily cleared and degraded monomers (**Figure 2b**). This idea is supported by in vitro data demonstrating that anti-Aβ antibodies mixed with synthetic Aβ could prevent (Solomon et al. 1996, Legleiter et al. 2004) or reverse (Solomon et al. 1997) aggregation of Aβ fibrils. Although most antibodies that bind to insoluble Aβ aggregates and clear plaques have epitopes directed against the N-terminus of Aβ, many N-terminal antibodies bind to both soluble and insoluble Aβ and may thus have more than one mechanism of action (see below).

Interestingly, direct injection of F(ab')$_2$ fragments of a different antibody (2286) resulted in diffuse plaque clearance but did not affect fibrillar Aβ deposit clearance (Wilcock et al. 2004a). These and other results suggest that there may be multiple, possibly sequential mechanisms of direct antibody-mediated plaque removal. Application of intact anti-Aβ antibody was reported to clear diffuse Aβ deposits within 24 h but clear fibrillar deposits only later at 72 h (Wilcock et al. 2003). Only the later phase was accompanied by a surge in the numbers of CD45-positive stained microglial cells. Fc-independent mechanisms appear capable of mediating the effects of active vaccination (Das et al. 2003) and systemic passive vaccination (Tamura et al. 2005) as well.

As noted above, dissociations between plaque clearance or Aβ levels measured by ELISA and behavioral effects have been well documented (Janus et al. 2000, Dodart et al. 2002, Kotilinek et al. 2002). The findings that exogenous administration of some specific forms of soluble Aβ could rapidly induce abnormalities in synaptic plasticity and in behavioral performance (Walsh et al. 2002, Cleary et al. 2005, Lesne et al. 2006) suggested that therapeutics that target specific soluble Aβ species may underlie these effects (**Figure 2c**). Experiments that unequivocally test the importance of this class of mechanisms in vivo have not been reported to our knowledge. There is, however, some supporting experimental evidence (Klyubin et al. 2005, Lee et al. 2006, Oddo et al. 2006a,b), and this is an area of active inquiry. A central question is still which if any of these protofibrils, oligomers, ADDLs, Aβ* species, etc., are present in the human brain, at which concentrations, and at which stages of disease. Oligomeric species of Aβ are challenging to measure, and detergents such as those used in extraction procedures or gel electrophoretic analyses lead to artificial assemblies (Hepler et al. 2006). This adds further uncertainty to the interpretation of these results.

Some evidence indicates that the effects of certain anti-Aβ antibodies may not actually require that these antibodies enter the brain. For example, the monoclonal antibody m266 has a very high affinity for soluble Aβ, can remove all free Aβ present in the blood of PDAPP mice, and does not bind Aβ plaques when given peripherally (DeMattos et al. 2001) or even when applied directly to AD brain sections (Racke et al. 2005). This and potentially other antibodies that bind to soluble Aβ may act as a "peripheral sink," preventing transport of Aβ from blood back into the brain (**Figure 2d**) and thereby enhancing net clearance of Aβ (DeMattos et al. 2001). Much work has gone into investigations of the mechanisms of Aβ clearance from the brain into the blood (Deane et al. 2003), and net clearance can involve not only a peripheral sink effect but also an active Fc receptor-mediated process at the blood-brain barrier (Deane et al. 2005). This process could further favor Aβ clearance without requiring antibody persistence in the brain.

Although most of the above discussion has focused on antibody-mediated effects, antibody-independent, cell-mediated effects may underlie some of the benefits seen following active vaccination (**Figure 2e**). This could also be a mechanism by which nonspecific immunization results in plaque clearance (Frenkel et al. 2005). Cell-based immunotherapeutic approaches have demonstrated benefits in animal models of Parkinson disease (Benner et al. 2004) and amyotropic lateral sclerosis (Angelov et al. 2003).

Overall, there may be many mechanisms that can affect Aβ plaques, soluble Aβ levels, and/or behavioral performance. These may act concomitantly or sequentially, be differentially important depending on the stage of disease or type of antibody, and may be more or less relevant depending on factors such as the specific animal model system used and the animals' ages. Other mechanisms not discussed here in depth such as effects on Aβ-mediated vasoconstriction or modulation of CNS cytokine production may also be involved. Some, all, or none of these mechanisms may be relevant to treatment of human disease. Substantial numbers of additional clinical trials in human patients with translational mechanistic studies incorporated into their designs will be required to sort this out.

ACTIVE AND PASSIVE VACCINATION STRATEGIES IN ANIMAL MODELS OF OTHER NEURODEGENERATIVE DISEASES

Several laboratories have investigated immunotherapeutic strategies targeting proteins known or suspected to be involved in the pathogenesis of other neurodegenerative diseases (see **Supplemental Table 3**). All these approaches are still in preclinical development to our knowledge.

Prion Diseases

Farthest advanced is work on transmissible spongiform encephalopathies, such as Creutzfeld-Jacob disease in humans and scrapie in other species. Because the native prion protein (PrP) is widely expressed throughout life and creates immune tolerance, it was initially challenging to produce effective active and passive vaccinations in experimental animals (reviewed in Bade & Frey 2007). Several clever strategies were used to overcome this hurdle, including genetically engineering mice to produce single-chain anti-PrP antibodies (Heppner et al. 2001), infusing monoclonal antibodies produced using PrP knockout mice (White et al. 2003), and actively vaccinating mice with highly immunogenic papilloma virus-derived particles displaying PrP epitopes (Handisurya et al. 2007), among many others. Overall, there has been substantial progress in preclinical models. For example, active mucosal vaccination of mice using a recombinant attenuated salmonella strain expressing mouse PrP delayed or prevented clinical disease following oral challenge with an otherwise universally fatal oral dose of a mouse scrapie strain (Goni et al. 2005). Likewise, passive vaccination of mice using intraperitoneal injection of monoclonal anti-PrP antibodies starting after inoculation with scrapie brain homogenates but before onset of symptoms prevented clinical disease development (White et al. 2003). To our knowledge, none of the experimental approaches has as yet demonstrated alleviation of symptomatic disease (Bade & Frey 2007).

Parkinson Disease and Amyotrophic Lateral Sclerosis

Several strategies similar to those used in AD have been tested in mouse models of Parkinson disease and amyotrophic lateral sclerosis (see **Supplemental Table 3**). Active vaccination with human α-synuclein in a familial Parkinson disease mouse model characterized by accumulation of α-synuclein aggregates provided some benefit in terms of pathological outcomes (Masliah et al. 2005b). Likewise, active vaccination with misfolded mutant superoxide dismutase 1 (SOD-1) in a familial amyotrophic lateral sclerosis mouse model carrying an SOD-1 mutation reduced loss of spinal cord neurons and caused a modest but statistically significant increase in life expectancy (Urushitani et al. 2007). Some benefits of nonspecific active vaccinations with Copaxone-based regimens have been reported (Angelov et al. 2003, Benner et al. 2004) in ALS and Parkinson models, although safety concerns have been raised (Haenggeli et al. 2007). The effects of passive vaccination approaches to these diseases have not been reported to our knowledge, but this clearly represents an avenue for future research.

Acute CNS Insults that May Trigger Neurodegeneration

Aβ may play a pathogenic role in other conditions besides AD, and some of the same immunotherapeutics under development for AD have been assessed as potential neuroprotectants for other central nervous system insults such as seizures and traumatic brain injury (TBI). For example, pretreatment with an anti-Aβ monoclonal antibody reduced seizure-related neuronal loss in the hippocampi of Tg2576 mice (Mohajeri et al. 2002). Likewise, our collaborative group treated young PDAPP mice with the anti-Aβ antibody m266 starting shortly before and continuing weekly after moderately severe experimental TBI. The m266-treated mice displayed markedly improved behavioral performance in the Morris water maze, reduced loss of hippocampal CA3 cells, and increased apparent neurogenesis in the dentate gyrus (Brody et al. 2005). An extensive epidemiological and pathological literature links TBI and the subsequent risk of AD (Jellinger 2004).

ONGOING CLINICAL DEVELOPMENT OF IMMUNOTHERAPEUTIC STRATEGIES TO COMBAT AD

Passive Vaccination

Farthest advanced at the time of this writing is the Elan/Wyeth trial of AAB-001 (Bapineuzumab), which was entering phase III testing in 2007. This approach involves passive immunization with an Aβ N-terminal directed, humanized monoclonal antibody. The murine version of this antibody binds to both soluble and aggregated Aβ (Bard et al. 2000, Racke et al. 2005). The N-terminal 8 amino acids of Aβ were the predominant epitope recognized by antibodies in the sera from 45 vaccinated patients in the AN1792 trial (Lee et al. 2005), which also bound to both soluble and aggregated Aβ. Thus, passive vaccination with an N-terminal antibody with similar features may be expected to mimic aspects of the antibody portion of the response to active vaccination. Phase I results have been reported only in abstract form (Ninth Int. Geneva/Springfield Symp. Alzeimer Ther. 2006), and the multiple-dose phase II trial has not yet been completed. A small amyloid imaging substudy is also part of this trial.

Next most advanced to our knowledge is the Eli Lilly and Co. phase II trial of LY2062430, which involves passive vaccination with an Aβ central domain directed, humanized monoclonal antibody. Systemic administration of the closely related central domain mouse monoclonal antibody m266 rapidly improved behavioral performance (Dodart et al. 2002, Bales et al. 2006) and decreased plaque formation (DeMattos et al. 2001) in preclinical studies. Thus far, this antibody has not been shown to reverse existing plaque pathology or bind directly to plaques. On the other hand, this antibody did not worsen intracerebral hemorrhage or vascular pathology in older APP transgenic mice (Racke et al. 2005). Unlike the AAB-001 trial, patients on anticoagulants or with contraindications to MRI are not specifically excluded from the LY2062430 trial (http://www.clinicaltrials.gov). The phase I results have been reported only in abstract form, and no safety concerns were raised (10th Int. Conf. Alzheimer's Dis. Relat. Disord. 2006).

The Pfizer/Rinat phase I trial of RN-1219 (PF-04360365), another humanized monoclonal antibody recognizing Aβ, excludes patients with hemorrhages, infarctions, or extensive white matter changes on MRI (http://www.clinicaltrials.gov). Detailed information about this antibody has not been publicly disclosed. Likewise, few details are available regarding the Hoffman-La Roche/MorphoSys phase I trial of R-1450 (http://www.centerwatch.com), a fully human monoclonal antibody developed using the MorphoSys HuCAL platform (Knappik et al. 2000). The protocol registry for this trial does not list MRI-based or anticoagulation-related exclusion criteria.

Active Vaccination

Refined active vaccination strategies for AD are also still considered viable options. The Elan/Wyeth ACC-001 phase I active vaccination trial started in the fall of 2005. The vaccine reportedly was designed to elicit specific antibody responses while minimizing the deleterious inflammatory responses seen in the AN1792 trial. This approach uses an Aβ fragment attached to a carrier protein (http://www.alzforum.com). Two other phase I active vaccination trials are underway: the CAD-106 trial led by a Novartis/Cytos collaboration and the V950 trial initiated by Merck.

FDA-Approved Immunotherapeutics

The safety concerns raised by AN1792, as well as several other new drugs in recent years, have spurred interest in using FDA-approved therapeutics with established safety records for new indications (e.g., Rothstein et al. 2005). Dodel et al. (2002) have found that human intravenous immunoglobulin (IVIg) contains some endogenously produced anti-Aβ antibodies, along with a great variety of other antibodies.

Two small phase I studies of IVIg infusions have been completed (Dodel et al. 2004), and a phase II study is underway as of this writing. An amyloid-imaging substudy with the PET ligand Pittsburg Compound-B (Klunk et al. 2004) is also part of this trial (http://www.clinicaltrials.gov).

FUTURE DIRECTIONS

Preclinical Predictors of Clinical Efficacy

One major question remains how to decide which preclinical therapeutics to move into clinical trials. The issue is that we do not have a solid understanding of which outcomes in preclinical models will best predict efficacy in human patients. Aβ plaques in many mouse models are very similar to those in human AD patients. However, it is not clear whether reducing Aβ plaque burden in the brains of patients with concomitant tangle pathology, reduced numbers of synapses, and extensive neuronal loss will have a meaningful benefit in terms of memory and other cognitive functions. In addition, the behavioral performance deficits in the mouse models are quite different in many respects from the clinical symptoms in human patients, thus the importance of targeting these deficits has been hotly debated. Specifically, many mouse models show early behavioral performance deficits on tests of memory and other cognitive function at young ages, often before substantial plaque pathology is observed (Smith et al. 1998, Lesne et al. 2006). In contrast, significant Aβ plaque pathology is likely present for a decade or longer before cognitive symptoms are apparent in humans (Braak & Braak 1997, Price & Morris 1999, Price et al. 2001).

Based on the data available in humans, cognitive symptoms correlate best with the onset and progression of neuronal and synaptic loss later in life (Masliah et al. 2001, Price et al. 2001, Scheff & Price 2003). However, no mouse model developed to date fully recapitulates these aspects of the disease.

Recently, attention from several groups has turned to Aβ-targeted immunotherapeutic effects on additional aspects of AD-related pathology seen in the mouse models. Preclinical studies have reported improvements in abnormalities of the geometry of neuronal processes (Lombardo et al. 2003), early tau pathology (Oddo et al. 2004), loss of synaptophysin immunoreactivity (Buttini et al. 2005), and neuritic dystrophy (Brendza et al. 2005). More advanced tau pathology was not affected by anti-Aβ antibody treatment (Oddo et al. 2004) nor did it appear that there was a clear reduction in tau pathology in the vaccinated human patients in the AN1792 trial.

Thus it is not clear at this time whether preclinical immunotherapeutics should be advanced to clinical trials on the basis of efficacy in (a) reducing Aβ plaque and/or tangle pathology, (b) blocking effects of soluble oligomeric Aβ species, (c) alleviating behavioral deficits, or (d) preventing neuronal and/or synaptic loss. An immunotherapeutic strategy may need to be successful at only some or all of these to offer a meaningful benefit to human patients. Although the panoply of available animal models may be useful, none currently recapitulates all aspects of human AD.

Because of the enormous resources being devoted to AD, effective clinical therapeutics for this condition may be developed by trial and error without a thorough mechanistic understanding of their actions. Although any effective therapeutic would be most welcome, the ability to generalize the approach to other, less common, or as-yet-undiscovered neurodegenerative conditions would be markedly enhanced if the mechanistic principles underlying efficacy were understood. Thus, further detailed investigation of the pathophysiology of neurodegenerative diseases in human patients and careful exploration of the effects of immunotherapeutic strategies in both mice and humans are clearly required.

Clinical Trial Design

A second major question revolves around the design of the clinical trials themselves. Should patients with relatively rare, heritable forms of the disorders be enrolled first? Nearly all the preclinical animal models are based on the genes implicated in these subsets of patients, and therefore one could argue that therapeutics developed using such animal models are most likely to work in the human diseases bearing the most resemblance to them. Or should the more common, sporadic, later-onset forms of the diseases be attacked first, in the hopes of bringing the greatest benefit soonest to the largest numbers of people?

Likewise, should symptomatic patients be enrolled, or should efforts focus on identifying presymptomatic patients with genetic, biomarker-based or imaging evidence of disease pathophysiology? An argument for presymptomatic treatment states that in preclinical studies, many therapeutic approaches have been more effective in younger mice with less advanced disease than in older mice with more advanced disease. Because the central nervous system has limited regenerative capacity, neuronal and synaptic loss is not likely to be reversible once it has occurred, and the most effective immunotherapeutic approaches will therefore require prevention of such losses. The analogy to treating coronary artery disease before the development of ischemic cardiomyopathy and consequent congestive heart failure is salient. Although none of our current approaches for detecting presymptomatic patients with nonautosomal dominant familial neurodegenerative disorders has been fully validated at present, encouraging progress along these lines has been made using noninvasive PET-based imaging of fibrillar Aβ structures and measurement of cerebrospinal fluid levels of Aβ42 and tau (Klunk et al. 2004, Fagan et al. 2006, Fagan et al. 2007). Thus this approach may be feasible in the near future.

Economic Issues

A third, but clearly important issue is an economic one. The expense of giving weekly or monthly monoclonal antibody infusions for years or decades could be considerable. The price of commonly used monoclonal antibodies on the market today is typically hundreds to thousands of dollars per dose, and this price does not include the costs associated with starting intravenous lines and monitoring infusions. This is not to say that the approach is not feasible; cancer treatments routinely involve monitored intravenous infusions of very expensive medications. However, cycles of chemotherapy usually last weeks to months, rather than years to decades. On the other hand, if these treatments can keep patients out of nursing homes longer, allow them to take better care of their other health issues, and potentially even keep them in the workforce longer, the economic effect could still be a net benefit. Nonetheless, the specter of effective therapeutics affordable only to an elite few has its harbinger in issues about the availability of Herceptin in certain health care systems (Hutchinson & DeVita 2005). Active immunization, despite the safety issues that have arisen, would likely be much more cost-effective than passive immunotherapy. Small molecules or other approaches unrelated to immunotherapeutics may prove still more economically efficient in the long run.

CONCLUSION

Overall, the field of active and passive immunotherapeutics for neurodegenerative disorders has been a vibrant one. However, a

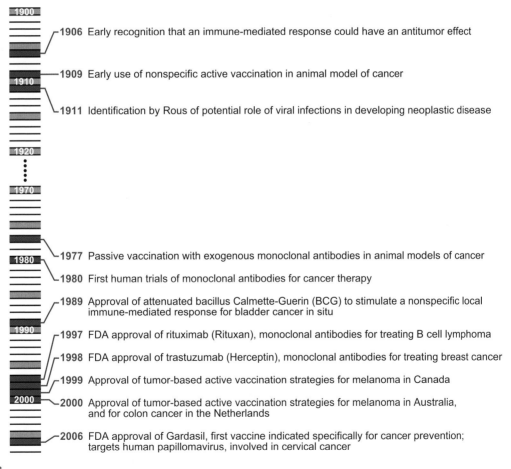

1900

1906 Early recognition that an immune-mediated response could have an antitumor effect

1910

1909 Early use of nonspecific active vaccination in animal model of cancer

1911 Identification by Rous of potential role of viral infections in developing neoplastic disease

1920

1970

1977 Passive vaccination with exogenous monoclonal antibodies in animal models of cancer

1980

1980 First human trials of monoclonal antibodies for cancer therapy

1989 Approval of attenuated bacillus Calmette-Guerin (BCG) to stimulate a nonspecific local immune-mediated response for bladder cancer in situ

1990

1997 FDA approval of rituximab (Rituxan), monoclonal antibodies for treating B cell lymphoma

1998 FDA approval of trastuzumab (Herceptin), monoclonal antibodies for treating breast cancer

1999 Approval of tumor-based active vaccination strategies for melanoma in Canada

2000

2000 Approval of tumor-based active vaccination strategies for melanoma in Australia, and for colon cancer in the Netherlands

2006 FDA approval of Gardasil, first vaccine indicated specifically for cancer prevention; targets human papillomavirus, involved in cervical cancer

Figure 3

Putting the timeline in perspective: immunotherapeutic strategies for neoplastic disorders.

review of the timeline in the field of immunotherapeutic development for neoplastic disorders suggests that we may still have a long way to go. To date, cancer research has subsumed an even greater net effort than has research in the neurodegenerative disorders. Yet a full 20 years elapsed from when the results of first trials of monoclonal antibodies in animal models were reported in 1977 until the first FDA approval of a monoclonal antibody-based therapeutic for a human cancer in 1997 (see **Figure 3**). Thus, far from being near the end of the road in terms of fundamental research on neurodegenerative disorders, it seems more likely that we are solidly in the middle of the journey, with many miles to go before we rest.

FUTURE ISSUES

1. How can patients with neurodegenerative disorders be identified in the earliest, even preclinical stages, and will immunotherapies initiated at these early stages prove especially effective?

2. Which pathophysiological events are targeted by effective immunotherapeutic strategies in preclinical animal models of disease? Do these same pathophysiological events occur and play important roles in human patients?

3. Will immunotherapeutic strategies specifically targeting rare but highly toxic soluble assemblies of the proteins believed to play central roles in neurodegenerative diseases offer therapeutic advantages over other targeting strategies?

4. How can clinical trials in neurodegenerative disorders—which often progress over the span of years to decades—be accelerated? Which biomarkers, clinical assessment strategies, or imaging approaches will make clinical trial design more efficient?

5. Can the essential aspects of the human immune response be modeled using experimental animals, including the tremendous variability caused by genetic and environmental factors? Will this approach lead to more rapid development of safe and effective therapeutics?

DISCLOSURE STATEMENT

David Holtzman is a coinventor on U.S. patent 7,195,761 entitled "Humanized antibodies that sequester abeta peptide" as well as other submitted patents on related topics. Washington University may receive income based on a license of these patents to Eli Lilly.

ACKNOWLEDGMENTS

This work was supported by NIH grants AG13956, AG20222, and NS049237, a Burroughs Wellcome Career Award in the Biomedical Sciences, The Cure Alzheimer's Fund, and the MetLife Foundation. We thank Ronald Demattos and Edgar Engleman for helpful comments.

LITERATURE CITED

Angelov DN, Waibel S, Guntinas-Lichius O, Lenzen M, Neiss WF, et al. 2003. Therapeutic vaccine for acute and chronic motor neuron diseases: implications for amyotrophic lateral sclerosis. *Proc. Natl. Acad. Sci. USA* 100:4790–95

Austin L, Arendash GW, Gordon MN, Diamond DM, DiCarlo G, et al. 2003. Short-term beta-amyloid vaccinations do not improve cognitive performance in cognitively impaired APP + PS1 mice. *Behav. Neurosci.* 117:478–84

Bacskai BJ, Kajdasz ST, McLellan ME, Games D, Seubert P, et al. 2002. Non-Fc-mediated mechanisms are involved in clearance of amyloid-beta in vivo by immunotherapy. *J. Neurosci.* 22:7873–78

Bade S, Frey A. 2007. Potential of active and passive immunizations for the prevention and therapy of transmissible spongiform encephalopathies. *Expert Rev. Vaccin.* 6:153–68

Bales KR, Tzavara ET, Wu S, Wade MR, Bymaster FP, et al. 2006. Cholinergic dysfunction in a mouse model of Alzheimer disease is reversed by an anti-Abeta antibody. *J. Clin. Invest.* 116:825–32

Bard F, Cannon C, Barbour R, Burke RL, Games D, et al. 2000. Peripherally administered antibodies against amyloid beta-peptide enter the central nervous system and reduce pathology in a mouse model of Alzheimer disease. *Nat. Med.* 6:916–19

Bard F, Barbour R, Cannon C, Carretto R, Fox M, et al. 2003. Epitope and isotype specificities of antibodies to beta-amyloid peptide for protection against Alzheimer's disease-like neuropathology. *Proc. Natl. Acad. Sci. USA* 100:2023–28

Bayer AJ, Bullock R, Jones RW, Wilkinson D, Paterson KR, et al. 2005. Evaluation of the safety and immunogenicity of synthetic Abeta42 (AN1792) in patients with AD. *Neurology* 64:94–101

Benner EJ, Mosley RL, Destache CJ, Lewis TB, Jackson-Lewis V, et al. 2004. Therapeutic immunization protects dopaminergic neurons in a mouse model of Parkinson's disease. *Proc. Natl. Acad. Sci. USA* 101:9435–40

Braak H, Braak E. 1997. Frequency of stages of Alzheimer-related lesions in different age categories. *Neurobiol. Aging* 18:351–57

Brendza RP, Bacskai BJ, Cirrito JR, Simmons KA, Skoch JM, et al. 2005. Anti-Abeta antibody treatment promotes the rapid recovery of amyloid-associated neuritic dystrophy in PDAPP transgenic mice. *J. Clin. Invest.* 115:428–33

Brody DL, Mac Donald CL, Fujimoto S, Conte V, Parsadanian M, et al. 2005. *Anti-Abeta antibody treatment attenuates cognitive impairment in a mouse model of experimental traumatic brain injury.* Presented at Annu. Natl. Neurotrauma Soc. Symp., 23rd, Washington DC

Buttini M, Masliah E, Barbour R, Grajeda H, Motter R, et al. 2005. Beta-amyloid immunotherapy prevents synaptic degeneration in a mouse model of Alzheimer's disease. *J. Neurosci.* 25:9096–101

Chen G, Chen KS, Kobayashi D, Barbour R, Motter R, et al. 2007. Active beta-amyloid immunization restores spatial learning in PDAPP mice displaying very low levels of beta-amyloid. *J. Neurosci.* 27:2654–62

Chishti MA, Yang DS, Janus C, Phinney AL, Horne P, et al. 2001. Early-onset amyloid deposition and cognitive deficits in transgenic mice expressing a double mutant form of amyloid precursor protein 695. *J. Biol. Chem.* 276:21562–70

Cleary JP, Walsh DM, Hofmeister JJ, Shankar GM, Kuskowski MA, et al. 2005. Natural oligomers of the amyloid-beta protein specifically disrupt cognitive function. *Nat. Neurosci.* 8:79–84

Das P, Howard V, Loosbrock N, Dickson D, Murphy MP, et al. 2003. Amyloid-beta immunization effectively reduces amyloid deposition in FcRgamma−/− knock-out mice. *J. Neurosci.* 23:8532–38

Das P, Murphy MP, Younkin LH, Younkin SG, Golde TE. 2001. Reduced effectiveness of Abeta1-42 immunization in APP transgenic mice with significant amyloid deposition. *Neurobiol. Aging* 22:721–27

Deane R, Du Yan S, Submamaryan RK, LaRue B, Jovanovic S, et al. 2003. RAGE mediates amyloid-beta peptide transport across the blood-brain barrier and accumulation in brain. *Nat. Med.* 9:907–13

Deane R, Sagare A, Hamm K, Parisi M, LaRue B, et al. 2005. IgG-assisted age-dependent clearance of Alzheimer's amyloid beta peptide by the blood-brain barrier neonatal Fc receptor. *J. Neurosci.* 25:11495–503

DeMattos RB, Bales KR, Cummins DJ, Dodart JC, Paul SM, et al. 2001. Peripheral anti-Abeta antibody alters CNS and plasma Abeta clearance and decreases brain Abeta burden in a mouse model of Alzheimer's disease. *Proc. Natl. Acad. Sci. USA* 98:8850–55

Dodart JC, Bales KR, Gannon KS, Greene SJ, DeMattos RB, et al. 2002. Immunization reverses memory deficits without reducing brain Abeta burden in Alzheimer's disease model. *Nat. Neurosci.* 5:452–57

Dodel R, Hampel H, Depboylu C, Lin S, Gao F, et al. 2002. Human antibodies against amyloid beta peptide: a potential treatment for Alzheimer's disease. *Ann. Neurol.* 52:253–56

Dodel RC, Du Y, Depboylu C, Hampel H, Frolich L, et al. 2004. Intravenous immunoglobulins containing antibodies against beta-amyloid for the treatment of Alzheimer's disease. *J. Neurol. Neurosurg. Psychiatry* 75:1472–74

Eng JA, Frosch MP, Choi K, Rebeck GW, Greenberg SM. 2004. Clinical manifestations of cerebral amyloid angiopathy-related inflammation. *Ann. Neurol.* 55:250–56

Fagan AM, Mintun MA, Mach RH, Lee SY, Dence CS, et al. 2006. Inverse relation between in vivo amyloid imaging load and cerebrospinal fluid Abeta42 in humans. *Ann. Neurol.* 59:512–19

Fagan AM, Roe CM, Xiong C, Mintun MA, Morris JC, et al. 2007. Cerebrospinal fluid tau/beta-amyloid(42) ratio as a prediction of cognitive decline in nondemented older adults. *Arch. Neurol.* 64:343–49

Ferrer I, Boada Rovira M, Sanchez Guerra ML, Rey MJ, Costa-Jussa F. 2004. Neuropathology and pathogenesis of encephalitis following amyloid-beta immunization in Alzheimer's disease. *Brain Pathol.* 14:11–20

Fox NC, Black RS, Gilman S, Rossor MN, Griffith SG, et al. 2005. Effects of Abeta immunization (AN1792) on MRI measures of cerebral volume in Alzheimer disease. *Neurology* 64:1563–72

Frenkel D, Maron R, Burt DS, Weiner HL. 2005. Nasal vaccination with a proteosome-based adjuvant and glatiramer acetate clears beta-amyloid in a mouse model of Alzheimer disease. *J. Clin. Invest.* 115:2423–33

Games D, Adams D, Alessandrini R, Barbour R, Berthelette P, et al. 1995. Alzheimer-type neuropathology in transgenic mice overexpressing V717F beta-amyloid precursor protein. *Nature* 373:523–27

Gilman S, Koller M, Black RS, Jenkins L, Griffith SG, et al. 2005. Clinical effects of Abeta immunization (AN1792) in patients with AD in an interrupted trial. *Neurology* 64:1553–62

Goni F, Knudsen E, Schreiber F, Scholtzova H, Pankiewicz J, et al. 2005. Mucosal vaccination delays or prevents prion infection via an oral route. *Neuroscience* 133:413–21

Haenggeli C, Julien JP, Lee Mosley R, Perez N, Dhar A, et al. 2007. Therapeutic immunization with a glatiramer acetate derivative does not alter survival in G93A and G37R SOD1 mouse models of familial ALS. *Neurobiol. Dis.* 26:146–52

Handisurya A, Gilch S, Winter D, Shafti-Keramat S, Maurer D, et al. 2007. Vaccination with prion peptide-displaying papillomavirus-like particles induces autoantibodies to normal prion protein that interfere with pathologic prion protein production in infected cells. *FEBS J.* 274:1747–58

Hartman RE, Izumi Y, Bales KR, Paul SM, Wozniak DF, et al. 2005. Treatment with an amyloid-beta antibody ameliorates plaque load, learning deficits, and hippocampal long-term potentiation in a mouse model of Alzheimer's disease. *J. Neurosci.* 25:6213–20

Hepler RW, Grimm KM, Nahas DD, Breese R, Dodson EC, et al. 2006. Solution state characterization of amyloid beta-derived diffusible ligands. *Biochemistry* 45:15157–67

Heppner FL, Musahl C, Arrighi I, Klein MA, Rulicke T, et al. 2001. Prevention of scrapie pathogenesis by transgenic expression of anti-prion protein antibodies. *Science* 294:178–82

Hock C, Konietzko U, Streffer JR, Tracy J, Signorell A, et al. 2003. Antibodies against beta-amyloid slow cognitive decline in Alzheimer's disease. *Neuron* 38:547–54

Hsiao K, Chapman P, Nilsen S, Eckman C, Harigaya Y, et al. 1996. Correlative memory deficits, Abeta elevation, and amyloid plaques in transgenic mice. *Science* 274:99–102

Hutchinson L, DeVita VT Jr. 2005. Herceptin: HERalding a new era in breast cancer care but at what cost? *Nat. Clin. Pract. Oncol.* 2:595

Janus C, Pearson J, McLaurin J, Mathews PM, Jiang Y, et al. 2000. Abeta peptide immunization reduces behavioural impairment and plaques in a model of Alzheimer's disease. *Nature* 408:979–82

Jellinger KA. 2004. Head injury and dementia. *Curr. Opin. Neurol.* 17:719–23

Jenner E. 1798. *An inquiry into the causes and effects of the Variolae-Vacciniae, a disease discovered in some of the western counties of England, particularly Gloucestershire, and known by the name of cow pox.* **http://www.bartleby.com/38/4/1.html**

Johnson-Wood K, Lee M, Motter R, Hu K, Gordon G, et al. 1997. Amyloid precursor protein processing and Abeta42 deposition in a transgenic mouse model of Alzheimer disease. *Proc. Natl. Acad. Sci. USA* 94:1550–55

Klunk WE, Engler H, Nordberg A, Wang Y, Blomqvist G, et al. 2004. Imaging brain amyloid in Alzheimer's disease with Pittsburgh Compound-B. *Ann. Neurol.* 55:306–19

Klunk WE, Price JC, Mathis CA, Tsopelas ND, Lopresti BJ, et al. 2007. Amyloid deposition begins in the striatum of presenilin-1 mutation carriers from two unrelated pedigrees. *J. Neurosci.* 27:6174–84

Klyubin I, Walsh DM, Lemere CA, Cullen WK, Shankar GM, et al. 2005. Amyloid beta protein immunotherapy neutralizes Abeta oligomers that disrupt synaptic plasticity in vivo. *Nat. Med.* 11:556–61

Knappik A, Ge L, Honegger A, Pack P, Fischer M, et al. 2000. Fully synthetic human combinatorial antibody libraries (HuCAL) based on modular consensus frameworks and CDRs randomized with trinucleotides. *J. Mol. Biol.* 296:57–86

Kotilinek LA, Bacskai B, Westerman M, Kawarabayashi T, Younkin L, et al. 2002. Reversible memory loss in a mouse transgenic model of Alzheimer's disease. *J. Neurosci.* 22:6331–35

Lee EB, Leng LZ, Zhang B, Kwong L, Trojanowski JQ, et al. 2006. Targeting amyloid-beta peptide (Abeta) oligomers by passive immunization with a conformation-selective monoclonal antibody improves learning and memory in Abeta precursor protein (APP) transgenic mice. *J. Biol. Chem.* 281:4292–99

Lee M, Bard F, Johnson-Wood K, Lee C, Hu K, et al. 2005. Abeta42 immunization in Alzheimer's disease generates Abeta N-terminal antibodies. *Ann. Neurol.* 58:430–35

Lee VM, Trojanowski JQ. 2006. Progress from Alzheimer's tangles to pathological tau points towards more effective therapies now. *J. Alzheimers Dis.* 9:257–62

Legleiter J, Czilli DL, Gitter B, DeMattos RB, Holtzman DM, et al. 2004. Effect of different anti-Abeta antibodies on Abeta fibrillogenesis as assessed by atomic force microscopy. *J. Mol. Biol.* 335:997–1006

Lesne S, Koh MT, Kotilinek L, Kayed R, Glabe CG, et al. 2006. A specific amyloid-beta protein assembly in the brain impairs memory. *Nature* 440:352–57

Lombardo JA, Stern EA, McLellan ME, Kajdasz ST, Hickey GA, et al. 2003. Amyloid-beta antibody treatment leads to rapid normalization of plaque-induced neuritic alterations. *J. Neurosci.* 23:10879–83

Masliah E, Hansen L, Adame A, Crews L, Bard F, et al. 2005a. Abeta vaccination effects on plaque pathology in the absence of encephalitis in Alzheimer disease. *Neurology* 64:129–31

Masliah E, Mallory M, Alford M, DeTeresa R, Hansen LA, et al. 2001. Altered expression of synaptic proteins occurs early during progression of Alzheimer's disease. *Neurology* 56:127–29

Masliah E, Rockenstein E, Adame A, Alford M, Crews L, et al. 2005b. Effects of alpha-synuclein immunization in a mouse model of Parkinson's disease. *Neuron* 46:857–68

Mohajeri MH, Saini K, Schultz JG, Wollmer MA, Hock C, et al. 2002. Passive immunization against beta-amyloid peptide protects central nervous system (CNS) neurons from increased vulnerability associated with an Alzheimer's disease-causing mutation. *J. Biol. Chem.* 277:33012–17

Morgan D, Diamond DM, Gottschall PE, Ugen KE, Dickey C, et al. 2000. Abeta peptide vaccination prevents memory loss in an animal model of Alzheimer's disease. *Nature* 408:982–85

Morris R. 1984. Developments of a water-maze procedure for studying spatial learning in the rat. *J. Neurosci. Methods* 11:47–60

Nicoll JA, Wilkinson D, Holmes C, Steart P, Markham H, et al. 2003. Neuropathology of human Alzheimer disease after immunization with amyloid-beta peptide: a case report. *Nat. Med.* 9:448–52

Oddo S, Billings L, Kesslak JP, Cribbs DH, LaFerla FM. 2004. Abeta immunotherapy leads to clearance of early, but not late, hyperphosphorylated tau aggregates via the proteasome. *Neuron* 43:321–32

Oddo S, Caccamo A, Tran L, Lambert MP, Glabe CG, et al. 2006a. Temporal profile of amyloid-beta (Abeta) oligomerization in an in vivo model of Alzheimer disease. A link between Abeta and tau pathology. *J. Biol. Chem.* 281:1599–604

Oddo S, Vasilevko V, Caccamo A, Kitazawa M, Cribbs DH, et al. 2006b. Reduction of soluble Abeta and tau, but not soluble Abeta alone, ameliorates cognitive decline in transgenic mice with plaques and tangles. *J. Biol. Chem.* 281:39413–23

Orgogozo JM, Gilman S, Dartigues JF, Laurent B, Puel M, et al. 2003. Subacute meningoencephalitis in a subset of patients with AD after Abeta42 immunization. *Neurology* 61:46–54

Patton RL, Kalback WM, Esh CL, Kokjohn TA, Van Vickle GD, et al. 2006. Amyloid-beta peptide remnants in AN-1792-immunized Alzheimer's disease patients: a biochemical analysis. *Am. J. Pathol.* 169:1048–63

Pfeifer M, Boncristiano S, Bondolfi L, Stalder A, Deller T, et al. 2002. Cerebral hemorrhage after passive anti-Abeta immunotherapy. *Science* 298:1379

Price JL, Ko AI, Wade MJ, Tsou SK, McKeel DW, et al. 2001. Neuron number in the entorhinal cortex and CA1 in preclinical Alzheimer disease. *Arch. Neurol.* 58:1395–402

Price JL, Morris JC. 1999. Tangles and plaques in nondemented aging and "preclinical" Alzheimer's disease. *Ann. Neurol.* 45:358–68

Racke MM, Boone LI, Hepburn DL, Parsadainian M, Bryan MT, et al. 2005. Exacerbation of cerebral amyloid angiopathy-associated microhemorrhage in amyloid precursor protein transgenic mice by immunotherapy is dependent on antibody recognition of deposited forms of amyloid beta. *J. Neurosci.* 25:629–36

Rothstein JD, Patel S, Regan MR, Haenggeli C, Huang YH, et al. 2005. Beta-lactam antibiotics offer neuroprotection by increasing glutamate transporter expression. *Nature* 433:73–77

Scheff SW, Price DA. 2003. Synaptic pathology in Alzheimer's disease: a review of ultrastructural studies. *Neurobiol. Aging* 24:1029–46

Schenk D, Barbour R, Dunn W, Gordon G, Grajeda H, et al. 1999. Immunization with amyloid-beta attenuates Alzheimer-disease-like pathology in the PDAPP mouse. *Nature* 400:173–77

Skovronsky DM, Lee VM, Trojanowski JQ. 2006. Neurodegenerative diseases: new concepts of pathogenesis and their therapeutic implications. *Annu. Rev. Pathol. Mech. Dis.* 1:151–70

Smith DH, Nakamura M, McIntosh TK, Wang J, Rodriguez A, et al. 1998. Brain trauma induces massive hippocampal neuron death linked to a surge in beta-amyloid levels in mice overexpressing mutant amyloid precursor protein. *Am. J. Pathol.* 153:1005–10

Solomon B, Koppel R, Frankel D, Hanan-Aharon E. 1997. Disaggregation of Alzheimer beta-amyloid by site-directed mAb. *Proc. Natl. Acad. Sci. USA* 94:4109–12

Solomon B, Koppel R, Hanan E, Katzav T. 1996. Monoclonal antibodies inhibit in vitro fibrillar aggregation of the Alzheimer beta-amyloid peptide. *Proc. Natl. Acad. Sci. USA* 93:452–55

Tamura Y, Hamajima K, Matsui K, Yanoma S, Narita M, et al. 2005. The F(ab)'2 fragment of an Abeta-specific monoclonal antibody reduces Abeta deposits in the brain. *Neurobiol. Dis.* 20:541–49

Towfighi A, Greenberg SM, Rosand J. 2005. Treatment and prevention of primary intracerebral hemorrhage. *Semin. Neurol.* 25:445–52

Urushitani M, Ezzi SA, Julien JP. 2007. Therapeutic effects of immunization with mutant superoxide dismutase in mice models of amyotrophic lateral sclerosis. *Proc. Natl. Acad. Sci. USA* 104:2495–500

Walsh DM, Klyubin I, Fadeeva JV, Cullen WK, Anwyl R, et al. 2002. Naturally secreted oligomers of amyloid beta protein potently inhibit hippocampal long-term potentiation in vivo. *Nature* 416:535–39

Walsh DM, Selkoe DJ. 2004. Deciphering the molecular basis of memory failure in Alzheimer's disease. *Neuron* 44:181–93

White AR, Enever P, Tayebi M, Mushens R, Linehan J, et al. 2003. Monoclonal antibodies inhibit prion replication and delay the development of prion disease. *Nature* 422:80–83

Wilcock DM, DiCarlo G, Henderson D, Jackson J, Clarke K, et al. 2003. Intracranially administered anti-Abeta antibodies reduce beta-amyloid deposition by mechanisms both independent of and associated with microglial activation. *J. Neurosci.* 23:3745–51

Wilcock DM, Munireddy SK, Rosenthal A, Ugen KE, Gordon MN, et al. 2004a. Microglial activation facilitates Abeta plaque removal following intracranial anti-Abeta antibody administration. *Neurobiol. Dis.* 15:11–20

Wilcock DM, Rojiani A, Rosenthal A, Subbarao S, Freeman MJ, et al. 2004b. Passive immunotherapy against Abeta in aged APP-transgenic mice reverses cognitive deficits and depletes parenchymal amyloid deposits in spite of increased vascular amyloid and microhemorrhage. *J. Neuroinflamm.* 1:24

Descending Pathways
in Motor Control

Roger N. Lemon

Sobell Department of Motor Neuroscience and Movement Disorders, Institute of
Neurology, University College London, London, WC1N 3BG, United Kingdom;
email: rlemon@ion.ucl.ac.uk

Annu. Rev. Neurosci. 2008. 31:195–218

First published online as a Review in Advance on
April 4, 2008

The *Annual Review of Neuroscience* is online at
neuro.annualreviews.org

This article's doi:
10.1146/annurev.neuro.31.060407.125547

Key Words

spinal cord, motoneuron, corticospinal, reticulospinal, tract

Abstract

Each of the descending pathways involved in motor control has a number of anatomical, molecular, pharmacological, and neuroinformatic
characteristics. They are differentially involved in motor control, a
process that results from operations involving the entire motor network rather than from the brain commanding the spinal cord. A given
pathway can have many functional roles. This review explores to what
extent descending pathways are highly conserved across species and
concludes that there are actually rather widespread species differences,
for example, in the transmission of information from the corticospinal
tract to upper limb motoneurons. The significance of direct, cortico-
motoneuronal (CM) connections, which were discovered a little more
than 50 years ago, is reassessed. I conclude that although these connections operate in parallel with other less direct linkages to motoneurons,
CM influence is significant and may subserve some special functions
including adaptive motor behaviors involving the distal extremities.

Contents

INTRODUCTION

The time is ripe for a review in this area, not only because of important new knowledge, but also because of the clinical importance attached to a fuller understanding of the role of descending pathways in motor control. There are two main areas of topical interest. In human spinal cord injury, trials are in progress in different parts of the world to test a variety of cell-based and drug-based therapeutic approaches (Ramer et al. 2005, Adams et al. 2006, Tator 2006, Courtine et al. 2007). In stroke, the physiological consequences of different types of therapy and rehabilitative approaches are being studied (Nudo 2007, Ward & Cohen 2004). For progress in these areas to continue, it is increasingly important to understand more fully the functional contribution of the different descending systems that have been injured. Whether these systems can regenerate or whether surviving systems can play compensatory roles and how these various processes can be boosted by appropriate therapy (Case & Tessier-Lavigne 2005, Deumens et al. 2005) (see **Figure 6**) are related clinical issues.

Scope and Structure

I focus very much on the comparative biology of the descending pathways and its relevance to the control of skilled hand movements. The review is structured to address the following points and issues:

- How are the descending pathways organized, and what are their defining characteristics?
- Does a descending pathway carry out single or multiple functions?
- Is the corticospinal tract a motor pathway?

- Is the organization of the descending pathways highly conserved across different mammalian species?
- What is the functional significance of direct connections with target motoneurons?

HOW ARE THE DESCENDING PATHWAYS ORGANIZED, AND WHAT ARE THEIR DEFINING CHARACTERISTICS?

Mammalian motor pathways involve a number of different descending systems, some of which are conserved from reptiles and other vertebrate species and others that have appeared much later in evolution. **Table 1** is a checklist of ten properties that characterize a descending pathway. They include key neuroanatomical features, including the origin, course, and pattern of pathway termination and also fiber number and size. We need to know much more about the molecular identity of each pathway and how this guides the pathway to cross or not to cross the midline, to find its target neurons, and to avoid others. The neuropharmacological features include the neurotransmitter and neuromodulators released at the pathway's terminals. Finally, we need to add the neuroinformatic features: the activity/information that the pathway transmits to those targets. In some cases one can inactivate or permanently lesion

a pathway in a selective manner that allows additional insight into function.

Unfortunately, a completed checklist of all these features is still not available for any of the major mammalian descending pathways. We now have advanced anatomical details for many of them, but the functional roles of each pathway and how they relate to these anatomical features are still unresolved. In particular, we lack evidence in the awake animal or human volunteer as to the nature of the information that these different pathways transmit to their spinal targets. For the generation of purposeful movements, target interneurons and motoneurons must integrate this information with that from other descending pathways and propriospinal and segmental inputs.

Mechanisms controlling features such as firing threshold, synaptic gain, and possible bistable properties could all play a major part in this integrative response (Hultborn et al. 2004). We need to understand that the descending pathways function as part of a large network rather than as separate controllers of the spinal cord. As Edgerton wrote, "the spinal cord functions as part of the brain, not as its servant." Descending pathways do not simply telegraph commands for movement to the spinal apparatus; hence we should abandon the use of terms such as "upper motoneuron" and "lower motoneuron," terms of undoubted convenience in the domain of clinical neurology but with

Table 1 Ten characteristics of a descending pathway

1	Origin	Location of cells of origin of the pathway
2	Synaptic input	Nature of the major inputs to these cells of origin
3	Fiber number and size	Numbers of fibers making up the pathway, and distribution of fiber diameters within the pathway
4	Course	Trajectory followed by fibers belonging to the pathway
5	Target/termination	Location and type of interneurons and motoneurons etc. receiving terminations from the pathway: defined both by level within the spinal cord and lamina within the gray matter
6	Collaterals	Other supraspinal targets innervated by axon collaterals from the same pathway
7	Molecular identity	Characteristic surface and other molecules important for axon guidance, target finding and synaptogenesis
8	Transmitter(s) and neuromodulators	Transmitters employed at synaptic and presynaptic targets of the descending pathway
9	Activity/information transmission	Timing, pattern and type of activity exhibited by neurons contributing to the pathway
10	Lesion/inactivation	Effects on behavior

no modern-day neuroanatomical or neuroscientific justification.

Grouping Descending Pathways According to Their Spinal Targets

Hans Kuypers's (1925–1989) major contributions to our understanding of the neuroanatomy of the descending pathways (see Kuypers 1981) have a continuing influence in the field. He worked first with silver degeneration methods, then with retrograde and anterograde labeling using radiolabeling, fluorescent, and other tracers, and finally viral transneuronal markers (Kuypers & Ugolini 1990).

Kuypers' study of descending pathways convinced him that the key to understanding their function was to examine their termination pattern within the spinal gray matter: to define the address to which descending activity is sent (Kuypers & Brinkman 1970). Using this approach, Kuypers identified three groups of brainstem pathways.

Group A: ventromedial brainstem pathways. These include the interstitiospinal and tectospinal tracts arising from the midbrain, the lateral and medial vestibulospinal tracts (Sugiuchi et al. 2004), and the reticulospinal and bulbospinal projections arising from the pontine and medullary reticular formation (Matsuyama et al. 1997, 1999; brainstem areas colored green on the right side of **Figure 1**). These pathways descend in the ventral and ventrolateral funiculi of the spinal cord and have characteristic terminations in Rexed lamina VII and VIII, i.e., the ventromedial part of the intermediate zone (IZ), with many axons terminating bilaterally (region colored green in the spinal cord, **Figure 1**). This region of the IZ gives rise to long propriospinal neurons whose bilateral projections link widely separated parts of the spinal cord, including the cervical and lumbar enlargements. Kuypers considered this group of pathways as a bilateral postural control system for head, neck, trunk, and proximal limb movements (Lawrence &

Kuypers 1968b, Kuypers 1981). An important subdivision of this group includes those involved in the control of respiration (Monteau & Hilaire 1991, Boers et al. 2005, de Troyer et al. 2005).

Group B: dorsolateral brainstem pathways. These include the rubrospinal tract arising from the magnocellular red nucleus (brainstem area and fibers colored red on the right side of **Figure 1**) (Kennedy 1990, Muir & Whishaw 2000, Kuchler et al. 2002) and the pontospinal tract (arising from the ventrolateral pontine tegmentum); they descend contralaterally in the dorsolateral funiculus. They are characterized by their pattern of termination in the dorsal and lateral regions of the IZ (region colored red in the spinal cord, **Figure 1**); this region in turn gives rise mostly to short propriospinal neurons and has more local, mainly unilateral projections. Kuypers considered this group of pathways to provide additional capacity for flexion-biased movements involving more distal limb segments, the elbow and wrist.

Emotional motor system. A number of other pathways influence motor and other functions at the spinal level. Holstege (1998) grouped these together as the 'emotional motor system'. A medial component comprises a diffuse system of pathways originating from the lower brainstem (raphespinal tract; Mason 1997), the ventromedial medullary tegmentum, the locus coeruleus, and the subcoeruleus (Tanaka et al. 1997). These fibers terminate widely at all spinal levels in the dorsal horn and among autonomic and somatic motoneuronal cell groups. Transmitters and neuromodulators in these pathways include serotonin (5-HT) and noradrenaline; the 5-HT pathways exert a major level-setting influence on spinal reflexes and motoneuronal membrane properties and excitability (Heckman & Lee 1999).

A second, lateral component originates from cell groups in the mid- and forebrain, which are involved in a number of specific motor activities, including defensive reactions, pupil

IZ: intermediate zone of the spinal cord gray matter

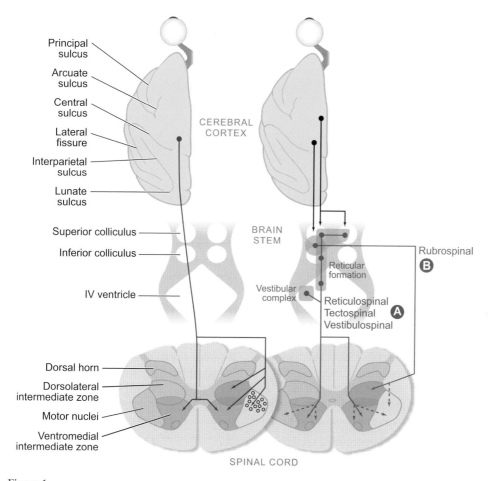

Figure 1

Schematic representations of the distributions of the corticospinal fibers and the fibers belonging to group A (ventromedial) and group B (dorsolateral) brainstem pathways, according to the scheme proposed by Kuypers (1981). On the right, Group A fibers (reticulospinal, tectospinal, vestibulospinal) are shown in green, arising from the brainstem reticular formation, superior colliculus, and vestibular complex. These fibers terminate bilaterally in the ventromedial part of the intermediate zone (IZ) shown as a green area in the spinal sections, with some direct projections to motoneurons supplying trunk and girdle muscles (dashed green lines). Group B fibers (rubrospinal) are shown in red, arising from the red nucleus. These fibers terminate contralaterally in the dorsolateral region of the IZ (shown in red in the spinal sections) with some projections to the lateral group of motor nuclei innervating the arm and hand (dashed red lines). These brainstem pathways receive significant cortical projections (black). On the left, corticospinal projections are shown in blue: Some parallel the group A fibers and terminate bilaterally in the ventromedial IZ (green area), whereas the majority parallel the group B system and terminate contralaterally in the dorsolateral IZ (red area) and directly on motoneurons innervating the arm and hand (blue region with small black circles).

dilation, cardiovascular changes, vocalization, micturition, and sexual behaviors. This lateral system influences the relevant motoneuronal groups through other brainstem pathways, such as the periaqueductal gray, which trig-

gers limbic vocalization during mating, fear, etc. (Vanderhorst et al. 2000).

The corticospinal and corticobular pathways. These pathways are present in all

CS: corticospinal

EMG:
electromyogram

CST: corticospinal
tract

mammals but to very different extents. Kuypers (1981) saw the corticospinal (CS; blue fibers, left side of **Figure 1**) and corticobulbar (black fibers, right side of **Figure 1**) projections as acting in parallel with brainstem systems. In the more primitive forms (edentates, marsupials, and lagomorphs) any overlap in areas of CS termination with brainstem pathways is largely restricted to group B (i.e., rubrospinal) pathways.

In a second group of mammals, including rodents, carnivores, and primates, there is a much more extensive CS projection, reaching all levels of the spinal cord (cervical, thoracic, lumbar, and sacral) and innervating all regions of the spinal gray matter, including, motoneurons (shown as black circles, left side of **Figure 1**) in the ventral horn in primates. In these forms, CS terminations overlap with those of both A and B brainstem pathways. **Figure 1** (*left*) shows that corticospinal projections (*blue fibers*) terminate both bilaterally within the ventromedial IZ (*zone colored green in the spinal cord*) and contralaterally within the dorsolateral IZ (*red zone*); some of these crossed fibers also terminate in the motor nuclei (lamina IX; *blue zone* in **Figure 1**).

How useful is this categorization of pathways? Overall, the Kuypers scheme still makes an impact. For example, in the cat, stimulation within the medullary reticular formation evokes responses in proximal limb extensors and flexors bilaterally, responses that are strongly gated by the locomotor cycle (Drew, 1991), whereas reticulospinal neurons originating from the ponto-medullary reticular formation (PMRF) facilitate or suppress electromyogram (EMG) activity in proximal fore- and hind-limb muscles, again bilaterally. Similar observations have been made in the monkey (Davidson & Buford 2004, 2006; Davidson et al. 2007). These findings of bilateral influences on trunk and proximal limb muscles generally agree with those described by Kuypers for the ventromedial pathways (group A). The reticulospinal system is involved both in the control of locomotion

and of the anticipatory postural changes needed to support these movements (Drew et al. 2004).

However, the divisions are certainly not absolute. Kuypers himself noted that some tectospinal and medullary reticulospinal fibers (group A pathways) also terminate in the lateral parts of the IZ (Alstermark et al. 1987, Holstege & Kuypers 1982) and therefore overlap with the terminations of both group B and CS projections, an overlap that is further enlarged by the extensive dendritic trees of both interneurons and motoneurons. New research in the macaque shows that stimulation within the PMRF and spike-triggered averaging from reticulospinal tract neurons show evidence of facilitation or suppression of EMG in more distal muscles, although these effects are far less common and weaker than for CS neurons (Baker & Riddle 2007).

Again, although the classical view of the corticospinal tract (CST) is that of a crossed pathway, a significant number of projections influence the ipsilateral spinal gray matter in the cervical (Galea & Darian-Smith 1997) (see **Figure 1**, *left*) and lumbosacral enlargement (Lacroix et al. 2004). These projections are of considerable potential significance for understanding the effect of cortical or spinal lesions.

The significance of collaterization. Descending pathways give off axon collaterals all along their route of descent toward and within the spinal cord. Kuypers noted that some pathways (e.g., group A) were more highly collateralized than others (e.g., the CST). He suggested that the latter exerted more focused actions and could mediate the more fractionated type of movements that characterize the use of the distal extremities, the hand and foot. The work of Shinoda and colleagues has since accumulated much evidence to show that highly collateralized vestibulospinal tract axon systems can actually subserve selective and coordinated head and neck movements in response to sensory input from the vestibular system (Sugiuchi et al. 2004).

DOES A DESCENDING PATHWAY CARRY OUT SINGLE OR MULTIPLE FUNCTIONS?

A single neuroanatomical pathway can mediate many different functions (Lemon & Griffiths 2005). The CST provides an excellent example; its functions include (*a*) descending control of afferent inputs, including these nociceptive inputs (Cheema et al. 1984, Wall & Lidierth 1997); (*b*) selection, gating, and gain control of spinal reflexes (Pierrot-Deseilligny & Burke 2005); (*c*) excitation and inhibition of motoneurons (Alstermark & Lundberg 1992, Porter & Lemon 1993, Maier et al. 1998); (*d*) autonomic control (Bacon & Smith 1993); (*e*) long-term plasticity of spinal cord circuits (Wolpaw 1997); and 6) trophic functions (Martin et al. 1999).

IS THE CORTICOSPINAL TRACT A MOTOR PATHWAY?

The anatomical characteristics of the CST support this multifunctional view. The CST originates from a wide variety of cortical areas, each with different functions, including, in the monkey, the primary motor cortex (M1), the dorsal and ventral premotor cortices, supplementary motor area (SMA), and cingulate motor areas (see Dum & Strick 2005). CS projections also extend from the parietal lobe (primary somatosensory cortex, S1), the posterior parietal cortex, and the parietal operculum. The origins of the CST from many different functional areas make it unlikely that the CST projection subserves a single role. In addition, the CST terminates widely within the spinal gray matter, presumably reflecting control of nociceptive, somatosensory, reflex, autonomic, and somatic motor functions. Although CS projections from different frontal cortical areas show a similar overall pattern (He et al. 1993), studies show marked quantitative differences in the projections, for example, from SMA vs. M1 (Maier et al. 2002; Boudrias et al. 2006).

The CST's involvement in controlling more than one function raises questions about its classical role as a motor pathway. Indeed, the CST projection to the dorsal horn is found in all mammals, and the early evolution of the CST to control afferent input may reflect the earliest evolved form of supraspinal control exerted by the CST (Kuypers, 1981, Jones 1986, Canedo 1997, Lemon & Griffiths 2005). In monkeys, the projection to the dorsal horn is derived from S1, not from M1 (Jones 1986, Armand et al. 1997). CST projections to the dorsal horn are probably involved in the descending control of proprioceptive inputs generated by movement or sensory reafference and the gating or filtering of such inputs to both local central pattern generators (CPGs) and supraspinal centers (Wolpert et al. 2001). CST projections to the dorsal horn are an important source of presynaptic inhibition of primary sensory afferent fibers (Canedo 1997, Wall & Lidierth 1997), and this mechanism could allow removal of predictable sources of afferent input associated with feedforward motor commands for voluntary movement. Lesions of the CST cause a breakdown in fine sensorimotor control, implying a deterioration not only in motor function, but also in the capacity to interrogate correctly the sensory feedback from the hand (Lemon & Griffiths 2005).

IS THE ORGANIZATION OF THE DESCENDING PATHWAYS HIGHLY CONSERVED ACROSS DIFFERENT MAMMALIAN SPECIES?

Many basic motor activities (e.g., breathing, swallowing, locomotion) are common to all mammalian species, and some studies suggest that the main function of the descending pathways is to modulate the CPGs that control each of these activities (Dietz 2003, Drew et al. 2004, Grillner & Wallen 2004, Yang & Gorassini 2006). Because the basic neuronal mechanisms underpinning these activities are present in all species, one can argue that the descending pathways should exhibit a highly conserved organization pattern across different species, and indeed, some CS and reticulospinal neurons are active during both locomotion and reaching

M1: primary motor cortex

SMA: supplementary motor area

CPG: central pattern generator

(Drew et al. 2004), as originally suggested in an influential paper by Georgopoulos & Grillner (1989).

However, Grillner & Wallen (2004) also note that the control of some movements, such as independent fine movements of the fingers, are exceptions to this scheme. Indeed, an emerging view of the descending pathways is that although an overall pattern is recognizable, considerable variations exist across species in the organization of the descending pathways, and particularly of the CST. Two examples of such differences are considered here: first, wide variation in the extent of direct cortico-motoneuronal (CM) connections in different primates and absence of this system in non-primates, and second, species differences in the organization of motor effects transmitted from the CST to motoneurons via propriospinal neurons.

The CM System is Developed to Different Extents across Species

The CM system, first discovered by Bernhard & Bohm (1954), represents a direct, monosynaptic projection from some CS fibers to spinal motoneurons.

Non-primates. Kuypers first showed that striking differences exist in the degree of CM system development across different species (Kuypers 1981). A number of studies have now confirmed that the CM system is a uniquely primate feature; there are no functional CM connections in the cat (Illert et al. 1976), rat (Yang & Lemon 2003, Alstermark et al. 2004), raccoon (Gugino et al. 1990), or mouse (Alstermark & Ogawa 2004) (see **Figure 2**).

Non-human primates. The CM system is developed to a variable extent in different primates (Lemon & Griffiths 2005). It is absent in tree

shrews, lemurs, and marmosets. It is well developed in some New World monkeys, such as the capuchin monkey, and in Old World monkeys, including the macaque (**Figure 2**); CM connections are generally most numerous in the great apes (Kuypers 1981). Bortoff & Strick (1993) made a direct comparison of the anterograde labeling in the cervical enlargement that resulted from injections into motor cortex of two New World species: the capuchin monkey (*Cebus appella*) and the squirrel monkey (*Saimiri sciureus*). Whereas the labeling of CST terminations in the IZ was similar in the two species, the labeling in the ventral horn was remarkably different. In *Cebus*, dense projections into lamina IX extended into the most lateral and dorsal motor nuclei that innervate the intrinsic hand muscles, whereas CST projections into the ventral horn in *Saimiri* were much more sparse.

When assessed electrophysiologically, the CM input to hand and forearm motoneurons in the squirrel monkey, *Saimiri*, is weak compared with that in the macaque (Nakajima et al. 2000), where there are strong functional CM connections to hand and digit muscles (**Figure 2**) as well as somewhat weaker connections to muscles acting more proximally (Porter & Lemon 1993, McKiernan et al. 1998). Evidence also demonstrates CM connections to foot and tail motoneurons (Jankowska et al. 1975; see Porter & Lemon 1993).

Humans. Anatomical evidence supports the existence of CM projections in human material (Kuypers, 1981) and monosynaptic effects on hand-muscle motoneurons in awake volunteers (**Figure 2**) (Palmer & Ashby 1992, Baldissera & Cavallari 1993, de Noordhout et al. 1999) as well as on many other upper limb muscles, even those acting at proximal joints (Colebatch et al. 1990). CM effects on lower limb muscles are also well defined (Brouwer & Ashby 1990, Rothwell et al. 1991, Nielsen et al. 1995).

Functional significance. Investigators are still debating the functional significance of these species differences for sensorimotor

---

[1]The term cortico-motoneuronal is quite specific to this particular monosynaptic connection and should not be used to refer to more general cortical influences over motoneurons mediated by oligosynaptic pathways.

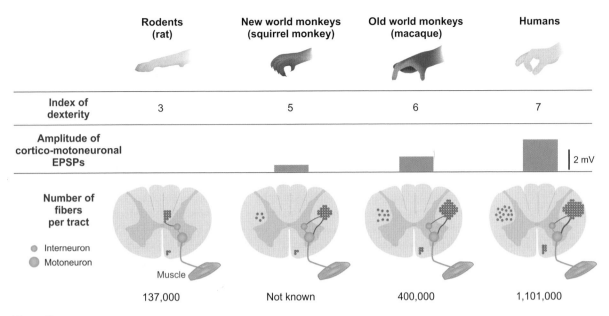

	Rodents (rat)	New world monkeys (squirrel monkey)	Old world monkeys (macaque)	Humans
Index of dexterity	3	5	6	7
Amplitude of cortico-motoneuronal EPSPs				
Number of fibers per tract	137,000	Not known	400,000	1,101,000

Figure 2

Relationship between the development of the CST and the emergence of fine motor control abilities. In rodents, there are no direct connections between CS neurons and the cervical motoneurons which innervate forelimb muscles—brainstem pathways and spinal interneurons relay cortical input to motor neurons. Most of the CST fibers in rodents travel in the dorsal columns. In non-human primates and humans, direct CS connections with motoneurons have evolved, together with an increase in the size and number of the CS fibers. This is reflected in an increase in the size of the excitatory postsynaptic potential (EPSP) elicited by cortical neurons in hand motoneurons. The primate CST is located mostly in the lateral columns, and a significant proportion of CS fibers (~10%) descend ipsilaterally. Development of the CST correlates with the improvement in the index of dexterity, particularly in the ability to perform finger-thumb precision grip (with permission from Courtine et al. 2007).

control of the hand (Lemon & Griffiths 2005, Iwaniuk et al. 1999) (see **Figure 2**). Heffner & Masterton (1983) already established the relationship across species between the extent of CST projections among the motoneuron pools of the spinal gray matter and the index of dexterity, a linear score of hand function that reflects the use of digits for prehensile purposes and for manipulation. For example, the capuchin monkey, *Cebus*, with many CM projections to hand motoneuron pools, is a very dexterous monkey and uses a pseudo-opposition between the tip of the thumb and the side of the index to grasp and manipulate small objects (Fragaszy 1998) and to use tools (Phillips 1998). In contrast, the squirrel monkey, *Saimiri*, with much weaker CM projections, has no precision grip and limited manipulatory skills.

The available data suggest that the CM system is a recently evolved feature and that it subserves evolutionarily new aspects of motor behavior, including voluntary control of relatively independent finger movements. Although such movements are not exclusive to species with CM connections, they are far better developed in primates than in non-primates.

Variation across Species in the Development of Propriospinal Transmission to Motoneurons

Descending excitation from the cerebral cortex is also transmitted from the CST to forelimb motoneurons (C6-Th1) through two rather different interneuronal routes: one via segmental interneurons (**Figure 2**) located in the same lower cervical segments (C6-T1) as the forelimb motoneurons, and the other via propriospinal neurons (PNs; i.e., interneurons located outside the forelimb segments),

PN: propriospinal neuron

PT: pyramidal tract

including those located in upper cervical segments (C3–C4). These are part of a propriospinal system that is present throughout the spinal cord (Molenaar & Kuypers 1978). After spinal injury, PN mechanisms may provide a means of reestablishing voluntary motor control without the need to regenerate long fiber tracts over the full length of the spinal cord (Fouad et al. 2001).

Cat. A large number of electrophysiological, anatomical, and lesion studies showed that the C3–C4 PNs are involved primarily in mediating the descending command for forelimb target reaching, whereas the local segmental system mediates the descending command for food taking (grasping a morsel of food with the digits) (Alstermark & Lundberg 1992; see Isa et al. 2007).

Non-human primates. Since the pioneering studies were carried out in the cat, considerable attention has been focused on whether this type of organization is seen in other species. The overall conclusion at present indicates that there are quite striking variations in the organization of propriospinal transmission of CS excitation to motoneurons (Isa et al. 2007). Once again, studies show differences between different primates: In the New World squirrel monkey (*Saimiri*), effects in upper limb motoneurons consistent with transmission through a C3-C4 PN system were observed, although these effects are significantly weaker than in the cat (Nakajima et al. 2000).

In the intact macaque monkey, such effects are rare (Maier et al. 1998, Alstermark et al. 1999, Olivier et al. 2001). Maier et al. (1998) speculated that this was because transmission through the C3-C4 system had been replaced by excitation mediated by the CM system. Alstermark et al. (1999) subsequently demonstrated that C3-C4 transmission can occur in the macaque monkey but only after the systemic administration of strychnine (Alstermark et al. 1999), which is a glycinergic antagonist and which therefore produces widespread abolition of inhibition in the spinal cord. Since then,

studies have shown that in the intact macaque monkey, stimulation of the pyramidal tract (PT) evokes feedforward and feedback inhibition of C3-C4 PNs, which may prevent onward transmission of excitation to upper limb motoneurons when the whole PT is stimulated (Isa et al. 2006, 2007).

Thus several striking differences exist between cat and macaque monkey: First, studies show more CS inhibition of C3-C4 PNs in the monkey than in the cat. A strychnine study has not yet been published for the cat, so we do not know how big the differences are. Second, although 84% of C3-C4 PNs in the cat project an ascending axon collateral to the lateral reticular nucleus (hypothesized to be an ascending readout of PN activity; Alstermark & Lundberg 1992), this projection was found for only 30% of PNs in the macaque (Isa et al. 2006). Finally, even after strychnine, it requires three or four high-frequency shocks to the PT to discharge the sampled PNs, which is not consistent with a powerful excitatory linkage between the CST and these PNs.

Rodents. A recent study by Alstermark et al. (2004) found no evidence of C3-C4 transmission of CS excitation to forelimb motoneurons in the rat; instead, they found that the cortex influenced motoneurons through corticoreticulospinal pathways. In the mouse, few motor effects on forelimb motoneurons are evoked from the CST, and reticulospinal control seems to be more important in this species (Alstermark & Ogawa 2004).

Conclusion: species differences in organization of descending pathways for motor control. Differences in the organization of CS projections across species may well reflect differences in the functional contributions made by this system and may explain species-dependent effects of CST lesions. Therefore, one should exercise caution in selecting which animal model should be adopted when interpreting the very considerable amount of indirect evidence for C3-C4 transmission in humans (Pierrot-Deseilligny and Burke 2005), and also in understanding changes that occur after spinal cord injury, stroke, and other disorders (Lemon & Griffiths 2005, Courtine et al. 2007).

WHAT IS THE FUNCTIONAL SIGNIFICANCE OF DIRECT CONNECTIONS WITH TARGET MOTONEURONS?

In addition to terminating in the IZ, several descending pathways produce monosynaptic actions on motoneurons, including fibers running in the rubro-, reticulo-, vestibulo-, and corticospinal tracts. These monosynaptic connections allow a direct contribution to motoneuron control, which operates in parallel with that from more indirect connections, possibly by adding the final spatiotemporal excitation patterns that would induce appropriate levels of motoneuronal recruitment and discharge (Lemon et al. 2004).

Some reviewers have taken the view that the CM input is unimportant because its excitatory input is rather small, rather like that of other monosynaptic inputs from other descending pathways (Baldissera et al. 1981, Grillner & Wallen 2004). An informed answer to this question requires the following issues to be addressed: (*a*) How extensive is the CM projection? (*b*) How many CM cells project to a given motoneuron or to a given muscle? And (*c*) how large are the postsynaptic CM effects in a given motoneuron?

New Details on the Origin of the CM Input from Viral Retrograde Tracer Studies

Fresh insights have come from recent work by Rathelot & Strick (2006) on the CM projection in the macaque monkey, using transneural labeling with rabies virus. Their approach has, for the first time, made it possible to identify anatomically the cells of origin of the CM projection to hand muscles in the macaque monkey. The rabies virus, when injected into the test muscle, is transported retrogradely and in-

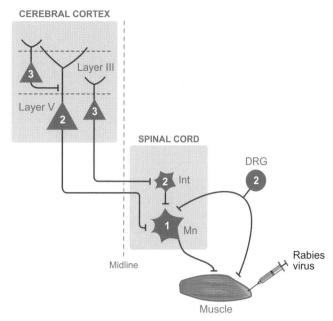

Figure 3

Retrograde viral tracing method for labeling CM cells in the macaque monkey. When the rabies virus is injected into a single digit muscle, it is transported in the retrograde direction to infect the motoneurons (i.e., first-order neurons, 1) that innervate the muscle. Then the virus is transported transneuronally in the retrograde direction to label all the second-order neurons (2) that synapse on the infected motoneurons. These include dorsal root ganglion cells that supply group Ia muscle spindle afferents, spinal cord interneurons, and cortical neurons in layer V (CM cells). At longer survival times, the virus can undergo another stage of retrograde transneuronal transport and label all the third-order neurons (3) that synapse on the infected second-order neurons. For example, the virus can move from second-order neurons in layer V to third-order neurons in layer III. Similarly, the virus can move from second-order interneurons in the spinal cord to third-order cortical neurons in layer V. DRG, dorsal root ganglion cell; Int, interneuron; Mn, motoneuron; 1, first-order neuron; 2, second-order neuron; 3, third-order neuron (from Rathelot & Strick 2006 with permission).

fects all the motoneurons of the injected muscle (first-order labeling; see **Figure 3**). The virus is subsequently transferred trans-synaptically to all the neurons that project directly to the infected motoneurons. This second-order labeling therefore involves different classes of neurons in the spinal cord (e.g., segmental interneurons, primary afferents), brainstem (e.g., cells of origin descending pathways), and cortex (CM cells). Subsequent third-order transneuronal transfer labels all the neurons, which project to these different classes of second-order neurons. In a typical experiment, first-order labeling of motoneurons is seen three days after muscle injection, with second-order labeling 3.5–5 days after injection and third order labeling after still longer survival times. By carefully timing their experiments, Rathelot & Strick (2006) were able to ensure that the transneuronal labeling process had not gone beyond the second-order stage so that any labeled cortical neurons could be confidently identified as CM.

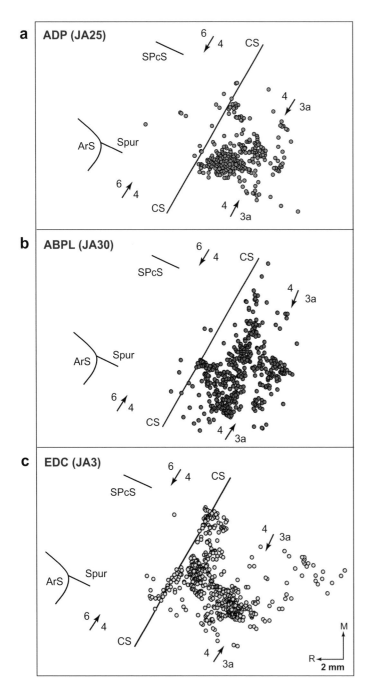

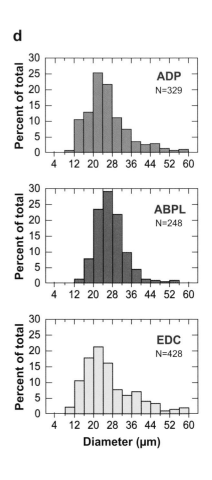

The labeled CM cells were distributed over a broad territory located in the classical arm and hand area of M1 (see **Figure 4a–c**). Most of the labeled CM neurons were found in the caudal part of classical M1, in the rostral bank of the central sulcus, and very few were found on the convexity of the precentral gyrus. The CM projection was therefore not coextensive with the total CS projection from M1, in keeping with older degeneration studies of the CM projection (Kuypers & Brinkman 1970).

The area of M1 that contained the entire population of CM cells innervating an individual muscle was quite large and measured between 42 and 54 square mm. It overlapped heavily with representations of other hand muscles (see **Figure 4a–c**). These findings finally rule out the existence of a fine, nonoverlapping, somatatotopic mapping of hand muscles within M1, thus concluding a long-running debate (Asanuma 1975, Andersen et al. 1975, Phillips 1975, Lemon 1988, Schieber & Hibbard 1993). The extensive representation of a single hand muscle reaches into regions of M1 concerned with elbow and shoulder movements (Kwan et al. 1978, Park et al. 2001) and suggests that this "could provide the neural substrate to create a broader range of functional synergies" involving hand and digit muscles as well as those acting at more proximal joints (see McKiernan et al. 1998).

The number of CM cells labeled from the three muscles, belonging to the cortical colony of that muscle (Andersen et al. 1975), varied between 248 (AbPL) and 428 (EDC); EDC is a larger muscle than AbPL. If a significant proportion of the CM cell colony converging on a hand muscle was active, and the unitary excitatory postsynaptic potential (EPSPs) from each CM cell motoneuron contact were within the range of other descending systems, this would produce a significant total excitatory drive (see below).

CM Cell Activity during Voluntary Movement

I cannot review here the very extensive literature that shows that CM cells in monkey M1 cortex are highly modulated during skilled hand and digit movements and the studies that show patterns of activity that are congruent with their connectivity as established by spike-triggered average (STA) (Bennett & Lemon 1996, Jackson et al. 2003, Schieber & Rivlis 2007). Information on how this activity is transformed into appropriate patterns of motoneuron activity is only just coming to light (Yanai et al. 2007). CM cells may also be involved in the generation of transcortical reflexes (Cheney & Fetz 1984), which could underpin fast and powerful responses to peripheral perturbations (Johansson et al. 1994).

CM Projections from Secondary Cortical Motor Areas?

Rathelot & Strick (2006) have not yet reported any labeled CM neurons in the

EPSP: excitatory postsynaptic potential

STA: spike-triggered average

Figure 4

(a–c) Cortical surface maps of CM cells that innervate the motoneurons for digit muscles. Each panel (a–c) displays a flattened or unfolded map of the central sulcus and precentral gyrus showing location of CM cells (*small round symbols*) labeled after injections of the rabies virus into ADP (adductor pollicis), ABPL (abductor pollicis longus) or EDC (extensor digitorum communis). Each muscle was injected in a different monkey (JA25, JA30, and JA3, respectively). Small arrows are placed at the area 4–6 border and the area 4–3a border. ArS, arcuate sulcus; CS, central sulcus; M, medial; R, rostral; SPcS, superior precentral sulcus. Note the large area occupied by the CM representation or colony belonging to each muscle and the extensive overlap between the three colonies. (d) Size distribution of the somata of CM cells. The graphs show the size of CM cells in M1 labeled after virus transport from ADP (*top*), ABPL (*middle*), and EDC (*bottom*). Cell size in these graphs represents the average of a cell's maximum and minimum diameter. Note the large proportion of small cells comprising the CM colony (from Rathelot & Strick 2006 with permission).

secondary motor areas: SMA, cingulate motor areas, or dorsal and ventral premotor areas, all of which give rise to CS projections (Dum & Strick 2005). In keeping with this, Maier et al. (2002) were unable to find evidence of CM-EPSPs in hand motoneurons evoked from intracortical stimulation in the SMA. In contrast, when the M1 hand area in the rostral bank of the central sulcus was stimulated, such EPSPs were evoked in most motoneurons tested. Stimulation of ventral premotor cortex (PMv) also failed to evoke CM-EPSPs in hand motoneurons (Shimazu et al. 2004). However, PMv stimulation could produce powerful modulation of M1 CS outputs (**Figure 5**), which may represent an important parallel route through which

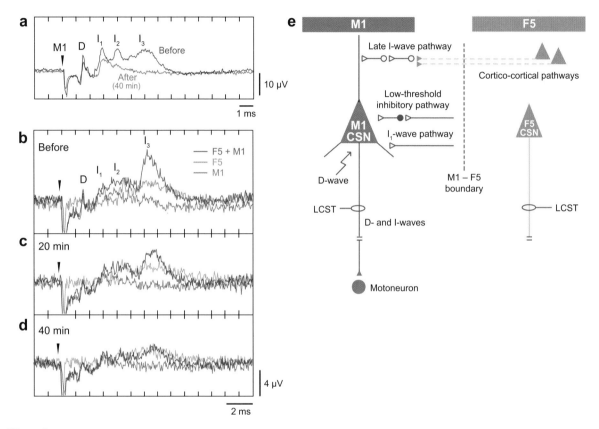

Figure 5

Facilitation of CS outputs from M1 from ventral premotor cortex (area F5). (*a*) Intracortical stimulation with a single pulse in the hand area of M1 evokes a series of descending CS volleys: the D (direct) wave and a number of indirect (I) waves (*blue trace*). Forty minutes after microinjection of the GABA$_a$ agonist, muscimol, close to the M1 stimulation site the late I waves (I$_2$, I$_3$) are mostly abolished (*red trace*); there is a small reduction in the I$_1$ wave and no obvious effect on the D wave. Averages of 150 sweeps, volleys recorded from the C3 spinal level. (*b–d*) Effects of muscimol injection in M1 on F5–M1 interaction. A single test (T) M1 shock was conditioned (C) by a single shock to the F5 division of PMv (*green*). Responses to the F5 shock alone are *orange* and to the M1 shock alone are *purple*. With the C–T interval of 0 ms, there was a marked facilitation of the I$_3$ wave. Twenty minutes after muscimol injection (C), this facilitation was considerably reduced, and it was abolished after 40 min (D), suggesting that F5-M1 interaction occurred within M1 itself. Averages of 100 sweeps. (*e*) Possible mechanisms explaining facilitation of late I waves. The elements (*in gray, on the left side of the diagram*) represent the three classes of excitatory inputs to CSNs: the D wave, the I$_1$ wave, and the later (I$_2$ and I$_3$) waves. A low-threshold inhibitory input pathway is also shown. These four inputs are all excited by stimulation within M1. The model proposes that the main excitatory input from F5 to M1 converges on the late interneuronal pathways in M1 giving rise to the I$_2$ and I$_3$ waves, thereby allowing F5 to influence I wave generation in M1 CS neurons. Note that although M1 CS neurons project directly to hand motoneurons, this is not the case for those located in F5 (from Shimazu et al. 2004, with permission).

a secondary motor area could exert its motor effects (Cerri et al. 2003, Shimazu et al. 2004). Reversible inactivation of M1 hand area can reduce or abolish motor responses evoked from PMv (Schmidlin et al. 2006).

Estimating the Size of the CM Input

Estimates of the amplitude of the CM input to a motoneuron or muscle have been derived from their responses to electrical stimulation of the CST or M1, or from spike-triggered averages derived from single CM cells. Both approaches suggest that CM effects can be substantial, particularly for motoneurons supplying muscles acting on the hand and digits. In the anaesthetized macaque monkey, stimulation of the pyramidal tract evokes monosynaptic EPSPs in most upper limb motoneurons, with a wide range in the size of these effects from a few hundred microvolts up to 5–7 mV; the largest effects occur in motoneurons supplying intrinsic hand muscles (Porter & Lemon 1993). The mean value for the compound EPSP (representing the summed excitatory input from all fast-conducting CM axons terminating on a motoneuron) was estimated at ∼2 mV. This figure is probably an underestimate because disynaptic inhibition often truncates the CM EPSP before it reaches its maximum (Maier et al. 1998). In the awake monkey, PT stimulation with single shocks is capable of discharging active motor units at monosynaptic latency with high probability (Olivier et al. 2001). In awake human volunteers, using transcranial electrical stimulation, the size of the compound CM-EPSP in hand muscle motoneurons has been estimated as at least 4–5 mV (de Noordhout et al. 1999).

Investigators find some drawbacks in using electrical stimulation of the cortex and PT, not the least of which is the unnatural and synchronous activation of many fibers together, which may well exert mixed excitatory and inhibitory effects on target neurons. The STA represents a more natural approach to the problem and, in addition, allows estimates of the excitatory CM influence of single motor cortex cells. CM cells can be identified in the awake animal and exert facilitation of motoneuron pools that is consistent with CM action (Fetz & Cheney 1980, Lemon et al. 1986). Although researchers now generally accept that the STA method reveals a spectrum of postspike effects in EMG (Baker and Lemon 1998; Schieber and Rivlis 2005), the strength of CM synapses is again underlined by STA findings. Even single CM cells can produce detectable changes in the discharge of both single motor units and multiunit EMGs recorded from hand and forelimb muscles, and the total CM input to motoneurons could provide a significant proportion of the facilitatory drive needed to maintain its steady discharge (Cheney et al. 1991).

CM Effects from Small, Slowly Conducting PTNs

Rathelot & Strick (2006) reported that labeled CM neurons had soma diameters ranging from 12 to 60 μM, the majority being small in size (see **Figure 4d**). Thus the CM output is not restricted to the large pyramidal cells in layer V (the classical Betz cells). STA studies have also demonstrated CM effects from small, slower-conducting CST neurons (Fetz & Cheney 1980; see Porter & Lemon 1993). Recording and stimulation approaches are both biased toward larger neurons with fast-conducting axons, so the contribution from the much more numerous smaller CM neurons has likely been seriously overlooked (Maier et al. 1998).

What Can We Learn from Lesion Studies?

Implicit in modern concepts of a distributed motor network is the acceptance that the behavioral and other changes that result from lesions to a descending pathway cannot be interpreted any longer as simply being due to the removal of the lesioned pathway. There are fast, activity-dependent, plastic changes that occur soon after a lesion is placed in a descending pathway (Steward et al. 2003, Bareyre et al. 2004, Deumens et al. 2005, Vavrek et al. 2006). The behavioral outcome is a consequence of compensatory changes of the motor network as a whole, including the response

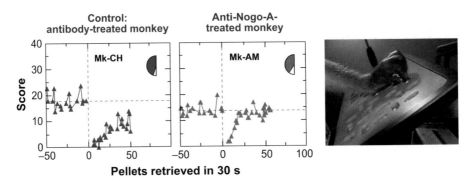

Figure 6

Effects of lesions to the CST at the cervical level on skilled independent finger movements in the macaque monkey, and partial recovery following treatment with anti-NOGO antibody. Data are shown for quantitative assessment of manual dexterity before and after lesion in a representative pair of monkeys (Mk-CH and Mk-AM), with similar extent of lesion in the dorsolateral funiculus at the C7-C8 border in the cervical spinal cord (the extent of the blue and red zones in the semicircular figures represents the extent of the hemicord lesion for each monkey). Monkeys were assessed on a modified Brinkman board (*right*) using the precision grip (opposition of thumb and index finger) to retrieve food pellets from the board. The untreated monkey (*blue*, Mk-CH) showed some limited recovery in its performance (score represents number of pellets retrieved in 30 s), but the monkey treated with the antibody to neutralize the neurite growth inhibitor Nogo-A (*red*, Mk-AM) rapidly returned to control levels of dexterity (from Freund et al. 2006, with permission).

of uninjured fibers, rather than a consequence of the removal of a single descending pathway or component. For example, lesions of the CST cause long-term changes in the influence of the rubrospinal tract (Belhaj-Saif & Cheney 2000). The exciting prospect is that this compensatory response can be boosted by appropriate therapies (see Case & Tessier-Lavigne 2005, Deumens et al. 2005, Buchli et al. 2007) (see **Figure 6**). Of course, spinal injury in humans results in damage to multiple ascending and descending pathways, and selective lesions of individual tracts are a poor model for such injury; however, they do provide important insights into the regrowth and possible regenerative capacities of different pathways.

Cortical and capsular lesions. These probably affect all the descending pathways because the cortex provides a major input to the cells of origin of the brainstem pathways (Kuypers 1981; see Matsuyama & Drew 1997, Matsuyama et al. 2004) (see **Figure 1**, *right*). Such lesions affect not only the CS and corticobulbar projections, but also corticostriatal, cortico-

pontine, and other systems related to motor function. Indeed it is remarkable that clinically, capsular lesions give rise to the classical pyramidal signs of contralateral weakness, and especially those affecting the distal musculature (Colebatch & Gandevia 1989), despite the fact that CS fibers probably constitute only a few percent of the total number of ascending and descending fibers within the capsule. The importance of the integrity of the CS tract has recently been reemphasized by studies of recovery of precision finger movements after subcortical stroke (Lang & Schieber 2003, Ward et al. 2007); good recovery of function and a normal pattern of brain activation were found in those patients in whom transcranial magnetic stimulation demonstrated an intact CS projection from the damaged hemisphere. During human spinal surgery, continuous intraoperative monitoring, below the surgery level, of the CS D-wave evoked from the motor cortex can be very useful in preventing poor outcomes in terms of locomotor and other functions; loss of the D-wave is a strong indicator for paraplegia following surgery (Sala et al. 2006).

D-wave:
Corticospinal activity reflecting direct activation of CS neurons

PYRAMIDAL AND CST LESIONS

Lawrence & Kuypers (1968a) originally showed that a complete pyramidotomy in the macaque monkey caused permanent deficits in skilled hand function, but we do not know how much of the deficit was due to interruption of CM projections (we must await a selective means of lesioning or inactivating the CM projection). Compared with the effects of complete pyramidotomy, a rather different picture emerges with subtotal lesions of the CST (Lawrence & Kuypers 1968a; see Porter & Lemon 1993). This applies to subtotal or unilateral lesions of the pyramidal tract and to funicular lesions or hemisections at the spinal level. Several recent studies have investigated the effect of unilateral surgical interruption of the fibers in the lateral CST at the cervical (Galea & Darian-Smith 1997, Sasaki et al. 2004, Freund et al. 2006) and thoracic levels (Courtine et al. 2005). The initial effects are similar to a complete CST lesion, with a striking deterioration in the speed, force, and accuracy of more distal hand (**Figure 6**) and/or foot movements, especially those involved in tasks demanding high levels of dexterity (e.g., catching with the hand or foot, removing food from small wells; **Figure 6**). The initial deficits must reflect the loss of CS connections to the spinal segments controlling the hand (or foot), including CM connections (Nishimura et al. 2007). The fast CM-EPSPs that are normally evoked in most hand motoneurons were absent after a C5 lesion by Sasaki et al. (2004). However, such lesions also sever the CS projections to segmental interneurons below the lesion (**Figure 2**), including those mediating inhibition of motoneurons.

Thereafter, these deficits often show a substantial degree of recovery, the extent of which seems to be related to the amount of CST fiber sparing; however, there is usually a permanent deficit of independent digit movements, and especially of the thumb. The subsequent recovery of grasping function (Sasaki et al. 2004, Freund et al. 2006) or of locomotor function (Courtine et al. 2005, Nishimura et al. 2007) after spinal CS lesions probably reflects the plastic changes induced elsewhere in the CNS, and we now know that there is significant sprouting of these fibers above the lesion (Fouad et al. 2001, Bareyre et al. 2004). Importantly, these changes can be further promoted by reducing the levels of inhibitory factors that suppress sprouting and regeneration, for example, by administration of antibody to the myelin inhibitory factor, NOGO (**Figure 6**) (Freund et al. 2007). The molecular identity of a given pathway must determine the capacity for sprouting, growth, and development of new synapses. It may also determine whether compensatory plasticity comes from uninjured fibers belonging to the same pathway or whether others can contribute, a key factor that might explain the substantial differences in the outcomes produced by total vs. subtotal lesions of a particular pathway.

Stimulation of the PT after a subtotal CST lesion evokes EPSPs in hand motoneurons with latencies longer than those for normal monosynaptic CM responses. These EPSPs could be disynaptic and originate from C3-C4 propriospinal projections (as suggested by Sasaki et al. 2004) but could also result from monosynaptic CM projections from slow uninjured CS fibres (as shown by Maier et al. 1998; see Lemon et al. 2004). This is potentially significant because of the large numbers of slow- vs. fast-conducting CS fibers and because fast fibers may be more susceptible to injury (Blight 1983, 1991; Quencer et al. 1992).

SUMMARY POINTS

1. Each descending pathway has specific characteristics that determine its neuronal targets within the spinal cord and therefore its functional role.

2. Each descending pathway may carry out a number of functional roles; these are linked together by the need for a coordinated set of operations underpinning performance of basic functions such as balance, posture, locomotion, and reaching.

3. Although some general principles underlie the organization of the descending pathways, striking differences can be found between species, and research suggests that, in the case of the corticospinal tract (CST), these differences reflect species-specific features of sensorimotor behavior, including differences in skilled use of the hand and digits.

4. The cortico-motoneuronal (CM) system provides the capacity for fractionation of movements and the control of small groups of muscles in a highly selective manner, an important feature of skilled voluntary movements in the acquisition of new motor skills. The CM system may provide an efferent pathway that is accessed by the extensive motor network of the primate brain for development of adaptive motor programs.

5. The CM system works in parallel with other indirect corticospinal (CS) influences, transmitted through segmental interneurons and propriospinal neurons; in some primates, the contribution of the CM inputs can be considerable.

FUTURE ISSUES

1. We need to understand the role of the CST in controlling or gating sensory afferent input. How exactly does it function to remove unwanted reafference?

2. We also need to understand how the movement-related information generated by the cells of origin of the descending pathways is integrated by spinal interneurons and motoneurons.

3. We need a better understanding of the compensatory mechanisms that are triggered by damage to different motor pathways.

DISCLOSURE STATEMENT

The author is not aware of any biases that might be perceived as affecting the objectivity of this review.

ACKNOWLEDGMENTS

I acknowledge the help and cooperation of all my colleagues and students who collaborated on much of the work cited in this article, which was funded by research grants from The Wellcome Trust, MRC and Brain Research Trust. I particularly thank Peter Kirkwood for his comments on an earlier draft.

LITERATURE CITED

Adams M, Carlstedt T, Cavanagh J, Lemon RN, McKernan R, et al. 2006. International spinal research trust research strategy. III: A discussion document. *Spinal Cord* 45:2–14

Alstermark B, Isa T, Ohki Y, Saito Y. 1999. Disynaptic pyramidal excitation in forelimb motoneurons mediated via C3-C4 propriospinal neurons in the macaca fuscata. *J. Neurophysiol.* 82:3580–85

Alstermark B, Kummel H, Tantisera B. 1987. Monosynaptic raphespinal and reticulospinal projection to forelimb motoneurones in cats. *Neurosci. Lett.* 74:286–90

Alstermark B, Lundberg A. 1992. The C3-C4 propriospinal system: target-reaching and food-taking. In *Muscle afferents and spinal control of movement*, ed. L Jami, E Pierrot-Deseilligny, D Zytnicki, pp. 327–54. Oxford, UK: Pergamon

Alstermark B, Ogawa J. 2004. In vivo recordings of bulbospinal excitation in adult mouse forelimb motoneurons. *J. Neurophysiol.* 92:1958–62

Alstermark B, Ogawa J, Isa T. 2004. Lack of monosynaptic corticomotoneuronal EPSPs in rats: disynaptic EPSPs mediated via reticulospinal neurones and polysynaptic EPSPs via segmental interneurones. *J. Neurophysiol.* 91:1832–39

Andersen P, Hagan PJ, Phillips CG, Powell TPS. 1975. Mapping by microstimulation of over-lapping projections from area 4 to motor units of the baboon's hand. *Proc. R. Soc. London Ser. B* 188:31–60

Armand J, Olivier E, Edgley SA, Lemon RN. 1997. The postnatal development of corti-cospinal projections from motor cortex to the cervical enlargement in the macaque monkey. *J. Neurosci.* 17:251–66

Asanuma H. 1975. Recent developments in the study of the columner arrangement of neurons within the motor cortex. *Physiol. Rev.* 55:143–56

Bacon SJ, Smith AD. 1993. A monosynaptic pathway from an identified vasomotor centre in the medial prefrontal cortex to an autonomic area in the thoracic spinal cord. *Neuroscience* 54:719–28

Baker SN, Lemon RN. 1998. Computer simulation of postspike facilitation in spike-triggered averages of rectified EMG. *J. Neurophysiol.* 80:1391–406

Baker SN, Riddle CN. 2007. The macaque reticulospinal tract forms monosynaptic connections with motoneurons in the cervical spinal cord controlling distal arm and hand muscles. *Soc. Neurosci. Abstr.* 191.3

Baldissera F, Cavallari P. 1993. Short-latency subliminal effects of transcranial magnetic stimula-tion on forearm motoneurones. *Exp. Brain Res.* 96:513–18

Baldissera F, Hultborn H, Illert M. 1981. Integration in spinal neuronal systems. See Brookhart & Mountcastle 1981, pp. 509–95

Bareyre FM, Kerschensteiner M, Raineteau O, Mettenleiter TC, Weinmann O, Schwab ME. 2004. The injured spinal cord spontaneously forms a new intraspinal circuit in adult rats. *Nat. Neurosci.* 7:269–77

Belhaj-Saif A, Cheney PD. 2000. Plasticity in the distribution of the red nucleus output to forearm muscles after unilateral lesions of the pyramidal tract. *J. Neurophysiol.* 83:3147–53

Bennett KMB, Lemon RN. 1996. Corticomotoneuronal contribution to the fractionation of mus-cle activity during precision grip in the monkey. *J. Neurophysiol.* 75:1826–42

Bernhard CG, Bohm E. 1954. Cortical representation and functional significance of the cortico-motoneuronal system. *Arch. Neurol. Psychiat.* 72:473–502

Blight AR. 1983. Cellular morphology of spinal cord injury in the cat: analysis of myelinated axons by line sampling. *Neuroscience* 10:521–43

Blight AR. 1991. Morphometric analysis of a model of spinal cord injury in guinea pigs, with behavioural evidence of delayed secondary pathology. *J. Neurol. Sci.* 103:156–71

Boers J, Ford TW, Holstege G, Kirkwood PA. 2005. Functional heterogeneity among neurons in the nucleus retroambiguus with lumbosacral projections in female cats. *J. Neurophysiol.* 94:2617–29

Bortoff GA, Strick PL. 1993. Corticospinal terminations in two New-World primates: further ev-idence that corticomotoneuronal connections provide part of the neural substrate for manual dexterity. *J. Neurosci.* 13:5105–18

Boudrias MH, Belhaj-Saif A, Park MC, Cheney PD. 2006. Contrasting properties of motor output from the supplementary motor area and primary motor cortex in rhesus macaques. *Cereb. Cortex* 16:632–38

Brookhart JM, Mountcastle VB, eds. 1981. *Handbook of Physiology—The Nervous System II*. Bethesda, MD: Am. Physiol. Soc.

Brouwer B, Ashby P. 1990. Corticospinal projections to upper and lower limb spinal motoneurons in man. *Electroenceph. Clin. Neurophysiol.* 76:509–19

Buchli AD, Rouiller E, Mueller R, Dietz V, Schwab ME. 2007. Repair of the injured spinal cord. A joint approach of basic and clinical research. *Neurodegener. Dis.* 4:51–56

Canedo A. 1997. Primary motor cortex influences on the descending and ascending systems. *Prog. Neurobiol.* 51:287–335

Case LC, Tessier-Lavigne M. 2005. Regeneration of the adult central nervous system. *Curr. Biol.* 15:R749–53

Cerri G, Shimazu H, Maier MA, Lemon RN. 2003. Facilitation from ventral premotor cortex of primary motor cortex outputs to macaque hand muscles. *J. Neurophysiol.* 90:832–42

Cheema SS, Rustioni A, Whitsel BL. 1984. Light and electron microscopic evidence for a direct corticospinal projection to superficial laminae of the dorsal horn in cats and monkeys. *J. Comp. Neurol.* 225:276–90

Cheney PD, Fetz EE. 1984. Corticomotoneuronal cells contribute to long-latency stretch reflexes in the rhesus monkey. *J. Physiol.* 349:249–72

Cheney PD, Fetz EE, Mewes K. 1991. Neural mechanisms underlying corticospinal and rubrospinal control of limb movements. *Prog. Brain Res.* 87:213–52

Colebatch JG, Gandevia SC. 1989. The distribution of muscular weakness in upper motor neuron lesions affecting the arm. *Brain* 112:749–63

Colebatch JG, Rothwell JC, Day BL, Thompson PD, Marsden CD. 1990. Cortical outflows to proximal arm muscles in man. *Brain* 113:1843–56

Courtine G, Bunge MB, Fawcett JW, Grossman RG, Kass JH, et al. 2007. Can experiments in nonhuman primates expedite the translation of treatments for spinal cord injury in humans? *Nat. Med.* 13:561–66

Courtine G, Roy RR, Raven J, Hodgson J, McKay H, et al. 2005. Performance of locomotion and foot grasping following a unilateral thoracic corticospinal tract lesion in monkeys (*Macaca mulatta*). *Brain* 128:2338–58

Davidson AG, Buford JA. 2004. Motor outputs from the primate reticular formation to shoulder muscles as revealed by stimulus triggered averaging. *J. Neurophysiol.* 92:83–95

Davidson AG, Buford JA. 2006. Bilateral actions of the reticulospinal tract on arm and shoulder muscles in the monkey: stimulus triggered averaging. *Exp. Brain Res.* 173:25–39

Davidson AG, Schieber MH, Buford JA. 2007. Bilateral spike-triggered average effects in arm and shoulder muscles from the monkey pontomedullary reticular formation. *J. Neurosci.* 27:8053–58

de Noordhout AM, Rapisarda G, Bogacz D, Gerard P, de Pasqua V, et al. 1999. Corticomotoneuronal synaptic connections in normal man. An electrophysiological study. *Brain* 122:1327–40

de Troyer A, Kirkwood PA, Wilson TA. 2005. Respiratory action of the intercostal muscles. *Physiol. Rev.* 85:717–56

Deumens R, Koopmans GC, Joosten EA. 2005. Regeneration of descending axon tracts after spinal cord injury. *Prog. Neurobiol.* 77:57–89

Dietz V. 2003. Spinal cord pattern generators for locomotion. *Clin. Neurophysiol.* 114:1379–89

Drew T. 1991. Functional organization within the medullary reticular formation of the intact anaesthetized cat. III. Microstimulation during locomotion. *J. Neurophysiol.* 66:919–38

Drew T, Prentice S, Schepens B. 2004. Cortical and brainstem control of locomotion. *Prog. Brain Res.* 143:251–61

Dum RP, Strick PL. 2005. Frontal lobe inputs to the digit representations of the motor areas on the lateral surface of the hemisphere. *J. Neurosci.* 25:1375–86

Fetz EE, Cheney PD. 1980. Postspike facilitation of forelimb muscle activity by primate cortico-motoneuronal cells. *J. Neurophysiol.* 44:751–72

Fouad K, Pedersen V, Schwab ME, Brosamle C. 2001. Cervical sprouting of corticospinal fibers after thoracic spinal cord injury accompanies shifts in evoked motor responses. *Curr. Biol.* 11:1766–70

Fragaszy DM. 1998. How non-human primates use their hands. In *The Psychobiology of the Hand*, ed. KJ Connolly, pp. 77–96. London: Mac Keith

Freund P, Schmidlin E, Wannier T, Bloch J, Mir A, et al. 2006. Nogo-A-specific antibody treat-ment enhances sprouting and functional recovery after cervical lesion in adult primates. *Nat. Med.* 12:790–92

Freund P, Wannier T, Schmidlin E, Bloch J, Mir A, et al. 2007. Anti-Nogo-A antibody treatment enhances sprouting of corticospinal axons rostral to a unilateral cervical spinal cord lesion in adult macaque monkey. *J. Comp. Neurol.* 502:644–59

Galea MP, Darian-Smith I. 1997. Corticospinal projection patterns following unilateral section of the cervical spinal cord in the newborn and juvenile macaque monkey. *J. Comp. Neurol.* 381:282–306

Georgopoulos AP, Grillner S. 1989. Visuomotor coordination in reaching and locomotion. *Science* 245:1209–10

Grillner S, Wallen P. 2004. Innate versus learned movements—a false dichotomy? *Prog. Brain Res.* 143:3–12

Gugino LD, Rowinski MJ, Stoney SD. 1990. Motor outflow to cervical motoneurons from racoon motorsensory cortex. *Brain Res. Bull.* 24:833–37

He SQ, Dum RP, Strick PL. 1993. Topographic organization of corticospinal projections from the frontal lobe: motor areas on the lateral surface of the hemisphere. *J. Neurosci.* 13:952–80

Heckman CJ, Lee RH. 1999. Synaptic integration in bistable motoneurons. *Prog. Brain Res.* 123:49–56

Heffner RS, Masterton RB. 1983. The role of the corticospinal tract in the evolution of human digital dexterity. *Brain Behav. Evol.* 23:165–83

Holstege G. 1998. The emotional motor system in relation to the supraspinal control of micturition and mating behavior. *Behav. Brain Res.* 92:103–9

Holstege G, Kuypers HGJM. 1982. The anatomy of brain stem pathways to the spinal cord in cat. A labeled amino acid tracing study. *Progr. Brain Res.* 57:145–75

Hultborn H, Brownstone RB, Toth TI, Gossard J-P. 2004. Key mechanisms for setting the input-output gain across the motoneuron pool. *Prog. Brain Res.* 143:77–95

Illert M, Lundberg A, Tanaka R. 1976. Integration in descending motor pathways controlling the forelimb in the cat. 1. Pyramidal effects on motoneurones. *Exp. Brain Res.* 26:509–19

Isa T, Ohki Y, Alstermark B, Pettersson LG, Sasaki S. 2007. Direct and indirect cortico-motoneuronal pathways and control of hand/arm movements. *Physiology* 22:145–52

Isa T, Ohki Y, Seki K, Alstermark B. 2006. Properties of propriospinal neurons in the C3-C4 segments mediating disynaptic pyramidal excitation to forelimb motoneurons in the macaque monkey. *J. Neurophysiol.* 95:3674–85

Iwaniuk AN, Pellis SM, Whishaw IQ. 1999. Is digital dexterity really related to corticospinal projections? A reanalysis of the Heffner and Masterton data set using modern comparative statistics. *Behav. Brain Res.* 101:173–87

Jackson A, Gee VJ, Baker SN, Lemon RN. 2003. Synchrony between neurons with similar muscle fields in monkey motor cortex. *Neuron* 38:15–125

Jankowska E, Padel Y, Tanaka R. 1975. Projections of pyramidal tract cells to p-motoneurones innervating hind-limb muscles in monkey. *J. Physiol.* 249:637–67

Johansson RS, Lemon RN, Westling G. 1994. Time varying enhancement of human excitability mediated by cutaneous inputs during precision grip. *J. Physiol.* 481:761–75

Jones EG. 1986. Connectivity of the primate sensory-motor cortex. In *Cerebral Cortex*, ed. EG Jones, A Peters, pp. 113–83. New York: Plenum

Kennedy PR. 1990. Corticospinal, rubrospinal and rubro-olivary projections: a unifying hypothesis. *Trends Neurosci.* 13:474–79

Kuchler M, Fouad K, Weinmann O, Schwab ME, Raineteau O. 2002. Red nucleus projections to distinct motor neuron pools in the rat spinal cord. *J. Comp. Neurol.* 448:349–59

Kuypers HGJM. 1981. Anatomy of the descending pathways. See Brookhart & Mountcastle 1981, pp. 597–666

Kuypers HGJM, Brinkman J. 1970. Precentral projections to different parts of the spinal intermediate zone in the rhesus monkey. *Brain Res.* 24:29–48

Kuypers HGJM, Ugolini G. 1990. Viruses as transneuronal tracers. *Trends Neurosci.* 13:71–75

Kwan HC, MacKay WA, Murphy JT, Wong YC. 1978. Spatial organization of precentral cortex in awake primates. II. Motor outputs. *J. Neurophysiol.* 41:1120–31

Lacroix S, Havton LA, McKay H, Yang H, Brant A, et al. 2004. Bilateral corticospinal projections arise from each motor cortex in the macaque monkey: a quantitative study. *J. Comp. Neurol.* 473:147–61

Lang CE, Schieber MH. 2003. Differential impairment of individuated finger movements in humans after damage to the motor cortex or the corticospinal tract. *J. Neurophysiol.* 90:1160–70

Lawrence DG, Kuypers HGJM. 1968a. The functional organization of the motor system in the monkey. I. The effects of bilateral pyramidal lesions. *Brain* 91:1–14

Lawrence DG, Kuypers HGJM. 1968b. The functional organization of the motor system in the monkey. II. The effects of lesions of the descending brain-stem pathway. *Brain* 91:15–36

Lemon R. 1988. The output map of the primate motor cortex. *Trends Neurosci.* 11:501–6

Lemon RN, Griffiths J. 2005. Comparing the function of the corticospinal system in different species: organizational differences for motor specialization? *Muscle Nerve* 32(3):261–79

Lemon RN, Kirkwood PA, Maier MA, Nakajima K, Nathan P. 2004. Direct and indirect pathways for corticospinal control of upper limb motoneurones in the primate. *Prog. Brain Res.* 143:263–79

Lemon RN, Mantel GWH, Muir RB. 1986. Corticospinal facilitation of hand muscles during voluntary movement in the conscious monkey. *J. Physiol.* 381:497–527

Maier MA, Armand J, Kirkwood PA, Yang HW, Davis JN, Lemon RN. 2002. Differences in the corticospinal projection from primary motor cortex and supplementary motor area to macaque upper limb motoneurons: an anatomical and electrophysiological study. *Cereb. Cortex* 12:281–96

Maier MA, Illert M, Kirkwood PA, Nielsen J, Lemon RN. 1998. Does a C3-C4 propriospinal system transmit corticospinal excitation in the primate? An investigation in the macaque monkey. *J. Physiol.* 511:191–212

Martin JH, Kably B, Hacking A. 1999. Activity-dependent development of cortical axon terminations in the spinal cord and brain stem. *Exp. Brain Res.* 125:184–99

Mason P. 1997. Physiological identification of pontomedullary serotonergic neurons in the rat. *J. Neurophysiol.* 77:1087–98

Matsuyama K, Drew T. 1997. Organisation of the projections from the pericruciate cortex to the pontomedullary brainstem of the cat: a study using the anterograde tracer Phaseolus vulgaris-leucoagglutinin. *J. Comp. Neurol.* 389:617–41

Matsuyama K, Mori F, Kuze B, Mori S. 1999. Morphology of single pontine reticulospinal axons in the lumbar enlargement of the cat: a study using the anterograde tracer PHA-L. *J. Comp. Neurol.* 410:413–30

Matsuyama K, Mori F, Nakajima K, Drew T, Aoki M, Mori S. 2004. Locomotor role of the corticoreticular-reticulospinal-spinal interneuron system. *Prog. Brain Res.* 143:239–49

Matsuyama K, Takakusaki K, Nakajima K, Mori S. 1997. Multisegmental innervation of single pontine reticulospinal axons in the cervico-thoracic region of the cat: anterograde PHA-L tracing study. *J. Comp. Neurol.* 377:234–50

McKiernan BJ, Marcario K, Karrer JH, Cheney PD. 1998. Corticomotoneuronal postspike effects in shoulder, elbow, wrist, digit, and intrinsic hand muscles during a reach and prehension task. *J. Neurophysiol.* 80:1961–80

Molenaar I, Kuypers HGJM. 1978. Cells of origin of propriospinal fiber and fibers ascending to supraspinal levels. An HRP study in cat and rhesus monkey. *Brain Res.* 152:429–50

Monteau R, Hilaire G. 1991. Spinal respiratory motoneurons. *Prog. Neurobiol.* 37:83–144

Muir GD, Whishaw IQ. 2000. Red nucleus lesions impair overground locomotion in rats: a kinetic analysis. *Eur. J. Neurosci.* 12:1113–22

Nakajima K, Maier MA, Kirkwood PA, Lemon RN. 2000. Striking differences in the transmission of corticospinal excitation to upper limb motoneurons in two primate species. *J. Neurophysiol.* 84:698–709

Nielsen J, Petersen N, Ballegaard M. 1995. Latency of effects evoked by electrical and magnetic brain stimulation in lower limb motoneurones in man. *J. Physiol. (London)* 484:791–802

Nudo RJ. 2007. Postinfarct cortical plasticity and behavioral recovery. *Stroke* 38:840–45

Olivier E, Baker SN, Nakajima K, Brochier T, Lemon RN. 2001. Investigation into nonmonosynaptic corticospinal excitation of macaque upper limb single motor units. *J. Neurophysiol.* 86:1573–86

Palmer E, Ashby P. 1992. Corticospinal projections to upper limb motoneurones in humans. *J. Physiol.* 448:397–412

Park MC, Belhaj-Saif A, Gordon M, Cheney PD. 2001. Consistent features in the forelimb representation of primary motor cortex in rhesus macaques. *J. Neurosci.* 21:2784–92

Phillips CG. 1975. Laying the ghost of "muscles versus movements". *Can. J. Neurol. Sci.* 2:209–18

Phillips KA. 1998. Tool use in wild capuchin monkeys. *Am. J. Primatol.* 46:259–61

Pierrot-Deseilligny E, Burke D. 2005. *The Circuitry of the Human Spinal Cord: Its Role in Motor Control and Movement Disorders.* Cambridge, UK: Cambridge Univ. Press

Porter R, Lemon RN. 1993. Corticospinal function and voluntary movement. In *Physiological Society Monograph*, p. 428. Oxford, UK: Oxford Univ. Press

Quencer RM, Bunge RP, Egnor M, Green BA, Puckett W, et al. 1992. Acute traumatic central cord syndrome: MRI-pathological correlations. *Neuroradiology* 34:85–94

Ramer LM, Ramer MS, Steeves JD. 2005. Setting the stage for functional repair of spinal cord injuries: a cast of thousands. *Spinal Cord* 43:134–61

Rathelot JA, Strick PL. 2006. Muscle representation in the macaque motor cortex: an anatomical perspective. *Proc. Natl. Acad. Sci. USA* 103:8257–62

Rothwell JC, Thompson PD, Day BL, Boyd S, Marsden CD. 1991. Stimulation of the human motor cortex through the scalp. *Exp. Physiol.* 76:159–200

Sala F, Palandri G, Basso E, Lanteri P, Deletis V, et al. 2006. Motor evoked potential monitoring improves outcome after surgery for intramedullary spinal cord tumors: a historical control study. *Neurosurgery* 58:1129–43

Sasaki S, Isa T, Pettersson LG, Alstermark B, Naito K, et al. 2004. Dexterous finger movements in primate without monosynaptic corticomotoneuronal excitation. *J. Neurophysiol.* 92:3142–47

Schieber MH, Hibbard LS. 1993. How somatotopic is the motor cortex hand area? *Science* 261:489–92

Schieber MH, Rivlis G. 2005. A spectrum from pure postspike effects to synchrony effects in spike-triggered averages of electromyographic activity during skilled finger movements. *J. Neurophysiol.* 94:3325–41

Schieber M, Rivlis G. 2007. Partial reconstruction of muscle activity from a pruned network of diverse motor cortex neurons. *J. Neurophysiol.* 97:70–82

Schmidlin EO, Brochier T, Maier MA, Lemon RN. 2006. Cortico-cortical interactions in macaque motor cortex investigated with intracortical microstimulation (ICMS) and transient inactivation with muscimol. *Soc. Neurosci. Abstr.* 255.17

Shimazu H, Maier MA, Cerri G, Kirkwood PA, Lemon RN. 2004. Macaque ventral premotor cortex exerts powerful facilitation of motor cortex outputs to upper limb motoneurons. *J. Neurosci.* 24:1200–11

Steward O, Zheng B, Tessier-Lavigne M. 2003. False resurrections: distinguishing regenerated from spared axons in the injured central nervous system. *J. Comp. Neurol.* 459:1–8

Sugiuchi Y, Kakei S, Izawa Y, Shinoda Y. 2004. Functional synergies among neck muscles revealed by branching patterns of single long descending motor-tract axons. *Prog. Brain Res.* 143:411–21

Tanaka H, Takahashi S, Oki J. 1997. Developmental regulation of spinal motoneurons by monoaminergic nerve fibers. *J. Peripher. Nerv. Syst.* 2:323–32

Tator CH. 2006. Review of treatment trials in human spinal cord injury: issues, difficulties, and recommendations. *Neurosurgery* 59:957–82

Vanderhorst VG, Terasawa E, Ralston HJ 3rd, Holstege G. 2000. Monosynaptic projections from the lateral periaqueductal gray to the nucleus retroambiguus in the rhesus monkey: implications for vocalization and reproductive behavior. *J. Comp. Neurol.* 424:251–68

Vavrek R, Girgis J, Tetzlaff W, Hiebert GW, Fouad K. 2006. BDNF promotes connections of corticospinal neurons onto spared descending interneurons in spinal cord injured rats. *Brain* 129:1534–45

Wall PD, Lidierth M. 1997. Five sources of a dorsal root potential: their interactions and origins in the superficial dorsal horn. *J. Neurophysiol.* 78:860–71

Ward NS, Cohen LG. 2004. Mechanisms underlying recovery of motor function after stroke. *Arch. Neurol.* 61:1844–48

Ward NS, Newton JM, Swayne OB, Lee L, Frackowiak RS, et al. 2007. The relationship between brain activity and peak grip force is modulated by corticospinal system integrity after subcortical stroke. *Eur. J. Neurosci.* 25:1865–73

Wolpaw JR. 1997. The complex structure of a simple memory. *Trends Neurosci.* 20:588–94

Wolpert DM, Ghahramani Z, Flanagan JR. 2001. Perspectives and problems in motor learning. *Trends Cogn. Sci.* 5:487–94

Yanai Y, Adamit N, Harel R, Israel Z, Prut Y. 2007. Connected corticospinal sites show enhanced tuning similarity at the onset of voluntary action. *J. Neurophysiol.* 27:12349–57

Yang H-W, Lemon RN. 2003. An electron microscopic examination of the corticospinal projection to the cervical spinal cord in the rat: lack of evidence for cortico-motoneuronal synapses. *Exp. Brain Res.* 149:458–69

Yang JF, Gorassini M. 2006. Spinal and brain control of human walking: implications for retraining of walking. *Neuroscientist* 12:379–89

Nishimura Y, Onoe H, Morichika Y, Perfiliev S, Tsukada H, Isa T. 2007. Time-dependent central compensatory mechanisms of finger dexterity after spinal cord injury. *Science* 318:1150–55

Task Set and Prefrontal Cortex

Katsuyuki Sakai

Department of Cognitive Neuroscience, Graduate School of Medicine, The University of Tokyo, 7-3-1 Hongo, Bunkyo-ku, Tokyo 113-0033, Japan; email: ksakai@m.u-tokyo.ac.jp

Annu. Rev. Neurosci. 2008. 31:219–45

First published online as a Review in Advance on April 2, 2008

The *Annual Review of Neuroscience* is online at neuro.annualreviews.org

This article's doi:
10.1146/annurev.neuro.31.060407.125642

Key Words

task rule, task switch, conflict, functional brain imaging, single-unit recording

Abstract

A task set is a configuration of cognitive processes that is actively maintained for subsequent task performance. Single-unit and brain-imaging studies have identified the neural correlates for task sets in the prefrontal cortex. Here I examine whether the neural data obtained thus far are sufficient to explain the behaviors that have been illustrated within the conceptual framework of task sets. I first discuss the selectivity of neural activity in representing a specific task. I then discuss the competitions between neural representations of task sets during task switch. Finally I discuss how, in neural terms, a task set is implemented to facilitate task performance. The processes of representing, updating, and implementing task sets occur in parallel at multiple levels of brain organization. Neural accounts of task sets demonstrate that the brain determines our thoughts and behaviors.

Contents

INTRODUCTION

In cognitive neuroscience studies, investigators usually begin experiments by giving task instructions to the subjects. The subjects heed the instructions and prepare for the experiment. The subjects may remember the instructions by verbally rehearsing them, but after practice for several trials, the task information is maintained as a configuration of perceptual, attentional, mnemonic, and motor processes necessary to perform the task. A "task set" refers to this prospectively configured task in an abstract form. When subjects are then asked to perform another task, they have to establish a new task set in a form distinct from the previous one. The stimulus set and response set are the same between the two tasks, but the rules of association between the stimuli and responses differ. For example, subjects judge whether a visually presented digit is even or odd in one task, and in the other, they judge whether the digit is larger or smaller than five. Thus a task set has to be specific in the sense that it represents a rule of a specific task to be performed. However, a task set is nonspecific in the sense that it can be applied to any stimulus as long as it belongs to a task-relevant stimulus set.

A number of behavioral experiments have provided conceptual frameworks to explain how the task sets are established (Meiran et al. 2000, Monsell 2003). Neurophysiological experiments have also identified the neural correlates of the cognitive components associated with task sets (Bunge & Wallis 2007). The aim of this review is not just to provide a list of brain areas with labels of psychological terminologies, but to examine whether the neural data obtained thus far are sufficient to explain the behaviors that have been explained within the framework of task sets. By focusing on the data obtained from the prefrontal cortex, I discuss the neural mechanisms involved in representing, updating, and implementing task sets. I specifically highlight the differences in the inferences that one can draw from single-unit recording studies on monkeys and brain-imaging studies on human subjects.

REPRESENTATION OF TASK SET

Rule-Specific Neural Activity

A task set is a psychological construct. Therefore, a fundamental question is whether such a construct exists in the brain. The neural correlates of a task set are considered to be the neural activity specific to the rule of a particular task. The activity should be independent of the task items (hereafter, called targets) that the

a   Task-cueing paradigm

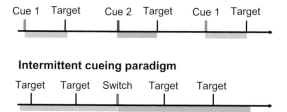

Intermittent cueing paradigm

b   Domain-specific rules

Location-match rule

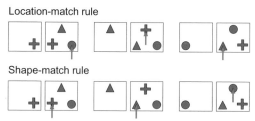

Shape-match rule

Operation-specific rules

Match rule

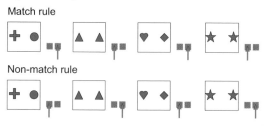

Non-match rule

Stimulus-response mapping rules

Shape-based mapping rule

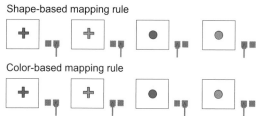

Color-based mapping rule

Phonological rule (one or two syllables)

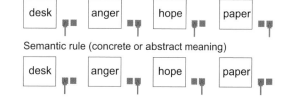

| desk | anger | hope | paper |

Semantic rule (concrete or abstract meaning)

| desk | anger | hope | paper |

Figure 1

Task paradigms used to study task sets. (*a*) In a task cuing paradigm, a cue specifies the task to be performed for each trial and subjects process the task item (target) accordingly. In an intermittent cuing paradigm, subjects perform one task over a block of trials and, after a switch cue, change to another task. The horizontal bars (*pink and blue*) indicate the activity associated with task sets that are maintained during the cue-target interval or during a block of trials. (*b*) Examples of tasks that are compared in a study. The tasks can differ in terms of the sensory domains of the task, the association between the stimulus and response (stimulus-response mapping), and cognitive operations such as comparative and linguistic operations. *Pink* arrows indicate correct responses.

subjects are to process using the rule. Also the task set activity should be independent of task cues as long as they instruct the same task.

To identify the rule-specific neural activity, a task-cuing paradigm has been used: For each trial, a cue instructing the to-be-performed task is presented, followed by a target. The neural correlates of a task set can be identified as the sustained neural activity during the cue-target interval in a particular task

(**Figure 1a**). Another paradigm is an intermittent cuing paradigm, in which subjects perform one task for a block of trials and then switch to another task on the basis of a switching cue or feedback. In this setting the neural correlates of a task set can be identified as neural activity that is sustained across trials of a particular task block.

Recent studies have identified activity of a single neuron that satisfies these criteria. Also

GO-NOGO task: a special case of a stimulus-response mapping in which a stimulus is associated with making a response (go) or not responding (no-go)

DLPFC: dorsolateral prefrontal cortex

VLPFC: ventrolateral prefrontal cortex

Stimulus-response (SR) mapping: a paradigm in which subjects respond to each stimulus in a particular set with a prespecified response

several imaging studies have successfully identified the rule-specific regional activity. In these studies, neural activity is compared between tasks that differ in a particular aspect of the cognitive processes, such as sensory domains of the tasks, rules of specific associations between a target and a response, and rules of cognitive operations applied to the targets (**Figure 1*b***).

I first examine the spatial distribution of the rule-specific neural activity within the prefrontal cortex. The prefrontal cortex is known to be constituted with multiple areas that differ in terms of cytoarchitectures and anatomical connections with other areas (Petrides & Pandya 1999, 2002) (**Figure 2**). Each of these prefrontal areas has its unique functional fingerprint, that is, a unique pattern of the degree of involvement across tasks (Passingham et al. 2002). The question here is how the representations of rules that differ in a particular cognitive component are associated with the regional segregation within the prefrontal cortex.

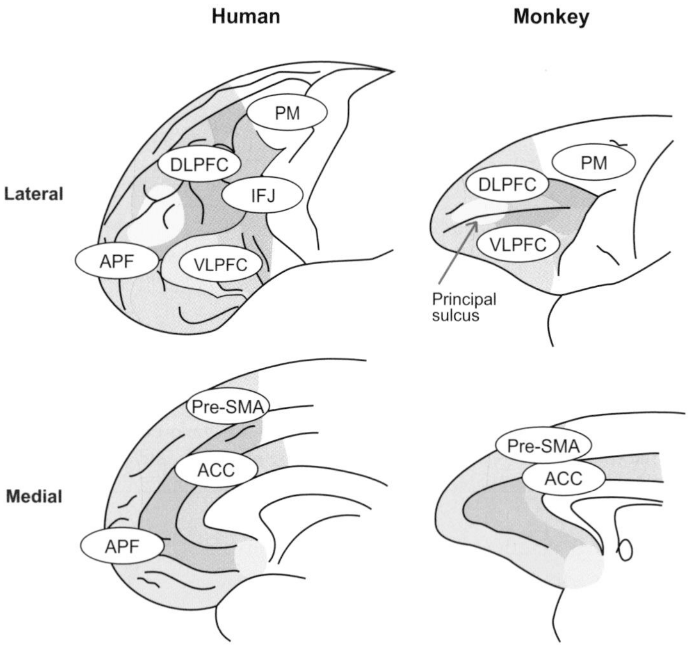

Figure 2

Schematic drawings of the lateral (*top*) and medial (*bottom*) prefrontal and other frontal regions in human and monkey. Regions defined by colors are Brodmann's areas based on Petrides & Pandya (1999). ACC, anterior cingulated cortex; APF, anterior prefrontal cortex; DLPF, dorsolateral prefrontal cortex; IFJ, inferior frontal junction; PM, premotor cortex; Pre-SMA, presupplementary motor area; VLPFC, ventrolateral prefrontal cortex.

Domain-Specific Rules

Single-unit studies have shown a difference in the distribution of rule-specific neurons when tasks are compared that differ in terms of the sensory domain of the task. For example, neurons that are active on a motion-based GO-NOGO task were found in the dorsolateral prefrontal cortex (DLPFC), whereas neurons that are active on a color-based GO-NOGO task were found in the ventrolateral prefrontal cortex (VLPFC) (Sakagami & Tsutsui 1999). In another study, neurons that are active in a location-match task were found in the more posterior part of the principal sulcus region relative to neurons that are active in a shape-match task (Hoshi et al. 2000). However, inactivation of this posterior region did not affect the performance of the location-match task, whereas inactivation of the more anterior region caused an increase in the number of errors for both the location- and the shape-matching tasks. Thus the difference in the distribution of task-specific neurons may not be clear when the functional relevance of these neurons is examined. The difference may also depend on whether the task rules are associated with specific features of the target items, as in a fixed stimulus-response mapping (SR mapping) task, or with sensory domains of the target items regardless of their specific visual features.

Human brain-imaging studies have also shown that task rules are represented in different prefrontal regions depending on the sensory domain of the task. The posterior part of the DLPFC and VLPFC is active during the cue-target interval for a location and a letter memory task, respectively (Sakai & Passingham 2003). Representations of different SR-mapping rules have also been associated with activation in different prefrontal regions when the rules are associated with different stimulus features. For example Yeung et al. (2006) found that, in addition to the feature-specific extrastriate areas, the right VLPFC was active when subjects responded on the basis of the gender of a face, whereas the left VLPFC was active when subjects responded on the basis of the number of syllables in a word.

These studies suggest that tasks that differ in the sensory domains are represented in different posterior prefrontal regions. However, the activation in these areas may simply reflect attentional sets for different sensory domains rather than task sets for different SR mapping rules (see sidebar on Attentional Set). More convincing evidence for the neural correlates of a task set can be obtained by comparing tasks that differ in terms of the way subjects process the same task items, i.e., operations of the task.

Operation-Specific Rules

Judgment of whether the sample and test objects are the same requires comparative operations between the two objects. This object-match task can be thought of as an abstract task because the task is not based on associations between specific stimuli and responses. The type of comparative operations differs in match vs. nonmatch tasks. Cognitive operations also differ in object-match vs. object-response association tasks. Single-unit studies have shown that there exist neurons that are active in a specific cognitive operation, and the spatial distribution of these neurons within the prefrontal cortex does not differ depending on the type of operations (Asaad et al. 2000, White & Wise 1999, Wallis et al. 2001) (**Figure 3a**). These operation-specific neurons have been found in the extensive prefrontal regions including the DLPFC, VLPFC, and orbitofrontal cortex.

Neurons representing different strategies are also represented by single-neuron activity within the same region in the DLPFC (Genovesio et al. 2005). Strategies are like special kinds of abstract rules that are acquired on the basis of past task performance history. In this study monkeys were trained to make a saccade to the left, right, or upward direction in response to a visual object. The monkeys could not learn any fixed SR mappings

ATTENTIONAL SET

We can respond to a stimulus faster when we have advanced knowledge about the features of the stimulus or the types of the movement we are to make. Such facilitation of behavior depends on the ability to represent the advanced information prior to the onset of the stimulus or movement. An attentional set is a definition of the representations of the advanced information involved in selecting task-relevant stimuli and responses (Corbetta & Shulman 2002). An attentional set is mediated by the sustained activity in the dorsal fronto-parietal network prior to the onset of the stimulus or movement. The network sends top-down signals to regions specifically involved in the actual processing of sensory features or in execution of the movements and facilitates neural processing in those regions. A task set reviewed in this article is regarded as an extension of an attentional set because it refers to representations of advanced information about the to-be-performed task.

because the response suggested by the same stimulus can differ depending on the previous trial. Instead the monkeys adopted freely between repeat-stay and change-shift strategies; that is, if the object repeated from a previous, successful trial, the monkeys repeated the response, and if the object changed, the monkeys shifted to a different response. In addition to the strategy-specific neurons, neurons in the same region in the DLPFC also represented fixed SR mapping rules that the monkeys learned during different sessions. The result suggests that task rules at different levels of abstraction, namely, exemplar-based fixed SR mappings and abstract response strategies, are represented by separate sets of neurons within the DLPFC.

By contrast, using brain imaging on human subjects, it is difficult to distinguish among different task operations. The conventional analysis of brain imaging such as functional magnetic resonance imaging (fMRI) is to compare, for each voxel of several millimeters, the magnetic resonance (MR) signals between task conditions, but neurons representing different operation-specific rules are likely to coexist within the same voxel.

fMRI: functional magnetic resonance imaging

A recent study, instead, used a multivoxel pattern of the MR signals to distinguish between the two arithmetic operation rules (Haynes et al. 2007). Human subjects were asked to choose between addition and subtraction and then, after a delay, were shown two digits to which they were to apply the selected arithmetic operation. The medial and

a Rule-specific activity of a single neuron

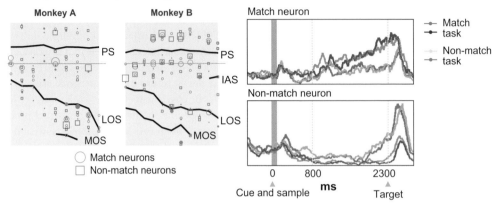

Match neuron

Non-match neuron

Match neurons
Non-match neurons

Match task
Non-match task

Cue and sample   ms   Target

b Rule-specific pattern of multi-voxel MR signals

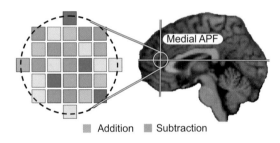

Addition   Subtraction

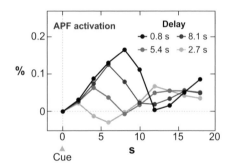

APF activation   Delay
0.8 s 8.1 s
5.4 s 2.7 s

Cue

c Rule-specific pattern of inter-regional interactions

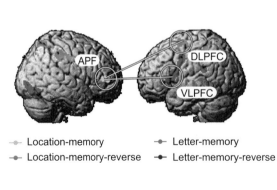

APF   DLPFC
VLPFC

Location-memory   Letter-memory
Location-memory-reverse   Letter-memory-reverse

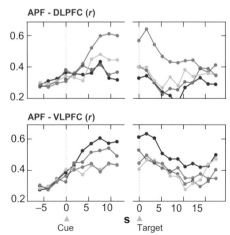

APF - DLPFC (r)

APF - VLPFC (r)

Cue   Target

lateral part of the anterior prefrontal cortex (APF) showed sustained activity before the actual arithmetic calculation. Conventional voxel-based comparison of the MR signals failed to distinguish between addition and subtraction, but the spatial pattern of the MR signals across multiple voxels within these prefrontal regions successfully distinguished between the arithmetic operations before the subjects actually performed the calculation (**Figure 3b**). The signal pattern may reflect slight differences in the proportion of addition and subtraction neurons across voxels. Although the precise neural mechanism of coding arithmetic operations remains open, this new analysis technique allows us to overcome the limitation of the spatial resolution of functional brain imaging and opens up the possibility to decode abstract human thought.

Imaging studies have also shown that the pattern of interregional interaction changes according to the rule of the upcoming task. The correlation of activity between the lateral APF and posterior DLPFC was higher when subjects were cued to remember a sequence of locations in reverse order, whereas the correlation between the same region in the APF and posterior VLPFC was higher when subjects were cued to remember a sequence of letters in reverse order (Sakai & Passingham 2003) (**Figure 3c**). The results suggest that the strength of functional connectivity changes depending on the requirement to perform a reversal operation on the sequence of remembered items. The changes were observed before the presentation of targets, that is, before the reversal operation was actually performed. The interregional interactions observed in this study reflect a preparatory process to perform a specific cognitive operation. Another study showed that the lateral APF interacted with the ventral premotor cortex (PM) and anterior VLPFC when subjects were cued to perform phonological judgment and semantic judgment for a visually presented word, respectively (Sakai & Passingham 2006). These interregional interactions reflect preparation for specific linguistic operations. Different sets of neurons within the APF may project to posterior frontal regions that are specifically involved in different tasks, and selective activation of each set of the APF neurons results in the rule-specific patterns of interregional interaction (see sidebar on Task Set for Memory Retrieval).

APF: anterior prefrontal cortex

PM: premotor cortex

Figure 3

(*a*) Rule-specific activity of a single neuron. *Left*: Neurons that are preferentially active in match task and nonmatch task are distributed in the same region of the prefrontal cortex in monkeys. IAS, inferior arcuate sulcus; LOS, lateral orbital sulcus; MOS, medial orbital sulcus; PS, principal sulcus. *Right*: Time course of activity of a match-selective neuron and a nonmatch-selective neuron. The neurons show a selective activity increase for the preferred task before a target is presented. Plots shown in dark and light *pink* and *blue* indicate trials in which different cues are used to instruct the same task. The original figures are kindly provided by J. Wallis based on Wallis et al. (2001). (*b*) Rule-specific pattern of multi-voxel signals. *Middle*: The medial part of the anterior prefrontal cortex (APF), as indicated by *yellow cross hairs and a circle*, is active when human subjects select between addition and subtraction. *Right*: The duration of the activation in the APF expands and contracts according to the length of the cue-target interval, indicating sustained activation. Adapted with permission from Haynes et al. (2007). *Left*: The magnetic resonance signals across multiple voxels in this region show a spatial pattern specific to whether the subjects intend to perform addition or subtraction as in the schematic drawing. The squares in *pink* and *blue* indicate voxels within the medial APF that are active in addition and subtraction, respectively. The lightness of the color indicates the strength of selectivity to a preferred task rule. The consistency of the signals' spatial patterns is confirmed across experimental sessions. (*c*) Rule-specific pattern of interregional interactions. *Left*: The anterior prefrontal cortex (APF), dorsolateral prefrontal cortex (DLPFC), and ventrolateral prefrontal cortex (VLPFC) are active when subjects are cued to perform a location and letter memory task. *Right*: Time course of the correlation coefficients (*r*) of magnetic resonance signals between the APF and the DLPFC (*top*) and between the APF and the VLPFC (*bottom*). The correlation of the signals between the APF and the DLPFC is significantly higher when human subjects remember a sequence of locations in reverse order. By contrast, the correlation of the signals between the APF and VLPFC is significantly higher when human subjects remember a sequence of letters in reverse order. The selective increase of correlation coefficients is observed during the cue-target interval as well as during task execution. Adapted with permission from Sakai & Passingham (2003).

TASK SET FOR MEMORY RETRIEVAL

Neural activity during preparation for memory retrieval differs depending on the kinds of information to be retrieved. Duzel et al. (1999) observed a sustained change of event-related potential in the right fronto-polar cortex during the cue-target interval of an episodic memory task irrespective of whether verbal or nonverbal items were used for the memory test, but the activity was not observed in a semantic task. This activity reflects retrieval mode, which refers to a tonically maintained cognitive set for episodic memory retrieval (Rugg & Wilding 2000). The sustained preparatory activity in the frontal region also differs within the episodic memory domain depending on the kinds of information to be retrieved, known as retrieval orientation (Herron & Wilding 2006). Retrieval mode and retrieval orientation can be considered as task sets for memory retrieval.

Hierarchical Organization of Rule Representations

Imaging studies suggest a rostro-caudal gradient within the prefrontal cortex in representing task sets according to the level of the abstractness of the task rule. When tasks with different sensory domains are compared, posterior prefrontal areas show differential activation depending on the task (Sakai & Passingham 2003). When tasks with different cognitive operations are compared, the amount of activation in these posterior areas does not differ significantly, but the interaction pattern with the anterior prefrontal area differs significantly. The multi-voxel signal pattern within a region also represents abstract arithmetic operations more strongly in the anterior prefrontal regions than in the posterior prefrontal regions (Haynes et al. 2007).

However, a regional difference in rule representation has rarely been demonstrated in single-unit studies. In single-unit recording, one can sample only two or three neurons in one electrode track, which is separated from another track by 1 or 2 mm, and regional difference is examined by comparing the number of rule-selective neurons across coronal sections of the brain or across regions defined by gross anatomical landmarks. Therefore, regional difference may be underestimated in single-unit studies. The advantage of a single-unit study is that it can clarify the local neural mechanisms of the rules represented at different levels of abstraction, which is the topic of future studies.

In addition to regions that show rule-specific activation, other regions are commonly involved in different task sets. Dosenbach et al. (2006) asked 183 subjects to perform 10 tasks that involve various kinds of cognitive processing, and they found that the anterior cingulated cortex (ACC) and anterior insula/operculum showed sustained activation during the cue-target interval as well as during the task performance, consistently across all these tasks. These results suggest a hierarchical organization of task set representation, in which ACC and anterior insula/operculum constitute a core task set system and play roles in regulating task-specific representations in the posterior frontal and other association areas. Whether these core areas contain multiple sets of task-specific neurons or whether the neurons in these areas are highly adaptive such that they can code different rules depending on experimental contexts are open issues.

Rule Selectivity

The rule specificity of neural activity is relative rather than absolute. In single-unit studies, the strength of rule selectivity has been estimated using receiver-operating characteristic (ROC) values, which are equivalent to the probability that an independent observer could identify the task condition solely on the basis of the neuron's firing rate. For example, in comparison between match and nonmatch tasks, the mean ROC value for the entire population of prefrontal neurons recorded, which include neurons nonspecific to the task condition, was 0.54 during the cue-target interval (Wallis & Miller 2003). This ROC value suggests that the task rules can be predicted on the basis of the neurons' firing rate with a probability significantly larger than chance, but at first glance the value seems to be very small. In fact, only 5% of the neurons exceeded the ROC value of

ACC: anterior cingulated cortex

ROC: receiver-operating characteristic

0.70. Because the monkeys actually performed a particular task, the ROC value should ideally be 1.0. The ROC value for the repeat-stay and change-shift strategies was higher than this, at 0.60, and the proportion of neurons exceeding the ROC value of 0.70 was ~20% (Genovesio et al. 2005). Rather surprisingly, the predictability of task rules can be higher in human brain imaging. In Haynes et al. (2007), the predictability of the intended arithmetic operators was 0.71 when multi-voxel pattern of the MR signals was examined in the medial APF.

However, a limitation exists in the inference that we can make regarding the rule specificity of the neural activity. Because rule specificity has been examined by comparing only two or three tasks in one experiment, whether the rule specificity can be generalized outside the context of that experiment remains to be determined. This is especially the case in experiments using monkeys because in most cases they had been trained only for the two tasks examined in the experiment. Also, because the neural coding is known to be adaptive depending on the experimental context (Duncan 2001), the neurons may be tuned to discriminate between the two tasks examined in a particular experiment; however, the same neurons can also be active for a completely different task when the monkeys were trained on a different task set. In fact, selectivity of neural coding for categorical sensory information changes depending on the task requirement (Freedman et al. 2002). Also, in single-unit studies, neurons are categorically classified as belonging to one task or the other on the basis of arbitrary chosen statistical threshold. In reality, however, the rule specificity of neurons changes gradually as indicated by the different ROC values across neurons. To what extent the categorical coding of task rules is part of a default mechanism of the brain based on its anatomical connectivity patterns remains an open issue.

Orthogonality of Rules

Another open issue is the orthogonality of the rules that are compared in a study. Rules are considered orthogonal when subjects switched between two SR mapping rules where the same stimulus can produce the same or a different response. However, in SR mapping reversal, the same stimulus always produces a different response, and subjects can use one rule as a default and perform the other task by first performing the default task and then selecting an alternative response: The rules are not orthogonal. Pro- and antisaccade tasks and match and nonmatch tasks are also nonorthogonal rule sets.

When match and nonmatch tasks were tested in human subjects, APF activity increased when subjects prepared to perform a nonmatch task compared with when they prepared for a match task (Bunge et al. 2003). The subjects may have coded the match task as a default and conceptualized the nonmatch task as the reverse of the match. The APF activity can be indicative of the reversing operation. By contrast, in single-unit studies on monkeys, investigators saw no indication of such default strategies in terms of both behavior and neural activity (Wallis et al. 2001). It may not be surprising, however, to find that different sets of neurons code either a match or a nonmatch rule because one rule is the reverse of the other.

Buschman et al. (2006) recently trained monkeys to perform logical OR, AND, and XOR operations on visual images associated with go or no-go responses. The advantage of this paradigm is that each abstract task rule is orthogonal to the other. The study also gives researchers an opportunity to investigate the neural mechanisms involved in the representation of hierarchically organized rules because the XOR rule is, in theory, built on AND and OR rules. The behavioral data were consistent with hierarchical rule organization.

TASK SET UPDATING

Building Up a Task Set

To adopt a task set is to select, link, and configure the elements of cognitive processes necessary to accomplish the task (Rogers & Monsell 1995). Thus establishing a task set is time consuming because it requires higher-order neural interactions between regions in the prefrontal and posterior association cortices that represent the elements of the task (Rougier et al. 2005). Preparation effect indicates the time required to establish a task set. However, merely showing a benefit of longer cue-target intervals is not sufficient to demonstrate that a task set has been established during this period because faster performance on trials with a long cue-target interval may be due to facilitation of processes nonspecific to the task, such as interpretation of task cues or general readiness for the presentation of targets. One must show the task-specific benefit of increased preparation time independently of task cues (Stoet & Snyder 2003) (**Figure 4a**).

The differential effect of preparation time across tasks is associated with the complexity of the neural mechanisms involved in representing the task rules. When human subjects performed phonological or semantic judgment for a visually presented word, reaction time (RT) on trials with a 300-ms preparation time increased compared with those with longer preparation times (Sakai & Passingham 2006). By contrast the RT of the case judgment was shorter than these and did not differ significantly between trials with preparation times of 300 ms and those

with longer preparation times. These behavioral data suggest an extra element of cognitive processes involved in preparation for phonological and semantic tasks compared with case judgment tasks, which may be the transformation of visually presented words into phonological and conceptual codes. Correspondingly APF activity was sustained during the preparation period as well as during the task performance in the phonological and semantic tasks but not in the case judgment task (**Figure 5a**). A 300-ms preparation time may be too short to recruit the APF, and on these trials, establishing the task set was carried over after the presentation of a target, thereby increasing RT (**Figure 4a**).

In single-unit recording studies, task establishment can be shown as an increment of task-selective activity. When monkeys were cued to perform a match or nonmatch task, the ROC value of the rule-selective prefrontal neurons gradually increased and exceeded 0.60 at 415 ms after the presentation of a task cue (Wallis & Miller 2003). This ROC value can be used to examine the order of recruitment of rule-selective neurons across brain areas. The ROC value in the PM exceeded 0.60 at 280 ms, significantly earlier than that in the prefrontal cortex (Muhammad et al. 2006, Wallis & Miller

Figure 4

Schemes to explain behavioral phenomenon associated with task switching. (*a*) Preparation effect. *Top*: On trials with long cue-target intervals (CTI), the process of preparation for both task A and B, indicated by the slope of the colored bars, is complete during the CTI. The difference in reaction time (RT) reflects the difference in the time involved in task execution between the two tasks, as indicated by light color shading. On trials with short CTI, the preparation for task A is complete during the CTI, and there is no change in the RT compared with that on long CTI trials, whereas the preparation for task B is incomplete during the CTI, and the remaining preparatory process is carried over after target presentation, thus resulting in an increase in the RT. (*b*) Switch cost. When subjects switch between two tasks (*pink and blue*), the establishment of a new task takes time, as indicated by the longer slope on the first trial after a task switch. On trials with long CTI, this effect is eliminated because the preparation of a new task is complete during the CTI. (*c*) Residual cost. Even with ample preparation time, there still remains a cost in switching from one task to the other. This may be due to an interference effect from a previous task set (*green arrow*, the residual of a previous task is indicated in *pink*), which causes an increase in target processing. (*d*) Mixing cost. The RT is longer on repeat trials in a mixed task block (the first, second, and fourth trials in the upper panel) than on trials in a single task block (all trials in the lower panel). This finding is likely due to the extra time needed to select between two tasks or due to an interference effect from an irrelevant task. Still to be determined is whether both of the two tasks, or neither of them (not shown), are simultaneously represented during the intertrial interval (*green arrows*).

a **Task-specific preparation effect**

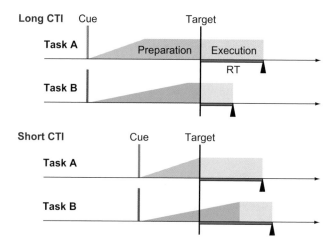

b **Switch cost and preparation effect**

Short CTI

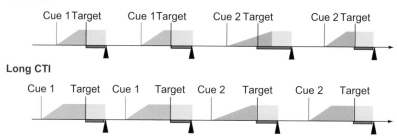

Long CTI

c **Residual cost**

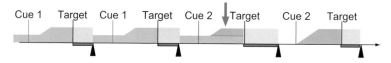

d **Mixing cost**

Mixed task block

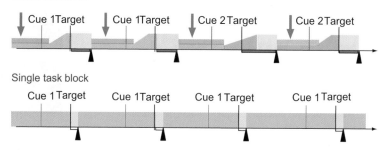

Single task block

a Rule–specific activity of a single neuron

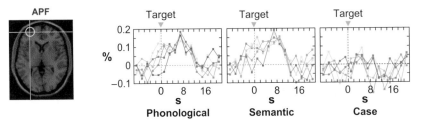

APF

Target Target Target

Phonological   Semantic   Case

b Activity associated with irrelevant task

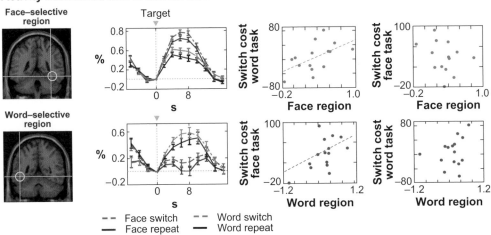

Face–selective region

Word–selective region

Target

Switch cost word task   Face region

Switch cost face task   Face region

Switch cost face task   Word region

Switch cost word task   Word region

- - - Face switch   - - - Word switch
—— Face repeat   —— Word repeat

c Activity associated with conceptual and response conflict

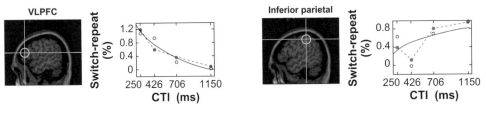

VLPFC   Inferior parietal

Switch-repeat (%)   CTI (ms)

Switch-repeat (%)   CTI (ms)

d Neuronal activity changes after task switch

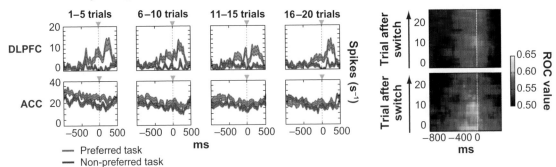

1–5 trials   6–10 trials   11–15 trials   16–20 trials

DLPFC

ACC

Spikes (s⁻¹)

ms

Trial after switch   Trial after switch

ROC value

ms

—— Preferred task
—— Non-preferred task

2003). It should be noted, however, that the time needed to establish a task set may change depending on the cue-target interval especially when a fixed interval is used, as in these studies. Also ROC values show only the discriminability between the two tasks tested in an experiment and do not provide the estimate of the time at which a particular task set is established outside the context of the experiment.

The time needed to establish task sets may also be one of the components that produce switch costs, which reflect the difference in performance between trials immediately after a task switch and trials that repeat the same task (**Figure 4b**). The switch costs are larger when switching between tasks with high rule complexity than when switching between tasks with low rule complexity (Rubinstein et al. 2001). Increasing costs when switching between complex tasks reflect increased demand for setting up these tasks, which requires incorporation of extra cognitive elements. The degree of stimulus discriminability or task difficulty did not affect the switch cost, suggesting dissociation between task set establishment and actual processing of task items.

Interference from a Previous Task Set

Switching costs cannot be eliminated completely even when ample preparation time of more than 1 s is allowed to configure a new task set. This residual cost supports the notion that a previous task set may interfere with the present task set (Allport et al. 1994, Meiran 1996; but see Rogers & Monsell 1995 for other accounts) (**Figure 4c**).

In neural terms, the residual effect of a previous task set is shown as a neural activity increase in areas involved in the previously relevant, now irrelevant task. Yeung et al. (2006) used overlapping face and word stimuli and asked human subjects to switch between face-based and word-based SR mapping tasks. Face-selective and word-selective areas in the extrastriate cortex were more active on trials immediately after task switch than on task repetition trials, regardless of whether the task currently performed was a face task or a word task (**Figure 5b**). The behavioral switch cost was positively correlated with the differential activity between switch and repeat trials in areas representing currently irrelevant stimulus feature. The

---

Figure 5

(*a*) Task-specific preparatory activity observed in the APF (*yellow cross hairs and a circle on the left*). *Right panels*: When human subjects are cued to perform a phonological or semantic judgment task, the APF shows sustained activation during the cue-target interval (CTI). The five levels of the lightness in color shading of the plots indicate trials with CTI of 0.3, 2, 4, 6, and 8 s. Note significant activation occurs before target presentation on trials with long CTIs. The APF does not show significant activation when subjects perform a visual case judgment task (*rightmost panel*). Based on data from Sakai & Passingham (2006). (*b*) Activity associated with an irrelevant task in task switching. When human subjects switched between stimulus-response mapping tasks on the basis of a face and a word in the stimulus, the face-selective and word-selective regions in the extrastriate cortex (*yellow cross hairs and a circle on the left panel*) were more active on trials immediately after task switch than on repetition trials, regardless of whether the task currently performed was a preferred or nonpreferred task for that area (*second panel from the left*). The switch cost in reaction time was positively correlated with activity in these regions when the task currently performed was a nonpreferred task for that region (*third panel from the left*) but was not significantly correlated when the task was a preferred task (*right*). Adapted with permission from Yeung et al. (2006). (*c*) Activity associated with conflicts at a conceptual level and response level during task switch. *Left*: The activation in the VLPFC changes depending on the CTI (*circles in blue; a curved line in blue is a fitted line*), consistent with a model estimate of conceptual conflict (*filled circles and dotted lines in red*). *Right*: The activation in the inferior parietal cortex changes depending on the CTI (*circles in blue, a curved line in blue is a fitted line*), consistent with a model estimate of cumulative response conflict (*filled circles and dotted lines in green*). The plots are adapted with permission from Badre & Wagner (2006). (*d*) Activity of single neurons during task switching. *Left four panels*: Activity of a single neuron in the prefrontal cortex (*top*) and ACC (*bottom*) when monkeys switched between pro- and antisaccade tasks. Activity during performance of the neurons' preferred (*red*) and nonpreferred task (*blue*) is shown in four five-trial sub-blocks after the task switch. *Right*: The change in the receiver-operating characteristics (ROC) values for a population of prefrontal and ACC neurons. Trials are sorted from bottom to top after the task switch, and time zero on the abscissa indicates the target presentation. In the ACC, the ROC values increase earlier and higher on trials immediately after the task switch, but on the following trials the onset becomes later and the peak becomes lower. The ROC values in the DLPFC remain unchanged after the task switch. Adapted with permission from Johnston et al. (2007).

Incongruent trials: a trial in which a stimulus is associated with different responses depending on the task rules

Pre-SMA: presupplementary motor area

IFJ: inferior frontal junction

Stroop task: tests a subject's ability to counter interference, such as naming the color of a visually presented word indicating a different color

correlation between the switch cost and the activity in areas representing a currently relevant stimulus feature was not significant. The results are consistent with the idea that switch costs are due to the interference from a previous task set.

The interference effect of a previous task endures several trials after a task switch (Wylie & Allport 2000). Consistent with this finding, an imaging study demonstrated that the APF activity increased on the third trial after a cognitive set shift in a Wisconsin card sorting task when the trial was incongruent with the previous set compared with when it was congruent (Konishi et al. 2005). The APF activity did not differ between the first trial on the set shift and task repeat trials. By contrast, the VLPFC activity was higher on set shift trials than on repeat trials and also on the later incongruent trials than on congruent trials. These results suggest that at least two separate mechanisms are involved in inhibition of a previous task set depending on temporal contexts.

Conflict Between Task Sets

Task switching is the process of selecting between the two competing task sets. Because a task set is a configuration of perceptual, attentional, mnemonic, and motor processes, the conflict can occur at various stages of cognitive processes depending on the differences between the two tasks, which could be associated with activation in different brain areas. When subjects switch between two tasks during which the same stimulus can produce a different response, conflicts occur at the response-selection stage, which has been associated with an increase in activation in the presupplementary motor area (pre-SMA) and possibly in the ACC in humans (Crone et al. 2006; Liston et al. 2006). By contrast the activity in the posterior parietal cortex and DLPFC was higher when the irrelevant stimulus dimension was salient, suggesting that these regions are sensitive to the conflict at a stimulus representation stage (Liston et al. 2006).

Compared with conflicts at stimulus- and response-processing stages, conflicts at a conceptual rule-processing stage occur regardless of the overlap among stimuli and among responses. Badre & Wagner (2006) used a three-layer connectionist model, in which task item, rule, and response layers are reciprocally connected. Activation in the human left VLPFC paralleled the model estimate of conceptual conflicts, whereas activation in the left inferior parietal cortex paralleled the model estimate of response conflict (**Figure 5c**).

Aside from these studies above, several studies have been conducted to identify the brain regions responsible for monitoring and resolving conflicts at different cognitive processing stages, but the results are far from clear (Barber & Carter 2005, Bunge 2004). Because neurons representing targets, responses, and rules coexist within the same regions, the stimulus processing, response selection, and task-set reconfiguration may take place in parallel across multiple brain regions, and the clear-cut regional segregation may be merely a product of statistical threshold.

In contrast with the areas associated with a conflict at a specific cognitive processing stage, a specific region in the posterior part of the VLPFC, termed the inferior frontal junction (IFJ), is involved in updating task representations regardless of the specific task features (Brass et al. 2005). The IFJ was more active on switch trials than on repeat trials in a task-switching paradigm, more active on incongruent trials than on congruent trials in Stroop tasks, and also more active in verbal working memory tasks that require updating of memory than in target-detection tasks.

One of the controversial topics is the area involved in monitoring response conflicts. One idea is that the ACC plays a role in this process (Botvinick et al. 2001, Ridderinkhof et al. 2004). In particular, the ACC is involved in anticipatory preparation of conflicts that may arise among incompatible SR mappings. Imaging studies also suggest that activation in the ACC engages regulatory processes in the lateral prefrontal cortex. Other studies are inconsistent with this idea (Rushworth et al. 2005). Single-unit studies have shown that neurons in the pre-SMA, but not in the ACC, are more active in conflicting situations. Several imaging studies also support the pre-SMA's role in conflict resolution, especially at a response-selection stage.

Evidence for Inhibition

A mechanism may exist that actively suppresses a previous task set (Allport et al. 1994). However, it is difficult to demonstrate that a neuron or brain region that is active on switch trials inhibits irrelevant task sets rather than facilitating a relevant task set. One way is to demonstrate that a region active on task switch trials is also active when inhibition alone should occur. The same region in the right VLPFC was active on set-shifting trials in a Wisconsin card sorting task and on no-go trials in a GO-NOGO task (Konishi et al. 1999). Patient studies consistently demonstrate that the right VLPFC inhibits irrelevant task sets and motor responses (Aron et al. 2004, Mayr et al. 2006). By contrast, the left VLPFC plays a role in endogenous control of establishing task sets regardless of switch or repeat.

Single-unit evidence for previous task set inhibition has been obtained from the pre-SMA. When monkeys performed a color-based selection of a saccade target, neurons in the pre-SMA showed phasic activity on trials after the color of a to-be-selected target changed (Isoda & Hikosaka 2007). When the same neurons were tested on a GO-NOGO task, the onset of the neural discharge was earlier for the neurons preferentially active on no-go trials than those active on go trials. The pre-SMA may have enabled the task switching by first suppressing automatic responses using a previous task set and then boosting a controlled response using a newly established task set.

Change in Task Representation after Task Switch

In imaging studies, frontal regions including the ACC, pre-SMA, and VLPFC have often been labeled as sites for general control mechanisms, whereas posterior regions have been labeled as targets for the control. Here the absence of task-specific activity in the frontal regions has been taken to indicate their involvement in the general control mechanism (see sidebar on Cognitive Control).

However, a recent single-unit recording study showed that the ACC, which has been regarded as the key area for cognitive control, also contains rule-specific neurons. Switching between two tasks was associated with a change in the strength and onset of rule-selective neuronal activity rather than with the recruitment of switch-selective neurons (Johnston et al. 2007) (**Figure 5d**). In this study, monkeys performed a pro- or antisaccade task for a block of 30 trials and then switched to another task on the basis of feedback. The rule selectivity of ACC neurons increased on trials immediately after switching tasks and then decreased on subsequent task repetition trials. By contrast, the rule selectivity of prefrontal neurons remained unchanged between switch and repetition trials. On trials immediately after the switch, the rule selectivity in the ACC also appeared earlier than that in the DLPFC, but on subsequent repetition trials, it appeared later than that of the DLPFC. The switch-related increase in the ROC value observed in this study may be due to using feedback as the signal to switch between tasks. The ROC value may decrease on switch trials in a task-cuing paradigm as a result of residual neural activity reflecting the rule of a previous task.

As discussed above, the distinction between controlling and controlled regions is becoming increasingly more obscure. The updating of a task set may be a self-regulatory mechanism emerging from interactions and competitions among rule representations. A regional difference in the strength and onset of representations for the to-be-performed task may provide clues to examine the mechanism of cognitive control.

Holding Multiple Task Sets

When subjects switch between two tasks, they are slower to respond even on task repetition trials compared with when they perform only one task. Mixing costs, which represent the difference in performance between task repeat trials in a mixed task block and trials in a single task block, reflect the demand to select among the task rules (**Figure 4d**). One idea is that, for each trial, subjects return to a neutral state immediately after executing a task, and establishing a task set is necessary even on task repetition trials. In fact, the same region in the VLPFC was active on switch and repetition trials in a mixed task block when the frequencies of the switch and stay trials were similar. Another possibility suggests that the two task sets are maintained with equal strength on each trial, and a target item activates the processing pathways for both tasks in parallel. According to this account, the mixing costs reflect competitions at the response-selection stage (Gilbert & Shallice 2002). Consistent with this idea, multiple motor sets are represented simultaneously in the PM during motor preparation (Cisek 2006).

The mixing costs were associated with activation in the lateral APF. An increase in APF activity was sustained throughout a mixed task block compared with a single task block, and larger activation in this area was associated with faster response in a mixed task block but not in a single task block (Braver et al. 2003). Consistent with this area's role in keeping multiple task sets at a relatively high level of activation, the APF was also active in a branching task, where subjects maintain an intention to perform a secondary task while continuously performing a baseline task (Koechlin et al. 1999).

TASK SET IMPLEMENTATION

Task Set Activity Facilitates Task Performance

The benefit of establishing a task set is facilitation of subsequent task performance, as shown in the preparation effect. Imaging studies have shown that higher activation during the pretask period is associated with faster performance. When subjects were cued to perform semantic judgment tasks, the VLPFC activity during the cue-target interval was larger on faster response trials, regardless of whether these were switch or repeat trials in a mixed task block or trials in a single task block (Braver et al. 2003). The APF also showed larger pretask activity on faster trials, but this correlation was observed on switch and repeat trials in a mixed task block but not on trials in a single task block. The results suggest that separate brain areas are involved in facilitating task performance depending on whether subjects perform multiple tasks or a single task. Another study showed that larger pretask APF activity was associated with a faster response on phonological and semantic tasks but not on a visual case judgment task (Sakai & Passingham 2006). These results indicate that the task set activity in a specific region is associated with the subject's performance on certain kinds of tasks.

Although human studies suggest a role of the APF in facilitating performance especially on tasks that require cognitive operations, no single-unit study has been performed on the APF because of the technical difficulty in recording unit activity from this region in monkeys. There is also an issue of whether the monkey APF is equivalent to the human APF (Ramnani & Owen 2004). Which subset of rule-specific neurons found in extensive prefrontal regions in monkeys contributes to the facilitation of task performance remains to be determined. The neurons close to the output of the rule-configuring network may play roles in task performance. Consistent with this idea, when monkeys performed pro- and antisaccade tasks, task-selective activity of the DLPFC neurons, which had been identified as directly projecting to the superior colliculus, was inversely correlated with saccade RT (Johnston & Everling 2006).

Task Set Activity Determines Task Execution Activity

Task set activity likely determines the activity in areas involved in task execution. Phonological and semantic processes are known to be subserved, at least in part, by the ventral PM and anterior VLPFC, respectively. Larger pretask activity in the lateral APF was associated with a decrease in PM activity during performance on a phonological task. By contrast, larger pretask activity in the lateral APF was associated with a decrease in anterior VLPFC activity during performance on a semantic task (Sakai & Passingham 2006) (**Figure 6a**). During the pretask period, the PM and anterior VLPFC may have been primed by the top-down task-specific signals from the APF, and a part of the task-related process was established in these areas. As shown in this study, implementation of a task rule is mediated by interactions between areas involved in representation of a task rule such as APF and areas involved in actual task performance such as PM and anterior VLPFC.

Patient studies suggest dissociation between maintenance and implementation of a task rule. Patients with lesions in the APF were impaired in following task instructions but could report the task instructions correctly when they were asked to do so (Burgess et al. 2000). Similarly, patients with prefrontal lesions show perseveration errors on a Wisconsin card sorting task but could report the relevant rule correctly (Milner 1963).

Then, is the APF necessary to induce task set activation in the task-specific posterior areas? An imaging study on patients with left APF lesions shows that this may not be the case (Rowe et al. 2007). In these patients, the task-selective posterior frontal areas show sustained pretask activity at a level comparable to that in normal subjects. Rather, the APF may play a role in setting up interregional interaction between task-relevant areas. In these patients, the correlation of activity in the posterior regions was signifi-

cantly lower than that of the normal subjects. Efficient implementation of a task set seems to be mediated by the interaction between posterior regions under the APF's control, especially when subjects are required to switch between multiple tasks.

Top-Down and Bottom-Up Signals

In addition to the top-down, task-specific signals, task item information, when it is presented, is conveyed as bottom-up signals from the lower-order sensory areas to areas involved in task execution. In Sakai & Passingham (2006), activity in the ventral PM and anterior VLPFC during performance of phonological and semantic tasks was positively correlated with activity in the fusiform gyrus, which is involved in processing of a visually presented word (**Figure 6a**). This correlation was observed for both the ventral PM and anterior VLPFC regardless of the kinds of tasks. Thus the neural activity associated with task execution can be regarded as a product of task-specific and region-specific top-down signals from anterior prefrontal regions and nonspecific bottom-up signals from lower-order sensory regions.

A recent study, however, suggests that the bottom-up process can also be modulated by a task set (Nakamura et al. 2006). A brief interruption of neural processing in the lateral temporal and inferior parietal cortex abolished the effects of a masked prime on a lexical and a pronunciation task, respectively. The results suggest that the information about the task item is conveyed through different routes depending on the task that the subjects intend to perform, even when the task item is not consciously perceived.

Whereas imaging studies suggest that focal brain areas are the sites at which the top-down and bottom-up signals interact, single-unit studies show that the interaction occurs at every region examined. When monkeys performed a location-match or shape-match task, neurons in the prefrontal cortex demonstrated activity specific to the sample stimuli, task rules,

Bottom-up signals: signals triggered by external environment thought to flow from lower-order to higher-order areas

Top-down signal: control signals thought to flow from higher-order to lower-order areas

and responses within the same prefrontal regions (Hoshi et al. 2000). Activity in these different sets of neurons is temporally structured in a way that is strategically appropriate for the rule-based selection of response (**Figure 6b**).

The information represented by each neuron also shows variable degrees of interaction between the rule and the response (Asaad et al. 2000, Genovesio et al. 2005, Hoshi et al. 1998, Wallis & Miller 2003). Furthermore the information represented by each single neuron can change during a single trial of task performance. Wallis & Miller (2003) found that one third of the prefrontal and premotor neurons that initially coded match or nonmatch rules demonstrated selectivity to a particular type of the matching outcome, i.e., whether the sample and test stimuli matched or did not match (**Figure 7**). Another subset of rule-selective

a   **Task set implementation across regions**

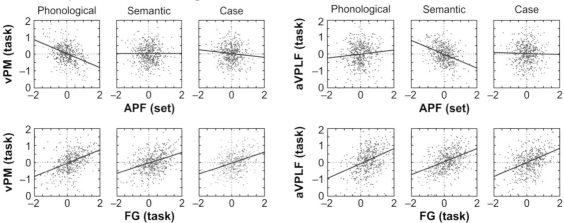

b   **Task set implementation across neurons**

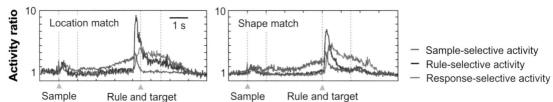

c   **Task set implementation within a single neuron**

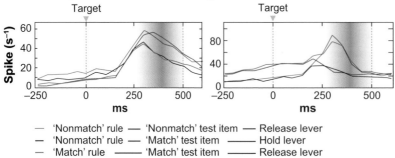

neurons demonstrated selectivity to a particular motor response, i.e., whether the monkeys released or held a response lever.

In sum, facilitation of task performance is mediated by interactions between representations of a task rule and a target item that occur across brain areas, across neurons within the same area, and within a single neuron. Future studies should be directed to understand how these neural processes at different levels are related to each other.

INTENTION AND TASK SET

Voluntary Selection of Task Set

One of the most controversial topics in studies on task sets is whether there exists a truly endogenous mechanism to prepare for a task (Logan et al. 2007, Monsell & Mizon 2006). One extreme idea is that the task cue automatically sets up the task and that there is no endogenous mechanism to prepare for the task. Neural data also provide mechanistic, self-regulatory accounts of task set mechanisms without the need to assume a "homunculus." However, we can freely select a task without an explicit task cue, suggesting the presence of some sort of "endogenous" mechanism (Arrington & Logan 2005).

When brain activation during voluntary task selection was compared with that during externally cued task selection, ACC activation increased (Forstmann et al. 2006). Other imaging studies have shown that the more anterior part of the medial prefrontal cortex is involved in voluntary selection. In a branching task, the lateral APF was active when the onset of the secondary task was unpredictable, whereas the medial APF was active when the onset of the secondary task was predictable (Koechlin et al. 2000). The medial APF was also active in a task-switching paradigm when the order of the two tasks and the timing of the task switch were predictable (Dreher et al. 2002). Performance of a predictable task sequence is regulated endogenously, thus the results suggest a role of the medial APF in voluntary task set selection.

Does this result also suggest that activation in the medial prefrontal cortex is the neural correlate of a cognitive homunculus? Haynes et al. (2007) have shown that, in a voluntary selection of arithmetic operations, medial APF activity reflects the specific task that the subjects intend to perform. Thus a task is not

Homunculus: a small, clever, imaginary person who lives in the brain and controls human thought and behavior

Figure 6

Neuronal activity associated with implementation of task sets. (*a*) *Left three panels*: The activity in the ventral premotor cortex (vPM) during task performance is plotted against anterior prefrontal cortex (APF) activity during the task preparation period (*top*) and the word-selective fugiform gyrus (FG) activity (*bottom*). The activity in the vPM and APF was negatively correlated only on the phonological task, whereas the activity in the vPM and FG was positively correlated in all task conditions. *Right three panels*: Activity in the anterior ventrolateral prefrontal cortex (aVLPFC) was inversely correlated with activity in the APF only in the semantic task, whereas it is positively correlated with FG activity in all task conditions. The values on the abscissa and ordinate are normalized values of the increase in magnetic resonance signals relative to the resting period. The blue oblique line on each panel indicates the estimate of linear regression. Adapted with permission from Sakai & Passingham (2006). (*b*) Time course of population activity of neurons showing selectivity to a particular sample stimulus, task rule, and response. Activity in location-match and shape-match tasks is shown separately. Sample-selective neurons showed gradually incrementing activity until the task instruction was presented (*green*). Rule-selective neurons then showed a sharp increase of activity peaking at 200 ms after the task instruction (*blue*), followed by an increase in activity of response-selective neurons, which peaked at 400 ms after the task instruction (*red*). The ordinates indicate activity ratio, which is calculated by dividing the neurons' discharge rate during the task by that during the intertrial interval. Adapted with permission from Hoshi et al. (2000). (*c*) Time course of activity of a single neuron when a monkey performs a match task and a nonmatch task. *Left*: A single neuron in the prefrontal cortex that initially shows higher activity in the nonmatch task than in the match task comes to show an increase of activity when the outcome of the actual comparison between the sample and test stimuli is the nonmatch. *Right*: A single neuron in the premotor cortex that initially shows more activity in the match task than in the nonmatch task showed an increase in activity when the monkey released a lever for a response. The ordinates indicate the discharge rate of the neuron (per second). Gradient shading in blue on each panel corresponds to the time window of the monkey's response. Adapted with permission from Wallis & Miller (2003).

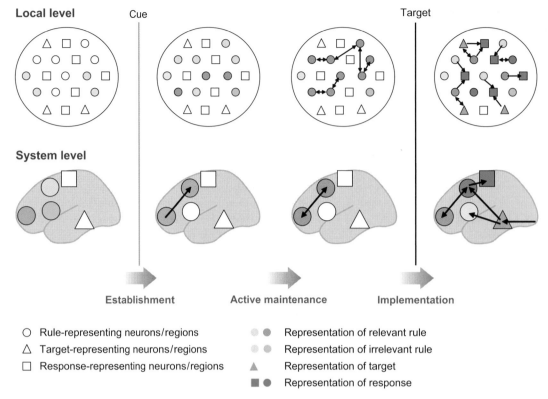

Local level

Cue

Target

System level

Establishment

Active maintenance

Implementation

○ Rule-representing neurons/regions
△ Target-representing neurons/regions
□ Response-representing neurons/regions

    Representation of relevant rule
    Representation of irrelevant rule
    Representation of target
    Representation of response

Figure 7

Scheme of neural mechanisms associated with task sets. *Top*: At a local level, establishment, maintenance, and implementation of a task set were mediated by a change in activity of a single neuron representing a task rule, target, and response (*indicated by a circle, triangle, and square, respectively*). After presentation of a task cue, activity and the number of neurons representing the rule of the relevant task increase (*pink*). Neuron activity representing the irrelevant task still remains, which causes a residual cost and a mixing cost. When target is presented, the rule-representing neurons interact with neurons representing the target item (*gray triangle*) and then interact with neurons representing a particular response, causing an increase in the activity of these neurons (*green square*). Some of the rule-representing neurons also show selectivity to the response (*green circle*). Interneuronal interactions likely play roles in these processes, but the mechanisms remain open (*arrows*). *Bottom*: At a system level, processes associated with task sets are mediated by interactions between different brain regions. Posterior frontal regions selectively represent a particular task (*pink and blue*), and anterior prefrontal regions represent both tasks (*purple*). When a task cue is presented, the anterior prefrontal region interacts with the posterior frontal region selectively involved in the cued task (*pink*). Subsequently, when a target is presented, the lower-order sensory region involved in target processing becomes active (*gray triangle*), and this region interacts nonspecifically with posterior frontal regions. The task-relevant posterior frontal region then sends signals to the region associated with response (*green square*). The regional segregation according to task rules, as indicated in *pink* and *blue* in the posterior frontal regions, is less clear when operations rules, rather than domain rules, differ.

selected by a homunculus in the medial prefrontal cortex; rather, the representation of an intended task emerges in this region. An interesting possibility is that specific information about the to-be-performed task is represented in different areas depending on whether the task is selected on the basis of free will or external cues.

Awareness of Intention to Select Task Set

It is not "us" but the self-regulatory mechanism of the brain that selects the task set. The question is how these neural events are associated with awareness of intention to perform a particular task. As the first step to answer the question, studies have been conducted to establish the boundaries between processes that can run subconsciously and those that require awareness (Mayr 2004).

Lau & Passingham (2007) asked subjects to perform a phonological or semantic task based on a cue. Before each cue, a masked prime that was similar in shape to the cue was also presented. When the prime and cue indicated different tasks, the subjects were slower to perform the task, and activation in the task-relevant brain areas was reduced. These results suggest that an invisible prime may have influenced the task set activation established by a visible task cue. This effect was observed only when the primes were invisible. The congruency between a prime and task cue affected the activation in different brain areas depending on whether the prime was visible. The ACC was more active on incongruent trials than on congruent trials when the prime was visible, whereas the DLPFC was more active on incongruent trials when the prime was invisible.

ACC activation was also associated with awareness of a conflict at a response-selection stage (Dehaene et al. 2003). In this study, schizophrenic patients, who show reduced ACC activation, were impaired in monitoring conscious conflicts. Also patients with lesions in the medial frontal lobe, including the ACC, sometimes showed an alien hand syndrome, during which one of the patient's hands moves automatically and performs a complex task such as picking up a cup and pouring extremely hot coffee into the patient's mouth. The patients try to stop this alien hand using the other hand. This suggests dissociation between the patients' intentions and the task set automatically established by environmental stimuli. This may also suggest that the activation in the medial frontal regions is necessary for awareness of intention to select a task set.

The most significant question pertains to identifying the causal mechanisms by which these medial frontal regions create awareness. Awareness of intention to act and attend is proposed to be mediated by efference copy signals (Driver & Frith 2000). Given similarities in the

mechanisms between attentional sets and task sets, one approach would be to test the possibility that efference copy signals are sent from task execution areas to the medial frontal regions during conscious establishment of a task set.

CONCLUSION

This review has attempted to provide neural accounts of task sets. Single-unit studies have shown that representing, updating, and implementing task sets are subserved by interactions among different sets of neurons in the same region of the prefrontal cortex. By contrast, imaging studies have shown that each of these processes is subserved by distinct regions in the prefrontal cortex and other areas. These results may suggest a parallel processing at different levels of brain organization: Competitions and interactions between representations of task rules, task items, and responses occur at the single-neuron level, within a single region, and across multiple regions during preparation and execution of a task (**Figure 7**).

Single-unit and imaging studies also provide different viewpoints for the neural mechanisms associated with task sets. Single-unit studies provide a representational view, by which brain functioning is considered on the basis of a set of specific information represented by single neurons (Wood & Grafman 2003). By contrast, imaging studies provide a processing-oriented view, by which the inferences are based on the correlation of regional brain activation with behavior changes. For example, the ACC activity during task switching is, from a processing-oriented view, thought to reflect conflict monitoring. From a representational point of view, the ACC activity reflects a change in the strength of task representations. Thus the two approaches, when considered together, provide comprehensive accounts of how the cognitive processes regulate our behavior and how these processes are configured by specific representations.

The key feature of a task set is its prospective and predictive nature. In this sense, the task set is a crucial concept in elucidating the causal mechanisms of the brain in creating complex behaviors and abstract thoughts. As discussed above, studies have shown significant progress in this attempt. This review also suggests the importance of identifying specific task information represented in each brain region and clarifying the relationship between representations and task-related behavioral changes.

SUMMARY POINTS LIST

1. A task set is maintained by sustained rule-specific activity of single neurons in the prefrontal cortex. Imaging studies have also identified the neural correlate of a task set as a rule-specific pattern of signals across multiple voxels within a region or a rule-specific pattern of interactions between multiple regions in the prefrontal cortex.

2. Establishing a task set takes several hundred milliseconds, which reflects the build-up of rule-specific activity of a single neuron or pattern of interactions between brain regions. This process is associated with preparation effects.

3. Task switching is associated with an increase of activity in areas selectively involved in the previous task, which results in residual costs. The right VLPFC in human and pre-SMA in monkeys are likely involved in inhibition of a previous task set.

4. Imaging studies have identified the VLPFC, ACC, and pre-SMA as involved specifically in monitoring and resolution of conflicts during task switch. Single-unit studies have shown that task switching is associated with changes in the strength and onset of task representation in ACC neurons.

5. Imaging studies have identified the ACC and anterior insula/operculum as commonly involved in active maintenance of task sets across different tasks. Imaging studies have also identified the IFJ as commonly involved in updating task sets across different tasks.

6. Imaging studies show that the APF is involved in preparation of cognitive operations. The APF is also involved in active maintenance of multiple task sets, and this demand is associated with mixing costs.

7. Activity in areas involved in task execution can be predicted on the basis of the rule-specific activity in higher-order prefrontal areas and the activity in areas involved in processing of task items, which in turn determines task performance. Active maintenance of a task rule can be thought of as a process to implement the rule for subsequent task performance.

8. The ACC and medial APF are active in voluntary task selection. The ACC is also associated with awareness of conflicts among task rules and responses.

FUTURE ISSUES

1. Many questions about the rule specificity of neural activity remain unanswered. First we need to test whether the rule-specific activity could be observed outside the context of a given experimental setting. We also need to investigate how the representations of rules at different levels of abstraction are associated with anatomically defined functional fingerprints of prefrontal regions. Experiments should be designed such that specific cognitive components are manipulated among the tasks that are compared.

2. The mechanisms of conflict monitoring and resolution need to be understood in terms of the competitions among representations of task-relevant information. More specifically, whether a winner-take-all mechanism can explain the neural processes associated with selection among task sets should be formally tested.

3. It is time to move beyond the analysis of correlation between brain activation and behavior. Future studies should be aimed at clarifying the causal relationship between them. For example, to demonstrate the causality of task set over subsequent task performance, it is necessary to manipulate the task set activity and examine its influence over behavior.

4. To obtain comprehensive accounts of the neural mechanisms associated with task sets, we need to clarify the dynamics in the relationships between the neural processes and representations at different levels of brain organization, from a single neuron to a network of multiple regions.

5. The core task set system including the ACC and anterior insula/operculum contains human-specific giant spindle neurons. Also, the human APF, which is involved in setting up cognitive operations and managing multiple task sets, has an expanded layer of synaptic connections. The IFJ, which is involved in updating tasks in human, may interact with neighboring language areas. An interesting possibility is that these evolutionary changes are associated with humans' capacity to manage a wide repertoire of task sets and represent highly abstract concepts.

DISCLOSURE STATEMENT

The author is not aware of any biases that might be perceived as affecting the objectivity of this review.

ACKNOWLEDGMENTS

The author is grateful to Richard E. Passingham for excellent collaboration, which forms the basis of this review. This work was supported by Grant-in-Aid for Scientific Research (A) and Grant-in-Aid for Young Scientists (S) from the JSPS.

LITERATURE CITED

Allport A, Styles EA, Hsieh S. 1994. Shifting intentional set: exploring the dynamic control of tasks. In *Attention and Performance XV: Conscious and Unconscious Information Processing*, ed. C Umilta, M Moscovitch, pp. 421–52. Cambridge: MIT Press

Aron AR, Robbins TW, Poldrack RA. 2004. Inhibition and the right inferior frontal cortex. *Trends Cogn. Sci.* 8:170–77

Arrington CM, Logan GD. 2005. Voluntary task switching: chasing the elusive homunculus. *J. Exp. Psychol. Learn. Mem. Cogn.* 31:683–702

Asaad WF, Rainer G, Miller EK. 2000. Task-specific neural activity in the primate prefrontal cortex. *J. Neurophysiol.* 84:451–59

Badre D, Wagner AD. 2006. Computational and neurobiological mechanisms underlying cognitive flexibility. *Proc. Natl. Acad. Sci. USA* 103:7186–91

Barber AD, Carter CS. 2005. Cognitive control involved in overcoming prepotent response tendencies and switching between tasks. *Cereb. Cortex* 15:899–912

Botvinick MM, Braver TS, Barch DM, Carter CS, Cohen JD. 2001. Conflict monitoring and cognitive control. *Psychol. Rev.* 108:624–52

Brass M, Derrfuss J, Forstmann B, von Cramon DY. 2005. The role of the inferior frontal junction area in cognitive control. *Trends Cogn. Sci.* 9:314–16

Braver TS, Reynolds JR, Donaldson DI. 2003. Neural mechanisms of transient and sustained cognitive control during task switching. *Neuron* 39:713–26

Bunge SA. 2004. How we use rules to select actions: a review of evidence from cognitive neuroscience. *Cogn. Affect Behav. Neurosci.* 4:564–79

Bunge SA, Kahn I, Wallis JD, Miller EK, Wagner AD. 2003. Neural circuits subserving the retrieval and maintenance of abstract rules. *J. Neurophysiol.* 90:3419–28

Bunge SA, Wallis JD, eds. 2007. *Neuroscience of Rule-Guided Behavior*. New York: Oxford Univ. Press

Burgess PW, Veitch E, de Lacy Costello A, Shallice T. 2000. The cognitive and neuroanatomical correlates of multitasking. *Neuropsychologia* 38:848–63

Buschman T, Machon M, Miller EK. 2006. *Comparison of AND, OR, and XOR rules in monkeys*. Presented at Annu. Meet. Soc. Neurosci., 36th, Atlanta

Cisek P. 2006. Integrated neural processes for defining potential actions and deciding between them: a computational model. *J. Neurosci.* 26:9761–70

Corbetta M, Shulman GL. 2002. Control of goal-directed and stimulus-driven attention in the brain. *Nat. Rev. Neurosci.* 3:201–15

Crone EA, Wendelken C, Donohue SE, Bunge SA. 2006. Neural evidence for dissociable components of task-switching. *Cereb. Cortex* 16:475–86

Dehaene S, Artiges E, Naccache L, Martelli C, Viard A, et al. 2003. Conscious and subliminal conflicts in normal subjects and patients with schizophrenia: the role of the anterior cingulate. *Proc. Natl. Acad. Sci. USA* 100:13722–27

Dosenbach NU, Visscher KM, Palmer ED, Miezin FM, Wenger KK, et al. 2006. A core system for the implementation of task sets. *Neuron* 50:799–812

Dreher JC, Koechlin E, Ali SO, Grafman J. 2002. The roles of timing and task order during task switching. *Neuroimage* 17:95–109

Driver J, Frith C. 2000. Shifting baselines in attention research. *Nat. Rev. Neurosci.* 1:147–48

Duncan J. 2001. An adaptive coding model of neural function in prefrontal cortex. *Nat. Rev. Neurosci.* 2:820–29

Duzel E, Cabeza R, Picton TW, Yonelinas AP, Scheich H, et al. 1999. Task-related and item-related brain processes of memory retrieval. *Proc. Natl. Acad. Sci. USA* 96:1794–99

Forstmann BU, Brass M, Koch I, von Cramon DY. 2006. Voluntary selection of task sets revealed by functional magnetic resonance imaging. *J. Cogn. Neurosci.* 18:388–98

Freedman DJ, Riesenhuber M, Poggio T, Miller EK. 2002. Visual categorization and the primate prefrontal cortex: neurophysiology and behavior. *J. Neurophysiol.* 88:929–41

Genovesio A, Brasted PJ, Mitz AR, Wise SP. 2005. Prefrontal cortex activity related to abstract response strategies. *Neuron* 47:307–20

Gilbert SJ, Shallice T. 2002. Task switching: a PDP model. *Cogn. Psychol.* 44:297–337

Haynes JD, Sakai K, Rees G, Gilbert S, Frith C, Passingham RE. 2007. Reading hidden intentions in the human brain. *Curr. Biol.* 17:323–28

Herron JE, Wilding EL. 2006. Neural correlates of control processes engaged before and during recovery of information from episodic memory. *Neuroimage* 30:634–44

Hoshi E, Shima K, Tanji J. 1998. Task-dependent selectivity of movement-related neuronal activity in the primate prefrontal cortex. *J. Neurophysiol.* 80:3392–97

Hoshi E, Shima K, Tanji J. 2000. Neuronal activity in the primate prefrontal cortex in the process of motor selection based on two behavioral rules. *J. Neurophysiol.* 83:2355–73

Isoda M, Hikosaka O. 2007. Switching from automatic to controlled action by monkey medial frontal cortex. *Nat. Neurosci.* 10:240–48

Johnston K, Everling S. 2006. Monkey dorsolateral prefrontal cortex sends task-selective signals directly to the superior colliculus. *J. Neurosci.* 26:12471–78

Johnston K, Levin HM, Koval MJ, Everling S. 2007. Top-down control-signal dynamics in anterior cingulate and prefrontal cortex neurons following task switching. *Neuron* 53:453–62

Koechlin E, Basso G, Pietrini P, Panzer S, Grafman J. 1999. The role of the anterior prefrontal cortex in human cognition. *Nature* 399:148–51

Koechlin E, Corrado G, Pietrini P, Grafman J. 2000. Dissociating the role of the medial and lateral anterior prefrontal cortex in human planning. *Proc. Natl. Acad. Sci. USA* 97:7651–56

Konishi S, Chikazoe J, Jimura K, Asari T, Miyashita Y. 2005. Neural mechanism in anterior prefrontal cortex for inhibition of prolonged set interference. *Proc. Natl. Acad. Sci. USA* 102:12584–88

Konishi S, Nakajima K, Uchida I, Kikyo H, Kameyama M, et al. 1999. Common inhibitory mechanism in human inferior prefrontal cortex revealed by event-related functional MRI. *Brain* 122:981–91

Lau HC, Passingham RE. 2007. Unconscious activation of the cognitive control system in the human prefrontal cortex. *J. Neurosci.* 27:5805–11

Liston C, Matalon S, Hare TA, Davidson MC, Casey BJ. 2006. Anterior cingulate and posterior parietal cortices are sensitive to dissociable forms of conflict in a task-switching paradigm. *Neuron* 50:643–53

Logan GD, Schneider DW, Bundesen C. 2007. Still clever after all these years: searching for the homunculus in explicitly cued task switching. *J. Exp. Psychol. Hum. Percept. Perform.* 33:978–94

Mayr U. 2004. Conflict, consciousness, and control. *Trends Cogn. Sci.* 8:145–48

Mayr U, Diedrichsen J, Ivry R, Keele SW. 2006. Dissociating task-set selection from task-set inhibition in the prefrontal cortex. *J. Cogn. Neurosci.* 18:14–21

Meiran N. 1996. Reconfiguration of processing mode prior to task performance. *J. Exp. Psychol. Learn. Mem. Cogn.* 22:1423–42

Meiran N, Chorev Z, Sapir A. 2000. Component processes in task switching. *Cogn. Psychol.* 41:211–53

Miller EK, Cohen JD. 2001. An integrative theory of prefrontal cortex function. *Annu. Rev. Neurosci.* 24:167–202

Milner B. 1963. Effects of different brain legions on card sorting. *Arch. Neurol.* 9:90–100

Monsell S. 2003. Task switching. *Trends Cogn. Sci.* 7:134–40

Monsell S, Mizon GA. 2006. Can the task-cuing paradigm measure an endogenous task-set reconfiguration process? *J. Exp. Psychol. Hum. Percept. Perform.* 32:493–516

Muhammad R, Wallis JD, Miller EK. 2006. A comparison of abstract rules in the prefrontal cortex, premotor cortex, inferior temporal cortex, and striatum. *J. Cogn. Neurosci.* 18:974–89

Nakamura K, Hara N, Kouider S, Takayama Y, Hanajima R, et al. 2006. Task-guided selection of the dual neural pathways for reading. *Neuron* 52:557–64

Passingham RE, Stephan KE, Kotter R. 2002. The anatomical basis of functional localization in the cortex. *Nat. Rev. Neurosci.* 3:606–16

Petrides M, Pandya DN. 1999. Dorsolateral prefrontal cortex: comparative cytoarchitectonic analysis in the human and the macaque brain and corticocortical connection patterns. *Eur. J. Neurosci.* 11:1011–36

Petrides M, Pandya DN. 2002. Comparative cytoarchitectonic analysis of the human and the macaque ventrolateral prefrontal cortex and corticocortical connection patterns in the monkey. *Eur. J. Neurosci.* 16:291–310

Ramnani N, Owen AM. 2004. Anterior prefrontal cortex: insights into function from anatomy and neuroimaging. *Nat. Rev. Neurosci.* 5:184–94

Ridderinkhof KR, Ullsperger M, Crone EA, Nieuwenhuis S. 2004. The role of the medial frontal cortex in cognitive control. *Science* 306:443–47

Rogers RD, Monsell S. 1995. The costs of a predictable switch between simple cognitive tasks. *J. Exp. Psychol. Gen.* 124:207–31

Rougier NP, Noelle DC, Braver TS, Cohen JD, O'Reilly RC. 2005. Prefrontal cortex and flexible cognitive control: rules without symbols. *Proc. Natl. Acad. Sci. USA* 102:7338–43

Rowe JB, Sakai K, Lund TE, Ramsoy T, Christensen MS, et al. 2007. Is the left prefrontal cortex necessary for establishing cognitive sets? *J. Neurosci.* 27:13303–10

Rubinstein JS, Meyer DE, Evans JE. 2001. Executive control of cognitive processes in task switching. *J. Exp. Psychol. Hum. Percept. Perform.* 27:763–97

Rugg MD, Wilding EL. 2000. Retrieval processing and episodic memory. *Trends Cogn. Sci.* 4:108–15

Rushworth MF, Kennerley SW, Walton ME. 2005. Cognitive neuroscience: resolving conflict in and over the medial frontal cortex. *Curr. Biol.* 15:R54–56

Sakagami M, Tsutsui K. 1999. The hierarchical organization of decision making in the primate prefrontal cortex. *Neurosci. Res.* 34:79–89

Sakai K, Passingham RE. 2003. Prefrontal interactions reflect future task operations. *Nat. Neurosci.* 6:75–81

Sakai K, Passingham RE. 2006. Prefrontal set activity predicts rule-specific neural processing during subsequent cognitive performance. *J. Neurosci.* 26:1211–18

Stoet G, Snyder LH. 2003. Task preparation in macaque monkeys (*Macaca mulatta*). *Anim. Cogn.* 6:121–30

Wallis JD, Anderson KC, Miller EK. 2001. Single neurons in prefrontal cortex encode abstract rules. *Nature* 411:953–56

Wallis JD, Miller EK. 2003. From rule to response: neuronal processes in the premotor and prefrontal cortex. *J. Neurophysiol.* 90:1790–806

White IM, Wise SP. 1999. Rule-dependent neuronal activity in the prefrontal cortex. *Exp. Brain Res.* 126:315–35

Wood JN, Grafman J. 2003. Human prefrontal cortex: processing and representational perspectives. *Nat. Rev. Neurosci.* 4:139–47

Wylie G, Allport A. 2000. Task switching and the measurement of "switch costs." *Psychol. Res.* 63:212–33

Yeung N, Nystrom LE, Aronson JA, Cohen JD. 2006. Between-task competition and cognitive control in task switching. *J. Neurosci.* 26:1429–38

RELATED RESOURCES

Multi-Voxel Pattern Analysis in fMRI

Haynes JD, Rees G. 2006. Decoding mental states from brain activity in humans. *Nat. Rev. Neurosci.* 7(7):523–34

Norman KA, Polyn SM, Detre GJ, Haxby JV. 2006. Beyond mind-reading: multi-voxel pattern analysis of fMRI data. *Trends Cogn. Sci.* 10(9):424–30

Functional Connectivity Analysis in Imaging

Friston KJ, Buechel C, Fink GR, Morris J, Rolls E, Dolan RJ. 1997. Psychophysiological and modulatory interactions in neuroimaging. *Neuroimage* 6(3):218–29

Horwitz B. 2003. The elusive concept of brain connectivity. *Neuroimage* 19(2):466–70

Multiple Sclerosis: An Immune or Neurodegenerative Disorder?

Bruce D. Trapp[1] and Klaus-Armin Nave[2]

[1]Department of Neurosciences, Lerner Research Institute, Cleveland Clinic, Cleveland, Ohio 44195; email: trappb@ccf.org

[2]Max Planck Institute of Experimental Medicine, Hermann-Rein-Strasse 3, D-37075 Göttingen, Germany; email: nave@em.mpg.de

Annu. Rev. Neurosci. 2008. 31:247–69

The *Annual Review of Neuroscience* is online at neuro.annualreviews.org

This article's doi: 10.1146/annurev.neuro.30.051606.094313

Key Words

myelin, demyelination, axonal degeneration, brain atrophy, cortical pathology, axon-glial interactions

Abstract

Multiple sclerosis (MS) is an inflammatory-mediated demyelinating disease of the human central nervous system. The clinical disease course is variable, usually starts with reversible episodes of neurological disability in the third or fourth decade of life, and transforms into a disease of continuous and irreversible neurological decline by the sixth or seventh decade. We review data that support neurodegeneration as the major cause of irreversible neurological disability in MS patients. We question whether inflammatory demyelination is primary or secondary in the disease process and discuss the challenges of elucidating the cause of MS and developing therapies that will delay or prevent the irreversible and progressive neurological decline that most MS patients endure.

Contents

INTRODUCTION

Multiple sclerosis (MS) is considered the prototype immune-mediated demyelinating disease of the human central nervous system (CNS). This concept that the immune system plays a central role in the pathogenesis of MS is indisputable. The pathological hallmark of MS is the presence of focal areas of inflammatory-mediated demyelination of the brain and spinal cord white matter. Increased susceptibility to MS is genetically linked with the immune molecule, major histocompatibility complex (MHC) class II. Brain-imaging studies have correlated breakdown of the brain-blood barrier and CNS inflammation with the neurological disability that typifies the initial clinical presentation and subsequent relapses that most MS patients endure. Antiinflammatory therapies are effective in slowing the progression of MS, but we have not reached the goal of stopping or preventing the disease process. To achieve this goal, it would help to know the cause of MS. Historically, investigators have generally assumed that MS is a disease in which our immune system is tricked first to see CNS myelin as foreign and then to destroy it. This concept was reinforced by experimental animal models in which immunization with myelin, myelin proteins, or myelin protein peptides induced immune-mediated destruction of CNS myelin. Although mechanisms of T cell–mediated myelin destruction have been described in detail in these animal models, the mechanism by which the immune system is tricked into seeing myelin as foreign in MS patients has eluded the MS community.

Are the immune systems of MS patients so easily fooled? Is there precedence for molecular mimicry and immune-mediated destruction within the nervous system? There is, and it took less than 10 years to identify *Campylobacter jejuni* as a causative agent in inducing molecular mimicry in acute motor axonal neuropathy (McKhann et al. 1991, Sheikh et al. 1998), a subtype of the Guillain-Barré syndrome that is seasonally prevalent in underdeveloped areas of the world including China and Mexico. The disease-causing agent is a glycoconjugate shared by the bacteria and motor axons of the peripheral nervous system (Ho et al. 1995). Destruction of axons occurs through an antibody and complement-mediated mechanism that is pathologically characterized by the presence of activated monocytes in the periaxonal space of the myelinated fibers (Griffin et al. 1995). Why has it been so difficult to identify the causative agent in MS? Is the immune response in MS secondary to a primary disease mechanism? This possibility is one focus of our review; we discuss the evolving concept that MS may be a primary neurodegenerative disease with secondary inflammatory demyelination.

CLINICAL DISEASE COURSE

Multiple sclerosis (MS) is the major cause of nontraumatic neurological disability in young adults in North America and Europe, where it affects more than 2.5 million individuals (Weinshenker 1996, Noseworthy et al. 2000,

---

Multiple sclerosis (MS): an immune-mediated demyelinating disease of the CNS

CNS: central nervous system

Hauser & Oksenberg 2006). MS is a chronic debilitating disease that reduces lifespan by seven to eight years on average. Fifty percent of MS patients are unable to perform household and employment responsibilities 10 years after disease onset, and 50% are nonambulatory 25 years after disease onset. Two aspects of MS significantly impact patients and clinicians. First, the neurological disease course is variable. Second, much of the disease process is clinically silent. Because MS afflicts individuals in their 20s and 30s, the uncertainty of disease progression influences patients' personal and professional decisions. More than 50% of MS patients experience depression, and suicide is 7–8 times more prevalent in individuals with MS when compared with aged-matched controls. Variability in disease progression impacts MS clinical trials, which require hundreds of patients to have sufficient statistical power to demonstrate treatment efficacy. Because much of the disease process is initially clinically silent, primary outcome measures of MS clinical trials rely heavily on brain imaging, which increases the trial costs and raises questions regarding the long-term predictive value of brain-imaging changes.

Approximately 85% of MS patients begin with a course of recurrent and reversible neurological deficits. This disease phase is termed relapsing-remitting MS (RRMS). Females are twice as likely to have RRMS as males. The reversible disability in RRMS is caused by focal areas of inflammatory demyelination in which myelin, myelin-forming cells (oligodendrocytes), and axons are destroyed. Much of the inflammation consists of blood-derived lymphocytes and monocytes. As these immune cells enter the brain, the blood-brain barrier is compromised and lesion areas become edematous. When the onset of neurological disability is rapid, axonal dysfunction probably results from nerve conduction block at nodes of Ranvier (caused by edema and influx of serum components). Relapses usually last no more than a few months and the patient regains neurological function. Resolution of the inflammation and edema, reorganization of axonal Na^+ chan-

nels on demyelinated axons, and remyelination help restore axonal conduction and contribute to the clinical recovery or remission. The relapse rate varies from patient to patient, averaging one to two episodes per year, and the RR stage of MS can last for years or decades. The axonal transection that occurs early in RRMS does not manifest clinically owing to the ability of the mammalian brain to compensate for neuronal loss. In addition, brain-imaging studies of RRMS patients indicate that inflammatory brain lesions can outnumber relapses by as much as 10 to 1 (Barkhof et al. 1992, Miller et al. 1993, Filippi et al. 1998), supporting the concept that much of the disease process is clinically silent. Relapses, therefore, are often a subjective readout of disease progression because their severity depends on whether lesions occur in an articulate area of the brain. After 8–20 years, the majority of RRMS patients enter a second disease phase characterized by continuous, irreversible neurological decline unassociated with relapses. Transition to this secondary progressive state of MS (SPMS) is an ominous event because therapeutics are not yet available to combat the physical, cognitive, and quality-of-life deterioration that most SPMS patients inevitably face. Prevention of SPMS, therefore, is a major therapeutic goal of MS research. Fifteen percent of MS patients have a primary progressive disease course (PPMS). Relapses are rare or nonexistent in PPMS. Clinical disease onset occurs later in life for PPMS patients than for RRMS patients (39 vs. 29 years), and in contrast to RRMS, the incidence of PPMS is similar in males and females.

Since the late 1990s, MS research has refocused on the role of axonal pathology and neurodegeneration as the cause of permanent neurological disability in MS patients. Axonal degeneration is now accepted as the major cause of irreversible neurological disability in MS patients. Axonal transection and degeneration occur in the setting of acute inflammatory demyelination (Trapp et al. 1998) and as a consequence of chronic demyelination (Ganter et al. 1999, Bjartmar et al. 2000, Lovas et al. 2000, Dutta et al. 2006). Current

Relapsing remitting MS (RRMS): common initial disease phase characterized by neurological episodes followed by recovery

Oligodendrocyte: myelin-forming cell in the CNS

Secondary progressive MS (SPMS): common late phase of continuous neurological disability, which follows RRMS

Primary progressive MS (PPMS): less common form in which disability progresses continuously without remissions

Experimental allergic encephalomyelitis (EAE): an inflammatory demyelinating disease used experimentally in rodents as a model for MS

dogma, therefore, describes MS as a primary demyelinating disease with secondary axonal degeneration. Recent studies also support the concept that demyelination may occur by different mechanisms in different subpopulations of MS patients (Lucchinetti et al. 2000). These mechanisms include T cell–mediated demyelination, antibody-mediated demyelination, and primary oligodendrocyte death.

Historical Perspective

Although it is often a controversial subject, axonal pathology was mentioned in the early MS literature (for review, see Kornek & Lassmann 1999). These reports include descriptions of axonal swellings, axonal transection, and Wallerian degeneration, as well as discussions regarding the functional consequences of such pathology. In their classical works, both Charcot and Marburg described MS pathology in terms of demyelination, reactive gliosis, and relative sparing of axons in MS lesions (Charcot 1868, Marburg 1906). In 1936, Putnam (1936) reported a 50% loss of axons in MS lesions from 11 patients. In contrast, Greenfield & King (1936) reported normal axon densities in more than 90% of MS lesions from 13 patients. The differences between these works were suggested to result from more sensitive axon staining in the latter. Subsequently, the axonal component of MS pathogenesis received less attention, and the question regarding axonal damage in MS remained unclear for a long time.

One of the major influences on MS pathogenesis concepts was rare side effects associated with rabies vaccinations in the 1930s and 1940s. Many aspects of human disease pathogenesis have been elucidated from the study of experimental animal models. The MS animal model, experimental allergic encephalomyelitis (EAE), has had a dramatic influence on shaping the current dogma of MS as a primary inflammatory demyelinating disease. EAE did not evolve as a model of MS, but as a model of postrabies vaccine encephalomyelitis. Fatal immune brain lesions following injection of attenuated rabies virus grown in rabbit brain were reported in

the United States (Rivers et al. 1933) and Japan (Uchimura & Shiraki 1957, Shiraki & Otani 1959). Subsequent studies suggested that the postvaccination brain lesions were due to antigens in the rabbit brain tissue and not due to the rabies vaccine. Brain tissue immunization was a weak and inconsistent inducer of inflammatory brain lesions in animal models until heat-killed bacteria was added as an adjuvant to emulsify the brain tissue. Emulsion of myelin, myelin proteins, or myelin protein peptides in complete Freunds adjavent now provides a number of reliable and reproducible animal models of immune-mediated CNS disease (for review, see Lassmann 2004). The pathological similarities among postvaccine encephalomyelitis, EAE, and MS provided a unifying hypothesis for their pathogenesis based on activated T cell destruction of myelin (Waksman 1959). This hypothesis dominated the MS research field between the late 1950s and the early 2000s when we learned much about T cell function in the CNS, but unfortunately the cause or essential aspects of the initiating factors in the human disease, MS, still remain a mystery. We discuss the possibility here that MS is a primary neurodegenerative disease with variable and secondary inflammatory demyelination. The study of MS patients provides much of the support for this concept.

NEURODEGENERATION IN MS

Histological demonstrations of axonal transection and loss in postmortem MS brains (Ferguson et al. 1997, Trapp et al. 1998, Ganter et al. 1999, Lovas et al. 2000, Bjartmar et al. 2001), progressive brain atrophy in MS patients (Rudick et al. 1999a, Simon et al. 1999, Miller et al. 2002), and reductions in the neuronal-specific marker, N-acetyl aspartic acid (Arnold et al. 1994, Matthews et al. 1996, De Stefano et al. 1998, Bjartmar et al. 2000), are abundant and unequivocal. Progressive axonal loss provided a logical explanation for the transition from RRMS to SPMS and for continuous and irreversible neurological decline in SPMS. Disruption of the myelin genome in mice results in late-onset axonal degeneration without significant myelin pathology (Griffiths et al. 1998, Yin et al. 1998, Lappe-Siefke et al. 2003) and strengthens the hypothesis that long-term axonal survival requires trophic support from oligodendrocytes, possibly independent of myelin, and provides proof of principle that chronically demyelinated axons in MS lesions may degenerate owing to loss of glial support. Recent studies also establish that the pathogenesis of MS involves demyelination and neurodegeneration of the cerebral cortex and deep gray matter.

Axons Are Transected During Inflammatory Demyelination

Because one function of myelin is to stabilize and organize the axonal cytoskeleton (de Waegh & Brady 1990, Brady et al. 1999), alterations in axonal cytoskeleton and fast axonal transport have been used to document pathology of demyelinated axons in MS lesions. The amyloid precursor protein (APP) is transported by fast axonal transport and detected in axons only when this transport is disrupted (Koo et al. 1990). Ferguson and colleagues (1997) described APP accumulation in axons located in active MS lesions and at the border of chronic active MS lesions. Some APP immunoreactive structures appeared to be terminal axonal swellings, suggesting axonal transection. The number of APP-labeled axonal swellings correlated with the degree of inflammation in the lesions (Ferguson et al. 1997). Myelination also increases axon diameter by increasing the neurofilament phosphorylation (Sanchez et al. 1996). Using confocal microscopy and computer-based three-dimensional reconstructions of nonphosphorylated neurofilaments, extensive axonal transection was demonstrated in cerebral white matter MS lesions from 11 patients with disease durations ranging from 2 weeks to 27 years (Trapp et al. 1998). Axonal ovoids were identified as terminal ends of transected axons in the confocal microscope (**Figure 1a**), and the degree of inflammation in the lesions was characterized by the presence of phagocytic macrophages and activated microglia. Acute, highly inflamed lesions contained more than 11,000 transected axons per mm^3 tissue and the edge of chronic active lesions contained more than 3000 transected axons per mm^3 tissue, whereas the core of chronic active lesions contained on average 875 transected axons per mm^3 tissue. In contrast, fewer than 1 transected axon/mm^3 tissue was found in control white matter. Together, these data demonstrate a positive correlation between axonal transection and degree of inflammation in cerebral white matter MS lesions undergoing demyelination. The presence of axonal ovoids in patients with short disease duration demonstrated that axonal transection begins at an early stage of MS (Trapp et al. 1998).

Mechanisms of Axonal Degeneration in Acute MS Lesions

Because axon pathology and the frequency of transected axons in MS lesions correlate with the degree of inflammation (Ferguson et al. 1997, Trapp et al. 1998), early axonal transection is thought to occur owing to vulnerability of demyelinated axons to inflammation (**Figure 1b**). Indeed, the inflammatory microenvironment contains a variety of substances that could injure axons, such as proteolytic enzymes, cytokines, oxidative products, and free radicals produced by activated immune and glial cells (Hohlfeld 1997). iNOS, one of the key enzymes involved in synthesis of nitric oxide (NO), is upregulated in acute inflammatory MS lesions (Bo et al. 1994, Liu et al. 2001, Smith & Lassmann 2002). Elevated levels of NO can have a detrimental effect on axonal survival (Smith & Lassmann 2002) by modifying the action of key ion channels, transporters, and glycolytic enzymes (McDonald & Moss 1993, Renganathan et al. 2002, Muriel et al. 2003). NO and its derivative, peroxynitrite, also inhibit mitochondrial respiration (Brown & Borutaite 2002) and limit the axon's ability to generate ATP. Recent data also indicate that cytotoxic CD8+ T cells can mediate axonal transection in active MS lesions (Babbe et al. 2000), in EAE

APP: amyloid precursor protein

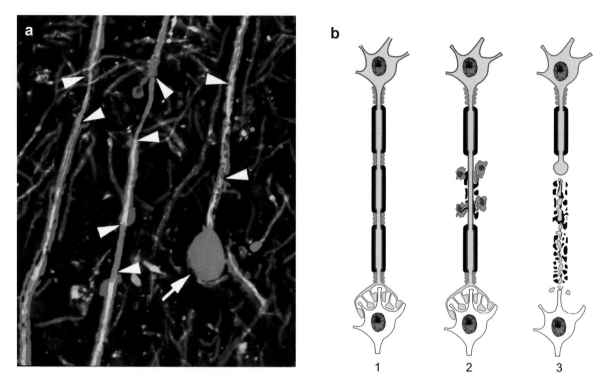

Figure 1

Axons are transected during inflammatory demyelination. (*a*) Confocal image of an actively demyelinating MS lesion stained for myelin protein (*red*) and axons (*green*). The three vertically oriented axons have areas of demyelination (*arrowheads*), which is mediated by microglia and hematogenous monocytes. The axon on the right ends in a large swelling (*arrowhead*), or axonal retraction bulb, which is the hallmark of the proximal end of a transected axon. Quantification of axonal retraction bulbs has established significant axonal transection in demyelinating lesions of MS. (*b*) Schematic summary of axonal response during and following transection. 1. Normal appearing myelinated axon. 2. Demyelination is an immune-mediated or immune cell–assisted process. 3. As many as 11,000 axons/mm³ of lesion area are transected during the demyelinating process. The distal end of the transected axon rapidly degenerates while the proximal end connected to the neuronal cell body survives. Following transection, the neuron continues to transport molecules and organelles down the axon, and they accumulate at the proximal site of the transection. These axon retraction bulbs are transient structures that eventually "die back" to the neuronal perikarya or degenerate. Panel *a* reproduced from Trapp et al. 1998 with permission.

mice (Huseby et al. 2001), and in vitro (Medana et al. 2001). In addition, directly or indirectly, inflammation reduces energy metabolism in demyelinated axons (Lassmann 2003). Inflammatory intermediates may act directly on the mitochondria, and local inflammatory edema may interfere with blood supply and induce an ischemic mechanism of axonal degeneration. Axoplasmic changes resulting from chronic demyelination also reduce ATP production and are discussed in detail below.

Glutamate-mediated excitotoxicity is observed in many acute and chronic neurodegen- erative conditions (Lipton & Rosenberg 1994). Support for glutamate excitoxicity during the pathogenesis of MS was first demonstrated by the observations that treatment with the AMPA/kainate glutamate receptor antagonist NBQX decreased neurological disability, increased oligodendrocyte survival, and reduced axonal damage in EAE (Pitt et al. 2000, Smith et al. 2000, Groom et al. 2003). When released in excess, glutamate activates ionotropic and metabotropic receptors resulting in toxic cytoplasmic Ca²⁺ accumulation and cell death. Investigators have shown several potential sources of glutamate in acute MS lesions including activated immune cells (Matute et al. 2001, Steinman 2001), axons (Lassmann 2003), and astrocytes (Ye et al. 2003, Parpura et al. 2004). Indeed, magnetic resonance spectroscopy studies of MS brains have detected elevated glutamate levels in acute MS lesions and in normal-appearing white matter (Srinivasan et al. 2005). Glutamate receptor expression excess glutamate may damage oligodendrocytes, myelin, and axons. Mature oligodendrocytes and astrocytes express AMPA and kainate receptors and are damaged by exogenous administration of agonist in vitro (Matute et al. 2006). In vitro studies have also demonstrated NMDA (N-methyl-D-aspartic acid) receptor–dependent signaling in oligodendrocytes (Karadottir et al. 2005) and their processes (Salter & Fern 2005). These receptors were also found in the mature myelin sheath and play an important role in the ischemic degeneration of this structure (Micu et al. 2006). Finally, activation of AMPA and/or kainate (but not NMDA) receptors can damage axons, and AMPA/kainate receptor antagonists can be axon-protective in hypoxic/ischemic conditions (Li & Stys 2000, Tekkok & Goldberg 2001). In light of the heterogeneous expression of glutamate receptors in white matter, development of therapeutics that target glutamate AMPA/kainate and NMDA receptors without producing side effects will be a challenge.

Although axonal loss is extensive in acute MS lesions, little permanent disability is associated with this axonal loss during early stages of RRMS because plasticity of the human CNS compensates for neuronal dysfunction and loss. Functional MRI studies have shown activation of new cortical regions participating in this compensation (Reddy et al. 2000, Pantano et al. 2002, Parry et al. 2003, Rocca et al. 2003, Morgen et al. 2007). Transition from RRMS to SPMS occurs when the CNS can no longer compensate for additional neuronal loss (Trapp et al. 1999). At later stages of MS, progressive and irreversible disability and brain atrophy often occur (Rudick et al. 1999b; Simon et al. 1998, 1999) in the absence of new inflammatory demyelinating lesions. Mechanisms other than inflammatory demyelination of white matter, therefore, must contribute to axonal degeneration.

MS Therapeutics

Inflammation, breakdown of the blood-brain barrier, demyelination, and axonal transection are pathological features of acute MS lesions. Therefore, aggressive antiinflammatory treatment during RRMS has, in addition to effects on inflammation, indirect effects in preventing axonal injury. In this sense, antiinflammatory therapies can also be considered neuroprotective. Two classes of therapeutics, interferon β (IFNβ) and glatiramer acetate (GA), are commonly used to treat RRMS. GA and three slightly different recombinant IFNβs reduce relapses, decrease MRI activity, and possibly slow progression of permanent neurological disability (Paty & Li 1993, Jacobs et al. 1996, ESGIB 1998, PRISMS 1998). Although IFNβ has a number of antiinflammatory effects including decreasing antigen presentation, apoptosis, and entry of immune cells into the CNS, its precise mode of actions in MS patients are unknown. GA is a polymer of amino acids that were designed to mimic myelin basic protein (MBP), a major component of CNS myelin. GA reduces antigen presentation and stimulates T cell secretion of cytokines associated with antiinflammatory or Th2 actions (Neuhaus et al. 2001). Delivery of GA and the IFNβs require frequent intramuscular or subcutaneous injections, which can produce many side effects including flu-like symptoms, dermal reactions, and in the case of IFNβ, neutralizing antibodies. Because some patients do not respond to individual treatments, micromanagement of therapeutic approaches is critical for maximizing the efficacy of these treatments. Identification of nonresponders and those with neutralizing antibodies is essential. In addition, recent studies support increased efficacy when these treatments are started as early as possible.

Multiple criteria, including relapses and MRI activity that separates new white matter lesions in space and time, are used for definitive diagnosis of MS (McDonald et al. 2001, Polman et al. 2005). Some investigators have relaxed the criteria for diagnosis in recent years. The availability of effective treatments and the concept that early treatment is key to greater efficacy have obviously influenced this trend. Although a number of new therapies are in clinical trials, it is beyond the scope of this review to describe these here. Developing an oral therapeutic that would be as effective as current therapies would be a major advance in MS therapeutics. This would ease the challenge of convincing young and seemingly invincible RRMS patients to take repeated injections when they are in remission.

Natalizumab (Tysabri), a humanized monoclonal antibody specific for α4 integrins, was developed to suppress the binding of leukocytes to vascular endothelia, a critical step for immune cell entry into the CNS (Yednock et al. 1992). In phase III trials, Tysabri showed great promise for the treatment of RRMS by reducing new gadolinium-enhancing MS lesions by 90% (O'Connor et al. 2004, Polman et al. 2006). The occurrence of progressive multifocal leukoencephalopathy (PML), a rare and most often fatal virus-induced demyelinating disease of immunocompromised individuals was reported in three MS patients receiving Tysabri (Kleinschmidt-DeMasters & Tyler 2005). This experience with Tysabri raises the question of how aggressive we can be in inhibiting the immune system of MS patients. Because we currently have no means to predict release of the PML-causing JC virus, which most of us harbor, future development and more effective inflammatory therapies will have to proceed with caution. Tysabri is currently back on the market and used for more aggressive disease courses that have failed INFβ and GA. The risk-benefit of Tysabri and other highly effective antiinflammatory therapies will likely be a matter of heated discussion until we have a better understanding of how the immune system controls the JC virus. Although no therapies are approved for PPMS or SPMS patients, neuroprotective therapies are currently being generated. Recombinant erythropoietin, for ex-

ample, is effective in animal models of MS (Li et al. 2004) and is clinically safe in MS patients (Konstantinopoulos et al. 2007).

CORTICAL DEMYELINATION IN MS

In addition to the more commonly described white matter locations, MS lesions can also involve gray matter (Brownell & Hughes 1962, Lumsden 1970, Kidd et al. 1999, Peterson et al. 2001, Bo et al. 2003b, Kutzelnigg & Lassmann 2005). Although recent reports of cortical demyelination were a surprise to many in the MS community, cortical demyelination has been reported in the MS literature for decades (Brownell & Hughes 1962, Lumsden 1970). Researchers have described three types of cortical lesions and are referred to as types I through III (**Figure 2a–c**). Type I lesions are leukocortical areas of demyelination that contiguously occupy subcortical white matter and cortex (Peterson et al. 2001). Type II lesions are small perivascular areas of demyelination that do not significantly contribute to cortical lesion load and are not discussed further. Type III lesions are strips of demyelination that extend from the pial surface of the cortex, often stop at cortical layer 3 or 4, and traverse several gyri. On the basis of current estimates, type I and III lesions contribute equally to cortical lesion load. Why do we know so little about cortical lesions? The reason may provide a major advance in elucidating whether peripheral immune cells play a primary or secondary role in demyelination in MS brains. Cortical demyelination occurs without significant influx of hematogenous leukocytes (Peterson et al. 2001, Bo et al. 2003a) (**Figure 2d**), and therefore, cortical lesions have intact blood-brain barriers and are not detected by conventional MRI modalities such as T2-weighted images (Kidd et al. 1999, Sharma et al. 2001). Gray matter lesions are also difficult to detect macroscopically in autopsy specimens because they do not change color, and they are not apparent from routine histological stains such as hematoxylin and eosin because they are not

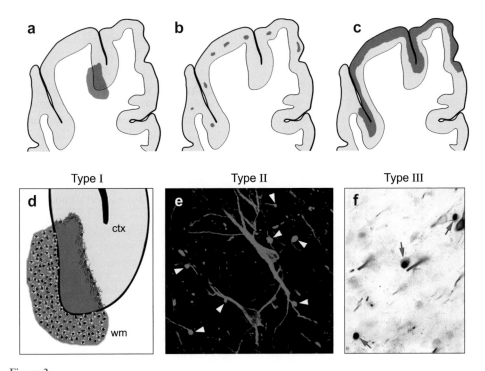

Figure 2

Cortical demyelination and neuronal pathology. Three types of cortical lesions have been described in MS brains (*a–c*, orange areas). (*a*) Type I lesions affect both white and gray matter. (*b*) Type II lesions are small perivascular areas of demyelination. (*c*) Type III lesions extend from the pial surface into the cortex and often demyelinate multiple gyri. (*d*) Cortical demyelination occurs without significant infiltration of hematogenous leukocytes, which is schematically depicted in a Type I lesion (ctx, cortex; wm, white matter). (*e*) Axons and dendrites are transected (*white arrowheads*) during cortical demyelination. (*f*) Apoptotic neurons (*red arrows*), identified by tunnel staining, are increased in demyelinated cortex. Reproduced from Peterson et al. 2005 (*a–c*) and Peterson et al. 2001 (*e, f*) with permission.

hypercellular. Estimates of cortical lesion load, therefore, have been restricted to immunocytochemical analysis of small samples of cortex. In one study, cortical lesion load was determined in 4 cortical areas from 20 patients with significant variation in disease duration, disability, and age (Bo et al. 2003b). The demyelinated cortical area exceeded 25% in these 80 cortical samples and varied between 2% and 75%. This small sample suggests that cortical lesions are a prominent feature in most MS patients. Furthermore, cortical lesion load may exceed white matter lesion load in some patients. The incidence, location, and dynamics of cortical demyelination await development of noninvasive imaging methods that reliably detect cortical demyelination during the course of the disease.

A study of postmortem MS brains using immunohistochemistry and confocal microscopy characterized neuronal pathology and immune cells in Type I gray matter lesions (Peterson et al. 2001). In cortical portions of acute type I lesions, transected neurites (axons and dendrites, see **Figure 2e**) averaged more than 3000/mm³ of lesion area, which is ~25%–30% of the number of transected axons in acute white matter lesions. Apoptotic neurons were also increased in cortical lesions (**Figure 2f**) when compared with myelinated cortex. Inflammatory cells were also quantified in gray and white matter portions of type I lesions, which should be of similar, if not identical, age. Gray matter lesions contained fewer inflammatory cells, no perivascular cuffs, and few phagocytic macrophages, but many activated microglia. Because T cells play a central role in the pathogenesis of MS, their density has been quantified in white matter and cortical MS lesions (Bo et al. 2003a). The highest density of lymphocytes was found in MS white matter lesions. Significantly fewer T cells were detected in cortical portions of type I gray matter lesions. The density of T cells in type III cortical lesions was equal to the lymphocyte density in myelinated cerebral cortex within the same tissue block.

How does cortical demyelination occur? A reasonable but unproven hypothesis proposes that microglia play a major role in cortical demyelination. Increased densities of activated microglia at the leading edge of active type I lesions (Peterson et al. 2001) support this possibility. Of equal interest, however, is the physical association between activated microglia and neuronal cell bodies and their proximal dendrites in acute gray matter lesions (Peterson et al. 2001). Type III lesions are most prominent in cortical regions bordering deep sulci that often have expanded Virchow-Robin spaces containing abundant immune cells. Because this is a cerebrospinal fluid space, a diffusible molecule may be secreted by these immune cells, penetrate the pial surface, and mediate or induce demyelination.

Why are cortical lesions not inflamed? One simple explanation is that myelin is less abundant in cortex, and peripheral immune cells are not needed for its removal following demyelination. It is likely more complex than this. The subcortical white matter of the human brain contains a population of interneurons that innervate vessels and help regulate blood flow. These neurons are destroyed during inflammatory demyelination in MS brains (Chang et al. 2007). Therefore, mechanisms that prevent inflammation of the gray matter may have evolved to protect neurons from destruction. Both explanations argue for more studies of cortical lesions because they may provide a means to elucidate primary mechanisms of demyelination in MS.

Gray matter demyelination may provide the pathological correlate to the executive and cognitive dysfunction that arises in 40%–70% of MS patients (Rao et al. 1991, Ron et al. 1991, Beatty et al. 1995, Foong et al. 1999, Bobholz & Rao 2003). Increased knowledge regarding the incidence of cortical and hippocampal lesions and extent of neuronal damage may contribute to our understanding of the cause of cognitive dysfunction in MS patients. Recent brain-imaging studies raise the possibility that a more global cortical pathology, partially independent of cortical lesions, occurs in MS patients. These studies have employed a variety of brain-imaging and postprocessing techniques to measure cortical thickness (Kutzelnigg & Lassmann 2005, Calabrese et al. 2007, Charil et al. 2007, Pirko et al. 2007). Investigators have reported several relatively consistent observations. Cortical thinning is an early event in MS pathogenesis, which can occur in mildly disabled patients (Calabrese et al. 2007, Charil et al. 2007). Cortical atrophy is progressive in MS patients (**Figure 3a–c**), is partially independent from white matter lesion load, and follows a pattern that is different from atrophy in normal aging (Charil et al. 2007). Cortical thinning varies regionally and is most prominent in areas of the brain that have extensive cortico-cortico connections, such as the cingulated gyrus and insular, frontal, temporal and parietal cortices (Charil et al. 2007). Primary motor, sensory, and visual cortices are less affected. These observations raise the possibility that MS pathogenesis includes an underlying cortical disease that manifests early in distinct brain regions. The limited atrophy in primary motor and sensory and visual cortex regions supports the concept that cortical atrophy is not necessarily dependent on white matter lesions, which are prominent in the cortical spinal and optic tracts. These studies have several limitations. They do not detect cortical lesions, and they have a relatively low measurement sensitivity. Imaging MS patients with higher strength magnets and application of functional MRI and PET should overcome many of these

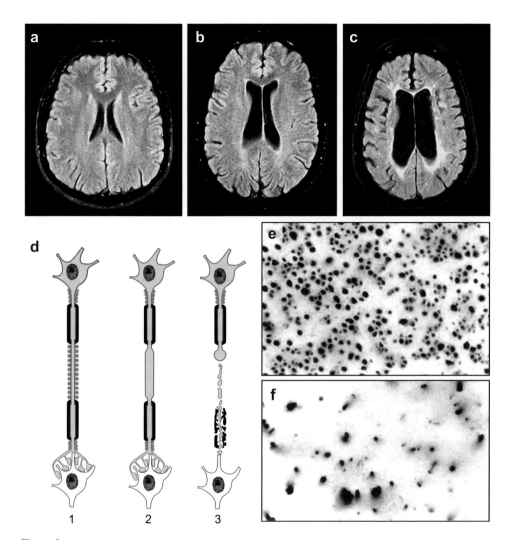

Figure 3

Chronically demyelinated axons degenerate owing to loss of myelin trophic support. A continuous and irreversible loss of brain tissue occurs during the chronic stages of MS despite a dramatic reduction or paucity in new demyelinating lesions. (*a*) Normal brain. (*b*) Brain of a relapsing-remitting MS patient. (*c*) Brain of a secondary progressive MS patient at end-stage disease. The progressive increase in ventricular volume highlights the brain atrophy that occurs as most MS patients age. (*d*) Degeneration of chronically demyelinated axons is a major contributor to neurological disability and brain atrophy. 1. Most demyelinated axons survive demyelination, redistribute Na$^+$ channels, and recover function. 2. Owing to loss of myelin trophic support, chronically demyelinated axons show signs of slowly progressive swelling and disorganization of the cytoskeleton. 3. These axons eventually degenerate. Quantification of total axonal loss is a challenge in MS brain because changes in axonal density and tissue atrophy must be measured. Panels *e* and *f* compare axonal density in control (*e*) and MS lesion (*f*) in spinal cord. Estimates of total axonal loss in chronic MS lesions approach 70%. Reproduced from Trapp et al. 1999 (*a–c*) and Bjartmar et al. 1999 (*e, f*) with permission.

Proteolipid protein
(PLP): the major
CNS myelin protein

limitations and provide more definitive correlations between cortical atrophy and specific disabilities including cognitive dysfunction, fatigue, and depression. The implications of cortical atrophy may be significant. If cortical pathology precedes white matter lesions in MS, it would indicate that white matter demyelination is a consequence of neuronal pathology.

Degeneration of Chronically Demyelinated Axons

Despite the large number of transected axons in acute MS lesions, most demyelinated axons initially survive, and the axonal changes associated with acute inflammatory demyelination are reversed. The MS brain, however, undergoes continuous atrophy in latter stages of the disease when new inflammatory demyelinating lesions are rare (**Figure 3**). Although atrophy is not a specific readout of any one pathological change, neuronal and axonal loss are major contributors. Thus, axonal degeneration must continue in environments other than the inflammatory demyelinating lesion. Evidence for degeneration of chronically demyelinated axons is derived from multiple sources. While the most definitive analysis, counting axons in fixed tissue sections from individual lesions cannot be performed at sequential time points, brain imaging, pathological, and animal model studies are most compelling. Brain-imaging studies play a fundamental and irreplaceable role in diagnosing MS and demonstrating treatment efficacy in clinical trials of MS therapeutics. Recent clinical trials of MS therapeutics have provided a valuable brain-imaging database during disease progression in MS and established brain atrophy as one of the most reliable predictors of clinical disease progression (Rudick et al. 1999a). Proof of principle for degeneration of chronically demyelinated axons comes from several lines of mice that harbor null mutations of individual myelin proteins. These studies were originally designed to investigate the function of myelin proteins during myelination. Myelination in mice null for the myelin-associated glycoprotein (MAG)

(Li et al. 1994, Yin et al. 1998) or for $2',3'$-cyclic nucleotide 3'-phosphodiesterase (CNP) (Lappe-Siefke et al. 2003) was surprisingly normal, whereas mice null for the major structural protein of CNS myelin, proteolipid protein (PLP), have altered compaction of CNS myelin (Klugmann et al. 1997, Griffiths et al. 1998). All three lines of mice, however, developed a late-onset, slowly progressing axonopathy and axon degeneration (Li et al. 1994, Klugmann et al. 1997, Griffiths et al. 1998, Yin et al. 1998, Lappe-Siefke et al. 2003). In addition, axonal degeneration was prominent in PLP-null mice when their compact myelin phenotype was rescued by the peripheral myelin protein, P_0 (Yin et al. 2006). These various lines of mice are described in detail elsewhere in this volume (Nave & Trapp 2008). These studies established that in addition to insulation, oligodendrocytes provide trophic support that is essential for long-term axonal survival. MAG- and CNP-dependent axonal trophic support can be segregated from their role in myelination. If removal of minor components of myelin, such as MAG and CNP, can cause axonal degeneration without dramatically affecting the structure of myelin, it should not be surprising that loss of large segments of myelin for decades, as occurs in MS, results in axonal degeneration.

Postmortem analysis of chronic inactive lesions (Kornek et al. 2000) provided direct histological evidence for degeneration of chronically demyelinated axons. Axonal retraction bulbs, the histological hallmark of axonal transection, are transient structures, and although only a few were detected in chronic lesions, the cumulative degeneration of demyelinated axons over decades would be substantial. Estimates of total axonal loss in spinal cords and corpora callosa obtained at autopsy from end-stage MS patients approach 70% (Ganter et al. 1999, Lovas et al. 2000, Bjartmar et al. 2000) (**Figure 3e,f**). Reports of total axonal loss in MS are limited because quantification of total axonal loss in the human brain is not easy to measure; one must account for reductions in axonal density and alterations in tissue volume.

Mechanisms of Degeneration of Chronically Demyelinated Axons

The MS community has seen considerable speculation regarding the mechanisms by which chronically demyelinated axons degenerate. Although several common themes have evolved, few have been directly tested because at present there is no animal model in which demyelinated axons persist for extended periods of time. The central postulates parallel mechanisms of axonal degeneration in ischemic/hypoxia insults of white matter and involve an imbalance between axonal energy demand and limited energy supply. Specifically, demyelination increases the energy demand of nerve conduction, and axoplasmic ATP production eventually becomes compromised in the chronically demyelinated axon. This double hit for the chronically demyelinated axon leads to an ionic imbalance that increases axoplasmic Ca^{2+}, which eventually destroys the axon. How does this happen? Myelination concentrates voltage-gated Na^+ channels at nodes of Ranvier, small unmyelinated axonal segments that separate individual myelin internodes (Trapp & Kidd 2004). The nerve impulse jumps from node to node by saltatory conduction. As soon as Na^+ enters the nodal axon, it is immediately pumped out by the Na^+/K^+ ATPase in an energy-dependent manner. This permits rapid and repetitive axon firing. Following demyelination, Na^+ channels are distributed along the entire length of the demyelinated axon (Bostock & Sears 1978, Felts et al. 1997). Although this action restores nerve conduction, it does so at the expense of increased energy demand to operate the Na^+/K^+ ATPase during axolemma repolarization. The other fundamental aspect of this hypothesis is reduced axoplasmic ATP production. Reduced ATP impairs Na^+/K^+ ATPase function, and the demyelinated axon cannot exchange axoplasmic Na^+ for extracellular K^+. As axonal Na^+ concentrations increase, the Na^+/Ca^{2+} exchanger, an energy-independent antiporter, is reversed and exchanges axoplasmic Na^+ for extracellular Ca^{2+} (Stys et al. 1992, Li et al. 2000). Ca^{2+} accumulates and initiates a variety of delete-

rious and eventually fatal effects on the axon. Excessive axoplasmic Ca^{2+} accumulation will cause a vicious cycle of activation of degradative enzymes, impaired mitochondrial operation, reduced energy production, compromised axonal transport, and more axoplasmic Ca^{2+}.

Recent studies support the concept that the mitochondria that reach chronically demyelinated axoplasm are compromised and have a reduced capacity for ATP production. This conclusion was based on an unbiased search for neuronal gene changes that may contribute to degeneration of chronically demyelinated axons (Dutta et al. 2006). In microarray comparisons of control and MS motor cortex, 26 nuclear encoded mitochondrial genes were decreased in MS cortex, and the function of mitochondrial complexes I and III was reduced by 40%–50% in mitochondrial-enriched preparations from MS cortex. These mitochondrial gene changes are restricted to neurons and were not detected in glia (Dutta et al. 2006). Neurons in chronic MS cortex, therefore, are likely to be sending defective mitochondria into chronically demyelinated axons. Reduced neuronal ATP production can render neurons less susceptible to noxious insults (Horiguchi et al. 2003). Therefore, reduced neuronal mitochondrial gene expression may prevent neuronal death in MS cortex, but at the expense of putting their demyelinated axonal segments at risk for degeneration. Recent studies indicate that chronically demyelinated axons eventually sustain an additional and probably fatal insult as the $\alpha1$, and $\alpha3$ Na^+/K^+ ATPase subunits, which are present in internodal axolemma of myelinated and acutely demyelinated axons, become virtually undetectable in many chronically demyelinated MS lesions (Young et al. 2008). Thus, these axons are chronically depolarized, inexcitable, and unable to sustain a number of critical homeostatic functions that depend on a healthy transaxolemmal Na^+ gradient. Loss of the Na^+/K^+ ATPase protein may be a key component of the vicious cycle of impaired energy production and Ca^{2+} accumulation and suggests that many chronically demyelinated axons are functionally dead before they degenerate.

Ultrastructural changes in chronically demyelinated spinal cord lesions (Dutta et al. 2006) also demonstrated Ca^{2+}-mediated destruction of chronically demyelinated axons. In the same spinal cord lesions that averaged 70% axonal loss, 50% of the remaining demyelinated axons contained fragmented neurofilaments and dramatically reduced numbers of mitochondria and microtubules. Fragmentation of neurofilaments is likely due to the activity of calpain, a Ca^{2+}-activated enzyme that is increased in chronic MS lesions (Shields et al. 1999). Another morphological feature of chronic MS lesions is axonal swelling. In a study of postmortem MS brains, Fisher et al. (2007) recently reported a correlation between demyelinated axonal swelling and T1 and magnetization transfer ratio (MTR) changes on MRI (but not T2-only MRI changes). Alterations in T1 and MTR sequences also correlated with histological features of more chronic MS lesions, whereas T2-only changes correlated with breakdown of the blood-brain barrier with or without acute demyelination. Therefore, axoplasmic swelling is a pathological hallmark of chronically demyelinated CNS axons that is likely to reflect, in part, increased axoplasmic Ca^{2+}.

Epidemiological Studies

The biphasic clinical course of MS with an initial relapsing-remitting phase followed by a secondary progressive phase has been interpreted to represent the dominance of two separate disease processes: inflammatory demyelination during RRMS and axonal degeneration during SPMS. In this model, secondary progression results from the delayed effects of previous demyelination that accumulated during the RRMS phase. The transition from RRMS to SPMS could be determined by a threshold for axonal loss beyond which the brain can no longer compensate. Recent epidemiological studies of the natural history of disease progression in MS patients (Compston 2006; Confavreux & Vukusic 2006a,b; Kremenchutzky et al. 2006), however, question the validity of this hypothesis. Knowing that the goal of MS research is to prevent the progression of irreversible disability in MS patients, these studies focus on variables that may contribute to onset and progression of permanent disability. Several important and somewhat surprising concepts are evolving from these longitudinal evaluations of large cohorts of MS patients. The chronological age at which MS patients reach permanent disability milestones is similar in patients with relapsing-remitting MS and primary progressive MS (Confavreux & Vukusic 2006a, Kremenchutzky et al. 2006). Relapses early in the disease process have limited effect on the accumulation of permanent disability in latter stages of the disease. For example, primary progressive MS patients have few or no relapses, and their clinical disease onset is ~10 years later than RRMS patients, but they reach disability milestones at an age similar to most RRMS patients. Thus much of the recent epidemiological data can be interpreted as indicating that all MS patients have a similar rate of primary neurodegeneration with variable secondary inflammatory demyelination. The reversible disability caused by inflammatory demyelination identifies the clinical disease at an earlier age in RRMS patients but has little effect on development and progression of permanent neurological disability that is age related. If relapses are secondary in MS pathogenesis, the progression of permanent neurological disability in MS patients supports a primary, age-related neurodegenerative disease process similar to that which occurs in Alzheimer disease, Parkinson disease, and amyotrophic lateral sclerosis.

Genetic Influences

Multiple sclerosis is considered a complex polygenic disease characterized by a modest inherited risk for disease susceptibility (Compston & Coles 2002, Hauser & Oksenberg 2006). The disease is not inherited, but one can inherit a greater susceptibility to acquiring MS. For example, the risk for MS in many parts of North America and Europe is 0.1%. If you have an affected parent, your risk rises to 2%–3%. Studies show increased MS incidence in families, relatively high risk in northern Europeans and relatively low risk in Africans, Asians, and American Indians (Rosati 2001). The only consistent MS-associated gene is the *HLA-DRB1* gene on chromosome 6p21 (Oksenberg et al. 2004); it accounts for 16%–60% of the genetic susceptibility in MS (Haines et al. 1998). Although this association has reinforced the immune etiology of MS, possibly by affecting antigen presentation, the mechanism by which *HLA-DRB1* contributes to disease susceptibility is unknown. HLA associations in the nonimmune disease narcolepsy are much greater than in MS (Mignot 1998, Tafti & Dauvilliers 2003). It is possible, therefore, that HLA-DR genotypes contribute to MS susceptibility by mechanisms other than immune cell function. The interleukin-7 receptor alpha chain and interleukin-2 receptor alpha chain have been identified as additional inheritable risk factors accounting for less than 0.4% of the variance in the risk for developing MS (Gregory et al. 2007, Hafler et al. 2007, Lundmark et al. 2007). MS, therefore, is a complex genetic disorder in which multiple interacting polymorphic genes have low penetrance, and each exerts a small effect on the overall disease risk. Many studies support a role for the environment in MS disease susceptibility. Data suggest that this environmental exposure is time sensitive and occurs before the age of 14. Although viruses are the most implicated environmental factor, no single environmental agent has been associated with the disease, and risk assessment studies identify genetic and not environmental factors as the major contributor to familiar clustering of MS (Ebers et al. 2000).

CONCLUSIONS AND FUTURE CHALLENGES

The major challenges for MS research are elucidating and understanding the cause of the disease. Although investigators have found genetic associations with disease risk, they are modest and not Mendelian. Gene linkage studies, therefore, are not likely to point easily to the cause of MS as has been the case in inherited CNS diseases. The concept that MS is an autoimmune disease induced by molecular mimicry has little direct support despite decades of searching for the initiating environmental agent. The past decade has refocused MS research on the symbiotic relationship of the axon and myelin sheath. This interaction may be the key to understanding the cause of MS. Answers to several critical questions will help address this issue.

The first question is whether inflammatory demyelination is primary or secondary in the disease process. How can this question be addressed? The major focus of MS therapeutics has been to reduce or stop new inflammatory demyelinating lesions. INFβ and GA have modest effects on reducing new gadolinium-enhancing lesions and relapses (Simon et al. 1998, Comi et al. 2001, Wolinsky et al. 2001) but do not stop disease progression. If we stop new white matter lesions, will we stop the disease? This may not be an easy experiment to run, considering the experience with Tysabri. The second question is whether primary-progressive and relapsing-remitting MS are the same disease. If so, one could argue that the inflammation is secondary to an underlying CNS disease process. The third question is what is the cause of MS. The lack of a disease-causing gene is a major obstacle in understanding or elucidating the cause of MS. Some lessons may be gleaned from recent success in identifying disease-causing genes in other CNS diseases, such as Alzheimer disease, Parkinson disease, Huntington disease, and amyotrophic lateral sclerosis. Without exception, the disease-causing gene came as a surprise to the neuroscience community: APP, presenilins, trinucleotide repeats, synuclein, and SOD were not a focus of research in these diseases prior to their discovery by gene linkage studies. Will this also be the case when the cause of MS is discovered? Is there a critical "missing link" that is responsible for the disease, MS? It may be prudent to consider the possibility that MS is not an autoimmune disease caused by an immune response to an environmental factor early in life that is reactivated and disease-causing later in life. Elucidating the cause of MS will require creative approaches combined with an investigative mindset rather than trying to support hypotheses that have proven so difficult to verify.

SUMMARY POINTS

1. Multiple sclerosis is an immune-mediated demyelinating disease of the human central nervous system.

2. Neurodegeneration is the major cause of permanent neurological disability in MS patients.

3. The cerebral cortex of MS patients is demyelinated without significant influx of immune cells.

4. Despite indirect evidence supporting an autoimmune etiology of MS, it remains to be determined if inflammation is primary or secondary to a degenerative process in the brain.

FUTURE ISSUES

1. If new inflammatory demyelinating lesions can be prevented in MS brains, will the disease process be halted?

2. Are relapsing-remitting MS and primary progressive MS the same disease or two different diseases?

3. Is there a global cortical-based disease in MS patients?

4. Investigators need to develop brain imaging modalities that can detect cortical demyelination.

DISCLOSURE STATEMENT

The authors are not aware of any biases that might be perceived as affecting the objectivity of this review.

ACKNOWLEDGMENTS

The authors wish to thank Dr. Grahame Kidd for assistance with the figures and manuscript production. This work was supported by NIH grants P01 NS38667 and R01 NS35058 (B.D.T.). K.A.N. is supported by grants from the Deutsche Forschungsgemeinschaft, the European Union (FP6), the Hertie Institute of MS Research, and the National MS Society.

LITERATURE CITED

Arnold DL, Riess GT, Matthews PM, Francis GS, Collins DL, et al. 1994. Use of proton magnetic resonance spectroscopy for monitoring disease progression in multiple sclerosis. *Ann. Neurol.* 36:76–82

Babbe H, Roers A, Waisman A, Lassmann H, Goebels N, et al. 2000. Clonal expansions of CD8(+) T cells dominate the T cell infiltrate in active multiple sclerosis lesions as shown by micromanipulation and single cell polymerase chain reaction. *J. Exp. Med.* 192:393–404

Barkhof F, Scheltens P, Frequin ST, Nauta JJ, Tas MW, et al. 1992. Relapsing-remitting multiple sclerosis: sequential enhanced MR imaging vs clinical findings in determining disease activity. *AJR Am. J. Roentgenol.* 159:1041–47

Beatty WW, Paul RH, Wilbanks SL, Hames KA, Blanco CR, Goodkin DE. 1995. Identifying multiple sclerosis patients with mild or global cognitive impairment using the Screening Examination for Cognitive Impairment (SEFCI). *Neurology* 45:718–23

Bjartmar C, Kidd G, Mork S, Rudick R, Trapp BD. 2000. Neurological disability correlates with spinal cord axonal loss and reduce *N*-acetyl aspartate in chronic multiple sclerosis patients. *Ann. Neurol.* 48:893–901

Bjartmar C, Kinkel RP, Kidd G, Rudick RA, Trapp BD. 2001. Axonal loss in normal-appearing white matter in a patient with acute MS. *Neurology* 57:1248–52

Bjartmar C, Yin X, Trapp BD. 1999. Axonal pathology in myelin disorders. *J. Neurocytol.* 28:383–95

Bo L, Dawson TM, Wesselingh S, Mork S, Choi S, et al. 1994. Induction of nitric oxide synthase in demyelinating regions of multiple sclerosis brains. *Ann. Neurol.* 36:778–86

Bo L, Vedeler CA, Nyland H, Trapp BD, Mork SJ. 2003a. Intracortical multiple sclerosis lesions are not associated with increased lymphocyte infiltration. *Mult. Scler.* 9:323–31

Bo L, Vedeler CA, Nyland HI, Trapp BD, Mork SJ. 2003b. Subpial demyelination in the cerebral cortex of multiple sclerosis patients. *J. Neuropathol. Exp. Neurol.* 62:723–32

Bobholz JA, Rao SM. 2003. Cognitive dysfunction in multiple sclerosis: a review of recent developments. *Curr. Opin. Neurol.* 16:283–88

Bostock H, Sears TA. 1978. The internodal axon membrane: electrical excitability and continuous conduction in segmental demyelination. *J. Physiol.* 280:273–301

Brady ST, Witt AS, Kirkpatrick LL, de Waegh SM, Readhead C, et al. 1999. Formation of compact myelin is required for maturation of the axonal cytoskeleton. *J. Neurosci.* 19:7278–88

Brown GC, Borutaite V. 2002. Nitric oxide inhibition of mitochondrial respiration and its role in cell death. *Free Radic. Biol. Med.* 33:1440–50

Brownell B, Hughes JT. 1962. Distribution of plaques in the cerebrum in multiple sclerosis. *J. Neurol. Neurosurg. Psychiatry* 25:315–20

Calabrese M, Atzori M, Bernardi V, Morra A, Romualdi C, et al. 2007. Cortical atrophy is relevant in multiple sclerosis at clinical onset. *J. Neurol.* 254:1212–20

Chang A, Smith M, Yin X, Staugaitis SM, Trapp B. 2007. Neurogenesis in the chronic lesions of multiple sclerosis. *J. Neuropathol. Exp. Neurol.* 66:431

Charcot M. 1868. Histologie de la sclerose en plaques. *Gaz Hosp.* 141:554-5-57-8

Charil A, Dagher A, Lerch JP, Zijdenbos AP, Worsley KJ, Evans AC. 2007. Focal cortical atrophy in multiple sclerosis: relation to lesion load and disability. *Neuroimage* 34:509–17

Comi G, Filippi M, Wolinsky JS. 2001. European/Canadian multicenter, double-blind, randomized, placebo-controlled study of the effects of glatiramer acetate on magnetic resonance imaging—measured disease activity and burden in patients with relapsing multiple sclerosis. European/Canadian Glatiramer Acetate Study Group. *Ann. Neurol.* 49:290–97

Compston A. 2006. Making progress on the natural history of multiple sclerosis. *Brain* 129:561–63

Compston A, Coles A. 2002. Multiple sclerosis. *Lancet* 359:1221–31

Confavreux C, Vukusic S. 2006a. Age at disability milestones in multiple sclerosis. *Brain* 129:595–605

Confavreux C, Vukusic S. 2006b. Natural history of multiple sclerosis: a unifying concept. *Brain* 129:606–16

De Stefano N, Matthews PM, Fu L, Narayanan S, Stanley J, et al. 1998. Axonal damage correlates with disability in patients with relapsing-remitting multiple sclerosis. Results of a longitudinal magnetic resonance spectroscopy study. *Brain* 121:1469–77

de Waegh S, Brady ST. 1990. Altered slow axonal transport and regeneration in a myelin-deficient mutant mouse: the trembler as an in vivo model for Schwann cell-axon interactions. *J. Neurosci.* 10:1855–65

Dutta R, McDonough J, Yin X, Peterson J, Chang A, et al. 2006. Mitochondrial dysfunction as a cause of axonal degeneration in multiple sclerosis patients. *Ann. Neurol.* 59:478–89

Ebers GC, Yee IM, Sadovnick AD, Duquette P. 2000. Conjugal multiple sclerosis: population-based prevalence and recurrence risks in offspring. Canadian Collaborative Study Group. *Ann. Neurol.* 48:927–31

ESGIB. 1998. Placebo-controlled multicentre randomised trial of interferon beta-1b in treatment of secondary progressive multiple sclerosis. European Study Group on interferon beta-1b in secondary progressive MS. *Lancet* 352:1491–97

Felts PA, Baker TA, Smith KJ. 1997. Conduction in segmentally demyelinated mammalian central axons. *J. Neurosci.* 17:7267–77

Ferguson B, Matyszak MK, Esiri MM, Perry VH. 1997. Axonal damage in acute multiple sclerosis lesions. *Brain* 120:393–99

Filippi M, Rocca MA, Martino G, Horsfield MA, Comi G. 1998. Magnetization transfer changes in the normal appearing white matter precede the appearance of enhancing lesions in patients with multiple sclerosis. *Ann. Neurol.* 43:809–14

Fisher E, Chang A, Fox R, Tkach JA, Svarovsky T, et al. 2007. Imaging correlates of axonal swelling in chronic multiple sclerosis brains. *Ann. Neurol.* 62:219–28

Foong J, Rozewicz L, Davie CA, Thompson AJ, Miller DH, Ron MA. 1999. Correlates of executive function in multiple sclerosis: the use of magnetic resonance spectroscopy as an index of focal pathology. *J. Neuropsychiatry Clin. Neurosci.* 11:45–50

Ganter P, Prince C, Esiri MM. 1999. Spinal cord axonal loss in multiple sclerosis: a postmortem study. *Neuropathol. Appl. Neurobiol.* 25:459–67

Greenfield JG, King LS. 1936. Observations on the histopathology of the cerebral lesions in disseminated sclerosis. *Brain* 59:445–58

Gregory SG, Schmidt S, Seth P, Oksenberg JR, Hart J, et al. 2007. Interleukin 7 receptor alpha chain (IL7R) shows allelic and functional association with multiple sclerosis. *Nat. Genet.* 39:1083–91

Griffin JW, Li CY, Ho TW, Xue P, Macko C, et al. 1995. Guillain-Barre syndrome in northern China. The spectrum of neuropathological changes in clinically defined cases. *Brain* 118:577–95

Griffiths I, Klugmann M, Anderson T, Yool D, Thomson C, et al. 1998. Axonal swellings and degeneration in mice lacking the major proteolipid of myelin. *Science* 280:1610–13

Groom AJ, Smith T, Turski L. 2003. Multiple sclerosis and glutamate. *Ann. N. Y. Acad. Sci.* 993:229–75

Hafler DA, Compston A, Sawcer C, Lander E, Daly MJ, et al. 2007. Risk alleles for multiple sclerosis identified by a genomewide study. *N. Engl. J. Med.* 357:851–62

Haines JL, Terwedow HA, Burgess K, Pericak-Vance MA, Rimmler JB, et al. 1998. Linkage of the MHC to familial multiple sclerosis suggests genetic heterogeneity. The Multiple Sclerosis Genetics Group. *Hum. Mol. Genet.* 7:1229–34

Hauser S, Oksenberg JR. 2006. The neurobiology of multiple sclerosis: genes, inflammation, and neurodegeneration. *Neuron* 52:61–76

Ho TW, Mishu B, Li CY, Gao CY, Cornblath DR, et al. 1995. Guillain-Barre syndrome in northern China. Relationship to *Campylobacter jejuni* infection and antiglycolipid antibodies. *Brain* 118:597–605

Hohlfeld R. 1997. Biotechnological agents for the immunotherapy of multiple sclerosis. Principles, problems and perspectives. *Brain* 120:865–916

Horiguchi T, Kis B, Rajapakse N, Shimizu K, Busija DW. 2003. Opening of mitochondrial ATP-sensitive potassium channels is a trigger of 3-nitropropionic acid-induced tolerance to transient focal cerebral ischemia in rats. *Stroke* 34:1015–20

Huseby ES, Liggitt D, Brabb T, Schnabel B, Ohlen C, Goverman J. 2001. A pathogenic role for myelin-specific CD8(+) T cells in a model for multiple sclerosis. *J. Exp. Med.* 194:669–76

Jacobs LD, Cookfair DL, Rudick RA, Herndon RM, Richert JR, et al. 1996. Intramuscular interferon beta-1a for disease progression in relapsing multiple sclerosis. *Ann. Neurol.* 39:285–94

Karadottir R, Cavelier P, Bergersen LH, Attwell D. 2005. NMDA receptors are expressed in oligodendrocytes and activated in ischaemia. *Nature* 438:1162–66

Kidd D, Barkhof F, McConnell R, Algra PR, Allen IV, Revesz T. 1999. Cortical lesions in multiple sclerosis. *Brain* 122:17–26

Kies M, Alvord EC, eds. 1959. *Allergic Encephalomyelitis*. Springfield, IL: Thomas

Kleinschmidt-DeMasters BK, Tyler KL. 2005. Progressive multifocal leukoencephalopathy complicating treatment with natalizumab and interferon beta-1a for multiple sclerosis. *N. Engl. J. Med.* 353:369–74

Klugmann M, Schwab MH, Puhlhofer A, Schneider A, Zimmermann F, et al. 1997. Assembly of CNS myelin in the absence of proteolipid protein. *Neuron* 18:59–70

Konstantinopoulos PA, Karamouzis MV, Papavassiliou AG. 2007. Selective modulation of the erythropoietic and tissue-protective effects of erythropoietin: time to reach the full therapeutic potential of erythropoietin. *Biochim. Biophys. Acta* 1776:1–9

Koo EH, Sisodia SS, Archer DR, Martin LJ, Weidemann A, et al. 1990. Precursor of amyloid protein in Alzheimer disease undergoes fast anterograde axonal transport. *Proc. Natl. Acad. Sci. USA* 87:1561–65

Kornek B, Lassmann H. 1999. Axonal pathology in multiple sclerosis. A historical note. *Brain Pathol.* 9:651–56

Kornek B, Storch MK, Weissert R, Wallstroem E, Stefferl A, et al. 2000. Multiple sclerosis and chronic autoimmune encephalomyelitis: a comparative quantitative study of axonal injury in active, inactive, and remyelinated lesions. *Am. J. Pathol.* 157:267–76

Kremenchutzky M, Rice GP, Baskerville J, Wingerchuk DM, Ebers GC. 2006. The natural history of multiple sclerosis: a geographically based study 9: observations on the progressive phase of the disease. *Brain* 129:584–94

Kutzelnigg A, Lassmann H. 2005. Cortical lesions and brain atrophy in MS. *J. Neurol. Sci.* 233:55–59

Lappe-Siefke C, Goebbels S, Gravel M, Nicksch E, Lee J, et al. 2003. Disruption of Cnp1 uncouples oligodendroglial functions in axonal support and myelination. *Nat. Genet.* 33:366–74

Lassmann H. 2003. Hypoxia-like tissue injury as a component of multiple sclerosis lesions. *J. Neurol. Sci.* 206:187–91

Lassmann H. 2004. Experimental autoimmune encephalomyelitis. In *Myelin Biology and Disorders*, ed. RA Lazzarini, pp. 1039–72. New York: Elsevier

Li C, Tropak MB, Gerial R, Clapoff S, Abramow-Newerly W, et al. 1994. Myelination in the absence of myelin-associated glycoprotein. *Nature* 369:747–50

Li S, Jiang Q, Stys PK. 2000. Important role of reverse Na$^{(+)}$-Ca$^{(2+)}$ exchange in spinal cord white matter injury at physiological temperature. *J. Neurophysiol.* 84:1116–19

Li S, Stys PK. 2000. Mechanisms of ionotropic glutamate receptor-mediated excitotoxicity in isolated spinal cord white matter. *J. Neurosci.* 20:1190–98

Li W, Maeda Y, Yuan RR, Elkabes S, Cook S, Dowling P. 2004. Beneficial effect of erythropoietin on experimental allergic encephalomyelitis. *Ann. Neurol.* 56:767–77

Lipton SA, Rosenberg PA. 1994. Excitatory amino acids as a final common pathway for neurologic disorders. *N. Engl. J. Med.* 330:613–22

Liu JS, Zhao ML, Brosnan CF, Lee SC. 2001. Expression of inducible nitric oxide synthase and nitrotyrosine in multiple sclerosis lesions. *Am. J. Pathol.* 158:2057–66

Lovas G, Szilagyi N, Majtenyi K, Palkovits M, Komoly S. 2000. Axonal changes in chronic demyelinated cervical spinal cord plaques. *Brain* 123:308–17

Lucchinetti C, Bruck W, Parisi J, Scheithauer B, Rodriguez M, Lassmann H. 2000. Heterogeneity of multiple sclerosis lesions: implications for the pathogenesis of demyelination. *Ann. Neurol.* 47:707–17

Lumsden CE. 1970. The neuropathology of multiple sclerosis. In *Handbook of Clinical Neurology*, ed. PJ Vinken, GW Bruyn, pp. 217–309. New York: Elsevier

Lundmark F, Duvefelt K, Iacobaeus E, Kockum I, Wallstrom E, et al. 2007. Variation in interleukin 7 receptor alpha chain (IL7R) influences risk of multiple sclerosis. *Nat. Genet.* 39:1108–13

Marburg O. 1906. Die sogenannte "akute multiple sklerose" (Encephalomyelitis peraxialis scleroticans). *Jahrb Neurol. Psych.* 27:211–312

Matthews PM, Pioro E, Narayanan S, De Stefano N, Fu L, et al. 1996. Assessment of lesion pathology in multiple sclerosis using quantitative MRI morphometry and magnetic resonance spectroscopy. *Brain* 119:715–22

Matute C, Alberdi E, Domercq M, Perez-Cerda F, Perez-Samartin A, Sanchez-Gomez MV. 2001. The link between excitotoxic oligodendroglial death and demyelinating diseases. *Trends Neurosci.* 24:224–30

Matute C, Domercq M, Sanchez-Gomez MV. 2006. Glutamate-mediated glial injury: mechanisms and clinical importance. *Glia* 53:212–24

McDonald LJ, Moss J. 1993. Stimulation by nitric oxide of an NAD linkage to glyceraldehyde-3-phosphate dehydrogenase. *Proc. Natl. Acad. Sci. USA* 90:6238–41

McDonald WI, Compston A, Edan G, Goodkin D, Hartung HP, et al. 2001. Recommended diagnostic criteria for multiple sclerosis: guidelines from the International Panel on the diagnosis of multiple sclerosis. *Ann. Neurol.* 50:121–27

McKhann GM, Cornblath DR, Ho TW, Li CY, Bai AY, et al. 1991. Clinical and electrophysiological aspects of acute paralytic disease of children and young adults in northern China. *Lancet* 338:593–97

Medana I, Martinic MA, Wekerle H, Neumann H. 2001. Transection of major histocompatibility complex class I-induced neurites by cytotoxic T lymphocytes. *Am. J. Pathol.* 159:809–15

Micu I, Jiang Q, Coderre E, Ridsdale A, Zhang L, et al. 2006. NMDA receptors mediate calcium accumulation in myelin during chemical ischaemia. *Nature* 439:988–92

Mignot E. 1998. Genetic and familial aspects of narcolepsy. *Neurology* 50:S16–22

Miller DH, Barkhof F, Frank JA, Parker GJ, Thompson AJ. 2002. Measurement of atrophy in multiple sclerosis: pathological basis, methodological aspects and clinical relevance. *Brain* 125:1676–95

Miller DH, Barkhof F, Nauta JJ. 1993. Gadolinium enhancement increases the sensitivity of MRI in detecting disease activity in multiple sclerosis. *Brain* 116:1077–94

Morgen K, Sammer G, Courtney SM, Wolters T, Melchior H, et al. 2007. Distinct mechanisms of altered brain activation in patients with multiple sclerosis. *Neuroimage* 37:937–46

Muriel P, Castaneda G, Ortega M, Noel F. 2003. Insights into the mechanism of erythrocyte Na$^+$/K$^+$-ATPase inhibition by nitric oxide and peroxynitrite anion. *J. Appl. Toxicol.* 23:275–78

Nave K-A, Trapp BD. 2007. Molecular biology of myelination. *Annu. Rev. Neurosci.* 31:535–61

Neuhaus O, Farina C, Wekerle H, Hohlfeld R. 2001. Mechanisms of action of glatiramer acetate in multiple sclerosis. *Neurology* 56:702–8

Noseworthy JH, Lucchinetti C, Rodriguez M, Weinshenker BG. 2000. Multiple sclerosis. *N. Engl. J. Med.* 343:938–52

O'Connor PW, Goodman A, Willmer-Hulme AJ, Libonati MA, Metz L, et al. 2004. Randomized multicenter trial of natalizumab in acute MS relapses: clinical and MRI effects. *Neurology* 62:2038–43

Oksenberg JR, Barcellos LF, Cree BA, Baranzini SE, Bugawan TL, et al. 2004. Mapping multiple sclerosis susceptibility to the HLA-DR locus in African Americans. *Am. J. Hum. Genet.* 74:160–67

Pantano P, Iannetti GD, Caramia F, Mainero C, Di Legge S, et al. 2002. Cortical motor reorganization after a single clinical attack of multiple sclerosis. *Brain* 125:1607–15

Parpura V, Scemes E, Spray DC. 2004. Mechanisms of glutamate release from astrocytes: gap junction "hemichannels", purinergic receptors and exocytotic release. *Neurochem. Int.* 45:259–64

Parry AM, Scott RB, Palace J, Smith S, Matthews PM. 2003. Potentially adaptive functional changes in cognitive processing for patients with multiple sclerosis and their acute modulation by rivastigmine. *Brain* 126:2750–60

Paty DW, Li DK. 1993. Interferon beta-1b is effective in relapsing-remitting multiple sclerosis. II. MRI analysis results of a multicenter, randomized, double-blind, placebo-controlled trial. UBC MS/MRI Study Group and the IFNB Multiple Sclerosis Study Group. *Neurology* 43:662–67

Peterson JW, Bo L, Mork S, Chang A, Trapp BD. 2001. Transected neurites, apoptotic neurons and reduced inflammation in cortical MS lesions. *Ann. Neurol.* 50:389–400

Peterson JW, Kidd GJ, Trapp BD. 2005. Axonal degeneration in multiple sclerosis: the histopathological evidence. In *Multiple Sclerosis as a Neuronal Disease*, ed. S Waxman, pp. 165–84. New York: Elsevier

Pirko I, Lucchinetti CF, Sriram S, Bakshi R. 2007. Gray matter involvement in multiple sclerosis. *Neurology* 68:634–42

Pitt D, Werner P, Raine CS. 2000. Glutamate excitotoxicity in a model of multiple sclerosis. *Nat. Med.* 6:67–70

Polman CH, O'Connor PW, Havrdova E, Hutchinson M, Kappos L, et al. 2006. A randomized, placebo-controlled trial of natalizumab for relapsing multiple sclerosis. *N. Engl. J. Med.* 354:899–910

Polman CH, Reingold SC, Edan G, Filippi M, Hartung HP, et al. 2005. Diagnostic criteria for multiple sclerosis: 2005 revisions to the "McDonald Criteria." *Ann. Neurol.* 58:840–46

PRISMS. 1998. Randomised double-blind placebo-controlled study of interferon beta-1a in relapsing/remitting multiple sclerosis. PRISMS (Prevention of Relapses and Disability by Interferon beta-1a Subcutaneously in Multiple Sclerosis) Study Group. *Lancet* 352:1498–504

Putnam TJ. 1936. Studies in multiple sclerosis. *Arch. Neurol. Psych.* 35:1289–308

Rao SM, Leo GJ, Bernardin L, Unverzagt F. 1991. Cognitive dysfunction in multiple sclerosis. I. Frequency, patterns, and prediction. *Neurology* 41:685–91

Reddy H, Narayanan S, Arnoutelis R, Jenkinson M, Antel J, et al. 2000. Evidence for adaptive functional changes in the cerebral cortex with axonal injury from multiple sclerosis. *Brain* 123:2314–20

Renganathan M, Cummins TR, Waxman SG. 2002. Nitric oxide blocks fast, slow, and persistent Na$^+$ channels in C-type DRG neurons by S-nitrosylation. *J. Neurophysiol.* 87:761–75

Rivers TM, Sprunt DH, Berry GP. 1933. Observations on attempts to produce acute disseminated encephalomyelitis in monkeys. *J. Exp. Med.* 58:39–53

Rocca MA, Mezzapesa DM, Falini A, Ghezzi A, Martinelli V, et al. 2003. Evidence for axonal pathology and adaptive cortical reorganization in patients at presentation with clinically isolated syndromes suggestive of multiple sclerosis. *Neuroimage* 18:847–55

Ron MA, Callanan MM, Warrington EK. 1991. Cognitive abnormalities in multiple sclerosis: a psychometric and MRI study. *Psychol. Med.* 21:59–68

Rosati G. 2001. The prevalence of multiple sclerosis in the world: an update. *Neurol. Sci.* 22:117–39

Rudick RA, Fisher E, Lee JC, Simon J, Jacobs L. 1999a. Use of the brain parenchymal fraction to measure whole brain atrophy in relapsing-remitting MS. Multiple Sclerosis Collaborative Research Group. *Neurology* 53:1698–704

Rudick RA, Goodman A, Herndon RM, Panitch HS. 1999b. Selecting relapsing remitting multiple sclerosis patients for treatment: the case for early treatment. *J. Neuroimmunol.* 98:22–28

Salter MG, Fern R. 2005. NMDA receptors are expressed in developing oligodendrocyte processes and mediate injury. *Nature* 438:1167–71

Sanchez I, Hassinger L, Paskevich PA, Shine HD, Nixon RA. 1996. Oligodendroglia regulate the regional expansion of axon caliber and local accumulation of neurofilaments during development independently of myelin formation. *J. Neurosci.* 16:5095–105

Sharma R, Narayana PA, Wolinsky JS. 2001. Grey matter abnormalities in multiple sclerosis: proton magnetic resonance spectroscopic imaging. *Mult. Scler.* 7:221–26

Sheikh KA, Ho TW, Nachamkin I, Li CY, Cornblath DR, et al. 1998. Molecular mimicry in Guillain-Barre syndrome. *Ann. N. Y. Acad. Sci.* 845:307–21

Shields DC, Schaecher KE, Saido TC, Banik NL. 1999. A putative mechanism of demyelination in multiple sclerosis by a proteolytic enzyme, calpain. *Proc. Natl. Acad. Sci. USA* 96:11486–91

Shiraki H, Otani S. 1959. Clinical and pathological features of rabies postvaccinal encephalomyelitis in man. See Kies & Alvord 1959, pp. 58–129

Simon JH, Jacobs LD, Campion M, Wende K, Simonian N, et al. 1998. Magnetic resonance studies of intramuscular interferon beta-1a for relapsing multiple sclerosis. The Multiple Sclerosis Collaborative Research Group. *Ann. Neurol.* 43:79–87

Simon JH, Jacobs LD, Campion MK, Rudick RA, Cookfair DL, et al. 1999. A longitudinal study of brain atrophy in relapsing multiple sclerosis. The Multiple Sclerosis Collaborative Research Group (MSCRG). *Neurology* 53:139–48

Smith KJ, Lassmann H. 2002. The role of nitric oxide in multiple sclerosis. *Lancet Neurol.* 1:232–41

Smith T, Groom A, Zhu B, Turski L. 2000. Autoimmune encephalomyelitis ameliorated by AMPA antagonists. *Nat. Med.* 6:62–66

Srinivasan R, Sailasuta N, Hurd R, Nelson S, Pelletier D. 2005. Evidence of elevated glutamate in multiple sclerosis using magnetic resonance spectroscopy at 3 T. *Brain* 128:1016–25

Steinman L. 2001. Multiple sclerosis: a two-stage disease. *Nat. Immunol.* 2:762–64

Stys PK, Waxman SG, Ransom BR. 1992. Ionic mechanisms of anoxic injury in mammalian CNS white matter: role of Na$^+$ channels and Na$^{(+)}$-Ca^{2+} exchanger. *J. Neurosci.* 12:430–39

Tafti M, Dauvilliers Y. 2003. Pharmacogenomics in the treatment of narcolepsy. *Pharmacogenomics* 4:23–33

Tekkök SB, Goldberg MP. 2001. AMPA/kainate receptor activation mediates hypoxic oligodendrocyte death and axonal injury in cerebral white matter. *J. Neurosci.* 21:4237–48

Trapp BD, Kidd GJ. 2004. Structure of the myelinated axon. In *Myelin Biology and Disorders*, ed. R Lazzarini, pp. 3–25. New York: Elsevier

Trapp BD, Peterson J, Ransohoff RM, Rudick R, Mork S, Bo L. 1998. Axonal transection in the lesions of multiple sclerosis. *N. Engl. J. Med.* 338:278–85

Trapp BD, Ransohoff RM, Fisher E, Rudick RA. 1999. Neurodegeneration in multiple sclerosis: relationship to neurological disability. *Neuroscientist* 5:48–57

Uchimura I, Shiraki J. 1957. A contribution to the classification and the pathogenesis of demyelinating encephalomyelitis; with special reference to the central nervous system lesions caused by preventive inoculation against rabies. *J. Neuropathol. Exp. Neurol.* 16:139–203

Waksman BH. 1959. Evidence favoring delayed sensitization as the mechansims underlying experimental "allergic" encephalomyelitis. See Kies & Alvord 1959, pp. 419–43

Weinshenker BG. 1996. Epidemiology of multiple sclerosis. *Neurol. Clin.* 14:291–308

Wolinsky JS, Narayana PA, Johnson KP. 2001. United States open-label glatiramer acetate extension trial for relapsing multiple sclerosis: MRI and clinical correlates. Multiple Sclerosis Study Group and the MRI Analysis Center. *Mult. Scler.* 7:33–41

Ye ZC, Wyeth MS, Baltan-Tekkok S, Ransom BR. 2003. Functional hemichannels in astrocytes: a novel mechanism of glutamate release. *J. Neurosci.* 23:3588–96

Yednock TA, Cannon C, Fritz LC, Sanchez-Madrid F, Steinman L, Karin N. 1992. Prevention of experimental autoimmune encephalomyelitis by antibodies against alpha 4 beta 1 integrin. *Nature* 356:63–66

Yin X, Baek RC, Kirschner DA, Peterson A, Fujii Y, et al. 2006. Evolution of a neuroprotective function of central nervous system myelin. *J. Cell Biol.* 172:469–78

Yin X, Crawford TO, Griffin JW, Tu P-H, Lee VMY, et al. 1998. Myelin-associated glycoprotein is a myelin signal that modulates the caliber of myelinated axons. *J. Neurosci.* 18:1953–62

Young EA, Fowler CD, Kidd GJ, Chang A, Rudick R, et al. 2008. Imaging correlates of decreased axonal Na^+/K^+ ATPase in chronic MS lesions. *Ann. Neurol.* In press

Multifunctional
Pattern-Generating Circuits

K.L. Briggman[1] and W.B. Kristan, Jr.[2]

[1]Department of Biomedical Optics, Max Planck Institute for Medical Research, Heidelberg, 69120 Germany; email: briggman@mpimf-heidelberg.mpg.de

[2]Neurobiology Section, Division of Biological Sciences, University of California, San Diego, La Jolla, California 92093-0357; email: wkristan@ucsd.edu

Annu. Rev. Neurosci. 2008. 31:271–94

First published online as a Review in Advance on April 2, 2008

The *Annual Review of Neuroscience* is online at neuro.annualreviews.org

This article's doi:
10.1146/annurev.neuro.31.060407.125552

Key Words

multistable, modular, behavior, dynamical systems

Abstract

The ability of distinct anatomical circuits to generate multiple behavioral patterns is widespread among vertebrate and invertebrate species. These multifunctional neuronal circuits are the result of multistable neural dynamics and modular organization. The evidence suggests multifunctional circuits can be classified by distinct architectures, yet the activity patterns of individual neurons involved in more than one behavior can vary dramatically. Several mechanisms, including sensory input, the parallel activity of projection neurons, neuromodulation, and biomechanics, are responsible for the switching between patterns. Recent advances in both analytical and experimental tools have aided the study of these complex circuits.

Contents

INTRODUCTION

Peter Getting introduced the term polymorphic neural networks more than 20 years ago to describe an anatomical network of neurons that can operate in more than one stable mode (Getting 1989, Getting & Deken 1985, Marder 1994). That is, a given anatomical connectivity pattern can switch between different functional activity patterns under various conditions. Discovering such circuits challenged (or at least complicated) the decades-old view that nervous systems select behaviors by activating dedicated pools of interneurons called "command neurons" (Kupfermann & Weiss 1978). At the same time, the highly distributed nature of networks driving even simple behaviors was beginning to be appreciated (Georgopoulos et al. 1986, Lockery & Kristan 1990). Polymorphic neural circuits have also been referred to as "reorganizing circuits" (Morton & Chiel 1994), or "multifunctional circuits" (Kristan et al. 1988). We choose the term multifunctional circuitry because it emphasizes the ability of adaptable, functional circuits to drive multiple behaviors.

Individual neurons can generate a wide variety of activity patterns depending on their state of modulation, synaptic input, and plasticity (Cymbalyuk & Shilnikov 2005, Frohlich et al. 2006, Guckenheimer et al. 1993, Kass & Mintz 2006). When these three dynamical elements are incorporated into a circuit, it is not surprising that the networks can also generate many different motor outputs. Most studies of multifunctional circuitry have focused on rhythmic behaviors and their underlying central pattern generators (CPGs) (Grillner 2006, Kiehn 2006, Marder 2000, Marder & Bucher 2001). The focus on invertebrate species derives from their relatively simple nervous systems and the close correlation between their neural output and behavior (Marder et al. 2005, Marder & Calabrese 1996; for historical perspective see Clarac & Pearlstein 2007, Leonard 2000). Indeed, many circuit mechanisms for generating neuronal rhythms, such as recurrent and reciprocal inhibition, were first discovered in invertebrate nervous systems (Friesen 1994, Getting 1989) and found later in the vertebrates (Marder 2000), demonstrating the extraordinary complexity of even "simple" invertebrate neuronal circuits. At the neuronal level, studies showed that a combination of the synaptic connectivity and intrinsic membrane properties generates the activity patterns in these circuits (Kristan et al. 2005, Marder & Bucher 2001, Marder & Calabrese 1996). Multifunctional circuits, in principle,

CPGs: central pattern
generators

contribute greatly to the versatility of central nervous systems for producing and modifying behaviors. This versatility, the ability to reconfigure anatomically defined circuits into many distinct functional circuits, is the focus of this review.

In particular, we focus on multifunctional networks generating recognizably discrete behaviors. We do not address the numerous interesting examples of networks that can modulate the magnitude or speed of a single behavior in a graded, continuous manner. For example, the activity level of descending interneurons in the goldfish spinal cord can modulate the strength of escape behavior in goldfish (Fetcho 1992) and zebrafish (Bhatt et al. 2007), and the number of recruited interneurons increases the strength of escape behaviors in fish and frogs (McLean et al. 2007, Sillar & Roberts 1993). Such gradation of behavior falls outside our definition of multifunctionalism.

Before reviewing the evidence for multifunctional circuits, we discuss several related principles that have crystallized over the past few decades regarding the function of circuits driving motor patterns. Key advances in understanding motor circuits include (*a*) a shift toward a general view of circuits operating as continuous dynamical systems rather than as discrete state machines; (*b*) the revelation that many behaviors are driven by the concerted actions of a large distributed network as opposed to simple, dedicated pathways; (*c*) the discovery that many behaviors, rhythmic and nonrhythmic, are generated using common building blocks, or behavioral modules; and (*d*) an increasing number of examples of multifunctional circuit architectures that can drive multiple behaviors.

Multifunctional Circuits as Continuous Dynamical Systems

Motor patterns are often divided into discrete behavioral phases, such as the flexion and extension phases of limb movements. Defining such discrete states, however, is only an approximation of a circuit's continuous underlying dynamics (Nowotny & Rabinovich 2007, Rabinovich et al. 2006, Schoner & Kelso 1988, Strogatz 2001). This richer set of capabilities can be captured by viewing the neural activity as a dynamical system in a phase space diagram (**Figure 1*a***). The coordinate axes in this diagram are two relevant parameters of the system plotted over time, representing variables such as the membrane potential of individual neurons (Selverston & Ayers 2006), neuron group activity (Briggman et al. 2006), or more abstract parameters such as limb motion (Chiel et al. 1999). A behavior can be traced by following the trajectories in the phase space, in this case along a stable orbit (limit cycle), representing an oscillating rhythm. One key insight is that stable behavioral patterns can produce low-dimensional trajectories in phase space, which can simplify their analysis. In the three-dimensional example in **Figure 1*a***, three neurons generate a stable oscillation, which lies on two-dimensional plane. Perturbations to the circuit (*blue trajectory*) may transiently disrupt or reset the oscillation, but the system resettles into its stable oscillatory state (Oprisan & Canavier 2002).

A circuit capable of generating more than one stable pattern is termed multistable, which is one form of multifunctionality. If a second pattern is also rhythmic, it can be represented as an additional stable orbit in the phase space (**Figure 1*b***). A perturbation to the system may now be sufficient to switch from one stable pattern to another (*blue trajectory*). Thus, the phase space diagram captures both the dimensionality of stable patterns (indicating which parameters must be measured) as well as the dynamics of the system in response to stimuli. The sensitivity of the stable states to external parameters, such as synaptic input or neuromodulator concentration, can also be viewed in parameter space, also called a bifurcation diagram (**Figure 1*c***) (Barak & Tsodyks 2007, Cymbalyuk et al. 2002). Within different parameter regimes, the circuit can attain additional stable (and unstable) states. For instance, at low concentrations of the neuromodulator, two stable states (I and III) are possible; at high

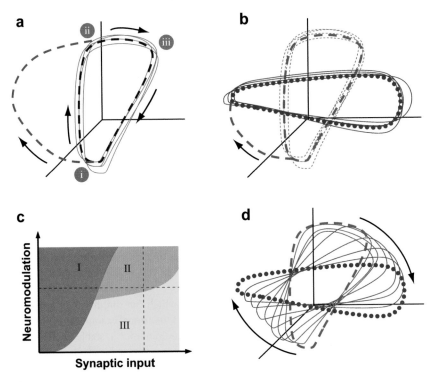

Figure 1

Multifunctional neuronal circuits viewed as dynamical systems. (*a*) Phase space plot of a unifunctional circuit generating a single rhythmic pattern, shown oscillating between three arbitrarily chosen phases (i, ii, iii); arrows indicate direction of time. The axes may represent the activity of individual neurons, groups of neurons, or more abstract values such as body movement. Light gray trajectories denote individual oscillation cycles; the black dashed trajectory is the average cycle. A perturbation to the system (e.g., sensory input) causes a transient deviation (*blue dashed trajectory*) from the steady-state oscillation, but the system recovers and resumes the stable oscillation cycle. (*b*) Phase space plot of a multifunctional system capable of generating two stable rhythmic behaviors (*purple and green cycles*). The green (*dashed*) cycle and the purple (*dotted*) cycle may oscillate at different frequencies. A perturbation (*blue dashed trajectory*) may be sufficient to switch the steady-state rhythm rapidly from the green cycle to the purple cycle. (*c*) A bifurcation diagram, or parameter space plot, for a circuit capable of operating in three distinct stable modes (I, II, III). The vertical dashed line indicates that at this level of synaptic drive, the circuit can operate in two modes (II, III), depending on the degree of neuromodulation. Likewise, the circuit is tristable (*horizontal dashed line*) at a constant modulation level as the synaptic input increases. (*d*) Phase space plot of the multifunctional circuit shown in part *b*. Instead of an abrupt switch between stable states, one stable rhythm (*green*) slowly evolves (*indicated by arrows*) into a second stable rhythm (*purple*).

concentrations, a different pair of stable states (I and II) exist; and at intermediate concentrations (*horizontal dashed line*) all three can occur, depending on the amount of synaptic input. Phase space plots can also indicate when transitions between stable states occur at different time scales, from abrupt transitions (**Figure 1*b***) to the slow evolution of one pattern into another (**Figure 1*d***).

These concepts have been successfully applied to the study of simplified behaviors such as one-dimensional walking (Beer et al. 1999, Chiel et al. 1999), which determined abstract dynamical modules that describe the alternation of the swing and stance phases of a leg. Such abstractions are necessary to distill the critical parameters driving a behavior, but are also necessary for practical reasons. It is impossible to view the activity of all the neurons in a nervous system, each plotted against each other, as a phase space plot. Determining exactly which are the relevant parameters necessary for describing the dynamics of a given circuit is a major challenge (Briggman et al. 2006). Viewing circuits as dynamical systems is important especially for multifunctional circuits in which the same groups of neurons can give rise to different stable states depending on the overall behavioral state of an animal, the environment, or neuromodulation.

Distributed Control

Many reflexive movements can be attributed to a direct pathway from sensory neurons, possibly via interneurons, to motor neurons (Hooper 2005). Such simple behaviors can potentially be understood by tracing the anatomical connections from the periphery to the central nervous system (CNS) and back out to the periphery. However, there are now many examples of far more distributed interactions at the premotor neuron level (Georgopoulos et al. 1986, Kristan & Shaw 1997, Sparks et al. 1997). A classic example is the gill withdrawal reflex in the sea hare, *Aplysia californica*. Studies using electrophysiological techniques had concluded that the sensory to motor pathway involved a handful of sensory interneurons (Leonard & Edstrom 2004). However, voltage-sensitive dye imaging of many neurons simultaneously found that a large fraction of neurons respond to stimuli eliciting the reflex (Wu et al. 1994) and that considering the concerted activity of this population was necessary to explain the position and velocity of the gill (Zochowski et al. 2000). Calcium imaging in the larval zebrafish recently yielded similar results: A large population of descending interneurons respond to stimuli that elicit escape behavior (Gahtan et al. 2002). In the leech CNS, the reflexive local bending behavior is generated by a population of interneurons (Lewis & Kristan 1998, Lockery & Kristan 1990). In addition to these reflexive behaviors, the phases of some rhythmic behaviors are generated by a distributed network of neurons oscillating in unison (Grillner 2006).

Electrode recordings, even with multielectrode techniques, are likely to underestimate greatly the degree of distributed control in circuits. In addition, traditional tests (such as the activation or ablation of single cells) to determine the sufficiency and/or necessity of a neuron in generating a behavior will most likely fail because of the functional redundancy of many systems. Thus the command neuron concept (DiDomenico & Eaton 1988, Eaton & DiDomenico 1985, Kupfermann & Weiss 1978) is of little use when studying distributed or multifunctional networks, although single neurons whose activation evokes complex behaviors continue to be found (Dembrow et al. 2003, Frost & Katz 1996, Hedwig 2000). Relying on correlations between network activity and behavior is often the only way to infer the function of neurons within a distributed network. Distributed circuitry creates the substrate—complex interactions among many cells connected by modifiable synapses and dynamic intrinsic membrane properties—to support multifunctional circuits. By applying modern imaging techniques (see New Techniques, below) to additional neuronal circuits, it will be possible to determine how distributed, multifunctional circuits produce behaviors.

Behavioral Modules

A second form of multifunctionality, in addition to multistable circuits, depends on the *modular organization* of many nervous systems (Kiehn & Kjaerulff 1998). Behavioral modules are populations of neurons that create distinct coordinated body movements (Buschges 2005). For example, modules can be used to coordinate different body segments, as for undulatory swimming (Grillner 2006, Kristan et al. 2005). Modules can also be a series of basic building blocks, made of neuronal pools dedicated to coordinating the activity of a muscle group. In their most basic form, modules serve dedicated functions, such as the rotation of a joint, but the dynamic

STG: stomatogastric
ganglion

coordination of different modules can generate a large variety of behaviors. Studies of directed limb movements have posited "muscle synergies," corresponding to motor modules in the vertebrate spinal cord (Bizzi et al. 1995, Bizzi et al. 2002, Tresch et al. 2002). These motor primitives can be combined to generate spatially and temporally precise force fields and the corresponding body and limb movements, for example, reaching, grasping, or kicking (Flash & Hochner 2005), that are linear combinations of the component muscle synergies (d'Avella et al. 2003, Kargo & Giszter 2000, Tresch et al. 1999). Repetitively activating a sequence of modules could also in principle generate rhythmic movements, although different modules may be responsible for rhythmic pattern generation (Saltiel et al. 1998). In this scheme, there must be control circuits—as yet unfound—that are capable of coordinating the temporal activation patterns of modules to produce multifunctionality (Bizzi et al. 2000).

Multifunctional Architectures

Defining a circuit as either multifunctional or unifunctional (i.e., dedicated) is often difficult because some neurons within a circuit may contribute to more than one behavior and other neurons to only a single behavior. In addition, the muscles that circuits control may be either unifunctional or multifunctional (Morton & Chiel 1994). To consider this point in detail, five basic architectures and behavioral examples are diagrammed in **Figure 2**. The first configuration illustrates truly dedicated pathways, with two dedicated neuronal pools driving two sets of unifunctional muscles (**Figure 2a**). For example, independent respiratory and locomotory circuits in mammals drive independent sets of muscles [although the circuits can receive common sensory input (Morin & Viala 2002)]. Any two independent behaviors belong to this category. Alternatively, two dedicated neuronal pools could drive a single multifunctional muscle set (**Figure 2b**). One example of this architecture comes from the flight and walking systems in the locust (Ramirez & Pearson

1988): One set of muscles is active in both behaviors but appears to be driven by two independent interneuron pools. In this example, the two behaviors are incompatible, so only one pool can drive the muscles at a given time.

Likewise, multistable neural circuits can be used to drive either unifunctional or multifunctional muscles. Crabs and lobsters use several different muscle groups during feeding, driven be a common multistable circuit, the stomatogastric ganglion (STG) (see Neurons Active During Multiple Behaviors, below). The sea slug, *Tritonia diomedea*, uses a common circuit to drive independent sets of muscles and cilia used for swimming and crawling, respectively (Popescu & Frost 2002). Multifunctional circuits implementing this architecture (**Figure 2c**) can potentially drive unifunctional muscles to generate two or more behaviors simultaneously, as in the crab STG (Bucher et al. 2006; Weimann et al. 1991). Alternatively, multifunctional circuits may drive a common set of multifunctional muscles (**Figure 2d**). For example, respiratory circuits in mammals can reconfigure to drive a variety of different inspiratory and expiratory rhythms (Lieske et al. 2000), and two forms of locomotion in the leech, swimming and crawling, are controlled by a multifunctional circuit activating multifunctional muscles (Briggman & Kristan 2006). Many of the behaviors generated using this architecture are mutually exclusive—a leech cannot both swim and crawl at the same time. The coordination of behavioral modules to drive muscle synergies is yet another form of multifunctionality (**Figure 2e**; see Behavioral Modules, above).

We use "neuronal pools" to be intentionally vague because separating the premotor interneuronal circuitry from the motor circuitry can be difficult (Staras et al. 1998). This distinction works only for extreme differences, but there is a large gray area between these extremes (Morton & Chiel 1994). For example, behaviors driven primarily by multifunctional circuits can also include apparently dedicated neurons (Berkowitz 2002, Briggman & Kristan 2006). Additionally, multistable

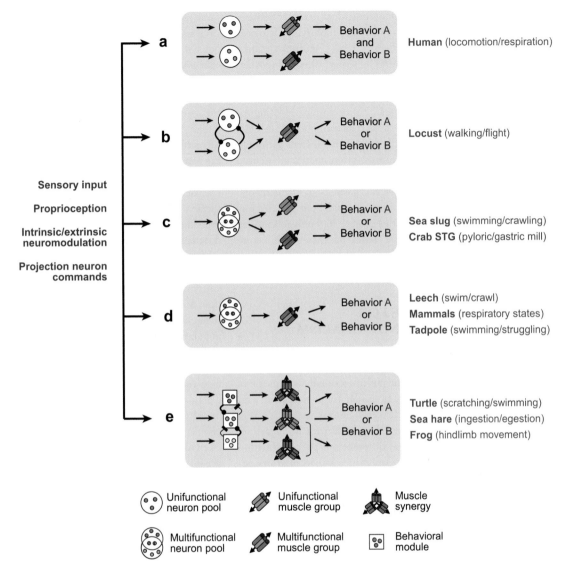

Figure 2

Multifunctional circuit architectures, including unifunctional and multifunctional neuron pools, uni/multifunctional muscle groups, behavioral modules, and muscle synergies (see the key). Inputs coordinating behaviors include sensory input, proprioception, neuromodulation, and command inputs. Multifunctional neuron pools represent multistable dynamical systems. They can be formed by neurons switching between two networks, the fusing of two networks, or the construction of a new network in response to input. (*a*) Dedicated neuron pools control distinct muscle groups and, consequently, distinct behaviors. (*b*) Dedicated neuron pools control multifunctional muscles that contribute to more than one behavior. Generally, these pools must inhibit each other to prevent simultaneous generation of incompatible behaviors. (*c*) A multifunctional neuron pool innervates two distinct muscle groups in parallel. This architecture allows for the simultaneous expression of two behaviors. (*d*) A multifunctional neuron pool excites a multifunctional muscle group. This organization leads to mutually exclusive behaviors. (*e*) Behavioral modules are coordinated by inputs and via intermodule connections denoting inhibition and excitation (*shown as filled circles and bold lines*) to activate muscle synergies. Muscle synergies are activated individually or in combinations to generate complex behaviors.

neuronal pools can arise by merging two circuits, switching individual neurons between pattern generators, or modifying intrinsic neuronal dynamics by neuromodulation (Dickinson 1995). Despite these important details, determining a general architecture can be useful for both didactic and experimental purposes. For example, ablating or inactivating dedicated neuron pools or muscles should not affect other dedicated pathways, whereas disrupting multifunctional neuron pools should affect several behaviors. Similarly, inducing short-term plasticity in elements of a multifunctional pool should alter all behaviors controlled by this pool. Ultimately, classifying circuits on the basis of their architecture requires knowing all relevant circuit elements and the behaviors they influence.

EVIDENCE FOR MULTIFUNCTIONALISM

The motor patterns produced by multifunctional circuits may show diverse dynamics. For instance, some multifunctional circuits drive two behaviors at similar frequencies but with different muscle coordination patterns, whereas others drive two behaviors with both distinct frequencies and different muscle coordination patterns. Consequently, a single neuron's activity pattern can change dramatically between two behaviors (**Figure 3**).

Neurons Active during Multiple Behaviors

Two different behaviors can be generated by changing the phase relationships of active motor neurons, such as those used during forward and backward stepping in humans (Grasso et al. 1998), or switching between synchronous and peristaltic beating of the leech heart (Norris et al. 2006). Individual interneurons, however, may produce very distinct activity patterns during different behaviors. The activity of many of the interneurons in the turtle spinal cord during three forms of hindlimb scratching is phase-locked to the scratching rhythm

(Berkowitz 2001, 2005; Stein 2005), but these same neurons can be rhythmically active, tonically active, or silent during swimming, another behavior involving the hindlimbs (**Figure 3a**) (Berkowitz 2002).

A change in the coordination of muscle activation can also be accompanied by a shift in the oscillation frequency of a multifunctional circuit. Struggling in frog tadpoles is produced by recruiting and recoordinating neurons from the same pool of neurons used for swimming (Soffe 1993); the level of excitation in the circuit determines which behavior is expressed (Soffe 1996). Swimming, a rostrocaudal progressing pattern, and struggling, a slower caudorostral pattern, are elicited by a common sensory pathway (Green & Soffe 1996; Soffe 1991, 1996). Neurons active during swimming oscillate rhythmically and typically produce one action potential per oscillation period. During struggling, these same neurons, plus a larger population, burst in phases during each oscillation period.

Swimming and crawling in the leech also involve some of the same muscle groups active in different phase relationships and at different frequencies (Kristan et al. 2005). Sampling a large proportion of the neurons in a leech segmental ganglion with voltage-sensitive dye imaging uncovered a distributed network of cells that oscillate in phase with both the crawling and swimming motor programs (**Figure 3b**) (Briggman & Kristan 2006). The swimming network contains fewer neurons, but the overlap between the two circuits is large (>90%). Stimulating just one of these multifunctional neurons can perturb both the swimming and the crawling rhythms. However, a neuron's activity phase during swimming did not predict its phase during crawling, suggesting a major reconfiguration of this multifunctional circuit to produce the two behaviors.

Although characterizing the activity of single neurons and relating their activity to two or more behaviors can lead to new insights about their multiple functions, knowing the complete anatomical circuit is invaluable. The stomatogastric nervous system of crabs

a Turtle

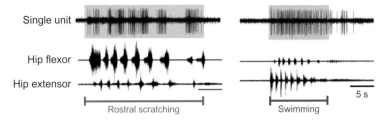

Single unit

Hip flexor

Hip extensor

Rostral scratching Swimming

5 s

b Leech

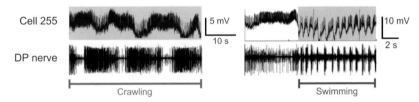

Cell 255

5 mV

10 s

10 mV

2 s

DP nerve

Crawling Swimming

c Crab

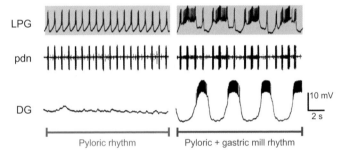

LPG

pdn

DG

10 mV

2 s

Pyloric rhythm Pyloric + gastric mill rhythm

Figure 3

Activity of single neurons within multifunctional circuits. In each preparation, a single neuron within the circuit was recorded during two behaviors (*blue and red*). (*a*) An extracellular single unit recorded from the turtle spinal cord during rostral scratching and swimming. The neuron bursts rhythmically during scratching, but is tonically active during swimming. (*b*) An interneuron (Cell 255) recorded from the isolated leech nerve cord during fictive crawling and swimming. The neuron oscillates with both rhythms but at very different frequencies. (*c*) A CPG neuron (LPG) recorded from the crab stomatogastric ganglion during the pyloric rhythm alone and the pyloric plus gastric mill rhythms. When both rhythms are expressed, the neuron fires in time with them. Modified with permission from Berkowitz (2002) (panel *a*), Briggman & Kristan (2006) (panel *b*), and Weimann et al. (1991) (panel *c*).

and lobsters provides such an advantage. The stomatogastric nervous system generates three different rhythms relating to the ingestion of food: the pyloric, gastric mill, and cardiac sac rhythms (Hooper & DiCaprio 2004, Marder & Bucher 2007). Individual neurons normally active during ongoing pyloric rhythms can switch to fire in time with either the gastric mill (Weimann & Marder 1994, Weimann et al. 1991) or the cardiac sac rhythms (Hooper & Moulins 1989, 1990; Hooper et al. 1990) when either of these two rhythms begins, or they can even oscillate with two rhythms simultaneously (**Figure 3c**). Switching the roles of neurons to support multiple simultaneous rhythms produces important behavioral consequences (Clemens et al. 1998, Heinzel et al. 1993, Thuma et al. 2003). Studies of this multifunctional system have been extremely influential in proposing general mechanisms for switching between multiple behaviors (see Mechanisms Driving Multifunctional Circuits, below). Switching neurons between rhythms is also observed during flight in the locust (Ramirez 1998). Interneurons normally active during abdominal respiratory movements switch to oscillate in time with the wing rhythm during flight. The consequent increase in respiration may play an important behavioral role, supporting the heightened metabolic requirements of flight.

Neurons Active During One Behavior May Be Inactive During Others

Neurons that are tonically or rhythmically active, or "on," during one behavior can be hyperpolarized during another behavior and are sometimes considered to be "off." Although it may be true that hyperpolarizing a neuron can remove it from influencing a motor pattern, this may not reflect its true function. For example, hyperpolarizing an inhibitory neuron may disinhibit other neurons. A complete account of all the elements involved in a circuit is required to observe such an effect (Selverston 1980). (The only truly independent neurons within a defined circuit are those that are active, either hyperpolarized or depolarized, during one behavior and inactive (i.e., remaining at its resting membrane potential) during other related behaviors.)

Investigators have found many examples of neurons that are excited during one behavior and inhibited during another. The following are examples from systems already discussed: Some interneurons rhythmically active in the locust during walking receive tonic or rhythmic inhibition during flight (Ramirez & Pearson 1988); at least one interneuron rhythmically active during crawling in the leech is tonically inhibited during swimming (Briggman & Kristan 2006); stimuli that elicit whole-body shortening in the leech inhibit some neurons involved

in the swim CPG network (Shaw & Kristan 1997); and interneurons that are excited during the hindlimb flexion reflex in the turtle are rhythmically inhibited during both scratching and walking (Berkowitz 2007). Both the mechanisms and the function of the inhibition remain to be found.

Behavioral Modules Generate Multiple Behaviors

The term behavioral module refers to single neurons or groups of neurons that coordinate a particular muscle synergy (e.g., the movement of a joint or a limb, or the coordinated movement of limbs) that is used in a variety of behaviors. For limbed animals, in addition to rhythmic locomotor movements, directed limb movements comprise a large fraction of their behavioral repertoire. One of the best characterized examples of behavioral modules comes from experiments on directed, nonrhythmic movements of frog limbs. The frog spinal cord contains behavioral modules to organize related muscle groups into coordinated behaviors, such as wiping or kicking (d'Avella et al. 2003). The individual modules are located at discrete locations along the spinal cord, rather than being distributed along the spinal cord (Saltiel et al. 1998). Focal iontophoretic application of the glutamate agonist N-methyl-D-aspartic acid (NMDA) stimulates local neuronal populations that elicit tonic motions of the frog hindlimbs, and activating other regions elicit rhythmic motions. Stimulated regions containing rhythmic modules are different from but spatially near regions that elicited the related tonic components. Functional imaging has also revealed that different brain regions are active during rhythmic and discrete movements (Schaal et al. 2004). The same behavioral modules and muscle synergies are coordinated, presumably by descending commands, into different temporal sequences to produce a variety of behaviors (Tresch et al. 1999, 2002).

This organization is likely shared with other vertebrate species, including humans. For instance, modules controlling the right and left leg during human stepping are autonomous (Yang et al. 2005). Remarkably, short-term adaptation of the right or left leg does not generalize between forward and backward stepping (Choi & Bastian 2007). This lack of generalization implies that the networks underlying walking are both leg and direction specific, consistent with the idea of discrete behavioral modules.

If a module (or single interneuron) is active during multiple behaviors, does it remain phase-locked to a certain motor neuron phase or switch to a new phase? It appears that modules can be either movement specific (phase-locked to a movement used in multiple behaviors) or behavior specific (phase-shifted depending on the behavioral context). Neurons in the turtle spinal cord during three forms of hindlimb scratching are often phase-locked (e.g., to hip flexion), regardless of changes in motor neuron coordination (**Figure 4a**) (Berkowitz 2001). Other behavior-specific interneurons dramatically alter their phase relationships to hip flexion during the different forms of scratching (**Figure 4b**) (Berkowitz 2001, 2005).

Neurons form both behavior-specific and movement-specific modules that contribute to feeding behavior in *Aplysia* (Jing et al. 2004; Jing & Weiss 2002, 2005). Ingestion and egestion feeding patterns are both generated by a protraction followed by a retraction of the food grasper, the radula. The neurons generating these phases are movement specific (**Figure 4c**). The patterns are distinguished by the phase in which the radula closes, which is driven by behavior-specific interneurons (**Figure 4c**) (Jing & Weiss 2002). Biting and swallowing are similar ingestive patterns; both require a closure of the food grasper, the radula, during the retraction phase of both behaviors. Despite this shared-phase relationship, different neurons elicit two closure phases, again indicating behavior-specific modules (Jing et al. 2004). Movement-specific neurons also control *Aplysia* neck movements in a variety of behaviors (Xin et al. 1996, 2000). Further studies of the vertebrate spinal cord are required to determine the extent to which such behavior-specific and movement-specific modules are used for limb movements.

MECHANISMS DRIVING MULTIFUNCTIONAL CIRCUITS

Many multifunctional networks are composed of neurons directly presynaptic to motor neurons. The ability of these networks to function in more than one mode raises the question of what causes the switch between behaviors. Likely candidates include (*a*) sensory or projection neurons providing input from the periphery or via descending and ascending inputs from higher-level networks, (*b*) the effects of neuromodulatory substances on intrinsic membrane properties and synapses, and (*c*) biomechanical constraints imposed by the body, detected by sensory feedback as the body moves.

Sensory and Projection Neurons Turn Behaviors On or Off Using Multifunctional Networks

Projection neurons can act in a parallel, combinatorial fashion to elicit different patterns from multistable circuits. The lobster STG can generate two distinct gastric mill rhythms (types I and II) depending on the parallel actions of two projection neurons. One of the projection neurons, CG, elicits the type I rhythm by its electrical coupling and a strong inhibitory synapse onto particular STG neurons (Simmers & Moulins 1988). The second projection neuron, GI, inhibits the gastric mill network (Combes et al. 1999b). When activated in parallel with CG, the GI activity disinhibits one of the gastric mill neurons and elicits the type II rhythm. Therefore, the parallel action of two descending projection neurons rapidly reconfigures a multistable network to generate two distinct rhythms. Sensory input from a mechanosensory neuron, AGR, modulates the activity in the two projection neurons (Combes et al. 1999a): Low AGR activity, coupled with a high intrinsic firing rate in CG, leads to the

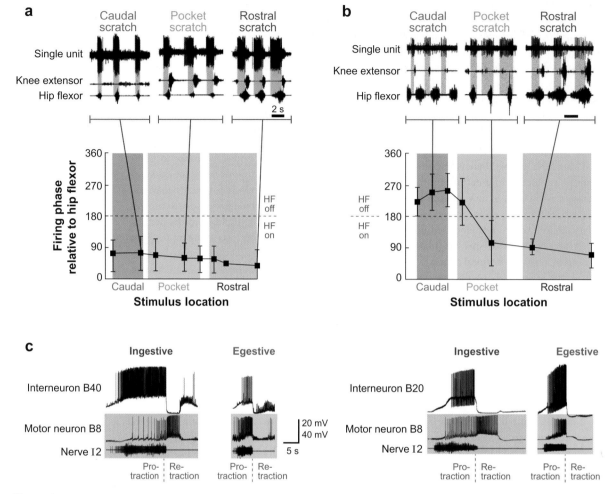

Figure 4

Behavioral modules can be behavior specific or movement specific. (*a*) A single unit recorded extracellularly from the turtle spinal cord during three forms of scratching. The phase of this neuron relative to the activity of the hip flexor remains constant in all three behaviors (*bottom panel*). This is an example of a movement-specific neuron. (*b*) A different single unit from the turtle spinal cord. The phase changes dramatically, by almost 180°, relative to hip flexor activity during different types of scratching. This neuron demonstrates behavior specificity. (*c*) *Aplysia* ingestion (*purple*) and egestion (*green*) feeding patterns are both characterized by protraction of the radula followed by retraction. Neurons generating these phases are movement specific. The phase during which the radula closes distinguishes the two patterns (motor neuron B8). The firing frequency of B40 is greater during ingestion, and it drives radula closing during the retraction phase of ingestion. B20 fires stronger during egestion, and it drives radula closing during the protraction phase of egestion. B40 and B20 play a behavior-specific role by controlling the phasing of radula closing, which distinguishes the two patterns. Modified with permission from Berkowitz (2001) (panels *a* and *b*) and Jing & Weiss (2002) (panel *c*).

type I rhythm; high AGR activity drives GI to fire faster than CG and generates the type II rhythm. Similarly, the activation of two distinct sensory inputs can also be used to activate different rhythms from the lobster STG (Beenhakker et al. 2004, 2005; Beenhakker & Nusbaum 2004; Blitz et al. 2004; Norris et al. 1994).

Projection neurons do not always act in parallel. In the tadpole of the frog, *Xenopus*, single intracellular stimuli delivered to Rohan-Beard sensory neurons elicit the swimming rhythm, and prolonged stimulation of the same neurons elicits a slower rhythm similar to struggling (Soffe 1997). Blocking ascending inhibition (Green & Soffe 1998) has no effect on swimming but altered the speed of struggling, which suggests that the swimming rhythm is insensitive to ascending inhibitory inputs owing to the kinetics and propagation of the swimming rhythm. Projection neurons that transmit behavior-specific information have also been demonstrated during swimming and escape behavior in the zebrafish spinal cord (Ritter et al. 2001). In vivo calcium imaging demonstrated two populations of descending interneurons: One population was active during Mauthner cell-mediated escapes but not during swimming, and the second population oscillated during swimming but was inactive during escapes. Because both of these behaviors involve bending the body, it appears that the two pathways activate local, segmental multifunctional circuits in different ways to generate the two behaviors.

Neuromodulatory Substances Switch Network Function

The release of neuromodulatory substances by nervous systems provides an alternative mechanism to modulate multifunctional networks. The nonsynaptic release and diffusion of neuromodulators allow the alteration of the intrinsic membrane properties and synaptic strengths of large populations of neurons causing short-term and long-lasting network effects (Dickinson 2006, Faumont et al. 2005, Katz 1998). Modulators may be released locally from individual projection neurons or in a more widespread pattern hormonally (Weimann et al. 1997, Willard 1981). They can alter single rhythms (Nusbaum & Marder 1989b), switch between rhythms (Jing et al. 2007), construct entirely new rhythms (Meyrand et al. 1994), or change the gain of neuromuscular transforms (Brezina et al. 2003). Much of what we know about the specific actions of a wide variety of modulatory substances is the culmination of a large body of work in the crustacean STG (Swensen & Marder 2000). The peptide proctolin, released from the projection neuron MPN, will start or speed up slow pyloric rhythms in the STG (Nusbaum & Marder 1989a,b). In contrast, MPN depresses the gastric mill rhythm by synaptically inhibiting other projection neurons, such as MCN1, that initiate the gastric mill rhythm (Blitz & Nusbaum 1997, 1999). Such corelease of transmitters and neuromodulatory compounds is widespread in invertebrate neurons (Blitz et al. 1999, Nusbaum et al. 2001, Vilim et al. 2000). The projection interneuron MCN1 also releases both proctolin and a tachykinin-related peptide (CabTRP Ia) onto STG neurons (Nusbaum 2002; Wood et al. 2000). CabTRP Ia is responsible for the excitation of the gastric mill rhythm (Wood et al. 2000). Like MPN, MCN1 also excites a pyloric rhythm, but the rhythm is distinct from the MPN-elicited pattern. This difference is partly explained by the existence of extracellular peptidases (Wood & Nusbaum 2002). Therefore, not only are individual neurons able to release a number of modulatory substances, but also the same substances can activate multifunctional networks in different ways.

Monoamines also play a role in multifunctional network activation. In the leech CNS, serotonin released from swim CPG neurons (Nusbaum & Kristan 1986) promotes swimming by altering voltage-gated conductances of another neuron in the swim CPG network (Angstadt & Friesen 1993a,b). Dopamine, on the other hand, terminates swimming and potentially promotes crawling, although the site of its action is unknown (Crisp & Mesce 2004). Amines can also significantly alter the strength of electrical synapses in the lobster STG (Johnson et al. 1993).

Most studies on the neuromodulation of in vitro vertebrate nervous systems use bath application of modulatory cocktails to elicit behaviors (Whelan et al. 2000). Exogenous application of neuromodulators does not necessarily mimic the effects of endogenously released substances because of heterogeneous cotransmission, receptor accessibility, peptidase activity, and concentration distribution (Nusbaum 1986). The identification of the neurons releasing modulators and the ability to stimulate these neurons selectively in a realistic manner allow one to test for differences between endogenous release and exogenous application (Nusbaum & Marder 1989b).

Biomechanical Constraints and Sensory Feedback Modulate Multifunctional Networks

The ability to maintain invertebrate nervous systems in vitro provides investigators almost unhindered access to large populations of neurons. The fact that isolated parts of these nervous systems continue to produce reasonably normal motor patterns has been extremely useful for characterizing rhythms generated by CPGs. But these pattern generators are only half of the picture. The biomechanical properties of musculature and joints and the transformation of motor neuron spikes into muscle contractions further constrain the behaviors that these patterns generate (Chiel & Beer 1997). Indeed, muscles that are context-sensitive add another layer of multifunctionality to nervous systems.

The feeding behavior of the sea hare *Aplysia californica* is generated by CPG-controlling muscles that drive both the ingestion and the rejection of food (Jing & Weiss 2005). The radula/odontophore complex (grasper) in the mouth alters its shape from spherical (when it is open during swallowing as it protracts to grab food) to ellipsoidal (when it is closed during rejection as it protracts to eject food) (Sutton et al. 2004b, Ye et al. 2006b). The formation of the ellipsoid stretches a protraction muscle, I2, which gives the ellipsoid a mechanical advantage to generate larger protractions during rejection (Sutton et al. 2004b). Larger protractions also pivot the grasping complex around a hinge muscle, causing a rotation of the rejected food (Novakovic et al. 2006, Sutton et al. 2004a). A biomechanical model of the system predicted that the activation of grasper closer muscles, I1/I3, must be delayed until the retraction phase of the rejection behavior to avoid pulling rejected food back into the mouth. Such a multifunctional neuron was found (Ye et al. 2006b). This complex interplay between muscle groups is an example of neuromechanical modulation: Coupling between muscles modulates their response to the same neural activity patterns under different conditions (Neustadter et al. 2007, Ye et al. 2006a). These studies clearly demonstrate the power of analyzing the biomechanics of a system both to yield predictions about the underlying neural activity and to understand the context-sensitive responses of muscles to incoming motor neuron commands.

Because of the nonlinear properties of individual muscles and variability between animals, predicting behavioral responses from motor neuron spike trains is difficult (Hooper et al. 2006, Hooper & Weaver 2000). The development of quantitative models of how neurons control muscles is necessary ultimately to determine the behavioral relevance of patterns produced by multifunctional circuits (Brezina et al. 2000, Brezina & Weiss 2000).

MODELING AND THEORETICAL ANALYSIS OF MULTIFUNCTIONAL CIRCUITS

Investigators have described several mechanisms to switch between patterns, e.g., the merging or fusion of multiple distinct networks and the switching of individual neurons or populations of neurons between networks (Dickinson 1995). However, do certain intrinsic properties of multistable networks allow these networks to support the generation of more than one pattern? Given the prevalence of circuits combining inhibitory chemical and electrical synapses, several models have been developed to study the multistable activity of such networks (Bem & Rinzel 2004, Chow & Kopell 2000, Lewis & Rinzel 2003, Pfeuty et al. 2003, Wang & Rinzel 1992). One such model of two neurons coupled by electrical synapses and reciprocal inhibition was used to probe the effects of voltage-gated conductances, the amplitude of external inputs, and the strengths and kinetics of the coupling (Pfeuty et al. 2005). Varying these parameters switched the neurons between stable synchronous and antisynchronous phase-locking. The parameters suggest biophysical mechanisms by which a single network can generate multiple output patterns. Therefore, in addition to driving the reconfiguration or fusion of circuits with external inputs, modifying the intrinsic properties of elements within a circuit can also switch between multiple states (Bem et al. 2005).

Transient inputs can also switch multistable circuits between patterns without any change in biophysical properties. For instance, rings of neurons connected by inhibitory synapses, a simple CPG network, are multistable (Luo et al. 2004). A single excitatory pulse delivered to one neuron initiates a cascade of events that delay and/or advance firing phases of neurons in the ring, which can rapidly switch the circuit into a different oscillation mode. (Modifying synaptic conductances can also switch the circuit into additional modes.) Switching between stable modes with transient inputs has also been demonstrated using dynamic clamp experiments (Manor & Nadim 2001) and in electrical circuits modeling multistable networks (Kier et al. 2006).

Can more elaborate network models, initially developed to produce a single behavior, be driven to produce alternate rhythms? The newt moves using one of two axial motor rhythms: swimming and terrestrial walking (Delvolve et al. 1997). A coupled oscillator model of lamprey swimming was modified to produce newt-like swimming and walking (Bem et al. 2003). Although the modifications to the lamprey model were substantial, including changing intersegmental connectivity and providing both tonic and phasic input to drive the network, the modified model successfully generated both rhythms. Stretch receptors active during locomotion could provide the phasic input, and the authors postulate an as-yet-

unidentified limb CPG will provide the tonic drive (Bem et al. 2003, Delvolve et al. 1997). The interaction between limb and trunk CPGs was also used to implement a model of swimming and walking in the salamander (Ijspeert 2001, Ijspeert et al. 2007).

BEHAVIORAL CHOICES

The study of how multifunctional networks operate naturally leads to the question of why a particular rhythm is expressed at any given time (Kristan & Gillette 2007). How do nervous systems choose an appropriate behavior in response to external stimuli? Behavior selection can involve a behavioral hierarchy, in which decisions are sequentially made, resulting in the selection of a motor program. The choice between swimming and crawling in the leech is mediated by descending interneurons that, when stimulated, elicit, first, an elongation followed either by swimming or crawling (Esch et al. 2002). When these neurons are stimulated in semi-intact preparations, the choice between swimming and crawling depends on the level of the surrounding saline bath. This finding suggests a hierarchy by which a sensory stimulus first prompts the leech to move and then the environment (saline depth) determines the appropriate behavior. At the level of segmental multifunctional circuits, the choice between swimming and crawling is determined by the covarying activity of a neuron population (Briggman et al. 2005). The choice between swimming and crawling in response to a sensory stimulus can be biased by manipulating the membrane potential of a neuron participating in the population. Putting together all the pieces—the influence of sensory pathways, descending commands, and multifunctional circuit pattern selection—is necessary to elucidate further the complex mechanisms of behavioral choice.

NEW TECHNIQUES

Our understanding of many mechanisms that underlly multifunctional networks comes from the numerous invertebrate nervous system studies (Hooper & DiCaprio 2004, Marder & Bucher 2007). The search for such mechanisms in larger nervous systems is limited by the technical difficulty in monitoring and manipulating large neuron populations. The crustacean STG multifunctional network is unique because the connectivity of the circuit is completely known, the output of many neurons can be monitored on nerves simultaneously, and single-cell ablations/manipulations can have profound consequences for pattern generation. In addition, the recent use of matrix-assisted laser desorption/ionization (MALDI) mass spectrometry has provided the ability to determine the complement of peptides within single neurons, a step necessary to understand fully the effects of neuromodulation within multifunctional circuits (Dickinson 2006, Hummon et al. 2006).

In larger circuits, the lack of detailed knowledge of anatomical connectivity can frustrate the efforts to identify all the elements potentially contributing to multifunctional pattern generation. New approaches to neural circuit reconstruction will help identify these elements (Briggman & Denk 2006). The rapid development of population-imaging techniques in recent years, including the bulk loading of calcium indicators and voltage-sensitive dyes, is necessary to characterize the extent to which a population is multifunctional (Bonnot et al. 2005, O'Donovan et al. 2005). Finally, the ability to excite and inhibit many specific neurons simultaneously may be the crucial step toward identifying the mechanisms of multifunctionality in larger nervous systems (Zhang et al. 2007).

CONCLUSIONS

Central to the discussion of multifunctional networks is the question of why these networks exist at all. Arguments can be made that the efficient use of a limited number of neurons for a large number of behaviors is evolutionarily advantageous. Alternatively, multifunctional networks may themselves reflect the evolutionary history of a species (Katz & Harris-Warrick 1999). The nearly complete overlap in the circuit producing swimming in the leech with that producing crawling, for instance, suggests that swimming evolved from crawling (Briggman & Kristan 2006), a possibility consistent with the molecular phylogeny of leeches (Siddall & Burreson 1996). It is tempting to compare multifunctional pattern-generating networks with the complex oscillatory dynamics of the brain, but such comparisons remain speculative (Grillner et al. 2005, Yuste et al. 2005). Indeed, demonstrating that the mechanisms shown to control multifunctional motor networks also control circuits in the central nervous system is a daunting, but extremely exciting, future goal. This goal may not be completely far-fetched given the apparent conservation of both multistable and modular multifunctional circuit mechanisms between vertebrate and invertebrate species. Most mechanisms described in this review were elucidated by meticulous experiments employing single-cell electrophysiology and pharmacology. The continued combination of new mathematical tools and experimental techniques is critical to bridge the gap between the knowledge of small-model organisms and vertebrate motor control.

DISCLOSURE STATEMENT

The authors are not aware of any biases that might be perceived as affecting the objectivity of this review.

LITERATURE CITED

Angstadt JD, Friesen WO. 1993a. Modulation of swimming behavior in the medicinal leech. I. Effects of serotonin on the electrical properties of swim-gating cell 204. *J. Comp. Physiol. [A]* 172:223–34

Angstadt JD, Friesen WO. 1993b. Modulation of swimming behavior in the medicinal leech. II. Ionic conductances underlying serotonergic modulation of swim-gating cell 204. *J. Comp. Physiol. [A]* 172:235–48

Barak O, Tsodyks M. 2007. Persistent activity in neural networks with dynamic synapses. *PLoS Comput. Biol.* 3:e35

Beenhakker MP, Blitz DM, Nusbaum MP. 2004. Long-lasting activation of rhythmic neuronal activity by a novel mechanosensory system in the crustacean stomatogastric nervous system. *J. Neurophysiol.* 91:78–91

Beenhakker MP, DeLong ND, Saideman SR, Nadim F, Nusbaum MP. 2005. Proprioceptor regulation of motor circuit activity by presynaptic inhibition of a modulatory projection neuron. *J. Neurosci.* 25:8794–806

Beenhakker MP, Nusbaum MP. 2004. Mechanosensory activation of a motor circuit by coactivation of two projection neurons. *J. Neurosci.* 24:6741–50

Beer RD, Chiel HJ, Gallagher JC. 1999. Evolution and analysis of model CPGs for walking: II. General principles and individual variability. *J. Comput. Neurosci.* 7:119–47

Bem T, Cabelguen JM, Ekeberg O, Grillner S. 2003. From swimming to walking: a single basic network for two different behaviors. *Biol. Cybern.* 88:79–90

Bem T, Le Feuvre Y, Rinzel J, Meyrand P. 2005. Electrical coupling induces bistability of rhythms in networks of inhibitory spiking neurons. *Eur. J. Neurosci.* 22:2661–68

Bem T, Rinzel J. 2004. Short duty cycle destabilizes a half-center oscillator, but gap junctions can restabilize the antiphase pattern. *J. Neurophysiol.* 91:693–703

Berkowitz A. 2001. Rhythmicity of spinal neurons activated during each form of fictive scratching in spinal turtles. *J. Neurophysiol.* 86:1026–36

Berkowitz A. 2002. Both shared and specialized spinal circuitry for scratching and swimming in turtles. *J. Comp. Physiol. A* 188:225–34

Berkowitz A. 2005. Physiology and morphology indicate that individual spinal interneurons contribute to diverse limb movements. *J. Neurophysiol.* 94:4455–70

Berkowitz A. 2007. Spinal interneurons that are selectively activated during fictive flexion reflex. *J. Neurosci.* 27:4634–41

Bhatt DH, McLean DL, Hale ME, Fetcho JR. 2007. Grading movement strength by changes in firing intensity versus recruitment of spinal interneurons. *Neuron* 53:91–102

Bizzi E, D'Avella A, Saltiel P, Tresch M. 2002. Modular organization of spinal motor systems. *Neuroscientist* 8:437–42

Bizzi E, Giszter SF, Loeb E, Mussa-Ivaldi FA, Saltiel P. 1995. Modular organization of motor behavior in the frog's spinal cord. *Trends Neurosci.* 18:442–46

Bizzi E, Tresch MC, Saltiel P, d'Avella A. 2000. New perspectives on spinal motor systems. *Nat. Rev. Neurosci.* 1:101–8

Blitz DM, Beenhakker MP, Nusbaum MP. 2004. Different sensory systems share projection neurons but elicit distinct motor patterns. *J. Neurosci.* 24:11381–90

Blitz DM, Christie AE, Coleman MJ, Norris BJ, Marder E, Nusbaum MP. 1999. Different proctolin neurons elicit distinct motor patterns from a multifunctional neuronal network. *J. Neurosci.* 19:5449–63

Blitz DM, Nusbaum MP. 1997. Motor pattern selection via inhibition of parallel pathways. *J. Neurosci.* 17:4965–75

Blitz DM, Nusbaum MP. 1999. Distinct functions for cotransmitters mediating motor pattern selection. *J. Neurosci.* 19:6774–83

Bonnot A, Mentis GZ, Skoch J, O'Donovan MJ. 2005. Electroporation loading of calcium-sensitive dyes into the CNS. *J. Neurophysiol.* 93:1793–808

Brezina V, Orekhova IV, Weiss KR. 2000. The neuromuscular transform: the dynamic, nonlinear link between motor neuron firing patterns and muscle contraction in rhythmic behaviors. *J. Neurophysiol.* 83:207–31

Brezina V, Orekhova IV, Weiss KR. 2003. Neuromuscular modulation in *Aplysia*. I. Dynamic model. *J. Neurophysiol.* 90:2592–612

Brezina V, Weiss KR. 2000. The neuromuscular transform constrains the production of functional rhythmic behaviors. *J. Neurophysiol.* 83:232–59

Briggman KL, Abarbanel HD, Kristan WB Jr. 2005. Optical imaging of neuronal populations during decision-making. *Science* 307:896–901

Briggman KL, Abarbanel HD, Kristan WB Jr. 2006. From crawling to cognition: analyzing the dynamical interactions among populations of neurons. *Curr. Opin. Neurobiol.* 16:135–44

Briggman KL, Denk W. 2006. Towards neural circuit reconstruction with volume electron microscopy techniques. *Curr. Opin. Neurobiol.* 16:562–70

Briggman KL, Kristan WB Jr. 2006. Imaging dedicated and multifunctional neural circuits generating distinct behaviors. *J. Neurosci.* 26:10925–33

Bucher D, Taylor AL, Marder E. 2006. Central pattern generating neurons simultaneously express fast and slow rhythmic activities in the stomatogastric ganglion. *J. Neurophysiol.* 95:3617–32

Buschges A. 2005. Sensory control and organization of neural networks mediating coordination of multisegmental organs for locomotion. *J. Neurophysiol.* 93:1127–35

Chiel HJ, Beer RD. 1997. The brain has a body: adaptive behavior emerges from interactions of nervous system, body and environment. *Trends Neurosci.* 20:553–57

Chiel HJ, Beer RD, Gallagher JC. 1999. Evolution and analysis of model CPGs for walking: I. Dynamical modules. *J. Comput. Neurosci.* 7:99–118

Choi JT, Bastian AJ. 2007. Adaptation reveals independent control networks for human walking. *Nat. Neurosci.* 10:1055–62

Chow CC, Kopell N. 2000. Dynamics of spiking neurons with electrical coupling. *Neural Comput.* 12:1643–78

Clarac F, Pearlstein E. 2007. Invertebrate preparations and their contribution to neurobiology in the second half of the 20th century. *Brain Res. Rev.* 54:113–61

Clemens S, Combes D, Meyrand P, Simmers J. 1998. Long-term expression of two interacting motor pattern-generating networks in the stomatogastric system of freely behaving lobster. *J. Neurophysiol.* 79:1396–408

Combes D, Meyrand P, Simmers J. 1999a. Dynamic restructuring of a rhythmic motor program by a single mechanoreceptor neuron in lobster. *J. Neurosci.* 19:3620–28

Combes D, Meyrand P, Simmers J. 1999b. Motor pattern specification by dual descending pathways to a lobster rhythm-generating network. *J. Neurosci.* 19:3610–19

Crisp KM, Mesce KA. 2004. A cephalic projection neuron involved in locomotion is dye coupled to the dopaminergic neural network in the medicinal leech. *J. Exp. Biol.* 207:4535–42

Cymbalyuk G, Shilnikov A. 2005. Coexistence of tonic spiking oscillations in a leech neuron model. *J. Comput. Neurosci.* 18:255–63

Cymbalyuk GS, Gaudry Q, Masino MA, Calabrese RL. 2002. Bursting in leech heart interneurons: cell-autonomous and network-based mechanisms. *J. Neurosci.* 22:10580–92

d'Avella A, Saltiel P, Bizzi E. 2003. Combinations of muscle synergies in the construction of a natural motor behavior. *Nat. Neurosci.* 6:300–8

Delvolve I, Bem T, Cabelguen JM. 1997. Epaxial and limb muscle activity during swimming and terrestrial stepping in the adult newt, *Pleurodeles waltl*. *J. Neurophysiol.* 78:638–50

Dembrow NC, Jing J, Proekt A, Romero A, Vilim FS, et al. 2003. A newly identified buccal interneuron initiates and modulates feeding motor programs in *Aplysia*. *J. Neurophysiol.* 90:2190–204

Dickinson PS. 1995. Interactions among neural networks for behavior. *Curr. Opin. Neurobiol.* 5:792–98

Dickinson PS. 2006. Neuromodulation of central pattern generators in invertebrates and vertebrates. *Curr. Opin. Neurobiol.* 16:604–14

DiDomenico R, Eaton RC. 1988. Seven principles for command and the neural causation of behavior. *Brain Behav. Evol.* 31:125–40

Eaton RC, DiDomenico R. 1985. Command and the neural causation of behavior: a theoretical analysis of the necessity and sufficiency paradigm. *Brain Behav. Evol.* 27:132–64

Esch T, Mesce KA, Kristan WB. 2002. Evidence for sequential decision making in the medicinal leech. *J. Neurosci.* 22:11045–54

Faumont S, Combes D, Meyrand P, Simmers J. 2005. Reconfiguration of multiple motor networks by short- and long-term actions of an identified modulatory neuron. *Eur. J. Neurosci.* 22:2489–502

Fetcho JR. 1992. Excitation of motoneurons by the Mauthner axon in goldfish: complexities in a "simple" reticulospinal pathway. *J. Neurophysiol.* 67:1574–86

Flash T, Hochner B. 2005. Motor primitives in vertebrates and invertebrates. *Curr. Opin. Neurobiol.* 15:660–66

Friesen WO. 1994. Reciprocal inhibition: a mechanism underlying oscillatory animal movements. *Neurosci. Biobehav. Rev.* 18:547–53

Frohlich F, Bazhenov M, Timofeev I, Steriade M, Sejnowski TJ. 2006. Slow state transitions of sustained neural oscillations by activity-dependent modulation of intrinsic excitability. *J. Neurosci.* 26:6153–62

Frost WN, Katz PS. 1996. Single neuron control over a complex motor program. *Proc. Natl. Acad. Sci. USA* 93:422–26

Gahtan E, Sankrithi N, Campos JB, O'Malley DM. 2002. Evidence for a widespread brain stem escape network in larval zebrafish. *J. Neurophysiol.* 87:608–14

Georgopoulos AP, Schwartz AB, Kettner RE. 1986. Neuronal population coding of movement direction. *Science* 233:1416–19

Getting PA. 1989. Emerging principles governing the operation of neural networks. *Annu. Rev. Neurosci.* 12:185–204

Getting PA, Deken MS. 1985. *Tritonia* swimming: a model system for integration within rhythmic motor systems. In *Model Neural Networks and Behavior*, ed. AI Selverston, pp. 3–20. New York: Plenum Press

Grasso R, Bianchi L, Lacquaniti F. 1998. Motor patterns for human gait: backward versus forward locomotion. *J. Neurophysiol.* 80:1868–85

Green CS, Soffe SR. 1996. Transitions between two different motor patterns in *Xenopus* embryos. *J. Comp. Physiol. [A]* 178:279–91

Green CS, Soffe SR. 1998. Roles of ascending inhibition during two rhythmic motor patterns in *Xenopus* tadpoles. *J. Neurophysiol.* 79:2316–28

Grillner S. 2006. Biological pattern generation: the cellular and computational logic of networks in motion. *Neuron* 52:751–66

Grillner S, Markram H, De Schutter E, Silberberg G, LeBeau FE. 2005. Microcircuits in action—from CPGs to neocortex. *Trends Neurosci.* 28:525–33

Guckenheimer J, Gueron S, Harris-Warrick RM. 1993. Mapping the dynamics of a bursting neuron. *Philos. Trans. R. Soc. London B Biol. Sci.* 341:345–59

Hedwig B. 2000. Control of cricket stridulation by a command neuron: efficacy depends on the behavioral state. *J. Neurophysiol.* 83:712–22

Heinzel HG, Weimann JM, Marder E. 1993. The behavioral repertoire of the gastric mill in the crab, *Cancer pagurus*: an in situ endoscopic and electrophysiological examination. *J. Neurosci.* 13:1793–803

Hooper SL. 2005. Movement control: dedicated or distributed? *Curr. Biol.* 15:R878–80

Hooper SL, DiCaprio RA. 2004. Crustacean motor pattern generator networks. *Neurosignals* 13:50–69

Hooper SL, Guschlbauer C, von Uckermann G, Buschges A. 2006. Natural neural output that produces highly variable locomotory movements. *J. Neurophysiol.* 96:2072–88

Hooper SL, Moulins M. 1989. Switching of a neuron from one network to another by sensory-induced changes in membrane properties. *Science* 244:1587–89

Hooper SL, Moulins M. 1990. Cellular and synaptic mechanisms responsible for a long-lasting restructuring of the lobster pyloric network. *J. Neurophysiol.* 64:1574–89

Hooper SL, Moulins M, Nonnotte L. 1990. Sensory input induces long-lasting changes in the output of the lobster pyloric network. *J. Neurophysiol.* 64:1555–73

Hooper SL, Weaver AL. 2000. Motor neuron activity is often insufficient to predict motor response. *Curr. Opin. Neurobiol.* 10:676–82

Hummon AB, Amare A, Sweedler JV. 2006. Discovering new invertebrate neuropeptides using mass spectrometry. *Mass Spectrom. Rev.* 25:77–98

Ijspeert AJ. 2001. A connectionist central pattern generator for the aquatic and terrestrial gaits of a simulated salamander. *Biol. Cybern.* 84:331–48

Ijspeert AJ, Crespi A, Ryczko D, Cabelguen JM. 2007. From swimming to walking with a salamander robot driven by a spinal cord model. *Science* 315:1416–20

Jing J, Cropper EC, Hurwitz I, Weiss KR. 2004. The construction of movement with behavior-specific and behavior-independent modules. *J. Neurosci.* 24:6315–25

Jing J, Vilim FS, Horn CC, Alexeeva V, Hatcher NG, et al. 2007. From hunger to satiety: reconfiguration of a feeding network by *Aplysia* neuropeptide Y. *J. Neurosci.* 27:3490–502

Jing J, Weiss KR. 2002. Interneuronal basis of the generation of related but distinct motor programs in *Aplysia*: implications for current neuronal models of vertebrate intralimb coordination. *J. Neurosci.* 22:6228–38

Jing J, Weiss KR. 2005. Generation of variants of a motor act in a modular and hierarchical motor network. *Curr. Biol.* 15:1712–21

Johnson BR, Peck JH, Harris-Warrick RM. 1993. Amine modulation of electrical coupling in the pyloric network of the lobster stomatogastric ganglion. *J. Comp. Physiol. [A]* 172:715–32

Kargo WJ, Giszter SF. 2000. Rapid correction of aimed movements by summation of force-field primitives. *J. Neurosci.* 20:409–26

Kass JI, Mintz IM. 2006. Silent plateau potentials, rhythmic bursts, and pacemaker firing: three patterns of activity that coexist in quadristable subthalamic neurons. *Proc. Natl. Acad. Sci. USA* 103:183–88

Katz PS. 1998. Comparison of extrinsic and intrinsic neuromodulation in two central pattern generator circuits in invertebrates. *Exp. Physiol.* 83:281–92

Katz PS, Harris-Warrick RM. 1999. The evolution of neuronal circuits underlying species-specific behavior. *Curr. Opin. Neurobiol.* 9:628–33

Kiehn O. 2006. Locomotor circuits in the mammalian spinal cord. *Annu. Rev. Neurosci.* 29:279–306

Kiehn O, Kjaerulff O. 1998. Distribution of central pattern generators for rhythmic motor outputs in the spinal cord of limbed vertebrates. *Ann. N.Y. Acad. Sci.* 860:110–29

Kier RJ, Ames JC, Beer RD, Harrison RR. 2006. Design and implementation of multipattern generators in analog VLSI. *IEEE Trans. Neural Netw.* 17:1025–38

Kristan WB, Gillette R. 2007. Behavioral choice. In *Invertebrate Neurobiology*, ed. G North, RJ Greenspan, pp. 533–54. New York: Cold Spring Harbor Lab. Press

Kristan WB Jr, Calabrese RL, Friesen WO. 2005. Neuronal control of leech behavior. *Prog. Neurobiol.* 76:279–327

Kristan WB Jr, Shaw BK. 1997. Population coding and behavioral choice. *Curr. Opin. Neurobiol.* 7:826–31

Kristan WB Jr, Wittenberg G, Nusbaum MP, Stern-Tomlinson W. 1988. Multifunctional interneurons in behavioral circuits of the medicinal leech. *Experentia* 44:383–89

Kupfermann I, Weiss KR. 1978. The command neuron concept. *Behav. Brain Sci.* 1:3–10

Leonard JL. 2000. Network architectures and circuit function: testing alternative hypotheses in multifunctional networks. *Brain Behav. Evol.* 55:248–55

Leonard JL, Edstrom JP. 2004. Parallel processing in an identified neural circuit: the *Aplysia californica* gill-withdrawal response model system. *Biol. Rev. Camb. Philos. Soc.* 79:1–59

Lewis JE, Kristan WB Jr. 1998. A neuronal network for computing population vectors in the leech. *Nature* 391:76–79

Lewis TJ, Rinzel J. 2003. Dynamics of spiking neurons connected by both inhibitory and electrical coupling. *J. Comput. Neurosci.* 14:283–309

Lieske SP, Thoby-Brisson M, Telgkamp P, Ramirez JM. 2000. Reconfiguration of the neural network controlling multiple breathing patterns: eupnea, sighs and gasps. *Nat. Neurosci.* 3:600–7

Lockery SR, Kristan WB Jr. 1990. Distributed processing of sensory information in the leech. II. Identification of interneurons contributing to the local bending reflex. *J. Neurosci.* 10:1816–29

Luo C, Clark JW Jr, Canavier CC, Baxter DA, Byrne JH. 2004. Multimodal behavior in a four neuron ring circuit: mode switching. *IEEE Trans. Biomed. Eng.* 51:205–18

Manor Y, Nadim F. 2001. Synaptic depression mediates bistability in neuronal networks with recurrent inhibitory connectivity. *J. Neurosci.* 21:9460–70

Marder E. 1994. Invertebrate neurobiology. Polymorphic neural networks. *Curr. Biol.* 4:752–54

Marder E. 2000. Motor pattern generation. *Curr. Opin. Neurobiol.* 10:691–98

Marder E, Bucher D. 2001. Central pattern generators and the control of rhythmic movements. *Curr. Biol.* 11:R986–96

Marder E, Bucher D. 2007. Understanding circuit dynamics using the stomatogastric nervous system of lobsters and crabs. *Annu. Rev. Physiol.* 69:291–316

Marder E, Bucher D, Schulz DJ, Taylor AL. 2005. Invertebrate central pattern generation moves along. *Curr. Biol.* 15:R685–99

Marder E, Calabrese RL. 1996. Principles of rhythmic motor pattern generation. *Physiol. Rev.* 76:687–717

McLean DL, Fan J, Higashijima S, Hale ME, Fetcho JR. 2007. A topographic map of recruitment in spinal cord. *Nature* 446:71–75

Meyrand P, Simmers J, Moulins M. 1994. Dynamic construction of a neural network from multiple pattern generators in the lobster stomatogastric nervous system. *J. Neurosci.* 14:630–44

Morin D, Viala D. 2002. Coordinations of locomotor and respiratory rhythms in vitro are critically dependent on hindlimb sensory inputs. *J. Neurosci.* 22:4756–65

Morton DW, Chiel HJ. 1994. Neural architectures for adaptive behavior. *Trends Neurosci.* 17:413–20

Neustadter DM, Herman RL, Drushel RF, Chestek DW, Chiel HJ. 2007. The kinematics of multifunctionality: comparisons of biting and swallowing in *Aplysia californica*. *J. Exp. Biol.* 210:238–60

Norris BJ, Coleman MJ, Nusbaum MP. 1994. Recruitment of a projection neuron determines gastric mill motor pattern selection in the stomatogastric nervous system of the crab, *Cancer borealis*. *J. Neurophysiol.* 72:1451–63

Norris BJ, Weaver AL, Morris LG, Wenning A, Garcia PA, Calabrese RL. 2006. A central pattern generator producing alternative outputs: temporal pattern of premotor activity. *J. Neurophysiol.* 96:309–26

Novakovic VA, Sutton GP, Neustadter DM, Beer RD, Chiel HJ. 2006. Mechanical reconfiguration mediates swallowing and rejection in *Aplysia californica*. *J. Comp. Physiol. A* 192:857–70

Nowotny T, Rabinovich MI. 2007. Dynamical origin of independent spiking and bursting activity in neural microcircuits. *Phys. Rev. Lett.* 98:128106

Nusbaum MP. 1986. Synaptic basis of swim initiation in the leech. III. Synaptic effects of serotonin-containing interneurones (cells 21 and 61) on swim CPG neurones (cells 18 and 208). *J. Exp. Biol.* 122:303–21

Nusbaum MP. 2002. Regulating peptidergic modulation of rhythmically active neural circuits. *Brain Behav. Evol.* 60:378–87

Nusbaum MP, Blitz DM, Swensen AM, Wood D, Marder E. 2001. The roles of cotransmission in neural network modulation. *Trends Neurosci.* 24:146–54

Nusbaum MP, Kristan WB Jr. 1986. Swim initiation in the leech by serotonin-containing interneurones, cells 21 and 61. *J. Exp. Biol.* 122:277–302

Nusbaum MP, Marder E. 1989a. A modulatory proctolin-containing neuron (MPN). I. Identification and characterization. *J. Neurosci.* 9:1591–99

Nusbaum MP, Marder E. 1989b. A modulatory proctolin-containing neuron (MPN). II. State-dependent modulation of rhythmic motor activity. *J. Neurosci.* 9:1600–7

O'Donovan MJ, Bonnot A, Wenner P, Mentis GZ. 2005. Calcium imaging of network function in the developing spinal cord. *Cell Calcium* 37:443–50

Oprisan SA, Canavier CC. 2002. The influence of limit cycle topology on the phase resetting curve. *Neural Comput.* 14:1027–57

Pfeuty B, Mato G, Golomb D, Hansel D. 2003. Electrical synapses and synchrony: the role of intrinsic currents. *J. Neurosci.* 23:6280–94

Pfeuty B, Mato G, Golomb D, Hansel D. 2005. The combined effects of inhibitory and electrical synapses in synchrony. *Neural Comput.* 17:633–70

Popescu IR, Frost WN. 2002. Highly dissimilar behaviors mediated by a multifunctional network in the marine mollusk *Tritonia diomedea*. *J. Neurosci.* 22:1985–93

Rabinovich MI, Varona P, Selverston AI, Abarbanel HDIA. 2006. Dynamical principles in neuroscience. *Rev. Mod. Phys.* 78:1213–65

Ramirez JM. 1998. Reconfiguration of the respiratory network at the onset of locust flight. *J. Neurophysiol.* 80:3137–47

Ramirez JM, Pearson KG. 1988. Generation of motor patterns for walking and flight in motoneurons supplying bifunctional muscles in the locust. *J. Neurobiol.* 19:257–82

Ritter DA, Bhatt DH, Fetcho JR. 2001. In vivo imaging of zebrafish reveals differences in the spinal networks for escape and swimming movements. *J. Neurosci.* 21:8956–65

Saltiel P, Tresch MC, Bizzi E. 1998. Spinal cord modular organization and rhythm generation: an NMDA iontophoretic study in the frog. *J. Neurophysiol.* 80:2323–39

Schaal S, Sternad D, Osu R, Kawato M. 2004. Rhythmic arm movement is not discrete. *Nat. Neurosci.* 7:1136–43

Schoner G, Kelso JA. 1988. Dynamic pattern generation in behavioral and neural systems. *Science* 239:1513–20

Selverston AI. 1980. Are central pattern generators understandable? *Behav. Brain Sci.* 3:535–71

Selverston AI, Ayers J. 2006. Oscillations and oscillatory behavior in small neural circuits. *Biol. Cybern.* 95:537–54

Shaw BK, Kristan WB Jr. 1997. The neuronal basis of the behavioral choice between swimming and shortening in the leech: control is not selectively exercised at higher circuit levels. *J. Neurosci.* 17:786–95

Siddall ME, Burreson EM. 1996. Leeches (Oligochaeta?: Euhirudinea), their phylogeny and the evolution of life-history strategies. *Hydrobiologia* 334:277–85

Sillar KT, Roberts A. 1993. Control of frequency during swimming in Xenopus embryos: a study on interneuronal recruitment in a spinal rhythm generator. *J. Physiol.* 472:557–72

Simmers J, Moulins M. 1988. A disynaptic sensorimotor pathway in the lobster stomatogastric system. *J. Neurophysiol.* 59:740–56

Soffe SR. 1991. Triggering and gating of motor responses by sensory stimulation: behavioural selection in *Xenopus* embryos. *Proc. Biol. Sci.* 246:197–203

Soffe SR. 1993. Two distinct rhythmic motor patterns are driven by common premotor and motor neurons in a simple vertebrate spinal cord. *J. Neurosci.* 13:4456–69

Soffe SR. 1996. Motor patterns for two distinct rhythmic behaviors evoked by excitatory amino acid agonists in the *Xenopus* embryo spinal cord. *J. Neurophysiol.* 75:1815–25

Soffe SR. 1997. The pattern of sensory discharge can determine the motor response in young *Xenopus* tadpoles. *J. Comp. Physiol. [A]* 180:711–15

Sparks DL, Kristan WB Jr, Shaw BK. 1997. The role of population coding in the control of movement. In *Neurons, Networks, and Motor Behavior*, ed. PS Stein, S Grillner, AI Selverston, DG Stuart, pp. 21–32. Cambridge, MA: MIT Press

Staras K, Kemenes G, Benjamin PR. 1998. Pattern-generating role for motoneurons in a rhythmically active neuronal network. *J. Neurosci.* 18:3669–88

Stein PS. 2005. Neuronal control of turtle hindlimb motor rhythms. *J. Comp. Physiol. A* 191:213–29

Strogatz SH. 2001. *Nonlinear Dynamics and Chaos: With Applications in Physics, Biology, Chemistry, and Engineering (Studies in Nonlinearity)*. Cambridge, MA: Perseus

Sutton GP, Macknin JB, Gartman SS, Sunny GP, Beer RD, et al. 2004a. Passive hinge forces in the feeding apparatus of *Aplysia* aid retraction during biting but not during swallowing. *J. Comp. Physiol. A* 190:501–14

Sutton GP, Mangan EV, Neustadter DM, Beer RD, Crago PE, Chiel HJ. 2004b. Neural control exploits changing mechanical advantage and context dependence to generate different feeding responses in *Aplysia*. *Biol. Cybern.* 91:333–45

Swensen AM, Marder E. 2000. Multiple peptides converge to activate the same voltage-dependent current in a central pattern-generating circuit. *J. Neurosci.* 20:6752–59

Thuma JB, Morris LG, Weaver AL, Hooper SL. 2003. Lobster (*Panulirus interruptus*) pyloric muscles express the motor patterns of three neural networks, only one of which innervates the muscles. *J. Neurosci.* 23:8911–20

Tresch MC, Saltiel P, Bizzi E. 1999. The construction of movement by the spinal cord. *Nat. Neurosci.* 2:162–67

Tresch MC, Saltiel P, d'Avella A, Bizzi E. 2002. Coordination and localization in spinal motor systems. *Brain Res. Brain Res. Rev.* 40:66–79

Vilim FS, Cropper EC, Price DA, Kupfermann I, Weiss KR. 2000. Peptide cotransmitter release from motorneuron B16 in *Aplysia californica*: costorage, corelease, and functional implications. *J. Neurosci.* 20:2036–42

Wang XJ, Rinzel J. 1992. Alternating and synchronous rhythms in reciprocally inhibitory model neurons. *Neural Comput.* 4:84–97

Weimann JM, Marder E. 1994. Switching neurons are integral members of multiple oscillatory networks. *Curr. Biol.* 4:896–902

Weimann JM, Meyrand P, Marder E. 1991. Neurons that form multiple pattern generators: identification and multiple activity patterns of gastric/pyloric neurons in the crab stomatogastric system. *J. Neurophysiol.* 65:111–22

Weimann JM, Skiebe P, Heinzel HG, Soto C, Kopell N, et al. 1997. Modulation of oscillator interactions in the crab stomatogastric ganglion by crustacean cardioactive peptide. *J. Neurosci.* 17:1748–60

Whelan P, Bonnot A, O'Donovan MJ. 2000. Properties of rhythmic activity generated by the isolated spinal cord of the neonatal mouse. *J. Neurophysiol.* 84:2821–33

Willard AL. 1981. Effects of serotonin on the generation of the motor program for swimming by the medicinal leech. *J. Neurosci.* 1:936–44

Wood DE, Nusbaum MP. 2002. Extracellular peptidase activity tunes motor pattern modulation. *J. Neurosci.* 22:4185–95

Wood DE, Stein W, Nusbaum MP. 2000. Projection neurons with shared cotransmitters elicit different motor patterns from the same neural circuit. *J. Neurosci.* 20:8943–53

Wu JY, Cohen LB, Falk CX. 1994. Neuronal activity during different behaviors in *Aplysia*: a distributed organization? *Science* 263:820–23

Xin Y, Weiss KR, Kupfermann I. 1996. An identified interneuron contributes to aspects of six different behaviors in *Aplysia*. *J. Neurosci.* 16:5266–79

Xin Y, Weiss KR, Kupfermann I. 2000. Multifunctional neuron CC6 in *Aplysia* exerts actions opposite to those of multifunctional neuron CC5. *J. Neurophysiol.* 83:2473–81

Yang JF, Lamont EV, Pang MY. 2005. Split-belt treadmill stepping in infants suggests autonomous pattern generators for the left and right leg in humans. *J. Neurosci.* 25:6869–76

Ye H, Morton DW, Chiel HJ. 2006a. Neuromechanics of coordination during swallowing in *Aplysia californica*. *J. Neurosci.* 26:1470–85

Ye H, Morton DW, Chiel HJ. 2006b. Neuromechanics of multifunctionality during rejection in *Aplysia californica*. *J. Neurosci.* 26:10743–55

Yuste R, MacLean JN, Smith J, Lansner A. 2005. The cortex as a central pattern generator. *Nat. Rev. Neurosci.* 6:477–83

Zhang F, Aravanis AM, Adamantidis A, de Lecea L, Deisseroth K. 2007. Circuit-breakers: optical technologies for probing neural signals and systems. *Nat. Rev. Neurosci.* 8:577–81

Zochowski M, Cohen LB, Fuhrmann G, Kleinfeld D. 2000. Distributed and partially separate pools of neurons are correlated with two different components of the gill-withdrawal reflex in *Aplysia*. *J. Neurosci.* 20:8485–92

Retinal Axon Growth at the Optic Chiasm: To Cross or Not to Cross

Timothy J. Petros, Alexandra Rebsam, and Carol A. Mason

Department of Pathology and Cell Biology, Department of Neuroscience, Columbia University, College of Physicians and Surgeons, New York, New York 10032; email: tjp2001@columbia.edu, adr2111@columbia.edu, cam4@columbia.edu

Annu. Rev. Neurosci. 2008. 31:295–315

First published online as a Review in Advance on April 2, 2008

The *Annual Review of Neuroscience* is online at neuro.annualreviews.org

This article's doi: 10.1146/annurev.neuro.31.060407.125609

Key Words

axon guidance, binocular vision, growth cone, Ephs/ephrins, patterning, transcription factors

Abstract

At the optic chiasm, retinal ganglion cell axons from each eye converge and segregate into crossed and uncrossed projections, a pattern critical for binocular vision. Here, we review recent findings on optic chiasm development, highlighting the specific transcription factors and guidance cues that implement retinal axon divergence into crossed and uncrossed pathways. Although mechanisms underlying the formation of the uncrossed projection have been identified, the means by which retinal axons are guided across the midline are still unclear. In addition to directives provided by transcription factors and receptors in the retina, gene expression in the ventral diencephalon influences chiasm formation. Throughout this review, we compare guidance mechanisms at the optic chiasm with those in other midline models and highlight unanswered questions both for retinal axon growth and axon guidance in general.

Contents

Optic chiasm:
X-shaped commissure on ventral brain surface where fibers from both eyes converge and decussate before projecting to higher visual targets

RGC: retinal ganglion cell

VT: ventrotemporal

Line of decussation: invisible demarcation in the retina of binocular species that separates ipsilaterally and contralaterally projecting retinal ganglion cells

INTRODUCTION

During development, growth cones must navigate through diverse cellular environments and decision regions, simultaneously integrate multiple cues along their pathway, then identify target regions and form synapses with appropriate target cells. Despite significant progress over the past several decades, axon guidance remains one of the most complex problems in developmental neuroscience (reviewed in Yu & Bargmann 2001). Axon pathfinding at the neuraxis midline constitutes an ideal model for studying growth cone behavior in response to guidance cues because a growth cone's decision to cross or not to cross the midline is crucial for establishing proper neuronal connectivity and circuitry of both sensory and motor pathways.

Divergence of retinal axons at the optic chiasm midline implements binocular vision—stereopsis and depth perception—in higher vertebrate species. To establish binocular pathways, information from the visual field of each retina must be transmitted to centers in the thalamus and the cerebral cortex on both sides of the brain. In binocular species, axons from the nasal retina of each eye project contralaterally, whereas axons from temporal retina project ipsilaterally. Animals with eyes near the front of their head have a higher degree of binocular overlap (thus binocular vision) and a higher percentage of ipsilaterally projecting retinal ganglion cells (RGCs) compared with animals whose eyes are displaced more laterally. Humans are highly binocular, with ~40% uncrossed RGC axons. Ferrets have ~15% uncrossed fibers, whereas mice are a poor binocular species with ~3%–5% uncrossed axons, arising exclusively from the most peripheral ventrotemporal (VT) crescent. In most fish and birds, the lateral location of the eyes does not provide for overlap in visual space, and thus the projections from both retinae are entirely crossed (**Figure 1**).

A distinctive aspect of the optic chiasm is that this pathway consists of axons from only one type of sensory neuron, RGCs, whose functions and targets are known. Within the retina, the uncrossed and crossed RGC populations are not intermingled and are separated by a rather sharp line of decussation. These aspects, combined with the accessibility of the retina and optic chiasm for both in vitro assays and in vivo analysis, make the retina-to-optic chiasm pathway a compelling model for analyzing the cellular and molecular mechanisms that guide axons at decision regions.

Here we discuss the development of retinal axon projections as the optic chiasm forms, highlighting older work on the spatiotemporal aspects of retinal axon growth and growth cone dynamics during divergence at the optic chiasm midline. We review guidance mechanisms for the uncrossed retinal projection during repulsion at the chiasm midline and candidate systems that might facilitate traversing

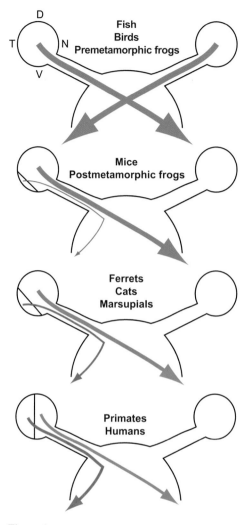

Figure 1

Retinal axon divergence at the optic chiasm of different vertebrate species. Depiction of the origins and relative proportions (but not the precise fiber trajectory) of the crossed and uncrossed retinal projections.

the midline. We then discuss transcription factors that specify RGC identity with respect to the laterality of projection and the development of the cellular terrain of the optic chiasm. Throughout this review, we discuss unresolved issues pertinent to RGC axon guidance at the optic chiasm, many of which are applicable to other midline models and axon guidance in general.

TIME COURSE OF OPTIC CHIASM FORMATION

RGC axon growth during the establishment of the optic chiasm can be divided into three stages. The earliest-born RGCs arise from the dorsocentral (DC) retina and enter the ventral diencephalon at E12–13.5, where they form both a crossed and an uncrossed projection (Colello & Guillery 1990, Godement et al. 1987b, Marcus & Mason 1995) (**Figure 2a**). Rather than orienting toward the midline, the uncrossed DC fibers extend directly into the ipsilateral optic tract. This early uncrossed projection is transient, but the fate of the uncrossed DC cells is unknown. These initial projections are considered pioneer axons because they appear to demarcate the future site of the chiasm and provide scaffolding for later-born axons, as observed in the zebrafish postoptic commissure (Bak & Fraser 2003). How the pioneer axons from DC retina penetrate and traverse the cellular terrain before the chiasmatic path is established remains unclear (Trousse et al. 2001).

During the peak phase of axon growth in mice from E14 to E17, axons from VT retina approach the chiasm midline and turn back to the ipsilateral optic tract, while axons from all other retinal regions (non-VT retina) traverse the chiasm and project into the contralateral optic tract (Guillery et al. 1995) (**Figure 2b**).

Crossed RGC projections arise from the expanding peripheral retina until birth. However, during this late phase of RGC axon extension (E17.5–P0), most newborn RGCs in the VT crescent project contralaterally rather than ipsilaterally (Drager 1985) (**Figure 2c**).

RETINAL AXON INTERACTIONS AND CELLULAR COMPONENTS AT THE OPTIC CHIASM

The insight that midline cues direct retinal axon divergence during optic chiasm formation came from observing the shape and trajectory of growth cones labeled with DiI during the peak growth phase. Whereas axons

DC: dorsocentral

Ventral diencephalon: ventral region of forebrain that includes the hypothalamus, where the optic chiasm forms

Ipsilateral: projecting to the same side of the midline

Pioneer axons: earliest-growing axons that navigate without the aid of preexisting tracts

Contralateral: projecting to the opposite side of the midline

DiI: 1,1′-dioactadecyl-3,3,3′,3′-tetramethylindocarbocyanine perchlorate

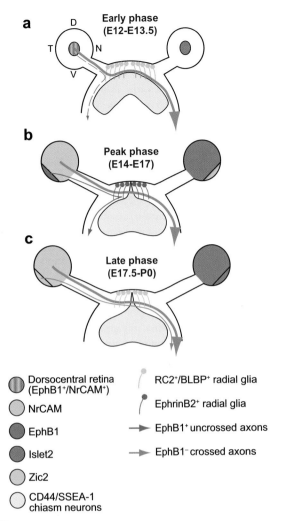

Dorsocentral retina
(EphB1⁺/NrCAM⁺)

NrCAM

EphB1

Islet2

Zic2

CD44/SSEA-1
chiasm neurons

RC2⁺/BLBP⁺ radial glia

EphrinB2⁺ radial glia

→ EphB1⁺ uncrossed axons

→ EphB1⁻ crossed axons

Figure 2

Phases of retinal axon extension during optic chiasm formation. View of the optic chiasm in approximate horizontal plane. Guidance molecules are indicated in the left retina; transcription factors are indicated in the right retina. (*a*) During the early phase (E12–E13.5), retinal fibers originating from the DC retina express *EphB1*, *NrCAM*, and *Islet2* (not *Zic2*) and project to both sides of the brain. Radial glial cells straddle the midline and express RC2 and BLBP. Note that crossing axons traverse the midline glia zone, but the transient uncrossed axons (*dotted line*) do not enter the glial palisade and instead turn directly into the ipsilateral optic tract. Both crossed and uncrossed axons follow the border of the CD44/SSEA-1 neurons. (*b*) During the peak phase (E14–E17), the radial glia cell palisade is more restricted than in earlier ages. Whereas ephrinB2 is weakly expressed in radial glia cells during the early and late phases, it is strongly upregulated in this peak phase. *Islet2* and *NrCAM* are expressed in non-VT (crossed) RGCs, whereas *Zic2* and *EphB1* are expressed in VT (uncrossed) RGCs. At this age, VT axons extend close to the midline before turning ipsilaterally. (*c*) During the late phase (E17.5–P0), *EphB1* and *Zic2* are downregulated in VT retina, and *NrCAM* and *Islet2* expression expands into the VT retina. Note that most late-born RGCs from VT retina project contralaterally.

arising from non-VT retina traverse the midline, uncrossed VT axons extend within several hundred microns of the chiasm midline and then turn sharply toward the ipsilateral optic tract (Godement et al. 1990). Video microscopy of DiI-labeled growth cones revealed the striking behaviors of ipsilaterally- and contralaterally-projecting retinal growth cones at the chiasm midline (Godement et al. 1994, Mason & Wang 1997, Sretavan & Reichardt 1993). All growth cones undergo saltatory growth in the optic nerve but extend rapidly when they advance. Upon entering the midline zone, RGC growth cones undergo cycles of spreading, pausing, and retraction, often lasting for several hours (Godement et al. 1994, Mason & Wang 1997) (**Figure 3**). While growth cones from non-VT retina quickly traverse the midline, the large spread growth cones of VT axons emit filopodia that resemble growth cone protrusions at the border of non-permissive substrates in vitro (Godement et al. 1990, Mason & Erskine 2000). Eventually, a backward-directed filopodium is consolidated, the main growth cone is progressively pruned, and the axon extends in the direction of the consolidated filopodium toward the ipsilateral optic tract (Godement et al. 1994) (**Figure 3**).

In the early 1990s, the advent of cell-specific markers made it possible to visualize and identify glial and neuronal cells at the chiasm midline. Radial glia cells at the base of the third ventricle extend processes that drape the midline of the chiasm (Marcus et al. 1995). Like radial glia in other regions of the immature brain (Lindwall et al. 2007), the chiasmatic midline radial glia express glial markers such as RC2, BLBP, and GLAST (but not GFAP) during the period of RGC axon growth, from E12 to P0 (Williams et al. 2003). In addition to these glial cells, a population of early-born neurons develops caudal to the chiasm and extends a raphe into the midline. The early-born neurons express epitopes such as SSEA-1, expressed by stem cells (Capela & Temple 2002), and CD44, expressed by cells of the immune system (Marcus & Mason 1995, Sretavan et al. 1994).

These immature glia and neurons are analogous to midline cells in the insect midline and vertebrate spinal cord floor plate (reviewed in Edenfeld et al. 2005, Mason & Sretavan 1997).

Colocalization of chiasm cells with DiI-labeled RGC axons revealed that both crossed and uncrossed RGC growth cones enter the glial palisade and intimately intertwine with glial processes (Marcus et al. 1995) (**Figure 3**). Uncrossed axons turn back at the outer edge of the palisade, and crossed axons traverse the palisade and midline raphe of the SSEA-1/CD44 neurons; both axon groups then extend caudally along the border of the chiasm neurons (Marcus et al. 1995, Mason & Sretavan 1997) (**Figure 2**). RGC axons cross the midline through the glial palisade at more dorsal levels, whereas the uncrossed fibers turn more ventrally (Colello & Coleman 1997, K.Y. Chung & C. Mason, unpublished observations). The interaction of axons with midline cells is thought to be important for crossing in other midline models, including forebrain commissures (Lindwall et al. 2007), but the precise cellular and molecular basis for this neuron-glial interaction has not been established.

In vitro studies in which retina explants were cocultured with cells from the chiasm midline indicated that the chiasmatic neurons and glia provide cues for axon divergence. When cocultured with chiasm explants in collagen gels, all retinal neurites display a reduction in outgrowth (Wang et al. 1996), indicating that chiasm cells express diffusible cues that are inhibitory to retinal axons. In addition, upon contacting chiasm cells, VT axons are repulsed while dorsotemporal (DT) axons extend relatively uninhibited, indicating that cells from the chiasm midline have differential contact-dependent effects on RGCs from specific retinal regions (Wang et al. 1995), modeling the divergent growth patterns in vivo. These results set the stage for investigating which family of guidance cues could underlie this differential response of crossed and uncrossed axons.

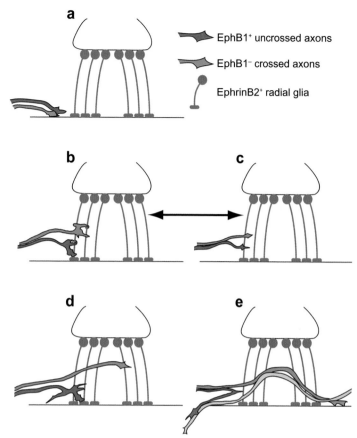

EphB1⁺ uncrossed axons

EphB1⁻ crossed axons

EphrinB2⁺ radial glia

Figure 3

Retinal growth cone behavior at the optic chiasm midline. Frontal view of the optic chiasm during the peak phase (**Figure 2b**). (*a*) After exiting the optic stalk, uncrossed and crossed retinal axons approach the radial glia palisade near the pial surface of the chiasm. (*b*) Both types of retinal axons interact with the midline glial processes and develop complex growth cones that wrap around midline glia. (*c*) All retinal axons withdraw their growth cone, then repeatedly extend, expand, and retract before advancing (*black double-headed arrow*). (*d*) Uncrossed growth cones spread extensively and emit protrusions in different axes. Crossed axons traverse the midline more dorsally within the radial glia palisade to the contralateral side. (*e*) Uncrossed axons consolidate a backward-oriented filopodium and the axon advances into the ipsilateral optic tract, possibly fasciculating with crossed axons from the contralateral retina (*light green*). Crossed axons descend toward the pia and enter the contralateral optic tract.

GUIDANCE FACTORS IMPORTANT FOR RETINAL AXON DIVERGENCE

The Usual Suspects

Soon after the behaviors of RGC growth cones and the character of the midline cells were

Filopodia: small, finger-like protrusions that extend from growth cones and probe the environment

Floor plate: an array of neuroepithelial cells at the ventral midline of the developing spinal cord; site through which commissural axons project

Fasciculation: axons contact each other and extend together in bundles

uncovered, the field witnessed the rapid identification of guidance factor families. Many of these cues were implicated in other midline-crossing models and were thus good candidates for retinal axon divergence at the chiasm midline (reviewed in Williams et al. 2004). Netrin-1 is a diffusible protein secreted by midline cells and acts as an attractant for axons expressing the deleted in colorectal cancer (DCC) receptor in the *Drosophila* ventral nerve cord and floor plate of the vertebrate spinal cord (reviewed in Kaprielian et al. 2001). However, *netrin-1* is not expressed at the optic chiasm, but is found at the optic disc where it is important for retinal fiber exit from the eye (Deiner et al. 1997). Although retinal axons in *netrin-1* and *DCC* mutant mice enter the chiasm at an altered angle, RGC divergence is not affected (Deiner & Sretavan 1999). Notably, although midline cells in other systems express an attractant cue, an analogous signal has not been identified at the optic chiasm.

Morphogens are now recognized to direct axon navigation (reviewed in Sanchez-Camacho et al. 2005). Sonic hedgehog (Shh) is expressed by floor plate cells where it induces cell differentiation in the vertebrate spinal cord (reviewed in Jessell 2000). Shh later plays a dual role in the floor plate, first as a chemoattractant for commissural axons toward the floor plate (Charron et al. 2003) and then as a repulsive cue for postcrossing axon projections into longitudinal tracts (Bourikas et al. 2005). Shh is expressed at the site of the future optic chiasm and appears to define the site of the chiasm (Torres et al. 1996), either by acting as a repulsive cue to retinal fibers (Trousse et al. 2001) and/or by regulating glial cell development and expression of guidance cues at the chiasm midline (Barresi et al. 2005). However, Shh appears to have no effect on axon divergence.

An obvious candidate for guiding RGC divergence is the Slit family of guidance cues and their receptors, Robos. Slits and Robos regulate the pathfinding of commissural axons across the midline in both the invertebrate ventral nerve cord and vertebrate spinal cord, where Slits act as repulsive guidance cues to Robo-expressing axons (reviewed in Dickson & Gilestro 2006). *Robo2* is expressed in RGCs, and *Slit1* and *Slit2* are present in the ventral diencephalon (Erskine et al. 2000). Whereas *Slit1* or *Slit2* single mutant mice show normal axon pathfinding at the optic chiasm, *Slit1/Slit2* double knockout mice display ectopic chiasm formation rostral to the normal chiasm, and retinal axons are misguided into the contralateral optic nerve and ventral diencephalon (Plump et al. 2002). However, even with these aberrations, a normal ipsilateral projection forms in these mice.

Similarly, in the zebrafish *robo2* mutant *astray*, RGC axons that are normally completely crossed display severe guidance errors at or after the midline. *astray* mutants also show ectopic chiasm formation and defasciculated axon growth (Fricke et al. 2001, Hutson & Chien 2002, Karlstrom et al. 1996), resembling the chiasm phenotype in *Slit1/Slit2* double knockout mice (Plump et al. 2002). Thus, Slit-Robo interactions channel RGC axons into the proper path and may regulate fasciculation rather than direct retinal axon divergence.

The Ephs and ephrins, particularly in the A family, mediate retinotectal topographic projections (reviewed in Lemke & Reber 2005, McLaughlin et al. 2003). Multiple EphAs and ephrinAs are expressed in gradients in the retina and optic tectum, and they are also expressed in chiasm midline cells. Whereas perturbation of EphA signaling in vitro blocks the normal inhibition caused by chiasm cells (Marcus et al. 2000), VT and non-VT axons show little difference in their response to ephrinAs. In chick, ephrinA2 or ephrinA5 overexpression increases the transitory ipsilateral projection (Dutting et al. 1999), but abnormalities in chiasm divergence have not been described in EphA or ephrinA mutant mice. So far, no experiments indicate that EphA-ephrinA signaling plays a significant role in retinal axon divergence at the chiasm midline.

The Uncrossed Retinal Projection

Studies on *Xenopus laevis* yielded the first hints of which molecular cues mediate retinal axon divergence at the optic chiasm. In *Xenopus* tadpoles, all retinal axons cross the midline, but during metamorphosis a subpopulation of RGCs from VT retina projects ipsilaterally, coinciding with the expression of ephrinBs at the optic chiasm (Nakagawa et al. 2000). Premature ectopic expression of ephrinB2 in the chiasm leads to a precocious ipsilateral projection, pointing to ephrinB as a cue for axon avoidance at the chiasm midline. Furthermore, EphB2 receptors are expressed in the ventral retina during metamorphosis. Because ephrinBs are not expressed at the optic chiasm in animals lacking an ipsilateral projection such as chick and fish, but are present in mice (Nakagawa et al. 2000), we asked whether EphB/ephrinB interactions might be important for retinal divergence in mice.

In situ hybridization experiments confirmed that ephrinB2 is indeed expressed at the mouse optic chiasm midline, specifically in the midline radial glia and most intensely during the peak growth phase of the ipsilateral projection (Williams et al. 2003) (**Figure 2b**). Moreover, ephrinB2 is selectively inhibitory to VT retinal axons in vitro, and blocking ephrinB2 in semi-intact chiasm preparations reduces the ipsilateral projection. Among the several EphB receptors that interact with ephrinB2, only EphB1 is expressed early in DC retina and then exclusively in VT retina at the time of peak midline divergence (**Figure 2b**). After E17.5, EphB1 expression is downregulated in VT RGCs, concurrent with the development of the late-born crossed projection from VT retina (**Figure 2c**). *EphB1*$^{-/-}$ mice show a decreased ipsilateral projection, with VT RGC axons ectopically crossing the midline, indicating that VT axons in *EphB1*$^{-/-}$ mice are not repelled by ephrinB2-expressing radial glial cells at the optic chiasm midline (Williams et al. 2003) (**Figure 4**).

Thus, EphB1 in the VT retina and ephrinB2 at the chiasm midline have emerged as the receptor-ligand system unique to the uncrossed RGC projection. However, several important questions remain. One curious finding is that EphB2 is highly expressed in the ventral retina

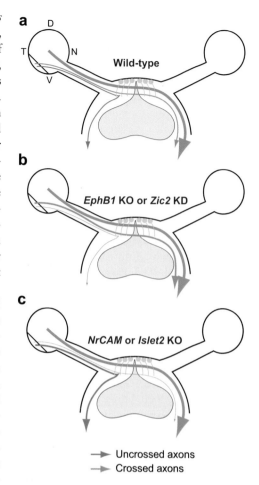

Figure 4

Abnormal divergence at the optic chiasm. (*a*) In wild-type mice, 3%–5% of retinal fibers project ipsilaterally, with uncrossed fibers arising from the VT retina during the peak phase and crossed fibers originating from the VT retina during the late phase. (*b*) In *EphB1* knockout (KO) and *Zic2* knockdown (KD) mice, the ipsilateral projection is reduced, with most VT retinal axons now projecting contralaterally. (*c*) In *NrCAM* KO and *Islet2* KO mice, the ipsilateral projection is increased because most late-born RGCs in VT retina project ipsilaterally rather than contralaterally. In all these mutants, note that the contralateral projection from non-VT retina is unchanged.

in both mice and *Xenopus* (Williams et al. 2003, Nakagawa et al. 2000). Even though EphB2 and EphB1 have a similar affinity to ephrinB2 (Flanagan & Vanderhaeghen 1998), EphB2 does not appear to play a role in directing the uncrossed projection (Williams et al. 2003). It will be interesting to determine why EphB1 is unique in its ability to induce repulsion upon contacting ephrinB2 at the chiasm midline.

An intriguing paradox of EphB1-ephrinB2 binding at the chiasm, as with all Eph-ephrin interactions, is that the receptor-ligand binding must be terminated for the growth cone to turn away from the midline and project ipsilaterally. Cleavage of the receptor-ligand complex is one mechanism to disengage the growth cone from the inhibitory cue. In *Xenopus*, matrix metalloproteinases are required for proper axon guidance at the optic chiasm (Hehr et al. 2005). Furthermore, both ephrin ligands (Hattori et al. 2000, Janes et al. 2005, Pascall & Brown 2004) and Eph receptors (Litterst et al. 2007) can be cleaved. Additionally, EphB/ephrinB complexes can be endocytosed bidirectionally by the receptor- or ligand-expressing cell (Mann et al. 2003, Marston et al. 2003, Zimmer et al. 2003), leading to cell detachment and axon retraction. EphrinAs and Sema3s can induce endocytosis (Fournier et al. 2000, Jurney et al. 2002), indicating that endocytosis may be a general mechanism for guidance cue regulation.

In addition to how an Eph/ephrin interaction is terminated, recent evidence indicates that ephrins can regulate Eph receptor signaling in *cis* (on the same cell) as well as in *trans* (Carvalho et al. 2006, Marquardt et al. 2005). Because RGCs express many Eph receptors and ephrins, *cis* interactions comprise a possible mechanism for adjusting the responsiveness of growth cones to various ephrins in the projection pathway. Along with *cis* interactions between members of the same guidance family, there can be receptor crosstalk between different guidance families. An example of this receptor crosstalk also comes from the vertebrate spinal cord midline, where Robo receptors are thought to silence the receptor DCC to netrin-1, such that axons that have crossed the midline are no longer attracted to netrin-1 (Stein & Tessier-Lavigne 2001).

Along with these forms of Eph receptor regulation, mRNA levels may be regulated by local translation of proteins necessary for guidance at the growth cone. There is ample evidence that mammalian axons contain mRNA and translational machinery (reviewed in Koenig & Giuditta 1999), and recent studies indicate that inhibiting local protein translation can alter growth cone response to guidance cues (Campbell & Holt 2001, Ming et al. 2002), likely by regulating the translation of downstream effectors such as β-actin or RhoA (Leung et al. 2006, Wu et al. 2005, Yao et al. 2006). However, only one study has shown that guidance receptors might be locally translated in the growth cone; EphA2 may undergo local translation in chick commissural neurons, but only after they cross the spinal cord midline (Brittis et al. 2002). The ability of growth cones to translate proteins locally on demand is a mechanism that allows for rapid response to guidance cues and represents a major advance in the field of axon guidance (reviewed in Lin & Holt 2007).

One important goal is to determine if and how these cellular mechanisms (cleavage and/or endocytosis, *cis-trans* Eph-ephrin interactions, and local translation) function in the context of the optic chiasm. Of note, EphB1 is expressed in the temporal retina and ephrinB2 at the optic chiasm in human fetal brain (Lambot et al. 2005), which suggests that EphB1-ephrinB2 may be an evolutionarily conserved mechanism for RGC guidance at the optic chiasm. Thus, unraveling how EphB1-ephrinB2 interactions lead to formation of the ipsilateral projection should illuminate how the binocular projection is patterned in higher vertebrates.

The Crossed Retinal Projection

The guidance factors described above are related to the uncrossed retinal projection. Which mechanisms underlie retinal axon crossing of the chiasm midline? One hypothesis proposes that traversing the chiasm occurs by default: Retinal axons project straight through the midline zone into the contralateral optic tract because they lack receptors to inhibitory cues. However, the suppression of neurite extension by chiasm cells (Wang et al. 1995, 1996) and the pausing and extension/retraction behaviors of all precrossing axons (Godement et al. 1994, Mason & Wang 1997) argue against this hypothesis.

To identify cues that mediate crossing, we focused on the immunoglobulin (Ig) superfamily of cell adhesion molecules (CAMs) because NrCAM plays a role in commissural axon midline crossing in chick spinal cord (Lustig et al. 1999, Stoeckli et al. 1997). Since NrCAM is expressed at the chiasm midline (Lustig et al. 2001), and CAMs can interact homotypically, we examined NrCAM expression in the retina. NrCAM is expressed in all retinal regions except for the VT crescent from E12 to E17 (**Figure 2a,b**). NrCAM expression then expands into the VT region from E17 to birth, corresponding to the time when crossing axons arise from VT retina (Williams et al. 2006) (**Figure 2c**). Thus, NrCAM is a strong candidate for mediating midline crossing. Indeed, blocking NrCAM function results in the decreased ability of fibers to cross the chiasm, and $NrCAM^{-/-}$ mice have an enhanced uncrossed RGC projection (**Figure 4**). Surprisingly, this increase arises strictly from the late-born VT RGCs. In contrast, the crossed projections from non-VT retina are unaltered despite their NrCAM expression profile (Williams et al. 2006).

The curious restriction in NrCAM function to the late-born crossed VT RGCs could be explained by other CAMs acting redundantly or in concert with NrCAM. Members of the CAM family such as *L1*, *neurofascin*, and *TAG-1* are strongly expressed in non-VT retina during the peak phase of the crossed RGC projection. During the late phase, *neurofascin* and *TAG-1* are upregulated in VT RGCs, similar to *NrCAM*. The exception is that *TAG-1* is strongly downregulated in non-VT retina at late stages (Williams et al. 2006). Thus, the L1

family of CAMs is expressed in all RGCs that cross the midline. While other CAMs may compensate for NrCAM function in non-VT retina, they cannot solely explain the restricted function of NrCAM to the late-born crossed VT projection because *neurofascin* and *TAG-1* are also upregulated in VT during the late phase. There are likely other as-yet-unidentified guidance factors or signaling molecules that make late-born VT RGCs distinct from crossed projection arising from non-VT retina.

Recently, the Semaphorins have been identified as playing a role in the formation of the crossed projection. In zebrafish, Sema3d is expressed around the chiasm in a pattern similar to Slit1 and Slit2 in mice, and perturbing Sema3d by overexpression or knockdown impairs crossing (Sakai & Halloran 2006). In this system, Sema3d is thought to act as a repulsive cue that guides retinal axons out of the chiasm into the contralateral optic tract. In mice, Sema5A channels RGC axons through the optic nerve (Oster et al. 2003), but the expression and function of other Semaphorins have gone unstudied. Of note, interactions between the L1 family of Ig CAMs and the semaphorin receptors, plexins and neuropilins, can change Semaphorin-induced repulsion to attraction (reviewed in Bechara et al. 2007). This interaction is critical for axon guidance in other commissural pathways (Falk et al. 2005). A role for semaphorins and their receptors, including potential interactions with CAMs, remains to be established in the mouse optic chiasm.

AXON ORGANIZATION AND FASCICULATION

A poorly understood aspect of RGC axon extension to central targets is how uncrossed and crossed fibers are organized when they enter, course within, and exit the optic chiasm. In eutherian species, RGC axons are organized in a grossly topographic manner as they exit the retina, but this topographic organization is lost as they approach the chiasm (reviewed in Guillery et al. 1995, Jeffery 2001). At the optic nerve–chiasm junction, RGC axon bundles are

Ig: immunoglobulin

CAM: cell adhesion molecule

NrCAM: Ng(neuron-glia)-CAM related cell adhesion molecule

no longer surrounded by glia and they defasciculate. As they progress through the chiasm, uncrossed axons separate from crossing axons, yet each component enters the optic tract again with crude retinotopic and age-related order (Jeffery 2001, Plas et al. 2005). This plan differs from that in marsupials, where the retinotopy and fasciculated bundles in the optic nerve are maintained as axons traverse the chiasm (Dunlop et al. 2000) and uncrossed fibers project directly into the optic tract, similar to the course of the first uncrossed RGC axons from DC retina in mice (reviewed in Guillery et al. 1995).

Which cellular and molecular mechanisms underlie this reorganization of fibers, and do such rearrangements play a role in axon guidance? For example, at the insect midline, commissural axon growth cones fasciculate with their contralateral homolog as they cross the midline, a rearrangement thought to ensure proper pathway formation (Myers & Bastiani 1993). Whether a similar interaction occurs in the vertebrate optic chiasm is not known. Another question is whether uncrossed axons fasciculate with new partners after executing the turn away from the midline back toward the ipsilateral optic tract. Insight into this issue comes from monocular enucleation experiments in eutherian species, in which a decrease in the uncrossed pathway is observed. One hypothesis for this decrease is that the uncrossed axons from one eye need to fasciculate with postcrossing axons from the other eye to project correctly into the ipsilateral optic tract (Chan et al. 1999, Godement et al. 1987a).

Genetic and cellular analyses of the fasciclins in the insect nervous system illustrate the importance of CAMs in axonal pathway formation (reviewed in Goodman 1996, Van Vactor 1998), but the molecular mechanisms directing axon reorganization (e.g., fasciculation and defasciculation) in commissural pathways are not yet understood. In the chick optic tract, enzymatic removal of polysialic acid (PSA) from neural CAM (NCAM) produces significant defasciculation of RGC axons (Yin et al. 1995). Changes in CAM expression on crossing axons within

the spinal cord midline were identified 20 years ago, with commissural neurons downregulating TAG-1 and upregulating L1 after crossing the midline (Dodd et al. 1988), although the function of these changes remains unknown. In the visual system, PSA-NCAM is similarly downregulated as axons enter the optic chiasm and upregulated in a subset of axons upon entering the optic tract (Chung et al. 2004). One function of these transitions may be to implement defasciculation, resorting, and subsequent refasciculation of axons at sites of decussation, potentially in collaboration with other guidance factors. For example, Sema3D can promote fasciculation by modulating L1 CAM receptor levels (Wolman et al. 2007). However, many CAM mutants do not display severe defects in fasciculation (Demyanenko & Maness 2003, Van Vactor 1998). Overlapping expression patterns and possible redundant function could explain this phenotype but also complicate unraveling the effects of CAMs on axon reorganization.

Other possible modifiers of guidance factors such as chondroitin sulfate (CSPGs) and heparan sulfate proteoglycans (HSPGs) are expressed in cells adjacent to the optic nerve–chiasm and chiasm–tract junctions, loci for axon fasciculation and defasciculation (Leung et al. 2003, Pratt et al. 2006, Reese et al. 1997). In support of this notion, heparan sulfate biosynthesis is perturbed in zebrafish mutants (*dak* and *box*) that display axon missorting in the optic tract (Lee et al. 2004). The well-documented interaction between CSPGs and CAMs (reviewed in Grumet et al. 1996) suggests a role for these families of molecules in axon rearrangement at the optic chiasm that may be independent of receptor-ligand signaling that mediates RGC divergence.

TRANSCRIPTION FACTORS THAT PATTERN THE RETINAL AXON PROJECTION

The link between transcription factor designation of neuron subpopulations and axon trajectory, through control of guidance cue expression, has been demonstrated in a number of models (reviewed in Butler & Tear 2007, Polleux et al. 2007). This has been especially well illustrated in the control of ephrins and netrins in motor neurons (Kania & Jessell 2003, Labrador et al. 2005). Here we describe regulatory genes that have been uncovered in the retina and ventral diencephalon that relate to optic chiasm formation.

Genes in the Retina

Zinc finger transcription factors (Zic1-5) are critical for early neural patterning and the midline of the body plan (reviewed in Merzdorf 2007). *Zic2* is expressed in the retina during eye cup formation (Nagai et al. 1997) and is then downregulated. Strikingly, *Zic2* is subsequently upregulated in the VT crescent between E14.5 and E17.5, precisely at the time when VT axons project ipsilaterally (**Figure 2**). Zic2 is required for formation of the uncrossed pathway, as demonstrated in *Zic2* knockdown mice in which the ipsilateral projection is significantly reduced (Herrera et al. 2003) (**Figure 4**). In addition, overexpression of *Zic2* in DT retinal explants induces these axons to be repulsed by chiasm cells, similar to the response of VT axons. *Zic2* expression in uncrossed RGCs appears to be evolutionarily conserved, as its expression in RGCs of animals such as *Xenopus* and ferrets correlates precisely with the proportion of ipsilaterally projecting RGCs (Herrera et al. 2003). The finding that Zic2 directs the uncrossed retinal projection heralded the first transcriptional program specifically directing an ipsilateral course of axon pathfinding at the midline in vertebrates. For the mouse retina, immediate questions arise: As both Zic2 and EphB1 are expressed in VT RGCs and are necessary for axon divergence at the optic chiasm, does Zic2 regulate EphB1, and are Zic2 and/or EphB1 sufficient to induce changes in the crossing behavior of non-VT RGCs? Recent findings indicate that ectopic expression of Zic2 in RGCs is sufficient to elicit avoidance of the midline, and this repulsion is largely EphB1-dependent (García-Frigola et al. 2008).

Irrespective of whether Zic2 directly or indirectly regulates EphB1 expression, these findings strengthen the notion that Zic2 and EphB1 are in the same pathway.

In contrast with Zic2 expression in VT retina, the LIM homeodomain transcription factor *Islet2* is expressed in non-VT retina and is absent from uncrossed VT RGCs from E13.5 to E15.5 (**Figure 2a,b**). However, *Islet2* is upregulated at E17.5 in late-born VT RGCs that cross the midline (Pak et al. 2004) (**Figure 2c**). Although Islet2 is not expressed in every RGC, the retinal domains containing Islet2$^+$ cells give rise exclusively to the crossed projection. Furthermore, the Islet2 expression pattern is remarkably similar to that of NrCAM, suggesting that Islet2 might regulate NrCAM, but this relationship has not been demonstrated. In *Islet2*^{tauLacZ} knockin mice, Islet2 is expressed only in contralateral projecting RGCs, and *Islet2*$^{-/-}$ mice display an increase in ipsilateral fibers. As with *NrCAM*$^{-/-}$ mice, this aberrant ipsilateral projection arises strictly from the VT retina. The increase in uncrossed axons in *Islet2*$^{-/-}$ mice is concurrent with an increase in *Zic2* and *EphB1* expression in the VT crescent (Pak et al. 2004) (**Figure 4**), but whether Islet2 suppresses Zic2 and/or EphB1 within the VT crescent remains to be established.

The differential expression of these transcription factors and guidance receptors in the retina points to one major difference in the patterning and implementation of axonal decussation at the optic chiasm midline compared with other midline models. In the retina, *Zic2* and *EphB1* are expressed solely in the uncrossed RGC population in the VT retina and are necessary for the formation of the ipsilateral pathway. A different set of genes, *Islet2* and *NrCAM*, are expressed in crossed RGC projections from non-VT and late-born VT retina. In contrast, at the vertebrate spinal cord midline or the *Drosophila* ventral nerve cord midline, both uncrossed and crossed axons express the same receptor, Robo, and the decision to cross or avoid the midline is achieved through fine regulation of this receptor at the growth cone membrane

(Dickson & Gilestro 2006). Crossing axons traverse the Slit$^+$ zone because the Commissureless protein prevents Robo trafficking to the membrane by sequestering it into endosomes (Keleman et al. 2005). In uncrossed axons, Robo expression remains elevated at all times, leading to avoidance of the midline. These findings have been extended to the vertebrate spinal cord commissural neurons, where Rig-1 causes Robo$^+$ axons to be insensitive to Slit prior to crossing the floor plate (Sabatier et al. 2004).

Which genes regulate Zic2 and Islet2? *Vax2* is expressed in a high-ventral to low-dorsal gradient (Barbieri et al. 2002, Mui et al. 2002). The effect of mutations in *Vax* genes in chiasm formation has been examined, but the expression of *Islet2* and *Zic2* in these mutants has not been reported. The ventral retina is further partitioned by *Foxd1* (previously known as *BF2*), which is confined to the VT quadrant in a zone extending more centrally than *Zic2*, whereas *Foxg1* (previously known as *BF1*) is expressed throughout the nasal retina (Herrera et al. 2004, Pratt et al. 2004). *Zic2* and *EphB1* expression are downregulated in *Foxd1* mutants, implicating *Foxd1* as a gene upstream of Zic2 (Herrera et al. 2004).

Zic2 and EphB1 comprise a guidance program for the uncrossed RGC projection from the VT retina. In mouse retina, as in other higher vertebrates, the line of decussation demarcates the sectors containing ipsilaterally and contralaterally projecting RGCs. Several questions then arise: Which molecules demarcate this distinct boundary between crossed and uncrossed RGCs? How is the VT crescent specified to have a unique spatial and temporal gene expression pattern for the uncrossed projection (*Zic2*, *EphB1*) during the peak phase and the crossed projection (*Islet2*, *NrCAM*) during the late phase? As outlined in the *Xenopus* retina, does the uniqueness of the VT region arise from the unusual embryonic derivation of these cells (Jacobson & Hirose 1978) or by differential regulation of cell division in this part of the retina at later stages (Marsh-Armstrong et al. 1999)? Is the uncrossed population of RGCs,

and thus the line of decussation, specified very early in development by cues from nonretinal tissues acting exogenously on the developing eye (Lambot et al. 2005), or is the line of decussation specified later by repulsive interactions, possibly between Zic2 and Islet2 and/or between Foxd1 and Foxg1?

Albinism is a unique condition that may provide clues to answer these questions on the development of the line of decussation. It is the only known mutation in which a decrease in the proportion of uncrossed axons is accompanied by a shift in the line of decussation toward the periphery (Guillery et al. 1995, Hoffmann et al. 2003, Jeffery & Erskine 2005). Albino mice show a reduction in the number of *Zic2*$^+$ cells, in agreement with the diminished ipsilateral projection (Herrera et al. 2003). It is unclear how perturbations in melanin biosynthesis and trafficking affect retinal gene expression and divergence at the midline, but one possibility is that factors in the melanin pathway could affect cell proliferation, perturbed in the albino (Rachel et al. 2002, Tibber et al. 2006), and in turn affect cell fate. Further studies are needed on the early stages of eye development to understand how the line of decussation is established.

Patterning of the Optic Chiasm Terrain

Previous studies have outlined regulatory gene expression in the ventral diencephalon where retinal axons converge and form the optic chiasm (Marcus et al. 1999, Wilson et al. 1993). Some genes are expressed in areas flanking the midline (e.g., *Foxd1*, *Dlx2*), whereas other genes are rostral or caudal to the chiasm with a raphe extending into the midline (e.g., *Foxg1*, *Nkx2.2*) (Herrera et al. 2004, Marcus et al. 1999) (**Figure 5a**). Strikingly, at the early stages of chiasm formation, the earliest-growing retinal axons follow the borders of these subdivisions (Marcus et al. 1999).

Many of these transcription factors are expressed in both the retina and the ventral diencephalon (e.g., *Foxd1* and *Zic2*), and mice lacking these genes can have severe

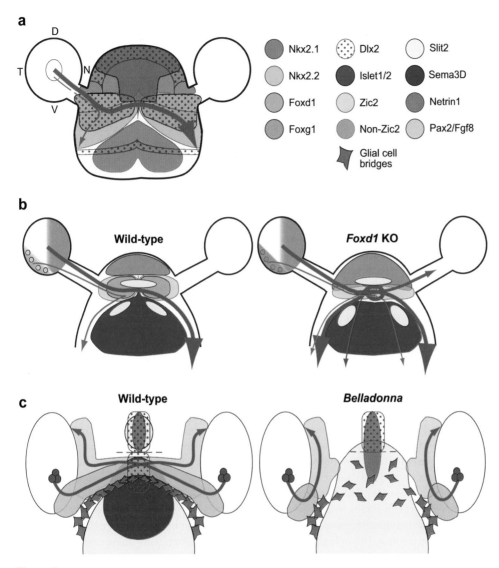

Figure 5

Patterning of the optic chiasm terrain. (*a*) At early stages of mouse optic chiasm formation (E11–E13), transcription factor expression delineates subregions within the ventral diencephalon. Retinal axons avoid *Nkx2.1* and *Foxg1* regions but project through the *Nkx2.2*, *Foxd1*, and *Dlx2* regions. Adapted from Marcus et al. 1999. (*b*) The retinal axon pathway with respect to patterns of gene expression in wild-type and *Foxd1* KO mice at E15.5. In wild-type mice, *Foxg1* is expressed in the nasal retina, whereas *Foxd1* is expressed in the VT quadrant, which includes the smaller *Zic2*+ zone (VT crescent). In *Foxd1* KO mice, *Foxg1* expression expands into the area normally occupied by *Foxd1* in both the retina and the ventral diencephalon, and *Zic2* is no longer expressed in the VT retina and is diminished near the chiasm. *Foxd1* KO mice display numerous projection errors at the optic chiasm due to the abnormal patterning of the ventral diencephalon, with an increased ipsilateral projection arising from all retinal regions. Adapted from Herrera et al. 2004. (*c*) In wild-type zebrafish, all retinal axons are crossed, extending through *Pax2*-, *fgf8*-, and *Zic2.1*-expressing regions to reach the contralateral optic tectum. *Belladonna* mutants are achiasmatic with an entirely uncrossed projection. *Sema3D* expression is completely abolished, *Pax2/fgf8* expression is strongly diminished, and *Slit2* and *netrin1* expression expands. In addition, the glial bridge is disrupted at the midline. These perturbations of the ventral diencephalon make this terrain refractory for all retinal axons. Adapted from Seth et al. 2006.

defects both in retinal axon trajectory and in the cellular organization of the optic chiasm (Bertuzzi et al. 1999, Herrera et al. 2003, 2004, Marcus et al. 1999, Pratt et al. 2004). In the retina of $Foxd1^{-/-}$ mice, $Zic2$ and $EphB1$ are missing, and $Foxg1$ expression extends into the temporal retina (Herrera et al. 2004) (**Figure 5b**). The absence of $Zic2$ and $EphB1$ in $Foxd1^{-/-}$ mice predicts a decrease in uncrossed projections. Unexpectedly, the uncrossed retinal projection increases dramatically, and these fibers arise from all regions of the retina. Such imbalances in RGC divergence result from a distortion in the architecture and regionalization of the ventral diencephalon: $Foxg1$ expands toward the $Foxd1$ region, $Zic2$ and $Islet1$ expression is diminished, and the $Slit2$ zone is extended and enhanced, resulting in abnormalities in chiasm shape (Herrera et al. 2004) (**Figure 5b**). Thus, alterations in the ventral diencephalon override defects in the retina to produce abnormalities in chiasm formation and retinal axon projection. Similar findings are observed in the *circletail*, *loop-tail*, and *Pax3* mutants (Rachel et al. 2000).

The zebrafish mutant *Belladonna* provides an additional example of the crucial role of molecular and cellular composition of the optic chiasm terrain. Like the Belgian sheepdog (Williams et al. 1994) and $Pax2^{-/-}$ mice (Torres et al. 1996), the *Belladonna* mutant is achiasmatic, having a totally uncrossed rather than the normal completely crossed retinal projection (Karlstrom et al. 1996) (**Figure 5c**). The mutated gene is $Lhx2$, and transcription factor and guidance cue expression is perturbed in the chiasm (Seth et al. 2006). $Zic2$ and $Dlx2$ zones are diminished medially and caudally, *netrin-1* and *Slit2* expand, and *Sema3D*, implicated in axon crossing (Sakai & Halloran 2006), is missing (**Figure 5c**). In addition, the glial bridge is disrupted, likely hindering crossing of the entire retinal axon cohort. This expansion of *Slit* expression and perturbed chiasm terrain is also observed in the Gli2 zebrafish mutant *you-too*, where Shh signaling is perturbed (Barresi et al. 2005). These examples illustrate the difficulty of unraveling a mutation's effect simply by examining the phenotypes of retinal axon trajectory alone.

CONCLUSIONS AND PERSPECTIVES

The specification of retinal ganglion cells and the receptor system for the uncrossed and late-forming crossed pathways from VT retina have begun to be uncovered. The program that regulates the crossed trajectory of RGCs that reside outside of the VT crescent remains to be identified. It will be especially challenging to dissect how various guidance families implement crossing through an apparently refractory midline and to illuminate the underpinnings of fasciculation during fiber reorganization in the chiasm. The field can look forward to the discovery of yet more regulatory genes that control navigation at the midline, and more importantly, explication of precisely how transcription factors regulate guidance molecule expression. Addressing these unsolved issues of axon guidance, with the optic chiasm and other midline scenarios as models, should enlighten the next discovery period.

SUMMARY POINTS

1. Retinal axon growth to the optic chiasm can be divided into three phases, with each cohort of retinal ganglion cell projections displaying different divergence patterns at the optic chiasm.

2. The transcription factor Zic2 and the guidance receptor EphB1 are expressed in the ventrotemporal retina during the peak phase of retinal ganglion cell axon outgrowth and regulate the uncrossed projection.

3. All growth cones enter and pause within the chiasm midline region, but only EphB1-expressing axons from ventrotemporal retina are repelled by ephrinB2-expressing midline radial glial cells and turn ipsilaterally.

4. The transcription factor Islet2 and the L1-family member NrCAM are expressed in non-ventrotemporal retina during the peak phase and are upregulated in the ventrotemporal retina during the late phase of retinal ganglion cell development, but Islet2 and NrCAM are required only for the late-born crossed projection from VT retina.

5. Transcription factors such as Foxd1 and Zic2 are crucial for patterning the ventral diencephalon as well as the retina and, in turn, affect the expression of cues for divergence at the optic chiasm.

FUTURE ISSUES

1. How is the retina patterned to produce two distinct sectors containing retinal ganglion cells that project ipsilaterally and contralaterally, and how is the line of decussation established?

2. Which molecular program(s) direct retinal ganglion cell projections arising from outside the ventrotemporal retina to cross the midline?

3. Which downstream signaling cascades become activated upon EphB1-ephrinB2 interaction, especially with respect to cytoskeletal reorganization during growth cone repulsion and turning?

4. What are the cellular interactions between growth cones and midline glia that instigate receptor trafficking, local translation, and termination of the receptor-ligand interaction during midline crossing and repulsion?

5. How do transcription factors regulate guidance factor expression in the retina and ventral diencephalon?

6. How do guidance programs for crossing or avoiding the midline relate to proper innervation of target regions in the thalamus and superior colliculus/tectum, and further distally, in the cortex?

DISCLOSURE STATEMENT

The authors are not aware of any biases that might be perceived as affecting the objectivity of this review.

ACKNOWLEDGEMENTS

We thank Dr. Takeshi Sakurai and past and present members of the Mason lab for their contributions to this work and comments on this manuscript. We also apologize to our colleagues whose research we could not cite due to space limitations. Our research is supported by the National

Institutes of Health (NEI, NINDS), Fondation pour la Recherche Medicale, the International Human Frontier Science Program Organization, and the Gatsby Foundation.

LITERATURE CITED

Bak M, Fraser SE. 2003. Axon fasciculation and differences in midline kinetics between pioneer and follower axons within commissural fascicles. *Development* 130:4999–5008

Barbieri AM, Broccoli V, Bovolenta P, Alfano G, Marchitiello A, et al. 2002. Vax2 inactivation in mouse determines alteration of the eye dorsal-ventral axis, misrouting of the optic fibres and eye coloboma. *Development* 129:805–13

Barresi MJ, Hutson LD, Chien CB, Karlstrom RO. 2005. Hedgehog regulated Slit expression determines commissure and glial cell position in the zebrafish forebrain. *Development* 132:3643–56

Bechara A, Falk J, Moret F, Castellani V. 2007. Modulation of semaphorin signaling by Ig superfamily cell adhesion molecules. *Adv. Exp. Med. Biol.* 600:61–72

Bertuzzi S, Hindges R, Mui SH, O'Leary DD, Lemke G. 1999. The homeodomain protein vax1 is required for axon guidance and major tract formation in the developing forebrain. *Genes Dev.* 13:3092–105

Bourikas D, Pekarik V, Baeriswyl T, Grunditz A, Sadhu R, et al. 2005. Sonic hedgehog guides commissural axons along the longitudinal axis of the spinal cord. *Nat. Neurosci.* 8:297–304

Brittis PA, Lu Q, Flanagan JG. 2002. Axonal protein synthesis provides a mechanism for localized regulation at an intermediate target. *Cell* 110:223–35

Butler SJ, Tear G. 2007. Getting axons onto the right path: the role of transcription factors in axon guidance. *Development* 134:439–48

Campbell DS, Holt CE. 2001. Chemotropic responses of retinal growth cones mediated by rapid local protein synthesis and degradation. *Neuron* 32:1013–26

Capela A, Temple S. 2002. LeX/ssea-1 is expressed by adult mouse CNS stem cells, identifying them as nonependymal. *Neuron* 35:865–75

Carvalho RF, Beutler M, Marler KJ, Knoll B, Becker-Barroso E, et al. 2006. Silencing of EphA3 through a cis interaction with ephrinA5. *Nat. Neurosci.* 9:322–30

Chan SO, Chung KY, Taylor JS. 1999. The effects of early prenatal monocular enucleation on the routing of uncrossed retinofugal axons and the cellular environment at the chiasm of mouse embryos. *Eur. J. Neurosci.* 11:3225–35

Charron F, Stein E, Jeong J, McMahon AP, Tessier-Lavigne M. 2003. The morphogen sonic hedgehog is an axonal chemoattractant that collaborates with netrin-1 in midline axon guidance. *Cell* 113:11–23

Chung KY, Leung KM, Lin CC, Tam KC, Hao YL, et al. 2004. Regionally specific expression of L1 and sialylated NCAM in the retinofugal pathway of mouse embryos. *J. Comp. Neurol.* 471:482–98

Colello RJ, Guillery RW. 1990. The early development of retinal ganglion cells with uncrossed axons in the mouse: retinal position and axonal course. *Development* 108:515–23

Colello SJ, Coleman LA. 1997. Changing course of growing axons in the optic chiasm of the mouse. *J. Comp. Neurol.* 379:495–514

Deiner MS, Kennedy TE, Fazeli A, Serafini T, Tessier-Lavigne M, Sretavan DW. 1997. Netrin-1 and DCC mediate axon guidance locally at the optic disc: loss of function leads to optic nerve hypoplasia. *Neuron* 19:575–89

Deiner MS, Sretavan DW. 1999. Altered midline axon pathways and ectopic neurons in the developing hypothalamus of netrin-1- and DCC-deficient mice. *J. Neurosci.* 19:9900–12

Demyanenko GP, Maness PF. 2003. The L1 cell adhesion molecule is essential for topographic mapping of retinal axons. *J. Neurosci.* 23:530–38

Dickson BJ, Gilestro GF. 2006. Regulation of commissural axon pathfinding by slit and its Robo receptors. *Annu. Rev. Cell. Dev. Biol.* 22:651–75

Dodd J, Morton SB, Karagogeos D, Yamamoto M, Jessell TM. 1988. Spatial regulation of axonal glycoprotein expression on subsets of embryonic spinal neurons. *Neuron* 1:105–16

Drager UC. 1985. Birth dates of retinal ganglion cells giving rise to the crossed and uncrossed optic projections in the mouse. *Proc. R. Soc. London B Biol. Sci.* 224:57–77

Dunlop SA, Tee LB, Beazley LD. 2000. Topographic order of retinofugal axons in a marsupial: implications for map formation in visual nuclei. *J. Comp. Neurol.* 428:33–44

Dutting D, Handwerker C, Drescher U. 1999. Topographic targeting and pathfinding errors of retinal axons following overexpression of ephrinA ligands on retinal ganglion cell axons. *Dev. Biol.* 216:297–311

Edenfeld G, Stork T, Klambt C. 2005. Neuron-glia interaction in the insect nervous system. *Curr. Opin. Neurobiol.* 15:34–39

Erskine L, Williams SE, Brose K, Kidd T, Rachel RA, et al. 2000. Retinal ganglion cell axon guidance in the mouse optic chiasm: expression and function of robos and slits. *J. Neurosci.* 20:4975–82

Falk J, Bechara A, Fiore R, Nawabi H, Zhou H, et al. 2005. Dual functional activity of semaphorin 3B is required for positioning the anterior commissure. *Neuron* 48:63–75

Flanagan JG, Vanderhaeghen P. 1998. The ephrins and Eph receptors in neural develoment. *Annu. Rev. Neurosci.* 21:309–45

Fournier AE, Nakamura F, Kawamoto S, Goshima Y, Kalb RG, Strittmatter SM. 2000. Semaphorin3A enhances endocytosis at sites of receptor-F-actin colocalization during growth cone collapse. *J. Cell Biol.* 149:411–22

Fricke C, Lee JS, Geiger-Rudolph S, Bonhoeffer F, Chien CB. 2001. Astray, a zebrafish roundabout homolog required for retinal axon guidance. *Science* 292:507–10

Godement P, Salaun J, Mason CA. 1990. Retinal axon pathfinding in the optic chiasm: divergence of crossed and uncrossed fibers. *Neuron* 5:173–86

Godement P, Salaun J, Metin C. 1987a. Fate of uncrossed retinal projections following early or late prenatal monocular enucleation in the mouse. *J. Comp. Neurol.* 255:97–109

Godement P, Vanselow J, Thanos S, Bonhoeffer F. 1987b. A study in developing visual systems with a new method of staining neurones and their processes in fixed tissue. *Development* 101:697–713

Godement P, Wang LC, Mason CA. 1994. Retinal axon divergence in the optic chiasm: dynamics of growth cone behavior at the midline. *J. Neurosci.* 14:7024–39

Goodman CS. 1996. Mechanisms and molecules that control growth cone guidance. *Annu. Rev. Neurosci.* 19:341–77

Grumet M, Friedlander DR, Sakurai T. 1996. Functions of brain chondroitin sulfate proteoglycans during developments: interactions with adhesion molecules. *Perspect. Dev. Neurobiol.* 3:319–30

Guillery RW, Mason CA, Taylor JS. 1995. Developmental determinants at the mammalian optic chiasm. *J. Neurosci.* 15:4727–37

Hattori M, Osterfield M, Flanagan JG. 2000. Regulated cleavage of a contact-mediated axon repellent. *Science* 289:1360–65

Hehr CL, Hocking JC, McFarlane S. 2005. Matrix metalloproteinases are required for retinal ganglion cell axon guidance at select decision points. *Development* 132:3371–79

Herrera E, Brown L, Aruga J, Rachel RA, Dolen G, et al. 2003. Zic2 patterns binocular vision by specifying the uncrossed retinal projection. *Cell* 114:545–57

A thoughtful, in-depth analysis of the mechanisms of Robo-Slit signaling in the insect and spinal cord midline.

Discovery that Zic2 is expressed specifically in VT retina and is required for the ipsilateral projection.

Herrera E, Marcus R, Li S, Williams SE, Erskine L, et al. 2004. Foxd1 is required for proper formation of the optic chiasm. *Development* 131:5727–39

Hoffmann MB, Tolhurst DJ, Moore AT, Morland AB. 2003. Organization of the visual cortex in human albinism. *J. Neurosci.* 23:8921–30

Hutson LD, Chien CB. 2002. Pathfinding and error correction by retinal axons: the role of astray/robo2. *Neuron* 33:205–17

Jacobson M, Hirose G. 1978. Origin of the retina from both sides of the embryonic brain: a contribution to the problem of crossing at the optic chiasm. *Science* 202:637–39

Janes PW, Saha N, Barton WA, Kolev MV, Wimmer-Kleikamp SH, et al. 2005. Adam meets Eph: an ADAM substrate recognition module acts as a molecular switch for ephrin cleavage in trans. *Cell* 123:291–304

Jeffery G. 2001. Architecture of the optic chiasm and the mechanisms that sculpt its development. *Physiol. Rev.* 81:1393–414

Jeffery G, Erskine L. 2005. Variations in the architecture and development of the vertebrate optic chiasm. *Prog. Retin. Eye. Res.* 24:721–53

Jessell TM. 2000. Neuronal specification in the spinal cord: inductive signals and transcriptional codes. *Nat. Rev. Genet.* 1:20–29

Jurney WM, Gallo G, Letourneau PC, McLoon SC. 2002. Rac1-mediated endocytosis during ephrin-A2- and semaphorin 3A-induced growth cone collapse. *J. Neurosci.* 22:6019–28

Kania A, Jessell TM. 2003. Topographic motor projections in the limb imposed by LIM homeodomain protein regulation of ephrin-A:EphA interactions. *Neuron* 38:581–96

Kaprielian Z, Runko E, Imondi R. 2001. Axon guidance at the midline choice point. *Dev. Dyn.* 221:154–81

Karlstrom RO, Trowe T, Klostermann S, Baier H, Brand M, et al. 1996. Zebrafish mutations affecting retinotectal axon pathfinding. *Development* 123:427–38

Keleman K, Ribeiro C, Dickson BJ. 2005. Comm function in commissural axon guidance: cell-autonomous sorting of Robo in vivo. *Nat. Neurosci.* 8:156–63

Koenig E, Giuditta A. 1999. Protein-synthesizing machinery in the axon compartment. *Neuroscience* 89:5–15

Labrador JP, O'Keefe D, Yoshikawa S, McKinnon RD, Thomas JB, Bashaw GJ. 2005. The homeobox transcription factor even-skipped regulates netrin-receptor expression to control dorsal motor-axon projections in Drosophila. *Curr. Biol.* 15:1413–19

Lambot MA, Depasse F, Noel JC, Vanderhaeghen P. 2005. Mapping labels in the human developing visual system and the evolution of binocular vision. *J. Neurosci.* 25:7232–37

Lee JS, von der Hardt S, Rusch MA, Stringer SE, Stickney HL, et al. 2004. Axon sorting in the optic tract requires HSPG synthesis by ext2 (dackel) and extl3 (boxer). *Neuron* 44:947–60

Lemke G, Reber M. 2005. Retinotectal mapping: new insights from molecular genetics. *Annu. Rev. Cell Dev. Biol.* 21:551–80

Leung KM, Taylor JS, Chan SO. 2003. Enzymatic removal of chondroitin sulphates abolishes the age-related axon order in the optic tract of mouse embryos. *Eur. J. Neurosci.* 17:1755–67

Leung KM, van Horck FP, Lin AC, Allison R, Standart N, Holt CE. 2006. Asymmetrical beta-actin mRNA translation in growth cones mediates attractive turning to netrin-1. *Nat. Neurosci.* 9:1247–56

Lin AC, Holt CE. 2007. Local translation and directional steering in axons. *EMBO J.* 26:3729–36

Lindwall C, Fothergill T, Richards LJ. 2007. Commissure formation in the mammalian forebrain. *Curr. Opin. Neurobiol.* 17:3–14

Litterst C, Georgakopoulos A, Shioi J, Ghersi E, Wisniewski T, et al. 2007. Ligand binding and calcium influx induce distinct ectodomain/gamma-secretase-processing pathways of EphB2 receptor. *J. Biol. Chem.* 282:16155–63

One of the first studies demonstrating a link between transcriptional specification of neural cell identity and regulation of guidance cues.

Genetic analysis of transcription factor coding of neuronal subtype and guidance factor expression in fly motor neurons.

Lustig M, Erskine L, Mason CA, Grumet M, Sakurai T. 2001. Nr-CAM expression in the developing mouse nervous system: ventral midline structures, specific fiber tracts, and neuropilar regions. *J. Comp. Neurol.* 434:13–28

Lustig M, Sakurai T, Grumet M. 1999. Nr-CAM promotes neurite outgrowth from peripheral ganglia by a mechanism involving axonin-1 as a neuronal receptor. *Dev. Biol.* 209:340–51

Mann F, Miranda E, Weinl C, Harmer E, Holt CE. 2003. B-type Eph receptors and ephrins induce growth cone collapse through distinct intracellular pathways. *J. Neurobiol.* 57:323–36

Marcus RC, Blazeski R, Godement P, Mason CA. 1995. Retinal axon divergence in the optic chiasm: uncrossed axons diverge from crossed axons within a midline glial specialization. *J. Neurosci.* 15:3716–29

Marcus RC, Mason CA. 1995. The first retinal axon growth in the mouse optic chiasm: axon patterning and the cellular environment. *J. Neurosci.* 15:6389–402

Marcus RC, Matthews GA, Gale NW, Yancopoulos GD, Mason CA. 2000. Axon guidance in the mouse optic chiasm: retinal neurite inhibition by ephrin "A"-expressing hypothalamic cells in vitro. *Dev. Biol.* 221:132–47

Marcus RC, Shimamura K, Sretavan D, Lai E, Rubenstein JL, Mason CA. 1999. Domains of regulatory gene expression and the developing optic chiasm: correspondence with retinal axon paths and candidate signaling cells. *J. Comp. Neurol.* 403:346–58

Marquardt T, Shirasaki R, Ghosh S, Andrews SE, Carter N, et al. 2005. Coexpressed EphA receptors and ephrin-A ligands mediate opposing actions on growth cone navigation from distinct membrane domains. *Cell* 121:127–39

Marsh-Armstrong N, Huang H, Remo BF, Liu TT, Brown DD. 1999. Asymmetric growth and development of the *Xenopus laevis* retina during metamorphosis is controlled by type III deiodinase. *Neuron* 24:871–78

Marston DJ, Dickinson S, Nobes CD. 2003. Rac-dependent trans-endocytosis of ephrinBs regulates Eph-ephrin contact repulsion. *Nat. Cell Biol.* 5:879–88

Mason C, Erskine L. 2000. Growth cone form, behavior, and interactions in vivo: retinal axon pathfinding as a model. *J. Neurobiol.* 44:260–70

Mason CA, Sretavan DW. 1997. Glia, neurons, and axon pathfinding during optic chiasm development. *Curr. Opin. Neurobiol.* 7:647–53

Mason CA, Wang LC. 1997. Growth cone form is behavior-specific and, consequently, position-specific along the retinal axon pathway. *J. Neurosci.* 17:1086–100

McLaughlin T, Hindges R, O'Leary DD. 2003. Regulation of axial patterning of the retina and its topographic mapping in the brain. *Curr. Opin. Neurobiol.* 13:57–69

Merzdorf CS. 2007. Emerging roles for zic genes in early development. *Dev. Dyn.* 236:922–40

Ming GL, Wong ST, Henley J, Yuan XB, Song HJ, et al. 2002. Adaptation in the chemotactic guidance of nerve growth cones. *Nature* 417:411–18

Mui SH, Hindges R, O'Leary DD, Lemke G, Bertuzzi S. 2002. The homeodomain protein Vax2 patterns the dorsoventral and nasotemporal axes of the eye. *Development* 129:797–804

Myers PZ, Bastiani MJ. 1993. Cell-cell interactions during the migration of an identified commissural growth cone in the embryonic grasshopper. *J. Neurosci.* 13:115–26

Nagai T, Aruga J, Takada S, Gunther T, Sporle R, et al. 1997. The expression of the mouse Zic1, Zic2, and Zic3 gene suggests an essential role for Zic genes in body pattern formation. *Dev. Biol.* 182:299–313

Nakagawa S, Brennan C, Johnson KG, Shewan D, Harris WA, Holt CE. 2000. Ephrin-B regulates the ipsilateral routing of retinal axons at the optic chiasm. *Neuron* 25:599–610

Oster SF, Bodeker MO, He F, Sretavan DW. 2003. Invariant Sema5A inhibition serves an ensheathing function during optic nerve development. *Development* 130:775–84

Describes the regional expression of transcription factors in the ventral diencephalon and their relation to chiasm formation.

The first demonstration that EphB/ephrinB guides RGC divergence at the optic chiasm.

Pak W, Hindges R, Lim YS, Pfaff SL, O'Leary DD. 2004. Magnitude of binocular vision controlled by islet-2 repression of a genetic program that specifies laterality of retinal axon pathfinding. *Cell* 119:567–78

> Discovered that the transcription factor Islet2 regulates the late-born crossed protection from VT retina.

Pascall JC, Brown KD. 2004. Intramembrane cleavage of ephrinB3 by the human rhomboid family protease, RHBDL2. *Biochem. Biophys. Res. Commun.* 317:244–52

Plas DT, Lopez JE, Crair MC. 2005. Pretarget sorting of retinocollicular axons in the mouse. *J. Comp. Neurol.* 491:305–19

Plump AS, Erskine L, Sabatier C, Brose K, Epstein CJ, et al. 2002. Slit1 and Slit2 cooperate to prevent premature midline crossing of retinal axons in the mouse visual system. *Neuron* 33:219–32

Polleux F, Ince-Dunn G, Ghosh A. 2007. Transcriptional regulation of vertebrate axon guidance and synapse formation. *Nat. Rev. Neurosci.* 8:331–40

Pratt T, Conway CD, Tian NM, Price DJ, Mason JO. 2006. Heparan sulphation patterns generated by specific heparan sulfotransferase enzymes direct distinct aspects of retinal axon guidance at the optic chiasm. *J. Neurosci.* 26:6911–23

Pratt T, Tian NM, Simpson TI, Mason JO, Price DJ. 2004. The winged helix transcription factor Foxg1 facilitates retinal ganglion cell axon crossing of the ventral midline in the mouse. *Development* 131:3773–84

Rachel RA, Dolen G, Hayes NL, Lu A, Erskine L, et al. 2002. Spatiotemporal features of early neuronogenesis differ in wild-type and albino mouse retina. *J. Neurosci.* 22:4249–63

Rachel RA, Murdoch JN, Beermann F, Copp AJ, Mason CA. 2000. Retinal axon misrouting at the optic chiasm in mice with neural tube closure defects. *Genesis* 27:32–47

Reese BE, Johnson PT, Hocking DR, Bolles AB. 1997. Chronotopic fiber reordering and the distribution of cell adhesion and extracellular matrix molecules in the optic pathway of fetal ferrets. *J. Comp. Neurol.* 380:355–72

> In zebrafish, demonstrated that Sema3D at the chiasm is important for RGC axon midline crossing.

Sabatier C, Plump AS, Le M, Brose K, Tamada A, et al. 2004. The divergent Robo family protein rig-1/Robo3 is a negative regulator of slit responsiveness required for midline crossing by commissural axons. *Cell* 117:157–69

Sakai JA, Halloran MC. 2006. Semaphorin 3d guides laterality of retinal ganglion cell projections in zebrafish. *Development* 133:1035–44

Sanchez-Camacho C, Rodriguez J, Ruiz JM, Trousse F, Bovolenta P. 2005. Morphogens as growth cone signalling molecules. *Brain. Res. Brain. Res. Rev.* 49:242–52

> Demonstrates that gene mutations in the ventral diencephalon perturb guidance factor expression, causing RGC projection errors.

Seth A, Culverwell J, Walkowicz M, Toro S, Rick JM, et al. 2006. Belladonna/(lhx2) is required for neural patterning and midline axon guidance in the zebrafish forebrain. *Development* 133:725–35

Sretavan DW, Feng L, Pure E, Reichardt LF. 1994. Embryonic neurons of the developing optic chiasm express L1 and CD44, cell surface molecules with opposing effects on retinal axon growth. *Neuron* 12:957–75

Sretavan DW, Reichardt LF. 1993. Time-lapse video analysis of retinal ganglion cell axon pathfinding at the mammalian optic chiasm: growth cone guidance using intrinsic chiasm cues. *Neuron* 10:761–77

Stein E, Tessier-Lavigne M. 2001. Hierarchical organization of guidance receptors: silencing of netrin attraction by slit through a Robo/DCC receptor complex. *Science* 291:1928–38

Stoeckli ET, Sonderegger P, Pollerberg GE, Landmesser LT. 1997. Interference with axonin-1 and NrCAM interactions unmasks a floor-plate activity inhibitory for commissural axons. *Neuron* 18:209–21

Tibber MS, Whitmore AV, Jeffery G. 2006. Cell division and cleavage orientation in the developing retina are regulated by l-DOPA. *J. Comp. Neurol.* 496:369–81

Torres M, Gomez-Pardo E, Gruss P. 1996. Pax2 contributes to inner ear patterning and optic nerve trajectory. *Development* 122:3381–91

Trousse F, Marti E, Gruss P, Torres M, Bovolenta P. 2001. Control of retinal ganglion cell axon growth: a new role for Sonic hedgehog. *Development* 128:3927–36

Van Vactor D. 1998. Adhesion and signaling in axonal fasciculation. *Curr. Opin. Neurobiol.* 8:80–86

Wang LC, Dani J, Godement P, Marcus RC, Mason CA. 1995. Crossed and uncrossed retinal axons respond differently to cells of the optic chiasm midline in vitro. *Neuron* 15:1349–64

Wang LC, Rachel RA, Marcus RC, Mason CA. 1996. Chemosuppression of retinal axon growth by the mouse optic chiasm. *Neuron* 17:849–62

Williams RW, Hogan D, Garraghty PE. 1994. Target recognition and visual maps in the thalamus of achiasmatic dogs. *Nature* 367:637–39

Williams SE, Grumet M, Colman DR, Henkemeyer M, Mason CA, Sakurai T. 2006. A role for Nr-CAM in the patterning of binocular visual pathways. *Neuron* 50:535–47

Williams SE, Mann F, Erskine L, Sakurai T, Wei S, et al. 2003. Ephrin-B2 and EphB1 mediate retinal axon divergence at the optic chiasm. *Neuron* 39:919–35

Williams SE, Mason CA, Herrera E. 2004. The optic chiasm as a midline choice point. *Curr. Opin. Neurobiol.* 14:51–60

Wilson SW, Placzek M, Furley AJ. 1993. Border disputes: Do boundaries play a role in growth-cone guidance? *Trends Neurosci.* 16:316–23

Wolman MA, Regnery AM, Becker T, Becker CG, Halloran MC. 2007. Semaphorin3D regulates axon-axon interactions by modulating levels of L1 cell adhesion molecule. *J. Neurosci.* 27:9653–63

Wu KY, Hengst U, Cox LJ, Macosko EZ, Jeromin A, et al. 2005. Local translation of RhoA regulates growth cone collapse. *Nature* 436:1020–24

Yao J, Sasaki Y, Wen Z, Bassell GJ, Zheng JQ. 2006. An essential role for beta-actin mRNA localization and translation in Ca^{2+} dependent growth cone guidance. *Nat. Neurosci.* 9:1265–73

Yin X, Watanabe M, Rutishauser U. 1995. Effect of polysialic acid on the behavior of retinal ganglion cell axons during growth into the optic tract and tectum. *Development* 121:3439–46

Yu TW, Bargmann CI. 2001. Dynamic regulation of axon guidance. *Nat. Neurosci.* 4:1169–76

Zimmer M, Palmer A, Kohler J, Klein R. 2003. EphB-ephrinB bi-directional endocytosis terminates adhesion allowing contact mediated repulsion. *Nat. Cell Biol.* 5:869–78

García-Frigola C, Carreres MI, Vegar C, Mason C, Herrera E. 2008. Zic2 promotes axonal divergence at the optic chiasm midline by EphB1-dependent and independent pathways. *Development.* In press

Established that EphB1/ephrinB2 in mouse direct the ipsilateral projection from VT retina.

RELATED RESOURCES

Chalupa LM, Williams RW, eds. 2007. *Eye, Retina, and Visual System of the Mouse.* Cambridge, MA: MIT Press. In press

Erskine L, Herrera E. 2007. The retinal ganglion cell axon's journey: insights into molecular mechanisms of axon guidance. *Dev. Biol.* 308:1–14

Pasquale EB. 2005. Eph receptor signalling casts a wide net on cell behaviour. *Nat. Rev. Mol. Cell Biol.* 6:462–75

Van Horck FPG, Weinl C, Holt CE. 2004. Retinal axon guidance: novel mechanisms for steering. *Curr. Opin. Neurobiol.* 14:61–66

Brain Circuits for the Internal Monitoring of Movements*

Marc A. Sommer[1] and Robert H. Wurtz[2]

[1] Department of Neuroscience, the Center for the Neural Basis of Cognition, and the Center for Neuroscience, University of Pittsburgh, Pittsburgh, Pennsylvania 15260; email: masommer@pitt.edu

[2] Laboratory of Sensorimotor Research, National Eye Institute, National Institutes of Health, Bethesda, Maryland 20892; email: bob@lsr.nei.nih.gov

Annu. Rev. Neurosci. 2008. 31:317–38

First published online as a Review in Advance on April 2, 2008

The *Annual Review of Neuroscience* is online at neuro.annualreviews.org

This article's doi:
10.1146/annurev.neuro.31.060407.125627

Key Words

vision, saccadic eye movements, corollary discharge, efference copy, perception

Abstract

Each movement we make activates our own sensory receptors, thus causing a problem for the brain: the spurious, movement-related sensations must be discriminated from the sensory inputs that really matter, those representing our environment. Here we consider circuits for solving this problem in the primate brain. Such circuits convey a copy of each motor command, known as a corollary discharge (CD), to brain regions that use sensory input. In the visual system, CD signals may help to produce a stable visual percept from the jumpy images resulting from our rapid eye movements. A candidate pathway for providing CD for vision ascends from the superior colliculus to the frontal cortex in the primate brain. This circuit conveys warning signals about impending eye movements that are used for planning subsequent movements and analyzing the visual world. Identifying this circuit has provided a model for studying CD in other primate sensory systems and may lead to a better understanding of motor and mental disorders.

Contents

INTRODUCTION

A major function of nervous systems is to process sensory input to detect changes in the environment. Yet animals have evolved not only to analyze but also to act, and actions have a multifold impact on sensory processing. A cat may notice a movement in a field and pounce toward a mouse. The cat's locomotion causes its legs to brush against grass, evoking touch sensations that must be ignored to avoid withdrawal reflexes. At the same time, the cat's head and eye movements disrupt its percept of auditory and visual space. It seems rather remarkable that a mouse is ever caught. Even at rest, if a cat perks up its ears to hear something better, the action momentarily disrupts the very sense it is intended to help. Nervous systems require some way of addressing these myriad, movement-related disruptions. One major mechanism is to provide warning signals to sensory systems from within the movement-generating systems themselves.

The importance of motor-to-sensory feedback may be emphasized by considering what happens if there is no such link. Imagine a security guard watching images of a room from a video camera. If a suspicious person enters, the guard can quickly recenter the camera by moving a joystick. This mobility comes at the price of momentary disorientation, however, because the entire image leaps across the monitor. An animal's eyes are its security cameras, and the brain encounters problems similar to those faced by the security guard. Approximately twice per second we move our eyes to inspect a new part of the visual scene, but we have no overt awareness that our visual input is constantly lurching about. We perceive the world as stable. The fundamental difference between the security camera and the eye is internal feedback of movement information. The visual system is not surprised by eye movements because it receives a warning signal about each one. This warning signal is corollary discharge (CD).

The hypothesis that CD exists in the service of vision is a venerable one (Grüsser 1995) and was discussed most influentially by von Helmholtz in the nineteenth century (1925), who argued for the importance of an "effort of will." The first strong experimental evidence for the hypothesis was presented in the twentieth century by two groups working independently. Sperry (1950) coined the term corollary discharge, and his contemporaries von Holst & Mittelstaedt (1950) referred to essentially the same principle as efference copy. We use the term CD to refer to the general concept of neuronal signals that provide internal information about movements.

In this review we concentrate on the demonstration of CD in the primate brain and its contribution to perception and motor planning.

Corollary discharge (CD): a copy of a motor command sent to brain areas for interpreting sensory inflow (also called efference copy)

Efference: motor outflow that causes movements

Our discussion focuses on the visual and oculo-motor system because that is where the physiological basis of CD is best understood. We omit some phenomena that likely depend on CD because little circuit-level data exist, and we do not cover equally important aspects of CD revealed in studies of invertebrates and simpler vertebrate systems. We briefly consider the relevance of CD to other systems in the brain and to human disease.

IDENTIFYING CD IN A COMPLEX BRAIN

A Model System for Studying CD: Saccades and Vision

In our example of the security camera, we emphasized the disruption of vision as the camera shifted rapidly from one point of the scene to another. For the eyes, such rapid movements are called saccades. Saccades play the essential role in vision of moving the fine grain analyzer of the retina, the fovea, onto salient objects in the visual scene. The generation of saccades is among the more thoroughly understood motor processes of the primate brain. The saccadic network therefore provides an attractive model system for investigating the sources of CD and the pathways that distribute CD to sensory areas.

The visuosaccadic network extends from retinal registration of visual input to muscle contraction for movement output (**Figure 1a**). Visual input arrives at area V1 in cerebral cortex, and movement output descends from the superior colliculus (SC), through the midbrain and pontine reticular formation, to the oculomotor nuclei that innervate each of the six eye muscles. In between these primary input and output pathways lie cortical processing networks that range from V1 through subsequent visual processing areas (**Figure 1a**, yellow arrows), and which include the lateral intraparietal (LIP) area of parietal cortex and the frontal eye field (FEF) of frontal cortex. These parietal and frontal areas contribute to the generation of voluntary saccades (see reviews by Andersen

& Buneo 2002, Colby & Goldberg 1999, Schall 2002). LIP, FEF, and associated cortical areas in turn project to the intermediate layers of the SC to influence lower motor pathways (**Figure 1a**, red arrows). Ancillary circuits (including one from FEF through the basal ganglia to SC) and a second visual pathway (from SC superficial layers through pulvinar to visual cortex) are not shown.

A CD Path to Frontal Cortex

Given this background about the saccadic system, we can now ask what a saccadic CD circuit should look like. In principle, a CD pathway should run counter to the normal downward flow of sensorimotor processing. One such pathway (**Figure 1b**, *red arrows*) ascends from the intermediate layers of the SC to the medial dorsal nucleus (MD) of the thalamus and then to FEF in frontal cortex. Other pathways may convey CD as well (*gray arrows* in the figure), but they have not yet been studied physiologically.

The pathway in **Figure 1b** was elucidated anatomically (as discussed below), but nothing was known about it functionally. The key question was, does it carry CD? Pathways for CD have been studied at the single neuron level in simpler nervous systems (Poulet & Hedwig 2007). In the highly complex primate brain, it is likely more of a challenge to establish that a neuron carries CD as opposed to a movement command. By definition both the command and the corollary should look the same. We therefore developed a set of criteria (Wurtz & Sommer 2004) that, if met, would satisfy us that we were studying a CD pathway.

Four Criteria for Identifying CD

First, the CD must originate from a brain region known to be involved in movement generation, but it must travel away from the muscles. The SC-MD-FEF pathway met this criterion. The pathway originates in the intermediate layers of the SC, an area clearly

Saccade: eye movement that rapidly redirects the fovea

SC: superior colliculus (of midbrain)

LIP: lateral intraparietal area (of parietal cortex)

FEF: frontal eye field (of frontal cortex)

MD: medial dorsal nucleus (of thalamus)

a

Visual-to-motor pathways

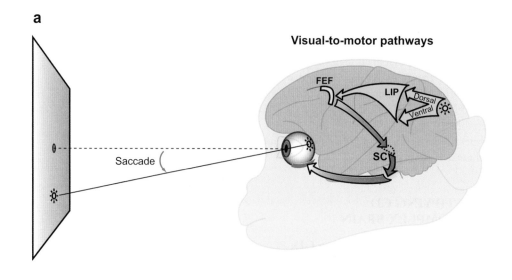

b   **Motor-to-visual pathways**   c   **CD signal**

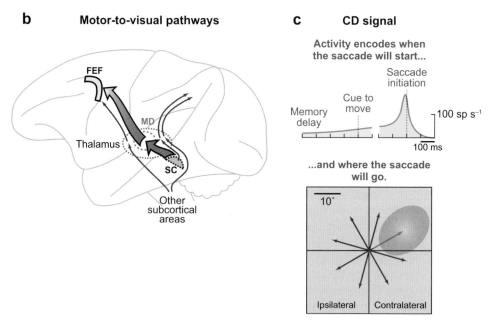

Figure 1

Visual and saccadic circuits in the monkey brain. (*a*) Outline of the major components of the circuit for generating visually guided saccades. Visual input arrives at area V1 of visual cortex (*far right*) through a retinal-thalamic pathway (not shown). Visual signals (*yellow*) then are processed by dorsal and ventral cortical streams. A major intermediary region is LIP. Visual information from both streams converges in the FEF. Signals are sent downstream (*red*) to influence the SC and deeper motor areas that trigger a saccade. (*b*) Candidate pathways for conveying saccade-related CD to cerebral cortex. The SC-MD-FEF pathway has been studied in detail. Other routes may exist as well (*gray arrows*). (*c*) The CD signal in MD thalamic relay neurons that project to FEF. In a task used to characterize behavior-related activity of neurons, monkeys prepare saccades to the location of a remembered visual target. During the memory delay and cue to move periods, MD neurons display an increase in firing rate. This increase culminates in a burst of activity just before saccade initiation (*top*). Spatially, the neurons fire only for specific ranges of contralateral saccades (*bottom*).

involved in saccade generation. Neurons in the area discharge before saccades, electrical stimulation in the area produces saccades, and inactivation in the area disrupts saccades (Sparks & Hartwich-Young 1989). Anatomical evidence for the pathway came from retrograde labeling and anterograde degeneration studies (Benevento & Fallon 1975, Goldman-Rakic & Porrino 1985) as well as trans-synaptic retrograde labeling using the herpes virus (Lynch et al. 1994). By combining the classic physiological techniques of antidromic and orthodromic stimulation, individual MD relay neurons were identified that received input from the SC and projected to the FEF (Sommer & Wurtz 1998, 2004a). Taken together, these results confirmed that the pathway originates in a strongly saccade-related area but travels away from the eye muscles to innervate the frontal cortex.

Second, the CD signal should occur just prior to the movement and represent its spatial and temporal parameters. Approximately three quarters of the MD neurons identified as relay neurons increased their activity just before the saccade (**Figure 1c**, *top*) (Sommer & Wurtz 2004a). Most MD neurons had peak activity for a limited range of saccadic amplitudes and directions, with strongest activity always occurring for directions into the contralateral visual field (**Figure 1c**, *bottom*). Thus the second criterion, a functional one, was met; neurons in this pathway are active before saccades and convey information about the parameters of the upcoming saccade.

Third, the neurons should not contribute to producing the movement. Silencing the pathway should not affect movements in simple tasks that do not require a CD. In accord with this criterion, inactivation of MD relay neurons did not alter saccadic accuracy and speed (Sommer & Wurtz 2002, 2004b).

The fourth criterion was the most critical: It required that perturbing the putative CD pathway does disrupt performance on a task that depends on a CD. Perturbing the CD pathway did disrupt performance in such a task. This evidence was so central to illuminating the nature of the CD that we devote the next section to these experiments.

CD AND SEQUENTIAL SACCADES

Influence of CD on Behavior: Double-Step Saccades

In order to evaluate whether CD influences behavior, other factors affecting that behavior must be eliminated. First, the behavior under study must prohibit visual feedback about performance because primates use visual information whenever it is available. Elimination of visual feedback is achieved in the double-step task (Hallett & Lightstone 1976), which has become standard in both the laboratory and the clinic. In this task, two targets are flashed sequentially. The subject's task is to make a saccade to the first target and then to the second. The first target may be flashed to the right, for example, and then the second target straight upward from there (**Figure 2a**, *yellow dots*). In correct performance, saccades would follow the targets as depicted by the solid arrows in the figure. No visual feedback is available during the task because the saccades are performed in total darkness; the fixation point and the two flashed targets are gone before the saccades begin. If the second saccade were made only on the basis of remembered visual information, it would go up and 45° to the right (**Figure 2a**, *dashed arrow*) because that was the target's location when the eye was originally fixating. To make the correct second saccade, the memory of the target location must be adjusted to account for where the eye lands at the end of the first saccade. Internal information about the execution of the first saccade is needed. Because such information is not provided by visual feedback and is unlikely to come from proprioception (see Proprioception in sidebar), CD must be the source.

Quantifying the influence of CD on behavior requires an observable measure of that influence. CD itself is an internal signal and hidden from direct psychophysical assessment. In the double-step task, the observable event is the second saccade, and the parameter of interest is

PROPRIOCEPTION

When visual feedback is absent or unreliable, extraretinal signals are required to monitor eye position. CD is one such signal, and proprioception of eye muscles is another (Wang et al. 2007). Although CD is of central origin and reaches cerebral cortex before a saccade, proprioception arises peripherally and reaches the cortex after a saccade. A series of experiments has shown that proprioception provides negligible information for everyday saccadic behavior. The common approach was to use microstimulation to displace a monkey's eye position while it prepared a voluntary saccade to a visual target. The question was, could a monkey internally monitor the displacement and compensate for it? The extent of compensation indicated the availability of extraretinal information. Stimulating the muscle-innervating motor neurons caused an eye displacement but no compensation, even though proprioception was available (Mays et al. 1987, Schiller & Sandell 1983, Sparks & Mays 1983). But stimulating higher in the circuitry, e.g., in the SC, caused a saccade that was immediately followed by full compensation, presumably because a concomitant CD signal was evoked. Compensation following SC stimulation persists even after the proprioceptive nerves are cut (Guthrie et al. 1983). Proprioception therefore seems insufficient as an extraretinal signal for making sequential saccades; instead, it may contribute to longer-term adjustments (Lewis et al. 2001, Steinbach 1987).

its vector. A second saccade vector that lands at the second target location, fully compensating for the first saccade, implies perfect CD. Inaccuracies of the second saccade reveal inaccuracies of CD.

Inactivation of the CD Pathway

The hypothesis that the pathway to frontal cortex provides CD information about the first saccade was tested by reversibly inactivating the MD relay node (**Figure 2b**) in the pathway. This left both the SC and FEF untouched while severing the link from SC to FEF (Sommer & Wurtz 2002, 2004b). If inactivation totally eliminates CD, the monkey should make the first saccade correctly but should not have internal information that it did so. In the absence of information about this first saccade, the sec-

ond saccade should be inaccurate; it should be directed toward the initial, retinotopic location of the second visual target. The average end points of second saccades were analyzed to see how they changed during inactivation.

Inactivation of MD did cause second saccade end points to shift as if CD of first saccades was impaired, as long as those first saccades were directed into the contralateral hemifield (**Figure 2c**). The impairment was significant but smaller than expected from total CD elimination. This partial deficit suggests that we did not silence the path completely or that other CD pathways exist. Both hypotheses are reasonable but have yet to be tested.

What makes the results more compelling is that the deficit was limited to trials in which the first saccades were made into the contralateral visual field (**Figure 2c**, *right*). This direction was the same preferred direction of the MD relay neurons. Double-step configurations in which first saccades went into the ipsilateral visual field were randomly interleaved, and these trials were unaffected (**Figure 2c**, *left*). In sum, there was a modest deficit in the average accuracy of second saccades during inactivation, but the deficit matched the direction of saccades represented in the affected MD nucleus.

Similar double-step deficits have been found in patients subsequent to thalamic injury (Bellebaum et al. 2005, Gaymard et al. 1994), parietal injury (Duhamel et al. 1992b, Heide et al. 1995), and parietal transcranial magnetic stimulation (Morris et al. 2007). Such deficits have not been reported for human FEF lesions, but the major confound that FEF damage causes impairment to saccade generation complicates analysis of CD deficits (discussed in Sommer & Wurtz 2004b).

Successive Saccades Depend on CD

A finer-grained analysis of the data, focusing on the precision of individual saccades, revealed a more striking deficit (Sommer & Wurtz 2004b). During normal behavior, monkeys made first saccades that landed at slightly different places

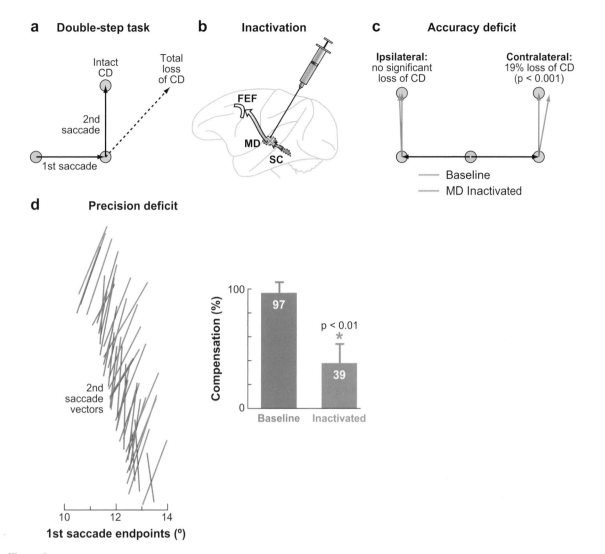

a **Double-step task**

Intact CD
Total loss of CD
2nd saccade
1st saccade

b **Inactivation**

FEF
MD
SC

c **Accuracy deficit**

Ipsilateral: no significant loss of CD
Contralateral: 19% loss of CD (p < 0.001)

—— Baseline
—— MD Inactivated

d **Precision deficit**

2nd saccade vectors

100

Compensation (%)

97

p < 0.01
*

39

0

Baseline   Inactivated

10   12   14

1st saccade endpoints (°)

Figure 2

Double-step task used to evaluate the influence of CD. (*a*) The pattern of double-step saccades (*arrows*) that would be expected with intact CD vs. total loss of CD. (*b*) Inactivation of the putative CD pathway: the GABA agonist muscimol was injected at the site of MD relay neurons. (*c*) Average accuracy deficit, measured as the horizontal shift in second saccade endpoints. Trials with contralateral first saccades (*right*) showed a clear deficit, but ipsilateral first saccades did not produce a deficit (*left*). (*d*) Precision deficit in compensating for trial-by-trial variability in first saccades. Compensation is illustrated for second saccade vectors in an example session. The vectors are spread out in the vertical dimension for clarity, but their relative horizontal positions are maintained (as indicated by axis at bottom). Deficit across all sessions (*n* = 22) is summarized with the bar graph (*inset*). Bars show means and standard errors; on average there was a 58% deficit ([97%-39%]/97%).

from trial to trial: sometimes short of where the first flashed target had been, sometimes beyond it. It therefore seemed likely that if a CD occurs with each saccade as predicted from the neuronal activity found in MD, a CD-driven compensation should also occur that is unique to each trial. As predicted, the individual second saccade vectors (**Figure 2d**, *blue lines*) rotated forward or backward depending on where they started (i.e., where the first saccades landed).

RF: receptive field

FF: future field

Hence in its normal state, the brain precisely monitors even small fluctuations in saccade amplitude and adjusts subsequent saccades accordingly.

Now if the CD were altered by inactivation, the compensation should be altered. This was, in fact, the case: the second saccades during inactivation (**Figure 2d**, *orange lines*) failed to rotate as a function of where the first saccades ended. Compensation was reduced by more than half (**Figure 2d**, *right*). Monkeys could still compensate slightly after MD inactivation, so the deficit was partial, as it was for the average accuracy deficit. And once again, the effect occurred only for saccades into the contralateral visual field.

These inactivation experiments suggest that activity in the SC-MD-FEF pathway not only correlates with what is expected of a CD signal, but actually functions as a CD signal. With this demonstration, the signals in the pathway from SC to FEF via MD satisfy the four criteria we regard as essential for identifying CD. The signals originate from a known saccade-related region, they encode the timing and spatial parameters of upcoming saccades, their removal does not affect the generation of saccades for which CD is not required, and their removal does disrupt saccades for which CD is required.

CD AND VISUAL PERCEPTION

Influence of CD on Vision: Shifting Receptive Fields

Having reviewed evidence that the information conveyed upward through the SC-MD-FEF pathway is a CD signal, we now return to the fundamental hypothesis: that the world appears stable across eye movements because CD informs the visual system about each imminent saccade. Whether cortical visual neurons truly use a CD signal was unknown until landmark experiments on parietal cortical area LIP by Duhamel et al. (1992a) revealed changes in neuronal activity that must result from CD input.

Neurons in the visual system have a restricted view of the world known as a receptive field (RF). The neurons respond only if light falls on a restricted zone of the retina—the RF—and conventional thought was that the zone's location changes little, if at all, when the eyes move. Duhamel et al. (1992a) found, however, that some neurons in LIP behave differently. A new zone of retinal sensitivity emerges at a location dependent on the saccade that is about to occur. Such a shifting RF is possible only if the neuron receives information about the imminent saccade.

The fundamental nature of a shifting RF is depicted in **Figure 3a**. Suppose that during fixation (*left panel*), a neuron has an RF (*blue ellipse*) aligned directly below the fovea (*orange dot*). The neuron would respond to an apple in its RF but not to a distant pepper. If the monkey plans to make a saccade to the point just above the pepper, however (*middle panel*), a neuron with a shifting RF shows a crucial change. It suddenly becomes sensitive to the pepper (*magenta ellipse*) because the pepper lies at the location where the RF will reside after the saccade—referred to as the future field (FF). Then the eyes move (*right panel*), and the presaccadic FF becomes the RF again. Such shifting RFs (also called remapped RFs) have been studied in many regions including LIP (Batista et al. 1999, Colby et al. 1996, Duhamel et al. 1992a, Heiser & Colby 2006, Kusunoki & Goldberg 2003), FEF (Sommer & Wurtz 2006; Umeno & Goldberg 1997, 2001), extrastriate visual areas (Nakamura & Colby 2002, Tolias et al. 2001), and the SC (Walker et al. 1994). They are likely present in the human brain as indicated by fMRI studies (Medendorp et al. 2003; Merriam et al. 2003, 2007), and their relation to corpus callosum function has been examined (Berman et al. 2005, 2007; Heiser et al. 2005).

Shifting RFs are not studied in the laboratory with apples and peppers, of course, but with well-controlled flashes of light (probes). The activity of a neuron with a shifting RF demonstrates its presaccadic change in visual sensitivity across space (**Figure 3b**, an FEF neuron). During initial fixation, the neuron responded

a   **Shifting RFs: concept**

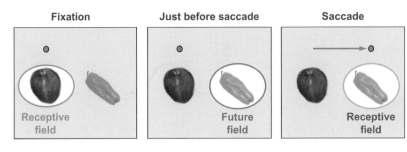

| Fixation | Just before saccade | Saccade |

Receptive field — Future field — Receptive field

b   **Neuron**

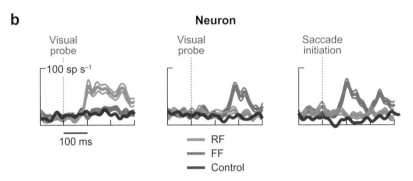

Visual probe — Visual probe — Saccade initiation

100 sp s^{-1}

100 ms

RF
FF
Control

Figure 3

Shifting visual receptive fields. (*a*) Concept shown schematically. (*b*) Example of an FEF neuron that shows the shift. A visual probe was flashed in the RF in some trials (*blue*) or in the FF in other trials (*magenta*). No probe appeared in control trials (*black*). Curves show mean and SE of average neuronal activity aligned to probe onset (*left and center panels*) or to saccade initiation (*right panel*).

only to probes flashed at the RF (*left*) and just before the saccade only to probes flashed at the FF (*middle*). Because the visual activity associated with the FF stimulation appeared around the time of the saccade (*right*), a fair question is whether the activity was related to making the saccade. This possible artifact was tested with interleaved trials in which no probe appeared, but the same saccade was made (*black traces*). Such control trials elicited no activity. The activity required a visual stimulus in the FF.

Our hypothesis was that, in FEF neurons, shifting RFs are caused by CD input from the MD-mediated pathway. This hypothesis led to three predictions. First, the spatial characteristics of the shifting RF should reflect those of the CD. Second, the temporal dynamics of the shift should match those of the CD. If both of these first two predictions were confirmed, the CD would be established as appropriate for driving

the shift. The third prediction was that the shift should be reduced if the CD were interrupted. Such a result would show that the CD is necessary for the shift.

Spatial Match Between CD and Shifting RFs

The first prediction about shifting RFs proposed that RFs should have spatial properties similar to the CD itself. The CD in the pathway to FEF originates in the SC (**Figure 4a**), a structure with exquisite topographic organization of saccade vectors. If SC activity provides the CD that causes the FEF shift, the shift should move as if prompted by a vector input. It should jump to its new location (**Figure 4b**) rather than spreading continuously from RF to FF. The prediction was testable by flashing probes not only at the RF and FF, but

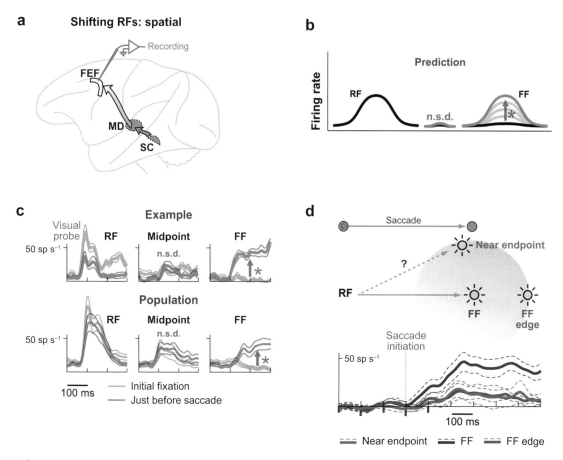

Figure 4

Shifting RFs: spatial. (*a*) Neurons with shifting RFs were recorded in the FEF. (*b*) It was predicted that shifting RFs would jump, as expected from a vector CD input. There should be a significant change in activity (*asterisk*) at the FF but no significant difference (n.s.d.) at the midpoint. (*c*) (*Top*) Example FEF neuron shows a jump, not a spread of activity. Just before a saccade, there was a decrease in activity at the RF (*left*), no change in activity at the midpoint (*middle*), and a large increase in activity at the FF (*right*). (*Bottom*) The same result was found in the average of the FEF population (*n* = 13 neurons). (*d*) Experiment to see if shifting RFs travel toward the saccadic endpoint ("?" trajectory). Shifting RF activity is shown for probes flashed near the endpoint (*green*), at the edge of the FF as defined by a shift parallel to the saccade (*purple*), and at the middle of the FF so defined (*black*). The data indicate that shift travels parallel to the saccade, not toward the saccadic endpoint (average data from *n* = 12 neurons).

also at the midpoint between them. If the RFs jump, the midpoint would exhibit little or no activity; if they spread, the midpoint would become highly active.

Data from an example neuron (**Figure 4c**, *top*) revealed that while activity increased dramatically at the FF just before a saccade, no change occurred at the midpoint. The result from this example neuron was reproduced in the overall population (**Figure 4c**, *bottom*). The data supported a jump, not a spread, of activ-

ity, compatible with a vector CD input. The small amount of activity at the midpoint, which did not change significantly before the saccade, was caused by the probe sometimes falling on the edges of the RF and FF (Sommer & Wurtz 2006).

In this analysis the implicit assumption has been that the shifting RFs move parallel to the saccade as implied by the horizontal shift with a horizontal saccade in **Figure 3a**. Shifting RFs in visual area V4, however, move toward the saccadic end point (Tolias et al. 2001). The direction of the shift is crucial because the FF location provides clues about the perceptual function of the shift, as discussed below. To examine the issue in FEF neurons, probes were introduced near the saccadic end point (**Figure 4d**, *top*). The shift at that location, however, was minimal (**Figure 4d**, *bottom*). It was no different from activity at the edge of the FF as defined by a parallel shift and much less than activity at the center of the FF so defined. The data supported our hypothesis that shifting RFs in the FEF move parallel to the saccade.

Temporal Match Between CD and Shifting RFs

The next prediction was that the timing of shifting RFs should match the timing of the CD signal in the SC-MD-FEF pathway. The recording study (Sommer & Wurtz 2004a) demonstrated that the pathway conveys a gradual increase in activity and then a burst of activity aligned with saccade onset (see **Figure 1c**). If CD from the pathway causes the shift, the shift should be time-locked with the saccade.

We analyzed whether shifting RF activity was better related to stimulus onset or to saccade initiation. Shifting RF activity requires visual stimulation in the FF, and **Figure 5a** shows for one neuron the temporal profile of visual probe presentation in the FF and the neuronal activity that it evoked. The probe appeared at time 0 and disappeared after 50 ms. The neuron showed no activity at its normal visual latency of 80 ms. Rather, activity appeared much later, and with appreciable variability. In individual trials, the start of the activity seemed closely linked to the start of each saccade. The correlation between activity onset and saccade onset was highly significant for this neuron and for the population (Sommer & Wurtz 2006). Alignment of the average visual activity improved when the same data were plotted relative to saccade initiation (**Figure 5b**, higher peak and narrower SEs). Thus a shifting RF is visual activity synchronized to a motor act.

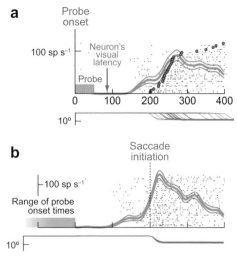

Shifting RFs: timing

Figure 5

Shifting RFs: timing. (*a*) Shifting RF activity aligned to probe onset for an example neuron. Each *raster* shows action potentials from a single trial, and the *bold orange trace* is average firing rate (surrounded by *thin traces* showing standard errors). Start of the shift is much later than the neuron's normal visual latency (*purple arrow*). Also shown are times of saccade onset in each trial (*green dots*) corresponding to the eye position data (*traces at bottom*). (*b*) Shifting RF activity aligned to saccade initiation. For this neuron, shift onset time and saccadic onset time were correlated with Pearson R = 0.97; for the population, R = 0.50 (both p < 0.01).

Although saccade-gated visual activity may seem bizarre, it makes sense; if the shifting RF is concerned with maintaining perceptual visual stability, it would be important for it to occur only if the generation of the saccade were inevitable. Saccades can be cancelled up to 100 ms before their initiation (Hanes & Schall 1995). The only way to ensure that shifts are linked to saccades, therefore, is to delay their onset to begin only after the "point of no return" for moving, i.e. 100 ms or less before saccade initiation.

Shifts were time-locked to saccades, but their exact onset times varied from neuron to neuron as indicated by four examples (**Figure 6a**). The shift started before the saccade began for some neurons and after the

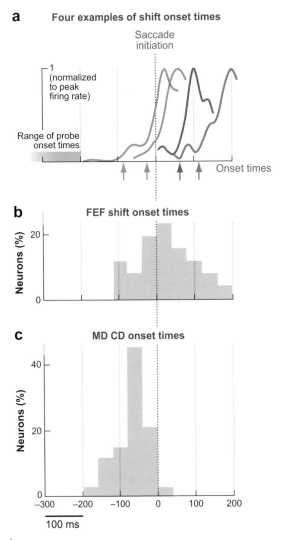

Shifting RFs: timing of CD influence

a   Four examples of shift onset times

Saccade
initiation

1
(normalized
to peak
firing rate)

Range of probe
onset times

Onset times

b   FEF shift onset times

Neurons (%)

20

0

c   MD CD onset times

Neurons (%)

40

20

0

-300   -200   -100   0   100   200

100 ms

Figure 6

Shifting RFs: timing of CD influence. (*a*) Shift onset times for four example
FEF neurons. Some neurons started their shift before saccade initiation (*orange*
and *blue* traces), while others started their shift after saccade initiation (*red* and
green traces). The orange trace is from the neuron in **Figure 5*a***, and the red
trace is from the neuron in **Figure 3*b***. The blue and green traces show data
from two other neurons. Shift magnitudes are normalized to each other to
facilitate comparison of timing. Note that the probe onset time always
preceded saccade onset. (*b*) Distribution of shift onset times for the sample of
FEF neurons (*n* = 26). (*c*) Distribution of CD onset times (the start of the
saccade-related bursts) in the population of MD relay neurons that project to
FEF (*n* = 33).

saccade began for others. Yet for all the neurons,
probes were flashed at identical times (*yellow
range* to left) and always appeared and disappeared in the neuron's FF. Even though some
shifts began after saccade initiation, they were
caused by a visual event happening before saccade initiation and outside the classic RF.

Overall, the shift onset times for the neuronal population were distributed approximately normally, over a range of 100 ms
before to 200 ms after saccade initiation
(**Figure 6*b***). The average shift onset time was
24 ms after the saccade started, which is in mid-
flight of a typical saccade. How does this distribution of shift onset times compare with the
CD signal that we posit is causing the shift?
Figure 6*c* shows the range of onset times for
the CD signal that MD relay neurons provide
to the FEF. Just as shift onset times in the FEF
were measured by a sudden increase in visual
activity of shifting neurons, CD onset times
were measured by a sudden increase in MD relay neuron saccade-related activity. The average
CD onset time was 72 ms before the saccade,
demonstrating that the CD signal could be a
causal trigger for the shift. It takes ∼100 ms
(72 + 24 = 96 ms) for CDs arriving at the
FEF to trigger a shifting RF. This delay implies
that shifts are not constructed by single, MD-
recipient FEF neurons, but rather that they develop through a multisynaptic network.

Necessity of CD Signal
for Shifting RFs

The final prediction was that interrupting the
CD signal should reduce the shift. Up to this
point the comparison of CD to shift was one
of spatial and temporal correlation; perturbing the CD allows us to move from correlation
to causation. The CD signal was inactivated at
the MD relay neuron level during continuous
recording of an FEF neuron with a shifting RF
(**Figure 7*a***) (Sommer & Wurtz 2006). If the
prediction were correct, the FEF neuron would
retain its classic visual response (in the RF) but
lose its ability to shift the response (into the FF).

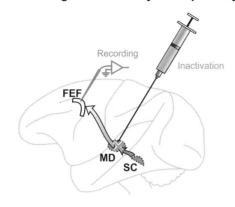

a   **Shifting RFs: necessity of CD pathway**

b

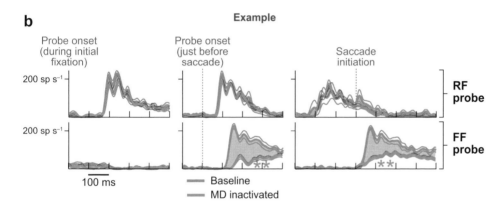

Example

Probe onset (during initial fixation)

Probe onset (just before saccade)

Saccade initiation

200 sp s⁻¹

RF probe

200 sp s⁻¹

FF probe

100 ms

— Baseline
— MD inactivated

c

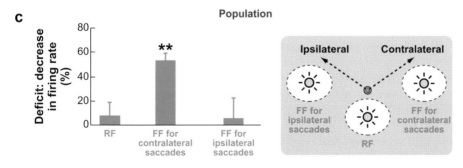

Population

Deficit: decrease in firing rate (%)

RF | FF for contralateral saccades | FF for ipsilateral saccades

Ipsilateral | Contralateral

FF for ipsilateral saccades | FF for contralateral saccades

RF

Figure 7

Shifting RFs: necessity of CD pathway. (*a*). Inactivation of the CD pathway while an FEF neuron with a shifting RF is recorded. (*b*) Example of a shifting RF that was impaired by MD inactivation. Activity before inactivation is depicted with *grey traces*. The neuron had a visual response in the RF during fixation (*left panels*), a shift to the FF just before the saccade (*lower middle panel*), and tight alignment of this shift with saccadic onset (*right panel*). The neuron maintained its activity in the RF even as it gained sensitivity at the FF, as many neurons did (*upper middle panel*). During inactivation (*orange traces*), the FF response was drastically reduced (*bottom row*). The classic RF response, however, was unchanged (*top row*). (*c, left*) The deficit in the population of neurons. Reductions in activity were seen only in the FF, not the RF, and only for contralateral, not ipsilateral, saccades. (*c, right*) Task configurations for testing shifting RFs that accompanied contralateral saccades vs. ipsilateral saccades. **, significant changes at p < 0.0001 level.

Data from an example FEF neuron are shown in **Figure 7b**. Before inactivation, the neuron had a strong visual response in the RF and an intense shift of visual sensitivity into the FF that was synchronized with the saccade. During inactivation, the FF response was drastically reduced even though the classic visual response in the RF was unchanged. In sum, when MD was inactivated, the FEF neuron retained its visual sensitivity and the monkey retained its ability to make a saccade, but the communication between saccade generation and visual processing was gone; the neuron lost advanced information about the saccade (i.e., CD), severely impairing its ability to shift visual sensitivity to the FF location.

The deficits found in the example neuron were replicated individually in most neurons tested (7 out of 8) and were highly significant in the pooled data (**Figure 7c**, *left*). Overall, the shift was reduced by an average of ∼50% when it accompanied contralateral saccades, those represented by the SC-MD-FEF pathway. Shifts that accompanied ipsilateral saccades were unimpaired.

In summary, our predictions were supported: Shifting RFs had spatial and temporal dynamics consistent with the CD signal, and the shifts depended on the CD signal. This provided systematic evidence for a functional link between the CD arising from the midbrain and visual processing in the FEF. This result produces an intriguing further implication when considering the FEF's position in the broader network of cortical visual areas. The FEF may play a role in relaying CD information, or shifting RFs themselves, to other cortical areas such as LIP where shifting RFs are found. Functional, reciprocal connections do link FEF and parietal cortex (Chafee & Goldman-Rakic 2000).

Shifting RFs and Visual Stability

The previous section and **Figures 3–7** have described shifting RFs. Because of their unique characteristics, shifting RFs likely play a role in the perception of visual stability across saccades (Duhamel et al. 1992a). To see why this hypothesis is plausible, we must discuss in more detail how CD signals influence shifting RFs and how shifting RFs, in turn, could influence visual analysis.

The CD signals produce a spatial effect in the retinotopic coordinate system used for visual processing. This effect is a new receptivity to visual stimulation at the FF. The FF's location is determined by the vector of the impending saccade conveyed by the CD. How the motor-to-sensory transformation between the saccadic system and the visual map could be made has been modeled by Quaia et al. (1998). For a neuron with a shifting RF, the CD signal enables visual responsiveness at a specific location on the visual map, and the information conveyed by the neuron is about visual events in visual coordinates.

In temporal terms, saccadic CD has a gating effect on visual activity. A normal visual response to a stimulus in the RF has a brisk onset and fixed latency relative to stimulus onset. But activity evoked by a visual stimulus in the FF has no such brisk onset; as was shown in **Figure 5**, it is tied not to stimulus onset but to saccade onset. The visual activity may be held in check for as long as 1000 ms until the saccade begins (Umeno & Goldberg 2001). Only around the time of the saccade is the visual activity released in a surge of action potentials. As noted above, saccadic gating of the FF visual activity is a strange but seemingly critical feature; it guarantees that a shifting RF occurs only if a saccade occurs.

The above description of shifting RFs confirms that the phenomenon has two phases: an early phase of visual sensitivity at the FF that can begin long before the saccade, and a later phase of unleashed visual activity that begins with the saccade. These two phases fit well with the nature of the CD signal (see **Figure 1c**, *top*), which has a low-frequency, prelude phase that could induce visual sensitivity at the FF and a high-frequency, burst phase that could trigger the saccade-aligned visual activity.

Considering these shifting RF properties, we can evaluate the hypothesis that they contribute to the perception of visual stability (Colby & Goldberg 1999, Sommer & Wurtz 2006). We think the hypothesis is compelling because of a specific, key feature of shifting RFs: The presaccadic FF location and the postsaccadic RF location sample the same absolute location in visual space. For an illustration of this overlap, see **Figure 3a** and note that the FF location (*middle box*) is identical to the postsaccadic RF location (*right box*). Information is available, therefore, about a single region of the visual world before and after the saccade. If the presaccadic image equals the postsaccadic image, objects must have remained stable across the saccade. If there is a presaccadic-postsaccadic difference, objects must have moved while the eyes moved. Thus, shifting RFs could allow single neurons to assess the stability of the world across saccades.

Confirmation or refutation of this hypothesis calls for two lines of research. One would address in a conceptual manner how transsaccadic information provided by shifting RFs could lead to a percept of visual stability. Which exact computational principles are involved? Must an overt representation of stable space emerge from the comparison of presaccadic FF with postsaccadic RF, or can it remain implicit at a network level? Answering these questions will almost certainly depend on developing more sophisticated models of how neuronal activity leads to perceptual experiences.

The other line of research would empirically test the link between shifting RFs and perceptual stability. On this issue, the experiments on the SC-MD-FEF pathway provide a new opportunity. Because we have established that at the neuronal level we can disrupt the shifting RFs in FEF by interrupting the ascending pathway, we can now test whether interrupting this pathway also disrupts visual stability when saccades occur. Investigators would train a monkey to report when an object moves in front of it as opposed to when an object seems to move as a result of its own saccades. We predict that during inactivation of the CD pathway, the monkey will erroneously report that an object moves during a saccade when, in fact, it is still.

Such a demonstration would provide the final link in the mechanistic chain from identified CD signal, to the shifting RF phenomenon, to the percept of visual stability across saccades.

CD BEYOND VISION

The Nature of the CD Signal

Until recent years, the mechanistic nature of CD in the primate has largely been a matter of speculation. Now, with insight into one example of the neuronal basis of CD, we can compare speculation with observation. Three salient concepts about how CD exerts its influences have been suppression, cancellation, and forward models.

The first CD concept is also the simplest: The CD simply suppresses visual information. Evidence for such suppression with saccades has been found in single neurons of the superficial visual layers in SC (Robinson & Wurtz 1976). The suppression resulted from a CD rather than from proprioceptive feedback because preventing the proprioception did not eliminate the suppression (Richmond & Wurtz 1980). This was the first demonstration of the CD's effect in the primate brain. The recent demonstration of an inhibitory input from the saccade-related intermediate layers of the SC to the visual superficial layers (Lee et al. 2007) suggests a mechanism for the effect. Clear suppression, which may also result from CD, has been demonstrated in the primate pulvinar (Robinson & Petersen 1985) and extrastriate area MT (Thiele et al. 2002); the lateral geniculate nucleus shows only modest suppression (Ramcharan et al. 2001, Reppas et al. 2002, Royal et al. 2006).

The second concept, cancellation, was put forward most forcefully by von Holst & Mittelstaedt (1950): "[T]he efference leaves an 'image' of itself somewhere in the CNS to which the reafference (i.e., sensory signs of the resulting movement) compares as the negative of a photograph compares to its print" (von Holtz quoted in MacKay 1972). Note that reafference is specifically due to self-movement,

Reafference: sensory inflow evoked by one's own movements

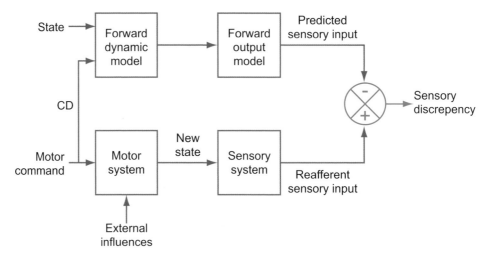

Figure 8

The forward model concept. CD (*left*) is combined with the current state of the system to generate a prediction of the sensory input (*upper right*). This prediction is compared with the actual, reafferent sensory input (*lower right*). Resultant discrepancies inform the brain about external influences and miscalibrations. Modified from Wolpert & Miall (1996) with permission from Elsevier.

in contrast to afference. Cancellation mechanisms have been found in some CD circuits, for example, in electric fish (Bell 1982), but have seemed less likely in the primate visual system (Bridgeman 2007). In our experiments, the activity representing the CD in MD (a vector encoding movement) and that representing visual activity in FEF (a response field on a retinotopic map) are strikingly different. Simple cancellation makes little sense, so we find no support for this interpretation of CD in the primate for the visual-saccadic system.

The third and most recently promoted concept, the forward model, has been useful particularly in skeletal movement analysis (Wolpert & Miall 1996) and has been invoked occasionally for the oculomotor system (Murthy et al. 2007, Robinson et al. 1986, Vaziri et al. 2006). **Figure 8** illustrates the concept. A copy of the motor command, or CD in our terminology, is routed into a forward dynamic model. This feedback about a new movement is used to create a feedforward model of the predicted sensory consequences of the movement: the forward output model. The prediction of the sensory input is compared with the actual, reafferent sensory input to determine the extent of

any sensory discrepancy—similar to the cancellation interpretation of CD. A major advance of forward models over previous cancellation hypotheses, however, is the explicit recognition that the copy of the motor command, in motor coordinates, must be transformed into sensory coordinates so that it can be compared directly with the sensory input (Webb 2004, Wolpert & Miall 1996). As we have noted above, such a motor-to-sensory transformation is exactly what we find in the FEF, which uses CD in saccadic motor coordinates to generate shifting receptive fields in visual coordinates. Hence the FEF may contain elements of a forward model (Crapse & Sommer 2008).

Results from a variety of studies in the primate visual-saccadic system therefore support two major concepts of CD: its role in suppression mechanisms and forward models. Little evidence supports a cancellation mechanism for saccades.

CD and Human Disease

For as long as CD has been recognized, researchers have speculated about the relation between CD and human disease. von Helmholtz

Afference: sensory inflow evoked by changes in the environment

(1925) described how eye muscle paralysis causes a mismatch between impaired saccades and intact "effort of will," as he called CD. He reported that the mismatch induces disturbing visual percepts, and subsequent work has refined his initial observation (Kornmueller 1931, Stevens et al. 1976). The reverse condition, in which muscles are intact but CD is lost, results from central nervous system lesions. Thalamic and parietal lesions cause CD deficits in double-step tasks, as discussed above, and parietal damage impairs the awareness of movement intentions (Sirigu et al. 2004). Frontal damage may cause CD-related deficits in perception (Teuber 1964), and visual cortex lesions can cause an illusion that the visual world is moving when the eyes move across a static background scene (Haarmeier et al. 1997).

Broader damage to CD networks, reaching beyond focal lesions, may have more devastating and long-term consequences. Schizophrenia may involve a CD disruption that could impair the self-monitoring of thoughts or actions so that they are misattributed to external sources (Feinberg 1978, Feinberg & Guazzelli 1999). CD deficits in schizophrenia could arise from the disrupted subcortical-thalamic-cortical circuits that are known to characterize the disease (Andreasen et al. 1999). Schizophrenics are impaired in tasks requiring CD (Malenka et al. 1986), and their auditory systems show particularly strong evidence for loss of CD-related modulation (Ford et al. 2001, 2007). Neurophysiological assessment of CD circuits (e.g., the SC-MD-FEF pathway) could be performed in monkey models of schizophrenia as induced with dopaminergic or glutamatergic drugs (Condy et al. 2005, Stone et al. 2007) or fetal irradiation (Selemon et al. 2005). While a contribution of impaired CD to schizophrenic positive symptoms, e.g., hallucinations, is the most obvious tie-in of the function to the disease, Frith (1987) argued that impaired CD could contribute to negative and cognitive symptoms as well.

CD deficits have also been implicated in disorders with a significant motor component such as Huntington's disease (Smith et al. 2000) and developmental coordination disorder (Katschmarsky et al. 2001). We expect that further insights into the possible role of CD impairments in disease will continue to develop as circuits for CD in the normal brain become better understood.

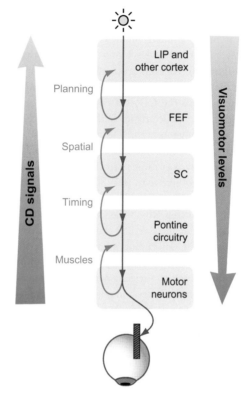

Figure 9

Hypothesized chain of saccadic CD in the primate brain. Descending visuomotor levels (*right*) extend from the cortical areas that detect a visual stimulus (*top*) to the brainstem motor neurons that innervate the muscles (*bottom*). Ascending CD signals may branch from any of these levels (*left*). The pathway that branches from SC up to the FEF conveys a CD signal rich with spatial vector information because that is the nature of the signal found in the SC and sent downstream from it. Other CD signals, branching from above and below the SC, would reflect other aspects of movement. It may be more accurate to consider CD as a set of signals representing the state of each saccade-related motor level, in contrast with a unitary CD signal representing the saccade itself.

CONCLUSION

We have described the specific issue of using CD from the saccadic system to influence visual processing and have indicated more generally how CD might be involved in other brain movement systems and in their pathology. We focus on a CD pathway that emanates from the midbrain and influences the cerebral cortex.

Although we have concentrated on a CD arising from one level of a sensorimotor pathway in the primate brain, CD could arise at any of several levels along such a pathway (**Figure 9**). At each level, the characteristics of the descending sensorimotor signals change, and because the CD is a copy of the motor signal, the characteristics of the ascending CD would change as well. The particular visual-saccadic CD that we consider branches off at a point where saccadic vectors are represented on a retinotopic map (the SC), and this spatial CD is well suited for visuomotor functions such as shifting RFs, which require input related to that map. A CD that branches from lower in the descending pathway, in the pontine circuitry, should convey information more tightly related to saccadic timing parameters. CD copies of velocity and position signals play an important role in low-level saccadic control

circuits (Robinson 1975) and may feed back to the SC (Soetedjo et al. 2002). CD signals that branch from the lowest levels would provide information about muscle innervation. At the upper extreme, CD signals may emerge from areas of cerebral cortex such as the FEF; they may convey more tentative plans to move rather than indicating distinct retinotopic vectors of the impending movement. Finally, the information at each level in the chain may diverge to influence multiple different targets. Which CD goes where would depend on the requirement of the target sensory area.

For any movement, there may be a set of CD signals; each CD conveys the state of the specific sensorimotor level from which it emerges. Although we may still speak of CD in loose terms as representing the state of a movement, it is more precise—and more beneficial to experimenters—to realize that in the primate brain, a CD signal represents the state of a movement structure such as the SC. The information content of the CD signal is limited by the signals available to the structure, and the potential targets of the CD signal are constrained by the anatomical reach of the structure's efferent pathways. Understanding these principles may be the key to identifying further CD pathways in the primate brain.

DISCLOSURE STATEMENT

The authors are not aware of any biases that might be perceived as affecting the objectivity of this review.

ACKNOWLEDGEMENTS

We thank M.E. Goldberg, C.L. Colby, R.A. Berman, T.B. Crapse, and J.P. Mayo for comments on an earlier draft of this review. The work was supported by the National Institutes of Health (EY017592) and the Alfred P. Sloan Foundation (M.A.S.), and the Intramural Research Program in the National Eye Institute, part of the National Institutes of Health (R.H.W.).

LITERATURE CITED

Andersen RA, Buneo CA. 2002. Intentional maps in posterior parietal cortex. *Annu. Rev. Neurosci.* 25:189–220

Andreasen NC, Nopoulos P, O'Leary DS, Miller DD, Wassink T, Flaum M. 1999. Defining the phenotype of schizophrenia: cognitive dysmetria and its neural mechanisms. *Biol. Psychiatry* 46:908–20

Batista AP, Buneo CA, Snyder LH, Andersen RA. 1999. Reach plans in eye-centered coordinates. *Science* 285:257–60

Bell CC. 1982. Properties of a modifiable efference copy in an electric fish. *J. Neurophysiol.* 47:1043–56

Bellebaum C, Daum I, Koch B, Schwarz M, Hoffmann KP. 2005. The role of the human thalamus in processing corollary discharge. *Brain* 128:1139–54

Benevento LA, Fallon JH. 1975. The ascending projections of the superior colliculus in the rhesus monkey (*Macaca mulatta*). *J. Comp. Neurol.* 160:339–61

Berman RA, Heiser LM, Dunn CA, Saunders RC, Colby CL. 2007. Dynamic circuitry for updating spatial representations. III. From neurons to behavior. *J. Neurophysiol.* 98:105–21

Berman RA, Heiser LM, Saunders RC, Colby CL. 2005. Dynamic circuitry for updating spatial representations. I. Behavioral evidence for interhemispheric transfer in the split-brain macaque. *J. Neurophysiol.* 94:3228–48

Bridgeman B. 2007. Efference copy and its limitations. *Comput. Biol. Med.* 37:924–29

Chafee MV, Goldman-Rakic PS. 2000. Inactivation of parietal and prefrontal cortex reveals interdependence of neural activity during memory-guided saccades. *J. Neurophysiol.* 83:1550–66

Colby CL, Duhamel JR, Goldberg ME. 1996. Visual, presaccadic, and cognitive activation of single neurons in monkey lateral intraparietal area. *J. Neurophysiol.* 76:2841–52

Colby CL, Goldberg ME. 1999. Space and attention in parietal cortex. *Annu. Rev. Neurosci.* 22:319–49

Condy C, Wattiez N, Rivaud-Pechoux S, Gaymard B. 2005. Ketamine-induced distractibility: an oculomotor study in monkeys. *Biol. Psychiatry* 57:366–72

Crapse TB, Sommer MA. 2008. The frontal eye field as a prediction map. *Prog. Brain. Res.* In press

Duhamel JR, Colby CL, Goldberg ME. 1992a. The updating of the representation of visual space in parietal cortex by intended eye movements. *Science* 255:90–92

Duhamel JR, Goldberg ME, Fitzgibbon EJ, Sirigu A, Grafman J. 1992b. Saccadic dysmetria in a patient with a right frontoparietal lesion. The importance of corollary discharge for accurate spatial behaviour. *Brain* 115(Pt. 5):1387–402

Feinberg I. 1978. Efference copy and corollary discharge: implications for thinking and disorders. *Schizophr. Bull.* 4:636–40

Feinberg I, Guazzelli M. 1999. Schizophrenia—a disorder of the corollary discharge systems that integrate the motor systems of thought with the sensory systems of consciousness. *Br. J. Psychiatry* 174:196–204

Ford JM, Gray M, Faustman WO, Roach BJ, Mathalon DH. 2007. Dissecting corollary discharge dysfunction in schizophrenia. *Psychophysiology* 44:522–29

Ford JM, Mathalon DH, Heinks T, Kalba S, Faustman WO, Roth WT. 2001. Neurophysiological evidence of corollary discharge dysfunction in schizophrenia. *Am. J. Psychiatry* 158:2069–71

Frith CD. 1987. The positive and negative symptoms of schizophrenia reflect impairments in the perception and initiation of action. *Psychol. Med.* 17:631–48

Gaymard B, Rivaud S, Pierrot-Deseilligny C. 1994. Impairment of extraretinal eye position signals after central thalamic lesions in humans. *Exp. Brain Res.* 102:1–9

Goldman-Rakic PS, Porrino LJ. 1985. The primate mediodorsal (MD) nucleus and its projection to the frontal lobe. *J. Comp. Neurol.* 242:535–60

Grüsser OJ. 1995. On the history of the ideas of efference copy and reafference. *Clio. Med.* 33:35–55

Guthrie BL, Porter JD, Sparks DL. 1983. Corollary discharge provides accurate eye position information to the oculomotor system. *Science* 221:1193–95

Haarmeier T, Thier P, Repnow M, Petersen D. 1997. False perception of motion in a patient who cannot compensate for eye movements. *Nature* 389:849–52

Hallett PE, Lightstone AD. 1976. Saccadic eye movements to flashed targets. *Vis. Res.* 16:107–14

Hanes DP, Schall JD. 1995. Countermanding saccades in macaque. *Vis. Neurosci.* 12:929–37

Heide W, Blankenburg M, Zimmermann E, Kompf D. 1995. Cortical control of double-step saccades: implications for spatial orientation. *Ann. Neurol.* 38:739–48

Heiser LM, Berman RA, Saunders RC, Colby CL. 2005. Dynamic circuitry for updating spatial representations. II. Physiological evidence for interhemispheric transfer in area LIP of the split-brain macaque. *J. Neurophysiol.* 94:3249–58

Heiser LM, Colby CL. 2006. Spatial updating in area LIP is independent of saccade direction. *J. Neurophysiol.* 95:2751–67

Katschmarsky S, Cairney S, Maruff P, Wilson PH, Currie J. 2001. The ability to execute saccades on the basis of efference copy: impairments in double-step saccade performance in children with developmental co-ordination disorder. *Exp. Brain Res.* 136:73–78

Kornmueller AE. 1931. Eine experimentelle anastesie der ausseren augenmusckeln am menschen und ihre auswirkungen. *J. Psychol. Neurol. Lpz.* 41:354–66

Kusunoki M, Goldberg ME. 2003. The time course of perisaccadic receptive field shifts in the lateral intraparietal area of the monkey. *J. Neurophysiol.* 89:1519–27

Lee PH, Sooksawate T, Yanagawa Y, Isa K, Isa T, Hall WC. 2007. Identity of a pathway for saccadic suppression. *Proc. Natl. Acad. Sci. USA* 104:6824–27

Lewis RF, Zee DS, Hayman MR, Tamargo RJ. 2001. Oculomotor function in the rhesus monkey after deafferentation of the extraocular muscles. *Exp. Brain Res.* 141:349–58

Lynch JC, Hoover JE, Strick PL. 1994. Input to the primate frontal eye field from the substantia nigra, superior colliculus, and dentate nucleus demonstrated by transneuronal transport. *Exp. Brain Res.* 100:181–86

MacKay DM. 1972. Voluntary eye movements as questions. *Bibl. Ophthalmol.* 82:369–76

Malenka RC, Angel RW, Thiemann S, Weitz CJ, Berger PA. 1986. Central error-correcting behavior in schizophrenia and depression. *Biol. Psychiatry* 21:263–73

Mays LE, Sparks DL, Porter JD. 1987. Eye movements induced by pontine stimulation: interaction with visually triggered saccades. *J. Neurophysiol.* 58:300–17

Medendorp WP, Goltz HC, Vilis T, Crawford JD. 2003. Gaze-centered updating of visual space in human parietal cortex. *J. Neurosci.* 23:6209–14

Merriam EP, Genovese CR, Colby CL. 2003. Spatial updating in human parietal cortex. *Neuron* 39:361–73

Merriam EP, Genovese CR, Colby CL. 2007. Remapping in human visual cortex. *J. Neurophysiol.* 97:1738–55

Morris AP, Chambers CD, Mattingley JB. 2007. Parietal stimulation destabilizes spatial updating across saccadic eye movements. *Proc. Natl. Acad. Sci. USA* 104:9069–74

Murthy A, Ray S, Shorter SM, Priddy EG, Schall JD, Thompson KG. 2007. Frontal eye field contributions to rapid corrective saccades. *J. Neurophysiol.* 97:1457–69

Nakamura K, Colby CL. 2002. Updating of the visual representation in monkey striate and extrastriate cortex during saccades. *Proc. Natl. Acad. Sci. USA* 99:4026–31

Poulet JF, Hedwig B. 2007. New insights into corollary discharges mediated by identified neural pathways. *Trends Neurosci.* 30:14–21

Quaia C, Optican LM, Goldberg ME. 1998. The maintenance of spatial accuracy by the perisaccadic remapping of visual receptive fields. *Neural Netw.* 11:1229–40

Ramcharan EJ, Gnadt JW, Sherman SM. 2001. The effects of saccadic eye movements on the activity of geniculate relay neurons in the monkey. *Vis. Neurosci.* 18:253–58

Reppas JB, Usrey WM, Reid RC. 2002. Saccadic eye movements modulate visual responses in the lateral geniculate nucleus. *Neuron* 35:961–74

Richmond BJ, Wurtz RH. 1980. Vision during saccadic eye movements. II. A corollary discharge to monkey superior colliculus. *J. Neurophysiol.* 43:1156–67

Robinson DA. 1975. Oculomotor control signals. In *Basic Mechanisms of Ocular Motility and Their Clinical Implications*, ed. G Lennerstrand, P Bach-y-Rita, pp. 337–74. Oxford: Pergamon

Robinson DA, Gordon JL, Gordon SE. 1986. A model of the smooth pursuit eye movement system. *Biol. Cybern.* 55:43–57

Robinson DL, Petersen SE. 1985. Response of pulvinar neurons to real and self-induced stimulus movement. *Brain Res.* 338:392–94

Robinson DL, Wurtz RH. 1976. Use of an extraretinal signal by monkey superior colliculus neurons to distinguish real from self-induced stimulus movement. *J. Neurophysiol.* 39:852–70

Royal DW, Sary G, Schall JD, Casagrande VA. 2006. Correlates of motor planning and postsaccadic fixation in the macaque monkey lateral geniculate nucleus. *Exp. Brain Res.* 168:62–75

Schall JD. 2002. The neural selection and control of saccades by the frontal eye field. *Philos. Trans. R. Soc. London Ser. B* 357:1073–82

Schiller PH, Sandell JH. 1983. Interactions between visually and electrically elicited saccades before and after superior colliculus and frontal eye field ablations in the rhesus monkey. *Exp. Brain Res.* 49:381–92

Selemon LD, Wang L, Nebel MB, Csernansky JG, Goldman-Rakic PS, Rakic P. 2005. Direct and indirect effects of fetal irradiation on cortical gray and white matter volume in the macaque. *Biol. Psychiatry* 57:83–90

Sirigu A, Daprati E, Ciancia S, Giraux P, Nighoghossian N, et al. 2004. Altered awareness of voluntary action after damage to the parietal cortex. *Nat. Neurosci.* 7:80–84

Smith MA, Brandt J, Shadmehr R. 2000. Motor disorder in Huntington's disease begins as a dysfunction in error feedback control. *Nature* 403:544–49

Soetedjo R, Kaneko CR, Fuchs AF. 2002. Evidence that the superior colliculus participates in the feedback control of saccadic eye movements. *J. Neurophysiol.* 87:679–95

Sommer MA, Wurtz RH. 1998. Frontal eye field neurons orthodromically activated from the superior colliculus. *J. Neurophysiol.* 80:3331–35

Sommer MA, Wurtz RH. 2002. A pathway in primate brain for internal monitoring of movements. *Science* 296:1480–82

Sommer MA, Wurtz RH. 2004a. What the brain stem tells the frontal cortex. I. Oculomotor signals sent from superior colliculus to frontal eye field via mediodorsal thalamus. *J. Neurophysiol.* 91:1381–402

Sommer MA, Wurtz RH. 2004b. What the brain stem tells the frontal cortex. II. Role of the SC-MD-FEF pathway in corollary discharge. *J. Neurophysiol.* 91:1403–23

Sommer MA, Wurtz RH. 2006. Influence of the thalamus on spatial visual processing in frontal cortex. *Nature* 444:374–77

Sparks DL, Hartwich-Young R. 1989. The neurobiology of saccadic eye movements. The deep layers of the superior colliculus. *Rev. Oculomot. Res.* 3:213–56

Sparks DL, Mays LE. 1983. Spatial localization of saccade targets. I. Compensation for stimulation-induced perturbations in eye position. *J. Neurophysiol.* 49:45–63

Sperry RW. 1950. Neural basis of the spontaneous optokinetic response produced by visual inversion. *J. Comp. Physiol. Psychol.* 43:482–89

Steinbach MJ. 1987. Proprioceptive knowledge of eye position. *Vis. Res.* 27:1737–44

Stevens JK, Emerson RC, Gerstein GL, Kallos T, Neufeld GR, et al. 1976. Paralysis of the awake human: visual perceptions. *Vis. Res.* 16:93–98

Stone JM, Morrison PD, Pilowsky LS. 2007. Glutamate and dopamine dysregulation in schizophrenia—a synthesis and selective review. *J. Psychopharmacol.* 21:440–52

Teuber HL. 1964. The riddle of the frontal lobe function in man. In *The Frontal Cortex and Behavior*. pp. 410–444. New York: McGraw-Hill

Thiele A, Henning P, Kubischik M, Hoffmann KP. 2002. Neural mechanisms of saccadic suppression. *Science* 295:2460–62

Tolias AS, Moore T, Smirnakis SM, Tehovnik EJ, Siapas AG, Schiller PH. 2001. Eye movements modulate visual receptive fields of V4 neurons. *Neuron* 29:757–67

Umeno MM, Goldberg ME. 1997. Spatial processing in the monkey frontal eye field. I. Predictive visual responses. *J. Neurophysiol.* 78:1373–83

Umeno MM, Goldberg ME. 2001. Spatial processing in the monkey frontal eye field. II. Memory responses. *J. Neurophysiol.* 86:2344–52

Vaziri S, Diedrichsen J, Shadmehr R. 2006. Why does the brain predict sensory consequences of oculomotor commands? Optimal integration of the predicted and the actual sensory feedback. *J. Neurosci.* 26:4188–97

von Helmholtz H. 1925 [1910]. *Helmholtz's Treatise on Physiological Optics.* Transl. JPC Southall. New York: Opt. Soc. Am. 3rd ed.

von Holst E, Mittelstaedt H. 1950. Das reafferenzprinzip. Wechselwirkungen zwischen central-nervensystem und peripherie. *Naturwissenschaften* 37:464–76

Walker MF, FitzGibbon EJ, Goldberg ME. 1994. Predictive visual responses in monkey superior colliculus. In *Contemporary Ocular Motor and Vestibular Research: A Tribute to David A. Robinson*, ed. U Büttner, T Brandt, A Fuchs, D Zee, pp. 512–19. Stuttgart, Germany: Thieme

Wang X, Zhang M, Cohen IS, Goldberg ME. 2007. The proprioceptive representation of eye position in monkey primary somatosensory cortex. *Nat. Neurosci.* 10:640–46

Webb B. 2004. Neural mechanisms for prediction: Do insects have forward models? *Trends Neurosci.* 27:278–82

Wolpert DM, Miall RC. 1996. Forward models for physiological motor control. *Neural Netw.* 9:1265–79

Wurtz RH, Sommer MA. 2004. Identifying corollary discharges for movement in the primate brain. *Prog. Brain Res.* 144:47–60

RELATED RESOURCES

Bell, CC. 2008. Cerebellum-like structures. *Annu. Rev. Neurosci.* 31:1–25

Bridgeman B. 2007. Efference copy and its limitations. *Comput. Biol. Med.* 37:924–29

Colby CL, Goldberg ME. 1999. Space and attention in parietal cortex. *Annu. Rev. Neurosci.* 22:319–49

Poulet JF, Hedwig B. 2007. New insights into corollary discharges mediated by identified neural pathways. *Trends Neurosci.* 30:14–21

Wnt Signaling in Neural Circuit Assembly

Patricia C. Salinas[1] and Yimin Zou[2]

[1] Department of Anatomy and Developmental Biology, University College London, London, WC1E 6BT, United Kingdom; email: p.salinas@ucl.ac.uk

[2] Division of Biological Sciences, Section of Neurobiology, University of California, San Diego, La Jolla, California; email: yzou@ucsd.edu

Annu. Rev. Neurosci. 2008. 31:339–58

The *Annual Review of Neuroscience* is online at neuro.annualreviews.org

This article's doi: 10.1146/annurev.neuro.31.060407.125649

0147-006X/08/0721-0339$20.00

Key Words

migration, polarity, axon guidance, dendrite, synapse formation

Abstract

The Wnt family of secreted proteins plays a crucial role in nervous system wiring. Wnts regulate neuronal positioning, polarization, axon and dendrite development, and synaptogenesis. These diverse roles of Wnt proteins are due not only to the large numbers of Wnt ligands and receptors but also to their ability to signal through distinct signaling pathways in different cell types and developmental contexts. Studies on Wnts have shed new light on novel molecular mechanisms that control the development of complex neuronal connections. This review discusses recent advances on how Wnt signaling influences different aspects of neuronal circuit assembly through changes in gene expression and/or cytoskeletal modulation.

Contents

INTRODUCTION

The function of the Wnt family of signaling proteins during embryonic development and disease has been well established. In recent years, it has become clear that Wnt signaling also plays a key role in the formation and modulation of neural circuits. Wnts regulate diverse cellular functions that include neuronal migration, neuronal polarization, axon guidance, dendrite development, and synapse formation, all of which are essential steps in the formation of functional neural connections. In addition, new emerging studies strongly suggest that Wnt signaling may also regulate synaptic function in the adult brain.

Wnt: Wingless Integration

PCP: planar cell polarity

Identifying the molecular principles of neuronal connectivity is central for understanding how complex neuronal circuits are formed and modulated but also how they can be repaired after brain injury. The Wnts' ability to stimulate neurite outgrowth and synaptic site formation suggests that Wnt activity modulation could be used to stimulate nerve regeneration and circuit repair. Elucidating the molecular mechanisms by which Wnts regulate diverse aspects of circuit formation will provide a basis for developing therapeutic approaches for nerve and brain regeneration after injury or disease.

This review focuses on recent findings on the function of Wnts in neuronal circuit assembly in different biological systems. Researchers have made substantial advances in identifying the Wnt molecular mechanisms of action during brain wiring. However, much of the progress on the role of Wnts in neuronal connectivity is relatively recent; therefore, many questions remain outstanding.

WNT LIGANDS SIGNAL THROUGH DIFFERENT SIGNALING CASCADES

Wnt proteins activate a number of signaling pathways, resulting in different cellular responses through binding to many receptors at the cell surface. Binding of Wnts to Frizzleds (Fzs), seven-pass transmembrane receptors, activates at least three different signaling pathways: the canonical or β-catenin pathway and the noncanonical pathways, which include the planar cell polarity (PCP) pathway and the calcium pathway (for review see Gordon & Nusse 2006, Logan & Nusse 2004). Different Wnt signaling pathways can be activated by different receptors. In addition to the Fz receptors (10 Fzs in mammals), the low-density lipoprotein receptor–related protein (LRP-5/6) is required as a coreceptor in the canonical pathway but not in the noncanonical pathways. Wnts also signal through two other receptors, Ryk/Derailed, a receptor tyrosine–like protein with an inactive kinase domain, and the receptor tyrosine kinase, Ror2 (Cadigan & Liu 2006). Although little is known about the signaling pathways downstream of these receptors, Ryk can form a complex with Frizzled and can signal through the canonical and noncanonical pathways depending on the cell or developmental context (Lu et al. 2004).

Binding Wnts to their receptors activates the cytoplasmic protein Dishevelled (Dvl), which brings together signaling components (Malbon & Wang 2006, Wallingford & Habas 2005). Downstream of Dvl, the pathway can branch into three different cascades. In the canonical pathway, Fz signals together with the LRP5/6 to activate Dvl. A key step in the canonical pathway is the inhibition of glycogen synthase kinase-3 (Gsk3), which phosphorylates the cytoplasmic protein β-catenin within a degradation complex containing Axin and adenomatous polyposis coli (APC). Canonical signaling increases the stability of β-catenin, which subsequently translocates to the nucleus, where it activates target gene transcription in association with transcription factors of the TCF/LEF family (Clevers 2006). A divergent canonical pathway in which transcription is not involved also operates in developing axons (Ciani et al. 2004) (see below). In the PCP pathway, Fz receptors are required, together with a number of core components such as Dvl, Flamingo, van Gogh, Diego, and Prickle, to activate Rac, Rho, and JNK (Jun N-terminal kinase) (for a review see Wang & Nathans 2007). This pathway regulates cell and tissue polarity through direct changes in the cytoskeleton. The calcium pathway, which also requires Fz receptors and Dvl, activates PKC (protein kinase C) and induces intracellular calcium mobilization and CAMKI activation (Kohn & Moon 2005). Documentation of cross-talk or interference between these pathways adds further complexity (Harris & Beckendorf 2007, Inoue et al. 2004, Mikels & Nusse 2006, Veeman et al. 2003, Wu et al. 2004). In neurons, both canonical and noncanonical pathways have been implicated in different aspects of embryonic patterning and also neuronal connectivity.

WNT-FRIZZLED SIGNALING AND DIRECTIONAL NEURONAL MIGRATION IN *CAENORHABDITIS ELEGANS*

Neuronal migration controls neuronal positioning in the nervous system, which is a vital step in nervous system wiring. Neurons undergo extensive migration to arrive at their final anatomical positions. Neuronal migration is typically highly directional and is controlled by several guidance systems (Hatten 2002).

Initial work in *Caenorhabditis* linked Wnt signaling to neuronal migration. While studying the guidance mechanisms controlling Q neuroblast migration, investigators found that *wnt/egl-20*, *frizzled/lin-17*, and *frizzled/mig-1* were involved in proper directional migration. Q neuroblasts are a pair of sensory neuron precursors that migrate along the anterior-posterior (A-P) body axis. The Q cell on the right side of the body, QR, migrates anteriorly, whereas the cell on the left side, QL, migrates posteriorly. In *egl-20* and *lin-17* mutants, both QR and QL migrate randomly along the A-P axis. Canonical Wnt signaling controls the posterior migration of QL, whereas an unknown noncanonical pathway mediates the anterior migration of QR cells (**Figure 1**).

Is EGL-20 a directional cue for Q cell migration? Moving the source of EGL-20 from posterior to anterior did not reverse the direction of Q cell migration (Harris et al. 1996, Maloof et al. 1999, Whangbo & Kenyon 1999). Therefore, EGL-20 may not be a directional cue for Q cells because the QR descendents do not require graded EGL-20 protein. However, one could not exclude the possibility that other Wnts may still be directional cues for Q cells because other Wnts may contribute to A-P guidance of Q cell migration. Indeed, two other Wnts are expressed in a low to high A-P gradient (CWN-1 and LIN-44). More recent work showed that Wnt signaling is also required for the anterior migration of a different neuron class, the hermaphrodite-specific neurons (HSNs). Genetic analyses showed that all Wnts and Frizzleds play a role in HSN migration (**Figure 1**). Misexpression experiments showed

APC: adenomatous polyposis coli

TCF: T-cell factor

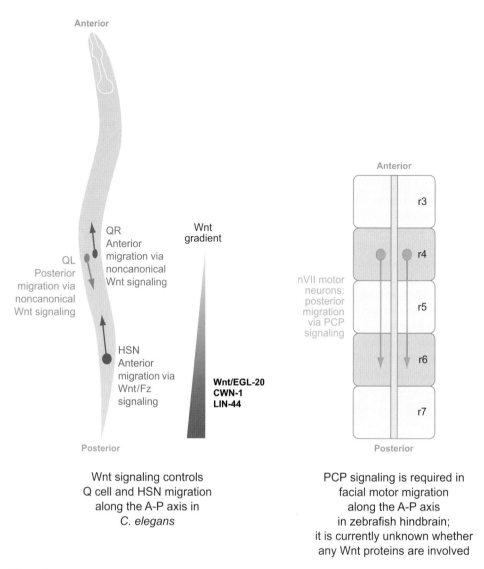

Figure 1

Neuronal migration along the anterior-posterior axis.

that at least Wnt/EGL-20 can act as a repellent for HSN neurons along the A-P axis (Pan et al. 2006). Therefore, Wnts can clearly act as directional cues in this case. Frizzleds likely mediate Wnt function in neuronal migration. *C. elegans* studies revealed that Frizzled signaling is rather complex. Frizzled/MIG-1 single mutation already results in HSN migration defects. Two other Frizzleds, MOM-5 and CFZ2, had no effect by themselves but enhanced the phenotypes of MIG-1, suggesting that they interact with MIG-1 somehow and assist the MIG-1 function. Another Frizzled, LIN-17, antagonized MIG-1 signaling. *Cfz-2* is also required cell nonautonomously for the anterior migration of anterior lateral microtubule (ALM) neurons (Zinovyeva & Forrester 2005). Therefore, further studies in this area will shed light on the complex Frizzled signaling networks.

Although many examples show that Wnt signaling controls the migration of normal and cancer cells in vertebrates, still unknown is whether Wnts are involved in vertebrate neuronal migration. Recent work in zebrafish hindbrain showed that components of the PCP pathway, strabismus (Stbm)/Van Gogh (Vang), Prickle 1, Frizzled3a, and Celsr2 (Flamingo) are required for branchiomotor neuron migration along the A-P axis (Bingham et al. 2002, Carreira-Barbosa et al. 2003, Jessen et al. 2002, Wada et al. 2006). Facial motor neurons originate from rhombomere 4 and migrate caudally to rhombomere 6 of the developing hindbrain (**Figure 1**). Mutations of these PCP genes blocked the caudal migration, and the motor neurons either stay in rhombomere 4 or start abnormal radial migration. Wnts would be good candidates involved in this migration given that Wnt (Wnt11) is involved in vertebrate convergent extension, which also requires the same PCP components (Heisenberg et al. 2000). However, currently there is no direct evidence for or against a Wnt involvement in this process.

WNTS AND FRIZZLEDS IN NEURONAL POLARITY

A fundamental feature of a vertebrate neuron is its highly polarized arrangement of axonal and dendritic processes. These polarized compartments are made of distinct components that mediate different cellular and physiological functions. Proper polarization of neuronal processes is essential for the development of neuronal connections and nervous system wiring. The mechanisms of neuronal polarity are beginning to be unveiled.

C. elegans neurons often do not have clear axon-dendrite distinction on a process having neighboring axonal and dendritic segments within the same process. In some neurons, such as the PLM neurons, however, the anterior and posterior processes have distinct functions and, therefore, have clear axon-dendrite polarity. Only the anterior process makes gap junctions and synapses, and there are no synapses in the posterior process (**Figure 2**). *Wnt/lin-44* and *frizzled/lin-17* are required for proper polarization of the PLM neurons (Hilliard & Bargmann 2006, Prasad & Clark 2006). In *lin-44* and *lin-17* mutants, the polarity of PLM neurons is completely reversed, with synapses formed in the posterior process. LIN-17 is asymmetrically localized only to the posterior process and requires LIN-44 for that localization.

Is a Wnt gradient important for neuronal polarity? Some evidence demonstrates that Wnts provide directional information for neuronal polarity of some neurons such as the ALM, although research is still unclear about others such as the PLM (Hilliard & Bargmann 2006). A localized source of LIN-44 is apparently not essential for proper PLM polarization because LIN-44 can partially rescue PLM when expressed uniformly from a heat-shock promoter. However, the directional role for Wnts is still possible because two other Wnts, EGL-20 and CMN-1, may also regulate PLM polarity. The reason that LIN-44 can rescue A-P polarity of PLM may be that LIN-44 may sensitize its response to these two other Wnts and plays a permissive role; these two other Wnts are still present in an A-P expression gradient and are present nearby. This latter possibility is consistent with the observation that when EGL-20 was ubiquitously expressed, the polarity of another neuron, ALM, was reversed (Hilliard & Bargmann 2006). Different from PLM, ALM is located anteriorly, far away from posterior sources of the other two Wnts (CWN-1 and LIN-44), and perhaps EGL-20 misexpression is sufficient to alter the gradient. It is currently unknown which signaling pathway mediates cell polarization along the A-P axis. A good candidate would be a pathway similar to the PCP, although a Wnt protein has not been implicated in fly PCP.

Wnt signaling through Dvl may regulate neuronal polarity through the PAR (partitioning defective) polarity pathway in vertebrates. Studies using hippocampal cultures have suggested that the PAR3-PAR6-aPKC pathway is essential for axonal differentiation and neuronal polarity (Nishimura et al. 2004; Shi et al. 2003).

Neuronal polarity: Neurons are highly polarized cells usually with axons and dendrites compartmentalized in distinct areas; in vertebrates, they typically grow out of opposite sides of the cell body

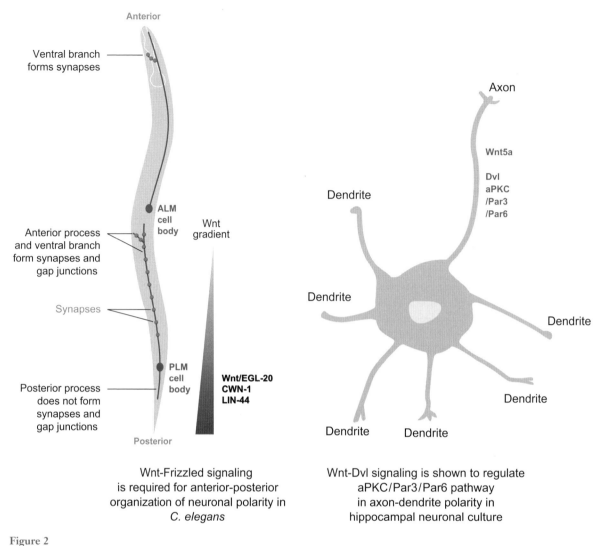

Figure 2

Wnt signaling and neuronal polarity.

Within the figure:

Anterior

Ventral branch forms synapses

Anterior process and ventral branch form synapses and gap junctions

Synapses

Posterior process does not form synapses and gap junctions

Posterior

ALM cell body

PLM cell body

Wnt gradient

Wnt/EGL-20 CWN-1 LIN-44

Wnt-Frizzled signaling is required for anterior-posterior organization of neuronal polarity in *C. elegans*

Axon

Wnt5a

Dvl aPKC /Par3 /Par6

Dendrite

Dendrite

Dendrite

Dendrite

Dendrite

Dendrite

Dendrite

Wnt-Dvl signaling is shown to regulate aPKC/Par3/Par6 pathway in axon-dendrite polarity in hippocampal neuronal culture

Upstream of this complex, both PAR1 and LKB-1 are required for neuronal polarization in mammals *in vivo*. *In vivo* evidence of PAR3-PAR6-aPKC regulating vertebrate neuronal polarity is still lacking (Barnes et al. 2007, Kishi et al. 2005, Shelly et al. 2007). In *Drosophila*, aPKC is not required for axon-dendrite polarity (Rolls & Doe 2004). A recent study using the same hippocampal culture system showed that Dvl may be upstream of this polarity pathway and that Wnt5a can stimulate the Dvl-mediated aPKC stabilization (**Figure 2**)

(Zhang et al. 2007). Overexpression of Dvl results in multiple axons, and aPKC mediates the Dvl-induced axon differentiation. This action potentially connects Wnt signaling to the PAR polarity pathway. The *in vivo* relevance of Wnts activating aPKC to regulate neuronal polarization awaits further investigation.

WNT SIGNALING IN AXON GUIDANCE

A remarkable part of brain wiring is the highly controlled process of axon growth and guidance. Neurons typically send out axons, which often navigate long distances to find their target area and then seek out their specific post-synaptic partners to make functional synapses. Many classes of axon guidance molecules have been discovered, including embryonic morphogens (Dickson 2002, Tessier-Lavigne & Goodman 1996, Zou & Lyuksyutova 2007). A large body of evidence from both in vivo and in vitro studies now demonstrates that Wnt proteins are guidance cues for axon pathfinding and target selection. In many cases, their biological functions and mechanisms are highly conserved between vertebrates and invertebrates.

Overexpression of *DWnt5* (previously known as *DWnt3*) disrupting commissural axon tracts (Fradkin et al. 1995) hinted that Wnts may affect axon growth in the *Drosophila* midline. Subsequent studies then showed that DWnt5 is a repulsive cue through Derailed, a receptor-tyrosine-like protein that controls the crossing of commissural axons in the *Drosophila* embryonic central nervous system (Yoshikawa et al. 2003). DWnt5 protein is enriched in the posterior commissure and prevents the Derailed-expressing axons from crossing the midline via the posterior commissure in each embryonic segment.

Studies in vertebrate spinal cord showed that Wnt signaling controls the direction of axon growth along the A-P axis of the spinal cord. *Wnt4* and *Wnt7b* are expressed in an A-P gradient in the ventral midline at midgestation and attract spinal cord commissural axons to turn anterior after midline crossing through an attractive guidance mechanism (Imondi & Thomas 2003, Lyuksyutova et al. 2003). In *Frizzled3* mutant mice, commissural axons turn randomly along the A-P axis. Recent studies suggest that atypical PKC signaling is required for Wnt attraction and that the anterior-posterior guidance of spinal cord commisural axons and PI3K may serve as an on switch for attractive Wnt reponsiveness upon midline cross (Wolf et al. 2008). Remarkably, two other Wnts, *Wnt1* and *Wnt5a*, are expressed in an A-P gradient in the dorsal spinal cord in neonatal rodents, and they repel descending corti-cospinal tract axons to grow down the spinal cord (Dickson 2005, Liu et al. 2005). Corticospinal tract axons are initially insensitive to Wnt repulsion. The onset of repulsion responsive to Wnts correlates with the timing of midline crossing (pyramidal decussation). After corticospinal tract axons have crossed the midline and entered the rostral spinal cord, they begin to respond to Wnts. Ryk, the vertebrate homolog of *Drosophila* Derailed, mediates this repulsion. The temporal expression pattern of Ryk correlates with the onset of Wnt repulsion. In the mammalian brain, Ryk mediates repulsive activity of Wnt5a to corpus collosum after they have crossed the midline and defines the tight fascicles of corpus collosum axons (Keeble & Cooper 2006). Ryk is also required for attractive Wnt response in dorsal root ganglion (DRG) cell axons. In Ryk-deficient mice created by an RNAi transgene, DRG axons failed to respond to Wnt3 attraction (Lu et al. 2004).

Wnt signaling plays highly conserved biological roles in axon guidance both in A-P guidance in vertebrate and in *C. elegans* axon pathfinding. Three *C. elegans* Wnts are expressed in an A-P increasing gradient. Wnt signaling is important for A-P guidance and neuronal polarity of several neurons as well as migration of neuroblasts as previously discussed (Hilliard & Bargmann 2006, Pan et al. 2006, Prasad & Clark 2006). Although Wnts provide A-P directional information in both vertebrates and nematodes, there are several differences, including the direction of Wnt gradients along the A-P axis and the role of Frizzleds in attractive and repulsive responses to Wnts (Zou 2006).

Wnts are also highly conserved topographic mapping cues in vertebrates and *Drosophila*. In the chick retinotectal system, Wnt signaling plays a role in medial-lateral topographic mapping by conveying positional information (Flanagan 2006, Luo 2006, Schmitt et al. 2006). Dorsal retina ganglia cell (RGC) axons target to lateral tectum, and ventral RGC axons project to medial tectum, forming a continuous spatial representation of the visual world along the dorsal-ventral axis. Computational

Morphogens: diffusible signaling proteins that form gradients and determine cell fate on the basis of concentration

Topographic mapping: spatial information is smoothly and continuous represented from one part of the nervous sytem to another through ordered projections

modeling suggested that for topographic maps to form, counterbalancing forces that oppose each other along each axis are necessary (Fraser & Hunt 1980, Fraser & Perkel 1990, Gierer 1983, Prestige & Willsaw 1975). Wnt3 is expressed in a medial-lateral decreasing gradient and Ryk is expressed in a dorsal-ventral increasing gradient. Wnt3 is a laterally directing mapping force, and the ventral axons are more sensitive to the lateral Wnt3 repulsion because of the higher Ryk expression level. The other opposing force is ephrinB1-EphB signaling, which is medially directed and whose activity is attractive. EphrinB1 is expressed at high medial and low lateral gradients in the tectum. EphB receptors are expressed by forming a low-dorsal and high-ventral gradient in the RGCs of the retina. Therefore, the more ventral RGCs are more sensitive to the medial ephrinB1 attraction, balancing the Wnt3 activity. The mechanisms by which these two molecular gradients act to balance each other are currently unknown. We also do not know whether the opposing gradient model is a general mechanism that applies to other maps. Wnt signaling also controls retinotopic mapping along the dorsal-ventral axis of the *Drosophila* visual system (Sato et al. 2006). The 750 ommatidia send axons from the fly compound eye to brain targets in a topographically organized way. Dorsal ommatidia axons project to the dorsal arm of the lamina, whereas ventral axons target to the ventral lamina, forming a continuous spatial map. DWnt4 is expressed in the ventral lamina and directs ventral axons toward the ventral lamina via DFz2 receptor. In *DWnt4*, *DFz2*, and *Dvl* mutants, ventral photoreceptor axons mistarget dorsally (Sato et al. 2006). Although the actual topographic organizations are different between vertebrates and fly, Wnt signaling plays essential roles in topographic organization along the same body axis. Moreover, the phenotype of a dorsal shift in the map when Wnt signaling is ablated suggests that an apposing dorsally directed force exists in the *Drosophila* visual system, which is yet to be identified (Zou & Lyuksyutova 2007).

WNT SIGNALING AND DENDRITE MORPHOGENESIS

Dendrite growth and branching are critical episodes in the formation of functional neuronal connections. A combination of intrinsic and environmental factors regulates the dendritic morphology of individual neurons through changes in the cytoskeleton (Parrish et al. 2007). Several extracellular cues that stimulate or inhibit dendritic growth and branching have been identified (McAllister 2000, Whitford et al. 2002). Intracellular molecules such as Rho GTPases have been implicated in dendritic development (Luo 2002, Van Aelst & Cline 2004), but little is known about how extracellular cues modulate these molecular switches.

Studies in hippocampal neurons led to the discovery that Wnt signaling regulates dendritic morphogenesis. Wnt7b, which is expressed during hippocampal dendritogenesis, increases the length and branching of cultured hippocampal neurons (Rosso et al. 2005). Wnt7b increases the number of secondary and tertiary branches, an effect that is blocked by the Wnt antagonist, Sfrp1. Sfrp1 on its own impairs dendritic development, which suggests that endogenous Wnts regulate this process (Rosso et al. 2005). Consistent with this finding, Yu & Malenka (2003) found that cultured hippocampal neurons release Wnt proteins into the conditioned media. These results indicate that endogenous Wnts contribute to the normal dendritic arborization of hippocampal neurons in culture.

A noncanonical Wnt pathway through Dvl and Rac regulates dendrite morphogenesis. Gain and loss of Dvl function studies showed that Dvl is required for dendrite outgrowth and branching (Rosso et al. 2005). Wnt7b and Dvl activate a noncanonical pathway through Rac. Activated Rac increases dendritic development, whereas dominant-negative Rac blocks the effect of Wnt or Dvl (Rosso et al. 2005). Wnt7b and Dvl also activate JNK to regulate dendritic morphogenesis. In hippocampal neurons, dominant-negative JNK or exposure to JNK inhibitors blocks the dendritogenic effect

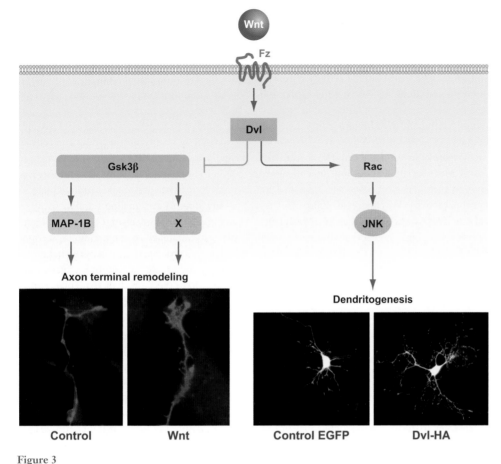

Figure 3

Wnts regulate the terminal arborization of axons and dendritic morphogenesis. Activation of the divergent canonical pathway regulates the terminal arborization of axons through inhibition of Gsk3 and changes in the phosphorylation of MAP1B and possibly other targets (X). In contrast, activation of the Wnt planar cell polarity pathway regulates dendritic morphogenesis. Wnt and Dvl activate Rac1 and JNK to increase dendrite length and branching.

of Dvl1. In contrast, JNK activation by low levels of anysomycin mimics the Wnt effect (Rosso et al. 2005). Thus Wnt signaling through a noncanonical pathway that requires Rac and JNK regulates dendritic development (**Figure 3**).

β-catenin, a component of the canonical pathway, also stimulates dendrite development. Expression of constitutively active β-catenin increases dendritic arborization in hippocampal neurons (Yu & Malenka 2003). However, this function is not dependent on β-catenin-mediated transcription. Instead, active β-catenin increases dendritic arborization

through its interaction with N-cadherin and αN-catenin (Yu & Malenka 2003). Moreover, a dominant-negative β-catenin that impairs TCF function does not block Dvl function on dendrites (Rosso et al. 2005). Thus, β-catenin stimulates dendritic morphogenesis through a pathway that is independent of canonical signaling.

Electrical activity regulates dendritic development by stimulating the expression of Wnts. It is well documented that neuronal activity stimulates the growth and maintenance of complex dendritic arbors (Libersat & Duch 2004, Scott & Luo 2001, Yuste & Bonhoeffer

NMDA: N-methyl-
D-aspartic acid

2001) and that activation of NMDA receptors and calcium release mediate this process (McAllister et al. 1996, Ruthazer et al. 2003, Sin et al. 2002, Wu et al. 1996). Recent studies showed that neuronal activity regulates dendrite growth by modulating the expression and/or release of Wnts. Indeed, conditioned media from depolarized hippocampal neurons contain higher levels of Wnt activity than does media from nonstimulated cells (Yu & Malenka 2003). More recently, Soderling and colleagues showed that Wnt2 expression is enhanced by neuronal activity (Wayman et al. 2006). Electrical activity activates CaMKK and CaMK1, triggering a signaling cascade that culminates with CREB-mediated transcriptional activation of Wnt2 (Wayman et al. 2006). Wnt2 is essential for activity-dependent dendritic growth and branching. Thus electrical activity enhances dendrite development through Wnt factors.

TERMINAL AXON REMODELING BY WNT SIGNALING

Upon contact with their appropriate targets, axons decelerate their growth and remodel extensively to form synaptic boutons, presynaptic structures containing the machinery responsible for neurotransmitter release (for review see Goda & Davis 2003, McAllister 2007, Waites et al. 2005). In some neurons, remodeling occurs at the terminal growth cones, whereas in other neurons, the axon shaft remodeling results in the formation of *en passant* synapses. Although axon remodeling is an essential early step in synapse formation, little is known about the mechanisms that regulate this process (Prokop & Meinertzhagen 2006, Ziv & Garner 2004).

In the central nervous system, Wnt factors enhance the terminal axon remodeling. *Wnt7a* is expressed in cerebellar granule cells when these neurons contact their presynaptic targets, the mossy fibers (Lucas & Salinas 1997). Upon contact with granule cells, mossy fiber axons are extensively remodeled, becoming larger and irregular in shape, and spread areas appear along the axon shaft where *en passant* synapses as-

semble (Hamori & Somogyi 1983, Mason & Gregory 1984). Gain and loss of function studies using a combination of cultured neurons and *Wnt7a* and double *Wnt7a/Dvl1* mutant mice have demonstrated a key role for Wnt7a and Dvl1 in mossy fiber remodeling (Ahmad-Annuar et al. 2006, Hall et al. 2000). Studies also found a similar function for Wnt3, which is released by lateral motoneurons and remodels proprioceptive DRG neurons (Krylova et al. 2002). In both systems, Wnts act as retrograde signals to regulate presynaptic remodeling.

How does Wnt signaling regulate axon remodeling? Profound changes in the organization and stability of microtubules seem to mediate this process. In actively growing axons, dynamic and stable microtubules form large bundles along the axon shaft that splay as they enter the growth cone (for review see Dent & Gertler 2003, Zhou & Cohan 2004). Growth cones contain many dynamic but few stable microtubules (Gordon-Weeks 2004, Dent & Gertler 2003). In the presence of Wnts, thicker bundles of microtubules form along the axon shaft with some unbundling at spread areas. At growth cones, both stable and dynamic microtubules form large loops in the central region (Hall et al. 2000). These findings suggest that Wnt signaling induces changes in microtubule stability and organization. Indeed, Dvl expression or Gsk3 inhibition mimics the Wnt remodeling effect (Ciani et al. 2004, Krylova et al. 2000, Lucas et al. 1998). Consistent with a role for the Wnt-Gsk3 pathway in terminal axon arborization, Gsk3 activity is required for proper arborization of retinotectal projections in the zebrafish (Tokuoka et al. 2002). Epistatic analyses revealed that Wnt-Dvl signaling regulates the microtubule cytoskeleton through a pathway that is independent of transcription (Ciani et al. 2004) (**Figure 3**). Axin, which functions as a negative regulator in the canonical Wnt pathway, collaborates with Dvl to regulate microtubule stability, and both bind very tightly to microtubules (Ciani et al. 2004, Krylova et al. 2000). These results suggest that the Wnt pathway directly signals to microtubules. Indeed, Wnt-mediated inhibition of Gsk3 changes the level of MAP1B phosphorylation (Ciani et al. 2004, Lucas et al. 1998), a microtubule-associated protein that regulates microtubule dynamics (Goold et al. 1999, Tint et al. 2005) (**Figure 3**). Thus Wnt signaling increases microtubule stability by decreasing MAP1B phosphorylation through Gsk3.

More recent studies showed that Wnt signaling also regulates microtubule stability through JNK. Expression of Dvl activates endogenous JNK, whereas JNK inhibition blocks Dvl-mediated microtubule stability. Interestingly, Rac and Rho do not seem to be involved in JNK activation (Ciani & Salinas 2007), suggesting the involvement of an alternative noncanonical Wnt pathway. The findings are consistent with the view that JNK collaborates with Gsk3 to increase microtubule stability. As JNK phosphorylates a number of microtubule-associated proteins, including MAP1B (Chang et al. 2003), these findings collectively suggest that Wnt-Dvl signaling modulates microtubule dynamics during axon remodeling through both JNK activation and Gsk3 inhibition.

WNT SIGNALING REGULATES THE FORMATION OF CENTRAL AND PERIPHERAL SYNAPSES

Wnts Stimulate Central Synaptogenesis

The first indication that Wnts modulate presynaptic differentiation came from studies using cultured cerebellar neurons (Lucas & Salinas 1997). However, loss-of-function studies looking at the mossy fiber–granule cell synapse provided the first clear demonstration of a role for Wnt signaling in synaptogenesis (Hall et al. 2000). In this system, Wnt7a from granule cells induces a significant increase in presynaptic protein clustering, a hallmark of presynaptic assembly on mossy fiber axons (Hall et al. 2000) (**Figure 4a**). The cerebellum of the *Wnt7a* mutant mouse exhibits defects in synapse formation, manifested by decreased accumulation of presynaptic proteins (Ahmad-Annuar et al. 2006, Hall et al. 2000). These findings collec-

tively demonstrate that Wnt7a acts as a retrograde signal to regulate presynaptic differentiation in the cerebellum.

Which pathway regulates synaptic differentiation? Dvl1 is required for Wnt7a function on synapse formation. Expression of Dvl1 mimics Wnt effects, whereas Dvl1 deficiency results in presynaptic differentiation defects manifested by the presence of fewer presynaptic sites in cultured mossy fibers. In vivo, the *Dvl1* mutant exhibits defects in the accumulation of presynaptic proteins as observed in the *Wnt7a* mutant. The double *Wnt7a/Dvl1* mutant mice exhibit more severe defects than do single-mutant mice (Ahmad-Annuar et al. 2006), demonstrating a requirement for Dvl1 function. Wnt7a-Dvl signals through Gsk3β, as inhibitors of Gsk3β mimic the effect of Wnt in vitro (Hall et al. 2000, 2002). Although the signaling pathway downstream of Gsk3β remains poorly understood, conditional knockout of β-catenin in hippocampal neurons indicates that β-catenin is required for the proper localization of synaptic vesicles along the axon. However, this function of β-catenin is independent of TCF-mediated transcription (Bamji et al. 2003). In agreement, loss or gain of function of Wnt signaling does not affect the levels of many presynaptic proteins (Ahmad-Annuar et al. 2006). It is worth noting that a similar Wnt-Gsk3 pathway operates presynaptically to regulate remodeling and presynaptic assembly, suggesting that both processes could be interconnected.

Which aspect of synapse formation is regulated by Wnt signaling? Waites et al. (2005) proposed that Wnt signaling regulates synaptogenesis indirectly by stimulating neuronal maturation through gene transcription. However, several findings do not support this suggestion. In cultured neurons, Wnt signaling rapidly increases the number and the size of synaptic vesicle recycling sites without affecting synaptic protein expression. In vivo, Wnt deficiency also affects the localization of synaptic proteins without affecting their levels (Ahmad-Annuar et al. 2006). Taken together, these findings strongly support the view that Wnt signaling regulates synaptic assembly.

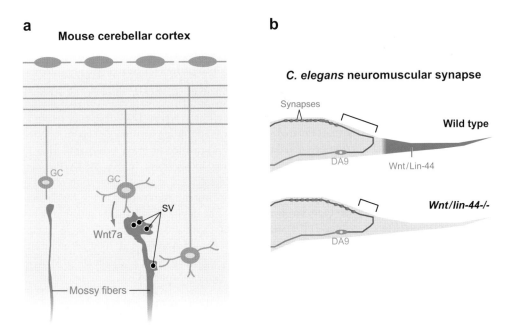

a **Mouse cerebellar cortex**

GC

GC

SV

Wnt7a

Mossy fibers

b **C. elegans neuromuscular synapse**

Synapses

DA9

Wnt/Lin-44

Wild type

DA9

Wnt/lin-44-/-

Figure 4

Wnts can function as pro- and antisynaptogenic factors. (*a*) In the mammalian nervous system, Wnts act as target-derived signals that regulate the assembly of synapses. In the cerebellum, Wnt7a induces the terminal remodelling of mossy fibre axons followed by the recruitment of synaptic components to remodeled axon areas. (*b*) In *C. elegans*, cholinergic DA9 motoneurons do not assemble synapses at the most posterior domain of the axon owing to the presence of Wnt/Lin-44, which is expressed in the posterior hypodermis.

Electrophysiological recordings at the mossy fiber–granule cell synapse using brain slices revealed that Wnt signaling could also regulate synaptic function. The *Wnt7a/Dvl1* double mutant exhibits a significant decrease in the frequency but not amplitude of miniature excitatory postsynaptic currents (mEPSCs), suggesting a presynaptic defect (Ahmad-Annuar et al. 2006). This mutant does not exhibit defects in the number or structures of active zones, suggesting that some aspects of synaptogenesis are normal. However, the decreased frequency of mEPSCs suggests a defect in neurotransmitter release (Ahmad-Annuar et al. 2006).

Wnt Signaling at the Neuromuscular Synapse

In *Drosophila*, the Wg/Wnt signaling pathway through Gsk3 regulates the assembly of the

neuromuscular junction (NMJ). The finding that Wg protein is present at motoneuron terminals led Budnik and collaborators to test the function of Wg in synaptogenesis using a temperature-sensitive allele to bypass any patterning defects (Packard et al. 2002). Wg is released from motoneurons but signals to both pre- and postsynaptic terminals. Loss of *wg* results in severe defects in the number and shape of synaptic boutons. At the ultrastructural level, active zones are not properly assembled, and pre- and postsynaptic markers fail to colocalize properly. Postsynaptically, Wg binds to DFz2 receptors, resulting in its internalization and cleavage at its C-terminus portion. This cleaved receptor is subsequently transported to the nucleus (Ataman et al. 2006). Whether the nuclear localization of the DFz2 is important for transcription is unclear; however, disruption of this receptor pathway affects synaptic growth (Ataman et al. 2006, Mathew et al. 2005). Thus

mEPSCs: miniature excitatory postsynaptic currents

NMJ: neuromuscular junction

Wg signaling is required at both pre- and post-synaptic sides of the neuromuscular synapse.

Wg signals through shaggy/gsk3 to regulate the shape of boutons through changes in synaptic microtubules. An interesting feature of NMJ boutons is the presence of looped microtubules, which are decorated with the microtubule binding protein Futsch, the homolog of MAP1B (Hummel et al. 2000, Roos et al. 2000). Loss of *wg* function leads to changes in the localization of Futsch and to an increase in the number of boutons without loops or containing splayed microtubules (Packard et al. 2002). Similarly, *shaggy* mutants exhibit defects in bouton size and distribution (Franco et al. 2004). These findings demonstrate that the organization of presynaptic microtubules contributes to the shape and size of boutons. The great similarity between the function of Wnt7a and Wg on presynaptic microtubules suggests that a common mechanism regulates presynaptic remodeling in central and peripheral synapses.

No clear evidence has been presented to support a role of Wnts at the NMJ. However, some components of the pathway are associated with the postsynaptic machinery. Dvl1 interacts with MuSK (Luo et al. 2002), a receptor for Agrin, a secreted molecule required for vertebrate NMJ formation (for review see Burden 2002, Kummer et al. 2006, Strochlic et al. 2005). Antisense constructs for the three *Dvl* mouse genes block Agrin's ability to cluster the postsynaptic acetylcholine receptors (AChRs) in cultured myotubules. In this system, Dvl does not signal through the canonical pathway. Instead, Dvl interacts with the p21 kinase PAK1. Moreover, Agrin activates PAK, and this process depends on Dvl (Luo et al. 2002). Another component of the Wnt pathway, APC, colocalizes with AChRs at the mature vertebrate NMJ (Wang et al. 2003), and disruption of APC-AChR interaction decreases Agrin's ability to induce AChR clustering. Unfortunately, no Wnt ligand has been identified to regulate NMJ development. Further studies are necessary to elucidate a possible role for Wnt signaling at the vertebrate NMJ.

Wnts as Antisynaptic Factors

In addition to promoting synaptogenesis, Wnts also inhibit synapse formation. Recent studies in *C. elegans* showed that Lin-44 (Wnt) (Herman et al. 1995), through its receptor Lin-17 (Frizzled) (Sawa et al. 1996), acts as an antisynaptogenic signal (Klassen & Shen 2007). The DA9 motoneuron, located on the ventral side of the animal, forms *en passant* synapses with muscles along the dorsal-most anterior regions of the axon. In contrast, a restricted area of the axon is completely devoid of synapses (Hall & Russell 1991, White et al. 1976). In *lin-44* and *lin-17* mutants, however, presynaptic puncta are present in this asynaptic area. The number of synaptic sites does not change, suggesting that Wnt signaling negatively regulates the distribution of neuromuscular synapses (Klassen & Shen 2007) (**Figure 4b**). Another interesting feature is the restricted localization of the Lin-17 receptor to the asynaptic area, and its distribution is dependent on Lin-44 (Klassen & Shen 2007). Lin-44 is secreted by four hypodermal cells in the tail forming a posterior-to-anterior gradient (Goldstein et al. 2006). However, when *lin-44* is ectopically expressed, the Lin-17 receptor localizes to axon areas close to the Lin-44 source (Klassen & Shen 2007). Thus the Wnt ligand regulates the localization of its receptor at specific domains within the axon. But how does Lin-44 inhibit synaptic assembly? One of the three dishevelled genes in the worm, *dsh-1*, is required in this process. However, β-catenin, TCF, and genes implicated in the PCP pathway are not required (Klassen & Shen 2007). Further studies are necessary to determine the pathway that regulates synaptic localization.

The assembly and distribution of synapses are crucial events in the formation of complex and elaborate neuronal circuits. Therefore, the discovery that Wnts act as pro- and antisynaptogenic factors is of great significance. A combination of Wnts and their receptors could sculpt complex neuronal networks. Although an antisynaptogenic activity for Wnts in vertebrates has not been reported, the absence of secreted Wnt antagonists such as Sfrps and Dkk1 in the *C. elegans* and *Drosophila* genomes (Lee et al. 2006) suggests that the interplay between Wnts and their secreted antagonists regulates synaptic assembly and distribution in higher organisms.

CONCLUSIONS

Although the field has made great progress in understanding the function of Wnt signaling in different aspects of neuronal circuit assembly, many issues remain outstanding. For example, during early embryonic patterning, Wnts can function as morphogens, wherein different levels of Wnt proteins can elicit different cellular responses within identical cells. Wnt gradients are apparently important in axon pathfinding and neuronal migration at later stages. How is this gradient used to control direction? Studies on retinotectal projection indicate that the Wnt gradients may convey positional information for the formation of axon termination zones within their correct topographic positions, but the underlying molecular mechanisms are still poorly understood. What is the interplay of different levels of Wnts, their receptors, and their secreted antagonists in axon guidance?

Wnts are clearly important for neuronal migration in *C. elegans*, conveying directional information in some neurons. It will be interesting to test whether Wnt signaling also plays a role in controlling vertebrate neuronal migration. Wnt signaling is also important for neuronal polarity in *C. elegans*. In vitro results suggest that Wnt5a may be a polarizing cue for vertebrate neurons. It will be important to determine whether Wnt signaling controls vertebrate neuronal polarity in vivo.

The discovery that Wnts can function as pro- and antisynaptogenic factors raises the intriguing and exciting possibility that a combinatorial function of different Wnts and their secreted antagonists plays a crucial role in determining the distribution of synapses within a neuron, thereby influencing the formation of complex patterned neuronal circuits.

Identification of a novel role for Wnt signaling in synaptic modulation is emerging. Wnts and their receptors are expressed throughout life, suggesting that Wnt signaling plays a role not only during early neuronal connectivity but also in synaptic modulation in the adult. The finding that Wnt deficiency leads to electrophysiological defects strongly suggests that Wnts regulate synaptic function (Ahmad-Annuar et al. 2006). Supporting this notion, further electrophysiological studies suggest that Wnts modulate synaptic transmission by regulating neurotransmitter release and long-term potentiation (LTP) (Beaumont et al. 2007, Chen et al. 2006). Although the mechanism by which Wnt regulates neurotransmitter release remains unknown, the finding that Wnts can increase presynaptic receptors, which modulate calcium entry (Farias et al. 2007), suggests a possible mechanism. The discovery that electrical activity can regulate Wnt expression and/or release (Wayman et al. 2006, Yu & Malenka 2003) and that Wnts might regulate neurotransmitter release suggests that Wnts could contribute to a positive feedback mechanism for synaptic modulation. Future studies using conditional genetic approaches will provide important insights into the function of Wnt signaling during the formation, maintenance, and modulation of neuronal circuits throughout an organism's life.

DISCLOSURE STATEMENT

The authors are not aware of any biases that might be perceived as affecting the objectivity of this review.

LITERATURE CITED

Ahmad-Annuar A, Ciani L, Simeonidis I, Herreros J, Fredj NB, et al. 2006. Signaling across the synapse: a role for Wnt and Dishevelled in presynaptic assembly and neurotransmitter release. *J. Cell Biol.* 174:127–39

Ataman B, Ashley J, Gorczyca D, Gorczyca M, Mathew D, et al. 2006. Nuclear trafficking of *Drosophila* Frizzled-2 during synapse development requires the PDZ protein dGRIP. *Proc. Natl. Acad. Sci. USA* 103:7841–46

Bamji SX, Shimazu K, Kimes N, Huelsken J, Birchmeier W, et al. 2003. Role of beta-catenin in synaptic vesicle localization and presynaptic assembly. *Neuron* 40:719–31

Barnes AP, Lilley BN, Pan YA, Plummer LJ, Powell AW, et al. 2007. LKB1 and SAD kinases define a pathway required for the polarization of cortical neurons. *Cell* 129:549–63

Beaumont V, Thompson SA, Choudhry F, Nuthall H, Glantschnig H, et al. 2007. Evidence for an enhancement of excitatory transmission in adult CNS by Wnt signaling pathway modulation. *Mol. Cell Neurosci.* 35:513–24

Bingham S, Higashijima S, Okamoto H, Chandrasekhar A. 2002. The Zebrafish trilobite gene is essential for tangential migration of branchiomotor neurons. *Dev. Biol.* 242:149–60

Burden SJ. 2002. Building the vertebrate neuromuscular synapse. *J. Neurobiol.* 53:501–11

Cadigan KM, Liu YI. 2006. Wnt signaling: complexity at the surface. *J. Cell Sci.* 119:395–402

Carreira-Barbosa F, Concha ML, Takeuchi M, Ueno N, Wilson SW, Tada M. 2003. Prickle 1 regulates cell movements during gastrulation and neuronal migration in zebrafish. *Development* 130:4037–46

Chang L, Jones Y, Ellisman MH, Goldstein LS, Karin M. 2003. JNK1 is required for maintenance of neuronal microtubules and controls phosphorylation of microtubule-associated proteins. *Dev. Cell* 4:521–33

Chen J, Park CS, Tang SJ. 2006. Activity-dependent synaptic Wnt release regulates hippocampal long term potentiation. *J. Biol. Chem.* 281:11910–16

Ciani L, Krylova O, Smalley MJ, Dale TC, Salinas PC. 2004. A divergent canonical WNT-signaling pathway regulates microtubule dynamics: dishevelled signals locally to stabilize microtubules. *J. Cell Biol.* 164:243–53

Ciani L, Salinas PC. 2007. c-Jun N-terminal kinase (JNK) cooperates with Gsk-3beta to regulate Dishevelled-mediated microtubule stability. *BMC Cell Biol.* 8:27

Clevers H. 2006. Wnt/beta-catenin signaling in development and disease. *Cell* 127:469–80

Dent EW, Gertler FB. 2003. Cytoskeletal dynamics and transport in growth cone motility and axon guidance. *Neuron* 40:209–27

Dickson BJ. 2002. Molecular mechanisms of axon guidance. *Science* 298:1959–64

Dickson BJ. 2005. Wnts send axons up and down the spinal cord. *Nat. Neurosci.* 8:1130–32

Farias GG, Valles AS, Colombres M, Godoy JA, Toledo EM, et al. 2007. Wnt-7a induces presynaptic colocalization of alpha 7-nicotinic acetylcholine receptors and adenomatous polyposis coli in hippocampal neurons. *J. Neurosci.* 27:5313–25

Flanagan JG. 2006. Neural map specification by gradients. *Curr. Opin. Neurobiol.* 16:59–66

Fradkin LG, Noordermeer JN, Nusse R. 1995. The *Drosophila* Wnt protein DWnt-3 is a secreted glycoprotein localized on the axon tracts of the embryonic CNS. *Dev. Biol.* 168:202–13

Franco B, Bogdanik L, Bobinnec Y, Debec A, Bockaert J, et al. 2004. Shaggy, the homolog of glycogen synthase kinase 3, controls neuromuscular junction growth in *Drosophila*. *J. Neurosci.* 24:6573–77

Fraser SE, Hunt RK. 1980. Retinotectal specificity: models and experiments in search of a mapping function. *Annu. Rev. Neurosci.* 3:319–52

Fraser SE, Perkel DH. 1990. Competitive and positional cues in the patterning of nerve connections. *J. Neurobiol.* 21:51–72

Gierer A. 1983. Model for the retino-tectal projection. *Proc. R. Soc. London B Biol. Sci.* 218:77–93

Goda Y, Davis GW. 2003. Mechanisms of synapse assembly and disassembly. *Neuron* 40:243–64

Goldstein B, Takeshita H, Mizumoto K, Sawa H. 2006. Wnt signals can function as positional cues in establishing cell polarity. *Dev. Cell* 10:391–96

Goold RG, Owen R, Gordon-Weeks PR. 1999. Glycogen synthase kinase 3beta phosphorylation of microtubule-associated protein 1B regulates the stability of microtubules in growth cones. *J. Cell Sci.* 112(Pt. 19):3373–84

Gordon MD, Nusse R. 2006. Wnt signaling: multiple pathways, multiple receptors, and multiple transcription factors. *J. Biol. Chem.* 281:22429–33

Gordon-Weeks PR. 2004. Microtubules and growth cone function. *J. Neurobiol.* 58:70–83

Hall AC, Brennan A, Goold RG, Cleverley K, Lucas FR, et al. 2002. Valproate regulates GSK-3-mediated axonal remodeling and synapsin I clustering in developing neurons. *Mol. Cell Neurosci.* 20:257–70

Hall AC, Lucas FR, Salinas PC. 2000. Axonal remodeling and synaptic differentiation in the cerebellum is regulated by WNT-7a signaling. *Cell* 100:525–35

Hall DH, Russell RL. 1991. The posterior nervous system of the nematode *Caenorhabditis elegans*: serial reconstruction of identified neurons and complete pattern of synaptic interactions. *J. Neurosci.* 11:1–22

Hamori J, Somogyi J. 1983. Differentiation of cerebellar mossy fiber synapses in the rat: a quantitative electron microscope study. *J. Comp. Neurol.* 220:365–77

Harris KE, Beckendorf SK. 2007. Different Wnt signals act through the Frizzled and RYK receptors during *Drosophila* salivary gland migration. *Development* 134:2017–25

Harris J, Honigberg L, Robinson N, Kenyon C. 1996. Neuronal cell migration in *C. elegans*: regulation of Hox gene expression and cell position. *Development* 122:3117–31

Hatten ME. 2002. New directions in neuronal migration. *Science* 297:1660–63

Heisenberg CP, Tada M, Rauch GJ, Saude L, Concha ML, et al. 2000. Silberblick/Wnt11 mediates convergent extension movements during zebrafish gastrulation. *Nature* 405:76–81

Herman MA, Vassilieva LL, Horvitz HR, Shaw JE, Herman RK. 1995. The *C. elegans* gene lin-44, which controls the polarity of certain asymmetric cell divisions, encodes a Wnt protein and acts cell nonautonomously. *Cell* 83:101–10

Hilliard MA, Bargmann CI. 2006. Wnt signals and frizzled activity orient anterior-posterior axon outgrowth in *C. elegans*. *Dev. Cell* 10:379–90

Hummel T, Krukkert K, Roos J, Davis G, Klambt C. 2000. *Drosophila* Futsch/22C10 is a MAP1B-like protein required for dendritic and axonal development. *Neuron* 26:357–70

Imondi R, Thomas JB. 2003. Neuroscience. The ups and downs of Wnt signaling. *Science* 302:1903–4

Inoue T, Oz HS, Wiland D, Gharib S, Deshpande R, et al. 2004. *C. elegans* LIN-18 is a Ryk ortholog and functions in parallel to LIN-17/Frizzled in Wnt signaling. *Cell* 118:795–806

Jessen JR, Topczewski J, Bingham S, Sepich DS, Marlow F, et al. 2002. Zebrafish trilobite identifies new roles for Strabismus in gastrulation and neuronal movements. *Nat. Cell Biol.* 4:610–15

Keeble TR, Cooper HM. 2006. Ryk: a novel Wnt receptor regulating axon pathfinding. *Int. J. Biochem. Cell Biol.* 38:2011–17

Kishi M, Pan YA, Crump JG, Sanes JR. 2005. Mammalian SAD kinases are required for neuronal polarization. *Science* 307:929–32

Klassen MP, Shen K. 2008. A Wnt signalling pathway inhibits synapse formation in the *C. elegans* tail. *Cell.* In press

Kohn AD, Moon RT. 2005. Wnt and calcium signaling: beta-catenin-independent pathways. *Cell Calcium* 38:439–46

Krylova O, Herreros J, Cleverley K, Ehler E, Henriquez J, et al. 2002. WNT-3, Expressed by motoneurons, regulates terminal arborization of neurotrophin-3-responsive spinal sensory neurons. *Neuron* 35:1043

Krylova O, Messenger MJ, Salinas PC. 2000. Dishevelled-1 regulates microtubule stability: a new function mediated by glycogen synthase kinase-3beta. *J. Cell Biol.* 151:83–94

Kummer TT, Misgeld T, Sanes JR. 2006. Assembly of the postsynaptic membrane at the neuro-muscular junction: paradigm lost. *Curr. Opin. Neurobiol.* 16:74–82

Lee PN, Pang K, Matus DQ, Martindale MQ. 2006. A WNT of things to come: evolution of Wnt signaling and polarity in cnidarians. *Semin. Cell Dev. Biol.* 17:157–67

Libersat F, Duch C. 2004. Mechanisms of dendritic maturation. *Mol. Neurobiol.* 29:303–20

Liu Y, Shi J, Lu CC, Wang ZB, Lyuksyutova AI, et al. 2005. Ryk-mediated Wnt repulsion regulates posterior-directed growth of corticospinal tract. *Nat. Neurosci.* 8:1151–59

Logan CY, Nusse R. 2004. The Wnt signaling pathway in development and disease. *Annu. Rev. Cell Dev. Biol.* 20:781–810

Lu W, Yamamoto V, Ortega B, Baltimore D. 2004. Mammalian Ryk is a Wnt coreceptor required for stimulation of neurite outgrowth. *Cell* 119:97–108

Lucas FR, Goold RG, Gordon-Weeks PR, Salinas PC. 1998. Inhibition of GSK-3beta leading to the loss of phosphorylated MAP-1B is an early event in axonal remodelling induced by WNT-7a or lithium. *J. Cell Sci.* 111:1351–61

Lucas FR, Salinas PC. 1997. WNT-7a induces axonal remodeling and increases synapsin I levels in cerebellar neurons. *Dev. Biol.* 193:31–44

Luo L. 2002. Actin cytoskeleton regulation in neuronal morphogenesis and structural plasticity. *Annu. Rev. Cell Dev. Biol.* 18:601–35

Luo L. 2006. Developmental neuroscience: Two gradients are better than one. *Nature* 439:23–24

Luo ZG, Wang Q, Zhou JZ, Wang J, Luo Z, et al. 2002. Regulation of AChR clustering by Dishevelled interacting with MuSK and PAK1. *Neuron* 35:489–505

Lyuksyutova AI, Lu CC, Milanesio N, King LA, Guo N, et al. 2003. Anterior-posterior guidance of commissural axons by Wnt-frizzled signaling. *Science* 302:1984–88

Malbon CC, Wang HY. 2006. Dishevelled: a mobile scaffold catalyzing development. *Curr. Top. Dev. Biol.* 72:153–66

Maloof JN, Whangbo J, Harris JM, Jongeward GD, Kenyon C. 1999. A Wnt signaling pathway controls Hox gene expression and neuroblast migration in *C. elegans*. *Development* 126:37–49

Mason CA, Gregory E. 1984. Postnatal maturation of cerebellar mossy and climbing fibers: transient expression of dual features on single axons. *J. Neurosci.* 4:1715–35

Mathew D, Ataman B, Chen J, Zhang Y, Cumberledge S, Budnik V. 2005. Wingless signaling at synapses is through cleavage and nuclear import of receptor DFrizzled2. *Science* 310:1344–47

McAllister AK. 2000. Cellular and molecular mechanisms of dendrite growth. *Cereb. Cortex* 10:963–73

McAllister AK. 2007. Dynamic aspects of CNS synapse formation. *Annu. Rev. Neurosci.* 30:425–50

McAllister AK, Katz LC, Lo DC. 1996. Neurotrophin regulation of cortical dendritic growth requires activity. *Neuron* 17:1057–64

Mikels AJ, Nusse R. 2006. Purified Wnt5a protein activates or inhibits beta-catenin-TCF signaling depending on receptor context. *PLoS Biol.* 4:e115

Nishimura T, Kato K, Yamaguchi T, Fukata Y, Ohno S, Kaibuchi K. 2004. Role of the PAR-3-KIF3 complex in the establishment of neuronal polarity. *Nat. Cell Biol.* 6:328–34

Packard M, Koo ES, Gorczyca M, Sharpe J, Cumberledge S, Budnik V. 2002. The *Drosophila* Wnt, wingless, provides an essential signal for pre- and postsynaptic differentiation. *Cell* 111:319–30

Pan CL, Howell JE, Clark SG, Hilliard M, Cordes S, et al. 2006. Multiple Wnts and Frizzled receptors regulate anteriorly directed cell and growth cone migrations in *Caenorhabditis elegans*. *Dev. Cell* 10:367–77

Parrish JZ, Emoto K, Kim MD, Jan YN. 2007. Mechanisms that regulate establishment, maintenance, and remodeling of dendritic fields. *Annu. Rev. Neurosci.* 30:399–423

Prasad BC, Clark SG. 2006. Wnt signaling establishes anteroposterior neuronal polarity and requires retromer in *C. elegans*. *Development* 133:1757–66

Prestige MC, Willshaw DJ. 1975. On a role for competition in the formation of patterned neural connexions. *Proc. R. Soc. London B Biol. Sci.* 190:77–98

Prokop A, Meinertzhagen IA. 2006. Development and structure of synaptic contacts in *Drosophila*. *Semin. Cell Dev. Biol.* 17:20–30

Rolls MM, Doe CQ. 2004. Baz, Par-6 and aPKC are not required for axon or dendrite specification in *Drosophila*. *Nat. Neurosci.* 7:1293–95

Roos J, Hummel T, Ng N, Klambt C, Davis GW. 2000. *Drosophila* Futsch regulates synaptic microtubule organization and is necessary for synaptic growth. *Neuron* 26:371–82

Rosso SB, Sussman D, Wynshaw-Boris A, Salinas PC. 2005. Wnt signaling through Dishevelled, Rac and JNK regulates dendritic development. *Nat. Neurosci.* 8:34–42

Ruthazer ES, Akerman CJ, Cline HT. 2003. Control of axon branch dynamics by correlated activity in vivo. *Science* 301:66–70

Sato M, Umetsu D, Murakami S, Yasugi T, Tabata T. 2006. DWnt4 regulates the dorsoventral specificity of retinal projections in the *Drosophila melanogaster* visual system. *Nat. Neurosci.* 9:67–75

Sawa H, Lobel L, Horvitz HR. 1996. The *Caenorhabditis elegans* gene lin-17, which is required for certain asymmetric cell divisions, encodes a putative seven-transmembrane protein similar to the *Drosophila* Frizzled protein. *Genes Dev.* 10:2189–97

Schmitt AM, Shi J, Wolf AM, Lu CC, King LA, Zou Y. 2006. Wnt-Ryk signalling mediates medial-lateral retinotectal topographic mapping. *Nature* 439:31–37

Scott EK, Luo L. 2001. How do dendrites take their shape? *Nat. Neurosci.* 4:359–65

Shelly M, Cancedda L, Heilshorn S, Sumbre G, Poo MM. 2007. LKB1/STRAD promotes axon initiation during neuronal polarization. *Cell* 129:565–77

Shi SH, Jan LY, Jan YN. 2003. Hippocampal neuronal polarity specified by spatially localized mPar3/mPar6 and PI 3-kinase activity. *Cell* 112:63–75

Sin WC, Haas K, Ruthazer ES, Cline HT. 2002. Dendrite growth increased by visual activity requires NMDA receptor and Rho GTPases. *Nature* 419:475–80

Strochlic L, Cartaud A, Cartaud J. 2005. The synaptic muscle-specific kinase (MuSK) complex: new partners, new functions. *Bioessays* 27:1129–35

Tessier-Lavigne M, Goodman CS. 1996. The molecular biology of axon guidance. *Science* 274:1123–33

Tint I, Fischer I, Black M. 2005. Acute inactivation of MAP1b in growing sympathetic neurons destabilizes axonal microtubules. *Cell Motil. Cytoskeleton* 60:48–65

Tokuoka H, Yoshida T, Matsuda N, Mishina M. 2002. Regulation by glycogen synthase kinase-3beta of the arborization field and maturation of retinotectal projection in zebrafish. *J. Neurosci.* 22:10324–32

Van Aelst L, Cline HT. 2004. Rho GTPases and activity-dependent dendrite development. *Curr. Opin. Neurobiol.* 14:297–304

Veeman MT, Slusarski DC, Kaykas A, Louie SH, Moon RT. 2003. Zebrafish prickle, a modulator of noncanonical Wnt/Fz signaling, regulates gastrulation movements. *Curr. Biol.* 13:680–85

Wada H, Tanaka H, Nakayama S, Iwasaki M, Okamoto H. 2006. Frizzled3a and Celsr2 function in the neuroepithelium to regulate migration of facial motor neurons in the developing zebrafish hindbrain. *Development* 133:4749–59

Waites CL, Craig AM, Garner CC. 2005. Mechanisms of vertebrate synaptogenesis. *Annu. Rev. Neurosci.* 28:251–74

Wallingford JB, Habas R. 2005. The developmental biology of Dishevelled: an enigmatic protein governing cell fate and cell polarity. *Development* 132:4421–36

Wang J, Jing Z, Zhang L, Zhou G, Braun J, et al. 2003. Regulation of acetylcholine receptor clustering by the tumor suppressor APC. *Nat. Neurosci.* 6:1017–18

Wang Y, Nathans J. 2007. Tissue/planar cell polarity in vertebrates: new insights and new questions. *Development* 134:647–58

Wayman GA, Impey S, Marks D, Saneyoshi T, Grant WF, et al. 2006. Activity-dependent dendritic arborization mediated by CaM-kinase I activation and enhanced CREB-dependent transcription of Wnt-2. *Neuron* 50:897–909

Whangbo J, Kenyon C. 1999. A Wnt signaling system that specifies two patterns of cell migration in *C. elegans. Mol. Cell* 4:851–58

White JG, Southgate E, Thomson JN, Brenner S. 1976. The structure of the ventral nerve cord of *Caenorhabditis elegans. Philos. Trans. R. Soc. London B Biol. Sci.* 275:327–48

Whitford KL, Marillat V, Stein E, Goodman CS, Tessier-Lavigne M, et al. 2002. Regulation of cortical dendrite development by Slit-Robo interactions. *Neuron* 33:47–61

Wolf AM, Lyuksyutova AI, Fenstermaker AG, Shafer B, Lo CG, Zou Y. 2008. PI3K-atypical PKC signaling is required for Wnt attraction and anterior-posterior axon guidance. *J. Neurosci.* 28:3456–67

Wu J, Klein TJ, Mlodzik M. 2004. Subcellular localization of Frizzled receptors, mediated by their cytoplasmic tails, regulates signaling pathway specificity. *PLoS Biol.* 2:E158

Wu G, Malinow R, Cline HT. 1996. Maturation of a central glutamatergic synapse. *Science* 274:972–76

Yoshikawa S, McKinnon RD, Kokel M, Thomas JB. 2003. Wnt-mediated axon guidance via the *Drosophila* Derailed receptor. *Nature* 422:583–88

Yu X, Malenka RC. 2003. Beta-catenin is critical for dendritic morphogenesis. *Nat. Neurosci.* 6:1169–77

Yuste R, Bonhoeffer T. 2001. Morphological changes in dendritic spines associated with long-term synaptic plasticity. *Annu. Rev. Neurosci.* 24:1071–89

Zhang X, Zhu J, Yang GY, Wang QJ, Qian L, et al. 2007. Dishevelled promotes axon differentiation by regulating atypical protein kinase C. *Nat. Cell Biol.* 9:743–54

Zhou FQ, Cohan CS. 2004. How actin filaments and microtubules steer growth cones to their targets. *J. Neurobiol.* 58:84–91

Zinovyeva AY, Forrester WC. 2005. The *C. elegans* Frizzled CFZ-2 is required for cell migration and interacts with multiple Wnt signaling pathways. *Dev. Biol.* 285:447–61

Ziv NE, Garner CC. 2004. Cellular and molecular mechanisms of presynaptic assembly. *Nat. Rev. Neurosci.* 5:385–99

Zou Y. 2006. Navigating the anterior-posterior axis with Wnts. *Neuron* 49:787–89

Zou Y, Lyuksyutova AI. 2007. Morphogens as conserved axon guidance cues. *Curr. Opin. Neurobiol.* 17:22–28

RELATED RESOURCE

Nusse R. 2008. *The Wnt homepage.* **http://www.stanford.edu/~rnusse/wntwindow.html**

Habits, Rituals, and the Evaluative Brain

Ann M. Graybiel

Department of Brain and Cognitive Science and the McGovern Institute for Brain Research, Massachusetts Institute of Technology, Cambridge, Massachusetts 02139; email: Graybiel@mit.edu

Annu. Rev. Neurosci. 2008. 31:359–87

The *Annual Review of Neuroscience* is online at neuro.annualreviews.org

This article's doi: 10.1146/annurev.neuro.29.051605.112851

Key Words

striatum, reinforcement learning, stereotypy, procedural learning, addiction, automatization, obsessive-compulsive disorder

Abstract

Scientists in many different fields have been attracted to the study of habits because of the power habits have over behavior and because they invoke a dichotomy between the conscious, voluntary control over behavior, considered the essence of higher-order deliberative behavioral control, and lower-order behavioral control that is scarcely available to consciousness. A broad spectrum of behavioral routines and rituals can become habitual and stereotyped through learning. Others have a strong innate basis. Repetitive behaviors can also appear as cardinal symptoms in a broad range of neurological and neuropsychiatric illness and in addictive states. This review suggests that many of these behaviors could emerge as a result of experience-dependent plasticity in basal ganglia–based circuits that can influence not only overt behaviors but also cognitive activity. Culturally based rituals may reflect privileged interactions between the basal ganglia and cortically based circuits that influence social, emotional, and action functions of the brain.

Contents

Habit is the most effective teacher of all things.

—*Pliny*

We are what we repeatedly do. Excellence, then, is not an act, but a habit.

—*Aristotle*

Habit is second nature, or rather, ten times nature.

—*William James*

For in truth habit is a violent and treacherous schoolmistress. She establishes in us, little by little, stealthily, the foothold of her authority; but having by this mild and humble beginning settled and planted it with the help of time, she soon uncovers to us a furious and tyrannical face against which we no longer have the liberty of even raising our eyes.

—*Montaigne*

INTRODUCTION

Habit, to most of us, has multiple connotations. On the one hand, a habit is a behavior that we do often, almost without thinking. Some habits we strive for, and work hard to make part of our general behavior. And still other habits are burdensome behaviors that we want to abolish but often cannot, so powerfully do they control our behavior. Viewed from this broad and intuitive perspective, habits can be evaluated as relatively neutral, or as "good" (desirable) or as "bad" (undesirable). Yet during much of our waking lives, we act according to our habits, from the time we rise and go through our morning routines until we fall asleep after evening routines. Taken in this way, habits have long attracted the interest of philosophers and psychologists, and they have been alternatively praised and cursed.

Whether good, bad, or neutral, habits can have great power over our behavior. When deeply enstated, they can block some alternate behaviors and pull others into the habitual repertoire. In early accounts, habits were broadly defined. Mannerisms, customs, and rituals were all considered together with simple daily habits, and habituation or sensitization (the lessening or increase in impact of stimuli and events with repetition) were included. Much current work on habit learning in neuroscience has pulled away from this broad view in an effort to define habit in a way that makes it accessible to scientific study. Much insight can also be gained by extending such constructs of habit and habit learning to include the rich array of behaviors considered by ethologists, neuropharmacologists, neurologists, and psychiatrists, as well as by students of motor control. Below, I review some of the definitions of habit that have developed in cognitive neuroscience and psychology and how these views have been formalized in computational theories. I then point to work on extreme habits and compulsions, ritualistic behaviors and mannerisms, stereotypies, and social and cultural "habits" and suggest that these are critical behaviors to consider in a neuroscience of habit formation.

This proposal is based on mounting evidence that this broad array of behaviors can engage neural circuits interconnecting the neocortex with the striatum and related regions of the basal ganglia. Different basal ganglia–based circuits appear to operate predominantly in relation to different types of cognitive and motor actions, for example, in intensely social behaviors such as mating and in the performance of practiced motor skills. Remarkably, however, evidence suggests that many of these basal ganglia-based subcircuits participate during the acquisition of habits, procedures, and repetitive behaviors, and these may be reactivated or misactivated in disorders producing repetitive thoughts and overt behaviors.

A starting point is to consider defining characteristics of habits. First, habits (mannerisms, customs, rituals) are largely learned; in current terminology, they are acquired via experience-dependent plasticity. Second, habitual behaviors occur repeatedly over the course of days or years, and they can become remarkably fixed. Third, fully acquired habits are performed almost automatically, virtually nonconsciously, allowing attention to be focused elsewhere. Fourth, habits tend to involve an ordered, structured action sequence that is prone to being elicited by a particular context or stimulus. And finally, habits can comprise cognitive expressions of routine (habits of thought) as well as motor expressions of routine. These characteristics suggest that habits are sequential, repetitive, motor, or cognitive behaviors elicited by external or internal triggers that, once released, can go to completion without constant conscious oversight.

This description is familiar to many who study animal behavior and observe complex repetitive behaviors [fixed action patterns (FAPs)]. Some of these appear to be largely innate, such as some mating behaviors, but others are learned, such as the songs of some orcene birds. Repetitive behaviors and thoughts are also major presenting features in human disorders such as Tourette syndrome and obsessive-compulsive disorder (OCD). Stereotypies and repetitive behaviors appear in a range of other clinical disorders including schizophrenia and Huntington's disease, as well as in addictive states. I suggest that there may well be a common theme across these behavioral domains. Many of these repetitive behaviors, whether motor or cognitive, are built up in part through the action of basal ganglia–based neural circuits that can iteratively evaluate contexts and select actions and can then form chunked representations of action sequences that can influence both cortical and subcortical brain structures (**Figure 1**). Both experimental evidence and computational analysis suggest that a shift from

FAPs: fixed action patterns

OCD: obsessive-compulsive disorder

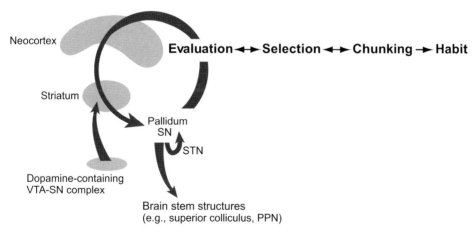

Figure 1

Schematic representation of the development of habits through iterative action of cortico–basal ganglia circuits. Circuits mediating evaluation of actions gradually lead to selection of particular behaviors that, through the chunking process, become habits. PPN, pedunculopontine nucleus; SN, substantia nigra; STN, subthalamic nucleus; VTA, ventral tegmental area.

S-R: stimulus-response

largely evaluation-driven circuits to those engaged in performance is a critical feature of habit learning. Chronic multielectrode recordings suggest that within the habit production system, as habits are acquired, neural activity patterns change dynamically and eventually settle into specific chunked patterns. This shift in neural activity from variable to repetitive matches the explore-exploit transition in behavioral output from a testing, exploratory mode to a focused, exploitive mode as habitual behaviors crystallize. This process may be critical to allow the emergence of habitual behaviors as entire structured entities once they are learned.

DEFINITIONS OF HABIT LEARNING IN COGNITIVE NEUROSCIENCE AND EXPERIMENTAL PSYCHOLOGY

Classic studies of habit learning distinguished this form of learning as a product of a procedural learning brain system that is differentiable from declarative learning brain systems for encoding facts and episodes. These definitions rest on findings suggesting that these two systems have different brain substrates (Knowlton et al. 1996, Packard & Knowlton 2002, Packard & McGaugh 1996). Deficits in learning facts contrast vividly with the preserved habits, daily routines, and procedural capabilities of patients with medial temporal lobe damage (Salat et al. 2006). By contrast, patients with basal ganglia disorders exhibit, in testing, procedural learning deficits and deficits in implicit (nonconsciously recognized) learning such as performance on mazes and probabilistic learning tasks in which the subject learns the probabilities of particular stimulus-response (S-R) associations without full awareness (Knowlton et al. 1996, Poldrack et al. 2001). The nonconscious acquisitions of S-R habits by amnesic patients has been documented most clearly by the performance of a patient who learned a probabilistic task with an apparent total lack of awareness of the acquired habit (Bayley et al. 2005).

Despite these distinctions, human imaging experiments suggest that both the basal ganglia (striatum) and the medial temporal lobe are active in such probabilistic learning tasks. When task conditions favor implicit learning, however, activity in the medial temporal lobe decreases as striatal activity increases, and when conditions favor explicit learning, the reverse is true (Foerde et al. 2006, Poldrack et al. 2001, Willingham et al. 2002). Moreover, in disease states involving dysfunction of the basal ganglia, medial temporal lobe activity can appear under conditions in which striatal activity normally would dominate (Moody et al. 2004, Rauch et al. 2006, Voermans et al. 2004). These findings demonstrate conjoint but differentiable contributions of both the declarative and the procedural memory systems to behaviors, as well as interactions between these two.

Comparable distinctions have been drawn for memory systems in experimental animals. The striatum is required for repetitive S-R or win-stay behaviors (for example, always turning right in a maze to obtain reward) as opposed to behaviors that can be flexibly adjusted when the context or rules change (for example, not just turning right, but turning toward the rewarded side even if it is now on the left). By contrast, the hippocampus is required for flexible (win-shift) behaviors (Packard & Knowlton 2002, Packard & McGaugh 1996). Nevertheless, the control systems for these behaviors cannot be simply divided into hippocampal and basal ganglia systems because both types of behavior can be supported by the striatum, depending on the hippocampal and sensorimotor connections of the striatal regions in question (Devan & White 1999, Yin & Knowlton 2004). Moreover, as conditional procedures are learned, neural activities in the striatum and hippocampus can become highly coordinated in the frequency domain (DeCoteau et al. 2007a).

In an effort to promote clearly interpretable experimentation on habit formation, Dickinson and his collaborators developed an operational definition of habits using characteristics of reward-based learning in rodents (Adams &

Dickinson 1981, Balleine & Dickinson 1998, Colwill & Rescorla 1985). In the initial stages of habit learning, behaviors are not automatic. They are goal directed, as in an animal working to obtain a food reward. But with extended training or training with interval schedules of reward, animals typically come to perform the behaviors repeatedly, on cue, even if the value of the reward to be received is reduced so that it is no longer rewarding (for example, if the animal is tested when it is sated or if its food reward has been repetitively paired with a noxious outcome). Dickinson defined the goal-oriented, purposeful, nonhabitual behaviors as action-outcome (A-O) behaviors and labeled the habitual behaviors occurring despite reward devaluation as S-R behaviors. Thus, in addition to habits being learned, repetitive, sequential, context-triggered behaviors, habits can be defined experimentally as being performed not in relation to a current or future goal but rather in relation to a previous goal and the antecedent behavior that most successfully led to achieving that goal.

The central finding from lesion work based on the reward-devaluation paradigm is that the transition from goal-oriented A-O to habitual S-R modes of behavior involves transitions in the neural circuits predominantly controlling the behaviors (**Figure 2**). Specifically, experiments suggest that different regions of the prefrontal cortex, the striatum, and the amygdala and other limbic sites critically influence these two different behavioral modes.

In rats, lesions in either the sensorimotor striatum (dorsolateral caudoputamen) or the infralimbic prefrontal cortex reduce the insensitivity to reward devaluation that defines habitual behavior in this paradigm. With such lesions, the animals exhibit sensitivity to

A-O: action-outcome

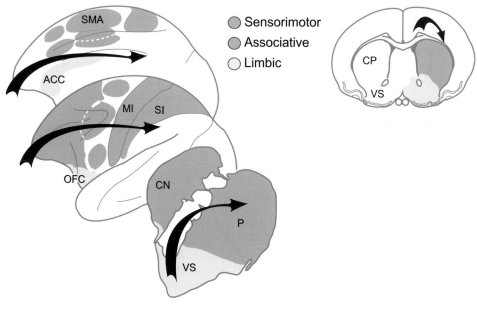

Figure 2

Dynamic shifts in activity in cortical and striatal regions as habits and procedures are learned. Sensorimotor, associative, and limbic regions of the frontal cortex (medial and lateral views) and striatum (single hemisphere) are shown for the monkey (*left*), and corresponding striatal regions are indicated for the rat (*right*). These functional designations are only approximate and are shown in highly schematic form. ACC, anterior cingulate cortex; CN, caudate nucleus; CP, caudoputamen; MI, primary motor cortex; OFC, orbitofrontal cortex; P, putamen; SI, primary somatosensory cortex; SMA, supplementary motor area; VS, ventral striatum.

reward value (A-O behavior) rather than habitual (S-R) behavior, even after overtraining (Killcross & Coutureau 2003, Yin & Knowlton 2004). By contrast, lesions of either the caudomedial (associative) striatum or the prelimbic prefrontal cortex reduce the sensitivity to reward devaluation that defines goal-oriented behavior in this paradigm; the animals are habit driven (Killcross & Coutureau 2003, Yin et al. 2005). The fact that lesions in either the striatum or the frontal cortex are disruptive suggests that the controlling systems represent neural circuits that have both cortical and subcortical components. In macaque monkeys, the basolateral amygdala and the orbitofrontal cortex are also required for sensitivity to reward devaluation (Izquierdo et al. 2004, Wellman et al. 2005). Thus multiple components of the goal-oriented system have been demonstrated across species, and these include regions strongly linked with the limbic system (Balleine et al. 2003, Corbit & Balleine 2005, Gottfried et al. 2003, Wellman et al. 2005).

Like the declarative vs. habit system distinction made in studies on humans, the distinction based on these experiments between action-outcome vs. stimulus-response systems is not absolute (Faure et al. 2005). Evidence suggests that these are not independent "systems." For example, after training that produces habitual behavior in rats, goal-oriented behavior can be reinstated if the infralimbic prefrontal cortex is inactivated (Coutureau & Killcross 2003). This finding suggests that the circuits controlling goal-directed behavior may be actively suppressed when behavior becomes habitual (Coutureau & Killcross 2003). The dichotomy between A-O and S-R behaviors also does not reflect the richness of behavior outside the narrow boundaries of their definitions in reward-devaluation paradigms (for example, when multiple choices are available or different reward schedules are used). The idea that there is a dynamic balance between control systems governing flexible cognitive control and more nearly automatic control of behavioral responses supports the long-standing view from clinical studies that frontal cortical inhibitory zones can suppress lower-order behaviors. This view has become important in models of such system-level interactions (Daw et al. 2005).

Most of these studies have been based on the effects of permanent lesions made in parts of either the dorsal striatum or the neocortex. The use of reversible inactivation procedures suggests that during early stages of instrumental learning, activity in the ventral striatum (nucleus accumbens) is necessary for acquisition of the behavior (Atallah et al. 2007, Hernandez et al. 2002, Hernandez et al. 2006, Smith-Roe & Kelley 2000). This requirement for the nucleus accumbens is apparently transitory: After learning, inactivation of the nucleus accumbens has less or no effect. Notably, inactivating the dorsolateral striatum during the very early stages of conditioning does not block learning and can even improve performance. This last result at first glance seems to conflict with the many reports concluding that the dorsolateral striatum is necessary for habit learning. However, these results fit well with the view, encouraged here, that the learning process is highly dynamic and engages in parallel, not simply in series, sets of neural circuits ranging from those most tightly connected with limbic and midbrain-ventral striatal reward systems to circuits engaging the dorsal striatum, neocortex, and motor structures such as the cerebellum.

Several groups have suggested that eventually the "engram" of the habit shifts to regions outside the basal ganglia, including the neocortex (Atallah et al. 2007, Djurfeldt et al. 2001, Graybiel 1998, Houk & Wise 1995, O'Reilly & Frank 2006). Evidence to settle this point is still lacking. There could be a competition between the early-learning ventral striatal system and the late-learning dorsal striatal system (Hernandez et al. 2002), an idea parallel to the proposal that, in maze training protocols that eventually produce habitual behavior, the hippocampus is required for learning early on, whereas later the dorsal striatum is required (Packard & McGaugh 1996). However, things are not likely to be so simple. The dorsal striatum can be engaged very early in the learning process (Barnes et al. 2005, Jog et al. 1999). And "the striatum" and "the hippocampus" each actually comprise a composite of regions that are interconnected with different functional networks.

COMPUTATIONAL APPROACHES TO HABIT LEARNING: HABIT LEARNING AND VALUE FUNCTIONS

Work on habit learning has been powerfully invigorated by computational neuroscience. A critical impetus for this effort came from the pioneering work of Sutton & Barto (1998), which explicitly outlined the essential characteristics of reinforcement learning (RL) and summarized a series of alternative models to account for such learning (RL models). For experimental neuroscientists, this work is of remarkable interest because neural signals and activity patterns are being identified that fit well with the essential elements of RL models (Daw et al. 2005, Daw & Doya 2006). The key characteristics of these models are that an agent (animal, machine, algorithm) undergoing learning starts with a goal and senses and explores the environment by making choices (selecting behaviors) in order to reach that goal optimally. The agent's actions are made in the context of uncertainty about the environment. The agent must explore the environment to reduce the uncertainty, but it must also exploit (for example, by selecting or deselecting an action) to attain the goal. Sequences of behaviors are seen as guided by subgoals, and the learning involves determining the immediate value of the state or state-action set (a reward function), the estimated (predicted) future value of the state in terms of that reward (a value function). To make this value estimate, the agent needs some representation of future actions (a policy). Then the choice can be guided by the estimated value of taking a given action in a given state with that policy (the action value). These value estimates are principal drivers of behavior. Most behaviors do not immediately yield primary reward, and so ordinarily they involve the generation of a model of the action space (environment) to guide future actions (planning) in the sense of optimal control. Thus the control of behavior crucially depends on value estimates learned through experience.

A pivotal convergence of RL models and traditional learning experiments came with two sets of findings based on conditioning experiments in monkeys (**Figure 3**). First, dopamine-containing neurons of the midbrain substantia nigra pars compacta and the ventral tegmental area (VTA) can fire in patterns that correspond remarkably closely to the properties of a positive reward prediction error of RL models such as in the temporal difference model (Montague et al. 1996, Romo & Schultz 1990, Schultz et al. 1997). Second, during such conditioning tasks, striatal neurons gradually acquire a response to the conditioning stimulus, and this acquired response depends on dopamine signaling in the striatum (Aosaki et al. 1994a,b). These two sets of findings suggested a teacher (dopamine)–student (striatum) sequence in which dopamine-containing nigral neurons, by coding reward-prediction errors, teach learning-related circuits in the striatum (Graybiel et al. 1994). The actor-critic architecture and its variants, in which the critic supplies value predictions to guide action selection by the actor, have been used to model these relationships (Schultz et al. 1997). Many studies have now focused on identifying signals corresponding to the parameters in models of this learning process.

The firing characteristics of midbrain dopamine-containing neurons suggest that they can signal expected reward value (reward probability and magnitude including negative reward prediction error) and motivational state in a context-dependent manner (Bayer & Glimcher 2005, Morris et al. 2004, Nakahara et al. 2004, Satoh et al. 2003, Tobler et al. 2005, Waelti et al. 2001), that they are specialized to respond in relation to positive but not aversive reinforcements, and that they may code uncertainty (Fiorillo et al. 2003, Hsu et al. 2005, Niv et al. 2005, Ungless et al. 2004) or salience (Redgrave & Gurney 2006). These characteristics may, among others, account for the remarkable capacity for placebo treatments to

RL: reinforcement learning

VTA: ventral tegmental area

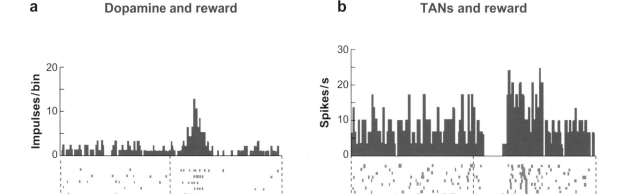

a   Dopamine and reward   **b**   TANs and reward

Figure 3

Reward-related activity of dopamine-containing neurons of nigral and striatal neurons. (*a*) Activity of nigral-VTA complex neurons (from Romo & Schultz 1990). (*b*) Activity of tonically active neurons in the striatum (from Aosaki et al. 1994b) Spike rasters (*below*) and histograms of those spikes (*above*) are aligned (*vertical lines*) at touch of the food reward (*a*) and at the conditional stimulus click sound indicating reward.

elicit dopamine release in the striatum (de la Fuente-Fernandez et al. 2001). Action value encoding was not detected by the original experimental paradigms used for recording from the dopaminergic neurons, which focused mostly on noninstrumental learning. Morris et al. (2006), using a decision task with a block design, have now shown that the action value of a future action can be coded in the firing of these neurons. This result is important in favoring computational models that take into account the value of a given action in a given state (the Q value). Remarkably, the dopaminergic neurons can signal which of two alternate actions will subsequently be taken in a given experimental task with a latency of less than 200 ms after the start of a given trial. This fast response suggests that another brain region has coded the decision and sent the information about the forthcoming action to the nigral neurons (Morris et al. 2006; compare Dommett et al. 2005).

Models that incorporate the value of chosen actions in a particular state include those known as the state-action-reward-state-action or SARSA models and advantage learning models.

Ironically, a main candidate for a neural structure that could deliver the action value signal to the midbrain dopamine-containing neurons is the striatum, the region originally thought to be the student of the dopaminergic substantia nigra. Many projection neurons in the striatum encode action value when monkeys perform in block design paradigms in which action values are experimentally manipulated (Samejima et al. 2005). Other structures projecting to the nigral dopamine-containing neurons are also candidates, including the pedunculopontine nucleus (one of the brain stem regions noted in **Figures 1** and **5**), the raphe nuclei including the dorsal raphe nucleus, the lateral habenular nucleus and forebrain regions including the amygdala and limbic-related cortex,

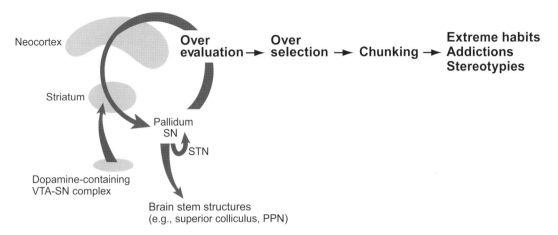

Figure 4

Schematic diagram suggesting the progression of functional activation in cortico-basal ganglia circuits as highly repetitive habits, addiction, and stereotypies emerge behaviorally. Note that in contrast to normal everyday habit learning (**Figure 1**), even the early stages of extreme habit formation involve steps that tend not to be readily reversible. PPN, pedunculopontine nucleus; SN, substantia nigra; STN, subthalamic nucleus; VTA, ventral tegmental area.

and also the striatum itself, including the striosomal system.

These findings highlight the difficulty of assigning an exclusive teaching function to any one node in interconnected circuits such as those linking the dopamine-containing midbrain neurons, the basal ganglia, and the cerebral cortex. Reinforcement-related signals of different sorts have been found in all of these brain regions (e.g., Glimcher 2003, Padoa-Schioppa & Assad 2006, Paton et al. 2006, Platt & Glimcher 1999, Sugrue et al. 2004), suggesting that signals related to reinforcement and motivation are widely distributed and can be used to modulate distributed neural representations guiding action. Reward-related activity has even been identified in the primary visual cortex (Shuler & Bear 2006) and the hippocampus (Suzuki 2007), neither of which is part of traditional reinforcement learning circuits. How these distributed mechanisms are coordinated is not yet clear.

Many of the ideas in reinforcement learning models and their close allies in neuroeconomics are now central to any consideration of habit learning. Experiments on goal-directed behavior in animals, including some with reward devaluation protocols, are increasingly being interpreted within the general framework of reinforcement learning (Daw & Doya 2006, Niv et al. 2006). For example, Daw et al. (2005) have proposed a model with two behavioral controllers. One (identified with the prefrontal cortex) uses a step-by-step, model-based reinforcement learning system to explore alternatives and make outcome predictions (their "tree-search" learning system for goal-oriented behaviors). The second, identified with the dorsolateral striatum, is a nonmodel-based cache system for determining a fixed value for an action or context that can be stored but that then is inflexible, corresponding to the habit system. The transition between behavioral control between the tree-search and cache systems is determined by the relative Bayesian uncertainty of the two systems. Allowing for interactions between these two systems would bring them into correspondence with the goal-oriented and habit systems of the reward devaluation literature. It seems unlikely, however, that there are only two learning systems or that these are dissociable as being exclusively cortical and subcortical.

The shift from ventral striatal to dorsal striatal activation during habit learning is also being incorporated explicitly into modified

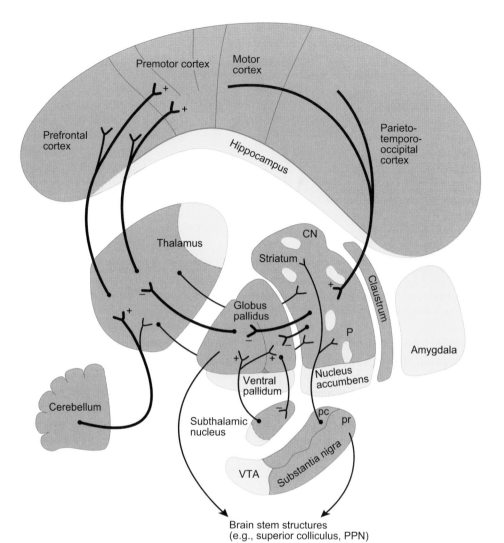

Figure 5

Basal ganglia circuit anatomy shown in simplified form with motor control pathways (*gray*) and regions related to limbic-system function (*yellow*). Descending pathways to brain stem structures (such as the superior colliculus and pedunculopontine nucleus) are not shown in detail. CN, caudate nucleus; P, putamen; pc, substantia nigra pars compacta; pr, substantia nigra pars reticulata; VTA, ventral tegmental area.

RL models. For example, evidence from human brain-scanning experiments has implicated the ventral striatum—active during the initial stages of learning—in reward-prediction error encoding whether or not responses by the subject are required. By contrast, reward-prediction error signaling that occurs during instrumental responding is differentially associated with activity in the dorsal striatum, especially dorsomedially (O'Doherty et al. 2004). These findings favor modified RL actor-critic models including advantage learning (Dayan 2002), in which the critic (ventral striatum) is

influenced by motivational state (e.g., hunger) as well as by the ongoing evaluation of the particular state, whereas the dorsal striatum (actor) is engaged when actions are instrumental in bringing reward (Dayan & Balleine 2002).

This reformulation of the actor-critic model adds the critical feature of motivation to RL treatments. Haruna & Kawato (2006) have drawn a contrast between the caudate nucleus (and ventral striatum) as being associated with reward-prediction error and the putamen as being associated with a stimulus-action-dependent reward-prediction error. The early requirement for ventral striatal activity in procedural learning has prompted the view that the ventral striatum is the director, guiding the dorsal striatal actor (Atallah et al. 2007). Imaging technologies cannot yet detect the neurochemically defined striosome and matrix compartments of the dorsal striatum, but the limbic-associated striosomes likely share some characteristics of the ventral striatum (Graybiel 1990, 1998). In line with this possibility, striosomes may represent state value in actor-critic architectures (Doya 2000, 2002) and thus may be critical components of striatum-based learning circuits.

The convergence of computational approaches with neuroimaging in humans has also led to new cognitive neuroscience models in which cognitive evaluation and choice depend on current reinforcement and expected future reinforcement and also on the value of behaviors not chosen so that choices themseves can be evaluated by comparing alternatives (Montague et al. 2006). The ventral-to-dorsal gradients found in these and other experiments (e.g., Tanaka et al. 2004, Zald et al. 2004) may in part reflect the predominance of immediate rewards (ventral striatum) and future rewards (dorsal striatum) in influencing behavior (Tanaka et al. 2004). This view emphasizes what I, below, call mental exploration.

EXTREME HABITS

Investigations of normal habit learning have immediate relevance for the field of addiction research. Habits induced by exposure to drugs of abuse can dominate behavior and produce states of craving and drug-seeking that persist for years. The dopamine-containing VTA and its striatal target region (the ventral striatum including the nucleus accumbens) have long been identified as forming a reward circuit essential for the initiation and expression of addictive behavior. Experiments on addiction have led to conclusions that are strikingly similar to those emerging from study of the normal transition from goal-directed to habit-driven behavior (**Figure 4**) (Everitt et al. 2001,

Hyman et al. 2006, Ito et al. 2002, Kalivas & Volkow 2005, Porrino et al. 2004, Wise 2004). (*a*) The addictive process involves plasticity in neural circuits, not just in a single brain region. (*b*) The circuits critical for addiction include midbrain dopamine-containing cell groups and the striatum (in particular, the ventral striatum) and the neocortex (especially the anterior cingulate cortex and orbitofrontal cortex), as well as limbic parts of the pallidum (ventral pallidum), the thalamus (mediodorsal nucleus), and the amygdala and extended amygdala (Everitt & Robbins 2005, Kalivas et al. 2005, Kalivas & Volkow 2005). (*c*) The neocortex (especially the anterior cingulate and orbitofrontal cortex) powerfully influences this circuit at multiple levels. (*d*) Within this cortico-subcortical circuitry, different subcircuits appear to be essential at different stages of acquisition: the VTA and shell of the nucleus accumbens for the initial learning process in addiction, and the core of the nucleus accumbens and neocortical regions for the expression of the learned behaviors. Thus, the predominant activity in a given learning context shifts over time as the addiction becomes fixed (Sellings & Clarke 2003).

A striking example of this dynamic patterning comes from imaging studies in which radiolabeled dopamine-receptor agonist ligands are given to addicted subjects. Given exposures to a drug (IV methylphenidate), increased dopamine-related signaling (displacement of the ligand) in cocaine addicts occurs in the ventral striatum (and anterior cingulate and orbitofrontal cortices). But when cocaine addicts view video tapes showing cues associated with cocaine use, it is the dorsal striatum (including the putamen) that exhibits the differentially heightened dopamine signal (Volkow et al. 2006, Wong et al. 2006).

These findings are supported by work on experimental animals. In macaque monkeys, cocaine self-administration alters metabolic activity mainly in the ventral striatum after a brief period of self-administration, but with extended self-administration, such changes occur increasingly in the dorsal striatum (Porrino et al. 2004). The gradients in these effects fit with anatomical evidence that the dopamine-containing inputs to the striatum, via indirectly recursive connections between the striatum and substantia nigra, follow such a ventral-to-dorsal gradient (Haber et al. 2000). Many other neurotransmitter-related compounds also follow such gradients, however, so that many aspects of striatal circuitry are modulated differently in ventral and dorsal striatum (and in medial and lateral striatal regions as well).

Differential release of dopamine in the nucleus accumbens in response to drug-associated cues has now been demonstrated directly in the rat by measuring extracellular dopamine by electrochemical detection with intrastriatal electrodes (Phillips et al. 2003). Cues produce dopamine release after the drug habit has been established (Ito et al. 2002). If the drug-taking behavior is extinguished and then is reinstated by exposing rodents to cues associated with the drug, as a model of relapse, dopamine release is also prominent in the neocortex and amygdala. These studies, in their entirety, suggest that as an individual moves from spaced, cognitively controlled experience with an addictive drug to increasingly repeated experience with the drug and then to the addicted state of compulsive drug use, major circuit-level changes occur at both cortical and subcortical levels. Within the striatum, there is a strong gradient from ventral to dorsal regions during the course of the addictive process.

Studies on normal habit learning and these studies on addictive habit learning both indicate a gradual change from ventral striatal control over the addictive behaviors to engagement of the dorsal striatum and neocortex (Everitt & Robbins 2005, Fuchs et al. 2006, Nelson & Killcross 2006). Such shared dynamic patterning (**Figure 2**) suggests that both in addiction and in the acquisition of nondrug-related habits, a similar set of coordinated changes in basic activity occurs at the circuit level. These dynamic patterns are a major clue for future experimental and computational work (Redish 2004).

A particular advantage of studies on addiction is that they can be used to uncover cellular and molecular mechanisms that promote transition to the addicted state (Hyman et al. 2006, Robinson & Kolb 2004, Self 2004). Many of these mechanisms are likely conserved and influence the learning processes that lead to repetitive, compulsive habits that are not triggered directly by drugs (Graybiel et al. 2000, Graybiel & Rauch 2000, Hyman et al. 2006). Links between the neurobiology of addictive and nonaddictive habits have already begun to be established by considering commonalities and interactions between drug effects and learning (e.g., Everitt & Robbins 2005, Fuchs et al. 2006, Nelson & Killcross 2006, Nordquist et al. 2007, Willuhn & Steiner 2006, Wise 2004).

LEARNING HABITS AND LEARNING PROCEDURES

How does habit learning relate to procedural or skill learning? Both involve learning sequential behaviors, but there are obvious differences: Learning how to ride a bicycle is quite different from having the habit of biking every evening after work. The skilled procedural performance of the baseball player is distinct from the habits and rituals for which baseball players are famous. And a regularly practiced habit may actually not be very skilled. Nevertheless, the learning of sequential actions to the point at which they can be performed with little effort of attention is common to both. So is the fact that, once consolidated, most acquired habits and procedures can be retained for long periods of time.

This transition from novice to expert is called proceduralization in the literature on human skill learning, most of which lies outside the reinforcement learning framework. These studies typically use reaction times (either simple or serial) as performance measures, along with a gradual reduction in interference if other tasks must be performed (Logan 1988, Nissen & Bullemer 1987). A large literature summarizes the varieties of motor sequence encoding in monkeys performing movement sequences (Georgopoulos & Stefanis 2007, Matsuzaka et al. 2007, Tanji & Hoshi 2007). This work demonstrates that neurons can have responses related to individual or multiple elements of movement sequences and that these sequence-selective activities can occur in widely distributed brain regions, both cortical and subcortical (**Figure 5**). Notably, imaging and other studies demonstrate dynamic changes in the patterns of neural activity in the neocortex and striatum that closely parallel those observed in studies of habit-learning (**Figure 2**). With practice, the activity shifts from anterior and ventral cortical regions to more posterior zones of the neocortex, and from more ventral and anterior striatal regions to more caudal zones in the striatum (Doyon & Benali 2005, Graybiel 2005, Isoda & Hikosaka 2007, Poldrack et al. 2005). These shifts have been interpreted as representing a shift in the coordinate frames of the early and late representations (Hikosaka et al. 1999, Isoda & Hikosaka 2007). Doyon et al. (2003) propose a key function for corticostriatal circuits in this process, contrasting these with cerebellum-based learning processes that allow adjustments of motor behavior to imposed changes (Kawato & Gomi 1992).

A function in online correction has been ascribed to the striatum (Smith & Shadmehr 2005), which would be consistent with evidence for the basal ganglia–based song system in birds (Brainard & Doupe 2000). The experiments of Haruno & Kawato (2006) suggest that the differential engagements of the more anterior striatum (caudate nucleus) and more posterior striatum (putamen) at different points during learning reflect a fundamental difference in the representations of learning-related signals in the caudate nucleus and ventral striatum (representing reward-prediction error) and the putamen (representing stimulus-action-reward association). These and related studies (Samejima et al. 2005) indicate the value of extending reinforcement learning and neuroeconomics models of habit learning to the study of the proceduralization process itself and underscore the dynamic shifts in focus of neural activity as learning proceeds. However, these mod-

els will likely be revised as more is learned about the concurrent neural activities in the neocortex, striatum, and cerebellum. For example, cerebellar input can reach the putamen via a cerebello-thalamo-striatal pathway (Hoshi et al. 2005). Activity in the putamen may in part reflect input from a cerebellar circuit. Perhaps the striatum evaluates the forward models of the cerebellum.

These findings may help in understanding the progression of symptoms in Parkinson's disease patients. The advance of dopamine denervation in Parkinson's disease follows a posterior-to-anterior gradient, the reverse of the gradient found for procedural learning and habit learning. The predominant difficulty of Parkinson's patients to perform even well-known procedures, like rising from a chair or trying to do two things at once, may reflect learning gradients seen during proceduralization.

HABITS, RITUALS, AND FIXED-ACTION PATTERNS

Work on habits and habit learning has clear connectivity with the study of repetitive behaviors that include species-specific, apparently instinctual action sequences: the fixed action patterns described by ethologists. Like habits, FAPs are regularly contrasted with voluntary behaviors and relegated to low-level behavioral control schemes in which a particular stimulus/context elicits an entire behavioral sequence, for example, a chick pecking the orange spot on the beak of its parent (Tinbergen 1953). Here, value functions are estimated over evolutionary time.

Rituals are common across animal species and can be either remarkably complex, as in the nest building of the bowerbird (Diamond 1986), or relatively simple, as in grazing animals following set routes to a water source. Rituals related to territoriality, mating, and social interactions of many types seem to dominate the lives of animals in the wild. These behaviors share cardinal characteristics of habits. They are repetitive, sequential action streams and can be triggered by particular cues.

Nearly all so-called fixed action patterns exhibit some flexibility (vulnerability to experience-based adjustment), and many rituals are acquired. A particularly instructive example of ritualistic behavior in animals is the use of song in mating behavior and other social behaviors in orcine birds. Their songs are composed of fixed sequences and, though in some species these appear to be genetically determined, in many species the songs are acquired by trial-and-error learning aimed at matching a tutor's template song. Bird song learning critically depends on a forebrain circuit that corresponds to a cortico–basal ganglia loop in mammals. Lesions within this circuit (called the anterior forebrain pathway) prevent the development of a mature song matching the tutor's song. The early song becomes prematurely stereotyped. At adulthood, such lesions impair the corrective adjustments of song required when abnormalities in singing behavior are induced experimentally (Brainard & Doupe 2000, Kao & Brainard 2006, Olveczky et al. 2005), and in some species the lesions can seriously degrade the mature song (Kobayashi et al. 2001). The song system is likely strongly influenced by dopamine, as the central core of the song circuit, area X (corresponding to the mammalian striatum and pallidum), receives a dopamine-containing fiber projection from the midbrain. Altogether, many aspects of avian song learning and song production bear a strong resemblance to the learning and performance of what we think of as habits and procedures in mammals: Both involve the acquisition of ordered sequences of behavior that are learned by balancing exploration and exploitation to attain a goal. And both require the activity of basal ganglia–based circuits as they are learned.

STEREOTYPIES

Very highly repetitive behavioral routines in animals are qualitatively distinguished as stereotypies on the basis of their apparent purposelessness and their great repetitiveness. In contrast to the habitual displays and rituals that are triggered in the course of normal behaviors, stereotypies (extremely rigid, repetitive sequences of behavior) are most prominent under aversive conditions, including stress, social isolation, and sensory deprivation (Ridley 1994). Exposure to psychomotor stimulants such as amphetamine also induces stereotypies, either of locomotion or of naturalistic behaviors such as sniffing, rearing, mouthing, and huddling. These categories can be likened to the route stereotypies and focused stereotypies of unmedicated animals. A distinction is often made between such highly repetitive, apparently purposeless behaviors and highly repetitive behaviors that appear goal directed, for example, repetitive grooming (barbering) of another animal or self-grooming.

Work on drug-induced stereotypies demonstrates that the basal ganglia are central to the development of these repetitive behaviors (Cooper & Dourish 1990). Low doses of drugs such as amphetamine (which releases dopamine and other biogenic amines) and cocaine (which blocks the reuptake of dopamine) induce prolonged and often repetitive bouts of locomotion. This behavior requires the functioning of the ventral striatum. The same drugs, given at higher doses, also induce bouts of highly stereotyped behavior, but these are more focused behaviors such as sniffing and grooming. These behaviors depend on the dorsal striatum (Joyce & Iversen 1984). Such dose-dependent drug effects ranging from locomotion to focused stereotypies follow the ventral striatal-to-dorsal striatal gradient that is thought to underlie both the acquisition of normal procedures and habits and the acquisition of addictive habits.

The strength of drug-induced stereotypic behaviors is correlated with differential activation of the striosomal compartment of the striatum, relative to activation of the matrix (**Figure 6**). This relationship has been demonstrated in rats, mice, and monkeys by the use of early-response gene assays (Canales 2005, Canales & Graybiel 2000, Graybiel et al. 2000, Saka et al. 2004). These findings tie the acquisition of at least one class of stereotypic

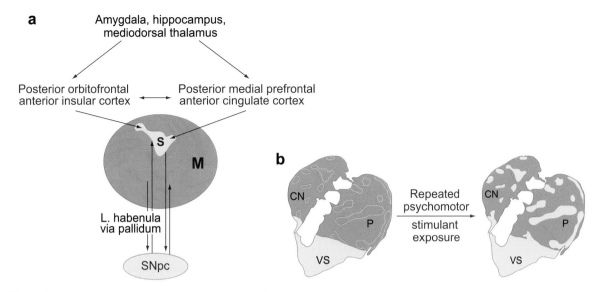

Figure 6

The striosomal system of the striatum. (*a*) Connections of striosomes with forebrain and brain stem regions including projections to nigral dopamine-containing neurons and/or nondopamine-containing nigral neurons near them. (*b*) Schematic illustration of the progressive accentuation of activity in striosomes that occurs as psychomotor stimulant-induced stereotypies develop. (Modified from Graybiel & Saka 2004). CN, caudate nucleus; M, matrix; P, putamen; S, striosome; SNpc, substantia nigra pars compacta; VS, ventral striatum.

behavior to a specialized set of striosome-predominant basal ganglia–based circuits. Striosomes in the anterior part of the striatum receive inputs from the anterior cingulate cortex and posterior orbitofrontal cortex and project to the substantia nigra both directly and indirectly via the pallidum and the limbic lateral habenular nucleus (Graybiel 1997). Through these connections, the striosomal system likely integrates and evaluates limbic information and affects the firing of dopamine-containing neurons in the nigral complex or immediately neighboring neurons. The striosomal system could thus alter the set point of subsets of the dopaminergic neurons and be a driving force behind some classes of stereotypic behaviors. We do not yet know what signal is transmitted by striosomes. One possibility is that they gate the magnitude of positive reward-prediction error or signal negative reward-prediction error through a connection with the pars compacta or nearby nigral neurons. They may also indirectly influence the lateral habenula, itself a source of negative reward-prediction error (Matsumoto & Hikosaka 2007).

A second line of evidence that the basal ganglia are required for some stereotypic behaviors comes from work on the highly repetitive movement sequences made by rodents during bouts of grooming called syntatic grooming (Berridge 1990). The grooming movements are strictly repetitive and temporally ordered, and lesion studies in the rat have demonstrated that a region in the mid-anterior dorsolateral caudoputamen must be intact for syntatic grooming to occur. Recording experiments suggest that some single striatal neurons in that region respond selectively to grooming movements that are part of a given syntatic grooming bout (Aldridge & Berridge 1998). The dopaminergic system can modulate the intensity of the grooming as well as the intensity of drug-induced stereotypies (Berridge et al. 2005).

How do stereotypies relate to habits? Evidently, they are different phenotypically. Stereotypies are more repetitive, are not as regularly elicited by triggers, and may not be governed in the same way by reinforcement learning. Yet many types of stereotypic behavior depend on basal ganglia–based circuits for their development and production, as do habits. Stereotypies, like addictions, seem to reflect an extreme state of functioning of these brain circuits, in which flexibility is minimal and repetitiveness is maximal. The well-learned song of a zebra finch, the product of a basal ganglia circuit analogue in the avian brain, is as stereotyped as a typical drug-induced stereotypy. The fact that one is naturally prompted and the other is not should not obscure a potential common neural basis of these behaviors in the operation of cortico–basal ganglia circuits and related brain networks.

HABITS, STEREOTYPIES, AND RITUALISTIC BEHAVIORS IN HUMANS

Habits and routines are woven into the fabric of our personal and social lives as humans. One can scarcely call to mind the events of a day without running up against them. In fact, "getting in a rut" is so easy to do for many of us that we must fight the tendency in order to get a fresh look at life. Yet, as William James and many others have argued, those ruts, fashioned carefully, are invaluable aids to making one's way through life and are critical in social order: famously, the flywheel of society (James 1890).

Helpful as habits can be in daily life, they can become dominant and intrusive in neuropsychiatric disorders, including obsessive-compulsive disorder (OCD), Tourette syndrome, and other so-called OC-spectrum disorders (Albin & Mink 2006, Graybiel & Rauch 2000, Leckman & Riddle 2000). They are major features of some autism spectrum disorders and can dominate in some of these conditions (for example, in Rett syndrome). Stereotypies are pronounced in unmedicated schizophrenics and in some patients with Huntington's disease or dystonia, and they appear in exaggerated form in some medicated Parkinson's disease patients. In these disorders—as in normal behavior—habits, stereotypies, and rituals can be cognitive as well as motor, and often, cognitive and motor acts and rituals are interrelated. This

commonality across cognitive and motor domains is a crucial attribute that suggests that there may be commonality also in the neural mechanisms underlying repetitive thoughts and actions.

In OCD, classified as an anxiety disorder, the repetitive behaviors that often occur do so in response to compelling and disturbing repetitive thoughts and felt needs (obsessions) that drive the repetitive behaviors (compulsions). The repetitive thoughts are dominant despite the fact that the person usually is aware that they are happening and does not want them to happen. Notably, the most common obsessions and compulsions are strongly cross-cultural. Checking ("Is it done? Did I do it just right?"), washing and cleaning due to obsessions about contamination, and repetitive ordering (lining up, straightening) are universal. OC-spectrum disorders include grooming and body-centered obsessions and compulsions, for example, extreme hair pulling or trichotillomania, for which barbering in animals is thought to be a model.

In Tourette syndrome, tics take the form of highly repetitive movements or vocalizations that range from abrupt single movements (twitches or sounds) to complex sequences of a behavior (including vocalizations) that appear as whole purposeful behaviors. They are exacerbated by stress, are sensitive to context, often occur in runs or bouts, and though suppressible for a time, often break through into overt expression. These actions are often prompted by an internal sensation or urge that builds up and produces stress; the urge can be relieved for a time by the expression of the tics. Nearly any piece of voluntary behavior can appear as a tic in Tourette syndrome. Sensory tricks, for example, touching a particular body part in a particular way, can also relieve the tics in some instances. It is as though a circuit goes into overdrive during a bout of tics but can be normalized by receiving new inputs.

At the behavioral level, obsessions and compulsions also can focus on a wide range of behaviors from grooming to eating to hoarding. Some specific genetic disorders involve compulsive engagement in one or more of these fundamental behaviors (for example, ritualistic hyperphagia and food-related obsessions in Prader-Willi syndrome). These disorders seem to accentuate one or another of a library of species-specific and culturally molded behaviors. This remarkable characteristic has led many investigators, from ethologists and neuroscientists to neurologists and psychiatrists, to make extensive comparisons between these human behaviors and the rituals, stereotypies, and FAPs of nonhuman species (Berridge et al. 2005, Graybiel 1997). These behaviors also have high relevance for the study of habit. Across the spectrum of OC disorders, the behaviors are repetitive and usually sequential; they are often performed as a whole after being triggered by external or internal stimuli or contexts (including thoughts); and they can be largely involuntary.

It is not yet known whether the brain states generated by these disorders are akin to the brain states generated as a result of habit learning. However, dysfunction of basal ganglia–based circuits appear to be centrally implicated in these disorders, from OCD-Tourette and other OC-spectrum disorders to eating disorders to the punding of overmedicated Parkinson's disease patients (e.g., Graybiel & Rauch 2000, Palmiter 2007, Voon et al. 2007). A disordered OCD circuit has been identified in imaging studies, and this circuit includes the regions most often implicated in drug addiction: the anterior cingulate cortex and the posterior orbitofrontal cortex and their thalamic correspondents, and the anterior part of the caudate nucleus and the ventral striatum. In Tourette syndrome, differential degeneration of parvalbumin-containing neurons in the striatum and pallidum has been reported (Kalanithi et al. 2005), suggesting that the basal ganglia also contribute to the symptomatology in this disorder. The focal nature of tics may reflect local abnormal activation of subdivisions in the matrix compartment of the striatum (matrisomes) (Mink 2001) or defects in particular cortico–basal ganglia subcircuits (Grabli et al. 2004). The efficacy of sensory tricks in some Tourette patients may also reflect this architecture, whereby inputs and outputs are matched in focal striatal domains (Flaherty & Graybiel 1994). Because of the limited spatial resolution of current imaging methods, it is not known whether the striosomal system is differentially involved; this would not be surprising, however. Differential vulnerability of striosomes has been reported for a subset of Huntington's disease patients in whom mood disorders, including OC-like symptoms, were predominant at disease onset (Tippett et al. 2007). Predominant striosomal degeneration has also been reported in X-linked dystonia-parkinsonism (DYT-3, Lubag), in which many affected individuals exhibit OCD-like symptoms and suffer from severe depression (Goto et al. 2005). Sites being targeted for deep brain stimulation therapy in OCD and Tourette syndrome are also within these basal ganglia–based circuits; anterior sites near the nucleus accumbens are favored for OCD, and intralaminar and internal pallidal sites are favored for Tourette syndrome (Mink et al. 2006).

HABITS AND RITUALS: THE BASAL GANGLIA AS A COMMON THEME

Habits, whether they are reflected in motor or cognitive activity, typically entail a set of actions, and these action steps typically are released as an entire behavioral episode once the habit is well engrained. Here, I have reviewed evidence that this characteristic expression of an entire sequential behavior is also typical of well-practiced procedures and extends to stereotypies and rituals, including personal and cultural rituals in humans. Neural mechanisms that could account for such extended, encapsulated behaviors are not yet understood, but clues are coming from experiments in which chronic electrophysiological recordings were made from ensembles of neurons of rats and monkeys performing repetitive tasks to receive reward (**Figure 7**) (Barnes et al. 2005; Fujii & Graybiel 2003, 2005; Graybiel et al. 2005; Jog et al. 1999).

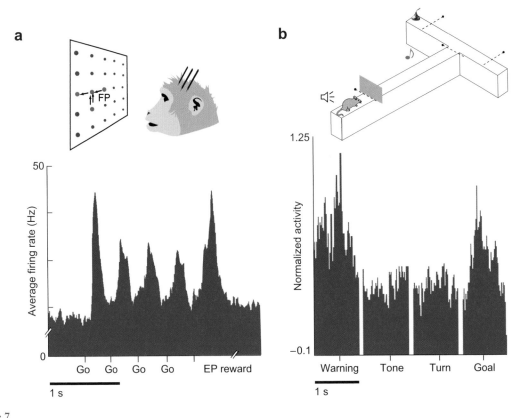

Figure 7

Heightened neural responses at action boundaries. Comparison between neural activity in prefrontal cortex accentuating the beginning and the end of saccade sequences in macaque monkeys performing sequential saccade tasks (*a*) and activity in sensorimotor striatum (dorsolateral caudoputamen) accentuating beginning and the end of maze runs in rats performing maze runs (*b*). Part *a* is modified from data in Fujii & Graybiel (2003), and part *b* is modified from data in Barnes et al. (2005).

In monkeys thoroughly trained to make sequences of saccadic eye movements, subsets of neurons in both the dorsolateral prefrontal cortex and the striatum have accentuated responses related to the first and last movements of the sequences, as though marking the boundaries of the action sequences (Fujii & Graybiel 2003, 2005). The accentuated beginning and end representations can be detected not only by looking at the task-related activity of the neurons, but also by looking at the temporal resolution of the neural representations (Graybiel et al. 2005). Time can be decoded with higher resolution at the beginning and the end of the movement sequences than during them. This last finding suggests that the neural representation of boundaries involves both action and time.

An explicit attempt has been made in rodents to track the neural activity patterns that occur in the striatum as habits are acquired (Barnes et al. 2005, Jog et al. 1999). Rats were trained to run a T-maze task in which they were cued to turn right or left to receive reward. The recordings were made in the dorsolateral caudoputamen, the region identified with S-R habits in reward devaluation studies and RL models. According to such studies, one would expect neural activity to be low during initial training but to increase during overtraining. The activity should be maintained during reward devaluation. This was not found. Instead, early in training, activity was strong throughout the maze runs. With extended overtraining, the activity did not increase. Instead, the activity changed in its pattern of distribution over the course of the maze runs. Task-related neural firing became concentrated at the beginning and end of the maze runs, in a pattern resembling the action boundary pattern found in the overtrained monkeys performing saccade sequences (**Figure 7**). Simultaneously, apparently nontask-responsive neural firing in the striatum was markedly reduced. When extinction training was then given (akin to reward devaluation), these acquired neural activity patterns were not maintained, but instead were gradually reversed. They were not lost or forgotten, however, because once reacquisition training began, the acquired activity patterns reappeared.

These findings suggest that one result of habit learning is to build in the sensorimotor striatum chunked, boundary-marked representations of the entire set of action steps that make up the behavioral habit. The changes in neural activity found in the maze experiments have been interpreted as representing the neural analog of explore-exploit behavior (Barnes et al. 2005, Graybiel 2005). During initial training, the animals shifted from an exploratory (variable) mode to a repetitive mode of running the maze. In parallel, the neural activity in the sensorimotor striatum measured across task time was initially variable (neural exploration), in that activity occurred throughout the maze runs. As the behavior became repetitive, the neural activity took on the accentuated beginning and end patterns (neural exploitation). This shift could represent part of the process by which action sequences are chunked for representation as a result of habit learning: When they are packaged as a unit ready for expression, the boundaries of the unit are marked and the behavioral steps unfold from the first to the last boundary marker (Graybiel 1998).

These shifts in neural activity during learning could be carried to the striatum by its input connections—notably, inputs from the cerebral cortex and the thalamus. Alternatively, they may be produced by intrastriatal network activity or by some combination of such circuit processing. Evidence favors a role for striatal interneurons in such circuit-level plasticity. For example, as

monkeys are taken through successive bouts of conditioning and extinction of conditioned eyeblink training, the tonically active neurons of the striatum (thought to be the cholinergic interneurons) successively gain and extinguish conditioned responses in parallel with the shifts in eyeblink behavior (Blazquez et al. 2002).

Further findings from the maze-learning experiments suggest that the formulation of the beginning and end pattern may depend, in part, on the rats learning the association between the instruction cues and the actions they instruct (T. Barnes, D. Hu, M. Howe, A. Dreyer, E. Brown, A. Graybiel, unpublished findings). The chunking process in sensorimotor tasks may thus occur as a result of S-R learning and be a hallmark of activity in the part of the striatum most directly engaged in affecting motor output. Work in progress also suggests that, at least in rats, the associative (dorsomedial) striatum exhibits different patterns during learning (Kubota et al. 2002a,b; C.A. Thorn, H. Atallah, Y. Kubota, A.M. Graybiel, unpublished findings). More work needs to be done to characterize these other learning-related striatal patterns along with those in other regions, including the neocortex; the locus coeruleus, identified as modulating exploration-exploitation balances (Aston-Jones & Cohen 2005); and serotonergic and cholinergic nuclei of the brainstem, which can modulate nigral neurons and reward functions (e.g., Doya 2002).

These results suggest that models epitomizing what physiological changes occur as habits and procedures are learned need to be revised to allow for repatterning of activity in different parts of the striatum and corresponding cortical and subcortical circuits. The gradients in activity traced from ventral to dorsal, anterior to posterior, and medial to lateral regions during habit learning do not necessarily mean that one region is transiently active and then becomes inactive as another region takes over, as if the habits are stored in just one site. The electrophysiological recordings suggest that we need dynamic models in which activity can occur simultaneously in multiple cortico-basal ganglia loops, not move in toto from one site to another, and models in which, as the learning process occurs, activity patterns change at all these sites. We need to capture how such simultaneous, dynamically changing activity patterns can become coordinated over time through the actions of plasticity mechanisms that act on neurotransmitter signaling systems that themselves are expressed in differential gradients and compartmental patterns across the striatum. An interesting possibility is that oscillatory activity helps to coordinate these activities (DeCoteau et al. 2007b, Thorn & Graybiel 2007). For example, even though different patterns of neuronal activity are found with simultaneous recordings in dorsomedial and dorsolateral regions of the striatum during maze learning, these two regions exhibit coordination of oscillatory activity in particular frequency ranges (Thorn & Graybiel 2007).

If this dynamic repatterning is a general function of cortico–basal ganglia loops, then it should occur for cognitive activity patterns as well, with a shift from mental exploration to mental exploitation as habits of thought are developed. Some evidence suggests that the striatum and associated cortico–basal ganglia loops are involved in such processing and chunking in human language (e.g., Crinion et al. 2006, Lieberman et al. 2004, Liegeois et al. 2003, Tettamanti et al. 2005). A system of executive control using such start and end states has been proposed for activity in Broca's area in the human, supporting the idea of boundary markers in hierarchically organized domains (Fujii & Graybiel 2003, Koechlin & Jubault 2006).

Could the chunking process also relate to the low levels of attention that we typically need to pay to a familiar behavior when performing it as a habit? The deaccentuation of neural activity in between the accentuated beginning and end activities in cortico–basal ganglia circuits could reflect this attribute of habits. The eventual chunking of action repertoires during habit learning is an endpoint of successive shifting of neural activity from regions more closely related to the limbic system to regions more closely related to motor and cognitive output. These shifts represent stages of evaluation, and

if the stages are successfully met—if the behavior is evaluated sufficiently positively—it is rerepresented in a chunked, readily releasable form. Thus, the relation between habits and the evaluative brain is that habits are an endpoint of the valuation process. Altogether, this process may engage a range of different cortico–basal ganglia loops and other neural circuits, potentially influencing different types of habit, from seemingly innocent mannerisms and rituals to dominating addictions. Studying this process should help investigators identify the neural systems underlying the shift from deliberative behavior controls to the nearly automatic, scarcely conscious control that we associate with acting through habit. Tracking this process may help us to understand the conscious state itself.

The power of social rituals may in part reflect an endpoint of this progressive evaluation process (**Figure 8**). The basal ganglia are strongly tied to the control and modulation of social behaviors. Human brain-imaging experiments have demonstrated strong activation of the dorsal striatum in experiments tracking activation for maternal love and romantic love (Aron et al. 2005, Bartels & Zeki 2004) and in social situations mimicked by interactive games and cost-benefit protocols (de Quervain et al. 2004, Elliott et al. 2004, Harbaugh et al. 2007, King-Casas et al. 2005, Montague & Berns 2002, O'Doherty et al. 2004, Tricomi et al. 2004, Zink et al. 2004). The nucleus accumbens and its dopamine receptors are necessary for monogamous pair bonding and for the maintenance of these bonds in the prairie vole (Aragona et al. 2006). Both language and song, with strong cortico–basal ganglia neural bases, serve social communication and are self-generated. They have the characteristic of agency, which heightens activation in the striatum in instrumental tasks relative to striatal activity in passive but otherwise corresponding tasks (Harbaugh et al. 2007, Zink et al. 2004). Finally, many of the rituals encountered in normal societies, and many of the ritualistic behaviors in neuropsychiatric disorders, have a strong social element, both in their content and

Figure 8

A ritual in humans (bull jumping in ancient Greece). Fresco from the East Wing of Knossos Palace, ~1500 B.C., Herakleion Museum, Crete.

in their likelihood for expression, and rituals and stereotypies in animals can also be strongly influenced by social context. Neural processing in circuits related to the basal ganglia, with their widespread interconnections with both limbic and sensorimotor systems, provides a common mechanistic theme across this large array of behaviors.

DISCLOSURE STATEMENT

The author is not aware of any biases that might be perceived as affecting the objectivity of this review.

ACKNOWLEDGMENTS

The author acknowledges the support for her laboratory from the National Institute of Mental Health MH60379, the National Institute of Neurological Disorders and Stroke NS25529, the National Eye Institute EY12848, the National Parkinson Foundation, and the Office of Naval Research N00014-04-1-0208. The author thanks H.F. Hall and Emily Romano for help with the figures; Clark Brayton for manuscript processing; and colleagues who read an earlier draft of this manuscript, including Russell Poldrack, Peter Dayan, Kyle Smith, and Hisham Atallah.

LITERATURE CITED

Adams CD, Dickinson A. 1981. Instrumental responding following reinforcer devaluation. *Q. J. Exp. Psychol.* 33:109–21

Albin RL, Mink JW. 2006. Recent advances in Tourette syndrome research. *Trends Neurosci.* 29:175–82

Aldridge JW, Berridge KC. 1998. Coding of serial order by neostriatal neurons: a "natural action" approach to movement sequence. *J. Neurosci.* 18:2777–87

Aosaki T, Graybiel AM, Kimura M. 1994a. Effects of the nigrostriatal dopamine system on acquired neural responses in the striatum of behaving monkeys. *Science* 265:412–15

Aosaki T, Tsubokawa H, Ishida A, Watanabe K, Graybiel AM, Kimura M. 1994b. Responses of tonically active neurons in the primate's striatum undergo systematic changes during behavioral sensorimotor conditioning. *J. Neurosci.* 14:3969–84

Aragona BJ, Liu Y, Yu YJ, Curtis JT, Detwiler JM, et al. 2006. Nucleus accumbens dopamine differentially mediates the formation and maintenance of monogamous pair bonds. *Nat. Neurosci.* 9:133–39

Aron A, Fisher H, Mashek DJ, Strong G, Li H, Brown LL. 2005. Reward, motivation, and emotion systems associated with early-stage intense romantic love. *J. Neurophysiol.* 94:327–37

Aston-Jones G, Cohen JD. 2005. An integrative theory of locus coeruleus-norepinephrine function: adaptive gain and optimal performance. *Annu. Rev. Neurosci.* 28:403–50

Atallah HE, Lopez-Paniagua D, Rudy JW, O'Reilly RC. 2007. Separate neural substrates for skill learning and performance in the ventral and dorsal striatum. *Nat. Neurosci.* 10:126–31

Balleine BW, Dickinson A. 1998. Goal-directed instrumental action: contingency and incentive learning and their cortical substrates. *Neuropharmacology* 37:407–19

Balleine BW, Killcross AS, Dickinson A. 2003. The effect of lesions of the basolateral amygdala on instrumental conditioning. *J. Neurosci.* 23:666–75

Barnes T, Kubota Y, Hu D, Jin DZ, Graybiel AM. 2005. Activity of striatal neurons reflects dynamic encoding and recoding of procedural memories. *Nature* 437:1158–61

Bartels A, Zeki S. 2004. The neural correlates of maternal and romantic love. *Neuroimage* 21:1155–66

Bayer HM, Glimcher PW. 2005. Midbrain dopamine neurons encode a quantitative reward prediction error signal. *Neuron* 47:129–41

Bayley PJ, Frascino JC, Squire LR. 2005. Robust habit learning in the absence of awareness and independent of the medial temporal lobe. *Nature* 436:550–53

Berridge KC. 1990. Comparative fine structure of action: rules of form and sequence in the grooming patterns of six rodent species. *Behaviour* 113:21–56

Berridge KC, Aldridge JW, Houchard KR, Zhuang X. 2005. Sequential superstereotypy of an instinctive fixed action pattern in hyperdopaminergic mutant mice: a model of obsessive compulsive disorder and Tourette's. *BMC Biol.* 3:4

Blazquez P, Fujii N, Kojima J, Graybiel AM. 2002. A network representation of response probability in the striatum. *Neuron* 33:973–82

Brainard MS, Doupe AJ. 2000. Interruption of a basal ganglia-forebrain circuit prevents plasticity of learned vocalizations. *Nature* 404:762–66

Canales JJ. 2005. Stimulant-induced adaptations in neostriatal matrix and striosome systems: transiting from instrumental responding to habitual behavior in drug addiction. *Neurobiol. Learn. Mem.* 83:93–103

Canales JJ, Graybiel AM. 2000. A measure of striatal function predicts motor stereotypy. *Nat. Neurosci.* 3:377–83

Colwill RM, Rescorla RA. 1985. Postconditioning devaluation of a reinforcer affects instrumental responding. *J. Exp. Psychol. Anim. Behav. Process.* 11:120–32

Cooper SJ, Dourish CT, eds. 1990. *Neurobiology of Stereotyped Behaviour.* Oxford, UK: Clarendon

Corbit LH, Balleine BW. 2005. Double dissociation of basolateral and central amygdala lesions on the general and outcome-specific forms of pavlovian-instrumental transfer. *J. Neurosci.* 25:962–70

Coutureau E, Killcross S. 2003. Inactivation of the infralimbic prefrontal cortex reinstates goal-directed responding in overtrained rats. *Behav. Brain Res.* 146:167–74

Crinion J, Turner R, Grogan A, Hanakawa T, Noppeney U, et al. 2006. Language control in the bilingual brain. *Science* 312:1537–40

Daw ND, Doya K. 2006. The computational neurobiology of learning and reward. *Curr. Opin. Neurobiol.* 16:199–204

Daw ND, Niv Y, Dayan P. 2005. Uncertainty-based competition between prefrontal and dorso-lateral striatal systems for behavioral control. *Nat. Neurosci.* 8:1704–11

Dayan P. 2002. Motivated reinforcement learning. In *Advances in Neural Information Processing Systems*, ed. TG Dietterich, S Becker, Z Ghahramans, pp. 11–18. San Mateo, CA: Morgan Kaufmann

Dayan P, Balleine BW. 2002. Reward, motivation, and reinforcement learning. *Neuron* 36:285–98

DeCoteau WE, Thorn CA, Gibson DJ, Courtemanche R, Mitra P, et al. 2007a. Learning-related coordination of striatal and hippocampal theta rhythms during acquisition of a procedural maze task. *Proc. Natl. Acad. Sci. USA* 104:5644–49

DeCoteau WE, Thorn C, Gibson DJ, Courtemanche R, Mitra P, et al. 2007b. Oscillations of local field potentials in the rat dorsal striatum during spontaneous and instructed behaviors. *J. Neurophysiol.* 97:3800–5

de la Fuente-Fernandez R, Ruth TJ, Sossi V, Schulzer M, Calne DB, Stoessl AJ. 2001. Expectation and dopamine release: mechanism of the placebo effect in Parkinson's disease. *Science* 293:1164–66

de Quervain DJ, Fischbacher U, Treyer V, Schellhammer M, Schnyder U, et al. 2004. The neural basis of altruistic punishment. *Science* 305:1254–58

Devan BD, White NM. 1999. Parallel information processing in the dorsal striatum: relation to hippocampal function. *J. Neurosci.* 19:2789–98

Diamond J. 1986. Animal art: variation in bower decorating style among male bowerbirds *Amblyornis inornatus*. *Proc. Natl. Acad. Sci. USA* 83:3042–46

Djurfeldt M, Ekeberg Ö, Graybiel AM. 2001. Cortex-basal ganglia interaction and attractor states. *Neurocomputing* 38–40:573–79

Dommett E, Coizet V, Blaha CD, Martindale J, Lefebvre V, et al. 2005. How visual stimuli activate dopaminergic neurons at short latency. *Science* 307:1476–79

Doya K. 2000. Complementary roles of basal ganglia and cerebellum in learning and motor control. *Curr. Opin. Neurobiol.* 10:732–39

Doya K. 2002. Metalearning and neuromodulation. *Neural. Netw.* 15:495–506

Doyon J, Benali H. 2005. Reorganization and plasticity in the adult brain during learning of motor skills. *Curr. Opin. Neurobiol.* 15:161–67

Doyon J, Penhune V, Ungerleider LG. 2003. Distinct contribution of the cortico-striatal and cortico-cerebellar systems to motor skill learning. *Neuropsychologia* 41:252–62

Elliott R, Newman JL, Longe OA, Deakin JFW. 2004. Instrumental responding for rewards is associated with enhanced neuronal response in subcortical reward systems. *Neuroimage* 21:984–90

Everitt BJ, Dickinson A, Robbins TW. 2001. The neuropsychological basis of addictive behaviour. *Brain Res. Brain Res. Rev.* 36:129–38

Everitt BJ, Robbins TW. 2005. Neural systems of reinforcement for drug addiction: from actions to habits to compulsion. *Nat. Neurosci.* 8:1481–89

Faure A, Haberland U, Conde F, El Massioui N. 2005. Lesion to the nigrostriatal dopamine system disrupts stimulus-response habit formation. *J. Neurosci.* 25:2771–80

Fiorillo CD, Tobler PN, Schultz W. 2003. Discrete coding of reward probability and uncertainty by dopamine neurons. *Science* 299:1898–902

Flaherty AW, Graybiel AM. 1994. Input-output organization of the sensorimotor striatum in the squirrel monkey. *J. Neurosci.* 14:599–610

Foerde K, Knowlton BJ, Poldrack RA. 2006. Modulation of competing memory systems by distraction. *Proc. Natl. Acad. Sci. USA* 103:11778–83

Fuchs RA, Branham RK, See RE. 2006. Different neural substrates mediate cocaine seeking after abstinence versus extinction training: a critical role for the dorsolateral caudate-putamen. *J. Neurosci.* 26:3584–88

Fujii N, Graybiel A. 2003. Representation of action sequence boundaries by macaque prefrontal cortical neurons. *Science* 301:1246–49

Fujii N, Graybiel A. 2005. Time-varying covariance of neural activities recorded in striatum and frontal cortex as monkeys perform sequential-saccade tasks. *Proc. Natl. Acad. Sci. USA* 102:9032–37

Georgopoulos AP, Stefanis CN. 2007. Local shaping of function in the motor cortex: motor contrast, directional tuning. *Brain Res. Rev.* 55:383–89

Glimcher PW. 2003. *Decisions, Uncertainty, and the Brain: The Science of Neuroeconomics.* Cambridge, MA: MIT Press

Goto S, Lee LV, Munoz EL, Tooyama I, Tamiya G, et al. 2005. Functional anatomy of the basal ganglia in X-linked recessive dystonia-parkinsonism. *Ann. Neurol.* 58:7–17

Gottfried JA, O'Doherty J, Dolan RJ. 2003. Encoding predictive reward value in human amygdala and orbitofrontal cortex. *Science* 301:1104–7

Grabli D, McCairn K, Hirsch EC, Agid Y, Feger J, et al. 2004. Behavioural disorders induced by external globus pallidus dysfunction in primates: I. Behavioural study. *Brain* 127:2039–54

Graybiel AM. 1990. Neurotransmitters and neuromodulators in the basal ganglia. *Trends Neurosci.* 13:244–54

Graybiel AM. 1997. The basal ganglia and cognitive pattern generators. *Schizophr. Bull.* 23:459–69

Graybiel AM. 1998. The basal ganglia and chunking of action repertoires. *Neurobiol. Learn. Mem.* 70:119–36

Graybiel AM. 2005. The basal ganglia: learning new tricks and loving it. *Curr. Opin. Neurobiol.* 15:638–44

Graybiel AM, Aosaki T, Flaherty AW, Kimura M. 1994. The basal ganglia and adaptive motor control. *Science* 265:1826–31

Graybiel AM, Canales JJ, Capper-Loup C. 2000. Levodopa-induced dyskinesias and dopamine-dependent stereotypies: a new hypothesis. *Trends Neurosci.* 23:S71–77

Graybiel AM, Fujii N, Jin DZ. 2005. Representation of time and states in the macaque prefrontal cortex and striatum during sequential saccade tasks. *Soc. Neurosci. Abstr. Viewer/Itiner.* 400.7

Graybiel AM, Rauch SL. 2000. Toward a neurobiology of obsessive-compulsive disorder. *Neuron* 28:343–47

Graybiel AM, Saka E. 2004. The basal ganglia and the control of action. In *The New Cognitive Neurosciences*, ed. MS Gazzaniga, pp. 495–510. Cambridge, MA: MIT Press. 3rd ed.

Haber SN, Fudge JL, McFarland NR. 2000. Striatonigrostriatal pathways in primates form an ascending spiral from the shell to the dorsolateral striatum. *J. Neurosci.* 20:2369–82

Harbaugh WT, Mayr U, Burghart DR. 2007. Neural responses to taxation and voluntary giving reveal motives for charitable donations. *Science* 316:1622–25

Haruno M, Kawato M. 2006. Different neural correlates of reward expectation and reward expectation error in the putamen and caudate nucleus during stimulus-action-reward association learning. *J. Neurophysiol.* 95:948–59

Hernandez PJ, Sadeghian K, Kelley AE. 2002. Early consolidation of instrumental learning requires protein synthesis in the nucleus accumbens. *Nat. Neurosci.* 5:1327–31

Hernandez PJ, Schiltz CA, Kelley AE. 2006. Dynamic shifts in corticostriatal expression patterns of the immediate early genes Homer 1a and Zif268 during early and late phases of instrumental training. *Learn. Mem.* 13:599–608

Hikosaka O, Nakahara H, Rand MK, Sakai K, Lu X, et al. 1999. Parallel neural networks for learning sequential procedures. *Trends Neurosci.* 22:464–71

Hoshi E, Tremblay L, Feger J, Carras PL, Strick PL. 2005. The cerebellum communicates with the basal ganglia. *Nat. Neurosci.* 8:1491–93

Houk JC, Wise SP. 1995. Distributed modular architectures linking basal ganglia, cerebellum, and cerebral cortex: their role in planning and controlling action. *Cereb. Cortex* 5:95–110

Hsu M, Bhatt M, Adolphs R, Tranel D, Camerer CF. 2005. Neural systems responding to degrees of uncertainty in human decision-making. *Science* 310:1680–83

Hyman SE, Malenka RC, Nestler EJ. 2006. Neural mechanisms of addiction: the role of reward-related learning and memory. *Annu. Rev. Neurosci.* 29:565–98

Isoda M, Hikosaka O. 2007. Switching from automatic to controlled action by monkey medial frontal cortex. *Nat. Neurosci.* 10:240–48

Ito R, Dalley JW, Robbins TW, Everitt BJ. 2002. Dopamine release in the dorsal striatum during cocaine-seeking behavior under the control of a drug-associated cue. *J. Neurosci.* 22:6247–53

Izquierdo A, Suda RK, Murray EA. 2004. Bilateral orbital prefrontal cortex lesions in rhesus monkeys disrupt choices guided by both reward value and reward contingency. *J. Neurosci.* 24:7540–48

James W. 1950 [1890]. *The Principles of Psychology.* New York: Dover

Jog M, Kubota Y, Connolly CI, Hillegaart V, Graybiel AM. 1999. Building neural representations of habits. *Science* 286:1745–49

Joyce EM, Iversen SD. 1984. Dissociable effects of 6-OHDA-induced lesions of neostriatum on anorexia, locomotor activity and stereotypy: the role of behavioural competition. *Psychopharmacology (Berl.)* 83:363–66

Kalanithi PS, Zheng W, Kataoka Y, DiFiglia M, Grantz H, et al. 2005. Altered parvalbumin-positive neuron distribution in basal ganglia of individuals with Tourette syndrome. *Proc. Natl. Acad. Sci. USA* 102:13307–12

Kalivas PW, Volkow N, Seamans J. 2005. Unmanageable motivation in addiction: a pathology in prefrontal-accumbens glutamate transmission. *Neuron* 45:647–50

Kalivas PW, Volkow ND. 2005. The neural basis of addiction: a pathology of motivation and choice. *Am. J. Psychiatry* 162:1403–13

Kao MH, Brainard MS. 2006. Lesions of an avian basal ganglia circuit prevent context-dependent changes to song variability. *J. Neurophysiol.* 96:1441–55

Kawato M, Gomi H. 1992. A computational model of four regions of the cerebellum based on feedback-error learning. *Biol. Cybern.* 68:95–103

Killcross S, Coutureau E. 2003. Coordination of actions and habits in the medial prefrontal cortex of rats. *Cereb. Cortex* 13:400–8

King-Casas B, Tomlin D, Anen C, Camerer CF, Quartz SR, Montague PR. 2005. Getting to know you: reputation and trust in a two-person economic exchange. *Science* 308:78–83

Knowlton BJ, Mangels JA, Squire LR. 1996. A neostriatal habit learning system in humans. *Science* 273:1399–402

Kobayashi K, Uno H, Okanoya K. 2001. Partial lesions in the anterior forebrain pathway affect song production in adult Bengalese finches. *NeuroReport* 12:353–58

Koechlin E, Jubault T. 2006. Broca's area and the hierarchical organization of human behavior. *Neuron* 50:963–74

Kubota Y, DeCoteau WE, Liu J, Graybiel AM. 2002a. Task-related activity in the medial striatum during performance of a conditional T-maze task. *Soc. Neurosci. Abstr. Viewer/Itiner.* 765.7

Kubota Y, DeCoteau WE, Liu J, Graybiel AM. 2002b. Task-related activity in the medial striatum during performance of a conditional T-maze task. *Soc. Neurosci. Abstr. Viewer/Itiner.* 765.7

Leckman JF, Riddle MA. 2000. Tourette's syndrome: when habit-forming systems form habits of their own. *Neuron* 28:349–54

Lieberman MD, Chang GY, Chiao J, Bookheimer SY, Knowlton BJ. 2004. An event-related fMRI study of artificial grammar learning in a balanced chunk strength design. *J. Cogn. Neurosci.* 16:427–38

Liegeois F, Baldeweg T, Connelly A, Gadian DG, Mishkin M, Vargha-Khadem F. 2003. Language fMRI abnormalities associated with *FOXP2* gene mutation. *Nat. Neurosci.* 6:1230–37

Logan GD. 1988. Toward an instance theory of automatization. *Psychol. Rev.* 95:492–527

Matsumoto M, Hikosaka O. 2007. Lateral habenula as a source of negative reward signals in dopamine neurons. *Nature* 447:1111–15

Matsuzaka Y, Picard N, Strick PL. 2007. Skill representation in the primary motor cortex after long-term practice. *J. Neurophysiol.* 97:1819–32

Mink JW. 2001. Basal ganglia dysfunction in Tourette's syndrome: a new hypothesis. *Pediatr. Neurol.* 25:190–98

Mink JW, Walkup J, Frey KA, Como P, Cath D, et al. 2006. Patient selection and assessment recommendations for deep brain stimulation in Tourette syndrome. *Mov. Disord.* 21:1831–38

Montague PR, Berns GS. 2002. Neural economics and the biological substrates of valuation. *Neuron* 36:265–84

Montague PR, Dayan P, Sejnowski TJ. 1996. A framework for mesencephalic dopamine systems based on predictive Hebbian learning. *J. Neurosci.* 16:1936–47

Montague PR, King-Casas B, Cohen JD. 2006. Imaging valuation models in human choice. *Annu. Rev. Neurosci.* 29:417–48

Moody TD, Bookheimer SY, Vanek Z, Knowlton BJ. 2004. An implicit learning task activates medial temporal lobe in patients with Parkinson's disease. *Behav. Neurosci.* 118:438–42

Morris G, Arkadir D, Nevet A, Vaadia E, Bergman H. 2004. Coincident but distinct messages of midbrain dopamine and striatal tonically active neurons. *Neuron* 43:133–43

Morris G, Nevet A, Arkadir D, Vaadia E, Bergman H. 2006. Midbrain dopamine neurons encode decisions for future action. *Nat. Neurosci.* 9:1057–63

Nakahara H, Itoh H, Kawagoe R, Takikawa Y, Hikosaka O. 2004. Dopamine neurons can represent context-dependent prediction error. *Neuron* 41:269–80

Nelson A, Killcross S. 2006. Amphetamine exposure enhances habit formation. *J. Neurosci.* 26:3805–12

Nissen MJ, Bullemer P. 1987. Attentional requirements of learning: evidence from performance measures. *Cognit. Psychol.* 19:1–32

Niv Y, Duff MO, Dayan P. 2005. Dopamine, uncertainty and TD learning. *Behav. Brain Funct.* 1:6

Niv Y, Joel D, Dayan P. 2006. A normative perspective on motivation. *Trends Cogn. Sci.* 10:375–81

Nordquist RE, Voorn P, de Mooij-van Malsen JG, Joosten RN, Pennartz CM, Vanderschuren LJ. 2007. Augmented reinforcer value and accelerated habit formation after repeated amphetamine treatment. *Eur. Neuropsychopharmacol.* 17:532–40

O'Doherty J, Dayan P, Schultz J, Deichmann R, Friston K, Dolan RJ. 2004. Dissociable roles of ventral and dorsal striatum in instrumental conditioning. *Science* 304:452–54

Olveczky BP, Andalman AS, Fee MS. 2005. Vocal experimentation in the juvenile songbird requires a basal ganglia circuit. *PLoS Biol.* 3:e153

O'Reilly RC, Frank MJ. 2006. Making working memory work: a computational model of learning in the prefrontal cortex and basal ganglia. *Neural Comput.* 18:283–328

Packard MG, Knowlton BJ. 2002. Learning and memory functions of the basal ganglia. *Annu. Rev. Neurosci.* 25:563–93

Packard MG, McGaugh JL. 1996. Inactivation of hippocampus or caudate nucleus with lidocaine differentially affects expression of place and response learning. *Neurobiol. Learn. Mem.* 65:65–72

Padoa-Schioppa C, Assad JA. 2006. Neurons in the orbitofrontal cortex encode economic value. *Nature* 441:223–26

Palmiter RD. 2007. Is dopamine a physiologically relevant mediator of feeding behavior? *Trends Neurosci.* 30:375–81

Paton JJ, Belova MA, Morrison SE, Salzman CD. 2006. The primate amygdala represents the positive and negative value of visual stimuli during learning. *Nature* 439:865–70

Phillips PE, Stuber GD, Heien ML, Wightman RM, Carelli RM. 2003. Subsecond dopamine release promotes cocaine seeking. *Nature* 422:614–18

Platt ML, Glimcher PW. 1999. Neural correlates of decision variables in parietal cortex. *Nature* 400:233–38

Poldrack RA, Clark J, Pare-Blagoev EJ, Shohamy D, Creso Moyano J, et al. 2001. Interactive memory systems in the human brain. *Nature* 414:546–50

Poldrack RA, Sabb FW, Foerde K, Tom SM, Asarnow RF, et al. 2005. The neural correlates of motor skill automaticity. *J. Neurosci.* 25:5356–64

Porrino LJ, Lyons D, Smith HR, Daunais JB, Nader MA. 2004. Cocaine self-administration produces a progressive involvement of limbic, association, and sensorimotor striatal domains. *J. Neurosci.* 24:3554–62

Rauch SL, Wedig MM, Wright CI, Martis B, McMullin KG, et al. 2006. Functional magnetic resonance imaging study of regional brain activation during implicit sequence learning in obsessive-compulsive disorder. *Biol. Psychiatry* 61:330–36

Redgrave P, Gurney K. 2006. The short-latency dopamine signal: a role in discovering novel actions? *Nat. Rev. Neurosci.* 7:967–75

Redish AD. 2004. Addiction as a computational process gone awry. *Science* 306:1944–47

Ridley RM. 1994. The psychology of perseverative and stereotyped behaviour. *Prog. Neurobiol.* 44:221–31

Robinson TE, Kolb B. 2004. Structural plasticity associated with exposure to drugs of abuse. *Neuropharmacology* 47(Suppl. 1):33–46

Romo R, Schultz W. 1990. Dopamine neurons of the monkey midbrain: contingencies of response to active touch during self-initiated arm movements. *J. Neurophysiol.* 63:592–606

Saka E, Goodrich C, Harlan P, Madras BK, Graybiel AM. 2004. Repetitive behaviors in monkeys are linked to specific striatal activation patterns. *J. Neurosci.* 24:7557–65

Salat DH, van der Kouwe AJ, Tuch DS, Quinn BT, Fischl B, et al. 2006. Neuroimaging H.M.: a 10-year follow-up examination. *Hippocampus* 16:936–45

Samejima K, Ueda Y, Doya K, Kimura M. 2005. Representation of action-specific reward values in the striatum. *Science* 310:1337–40

Satoh T, Nakai S, Sato T, Kimura M. 2003. Correlated coding of motivation and outcome of decision by dopamine neurons. *J. Neurosci.* 23:9913–23

Schultz W, Dayan P, Montague PR. 1997. A neural substrate of prediction and reward. *Science* 275:1593–99

Self DW. 2004. Regulation of drug-taking and -seeking behaviors by neuroadaptations in the mesolimbic dopamine system. *Neuropharmacology* 47(Suppl. 1):242–55

Sellings LH, Clarke PB. 2003. Segregation of amphetamine reward and locomotor stimulation between nucleus accumbens medial shell and core. *J. Neurosci.* 23:6295–303

Shuler MG, Bear MF. 2006. Reward timing in the primary visual cortex. *Science* 311:1606–9

Smith MA, Shadmehr R. 2005. Intact ability to learn internal models of arm dynamics in Huntington's disease but not cerebellar degeneration. *J. Neurophysiol.* 93:2809–21

Smith-Roe SL, Kelley AE. 2000. Coincident activation of NMDA and dopamine D1 receptors within the nucleus accumbens core is required for appetitive instrumental learning. *J. Neurosci.* 20:7737–42

Sugrue LP, Corrado GS, Newsome WT. 2004. Matching behavior and the representation of value in the parietal cortex. *Science* 304:1782–87

Sutton RS, Barto AG. 1998. *Reinforcement Learning: An Introduction.* Cambridge, MA: MIT Press

Suzuki WA. 2007. Associative learning signals in the monkey medial temporal lobe. Presented at *Cosyne 2007*, Salt Lake City, Utah

Tanaka SC, Doya K, Okada G, Ueda K, Okamoto Y, Yamawaki S. 2004. Prediction of immediate and future rewards differentially recruits cortico-basal ganglia loops. *Nat. Neurosci.* 7:887–93

Tanji J, Hoshi E. 2007. Role of the lateral prefrontal cortex in executive behavioral control. *Physiol. Rev.* 88:37–57

Tettamanti M, Moro A, Messa C, Moresco RM, Rizzo G, et al. 2005. Basal ganglia and language: phonology modulates dopaminergic release. *NeuroReport* 16:397–401

Thorn CA, Graybiel AM. 2007. Medial and lateral striatal LFPs exhibit task-dependent patterns of coherence in multiple frequency bands. *Soc. Neurosci. Abstr. Viewer/Itiner.* 622.14

Tinbergen N. 1953. *The Herring Gull's World: A Study of the Social Behaviour of Birds.* London: Collins

Tippett LJ, Waldvogel HJ, Thomas SJ, Hogg VM, van Roon-Mom W, et al. 2007. Striosomes and mood dysfunction in Huntington's disease. *Brain* 130:206–21

Tobler PN, Fiorillo CD, Schultz W. 2005. Adaptive coding of reward value by dopamine neurons. *Science* 307:1642–45

Tricomi EM, Delgado MR, Fiez JA. 2004. Modulation of caudate activity by action contingency. *Neuron* 41:281–92

Ungless MA, Magill PJ, Bolam JP. 2004. Uniform inhibition of dopamine neurons in the ventral tegmental area by aversive stimuli. *Science* 303:2040–42

Voermans NC, Petersson KM, Daudey L, Weber B, Van Spaendonck KP, et al. 2004. Interaction between the human hippocampus and the caudate nucleus during route recognition. *Neuron* 43:427–35

Volkow ND, Wang GJ, Telang F, Fowler JS, Logan J, et al. 2006. Cocaine cues and dopamine in dorsal striatum: mechanism of craving in cocaine addiction. *J. Neurosci.* 26:6583–88

Voon V, Potenza MN, Thomsen T. 2007. Medication-related impulse control and repetitive behaviors in Parkinson's disease. *Curr. Opin. Neurol.* 20:484–92

Waelti P, Dickinson A, Schultz W. 2001. Dopamine responses comply with basic assumptions of formal learning theory. *Nature* 412:43–48

Wellman LL, Gale K, Malkova L. 2005. GABA$_A$-mediated inhibition of basolateral amygdala blocks reward devaluation in macaques. *J. Neurosci.* 25:4577–86

Willingham DB, Salidis J, Gabrieli JD. 2002. Direct comparison of neural systems mediating conscious and unconscious skill learning. *J. Neurophysiol.* 88:1451–60

Willuhn I, Steiner H. 2006. Motor-skill learning-associated gene regulation in the striatum: effects of cocaine. *Neuropsychopharmacology* 31:2669–82

Wise RA. 2004. Dopamine, learning and motivation. *Nat. Rev. Neurosci.* 5:483–94

Wong DF, Kuwabara H, Schretlen DJ, Bonson KR, Zhou Y, et al. 2006. Increased occupancy of dopamine receptors in human striatum during cue-elicited cocaine craving. *Neuropsychopharmacology* 31:2716–27

Yin HH, Knowlton BJ. 2004. Contributions of striatal subregions to place and response learning. *Learn. Mem.* 11:459–63

Yin HH, Knowlton BJ, Balleine BW. 2005. Blockade of NMDA receptors in the dorsomedial striatum prevents action-outcome learning in instrumental conditioning. *Eur. J. Neurosci.* 22:505–12

Zald DH, Boileau I, El-Dearedy W, Gunn R, McGlone F, et al. 2004. Dopamine transmission in the human striatum during monetary reward tasks. *J. Neurosci.* 24:4105–12

Zink CF, Pagnoni G, Martin-Skurski ME, Chappelow JC, Berns GS. 2004. Human striatal responses to monetary reward depend on saliency. *Neuron* 42:509–17

Crittenden J, Sauvage M, Cepeda C, Andre V, Costa C, et al. 2007. *CalDAG-GEFI modulates behavioral sensitization to psycho motor stimulants and is required for cortico-striatal long-term potentiation.* Presented at Annu. Meet. Am. Coll. Neuropsychopharmacol., 46th, Boca Raton, FL

Welch J, Lu J, Rodriguez RM, Trotta NC, Peca J, et al. 2007. Cortico-striatal synaptic defects and OCD-like behaviours in Sapap3-mutant mice. *Nature* 448:894–900

NOTE ADDED IN PROOF

Valuable new mouse models of disorders involving action since this review was submitted include emerging Sapap-3 mutant mice and CalDAG-GEF I mutant mice (Crittenden et al. 2007, Welch et al. 2007).

Mechanisms of Self-Motion Perception

Kenneth H. Britten

Center for Neuroscience and Department of Neurobiology, Physiology, and Behavior, University of California, Davis, California 95616; email: khbritten@ucdavis.edu

Annu. Rev. Neurosci. 2008. 31:389–410

First published online as a Review in Advance on March 25, 2008

The *Annual Review of Neuroscience* is online at neuro.annualreviews.org

This article's doi:
10.1146/annurev.neuro.29.051605.112953

0147-006X/08/0721-0389$20.00

Key Words

sensorimotor, optic flow, heading perception, motion processing, visual cortex, nonhuman primate

Abstract

Guiding effective movement through the environment is one of the visual system's most important functions. The pattern of motion that we see allows us to estimate our heading accurately in a variety of environments, despite the added difficulty imposed by our own eye and head movements. The cortical substrates for heading perception include the medial superior temporal area (MST) and the ventral intraparietal area (VIP). This review discusses recent work on these two areas in the context of behavioral observations that establish the important problems the visual system must solve. Signals relevant to self motion are both more widespread than heretofore recognized and also more complex because they are multiplexed with other sensory signals, such as vestibular, auditory, and tactile information. The review presents recent work as a background to highlight important problems that remain unsolved.

Contents

Optic flow: the image deformation that results from motion through the world

INTRODUCTION

Freely moving animals navigate through a cluttered three-dimensional world with ease. Although we use a variety of information for this process, most primates rely on vision first and foremost. Gibson (1950) is usually credited with noting the usefulness of visual motion perception for the guidance of self motion, but the ideas go back at least as far as Helmholtz (Helmholtz & Southall 1924). Gibson coined the term optic flow to describe the pattern of motion on our retinae that occurs as we move, and he sets the stage for a hugely productive field of research that effectively combines perceptual, physiological, and theoretical approaches. When an animal moves, a large-scale pattern of relative motions is produced (**Figure 1**). This optic flow field depends on the speed and direction of observer motion and on the depth structure in the scene. The yellow arrows in **Figure 1** indicate the velocity of the different points on the image, and there are two main regularities. First, all the vectors emanate from the current direction of motion ("heading"), forming a focus of expansion at this point. Secondly, more distant image points move much more slowly, as a consequence of motion parallax.

Recovering heading from such an image would be trivial were it not for a thorny issue often called the "rotation problem." Our gaze is rarely still; we frequently track points in a moving image or follow independently moving objects in the scene using a combination of eye and head rotations, which greatly distort the optic flow field. Thus, to estimate heading accurately, the visual system must first decompose the resulting complex vector field into the component that is caused by the gaze shift and the component that results from our movement. How this is accomplished is not well understood, despite extensive study.

The field has been extensively reviewed (Andersen et al. 1990, Lappe 2000, Orban 2001, Warren 1998, Wurtz et al. 1993), so this review emphasizes the most recent work. Owing to space limitations, this review focuses almost entirely on work on nonhuman primates, despite the considerable work on optic flow processing in cats (Sherk & Fowler 2001) and in flying insects (Collett et al. 2006, Egelhaaf & Borst 1993, Srinivasan & Zhang 1997). Likewise,

Figure 1

The pattern of optic flow resulting from observer translation over a landscape. Yellow arrows indicate the local, instantaneous velocity of different points in the image. The black X denotes the current heading direction. Headings can vary in two dimensions: azimuth (*horizontally*) or elevation (*vertically*).

I must largely ignore two important areas of work: optical imaging in humans and theoretical work.

PERCEPTUAL STUDIES

Heading is normally used in the guidance of ongoing movement, but most of what we know about it comes from traditional psychophysical experiments. These typically use a forced-choice discrimination task (e.g., Warren et al. 1988) or sometimes the method of adjustment, in which the observer sets a cursor to indicate perceived heading (e.g., Royden et al. 1992). The stimulus used for these experiments is usually a field of random dots, simulating a flat plane (**Figure 2a**), a three-dimensional cloud of points (**Figure 2b**), or a ground plane (**Figure 2c**). These are importantly different, but heading perception is sufficiently robust (absent eye movements; see below) that results are comparable across stimuli. The accuracy of heading perception is remarkably good: thresholds average 1°–2° of heading angle (Warren et al. 1988). Psychometric func-

tions for two-alternative heading discrimination in human and monkey observers are nearly superimposable (**Figure 3**).

Heading thresholds are also stable across a variety of manipulations to degrade the optic flow field. They are asymptotically low with as few as 10 dots in the image (Warren et al. 1988) and across a range of scene geometries (Crowell & Banks 1993, te Pas 1996). Thresholds are also little affected by the speed of simulated motion, at least within the range encountered in normal locomotion (Warren et al. 1988). However, adding even as little as 20%–30% noise to the display has a substantial effect, especially for a stimulus simulating a cloud of points (**Figure 2c**). Thresholds are considerably more immune to noise for ground plane stimuli (van den Berg 1992) because perspective cues provide extra depth information.

The Rotation Problem

Primates have very active vision and are constantly exploring the scene. During self motion, these gaze changes (arising from a combination

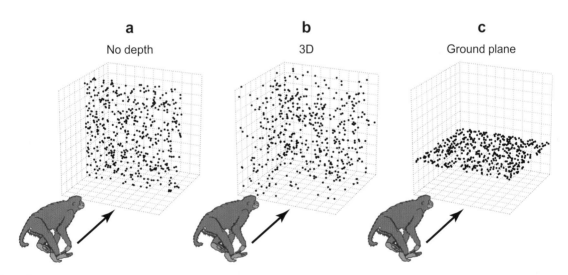

a
No depth

b
3D

c
Ground plane

Figure 2

Stimulus configurations simulated in typical heading studies. *a*: No depth in the scene, as if the dots were painted on a wall. In this version, the speed of the dot is directly related to its distance from the center of heading, and nearby dots will have similar speeds. *b*: Depth containing, with dots therefore moving at different velocities at any given point in the screen. This produces a vivid percept of depth in most observers. *c*: Ground plane. Depth is very evident in the percept from this display as well.

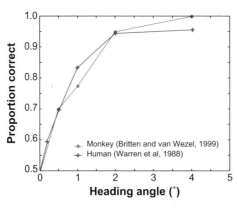

Figure 3

Psychometric functions from two-alternative heading discriminations in monkey and man. In each experiment, the subject is asked to decide if a simulated trajectory is to the right or left of a reference green X. In the Warren experiment, the reference point was a visible landmark on the simulated horizon; in the Britten & van Wezel study, it was the middle of the screen (directly ahead).

of eye-in-head, head-on-neck, and torso movements; Lappe & Hoffman 2000) are nearly omnipresent. Eye movements are very task dependent, differing greatly depending on the nature of the movement and behavioral goals in the task. Human studies have provided the most information about eye movement behavior during locomotion. When traveling straight ahead, drivers spend much of their time looking in a direction near where they are going (Land & Lee 1994). However, in a curve, drivers tend to look at the limiting point on the inside of the curve: the so-called tangent point. Walking subjects frequently fixate objects on the ground, especially if the objects are stepping targets (Hollands et al. 2002; Patla & Vickers 1997, 2003). Observers presented with ground-plane displays (**Figure 2c**), however, showed eye movements similar to optokinetic nystagmus, alternating slow phases following the local motion near the fovea and resetting saccades toward more distant points (Niemann et al. 1999). Any movement of the gaze has profound implications for the pattern of motion on the retina (**Figure 4**). In this case, the subject is fixating a point on the ground plane, tracking

it with a smooth-pursuit eye movement. The eye movement itself (absent self motion) would cause a nearly uniform pattern of vectors across frontal space (red arrows in **Figure 4**): the rotation component. The optic flow field from the observer's motion alone would cause a simple expanding pattern: the translation component (blue arrows). The visual system receives the vector sum of the two components (purple arrows). The flow field is greatly affected by the eye rotation: The center of expansion no longer aligns with the heading (which remains straight ahead) because the eye is instead zeroing the velocity of a point on the plane. If the eye were tracking an independently moving object in the scene, then there need not be a center of expansion at all. Nonetheless, observers are generally accurate at recovering heading in the presence of eye movements (Crowell et al. 1998; Royden et al. 1992, 1994; van den Berg 1996; Warren & Hannon 1990). Therefore, the visual system must have some way of recovering the translation component from the flow field. Two general classes of solution exist for this problem; both are effective in theoretical work. The first is to use a purely visual algorithm that exploits the depth dependence of the translation component and the depth invariance of the rotation component (Heeger & Jepson 1992; Longuet-Higgins & Prazdny 1980; Perrone & Stone 1994, 1998; Royden 1997). Obviously, this kind of approach will work only in a scene where there is visual depth. (See sidebar on Occlusion Boundaries and Objects.) The other general class of approach is to use an explicit (extraretinal) signal of eye velocity to correct one's heading estimate (Ben Hamed et al. 2003, Hanada 2005, Lappe et al. 1996, van den Berg & Beintema 1997).

Many experiments have asked which of these algorithms is actually used when observers discriminate heading in the presence of eye movements. The usual approach is to compare performance on trials containing a gaze shift (and therefore where an extraretinal signal was available) with performance on trials where the retinal consequences of eye movements are simulated (removing the extraretinal information).

With a ground-plane stimulus and low-speed pursuit eye movements (1–2 degrees per second, which simulates tracking of an object a few meters away at a normal walking pace) (Warren & Hannon 1990), performance is very similar on real and simulated eye movement trials, suggesting that visual information is sufficient. However, if depth is removed from the scene, or if higher-speed eye movements are used, then substantial errors are seen on simulated eye movement trials but not in the presence of real eye movements, which shows that extraretinal signals are employed under these conditions (Royden et al. 1992, 1994). The importance of depth in the perception of heading is particularly evident in the experiments of van den Berg and colleagues, who tested perception of heading with stimuli degraded by noise (van den Berg 1992). Perception is quickly degraded by noise, especially for three-dimensional cloud stimuli, which contain fewer cues to depth. For these stimuli, adding real eye movements again considerably improves heading perception. Also, although stereo depth cues do not normally improve heading thresholds, when these are degraded by noise, giving additional depth information restores performance.

From these and other experiments, it is clear that heading perception depends on the integration of many cues (optic flow; monocular depth; stereo, vestibular, and presumably proprioceptive cues). Furthermore, how these cues are weighted is not constant but depends on the experimental situation. How this is accomplished at a physiological level is just barely beginning to be understood.

PHYSIOLOGICAL MECHANISMS

One strength of this research field is a well-studied network of cortical areas that form a natural substrate for the analysis of optic flow information. The motion system of dorsal extrastriate cortex (for review, see Albright 1993, Andersen et al. 1993, Britten 2008, Zeki 1990) is characterized by a preponderance of neurons selective for the speed and direction of visual motion (**Figure 5**). Early and middle ar-

OCCLUSION BOUNDARIES AND OBJECTS

Occlusion boundaries have a special place in computational vision because they provide unambiguous depth order information. In an optic flow field, such boundaries produce abrupt speed changes in the image, which can be exploited in certain computational models (Rieger & Lawton 1985, Royden 1997). It is interesting that under normal circumstances, adding this information does not improve heading thresholds. More interesting, perhaps, is an asymmetry in the effects of attention. In a dual-response design, Royden and colleagues cued subjects to attend to either self-motion direction or the direction of a superimposed, independently moving object (Royden & Hildreth 1999). Attending to the object had no effect on the heading thresholds, but attending to self motion raised thresholds for object motion discrimination. This finding suggests a certain automaticity in the calculation of heading, which makes sense in natural behavior. In some situations, the presence of large moving objects causes small biases in heading judgments (Royden & Hildreth 1999), but these often dominate the MST neuron responses (Logan & Duffy 2006). Considering the responses of an MST neuron population—many being less influenced by the presence of the moving object—appears to resolve this apparent discrepancy. No one has yet measured the physiological effects in MST of directing attention to objects, but it could be very revealing.

eas in this pathway (yellow and orange tones in **Figure 5**) have small- to medium-sized receptive fields (RFs) and relatively simple, linear directional preferences [V1, Bair & Movshon 2004, De Valois et al. 1982, Livingstone & Conway 2003; middle temporal (MT), Albright 1984, Maunsell & Van Essen 1983, Zeki 1974; V2, Gattass et al. 1981, Levitt et al. 1994; V3, Gattass et al. 1988], seemingly nonoptimal for the analysis of whole-field optic flow. Upper visual areas, most notably the medial superior temporal area (MST) and the ventral intraparietal area (VIP), are much better suited to the problem. These areas are also directly connected to anterior areas that have been implicated in eye, limb, and body movements, both cortically and subcortically. This review emphasizes these two areas because of extensive recent work and their clear suitability for the analysis of optic flow patterns.

MT: middle temporal

MST: medial superior temporal

VIP: ventral intraparietal

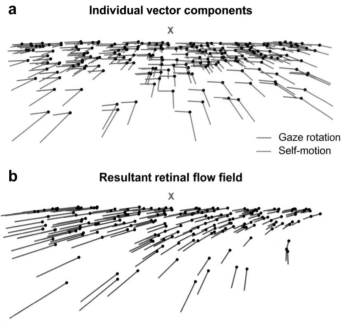

a **Individual vector components**

X

— Gaze rotation
— Self-motion

b **Resultant retinal flow field**

X

Figure 4

Optic and retinal flow vectors from observer movement (always toward the green X) in the presence of a modest horizontal smooth pursuit eye movement to the right. The blue vectors in *a* are the motions resulting from the observer movement alone, whereas the red vectors are the retinal motions resulting from the eye movement alone. In *b*, the vector sum is illustrated. Note how the red vectors are the same, independent of the distance to the dot, whereas the blue vectors depend on the depth of the point in the simulated scene. Therefore, the amount of distortion (shift of the center of expansion) depends on depth in a predictable way, with very small shifts close by (where the self-motion vectors are long) and very large ones far away, where the blue vectors are very small.

AREA MST

Area MST was the first area reported to have large-field, complex-motion-sensitive responses. It remains the most studied area in the context of self-motion. In macaques, it extends dorsally from the floor of the superior temporal sulcus up along the anterior bank (see **Figure 5**). Like antecedent areas, it contains a preponderance of direction selectivity but differs in important ways from earlier areas such as MT.

Basic Properties

MST has very large, frequently bilateral RFs, which differ from their afferent areas often by

having selectivity for nonuniform, large area motion (Duffy & Wurtz 1991; Graziano et al. 1994; Lagae et al. 1994; Tanaka et al. 1986, 1989; Tanaka & Saito 1989). MST consists of at least two subdivisions: the dorsal and lateral (MSTd and MSTl, respectively). Nearly all the work relating MST to heading has been performed in MSTd because this area is more responsive to large-field visual stimuli (Tanaka et al. 1986). Many physiologists have probed the properties of neurons in MST with motion stimuli containing motion patterns commonly found in optic flow from self motion: uniform motion, expansion, and rotation. Tested in this way, MST neurons usually display a mix of selectivities. More neurons in MST are selective for uniform translational motion than for complex motion (Duffy & Wurtz 1991). Among complex motion directions, the cell population shows a distinct bias for expansion over other directions (see sidebar on Anisotropies in Physiology and Perception).

Responses to Stimuli-Simulating Movement

The most revealing experiments about the role of MST in heading employ more natural stimuli—random dots moved so as to simulate observer motion. A very large fraction of MST neurons are selective for such stimuli (Duffy & Wurtz 1995, Lappe et al. 1996, Paolini et al. 2000, Pekel et al. 1996). Perhaps the most complete description of MST responses comes from DeAngelis and Angelaki (Gu et al. 2006b), who describe the complete tuning in all directions, rather than the reduced slices chosen by many experimenters for practical reasons. **Figure 6a** shows the full two-dimensional tuning surface of single MST neurons. The maximum response for the upper neuron is elicited by translations directly to the right. This study shows a distinct anisotropy in preferred headings across the MST population (**Figure 6b**), with a preponderance of neurons tuned to either left or right headings. Thus, the prevalence of expansion-preferring MST neurons (see above) does not produce a peak in forward-preferring neurons. This type of population, however, is ideal for making fine discriminations of forward headings.

These experiments used extremely rich heading stimuli, with simulated depth and also size cues to depth. Many experiments, however, have employed simpler stimuli lacking depth (**Figure 2a**). A particularly revealing experiment allows direct examination of the contribution of simulated depth to heading tuning in MST (Upadhyay et al. 2000). Heading tuning is significantly improved in the majority of MST neurons when multiple depth planes are included, compared with the no-depth condition. This finding is consistent with perceptual results (see above) and strongly suggests a role for the integration of multiple speeds in the generation of MST heading tuning (see Future Issues section, below).

If the heading signals are to be used for the real-time guidance of motion, it is critical that

ANISOTROPIES IN PHYSIOLOGY AND PERCEPTION

Physiological observations in MST, VIP, and 7a all point to an overrepresentation of expansion, compared with other directions of complex motion (Duffy & Wurtz 1991, Graziano et al. 1994, Heuer & Britten 2004, Siegel & Read 1997). This finding clearly suggests a role in the guidance of locomotion because expansion is a dominant component of such optic flow patterns. Although psychophysical measurements point to mechanisms tuned to complex motions (for review see Bex & Makous 1997, Vaina 1998), they typically do not show greater sensitivity to expansion over contraction (Snowden & Milne 1996). This contrast carries two interesting implications. First, there is likely to be an opponent step in discrimination, downstream from the sensory representations of complex motion. The interesting corollary question is how far downstream this opponent step is being pushed because sensory signals of optic flow are found in parietal and motor cortex.

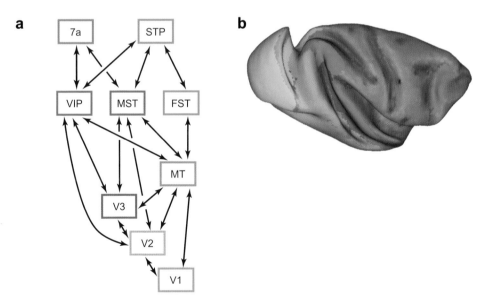

Figure 5

Anatomy of the motion system. *a*: Simplified schematic of the connections between areas known to play a part in motion analysis. *b*: Anatomical locations of the areas on a slightly "inflated" monkey brain to allow visualization of areas within sulci. The viewpoint is dorsal and lateral. Nomenclature and area boundaries after Felleman & Van Essen (1991); image generated with public domain software CARET (**http://brainmap.wustl.edu/caret**; Van Essen et al. 2001).

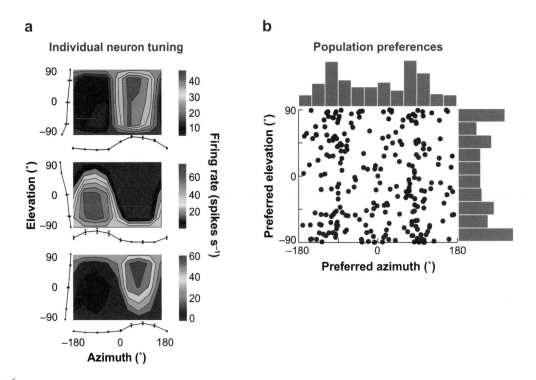

Figure 6

Heading tuning of MST neurons to visual stimuli. *a*: Responses of three individual neurons plotted as a function of azimuth (*horizontal*) and elevation (*vertical*) heading dimensions. Firing rate is illustrated by the color scale, with hotter colors indicating higher firing rates. *b*: Preferred headings of a sample of 255 MST neurons, with the azimuth and elevation of maximum response indicated by the location of each dot. After Gu et al. (2006b).

they be both rapid and linear in their temporal dynamics. Despite suggestions that complex temporal dynamics exist in the initial responses of MST neurons when simpler stimuli are used (Duffy & Wurtz 1997), MST neurons appear faithful to stimulus dynamics when tested with more natural stimuli (Paolini et al. 2000). When trajectories are not presented in short, uniform pulses but rather in a continuously varying manner, MST firing rates also vary smoothly and little hysteresis is observed. This result is interesting viewed against the backdrop of studies on MT and MST as potential substrates for the guidance of pursuit eye movements. The initial stages of pursuit require acceleration signals (Krauzlis & Lisberger 1994), which occur in area MT (Lisberger & Movshon 1999; Priebe et al. 2002; Priebe & Lisberger 2002). Because the

same signals are presumably being used for both purposes, and often at the same time, it remains an open question how the potentially different constraints for different behavioral purposes are jointly satisfied.

Extraretinal Signals in MST

Although vision is one strong signal for the guidance of self motion, it is not the only one. In normal locomotion, both proprioceptive and vestibular signals are also used. This has led several groups to investigate the contribution of vestibular inputs to the neuronal responses in MST. MST responds well to vestibular signals alone (Fetsch et al. 2007; Gu et al. 2006b, 2007; Page & Duffy 2003) (see **Figure 7**). A single MST neuron shows good tuning to visual, vestibular, and combined stimulation

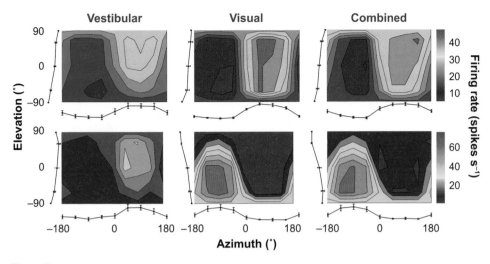

Figure 7

Visual and vestibular tuning of two representative MST neurons. Monkeys were seated in a hydraulically operated lift chair and moved in linear trajectories with peak accelerations of 0.1 g. The vestibular responses were measured in complete darkness; visual responses were measured in the usual way with the chair stationary, and the combined response was measured with visual and vestibular stimulation corresponding to the same trajectory. Note that in the lower cell, the visual and vestibular responses peaked in opposite directions ("incongruent"). After Gu et al. (2006b).

(**Figure 6**), but what is puzzling about these signals is true of half the neurons in MST. The tuning for visual and vestibular signals is approximately opposite in experiments using either linear or circular trajectories. The relationship between these vestibular signals and perception has been well studied and is discussed further below (see Relating MST to Perception, below).

Another important feature that distinguishes MST from antecedent areas is the presence of extraretinal signals concerning eye position or movement. A large fraction of MST neurons carry such signals in either of two forms. Some neurons carry explicit signals of ongoing smooth-pursuit eye movements, which persist even without any visual stimulation (Newsome et al. 1988). Additionally, MST neurons modulate their visual responses according to current eye position and speed ("gain field"; Squatrito & Maioli 1997). These responses are potentially important in interpreting the results of pursuit compensation experiments (see next section).

The Rotation Problem and MST

Many studies have explored how MST neurons respond to heading stimuli in the presence of eye and head movements. In seminal studies (using stimuli without simulated depth), the Andersen laboratory asked whether MST neurons would compensate for ongoing pursuit eye movements (Bradley et al. 1996, Shenoy et al. 1999). One advantage of using such stimuli is that it allows one to predict explicitly what the tuning would be without any compensation because the retinal flow field is simply shifted in the direction of pursuit. Some cells compensate not at all, others compensate nearly perfectly, and still others overcompensate. On average, MST cells undercompensate for pursuit, shifting ~70% of the amount needed to maintain a stable representation in head-centered or world coordinates (which are equivalent when the head is fixed). The amount of compensation scales approximately linearly with pursuit speed (Shenoy et al. 2002) and with different speeds of simulated translation (Lee et al. 2007). When

Retinal flow: image that is seen by the retina, and additionally includes the results of eye movements

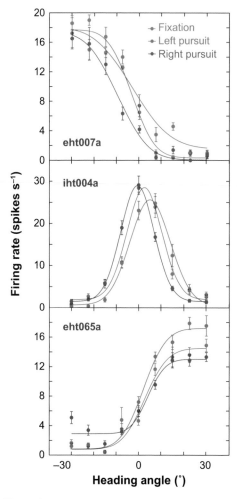

Figure 8

Tuning functions of three MST neurons as a function of azimuth (*horizontal*) heading directions, illustrating compensation for ongoing pursuit eye movements. Pursuit speed was 10°/s.

examined with eight directions of pursuit, apparently larger effects of pursuit are observed, leaving very few MST neurons truly tolerant of pursuit (Page & Duffy 1999).

The presence of simulated depth in the visual stimuli helps the pursuit compensation by MST neurons. **Figure 8a** shows frontal heading tuning of a single MST neuron measured in the author's laboratory (J. Maciokas & K. Britten, unpublished results). In this experiment, the visual stimulus simulates a three-dimensional cloud of points (**Figure 2b**). This

neuron shows nearly pursuit-invariant heading tuning and is fairly representative of the population (**Figure 8b**). This figure shows representative horizontal shifts of MST heading tuning functions, which average less than 10°. It is difficult, when using such stimuli, to know the amount of shift without compensation because the shift depends on depth, but the average in these stimuli was 30°. Similar results were obtained in the eight-direction pursuit experiments of Upadhay et al. (2000). Comparing these results with the original results from the Andersen laboratory strongly suggests that depth cues help in pursuit compensation by MST cells as well as in perception. The mechanism by which this occurs remains mysterious.

Relating MST to Perception

Several forms of evidence can suggest that a particular CNS structure directly contributes to perception or behavior (for review, see Parker & Newsome 1998). One involves overall sufficiency of neuronal signals to support threshold performance. To make neuronal and perceptual measurements quantitatively comparable, "neurometric function analysis" is typically used to allow direct estimation of the limits of neuronal performance. MST signals of optic flow have been analyzed by this approach in three different tasks, and the differences between the results are intriguing. When the sensitivity of MST neurons is compared with monkeys' discrimination of simple translational motion masked in noise, the neurons are on average slightly more sensitive than the monkey (Celebrini & Newsome 1994b). If, however, the discrimination is between opposite rotations, expansions, or spirals (e.g., contraction vs. expansion), then the threshold of MST neurons is worse relative to monkey thresholds (Heuer & Britten 2004). The best neurons in MST are still a bit better than the monkey, but the average neuron has a threshold ~3 times worse. The comparison between the results of these very similar experiments strongly argues against MST being specialized for complex motion pattern analysis.

In an experiment more directly related to heading perception (Gu et al. 2006a), monkeys were trained to discriminate small differences in heading (rather than opposites) in an apparatus that allowed both visual and vestibular heading signals. The visual signals were modestly degraded by noise to allow visual and vestibular signals to be comparable. Tested in this way, MST neurons are on average much worse than the monkey, and only the most sensitive neurons have thresholds similar to the monkeys'. This finding is consistent with Mountcastle's "lower envelope principle," in which the most sensitive neurons limit performance, but also with population pooling models in which the averaging of the signals from many insensitive neurons bestows lower thresholds to the monkey. This consensus result emerges from related experiments in a variety of cortical areas, when small differences in a stimulus are being discriminated (Krug et al. 2004, Purushothaman & Bradley 2005, Uka & DeAngelis 2001).

Another form of evidence that supports the involvement of candidate neuronal signals with perception is trial-by-trial correlations between neuronal responses and the monkey's subsequent choice on that trial. This correlation is frequently captured with a statistical measure termed choice probability (Britten et al. 1996). These correlations are typically small but nonetheless reveal an intimate connection between neuronal activity and perception. MST neurons show the largest and most consistent choice probabilities in two tasks: two-alternative opposed-motion discrimination (Celebrini & Newsome 1994a) and the vestibular-cued heading task (Gu et al. 2007). One curious finding is that even though visual responses are overall stronger than vestibular responses in MST, choice probabilities are stronger and more consistent for the vestibular version of the task. One also finds a revealing contrast with the results of Heuer & Britten (2004), in which monkeys discriminate more abstract spiral space patterns matched to neuronal preferences rather than stimuli simulating self motion. In this task, choice probabilities are not significant. Although one needs to be cautious in interpreting negative evidence, especially for a relatively subtle measurement such as choice probability, this finding suggests a greater perceptual role for MST in the context of more realistic discriminations designed to resemble actual self motion.

Perhaps the most compelling evidence for direct involvement of cortical signals in perception comes from electrical microstimulation experiments (**Figure 9**). MST is stimulated in the context of a two-alternative discrimination between nearby headings (Britten & Van Wezel 1998, 2002). Microstimulation in this case produces a significant bias in the monkey, consistent with the preferred heading of the cluster of neurons near the stimulating electrode. The results are a little inconsistent—many sites produce oppositely directed perceptual effects—but the signals directly contribute to heading discrimination. The results are stronger and more consistent in the presence of smooth pursuit eye movements, which suggests a larger role for MST signals when the animals actively solve the rotation problem.

Overall, then, this broad pattern of results clearly demonstrates a direct, but probably not obligate or one-to-one, connection between MST activity and heading perception. However, MST clearly does not act alone.

THE VENTRAL INTRAPARIETAL AREA

As its name suggests, the ventral intraparietal area (VIP) is found in the depths of the intraparietal sulcus (**Figure 5b**) and is one of several multimodal areas in this sulcus. It contains neurons that respond to visual, somatosensory, vestibular, and auditory stimuli (Bremmer et al. 2002b; Colby et al. 1993; Schlack et al. 2002, 2005). Of particular interest in this review, the visual signals are very similar in many ways to those in MST, despite the many other differences between the physiology in these two areas (Bremmer et al. 2002a, Schaafsma et al. 1997, Schaafsma & Duysens 1996). These two

Choice probability: the ROC area of two spike-count distributions sorted by the monkey's choice rather than by the stimulus, as in normal ROC analysis. Bounded between 0 and 1, with a value of 0.5 indicating no relationship

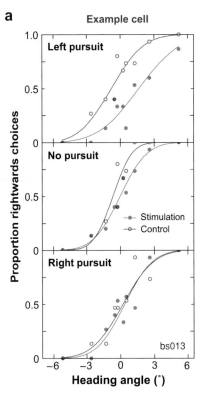

a

Example cell

Left pursuit

No pursuit

- ● Stimulation
- ○ Control

Right pursuit

bs013

Proportion rightwards choices

Heading angle (˚)

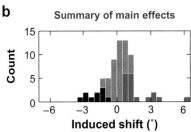

b

Summary of main effects

Count

Induced shift (˚)

Figure 9

Effects on heading perception of electrically microstimulating MST. *a*: Single example case. Each curve is an "unfolded" psychometric function showing the proportion of choices in favor of rightward headings as a function of the heading direction. The stimulation site for this experiment preferred left headings, and stimulation significantly increased the proportion of left choices. Three pursuit conditions are included, and the perceptual effects are larger during left pursuit. *b*: Summary across experiments, showing the average effect size, expressed as a horizontal shift of the psychometric function, across the different pursuit conditions. Purple bars denote significant effects where the perceptual effect was aligned with the site preference; black bars indicate significant reversed effects.

areas show many similarities in optic flow processing (Bremmer 2005, Bremmer et al. 2000), but this review focuses on recent findings. Investigators increasingly agree that VIP consists of at least two subdivisions (Lewis & Van Essen 2000); however, because this picture is still developing, and because most experiments did not track the distinction, I treat it as a single area.

VIP visual RFs are, on average, somewhat smaller than those in MST and are highly directional for visual motion (Colby et al. 1993). The visual and somatosensory receptive fields of VIP neurons closely correspond with each other (Duhamel et al. 1991). Thus, a neuron with a central, inferior visual RF might have a somatosensory RF on the nose or mouth, whereas one with a more peripheral visual RF would have an RF on the shoulder or arm. Somatosensory RFs are also often directionally selective, and the visual and somatosensory preferred directions are frequently aligned. Also, when tested with real object stimuli, the visual RFs of these cells are sometimes restricted in depth—such neurons stop responding once a stimulus is more than a few meters from the eyes (Colby et al. 1993). This suite of properties has led some investigators to suggest that VIP is particularly involved in near-field motion analysis—of objects being approached or avoided. More evidence supporting this notion comes from the electrical microstimulation experiments of Graziano and colleagues. When VIP is electrically activated with pulse trains of fairly long duration and moderately large intensities, stereotyped face, shoulder, and arm movements result, which strongly resemble normal defensive movements evoked by air puff stimuli (Cooke et al. 2003, Cooke & Graziano 2003). VIP neurons are also clearly implicated in the detection of the onset of coherent motion (Cook & Maunsell 2002b) and in attention to motion (Cook & Maunsell 2002a). In these experiments, VIP is clearly associated more closely with behavior (higher choice probability, greater attentional effects) than are neurons in the archetype motion area, MT.

Another interesting property of area VIP, which distinguishes it from MST, is the influence of eye position on visual receptive fields. Although MST has modulation by eye position, many neurons in VIP show RFs that shift on the retina when the eyes move, such that the receptive field is partially or completely transformed into a head-centered coordinate system (Avillac et al. 2005, 2007; Duhamel et al. 1997). Although one could orchestrate effective movements without such a coordinate system (Zipser & Andersen 1988), an explicitly head-centered representation is probably more directly useful because it could be easily "read out."

VIP AND HEADING

Because of its position in the motion system, and because of the many interesting properties described above, VIP has attracted considerable attention from researchers studying heading. Despite many other differences, the properties of VIP in the context of heading perception seem very similar to those in MST. Almost all VIP neurons are well tuned for heading, and a wide range of headings is represented by the population (Bremmer et al. 2002a, Zhang et al. 2004, Zhang & Britten 2004). Neurons integrate across at least visual and vestibular information to represent heading and produce good responses to either input alone. Just as in MST, there are two broad classes of neurons: those with very similar preferred directions for vestibular and visual stimuli (congruent) and those that prefer opposite directions for visual and vestibular stimuli. As in MST, the presence of large numbers of incongruently tuned neurons is very puzzling.

As in MST, there seems to be a preponderance of neurons tuned for headings to the left or to the right, causing firing rates to change rapidly over frontal headings, where discrimination thresholds are lowest. These signals are quite sensitive to small changes in heading, and as in MST, the most sensitive neu-rons in the population have thresholds approximately equal to the monkeys' (Zhang & Britten 2005). Also, many VIP neurons have significant choice probabilities in a heading discrimination task on the basis of visual information alone (**Figure 10**).

VIP has also been tested for involvement in heading discrimination using microstimulation (Zhang & Britten 2003). Microstimulation causes significant choice biases in most cases (**Figure 10c**), which strongly suggests a perceptually causal role for signals in this area. Even the most peculiar aspects of the data—the minority of backward stimulation effects—are quantitatively comparable in VIP and MST. Together, these data all clearly show parallel roles for VIP and MST in the discrimination of heading.

OTHER AREAS WITH OPTIC FLOW SIGNALS

The processing of large-scale image motion is not restricted to the classically defined visual motion system, which is generally considered to end in the parietal cortex. However, this is clearly not the end of the story because vigorous visual motion responses are found in many regions, including area 7a (Merchant et al. 2001; Siegel & Read 1997), PEc (caudal part of area PE of von Bonin and Bailey) (Raffi et al. 2002), STP (superior temporal polysensory) (Anderson & Siegel 1999, 2005; Hietanen & Perrett 1996), motor cortex (Merchant et al. 2001), and others.

The representation of optic flow may become more complex in these higher areas. In 7a, for instance, complex temporal dynamics strongly depend on the speed of the stimuli (Phinney & Siegel 2000). In STPa, which receives converging imput from both the dorsal and the ventral streams of visual processing (Boussaoud et al. 1990), many neurons prefer three-dimensional stimuli-simulating objects rotating in depth over stimuli-simulating observer translation (Anderson & Siegel 2005).

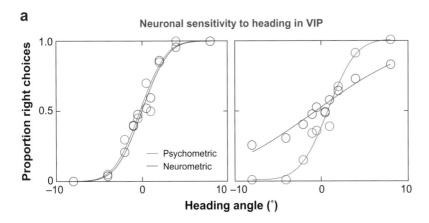

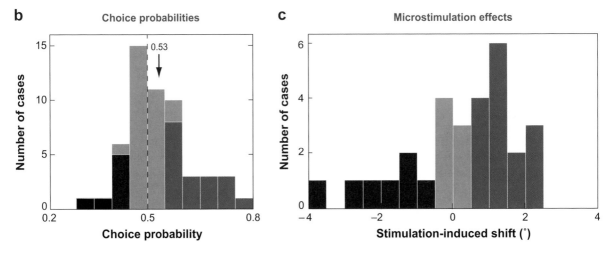

Figure 10

VIP related to heading perception. *a*: Neurometric functions showing sensitivity of individual neurons to heading direction, related to the simultaneously measured monkey performance. The best neuron is on the left, whereas a typical neuron on the right shows substantially lower sensitivity to heading. *b*: Choice probabilities of a sample of 54 neurons from two monkeys. Purple bars indicate significant "positive" choice probability (higher firing for preferred decisions); black bars denote the converse. *c*: Microstimulation effects in VIP; conventions as in **Figure 9b**.

CONCLUSION

Recovery of self motion is a very complex task, and recent discoveries about the cortical underpinnings have made it even more so. We now know that it involves visual-vestibular interactions and probably, under more natural circumstances, involves proprioception as well. One unfortunate aspect of the new complexity is to make the problem less tractable. The initial interest in heading was spurred in part by its relative simplicity, and this characteristic is vanish-

ing. However, our methodology and ideas are also becoming more sophisticated, and the next decade should see major advances. For these advances to occur, the first most important step will be to understand how multiple cortical areas work together in real time, when the exact details of the information in each area and cortical column differ in both large and small ways. The modern large-scale multiple-neuron recording systems have yet to be much applied in understanding self motion (in striking contrast with the related problem of reaching behavior) and will presumably be necessary to make the next advances. Our theoretical understanding of the problem is in good step with the physiological data we currently possess, but it must also be expanded to incorporate serial and parallel interactions between multiple areas working in concert. We can hope that self motion might be one of the first truly complex behavioral tasks, in which volition and cognition are key components, where we can truly close the loop through cortex.

SUMMARY POINTS

1. Heading perception is more complicated than many reduced experimental situations suggest. Many cues are in use beyond just optic flow.

2. Depth is a critical parameter for optic flow analysis, both perceptually and physiologically. Especially under conditions where performance is made suboptimal by adding noise both vestibular and visual cues assist performance.

3. In both VIP and MST, strong vestibular responses are found in addition to the well-studied visual responses. One peculiarity of these responses is that in about half the cells in both areas, the tuning for visual and vestibular headings are opposite rather than aligned.

4. Perceptual correlates in both MST and VIP strongly suggest a direct and causal relationship between cortical activity and the perception of heading.

5. Optic flow signals are widespread in higher cortical areas, so clearly this is a distributed function. A corollary of this observation is that individual areas must be serving multiple functions. One good example is VIP, which is also strongly implicated in the analysis of near-field movement.

FUTURE ISSUES

1. Which stimuli are preferred by optic-flow-sensitive neurons? Stimulus selection is limited by time, equipment, or simply tradition. Complex motion can be mathematically decomposed into at least five orthogonal dimensions (Koenderink & van Doorn 1987), which are often correlated. However, only three are commonly explored by physiologists. Stimuli simulating self motion contain complex mixtures of these components. Therefore, we do not know which motion patterns are maximally effective in MST (or other areas with optic flow responses). One approach is to use a full suite of orthogonal basis functions, which can get tedious but would be useful. Another approach is to use reverse correlation or some other system-identification method to efficiently find the stimulus dimensions that maximally modulate MST neurons.

2. How is speed integrated in MST? Despite the importance of depth in heading perception and in the generation of MST tuning, there has been little study of how multiple speeds interact within the large RFs of MST neurons. Theory (Perrone & Stone 1994) makes strong predictions for the nature of this interaction, that remain untested.

3. How are the representations of object- and self-motion related? Many experiments have targeted the physiological basis of self motion because of its importance and experimental tractability. The field of active vision has been working for some years on the visual parameters of reaching for moving objects, but much less is known about the underlying physiology. Given that the same areas are likely to be used for both kinds of motion analysis, these two fields need to be integrated.

4. How do parietal regions interact? Although investigators have given considerable attention to fronto-parietal interactions, different labs tend to focus on one or another parietal area (e.g., LIP or VIP). Given their proximity and the diversity of signals in the parietal cortex, useful for orchestration of many behaviors (self motion, reaching), it is critical to explore the joint activity (or necessity) of multiple parietal regions in the same monkeys on a single task.

5. How do sensory areas interact? The evidence is incontrovertible that multiple areas directly contribute to heading perception. The Newsome toolkit of modern methods for relating neuronal activity to performance was developed for, and has been applied mostly to, single areas. These techniques will need to be extended to handle a range of possible interactions between signals in multiple areas that contribute to perception at the same time.

6. What is the relationship between cortical, brainstem, and cerebellar signals in self-motion? Many kinds of visuomotor behavior (ocular following, postural adjustment, etc.) depend on large-scale retinal flow, and the accessory optic system and cerebellum are deeply connected to the cortical areas that receive most of the study. In natural locomotor behavior, the active, volitional components of locomotion must be closely connected to the more reflexive aspects that are presumably the province of these deep structures. More work needs to be directed to this interaction.

DISCLOSURE STATEMENT

The author is not aware of any biases that might be perceived as affecting the objectivity of this review.

ACKNOWLEDGMENTS

Thanks to T. Zhang, C.H. McCool, S.W. Egger, E.A. Marshall, and H.R. Engelhardt for providing useful comments on an earlier draft of the manuscript. Thanks to S.M. Aamodt for help with the proofreading and constructive criticism. Thanks to S.W. Egger for creating **Figure 4**. Support to K.H.B. provided by the National Eye Institute (EY10562 and Vision Core Center grant EY12576).

LITERATURE CITED

Albright TD. 1984. Direction and orientation selectivity of neurons in visual area MT of the macaque. *J. Neurophysiol.* 52:1106–30

Albright TD. 1993. Cortical processing of visual motion. In *Visual Motion and its Role in the Stabilization of Gaze*, ed. FA Miles, J Wallman, pp. 177–201. New York: Elsevier

Andersen R, Snowden R, Treue S, Graziano M. 1990. Hierarchical processing of motion in the visual cortex of monkey. In *Cold Spring Harbor Symposia on Quantitative Biology*, Vol. LV, *The Brain*, pp. 741–48. Plain View, NY: Cold Spring Harbor Lab. Press

Andersen RA, Treue S, Graziano M, Snowden RJ, Qian N. 1993. From direction of motion to patterns of motion: hierarchies of motion analysis in the visual cortex. See Ono et al. 1993, 284–199

Anderson KC, Siegel RM. 1999. Optic flow selectivity in the anterior superior temporal polysensory area, STPa, of the behaving monkey. *J. Neurosci.* 19:2681–92

Anderson KC, Siegel RM. 2005. Three-dimensional structure-from-motion selectivity in the anterior superior temporal polysensory area, STPa, of the behaving monkey. *Cereb. Cortex* 15:1299–307

Avillac M, Ben Hamed S, Duhamel JR. 2007. Multisensory integration in the ventral intraparietal area of the macaque monkey. *J. Neurosci.* 27:1922–32

Avillac M, Deneve S, Olivier E, Pouget A, Duhamel JR. 2005. Reference frames for representing visual and tactile locations in parietal cortex. *Nat. Neurosci.* 8:941–49

Bair W, Movshon JA. 2004. Adaptive temporal integration of motion in direction-selective neurons in macaque visual cortex. *J. Neurosci.* 24:7305–23

Ben Hamed S, Page W, Duffy C, Pouget A. 2003. MSTd neuronal basis functions for the population encoding of heading direction. *J. Neurophysiol.* 90:549–58

Bex PJ, Makous W. 1997. Radial motion looks faster. *Vis. Res.* 37:3399–405

Boussaoud D, Ungerleider LG, Desimone R. 1990. Pathways for motion analysis: cortical connections of the medial superior temporal and fundus of the superior temporal visual areas in the macaque. *J. Comp. Neurol.* 296:462–95

Bradley DC, Maxwell M, Andersen RA, Banks MS, Shenoy KV. 1996. Mechanisms of heading perception in primate visual cortex. *Science* 273:1544–47

Bremmer F. 2005. Navigation in space—the role of the macaque ventral intraparietal area. *J. Physiol.* 566:29–35

Bremmer F, Duhamel JR, Ben Hamed S, Graf W. 2000. Stages of self-motion processing in primate posterior parietal cortex. *Int. Rev. Neurobiol.* 44:173–98

Bremmer F, Duhamel JR, Ben Hamed S, Graf W. 2002a. Heading encoding in the macaque ventral intraparietal area (VIP). *Eur. J. Neurosci.* 16:1554–68

Bremmer F, Klam F, Duhamel JR, Ben Hamed S, Graf W. 2002b. Visual-vestibular interactive responses in the macaque ventral intraparietal area (VIP). *Eur. J. Neurosci.* 16:1569–86

Britten KH. 2008. Cortical processing of visual motion. In *The Senses: A Comprehensive Reference*, ed. TD Albright, RH Masland. London: Elsevier. In press

Britten KH, Newsome WT, Shadlen MN, Celebrini S, Movshon JA. 1996. A relationship between behavioral choice and the visual responses of neurons in macaque MT. *Vis. Neurosci.* 13:87–100

Britten KH, Van Wezel RJ. 2002. Area MST and heading perception in macaque monkeys. *Cereb. Cortex* 12:692–701

Britten KH, van Wezel RJA. 1998. Electrical microstimulation of cortical area MST biases heading perception in monkeys. *Nat. Neurosci.* 1:1–5

Celebrini S, Newsome WT. 1995. Microstimulation of extrastriate area MST influences performance on a direction discrimination task. *J. Neurophysiol.* 73:437–48

Celebrini S, Newsome WT. 1994b. Neuronal and psychophysical sensitivity to motion signals in extrastriate area MST of the macaque monkey. *J. Neurosci.* 14:4109–24

Colby CL, Duhamel JR, Goldberg ME. 1993. Ventral intraparietal area of the macaque: anatomic location and visual response properties. *J. Neurophysiol.* 69:902–14

Collett M, Collett TS, Srinivasan MV. 2006. Insect navigation: measuring travel distance across ground and through air. *Curr. Biol.* 16:R887–90

Cook EP, Maunsell JH. 2002a. Attentional modulation of behavioral performance and neuronal responses in middle temporal and ventral intraparietal areas of macaque monkey. *J. Neurosci.* 22:1994–2004

Cook EP, Maunsell JH. 2002b. Dynamics of neuronal responses in macaque MT and VIP during motion detection. *Nat. Neurosci.* 5:985–94

Cooke DF, Graziano MS. 2003. Defensive movements evoked by air puff in monkeys. *J. Neurophysiol.* 90:3317–29

Cooke DF, Taylor CS, Moore T, Graziano MS. 2003. Complex movements evoked by microstimulation of the ventral intraparietal area. *Proc. Natl. Acad. Sci. USA* 100:6163–68

Crowell JA, Banks MS. 1993. Perceiving heading with different retinal regions and types of optic flow. *Percept. Psychophys.* 53:325–37

Crowell JA, Banks MS, Shenoy KV, Andersen RA. 1998. Visual self-motion perception during head turns. *Nat. Neurosci.* 1:732–37

De Valois RL, Yund EW, Hepler N. 1982. The orientation and direction selectivity of cells in the macaque visual cortex. *Vis. Res.* 22:531–44

Duffy CJ, Wurtz RH. 1991. Sensitivity of MST neurons to optic flow stimuli. I. A continuum of response selectivity of large-field stimuli. *J. Neurophysiol.* 65:1329–45

Duffy CJ, Wurtz RH. 1995. Response of monkey MST neurons to optic flow stimuli with shifted centers of motion. *J. Neurosci.* 15:5192–208

Duffy CJ, Wurtz RH. 1997. Multiple temporal components of optic flow responses in MST neurons. *Exp. Brain Res.* 114:472–82

Duhamel J-R, Colby CL, Goldberg ME. 1991. Congruent representations of visual and somatosensory space in single neurons of monkey ventral intraparietal sulcus. In *Brain and Space*, ed. J Paillard, pp. 223–36. Oxford, UK: Oxford Univ. Press

Duhamel JR, Bremmer F, Ben Hamed S, Graf W. 1997. Spatial invariance of visual receptive fields in parietal cortex neurons. *Nature* 389:845–48

Egelhaaf M, Borst A. 1993. A look into the cockpit of the fly: visual orientation, algorithms, and identified neurons. *J. Neurosci.* 13:4563–74

Felleman D, Van Essen D. 1991. Distributed hierarchical processing in the primate cerebral cortex. *Cereb. Cortex* 1:1–47

Fetsch CR, Wang S, Gu Y, Deangelis GC, Angelaki DE. 2007. Spatial reference frames of visual, vestibular, and multimodal heading signals in the dorsal subdivision of the medial superior temporal area. *J. Neurosci.* 27:700–12

Gattass A, Sousa PB, Gross CG. 1988. Visuotopic organization and extent of V3 and V4 of the macaque. *J. Neurosci.* 8:1831–45

Gattass R, Gross CG, Sandell JH. 1981. Visual topography of V2 in the macaque. *J. Comp. Neurol.* 201:519–39

Gibson JJ. 1950. *Perception of the Visual World*. Boston: Houghton-Mifflin

Graziano MSA, Andersen RA, Snowden RJ. 1994. Tuning of MST neurons to spiral motions. *J. Neurosci.* 14:54–67

Green DM, Swets JA. 1966. *Signal Detection Theory and Psychophysics*. New York: Wiley

Gu Y, DeAngelis GC, Angelaki DE. 2007. A functional link between area MSTd and heading perception based on vestibular signals. *Nat. Neurosci.* 10:1038–47

Gu Y, Watkins PV, Angelaki DE, DeAngelis GC. 2006b. Visual and nonvisual contributions to three-dimensional heading selectivity in the medial superior temporal area. *J. Neurosci.* 26:73–85

Hanada M. 2005. An algorithmic model of heading perception. *Biol. Cybern.* 92:8–20

Heeger DJ, Jepson AD. 1992. Subspace methods for recovering rigid motion I: algorithm and implementation. *Int. J. Comp. Vis.* 7:95–117

Helmholtz Hv, Southall JPC. 1924. *Helmholtz's Treatise on Physiological Optics.* Rochester, NY: Opt. Soc. Am. 3 v. pp.

Heuer HW, Britten KH. 2004. Optic flow signals in extrastriate area MST: comparison of perceptual and neuronal sensitivity. *J. Neurophysiol.* 91:1314–26

Hietanen JK, Perrett DI. 1996. Motion sensitive cells in the macaque superior temporal polysensory area: response discrimination between self-generated and externally generated pattern motion. *Behav. Brain. Res.* 76:155–67

Hollands MA, Patla AE, Vickers JN. 2002. "Look where you're going!": gaze behaviour associated with maintaining and changing the direction of locomotion. *Exp. Brain Res.* 143:221–30

Koenderink JJ, van Doorn AJ. 1987. Facts on optic flow. *Biol. Cybern.* 56:247–54

Krauzlis RJ, Lisberger SG. 1994. A model of visually-guided smooth pursuit eye movements based on behavioral observations. *J. Comput. Neurosci.* 1:265–83

Krug K, Cumming BG, Parker AJ. 2004. Comparing perceptual signals of single V5/MT neurons in two binocular depth tasks. *J. Neurophysiol.* 92:1586–96

Lagae L, Maes H, Raiguel S, Xiao DK, Orban GA. 1994. Responses of macaque STS neurons to optic flow components: a comparison of MT and MST. *J. Neurophysiol.* 71:1597–626

Land MF, Lee DN. 1994. Where we look when we steer. *Nature* 369:742–44

Lappe M, ed. 2000. *Neuronal Processing of Optic Flow*, Vol. 44. San Diego: Academic

Lappe M, Bremmer F, Pekel M, Thiele A, Hoffmann KP. 1996. Optic flow processing in monkey STS: a theoretical and experimental approach. *J. Neurosci.* 16:6265–85

Lappe M, Hoffman K-P. 2000. Optic flow and eye movements. In *Neuronal Processing of Optic Flow*, ed. M Lappe, pp. 29–50. San Diego: Academic

Lee B, Pesaran B, Andersen RA. 2007. Translation speed compensation in the dorsal aspect of the medial superior temporal area. *J. Neurosci.* 27:2582–91

Levitt J, Kiper D, Movshon A. 1994. Receptive fields and functinal architecture of macaque V2. *J. Neurophysiol.* 71:2517–42

Lewis JW, Van Essen DC. 2000. Corticocortical connections of visual, sensorimotor, and multimodal processing areas in the parietal lobe of the macaque monkey. *J. Comp. Neurol.* 428:112–37

Lisberger SG, Movshon JA. 1999. Visual motion analysis for pursuit eye movements in area MT of macaque monkeys. *J. Neurosci.* 19:2224–46

Livingstone MS, Conway BR. 2003. Substructure of direction-selective receptive fields in macaque V1. *J. Neurophysiol.* 89:2743–59

Logan DJ, Duffy CJ. 2006. Cortical area MSTd combines visual cues to represent 3-D self-movement. *Cereb. Cortex* 16:1494–507

Longuet-Higgins HC, Prazdny K. 1980. The interpretation of a moving retinal image. *Proc. R. Soc. London Ser. B* 208:385–97

Maunsell JHR, Van Essen DC. 1983. Functional properties of neurons in the middle temporal visual area (MT) of the macaque monkey: I. Selectivity for stimulus direction, speed and orientation. *J. Neurophysiol.* 49:1127–47

Merchant H, Battaglia-Mayer A, Georgopoulos AP. 2001. Effects of optic flow in motor cortex and area 7a. *J. Neurophysiol.* 86:1937–54

Newsome WT, Wurtz RH, Komatsu H. 1988. Relation of cortical areas MT and MST to pursuit eye movements. II. Differentiation of retinal from extraretinal inputs. *J. Neurophysiol.* 60:604–20

Niemann T, Lappe M, Buscher A, Hoffmann KP. 1999. Ocular responses to radial optic flow and single accelerated targets in humans. *Vis. Res.* 39:1359–71

Ono T, Squire L, Raichle M, Perrett D, Fukuda M, eds. 1993. *Brain Mechanisms of Perception and Memory: From Neuron to Behavior.* Oxford, UK: Oxford Univ. Press

Orban GA. 2001. Imaging image processing in the human brain. *Curr. Opin. Neurol.* 14:47–54

Page WK, Duffy CJ. 1999. MST neuronal responses to heading direction during pursuit eye movements. *J. Neurophysiol.* 81:596–610

Page WK, Duffy CJ. 2003. Heading representation in MST: sensory interactions and population encoding. *J. Neurophysiol.* 89:1994–2013

Paolini M, Distler C, Bremmer F, Lappe M, Hoffmann KP. 2000. Responses to continuously changing optic flow in area MST. *J. Neurophysiol.* 84:730–43

Parker AJ, Newsome WT. 1998. Sense and the single neuron: probing the physiology of perception. *Annu. Rev. Neurosci.* 21:227–77

Patla AE, Vickers JN. 1997. Where and when do we look as we approach and step over an obstacle in the travel path? *Neuroreport* 8:3661–65

Patla AE, Vickers JN. 2003. How far ahead do we look when required to step on specific locations in the travel path during locomotion? *Exp. Brain Res.* 148:133–38

Pekel M, Lappe M, Bremmer F, Thiele A, Hoffmann KP. 1996. Neuronal responses in the motion pathway of the macaque monkey to natural optic flow stimuli. *Neuroreport* 7:884–88

Perrone JA, Stone LS. 1994. A model of self-motion estimation within primate extrastriate visual cortex. *Vis. Res.* 34:2917–38

Perrone JA, Stone LS. 1998. Emulating the visual receptive-field properties of MST neurons with a template model of heading estimation. *J. Neurosci.* 18:5958–75

Phinney RE, Siegel RM. 2000. Speed selectivity for optic flow in area 7a of the behaving macaque. *Cereb Cortex* 10:413–21

Priebe NJ, Churchland MM, Lisberger SG. 2002. Constraints on the source of short-term motion adaptation in macaque area MT. I. the role of input and intrinsic mechanisms. *J. Neurophysiol.* 88:354–69

Priebe NJ, Lisberger SG. 2002. Constraints on the source of short-term motion adaptation in macaque area MT. II. tuning of neural circuit mechanisms. *J. Neurophysiol.* 88:370–82

Purushothaman G, Bradley DC. 2005. Neural population code for fine perceptual decisions in area MT. *Nat. Neurosci.* 8:99–106

Raffi M, Squatrito S, Maioli MG. 2002. Neuronal responses to optic flow in the monkey parietal area PEc. *Cereb Cortex* 12:639–46

Rieger JH, Lawton DT. 1985. Processing differential image motion. *J. Opt. Soc. Am. A* 2:354–60

Royden CS. 1997. Mathematical analysis of motion-opponent mechanisms used in the determination of heading and depth. *J. Opt. Soc. Am. A* 14:2128–43

Royden CS, Banks MS, Crowell JA. 1992. The perception of heading during eye movements. *Nature* 360:583–85

Royden CS, Crowell JA, Banks MS. 1994. Estimating heading during eye movements. *Vis. Res.* 34:3197–214

Royden CS, Hildreth EC. 1999. Differential effects of shared attention on perception of heading and 3-D object motion. *Percept. Psychophys.* 61:120–33

Schaafsma SJ, Duysens J. 1996. Neurons in the ventral intraparietal area of awake macaque monkey closely resemble neurons in the dorsal part of the medial superior temporal area in their responses to optic flow patterns. *J. Neurophysiol.* 76:4056–68

Schaafsma SJ, Duysens J, Gielen CC. 1997. Responses in ventral intraparietal area of awake macaque monkey to optic flow patterns corresponding to rotation of planes in depth can be explained by translation and expansion effects. *Vis. Neurosci.* 14:633–46

Schlack A, Hoffmann KP, Bremmer F. 2002. Interaction of linear vestibular and visual stimulation in the macaque ventral intraparietal area (VIP). *Eur. J. Neurosci.* 16:1877–86

Schlack A, Sterbing-D'Angelo SJ, Hartung K, Hoffmann KP, Bremmer F. 2005. Multisensory space representations in the macaque ventral intraparietal area. *J. Neurosci.* 25:4616–25

Shenoy KV, Bradley DC, Andersen RA. 1999. Influence of gaze rotation on the visual response of primate MSTd neurons. *J. Neurophysiol.* 81:2764–86

Shenoy KV, Crowell JA, Andersen RA. 2002. Pursuit speed compensation in cortical area MSTd. *J. Neurophysiol.* 88:2630–47

Sherk H, Fowler GA. 2001. Neural analysis of visual information during locomotion. *Prog. Brain Res.* 134:247–64

Siegel RM, Read HL. 1997. Analysis of optic flow in the monkey parietal area 7a. *Cereb Cortex* 7:327–46

Snowden RJ, Milne AB. 1996. The effect of adapting to complex motions: position invariance and tuning to spiral motions. *J. Cogn. Neurosci.* 8:435–52

Squatrito S, Maioli MG. 1997. Encoding of smooth pursuit direction and eye position by neurons of area MSTd of macaque monkey. *J. Neurosci.* 17:3847–60

Srinivasan MV, Zhang SW. 1997. Visual control of honeybee flight. *Exs* 84:95–113

Tanaka K, Fukada Y, Saito H. 1989. Underlying mechanisms of the response specificity of expansion/contraction and rotation cells in the dorsal part of the medial superior temporal area of the Macaque monkey. *J. Neurophysiol.* 62:642–56

Tanaka K, Hikosaka H, Saito H, Yukie Y, Fukada Y, Iwai E. 1986. Analysis of local and wide-field movements in the superior temporal visual areas of the macaque monkey. *J. Neurosci.* 6:134–44

Tanaka K, Saito H. 1989. Analysis of motion of the visual field by direction, expansion/contraction and rotation cells clustered in the dorsal part of the medial superior temporal area of the Macaque monkey. *J. Neurophysiol.* 62:626–41

te Pas SF. 1996. *Perception of Structure in Optical Flow Fields*. Utrecht, The Neth.: Utrecht Univ.

Uka T, DeAngelis GC. 2001. Contribution of MT neurons to depth discrimination. I. Comparison of neuronal and behavioral sensitivity. *Soc. Neurosci. Abstr.* 680.12

Upadhyay UD, Page WK, Duffy CJ. 2000. MST responses to pursuit across optic flow with motion parallax. *J. Neurophysiol.* 84:818–26

Vaina LM. 1998. Complex motion perception and its deficits. *Curr. Opin. Neurobiol.* 8:494–502

van den Berg AV. 1992. Robustness of perception of heading from optic flow. *Vis. Res.* 32:1285–96

van den Berg AV. 1996. Judgements of heading. *Vis. Res.* 36:2337–50

van den Berg AV, Beintema JA. 1997. Motion templates with eye velocity gain fields for transformation of retinal to head centric flow. *Neuroreport* 8:835–40

Van Essen DC, Drury HA, Dickson J, Harwell J, Hanlon D, Anderson CH. 2001. An integrated software suite for surface-based analyses of cerebral cortex. *J. Am. Med. Inform. Assoc.* 8:443–59

Warren WH. 1998. The state of flow. In *High-Level Motion Processing*, ed. T Watanabe, pp. 315–58. Cambridge, MA: MIT Press

Warren WH, Hannon DJ. 1990. Eye movements and optical flow. *J. Opt. Soc. Am. A* 7:160–69

Warren WH, Morris MW, Kalish M. 1988. Perception of translational heading from optical flow. *J. Exp. Psychol. Hum. Percept. Perform.* 14:646–60

Wurtz RH, Duffy CJ, Roy J-P. 1993. Motion processing for guiding self-motion. See Ono et al. 1993, pp. 141–65

Zeki SM. 1974. Functional organization of a visual area in the posterior bank of the superior temporal sulcus of the rhesus monkey. *J. Physiol.* 236:549–73

Zeki SM. 1990. The motion pathways of the visual cortex. In *Vision: Coding and Efficiency*, ed. C Blakemore, pp. 321–45. Cambridge, UK: Cambridge Univ. Press

Zhang T, Britten KH. 2003. Microstimulation of area VIP biases heading perception in monkeys. *Soc. Neurosci. Abstr. Viewer/Itiner.* 339.9

Zhang T, Britten KH. 2004. Clustering of selectivity for optic flow in the ventral intraparietal area. *Neuroreport* 15:1941–45

Zhang T, Britten KH. 2005. Neuronal sensitivity and choice probability in the ventral intraparietal area during a heading discrimination task. *Soc. Neurosci. Abstr. Viewer/Itiner.* 390.16

Zhang T, Heuer HW, Britten KH. 2004. Parietal area VIP neuronal responses to heading stimuli are encoded in head-centered coordinates. *Neuron* 42:993–1001

Zipser D, Andersen RA. 1988. A back-propagation programmed network that simulates response properties of a subset of posterior parietal neurons. *Nature* 331:679–84

Mechanisms of
Face Perception

Doris Y. Tsao[1] and Margaret S. Livingstone[2]

[1]Centers for Advanced Imaging and Cognitive Sciences, Bremen University,
D-28334 Bremen, Germany; email: doris@nmr.mgh.harvard.edu

[2]Department of Neurobiology, Harvard Medical School, Boston, Massachusetts 02115;
email: mlivingstone@hms.harvard.edu

Annu. Rev. Neurosci. 2008. 31:411–37

First published online as a Review in Advance on
April 2, 2008

The *Annual Review of Neuroscience* is online at
neuro.annualreviews.org

This article's doi:
10.1146/annurev.neuro.30.051606.094238

Key Words

face processing, face cells, holistic processing, face recognition, face
detection, temporal lobe

Abstract

Faces are among the most informative stimuli we ever perceive: Even
a split-second glimpse of a person's face tells us his identity, sex, mood,
age, race, and direction of attention. The specialness of face processing
is acknowledged in the artificial vision community, where contests for
face-recognition algorithms abound. Neurological evidence strongly
implicates a dedicated machinery for face processing in the human
brain to explain the double dissociability of face- and object-recognition
deficits. Furthermore, recent evidence shows that macaques too have
specialized neural machinery for processing faces. Here we propose a
unifying hypothesis, deduced from computational, neurological, fMRI,
and single-unit experiments: that what makes face processing special is
that it is gated by an obligatory detection process. We clarify this idea
in concrete algorithmic terms and show how it can explain a variety of
phenomena associated with face processing.

Contents

INTRODUCTION

The central challenge of visual recognition is the same for both faces and objects: We must distinguish among often similar visual forms despite substantial changes in the image arising from changes in position, illumination, occlusion, etc. Although face identification is often singled out as demanding particular sensitivity to differences between objects sharing a com-

mon basic configuration, in fact such differences must be represented in the brain for both faces and nonface objects. Most humans can easily identify hundreds of faces (Diamond & Carey 1986), but even if one cannot recognize a hundred different bottles by name, one can certainly distinguish them in pairwise discrimination tasks. Furthermore, most of us can recognize tens of thousands of words at a glance, not letter by letter, a feat requiring expert detection of configural patterns of nonface stimuli. Thus, face perception is in many ways a microcosm of object recognition, and the solution to the particular problem of understanding face recognition will undoubtedly yield insights into the general problem of object recognition.

The system of face-selective regions in the human and macaque brain can be defined precisely using fMRI, so we can now approach this system hierarchically and physiologically to ask mechanistic questions about face processing at a level of detail previously unimaginable. Here we review what is known about face processing at each of Marr's levels: computational theory, algorithm, and neural implementation.

Computer vision algorithms for face perception divide the process into three distinct steps. First, the presence of a face in a scene must be detected. Then the face must be measured to identify its distinguishing characteristics. Finally, these measurements must be used to categorize the face in terms of identity, gender, age, race, and expression.

Detection

The most basic aspect of face perception is simply detecting the presence of a face, which requires the extraction of features that it has in common with other faces. The effectiveness and ubiquity of the simple T-shaped schematic face (eye, eye, nose, mouth) suggest that face detection may be accomplished by a simple template-like process. Face detection and identification have opposing demands: The identification of individuals requires a fine-grained analysis to extract the ways in which each face differs from the others despite the fact that all faces share the same basic T-shaped configuration, whereas detection requires extracting what is common to all faces. A good detector should be poor at individual recognition and vice versa.

Another reason why detection and identification should be separate processes is that detection can act as a domain-specific filter, ensuring that precious resources for face recognition [e.g., privileged access to eye movement centers (Johnson et al. 1991)] are used only if the stimulus passes the threshold of being a face. Such domain-specific gating may be one reason for the anatomical segregation of face processing in primates (it is easier to gate cells that are grouped together). A further important benefit of preceding identification by detection is that detection automatically accomplishes face segmentation; i.e., it isolates the face from background clutter and can aid in aligning the face to a standard template. Many face-recognition algorithms require prior segmentation and alignment and will fail with nonuniform backgrounds or varying face sizes.

Measurement and Categorization

After a face has been detected, it must be measured in a way that allows for accurate, efficient identification. The measurement process must not be so coarse as to miss the subtle features that distinguish one face from another. On the other hand, it must output a set of values that can be efficiently compared with stored templates for identification. There is a zero-sum game between measurement and categorization: The more efficient the measurement, the easier the classification; conversely, less efficient measurement (e.g., a brute force tabulation of pixel gray values) makes the classification process more laborious.

COMPUTER VISION ALGORITHMS

A comprehensive review of computer algorithms for face recognition can be found in Zhao et al. (2003) and Shakhnarovich &

Moghaddam (2004). Our goal here is to discuss algorithms that offer special insights into possible biological mechanisms.

Detection

How can a system determine if there is a face in an image, regardless of whose it is? An obvious approach is to perform template matching (e.g., search for a region containing two eyes, a mouth, and a nose, all inside an oval). In many artificial face-detection systems a template is swept across the image at multiple scales, and any part of the image that matches the template is scored as a face. This approach works, but it is slow.

To overcome this limitation, Viola & Jones (2004) introduced the use of a cascade of increasingly complex filters or feature detectors. Their reasoning was that the presence of a face can be ruled out most of the time with a very simple filter, thus avoiding the computational effort of doing fine-scale filtering on uninformative parts of the image. The first stage in their cascade consists of only two simple filters, each composed of a few rectangular light or dark regions (**Figure 1a**). Subsequent stages of filtering are performed only on regions scoring positive at any preceding stage. This cascade approach proved just as accurate, but 10 times faster, than single-step face-detector algorithms.

Sinha's face-detection algorithm (Sinha 2002a) is based on the observation that qualitative contrast relationships between different parts of a face are highly conserved, even under different lighting conditions (**Figure 1b**). Even though any single contrast relationship between two facial regions would be inadequate to detect a face, a set of such relationships could be adequate (because probabilities multiply). A subset of Sinha's directed contrasts ([r2, r3] and [r4, r5]) are equivalent to the first stage of the Viola-Jones face detector.

Effective primitives for face detection can also be computed using an information theory approach by identifying fragments (subwindows) of face images that are maximally

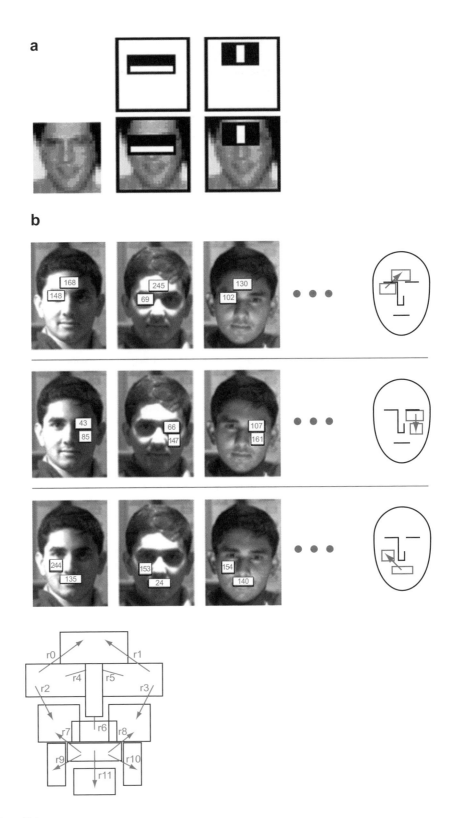

informative about the presence or absence of a face (Ullman et al. 2002). The resulting fragments consist of medium-resolution face parts, e.g., an eye, rather than the whole face, so in this algorithm, face detection is triggered by detection of a threshold number of such fragments.

All three algorithms discussed above use basic feature detectors much simpler than a whole face (rectangle features in the Viola-Jones algorithm, qualitative contrast ratios between pairs of face regions in the Sinha algorithm, and face parts in the Ullman algorithm). Yet, all three algorithms perform holistic detection, that is, they obligatorily detect faces as correctly arranged wholes. This is because all three algorithms detect overlapping constellations of elemental features that cover the whole face. The feature overlaps implicitly enforce the correct overall arrangement of features.

Measurement

Once a face has been detected, it may need to be identified or classified. Algorithms for the identification of individual faces are generally either feature-based or holistic. In feature-based methods, fiducial points (e.g., eyes, mouth, nose) are identified and used to compute various geometric ratios. As long as the features can be detected, this approach is robust to position and scale variations. In holistic methods, the entire face is matched to memory templates without isolating specific features or parts. One advantage of holistic methods is that all parts of the face are used, and no information is discarded.

The simplest holistic recognition algorithm is to correlate a presented image directly to a bank of stored templates, but having templates for every face is expensive in time and memory space. Turk & Pentland (1991) developed the eigenface algorithm to overcome

these limitations. The eigenface algorithm exploits the fact that all faces share a common basic structure (round, smooth, symmetric, two eyes, a nose, and a mouth). Thus the pixel arrays defining various faces are highly correlated, and the distinguishing characteristics of a face can be expressed more efficiently if these correlations are removed using principal components analysis (PCA). When PCA is performed on a large set of faces, the eigenvectors with largest eigenvalues all look like faces, and hence are called "eigenfaces" (**Figure 2a**). An arbitrary face can be projected onto a set of eigenfaces to yield a highly compressed representation; good face reconstructions can typically be obtained with just 50 eigenfaces and passable ones with just 25. In other words, something as ineffable as an identity can be reduced to 25 numbers (**Figure 2b**).

PCA on sets of faces varying in both expression and identity generates some principal components that are useful for only expression or only identity discrimination and others that are useful for both (Calder et al. 2001). This partial independence of PCs can successfully model the independent perception of expression and identity (Cottrell et al. 2002).

The eigenface algorithm does not perform well if the sample face is not accurately aligned in scale and position to the template eigenfaces. Human face perception, however, is tolerant to changes in both scale and position. Moreover, if a face is transformed further along the morph line representing the deviation of that face from the average face, the transformed face is easily recognized as the same individual; this is the basis of caricature (Leopold et al. 2001). The process of morphing one individual into another (Wolberg 1996) involves both an intensity transform (which eigenfaces model very well) and a simultaneous geometric transform (**Figure 3a**). Because eigenfaces represent axes

Eigenface: an eigenvector of the covariance matrix defined by a set of faces that allows a compressed representation

PCA: principal components analysis

Caricature: an artistic technique to enhance the recognizability of a face by exaggerating features distinguishing that face from the average face

---

Figure 1

(*a*) The two most diagnostic features defining a face comprise the first level of the detection cascade in the Viola-Jones algorithm for face detection. From Viola & Jones 2004. (*b*) The Sinha algorithm for face detection, showing the ratio-templates defining a face. From Sinha 2002a.

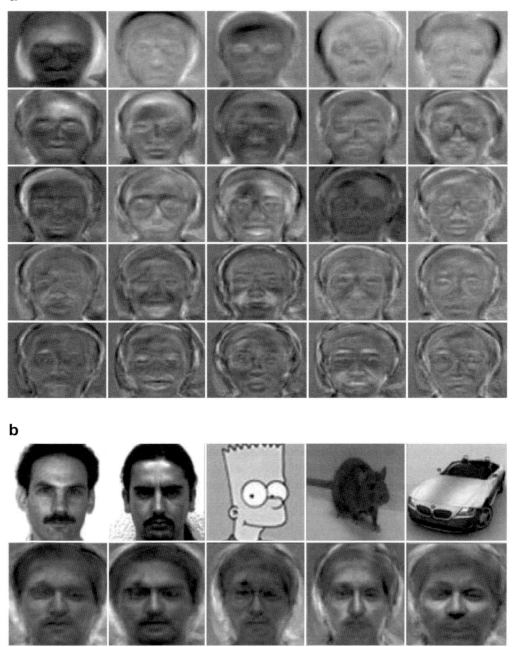

Figure 2

The eigenface algorithm for face recognition. (*a*) The first 25 eigenvectors computed from the Yale face database (a collection of 165 face images). (*b*) Eigenface reconstructions of 5 different images, using the 25 eigenfaces shown in panel *a*. Note that nonface images can have nontrivial projections onto eigenfaces. Courtesy of C. DeCoro.

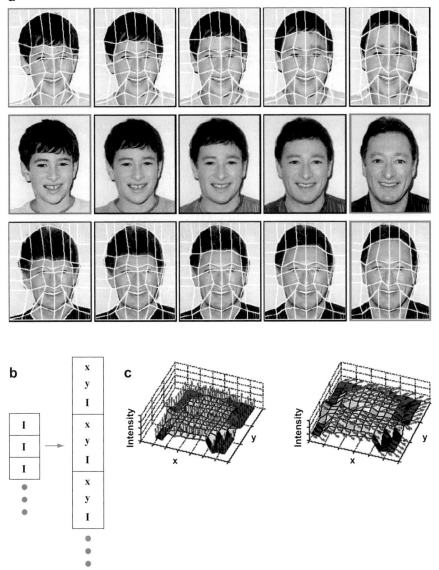

Figure 3

A computational approach that can represent both spatial and intensity variations. (*a*) The computer graphics technique of morphing, in which the identity of one individual can be continuously transformed into that of another, provides insights about the nature of the face template. In the middle row, the individual outlined in *red* is continuously morphed into the individual outlined in *green*, which requires both a geometric transform and an intensity transform. The top and bottom rows show pure geometric transforms (morphing of the mesh) of the same 2 faces (the top rows show the geometric distortion of the *red* face into the shape of the *green* face, and the bottom row shows the distortion of the *green* face into the shape of the *red* face). The middle row shows a weighted intensity average of the aligned meshes from the top and bottom rows. From Wolberg 1998. (*b*) Bags of Pixels variant on the eigenface algorithm. The (x, y) coordinate of each pixel is elevated to the same status as the intensity value. (*c*) Adding or subtracting traditional eigenfaces to an average face produces only intensity variations at each pixel. Adding or subtracting eigenfaces computed using Bags of Pixels, however, can produce geometric variations in addition to intensity variations. From Jebara 2003.

of intensity values on a fixed spatial basis, the eigenface approach does not interpret caricature transformations as the same individual.

Jebara (2003) proposed a clever way to get around the spatial rigidity of the original eigenface approach: Instead of performing PCA on the intensity values, the size of the representation is tripled, so each pixel conveys not only the image intensity value but also the intensity value's (x, y) location. Then PCA can be done on the triple-sized image containing a concatenation of (x, y, I) values (**Figure 3b,c**). The power of this approach is that spatial coordinates are treated just like intensity coordinates, and thus the resultant eigenfaces represent both geometric and intensity variations. The fact that this bags of pixels approach performs three orders of magnitude better than standard eigenface analysis on face sets with changes in pose, illumination, and expression is computational proof of the importance of representing geometric variations in addition to intensity variations.

Categorization

Turk and Pentland used a simple Euclidean distance metric on face eigen-coordinates to perform recognition. More powerful classifiers that have been applied to the problem of face recognition include Fisher linear discriminants (Belhumeur et al. 1997), Bayesian estimation (Moghaddam et al. 2000), and support vector machines (Shakhnarovich & Moghaddam 2004). These classification techniques can be regarded as second-tier add-ons to the basic eigenface measurement system. Measurement yields analog descriptions, whereas classification is nonlinear and yields discrete boundaries between descriptions.

Separating the process of measurement from the process of classification gives a computational system maximum flexibility because different categorizations (e.g., emotion, identity, gender) can all operate on the same set of basic eigenvector projections. Gender determination can be based on large eigenvalue eigenvectors, whereas identification of individuals relies on lower-value eigenvectors (O'Toole et al. 1993). Furthermore, because classifications are necessarily nonlinear, the independence of classification mechanisms from measurement mechanisms would be very exciting from an experimental point of view because the templates for measurement could thus be linear, and therefore their detailed structure could be mapped. We will return to the idea of linear measurement mechanisms when we discuss tuning properties of face cells.

Invariance

Developing position and scale invariant recognition is a huge challenge for artificial face-recognition systems. Initial attempts to compute a meaningful set of eigen-coordinates for a face required that the face be accurately aligned in scale, position, and rotation angle to the template eigenfaces. However, if, as we propose, face detection precedes measurement, the detector can determine the location, size, and rotation angle of the eyes and face outline and then use these to normalize the input to face-measurement units.

Summary

The main lesson we can extract from artificial systems for face processing is that detection and recognition are distinct processes, with distinct goals, primitives (coarse contrast relationships vs. detailed holistic templates), and computational architectures (filter cascade vs. parallel measurements). By preceding recognition, detection can act as a domain-specific filter to gate subsequent processing and can include alignment and segmentation, preparing the face representations for subsequent measurement. The effectiveness of the eigenface algorithm for face recognition shows that faces can be represented by their deviation from the average in a compressed subspace. To characterize faces most effectively, this subspace needs to include spatial variations as well as intensity variations.

Some machine vision models of recognition use common meta-algorithms to learn the primitives for both detection and recognition of faces (Riesenhuber & Poggio 2000, Ullman 2007). Thus the two processes may share core computational principles. Whether biological systems use discrete steps of detection, measurement, and classification to recognize faces is a question that can only be resolved empirically.

HUMAN BEHAVIOR AND FUNCTIONAL IMAGING

The extensive behavioral literature on face perception provides a rich source of clues about the nature of the computations performed in processing faces (**Figure 4**). One of the hallmarks of face processing is that recognition performance drops substantially when faces are presented upside down (**Figure 4a**) or in negative contrast, and both effects are much smaller for objects (Kemp et al. 1990, Yin 1969). We propose that both these properties can be explained if only upright, positive-contrast faces gain access to the face-processing system, i.e., if an upright, positive-contrast template is used for face detection. This template may be innate in humans, as evidenced by the tendency for newborns to track normal schematic faces longer

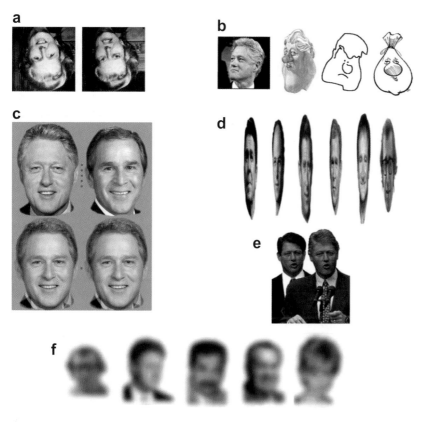

Figure 4

Behavioral observations on the nature of human face processing. (*a*) Flip the page upside down. The Thatcher Illusion shows that faces are obligatorily processed as wholes (an identical pair of features such as the upright and inverted mouth can appear similar or dramatically different depending on the surrounding context). From Thompson 1980. (*b*) Robustness of face identification to caricature. (*c*) Adaptation: Run your eyes along the 5 *red* dots for a minute, and then shift your gaze to the single *red* dot. From Afraz & Cavanagh 2008. (*d*) Robustness to compression. From Sinha et al. 2006. (*e*) The importance of external features. From Sinha & Poggio 1996. (*f*) Robustness to low resolution. From Sinha 2002b.

than scrambled schematic faces (Johnson et al. 1991, Simion et al. 1998).

Norm-Based Coding

Caricatures are remarkably powerful in evoking recognition (**Figure 4b**): Caricatured faces are often more identifiable than veridical photographs (Lee et al. 2000). This finding has led to the proposal that faces are represented in terms of their deviation from the norm, or average, face (Leopold et al. 2001, Rhodes et al. 1987). Furthermore, the existence of face aftereffects (**Figure 4c**) shows that the face norm is adaptable (Webster & MacLin 1999). Because such face aftereffects transfer across retinal positions (Leopold et al. 2001) and image sizes (Jeffery et al. 2006), they apparently do not reflect adaptation to specific low-level image features, but instead indicate adaptation of higher-level representations. This face identity aftereffect was interpreted as indicating that adaptation to a given face shifts the norm or average face in the direction of the adapting face, making faces on the opposite side of the norm more distinctive (i.e., more different from the norm). To explain these results Rhodes & Jeffrey (2006) propose that face identity is coded by pairs of neural populations that are adaptively tuned to above-average and below-average values along each dimension of face space.

Opposite adaptation can occur simultaneously for upright and inverted faces, consistent with the idea that distinct neural pathways underlie the coding (and adaptation to) upright versus inverted faces (Rhodes & Jeffery 2006). Finally, although norm-based coding can work only for classes of stimuli that have similar enough first-order shape that a norm can be defined, this situation may not be unique to faces. Rhodes & McLean (1990) showed evidence for norm-based coding for images of birds, and adaptation effects can also be observed for simple shapes such as taper and overall curvature (Suzuki & Cavanagh 1998). Thus adaptive norm-based coding may be a general feature of high-level form-coding processes.

Prosopagnosia: highly specific inability to recognize faces, due to either congenital brain miswiring ("developmental prosopagnosia") or focal brain lesions ("acquired prosopagnosia")

Detection

As argued in the modeling section, it is computationally efficient to separate detection and recognition and to have detection precede recognition because detection can act as a domain-specific filter to make the recognition process more efficient (by focusing recognition on regions actually containing faces). That there are also separate detection and recognition stages in human face processing fits with one of the most striking findings from the neuropsychology literature: Patient CK, who was severely impaired at object recognition, including many basic midlevel visual processes, was nonetheless 100% normal at face recognition (Moscovitch et al. 1997). His pattern of deficits indicated that face processing is not simply a final stage tacked onto the end of the nonface object recognition pathway but rather a completely different pathway that branches away from object recognition early in the visual hierarchy, and it is this branching off that we propose to equate with the detection process. CK's dissociation is illustrated by his perception of the painting of a face made up of vegetables by Arcimbaldo—CK sees the face but not the constituent vegetables.

CK's ability to recognize famous or familiar faces was at least as good as normal controls, until the faces were shown upside down, and then his performance became much worse than that of controls. Conversely, patients with prosopagnosia perform better than controls in recognizing inverted faces (Farah et al. 1995). This double dissociation of the inversion effect is consistent with the existence of a face-specific processing system that can be accessed only by upright faces, present in CK and absent in prosopagnosics. Presumably, CK can process objects using only the face-specific system, prosopagnosics have a general object-recognition system but not the face-specific system, and normal subjects have both systems. The general nonface object system is not as good at processing faces as the face-specific system (hence the inversion effect in normal subjects), is missing in CK (hence his disproportionate deficit for inverted faces), and is the only way prosopagnosics can process any face (hence their relatively superior performance with inverted faces because their general object system gets extra practice processing faces).

Holistic Processing of Faces

Face processing is said to be distinct from nonface object processing in that it is more holistic; that is, faces are represented as nondecomposed wholes rather than as a combination of independently represented component parts (eyes, nose, mouth) and the relations between them (Farah et al. 1998). Evidence for holistic processing of faces comes from a number of behavioral paradigms, of which the two most cited are the part-whole effect (Tanaka & Farah 1993) and the composite effect (Young et al. 1987). In the part-whole effect, subjects are better at distinguishing two face parts in the context of a whole face than in isolation. In the composite effect, subjects are slower to identify half of a chimeric face aligned with an inconsistent other half-face than if the two half-faces are misaligned (Young et al. 1987). As with the part-whole effect, the composite effect indicates that even when subjects attempt to process only part of the face, they suffer interference from the other parts of the face, suggesting a lack of access to parts of the face and mandatory processing of the whole face.

One interpretation of the uniqueness of face processing is that it uses special neural machinery not shared by other kinds of objects, an idea that is consistent with functional imaging studies, as described below. Another interpretation is that holistic processing is characteristic of any kind of object that must be distinguished on a subordinate level, especially objects with which the subject is highly trained or familiar (Diamond & Carey 1986). It is not yet clear what the perceptual phenomenology of holistic processing implies either mechanistically or computationally. We suggest that holistic face processing can be explained by an obligatory detection stage that uses a coarse upright template to detect whole faces (**Figure 5**). This model explains the composite effect because an aligned chimera would be detected as a whole face and therefore would be processed as a unit by subsequent measurement and classification stages.

However, we cannot rule out alternatives, such as one-stage models in which both face detection and identification are carried out by the same set of face-selective cells. In this case, to explain holistic properties of face processing, we would have to postulate that individual face cells, unlike nonface cells, are selective not just for local features but for whole faces or that the readout of face information must comprise all or most of the population code. Either or both of these models would produce the behavioral holistic effects, even without an antecedent detection gate. The key evidence favoring our early detection gating hypothesis over a single-stage system comes from the identification of a series of face-selective areas in the macaque (Pinsk et al. 2005, Tsao et al. 2003) and the finding that an area early in this hierarchy already consists entirely of face-selective cells (Tsao et al. 2006); both these results are discussed more extensively below.

Although faces are unique in the degree to which they are processed holistically, other nonface objects can also show holistic effects, especially well-learned categories; for review see Gauthier & Tarr (2002). Words may approach faces in the degree to which they are processed holistically: Coltheart et al. (1993) found that some acquired dyslexics can read whole words and understand their meanings but cannot distinguish individual letters making up the words. And Anstis (2005) showed that word recognition can show the composite effect, in that observers cannot tell whether two words have same or different top halves.

HUMAN FUNCTIONAL IMAGING

Positron emission tomography studies initially showed activation of the fusiform gyrus in a variety of face-perception tasks (Haxby et al. 1991, Sergent et al. 1992), and fMRI subsequently

Inversion effect: some objects are recognized better when they are upright than inverted, this is especially true for faces and words

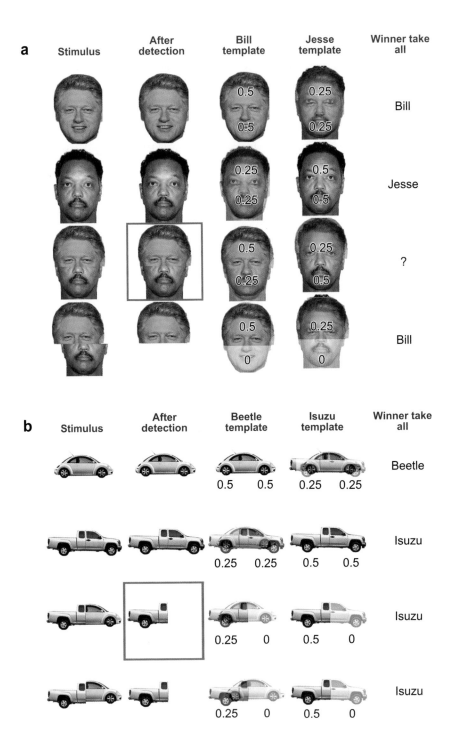

revealed more specificity in these cortical regions for faces with demonstrations of fusiform regions that responded more strongly to faces than to letter strings and textures (Puce et al. 1996), flowers (McCarthy et al. 1997), everyday objects, houses, and hands (Kanwisher et al. 1997). Although face-specific fMRI activation can also be seen in the superior temporal sulcus (fSTS) and in part of the occipital lobe [the occipital face area (OFA)], the most robust face-selective activation is consistently found on the lateral side of the right mid-fusiform gyrus, the fusiform face area (FFA) (Kanwisher et al. 1997) (**Figure 6**). The fact that this part of the brain is activated selectively in response to faces indicates that activity in this region must arise at or subsequent to a detection stage.

Many studies support the idea that the FFA is activated specifically by faces and not by the low-level stimulus features usually present in faces, that is, activity in the FFA indicates that stimuli have been detected as faces: The FFA shows increased blood flow in response to a wide variety of face stimuli: front and profile photographs of faces (Tong et al. 2000), line drawings of faces (Spiridon & Kanwisher 2002), and animal faces (Tong et al. 2000). Furthermore, the FFA BOLD signal to upright Mooney faces (low-information two-tone pictures of faces; Mooney 1957) is almost twice as strong as to inverted Mooney stimuli (which have similar low-level features but do not look like faces) (Kanwisher et al. 1998). Finally, for bistable stimuli such as the illusory face-vase, or for binocularly rivalrous stimuli in which a face is presented to one eye and a nonface is presented to the other eye, the FFA responds more strongly when subjects perceive a face than when they do not, even though the retinal stimulation is unchanged (Andrews et al. 2002, Hasson et al. 2001).

Although the FFA shows the strongest increase in blood flow in response to faces, it does also respond to nonface objects. Therefore, two alternative hypotheses have been proposed to the idea that activity in the FFA represents face-specific processing. First is the expertise hypothesis. According to this idea, the FFA is engaged not in processing faces per se, but rather in processing any sets of stimuli that share a common shape and for which the subject has gained substantial expertise (Tarr & Gauthier 2000). Second is the distributed coding hypothesis: In an important challenge to a more modular view of face and object processing, Haxby et al. (2001) argued that objects and faces are coded via the distributed profile of neuronal activity across much of the ventral visual pathway. Central to this view is the suggestion that nonpreferred responses, for example, to objects in the FFA, may form an important part of the neural code for those objects. The functional significance of the smaller but still significant response of the FFA to nonface objects will hopefully be unraveled by the combined assaults of higher-resolution imaging in humans and single-unit recordings in nonhuman primates.

Measurement and Categorization

Does the human brain use separate systems for face measurement and face classification? Some fMRI evidence suggests that it does. For

STS: superior temporal sulcus

FFA: fusiform face area

Blood-oxygen-level-dependent (BOLD) signal: hemodynamic signal measured in fMRI experiments. Active neurons consume oxygen, causing a delayed blood flow increase 1–5 s later

Expertise hypothesis: face-processing mechanisms are used to process any stimuli sharing a common shape and visual expertise

Distributed coding: representation scheme using distributed activity of coarsely tuned units. A key challenge for this idea is specifying how distributed codes can be read out

Figure 5

We propose that holistic (composite) effects of face processing can be explained by a detection stage that obligatorily segments faces as a whole. Subjects are asked to identify the top (faces) or left (car) part of each chimera (third and fourth rows) or simply to identify the object (first and second rows). Four face (*a*) and car (*b*) stimuli are detected, projected onto holistic templates, and then identified through a winner-take-all mechanism. The numbers in the third and fourth columns indicate the result of projecting each stimulus, after detection, onto the respective templates. Aligned faces are obligatorily detected as a whole, but misaligned faces and cars are not, and therefore their attended parts can be processed independently. According to this hypothesis, the essential difference between face (*a*) and nonface (*b*) processing occurs at the detection stage (*red boxes*). Subsequent measurement and classification could use similar mechanisms.

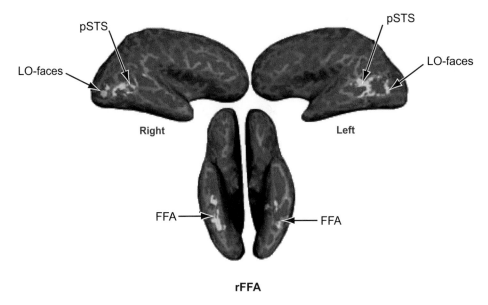

pSTS

LO-faces

pSTS

LO-faces

Right

Left

FFA

FFA

rFFA

Figure 6

Face-selective regions in one representative subject. Face-selective regions (*yellow*) were defined as regions that respond more strongly to faces than to houses, cars, and novel objects ($p < 10^{-4}$). From Grill-Spector et al. 2004.

example, in a study of morphing between Marilyn Monroe and Margaret Thatcher, adaptation strength in the OFA followed the amount of physical similarity along the morph line, while in the FFA it followed the perceived identity (Rotshtein et al. 2005), suggesting that the OFA performs measurement and the FFA performs classification. However, another study indicates that release from adaptation occurs in the FFA when physical differences are unaccompanied by changes in perceived identity (Yue et al. 2006).

According to Bruce & Young (1986), the processing of facial expression (one form of categorization) and facial identity (another form of categorization) takes separate routes. Haxby and colleageus (2000) proposed a neural basis for this model. According to this idea, the inferior occipital gyri are involved in early perception of facial features (i.e., measurement). The pathway then diverges, with one branch going to the superior temporal sulcus, which is proposed to be responsible for processing changeable aspects of faces including direction of eye gaze, view angle, emotional expression,

and lip movement. The other projection is to the lateral fusiform gyrus, which is responsible for processing identity. A recent review has challenged the Bruce and Young model, arguing that changeable aspects and invariant identity may instead be processed together and rely on partially overlapping visual representations (Calder & Young 2005).

Invariance

Several studies have used fMRI adaptation for face identity in the FFA and found invariance to image size (Andrews & Ewbank 2004) and spatial scale (Eger et al. 2004). Thus representations in the FFA are not tied to low-level image properties, but instead show at least some invariance to simple image transformations, though not to viewpoint (Pourtois et al. 2005).

Summary

Behavioral studies complement computational approaches by indicating that specialized

fMRI adaptation: controversial technique for deducing tuning properties of single cells from the magnitude of the BOLD signal, which averages activity of tens of thousands of cells

machinery may be used to process faces and that a face-detection stage gates the flow of information into this domain-specific module. The filters, or templates, used by this detection stage require an upright, positive contrast face, with the usual arrangement of features, and images that do not fit the template are analyzed only by the general object-recognition system. Even images that pass into the face-specific module are probably also processed in parallel by the general system, but the face module appears to process images differently from the general object system: Face processing is holistic in the sense that we cannot process individual face parts without being influenced by the whole face. We suggest that this difference arises early in the face processing pathway. The face-detection stage may, in addition to gating access, obligatorily segment faces as a whole for further processing by the face module. Finally, substantial recent evidence suggests that face identity is coded in an adaptive norm-based fashion.

Human imaging studies converge on the conclusion that faces are processed in specific locations in the temporal lobe, but the degree of specialization for faces within these locations is debated. The modular interpretation is consistent with neurological findings and, as described below, with single-unit recordings in macaques. The role of experience in determining both the localization of face processing and its holistic characteristics is also debated. And the relationship, if any, between modular organization and holistic processing is completely unexplored. Only a few visual object categories show functional localization in fMRI: faces, body parts, places, and words (for review see Cohen & Dehaene 2004, Grill-Spector & Malach 2004). Faces, bodies, and places are all biologically significant, and their neural machinery could conceivably be genetically programmed, but the use of writing arose too recently in human history for word processing to be genetically determined. Therefore, at least one kind of anatomical compartmentalization must be due to extensive experience. We have suggested that the existence of discrete brain regions dedicated to face processing implies an obligatory detection stage and that an obligatory detection stage results in holistic processing. What we know about word processing suggests that it too displays holistic properties, and it is localized, interestingly, in the left hemisphere in an almost mirror symmetric location to the position of the FFA in the right hemisphere (Cohen & Dehaene 2004, Hasson et al. 2002).

MONKEY fMRI AND SINGLE-UNIT PHYSIOLOGY

Detection

The seminal finding by Gross and his colleagues (1969, 1972) that there exist cells in inferotemporal cortex (IT) that are driven optimally by complex biologically relevant stimuli, such as hands or faces, was novel and initially not well accepted, despite the fact that Konorski (1967) had predicted the existence of face-selective cells, or gnostic units, and that they would be found in IT. Although IT cells do not generally appear to be detectors for complex objects, there are consistently observed populations of cells selectively responsive to faces, bodies, and hands, suggesting that faces, bodies, and hands are treated differently from other types of complex patterns, consistent with their also being among the only object categories, aside from words and numbers, that show localization in human fMRI. But the strong possibility remained that these cells were not really tuned to biologically relevant objects, but rather to some more abstract basis set, in which all possible shapes are represented by different cells and some cells were tuned to particular parameters that happened to fit the face or hand stimuli better than any of the other objects tested. Foldiak et al. (2004) recently provided evidence that face selectivity is not just an incidental property of cells tuned to an exhaustive set of image features: They presented 600–1200 stimuli, randomly chosen from several image archives, to cells recorded from both the upper and the lower bank of the STS and found

Inferotemporal cortex (IT): ventral temporal lobe, including the lower bank of the STS and outer convexity, specialized for visual object recognition

gnostic unit (or grandmother cell): a hypothetical cell responding exclusively to a single high-level percept in a highly invariant manner

that the distribution of tuning to these images showed bimodality, i.e., cells were either predominantly face selective or not face selective. It is not unprecedented to have specialized neural systems for socially important functions: Birds have evolved specialized structures for the perception and generation of song, and in humans there are specialized parts of the auditory and motor systems devoted specifically to language.

Direct evidence that some face cells are used for face detection comes from a microstimulation study by Afraz et al. (2006). Monkeys were trained to discriminate between noisy pictures of faces and nonface objects. Through systematic sampling, Afraz et al. identified cortical locations where clusters of face-selective cells could be reliably recorded. When they stimulated these regions and observed the monkeys' perceptual choices, they found a shift in the psychometric curve favoring detection of a face.

Holistic Processing of Faces

In general, face cells require an intact face and are not selective just for individual features (Bruce et al. 1981; Desimone et al. 1984; Kobatake & Tanaka 1994; Leonard et al. 1985; Oram & Perrett 1992; Perrett et al. 1982, 1984; Scalaidhe et al. 1999; Tsao et al. 2006). **Figure 7** shows nonlinear combinatorial response properties of a face-selective cell recorded in IT by Kobatake & Tanaka (1994). Out of a large number of three-dimensional objects, this cell responded best to the face of a toy monkey (panel *a*), and by testing various simplified two-dimensional paper stimuli, they determined that the cell would also respond to a configuration of two black dots over a horizontal line within a disk (panel *b*) but not in the absence of either the spots or the line (panels *c* and *d*) or the circular outline (panel *e*). The contrast between the inside and the outside of the circle was not critical (panel *g*), but the spots and the bar had to be darker than the disk (panel *h*). Thus the cell responded only when the stimulus looked like a face, no matter how simplified.

The response selectivity of face cells indicates that they must not only combine fea-

tures nonlinearly but also require them to be in a particular spatial configuration. However, such spatial-configuration selective responses and nonlinear combination of features are not restricted to face cells as such behavior has been reported for other kinds of complex object-selective cells in the temporal lobe (Baker et al. 2002, Kobatake & Tanaka 1994, Tanaka et al. 1991). Even earlier in the temporal pathway, nonlinear combinatorial shape selectivity can be seen (Brincat & Connor 2004).

Anatomical Specialization of Face Cells

Most studies on face cells reported face-selective cells scattered throughout the temporal lobe, though they tended to be found in clusters (Perrett et al. 1984). Because other kinds of shape selectivities also tend to be clustered (Desimone et al. 1984, Fujita et al. 1992, Tanaka et al. 1991, Wang et al. 1996), it was assumed that within the temporal lobe there was a columnar organization for shape, in which face columns represented just one of many shape-specific types of columns. However, this view was inconsistent with emerging evidence from human neurology and functional imaging that human face processing was localized to specific, reproducible regions of the temporal lobe. The apparent discrepancy was resolved by two recent studies by Tsao et al. (2003, 2006), who found that in monkeys, as in humans, face processing, as revealed by functional imaging, is localized to discrete regions of the temporal lobe, and they further showed that even at the single-unit level, face processing is highly localized (**Figure 8**; note also **Figure 7**, *top*).

Tsao et al. used functional imaging to localize regions in the macaque temporal lobe that were selectively activated by faces, compared with nonface objects, and then they recorded almost 500 single units within the largest of these face-selective regions in two monkeys. They found a remarkable degree of face selectivity within this region; 97% of the cells were face selective, on average showing almost 20-fold larger responses to faces than to nonface

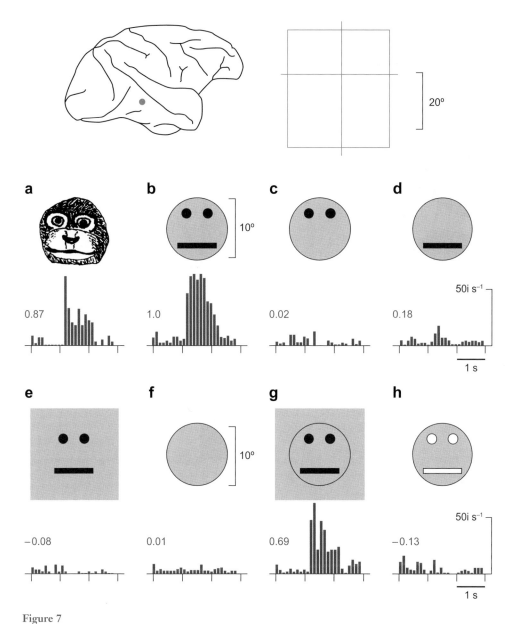

Figure 7

Holistic face detection. (*Top*) recording site and receptive-field location of a face cell. (*a-h*) Response
selectivity. From Kobatake & Tanaka (1994).

objects. The region where they recorded was
quite posterior in the temporal lobe (6 mm
anterior to the interaural canal, corresponding to posterior TE/anterior TEO). The fact
that an area consisting almost entirely of face-
selective cells exists so early in the ventral

stream provides strong support for the hypothesis that the face pathway is gated by an obligatory detection stage.

In light of the clear large-scale organization of face processing in macaques revealed by
Tsao et al. and recently by Pinsk et al. (2005),

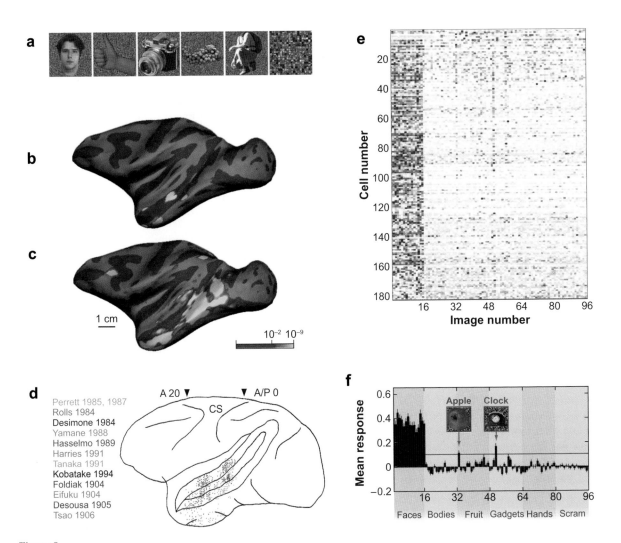

Figure 8

Mapping face and object selectivity in the monkey brain. (*a*) Five stimulus categories included faces, four nonface object categories (hands, gadgets, fruits, and bodies), and grid scrambled patterns. (*b*) Map of faces > objects. (*c*) Map of objects > scrambled. (*d*) Meta-analysis showing the location of physiologically identified face-selective cells; studies identified by first author and date. Five hundred face-selective cells were recorded by Tsao et al. 2006 at the location indicated by the pink asterisk. (*e*) Responses of 182 neurons from M1's middle face patch to 96 images of faces and nonface objects. (*f*) Average normalized population response to each image. Panels *a–c*, e, *f* are from Tsao et al. 2006.

we reexamined all previous physiological studies that mapped out locations of face-selective cells, and by remapping their face-cell localizations onto a common map, we found that, taken en masse, these studies do show a concentration of face selectivity in two major regions of the temporal lobe, regions that correspond to the middle and anterior face patches described by

Tsao and colleagues using functional imaging (**Figure 8*d***).

The Functional Significance of the Anatomical Localization of Face Processing

The cerebral cortex is functionally parcellated: Neurons concerned with similar things are organized into areas and columns, each having extensive interconnections and common inputs and outputs (Mountcastle 1997). It is not surprising that face processing, being an important, identifiable and discrete form of object recognition, is also organized into anatomically discrete processing centers. Individual neurons connect with only a small fraction of the rest of the neurons in the brain, usually to nearby cells, because longer axons delay neural transmission, are energetically expensive, and take up space. Barlow (1986) has noted that facilitatory interactions within a functional area or column could underlie Gestalt linking processes—clustering cells concerned with color or motion might facilitate interactions between parts of the visual field having common color or motion. However, enriched local inhibitory interactions and sharpening of tuning might be an even more important function of colocalization because inhibitory neurons are always local, and long-range intracortical connections are invariably excitatory (Somogyi et al. 1998). Wang et al. (2000) recorded responses in anterior IT to a set of complex stimuli before, during, and after applying the GABA antagonist bicuculline near the recording electrode. In many cases, for both face-selective and nonface-selective cells, blocking local inhibition revealed responses to previously nonactivating stimuli, which were often activating stimuli for neighboring cells. This suggests that neighboring cells refine each other's response selectivity by mutual inhibition.

Time Course of Feature-Combination Responses

Although a large fraction of the information about which face stimulus was shown is carried by the earliest 50 ms of the response of face-selective cells (Tovee et al. 1993), several studies have shown that the information carried by the early part of the response is different from the information carried by later spikes. In particular, the earliest spikes in a response are sufficient for distinguishing faces from other object categories, but information about in-

dividual facial identity does not develop until ~50 ms later (Sugase et al. 1999, Tsao et al. 2006).

Similarly, responses in IT to nonface stimuli also become more selective, or sparser, over time (Tanaka et al. 1991, Tamura & Tanaka 2001). Similar temporal dynamics indicative of early detection activity followed by later individual identification activity have been observed for face-selective MEG responses in human occipitotemporal cortex (Liu et al. 2002). The observations that global information precedes finer information are consistent with a role for local inhibition in sharpening tuning within a local cluster of cells having similar response properties. Such response dynamics suggest a feedback or competitive process, whereby cells that respond best to a given stimulus inhibit nearby cells, resulting in a winner-take-all situation.

Norm-Based Coding

Recently an idea has emerged for both face processing and general object coding in the temporal lobe—that firing rate represents the magnitude of deviation from a template or norm for that property. Cells in V4 can be tuned to curvature, but the optimal values for curvature are most often found at either extreme or zero curvature, with few cells tuned to intermediate curvature (Pasupathy & Connor 2001). Kayaert and colleagues (2005a) found norm-based tuning for shapes in IT; neurons tuned to different shapes tended to show monotonic tuning, with maximum responses to extreme values of those shapes. Lastly, Leopold et al. (2006) recorded from face-responsive cells in anterior IT and found that most cells were tuned around an identity-ambiguous average human face, showing maximum firing to faces farthest from an average face (i.e., tuning was V-shaped around the average). Freiwald et al. (2005), on the other hand, reported that many cells in the macaque middle face patch showed monotonic tuning curves to different feature dimensions in a large cartoon face space, with the maximum response at one extreme and the minimum response at

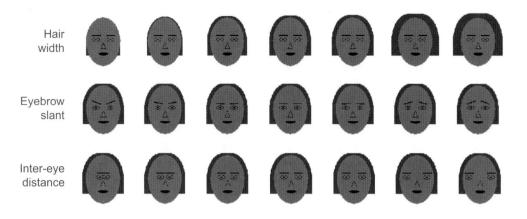

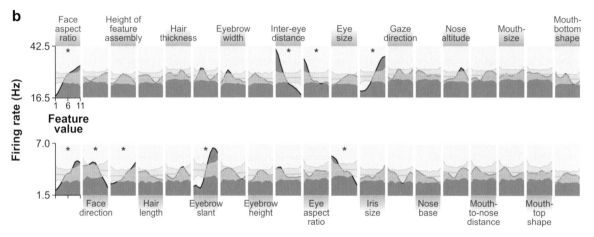

Figure 9

Tuning of face cells to a cartoon face space. (*a*) Three example dimensions of the 19-dimensional cartoon space. Each row shows example values for one parameter, with all other parameters fixed at their mean. (*b*) Tuning curves of two example cells to each of the 19 feature dimensions. Maximal, minimal, and mean values from shift predictor are shown in *gray*. Stars mark significant modulation. From Freiwald et al. 2005.

the opposite extreme (**Figure 9**). This ramp-shaped tuning is consistent with the model proposed by Rhodes et al. (2004) for explaining the face-adaptation effect (**Figure 5*b***)—that each face feature axis is coded by two opponent cell populations; thus the face norm would be implicitly represented as the virtual point of intersection between face cell populations with opponent ramp-shaped feature tuning curves. For both faces and nonface objects, many cells show tuning to several feature dimensions, and the tuning is separable, or independent, for the different tuning axes (Freiwald et al. 2005, Kayaert et al. 2005b).

Invariance

Face-selective cells in the temporal lobe are usually position and scale invariant in their ability to detect and distinguish faces, but they are seldom view and angle invariant (Desimone et al. 1984; Perrett et al. 1984, 1985, 1989, 1991; Rolls & Baylis 1986; Tanaka et al. 1991; Tovee et al. 1994; Tsao et al. 2006). The marked view selectivity of some IT cells may reflect a role in interpreting social gestures (who is looking at whom) (Argyle & Cook 1976, Bertrand 1969). De Souza et al. (2005) recently found a striking pattern of view selectivity in rostral versus caudal anterior STS. In caudal anterior STS, they found mirror-symmetric view-tuned cells, but in rostral anterior STS, view tuning was not mirror symmetric. Furthermore, view angle and gaze direction interacted, with neurons selective for a particular combination of face view and direction of gaze and often were strongly modulated by eye contact.

Recordings from the medial temporal lobe of human epilepsy patients have revealed the existence of cells that respond to familiar individuals in a highly invariant manner (Quiroga et al. 2005), as expected of a grandmother cell. For example, some cells responded to multiple pictures of a well-known individual as well as to a letter string of the person's name but were unresponsive to all other images. Such individual-specific cells have not been found in the lateral inferior temporal lobe, where most face cells in monkeys have been recorded, although as a population, cells in the anterior inferior temporal gyrus of the macaque can represent view-invariant identity (Eifuku et al. 2004).

Summary

The correlation between fMRI localization of face processing in macaques and the strong clustering of physiologically identified face-selective cells supports the idea of domain specificity, suggested by neurological findings and fMRI studies in humans. The strength and predominance of face selectivity within the middle face patch are not consistent with either the expertise hypothesis or the distributed coding model. The existence of neurons located at an early stage of form processing in the macaque brain that respond selectively to faces supports the idea that face processing begins with a detection stage, and the response properties of face cells indicate that this stage is highly nonlinear.

However, face cells seem to measure different face variables independently and linearly, so how does this reconcile with evidence that face perception in humans is holistic; i.e., how can we explain the composite effect and the part-whole effect neurally? We suggest that both these apparently nonlinear perceptual effects are consistent with a linear neural measuring stage if the preceding detection stage is holistic and nonlinear. One surprising result from physiological studies on face processing is the preponderance of view-selective units, but what role they play in face processing is still unclear.

FUTURE DIRECTIONS

1. Is face processing unique? We do not yet understand the details of how either faces or nonface objects are represented in the brain—perceptual studies have shown major differences in the ways that faces and objects are recognized, but there are nevertheless similarities in the response properties between face-selective cells and object-selective cells in IT. Both face- and object-selective cells in IT show tuning characteristics of a norm-based code. A variety of evidence suggest that our perception of faces is holistic, but processing of some nonface objects, like words, also shows important context effects. One fact is clear: The basic computational challenges to face processing are common to all object recognition (namely, detection, measurement, and classification). What is a face template in computational and neural terms, and how does it differ from a chair template? A truly satisfying answer to the question of whether face processing is unique will come only when we understand the precise neural mechanism underlying both face and nonface object recognition.

2. Is face processing modular? Perhaps the most striking result to come from the neurobiological research on face perception in the past decade is that specialized machinery is used for processing faces. Evidence reveals a fundamental specialization both at the gross anatomical level and at the level of single cells. It will be exciting to move forward along this pathway to understand how these face cells are used for different high-level percepts and behaviors; e.g., conveying invariant identity, expression, direction of attention, social dominance. But we believe that equally important new insights will come from looking back, asking how these cells acquire their face selectivity—undertaking a systematic study of the face-detection process.

3. What makes face processing special? We have proposed that what is special about face processing is that it is gated by an obligatory detection process. Such a design would be computationally elegant (by allowing for fast domain-specific filtering, segmentation, and alignment prior to fine-grained identification) and could explain the existence of face cells, face areas, prosopagnosia, and holistic processing. This detection-gating hypothesis naturally leads to the idea that there are two distinct classes of face cells: face-recognition cells, which encode different kinds of face templates, and face-detector cells, which (contrary to their name) could perform the triple function of detection, segmentation, and alignment. However, it is also possible that detection and discrimination are carried out by the same cells (either simultaneously or sequentially). Either way, we should at least be able to find out the answer. Because we know that face-selective cells are coding faces, we can distinguish detection-related activity from discrimination-related activity, which is impossible when one is studying a cell whose form specialization is unknown. Perhaps what is truly special about face processing is that it is now amenable to being understood. We have a beautiful hierarchy, a gift from nature, and we should exploit it in both directions.

DISCLOSURE STATEMENT

The authors are not aware of any biases that might be perceived as affecting the objectivity of this review.

LITERATURE CITED

Afraz SR, Cavanagh P. 2008. Retinotopy of the face aftereffect. *Vision Res.* 48:42–54

Afraz SR, Kiani R, Esteky H. 2006. Microstimulation of inferotemporal cortex influences face categorization. *Nature* 442:692–95

Andrews TJ, Ewbank MP. 2004. Distinct representations for facial identity and changeable aspects of faces in the human temporal lobe. *Neuroimage* 23:905–13

Andrews TJ, Schluppeck D, Homfray D, Matthews P, Blakemore C. 2002. Activity in the fusiform gyrus predicts conscious perception of Rubin's vase-face illusion. *Neuroimage* 17:890–901

Anstis S. 2005. Last but not least. *Perception* 34:237–40

Argyle M, Cook M. 1976. *Gaze and Mutual Gaze*. Cambridge, UK/New York: Cambridge Univ. Press. xi, 210 pp.

Baker CI, Behrmann M, Olson CR. 2002. Impact of learning on representation of parts and wholes in monkey inferotemporal cortex. *Nat. Neurosci.* 5:1210–16

Barlow HB. 1986. Why have multiple cortical areas? *Vision Res.* 26:81–90

Belhumeur PN, Hespanha JP, Kriegnam DJ. 1997. Eigenfaces vs Fisherfaces: recognition using class specific linear projection. *IEEE Trans. Patt. Anal. Mach. Intell.* 19:711–20

Bertrand M. 1969. *The Behavioural Repertoire of the Stumptail Macaque: A Descriptive and Comparative Study.* Basel, Switz.: Karger

Brincat SL, Connor CE. 2004. Underlying principles of visual shape selectivity in posterior inferotemporal cortex. *Nat. Neurosci.* 7:880–86

Bruce C, Desimone R, Gross CG. 1981. Visual properties of neurons in a polysensory area in superior temporal sulcus of the macaque. *J. Neurophysiol.* 46:369–84

Bruce V, Young A. 1986. Understanding face recognition. *Br. J. Psychol.* 77(Pt. 3):305–27

Calder AJ, Burton AM, Miller P, Young AW, Akamatsu S. 2001. A principal component analysis of facial expressions. *Vision Res.* 41:1179–208

Calder AJ, Young AW. 2005. Understanding the recognition of facial identity and facial expression. *Nat. Rev. Neurosci.* 6:641–51

Cohen L, Dehaene S. 2004. Specialization within the ventral stream: the case for the visual word form area. *Neuroimage* 22:466–76

Coltheart M, Curtis B, Atkins P, Heller M. 1993. Models of reading aloud: dual-route and parallel-distributed-processing approaches. *Psychol. Rev.* 100:589–608

Cottrell GW, Branson KM, Calder AJ. 2002. *Do expression and identity need separate representations?* Presented at Annu. Meet. Cogn. Sci. Soc., 24th, Fairfax, Va.

Desimone R, Albright TD, Gross CG, Bruce C. 1984. Stimulus-selective properties of inferior temporal neurons in the macaque. *J. Neurosci.* 4:2051–62

De Souza WC, Eifuku S, Tamura R, Nishijo H, Ono T. 2005. Differential characteristics of face neuron responses within the anterior superior temporal sulcus of macaques. *J. Neurophysiol.* 94:1252–66

Diamond R, Carey S. 1986. Why faces are and are not special: an effect of expertise. *J. Exp. Psychol. Gen.* 115:107–17

Eger E, Schyns PG, Kleinschmidt A. 2004. Scale invariant adaptation in fusiform face-responsive regions. *Neuroimage* 22:232–42

Eifuku S, De Souza WC, Tamura R, Nishijo H, Ono T. 2004. Neuronal correlates of face identification in the monkey anterior temporal cortical areas. *J. Neurophysiol.* 91:358–71

Farah MJ, Wilson KD, Drain HM, Tanaka JR. 1995. The inverted face inversion effect in prosopagnosia: evidence for mandatory, face-specific perceptual mechanisms. *Vision Res.* 35:2089–93

Farah MJ, Wilson KD, Drain M, Tanaka JN. 1998. What is "special" about face perception? *Psychol. Rev.* 105:482–98

Foldiak P, Xiao D, Keysers C, Edwards R, Perrett DI. 2004. Rapid serial visual presentation for the determination of neural selectivity in area STSa. *Prog. Brain Res.* 144:107–16

Freiwald WA, Tsao D, Tootell RB, Livingstone MS. 2005. Single-unit recording in an fMRI-identified macaque face patch. II. Coding along multiple feature axes. *Soc. Neurosci. Abstr.* 362.6

Fujita I, Tanaka K, Ito M, Cheng K. 1992. Columns for visual features of objects in monkey inferotemporal cortex. *Nature* 360:343–46

Gauthier I, Tarr MJ. 2002. Unraveling mechanisms for expert object recognition: bridging brain activity and behavior. *J. Exp. Psychol. Hum. Percept. Perform.* 28:431–46

Grill Spector K, Knouf N, Kanwisher N. 2004. The fusiform face area subserves face perception, not generic within-category identification. *Nat. Neurosci.* 7:555–62

Grill-Spector K, Kushnir T, Edelman S, Avidan G, Itzchak Y, Malach R. 1999. Differential processing of objects under various viewing conditions in the human lateral occipital complex. *Neuron* 24:187–203

Grill-Spector K, Malach R. 2004. The human visual cortex. *Annu. Rev. Neurosci.* 27:649–77

Gross CG, Bender DB, Rocha-Miranda CE. 1969. Visual receptive fields of neurons in inferotemporal cortex of the monkey. *Science* 166:1303–6

Gross CG, Rocha-Miranda CE, Bender DB. 1972. Visual properties of neurons in inferotemporal cortex of the macaque. *J. Neurophysiol.* 35:96–111

Hasson U, Hendler T, Ben Bashat D, Malach R. 2001. Vase or face? A neural correlate of shape-selective grouping processes in the human brain. *J. Cogn. Neurosci.* 13:744–53

Hasson U, Levy I, Behrmann M, Hendler T, Malach R. 2002. Eccentricity bias as an organizing principle for human high-order object areas. *Neuron* 34:479–90

Haxby JV, Gobbini MI, Furey ML, Ishai A, Schouten JL, Pietrini P. 2001. Distributed and overlapping representations of faces and objects in ventral temporal cortex. *Science* 293:2425–30

Haxby JV, Grady CL, Horwitz B, Ungerleider LG, Mishkin M, et al. 1991. Dissociation of object and spatial visual processing pathways in human extrastriate cortex. *Proc. Natl. Acad. Sci. USA* 88:1621–25

Haxby JV, Hoffman EA, Gobbini MI. 2000. The distributed human neural system for face perception. *Trends Cogn. Sci.* 4:223–33

Jebara T. 2003. *Images as bags of pixels*. Presented at IEEE Int. Conf. Comp. Vis. (ICCV'03), 9th, Nice, France

Jeffery L, Rhodes G, Busey T. 2006. View-specific coding of face shape. *Psychol. Sci.* 17:501–5

Johnson MH, Dziurawiec S, Ellis H, Morton J. 1991. Newborns' preferential tracking of face-like stimuli and its subsequent decline. *Cognition* 40:1–19

Kanwisher N, Tong F, Nakayama K. 1998. The effect of face inversion on the human fusiform face area. *Cognition* 68:B1–11

Kanwisher NG, McDermott J, Chun MM. 1997. The fusiform face area: a module in human extrastriate cortex specialized for face perception. *J. Neurosci.* 17:4302–11

Kayaert G, Biederman I, Op de Beeck H, Vogels R. 2005a. Tuning for shape dimensions in macaque inferior temporal cortex. *Eur. J. Neurosci.* 22:212–24

Kayaert G, Biederman I, Vogels R. 2005b. Representation of regular and irregular shapes in macaque inferotemporal cortex. *Cereb. Cortex* 15:1308–21

Kemp R, McManus C, Pigott T. 1990. Sensitivity to the displacement of facial features in negative and inverted images. *Perception* 19:531–43

Kobatake E, Tanaka K. 1994. Neuronal selectivities to complex object features in the ventral visual pathway of the macaque cerebral cortex. *J. Neurophysiol.* 71:856–67

Konorski J. 1967. *Integrative Activity of the Brain: An Interdisciplinary Approach*. Chicago: Univ. Chicago Press. xii, 531 pp.

Lee K, Byatt G, Rhodes G. 2000. Caricature effects, distinctiveness, and identification: testing the face-space framework. *Psychol. Sci.* 11:379–85

Leonard CM, Rolls ET, Wilson FA, Baylis GC. 1985. Neurons in the amygdala of the monkey with responses selective for faces. *Behav. Brain Res.* 15:159–76

Leopold DA, Bondar IV, Giese MA. 2006. Norm-based face encoding by single neurons in the monkey inferotemporal cortex. *Nature* 442:572–75

Leopold DA, O'Toole AJ, Vetter T, Blanz V. 2001. Prototype-referenced shape encoding revealed by high-level aftereffects. *Nat. Neurosci.* 4:89–94

Liu J, Harris A, Kanwisher N. 2002. Stages of processing in face perception: an MEG study. *Nat. Neurosci.* 5:910–16

McCarthy G, Luby M, Gore J, Goldman-Rakic P. 1997. Infrequent events transiently activate human prefrontal and parietal cortex as measured by functional MRI. *J. Neurophysiol.* 77:1630–34

Moghaddam B, Jebara T, Pentland A. 2000. Bayesian face recognition. *Pattern Recognit.* 33:1771–82

Mooney CM. 1957. Age in the development of closure ability in children. *Can. J. Psychol.* 11:219–26

Moscovitch M, Winocur G, Behrmann M. 1997. What is special about face recognition? Nineteen experiments on a person with visual object agnosia and dyslexia but normal face recognition. *J. Cogn. Neurosci.* 9:555–604

Mountcastle VB. 1997. The columnar organization of the neocortex. *Brain* 120(Pt. 4):701–22

Oram MW, Perrett DI. 1992. Time course of neural responses discriminating different views of the face and head. *J. Neurophysiol.* 68:70–84

O'Toole A, Abdi H, Deffenbacher K, Valentin D. 1993. Low dimensional representation of faces in high dimensions of the space. *J. Opt. Soc. Am. A* 10:405–10

Pasupathy A, Connor CE. 2001. Shape representation in area V4: position-specific tuning for boundary conformation. *J. Neurophysiol.* 86:2505–19

Perrett DI, Harries MH, Bevan R, Thomas S, Benson PJ, et al. 1989. Frameworks of analysis for the neural representation of animate objects and actions. *J. Exp. Biol.* 146:87–113

Perrett DI, Oram MW, Harries MH, Bevan R, Hietanen JK, et al. 1991. Viewer-centred and object-centred coding of heads in the macaque temporal cortex. *Exp. Brain Res.* 86:159–73

Perrett DI, Rolls ET, Caan W. 1982. Visual neurones responsive to faces in the monkey temporal cortex. *Exp. Brain Res.* 47:329–42

Perrett DI, Smith PA, Potter DD, Mistlin AJ, Head AS, et al. 1984. Neurones responsive to faces in the temporal cortex: studies of functional organization, sensitivity to identity and relation to perception. *Hum. Neurobiol.* 3:197–208

Perrett DI, Smith PA, Potter DD, Mistlin AJ, Head AS, et al. 1985. Visual cells in the temporal cortex sensitive to face view and gaze direction. *Proc. R. Soc. London B Biol. Sci.* 223:293–317

Pinsk MA, Desimone K, Moore T, Gross CG, Kastner S. 2005. Representations of faces and body parts in macaque temporal cortex: a functional MRI study. *Proc. Natl. Acad. Sci. USA* 102:6996–7001

Pourtois G, Schwartz S, Seghier ML, Lazeyras F, Vuilleumier P. 2005. View-independent coding of face identity in frontal and temporal cortices is modulated by familarity: an event-related fMRI study. *Neuroimage* 24:1214–24

Puce A, Allison T, Asgari M, Gore JC, McCarthy G. 1996. Differential sensitivity of human visual cortex to faces, letterstrings, and textures: a functional magnetic resonance imaging study. *J. Neurosci.* 16:5205–15

Quiroga RQ, Reddy L, Kreiman G, Koch C, Fried I. 2005. Invariant visual representation by single neurons in the human brain. *Nature* 435:1102–7

Rhodes G, Brennan S, Carey S. 1987. Identification and ratings of caricatures: implications for mental representations of faces. *Cogn. Psychol.* 19:473–97

Rhodes G, Jeffery L. 2006. Adaptive norm-based coding of facial identity. *Vision Res.* 46:2977–87

Rhodes G, Jeffery L, Watson TL, Jaquet E, Winkler C, Clifford CW. 2004. Orientation-contingent face aftereffects and implications for face-coding mechanisms. *Curr. Biol.* 14:2119–23

Rhodes G, McLean IG. 1990. Distinctiveness and expertise effects with homogeneous stimuli: towards a model of configural coding. *Perception* 19:773–94

Riesenhuber M, Poggio T. 2000. Models of object recognition. *Nat. Neurosci.* 3(Suppl.):1199–204

Rolls ET, Baylis GC. 1986. Size and contrast have only small effects on the responses to faces of neurons in the cortex of the superior temporal sulcus of the monkey. *Exp. Brain Res.* 65:38–48

Rotshtein P, Henson RN, Treves A, Driver J, Dolan RJ. 2005. Morphing Marilyn into Maggie dissociates physical and identity face representations in the brain. *Nat. Neurosci.* 8:107–13

Scalaidhe SP, Wilson FA, Goldman-Rakic PS. 1999. Face-selective neurons during passive viewing and working memory performance of rhesus monkeys: evidence for intrinsic specialization of neuronal coding. *Cereb. Cortex* 9:459–75

Sergent J, Ohta S, MacDonald B. 1992. Functional neuroanatomy of face and object processing. A positron emission tomography study. *Brain* 115:15–36

Shakhnarovich G, Moghaddam B. 2004. Face recognition in subspaces. In *Handbook of Face Recognition*, ed. SZ Li, AK Jain. Berlin: Springer-Verlag

Simion F, Valenza E, Umilta C, Dalla Barba B. 1998. Preferential orienting to faces in newborns: a temporal-nasal asymmetry. *J. Exp. Psychol. Hum. Percept. Perform.* 24:1399–405

Sinha P. 2002a. Qualitative representations for recognition. In *Lecture Notes in Computer Science*, pp. 249–62. Berlin: Springer-Verlag

Sinha P. 2002b. Recognizing complex patterns. *Nat. Neurosci.* 5(Suppl.):1093–97

Sinha P, Balas BJ, Ostrovsky Y, Russell R. 2006. Face recognition by humans: nineteen results all computer vision researchers should know about. *Proc. IEEE* 94(11):1948–62

Sinha P, Poggio T. 1996. I think I know that face. *Nature* 384:404

Somogyi P, Tamas G, Lujan R, Buhl EH. 1998. Salient features of synaptic organisation in the cerebral cortex. *Brain Res. Brain Res. Rev.* 26:113–35

Spiridon M, Kanwisher N. 2002. How distributed is visual category information in human occipito-temporal cortex? An fMRI study. *Neuron* 35:1157–65

Sugase Y, Yamane S, Ueno S, Kawano K. 1999. Global and fine information coded by single neurons in the temporal visual cortex. *Nature* 400:869–73

Suzuki S, Cavanagh P. 1998. A shape-contrast effect for briefly presented stimuli. *J. Exp. Psychol. Hum. Percept. Perform.* 24:1315–41

Tamura H, Tanaka K. 2001. Visual response properties of cells in the ventral and dorsal parts of the macaque inferotemporal cortex. *Cereb. Cortex* 11:384–99

Tanaka JW, Farah MJ. 1993. Parts and wholes in face recognition. *Q. J. Exp. Psychol. A Hum. Exp. Psychol.* 46A:225–45

Tanaka K, Saito H, Fukada Y, Moriya M. 1991. Coding visual images of objects in the inferotemporal cortex of the macaque monkey. *J. Neurophysiol.* 66:170–89

Tarr MJ, Gauthier I. 2000. FFA: a flexible fusiform area for subordinate-level visual processing automatized by expertise. *Nat. Neurosci.* 3:764–69

Thompson P. 1980. Margaret Thatcher: a new illusion. *Perception* 9:483–84

Tong F, Nakayama K, Moscovitch M, Weinrib O, Kanwisher N. 2000. Response properties of the human fusiform face area. *Cogn. Neuropsychol.* 17:257–79

Tovee MJ, Rolls ET, Azzopardi P. 1994. Translation invariance in the responses to faces of single neurons in the temporal visual cortical areas of the alert macaque. *J. Neurophysiol.* 72:1049–60

Tovee MJ, Rolls ET, Treves A, Bellis RP. 1993. Information encoding and the responses of single neurons in the primate temporal visual cortex. *J. Neurophysiol.* 70:640–54

Tsao DY, Freiwald WA, Knutsen TA, Mandeville JB, Tootell RB. 2003. Faces and objects in macaque cerebral cortex. *Nat. Neurosci.* 6:989–95

Tsao DY, Freiwald WA, Tootell RB, Livingstone MS. 2006. A cortical region consisting entirely of face-selective cells. *Science* 311:670–74

Turk M, Pentland A. 1991. Eigenfaces for recognition. *J. Cogn. Neurosci.* 3:71–86

Ullman S. 2007. Object recognition and segmentation by a fragment-based hierarchy. *Trends Cogn. Sci.* 11:58–64

Ullman S, Vidal-Naquet M, Sali E. 2002. Visual features of intermediate complexity and their use in classification. *Nat. Neurosci.* 5:682–87

Viola P, Jones M. 2004. Robust real-time face detection. *Int. J. Comp. Vision* 57:137–54

Wang G, Tanaka K, Tanifuji M. 1996. Optical imaging of functional organization in the monkey inferotemporal cortex. *Science* 272:1665–68

Wang Y, Fujita I, Murayama Y. 2000. Neuronal mechanisms of selectivity for object features revealed by blocking inhibition in inferotemporal cortex. *Nat. Neurosci.* 3:807–13

Webster MA, MacLin OH. 1999. Figural aftereffects in the perception of faces. *Psychon. Bull. Rev.* 6:647–53

Wolberg G. 1998. Image morphing: a survey. *Vis. Comput.* 14:360–72

Yin R. 1969. Looking at upside-down faces. *J. Exp. Psychol.* 81:141–45

Young AW, Hellawell D, Hay DC. 1987. Configurational information in face perception. *Perception* 16:747–59

Yue X, Tjan BS, Biederman I. 2006. What makes faces special? *Vision Res.* 46:3802–11

Zhao W, Chellappa R, Phillips PJ, Rosenfeld A. 2003. Face recognition: a literature survey. *ACM Comput. Surv.* 35:399–458

The Prion's Elusive Reason for Being

Adriano Aguzzi, Frank Baumann, and Juliane Bremer

Institute of Neuropathology, University of Zurich, CH-8091 Zurich, Switzerland, email: Adriano.Aguzzi@usz.ch

Annu. Rev. Neurosci. 2008. 31:439–77

First published online as a Review in Advance on April 2, 2008

The *Annual Review of Neuroscience* is online at neuro.annualreviews.org

This article's doi: 10.1146/annurev.neuro.31.060407.125620

Key Words

transmissible spongiform encephalopathy, prion diseases, PrP^C, PrP^{Sc}

Abstract

The protein-only hypothesis posits that the infectious agent causing transmissible spongiform encephalopathies consists of protein and lacks any informational nucleic acids. This agent, termed prion by Stanley Prusiner, is thought to consist partly of PrP^{Sc}, a conformational isoform of a normal cellular protein termed PrP^C. Scientists and lay persons have been fascinated by the prion concept, and it has been subjected to passionate critique and intense experimental scrutiny. As a result, PrP^C and its isoforms rank among the most intensively studied proteins encoded by the mammalian genome. Despite all this research, both the physiological function of PrP^C and the molecular pathways leading to neurodegeneration in prion disease remain unknown. Here we review the salient traits of those diseases ascribed to improper behavior of the prion protein and highlight how the physiological functions of PrP^C may help explain the toxic phenotypes observed in prion disease.

Contents

TRANSMISSIBLE PRION DISEASES AND NONTRANSMISSIBLE PRION-RELATED DISEASES

The origin of the word "prion" stems from the anagram of "proteinaceous infectious particle." Naturally, the latter is by no means a qualifying attribute because all conventional infectious agents—including all viruses and bacteria—are proteinaceous to some degree. What sets prions apart, as proposed by Prusiner, is that the actual infectious principle consists merely of protein and is capable of replicating and transmitting infections without the need for informational nucleic acids. This postulate counters much of the established molecular biological evidence, which predicates that nucleic acids are the basis for self-replicating biological information in all living beings, including even the most elementary infectious particles.

Prion diseases are generally characterized by widespread neurodegeneration and therefore exhibit clinical signs and symptoms of cognitive and motor dysfunction, in addition to propagating infectious prions and, in many instances, forming striking amyloid plaques. The latter plaques contain aggregates of PrP^{Sc}, a misfolded and beta-sheet-rich isoform of the protein PrP^C encoded by the *PRNP* gene. Further neuropathological features are neuronal loss, astrocytic activation (gliosis), and spongiform change. All prion diseases are progressive, fatal, and presently incurable.

Although the normal cellular prion protein PrP^C can easily be digested with proteinase K (PK), the beta-sheet-rich, misfolded form PrP^{Sc} is partially proteinase K (PK) resistant. A crucial piece of evidence demonstrating that PrP^C is a key player in prion disease came from experiments showing that mice lacking the prion protein gene are resistant to prions (Bueler et al. 1993).

Although the formation of PrP^{Sc} accompanies neurodegeneration in prion disease, many lines of evidence indicate that PrP^{Sc} is not intrinsically neurotoxic. PrP^C needs to be presented by host neurons for neurodegeneration to occur. Thus, when neurografts propagating PrP^{Sc} were implanted into *Prnp*$^{o/o}$ mice, the host mice did not develop prion disease (Brandner et al. 1996). Additionally, transgenic mice expressing only a secreted form of PrP^C, lacking its membrane attachment via glycosylphosphatidylinositol (GPI) anchor, have been reported to be refractory to develop clinical signs of prion diseases, although prion inoculation induces PrP^{Sc} formation and aggregation of amyloid plaques (Chesebro et al. 2005). This finding indicates that membrane attachment of PrP^C is a prerequisite for neurodegeneration to occur and that the presence of PrP^{Sc} alone does not cause disease. The importance of neuronal expression of PrP^C for prion disease development has been corroborated by the phenotype of mice with neuron-specific ablation of PrP^C eight weeks after prion

GPI: glycosylphosphatidylinositol

inoculation. Early spongiform changes were reversed, and clinical disease was prevented. This reversal occurred despite the accumulation of extraneuronal PrP^{Sc} (Mallucci et al. 2003).

Bona fide prion diseases are characterized by their transmissibility and are therefore also termed transmissible spongiform encephalopathies (TSE). Transmissibility is a defining, and hence indispensable, trait of all prion diseases. However, transmissibility has not been formally proven for all kinds of diseases thought to be caused by prions. In addition, some diseases are genetically associated with the prion protein, yet they are nontransmissible. These diseases are sometimes called prionopathies. Among these are rare genetic syndromes that cosegregate with point mutations in the open reading frame of the *PRNP* gene. In addition to these naturally occurring prionopathies, several transgenic mice have been used to gain insight into functional domains of PrP^C (Weissmann & Flechsig 2003). In these mice, deletion of parts of PrP^C caused prionopathies, characterized by a shortened life span and the development of white matter disease in the central nervous system (CNS) as well as neuronal cell death in the cerebellum (Baumann et al. 2007, Li et al. 2007, Shmerling et al. 1998). Overexpression of wild-type PrP^C also caused disease in transgenic mice (Westaway et al. 1994).

Prion Disease in Humans and Animals

Prion diseases have occurred in humans and animals for many years. A disease similar to scrapie was recorded in the mid eighteenth century, and scholars heavily debated its origin. A crucial experiment showing incontrovertible transmissibility of scrapie to goats was performed by Cuille & Chellè in the 1930s (Cuille & Chellè 1939). The first cases of human prion disease, Creutzfeldt-Jakob disease (CJD), were reported in the 1920s (Creutzfeldt 1920, Jakob 1921). The number of human and animal diseases recognized as TSEs has increased steadily and now includes Gerstmann-Sträussler-Scheinker syndrome (GSS), fatal familial insomnia (FFI),

and Kuru in humans; bovine spongiform encephalopathy (BSE) in cattle; chronic wasting disease (CWD) in deer and elk; and transmissible mink encephalopathy. BSE has been inadvertently transmitted to a variety of captive animals, causing feline spongiform encephalopathy (FSE) and a plethora of diseases in zoo animals including kudus, nyalas, and greater cats, for example.

Creutzfeldt-Jakob disease. CJD was initially described as a sporadic disease occurring for no known cause (sCJD). The incidence of CJD is low in all ethnicities and typically affects ~1 person in one million each year. Very rapid cognitive decline, causing dementia, is the main symptom. Cerebellar symptoms including ataxia and myoclonus are also frequent presenting symptoms. Death often occurs within few weeks of the first signs of disease, and a fulminant, "apoplectiform" course of disease has been documented in the past. Somatic mutations in the *PRNP* gene analogous to those in the germline of genetic CJD patients (see below) have been hypothesized to underlie sporadic CJD. Alternatively, Aguzzi & Glatzel (2006) suggested that some cases of alleged sCJD derive from heretofore unrecognized infections. Finally, PrP^C may possess a finite, albeit extremely low, propensity to self-assemble into ordered aggregates of PrP^{Sc}, thereby stochastically initiating prion replication and, ultimately, a sporadic form of disease. The latter scenario could be regarded as the bad-luck hypothesis. However, none of this has been proven, and therefore the cause of sCJD is still unknown.

Variant Creutzfeldt-Jakob disease and Bovine Spongiform Encephalopathy. Public understanding of prion disease remained limited for a long time: For example, we have heard neurologists saying that CJD is an essentially nonexistent disease. However, this mindset changed completely when BSE was first reported in the early 1980s (Wells et al. 1987). In the following years and until mid 2007, BSE affected ~190,000 cows

TSE: transmissible spongiform encephalopathies

CJD: Creutzfeldt-Jakob disease

BSE: bovine spongiform encephalopathy

CWD: chronic wasting disease

FSE: feline spongiform encephalopathy

sCJD: sporadic CJD

vCJD: variant CJD

(http://www.oie.int/). Some investigators suggested that BSE could cause a new variant form of CJD (vCJD) in humans. A direct experimental proof that vCJD represents transmission of BSE prions to humans cannot be produced. However, epidemiological, biochemical, neuropathological evidence and transmission studies strongly suggest that BSE has transmitted to humans in the form of vCJD (Aguzzi 1996, Aguzzi & Weissmann 1996, Bruce et al. 1997, Hill et al. 1997). The incidence of vCJD has been rising between 1994, when the first patients suffering from vCJD presented with their initial symptoms, and 2001, raising fears that a very large epidemic may be looming. At the time of this writing, vCJD has killed ∼200 individual victims worldwide (http://www.cjd.ed.ac.uk/). Most of the affected individuals lived in United Kingdom and France. Fortunately, in the United Kingdom the incidence appears to be decreasing from the year 2001 to 6 diagnosed cases yearly in 2005 and 2006. In contrast, in France the number of probable and definite cases of vCJD increased from 0 to 3 diagnosed cases per year in 1996–2004 to 6 per year in 2005 and 2006. In 2007, the number of cases was back to 3 again. (http://www.invs.sante.fr/publications/mcj/donnees_mcj.html). A 30+-year mean incubation time of BSE/vCJD in humans is not entirely implausible, and therefore some authors have predicted a multiphasic human BSE endemic with a second increase in the incidence of vCJD affecting people heterozygous at codon 129 (Collinge et al. 2006). Others, these authors included, regard the incidence of vCJD as subsiding (Andrews et al. 2003) (**Figure 1**).

It is important to note, however, that the above considerations apply primarily to the epidemiology of primary transmission from cows to humans. Although, by now a pool of preclinically infected humans may have been built. Human-to-human transmission may present with characteristics very different from those of primary cow-to-human transmission, including enhanced virulence, shortened incubation times, disrespect of allelic *PRNP* poly-

morphisms (129MM, MV or VV), and heterodox modes of infection including blood-borne transmission. If we account for the time it will take to eradicate these secondary transmissions in the population, vCJD is not likely to disappear entirely in the coming four decades.

Iatrogenic CJD. Iatrogenic CJD is accidentally transmitted during the course of medical or surgical procedures. The first documented case of iatrogenic prion transmission occurred in 1974 and was caused by corneal transplantation of a graft derived from a patient suffering from sCJD (Duffy et al. 1974). Iatrogenic CJD is also rare, most often observed in individuals that have received cadaveric dura mater implants and human growth hormone; some of these individuals received gonadotrophin extracted from human pituitary glands or had stereotactically placed electrodes in their brains (Will 2003). Four cases of vCJD transmission by blood transfusions have been reported recently in the United Kingdom (Llewelyn et al. 2004, Peden et al. 2004, Wroe et al. 2006) (see also http://www.cjd.ed.ac.uk/TMER/TMER.htm). The fact that preclinically infected individuals can transmit vCJD underscores the important medical need for sensitive diagnostic tools, which could be used for screening blood units prior to transfusion, for example.

Kuru. In the mid 1950s, when the remote parts of Papua New Guinea were first explored by Australians and Westerners, Kuru was first described in research (Gajdusek & Zigas 1957). Kuru was, at that time and at least since 1941, an endemic disease among some tribes of New Guinea aborigines, especially among the Fore linguistic group and neighboring tribes (Gajdusek & Reid 1961). *Kuru* in the Fore language means "to shiver," and along with other signs of cerebellar ataxia, shivering is a hallmark of the disease. The ritual consumption of dead relatives as a symbol of respect and mourning is the attributed route of transmission. As a consequence, the incidence has steadily fallen after cessation of cannibalism in Papua New

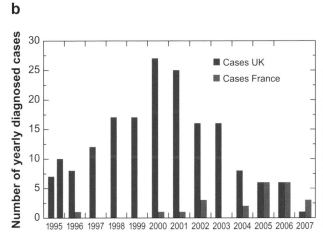

a

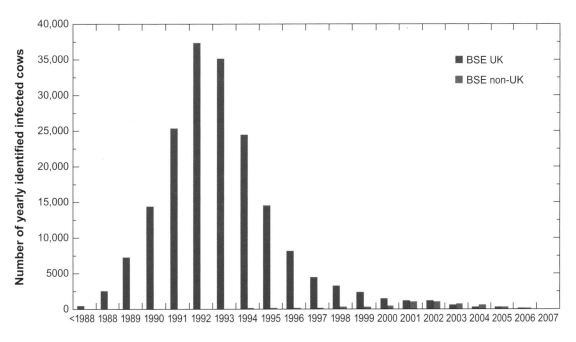

b

	vCJD cases (dead)	vCJD cases (still alive)	Secondary vCJD (blood transfusion)
UK	160	3	4
France	21	2	–
Republic of Ireland	4	0	–
Italy	1	–	–
USA	3	–	–
Canada	1	–	–
Saudi Arabia	–	1	–
Japan	1	–	–
Netherlands	2	–	–
Portugal	1	1	–
Spain	2	–	–

Figure 1

BSE and vCJD cases reported worldwide. (*a*) Reported cases of bovine spongiform encephalopathy (BSE) in the United Kingdom (UK) (*blue*), and in countries excluding the UK (*red*). Non-UK BSE cases include cases from countries both within and outside of the European Union (EU). Data are as of December 2006 (**http://www.oie.int**). (*b*) Reported cases of variant Creutzfeldt-Jakob disease (vCJD) in the UK (*blue*) and in countries outside the UK (*red*). Non-UK vCJD cases include those reported in France, Republic of Ireland, Italy, United States, Canada, Saudi Arabia, Japan, the Netherlands, Portugal, and Spain. Data are as of February 2008 and include cases of vCJD in patients who resided in the UK in the 1980s or 1990s [see the National Creutzfeldt-Jakob Disease Surveillance Unit Web site for vCJD data to July 2007 (**http://www.cjd.ed.ac.uk/**)].

gCJD: genetic CJD

OR: octarepeat region

Guinea (Collinge et al. 2006). In a concise and extremely clairvoyant observation published in 1959, Bill Hadlow noted the epidemiological, clinical, and neuropathological similarities between Kuru and scrapie (Hadlow 1959). These were taken up by Carleton Gajdusek who, in 1966, succeeded in transmitting Kuru to three chimpanzees (Gajdusek et al. 1966). Soon thereafter, serial passage of Kuru and of several other prion diseases was demonstrated in chimpanzees and other primates (Gajdusek et al. 1967, 1968). Investigators have since transmitted human prion disease to various species including laboratory rodents.

Genetic CJD and Gerstmann-Sträussler-Scheinker syndrome. Several mutations in the prion protein gene (*PRNP*) have been found in families with hereditary or genetic CJD (gCJD). **Figure 2** summarizes known mutations causing human TSEs. gCJD occurs with point mutations mostly affecting the region between the second and the third helix of the carboxy-terminus. However, insertions in the octarepeat region (OR) in the amino-terminus, and even one instance of a premature termination codon at position 145, have also been associated with human prion disease. The inheritance was, in all cases, autosomal dominant, often with very high penetrance. The clinicopathological disease phenotype varies depending on the actual mutation, as well as on polymorphisms at codon 129, and most likely on a plethora of yet unidentified modifiers and cofactors (Kovacs et al. 2002).

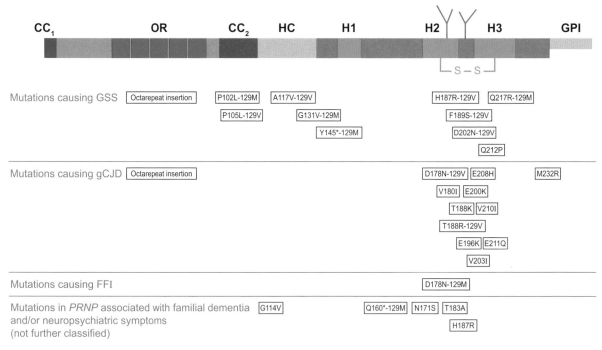

Figure 2

The human PrP^C protein and its mutants. The mature human PrP^C protein contains 208 amino acid residues. It features two positively charged amino acid clusters denoted CC$_1$ and CC$_2$ (*blue boxes*), an octapeptide repeat region (OR) (*green boxes*), a hydrophobic core (HC) (*gray box*), three α-helixes (H1-H3) (*red boxes*), one disulphide bond (S–S) between cysteine residues 179 and 214, and two potential sites for N-linked glycosylation (*red forks*) at residues 181 and 197. A glycosylphosphatidylinositol anchor (GPI) (*yellow box*) is attached to the C-terminus of PrP. This figure indicates in black framed boxes point mutations and insertions found in the human *PRNP* gene in patients with prion disease. The associated polymorphisms of codon 129 (methionine M or valine V) are indicated. Amino acids are given in single-letter code. The asterisk indicates a stop codon; therefore, this mutation results in a truncated protein.

The first descriptions of Gerstmann-Sträussler-Scheinker syndrome (GSS) originate from 1928 and 1936 in an Austrian family (Gerstmann 1928, Gerstmann et al. 1936). In the following years, analogous disorders have been described, but its classification as a TSE lagged until 1981, when Masters and colleagues (1981) reported that inoculation of brain tissue from three patients with GSS resulted in spongiform encephalopathy in nonhuman primates. The authors also defined clinical hallmarks of GSS (earlier age at onset, longer disease duration, and prominent cerebellar ataxia) differentiating the disease from CJD. Nowadays GSS is considered an autosomal-dominantly inherited TSE caused by mutations in the prion protein open reading frame, manifesting typically with progressive cerebellar ataxia or spastic paraparesis and cognitive decline. The known GSS-causing mutations are summarized in **Figure 2**. In addition to the regions affected in gCJD, mutations altering the sequence of the central domain can cause GSS. Its distinctive neuropathological feature is the presence of widespread large and multicentric amyloid plaques (Collins et al. 2001).

GSS is generally transmissible (Hsiao et al. 1989, Masters et al. 1981, Tateishi et al. 1988); therefore, its classification as a TSE is widely accepted. However, the overall experimental transmissibility of GSS to nonhuman primates and rodents is low. Only for the most common GSS-associated mutations (P102L), and only in approximately one third of the cases, were brain homogenates derived from patients reproducibly capable of inducing disease upon transmission (Tateishi et al. 1996a). The less frequent mutations causing GSS often failed to induce disease after experimental transmission to nonhuman primates and rodents, and in many cases transmissibility was never assessed (Brown et al. 1994, Tateishi et al. 1996a).

Fatal familial insomnia. Fatal familial insomnia (FFI) is the descriptive name given to a disease identified in 1986. Five members of an Italian family presented with insomnia and dysautonomia (Lugaresi et al. 1986). In 1992, the disease-causing mutation in the prion protein gene (D178N) was identified, thereby allowing the classification of FFI as a genetically determined prion disease (Medori et al. 1992). The final proof that FFI is a TSE was achieved when FFI was successfully transmitted to mice (Tateishi et al. 1995). FFI typically affects the thalamus, and accordingly, the core clinical features are disruption of the normal sleep-wake cycle, sympathetic overactivity, endocrine abnormalities, and impaired attention (Collins et al. 2001). In addition to the pathogenic point mutation D178N, the methionine-valine polymorphism at codon 129 of the *PRNP* gene controls the disease phenotype. Whereas D178N-129MM (homozygosity for methionine at codon 129) was associated with FFI, heterozygosity at codon 129 (D178N-129MV) segregated with the familial CJD subtype (Goldfarb et al. 1992). However, Zarranz et al. (2005) reported more recently that this genotype-phenotype association is not absolute. In one study, several patients have been identified with a CJD phenotype and a D178N-129MM genotype. The authors concluded that rather than being separate disease entities, prion disease phenotypes such as FFI and CJD represent two extreme manifestations of a continuous disease spectrum (Zarranz et al. 2005).

In addition to the familial form of fatal insomnia, a sporadic form of the disease, termed sporadic fatal insomnia, was described. Sporadic FFI is not associated with mutations in the *PRNP* gene (Mastrianni et al. 1999, Parchi et al. 1999).

The Nature of the Prion

Although formulated a century ago, Koch's postulates remain the bedrock of microbiology. According to Koch, three conditions must be met to identify a microbe as the causative agent of any given infection: (*a*) The microorganism must be detectable in all diseased tissues, (*b*) its isolation and growth must be achieved in pure culture, and (*c*) the culture-derived microorganisms must be able to induce disease after experimental infection of a subject, from which

GSS: Gerstmann-Sträussler-Scheinker syndrome

FFI: fatal familial insomnia

PMCA: protein misfolding cyclic amplification

a further round of reisolation of the microorganism should be possible. Although Koch's work was performed long before contemporary molecular biology, his postulates continue to serve remarkably well in defining conventional viral and bacterial agents.

However, as prions are thought to be infectious proteins that amplify in a self-catalytic misfolding process, their microbiological culture sensu strictiori is not possible. Therefore whether Koch's postulates can be meaningfully applied to prion disease is questionable. Furthermore, Koch's postulates account for the influence of host susceptibility, which is of utmost importance in prion disease. Prion disease development depends on the presence of PrP^C on host cells, and the species-specific amino acid sequence and polymorphism of codon 129 are important. Alternate postulates for infectious proteinaceous agents have recently been suggested (Walker et al. 2006), but it remains to be seen whether they will garner universal acceptance.

In the prion field, researchers generally accept that a reasonable surrogate for Koch's second postulate be fulfilled by the generation of synthetic prions in vitro, i.e., the recovery of perpetually transmissible infectivity from prion protein produced recombinantly or chemically from defined constituents. Major progress toward this end has been made in recent days. Purified PrP^{Sc} was used to generate PK-resistant PrP (PrP^{res}) in a cell-free system that could even reflect two typical features of prions: species barrier and strain specificity (Bessen et al. 1995, Kocisko et al. 1995).

Another approach used a method called PMCA (protein misfolding cyclic amplification), in which PrP^{res} can be amplified by incubating and sonicating PrP^{res}-containing brain homogenate diluted in normal brain homogenate. Soto and coworkers amplified PrP^{res} derived from scrapie-infected hamsters indefinitely by using PMCA in serial dilutions. Amplification of PrP^{res} was accompanied by amplification of infectivity (Castilla et al. 2005a). Certainly PMCA is a very sensitive method to detect PrP^{Sc} even in complex samples such as

blood and already in a presymptomatic disease state (Castilla et al. 2005b, Saa et al. 2006). The use of purified PrP^C instead of brain homogenate as a substrate decreased the efficiency of amplification, suggesting that additional cofactors may facilitate misfolding (Deleault et al. 2005). For a long time, all attempts to use recombinant PrP as a substrate for PMCA failed. However, Caughey and coworkers have now succeeded in carrying out PMCA using bacterially expressed hamster PrP as a substrate. While this represents a major advance in many ways, the sensitivity was not quite as high as that of the original PMCA (Aguzzi 2007, Atarashi et al. 2007).

Infectivity may not have been generated de novo in PMCA in these studies. Instead, prion-infected brain could have been inadvertently added in the beginning. In an fascinating study, Supattapone and coworkers identified the minimal components (PrP^C, copurified lipids, and single-stranded polyanionic molecules) required for amplification of PK-resistant PrP, and they convincingly showed that prion infectivity can be generated de novo in brain homogenates derived from healthy hamsters using PMCA. Inoculation of further healthy hamsters with the de novo–formed prions caused a transmissible prion disease (Deleault et al. 2007). This study might be regarded as the final proof of the prion hypothesis. However, it also acknowledges PMCA's limitation for diagnostic purposes because PK-resistant material and infectivity can be formed in the absence of prions, thereby risking the reporting of false positive results.

A second approach comprises de novo generation of infectivity by misfolding recombinant PrP^C and subsequently inoculating wild-type animals. In one attempt, a 55-residue peptide encompassing the GSS mutation P101L was refolded in vitro to a beta-sheet rich peptide and could induce disease similar to GSS in transgenic mice expressing PrP (P101L). Transmission to wild-type mice was not successful, and PrP (P101L) was not resistant to PK. Because transgenic mice expressing PrP (P101L) develop disease spontaneously, although later in life than those exposed to the peptide, Nazor et al. (2005) remarked that the misfolded peptide may have simply accelerated a spontaneously occurring disease.

Transmission to wild-type mice of an in vitro–generated misfolded part of the prion protein (amino acid residues 89–231) was achieved a few years later. Legname and coworkers produced PrP (89–231) recombinantly and generated amyloid fibrils in vitro. These fibrils induced prion disease in transgenic mice overexpressing PrP (89–231), which was subsequently transmissible to wild-type mice (Legname et al. 2004, 2005).

Neurotoxicity

Current knowledge about the mechanisms behind neurodegeneration in prion disease and prionopathies is limited. Apoptosis and oxidative stress certainly contribute to some stages of TSE pathology (Milhavet & Lehmann 2002), but little is known about damage causing primary events. Early pathologic changes that occur during prion disease involve synapses, yet the molecular underpinnings of these findings remain unknown.

It is still unclear whether the toxicity of PrP^{Sc} represents a gain of function or whether loss of function of PrP^C is responsible for neuropathological changes induced by prions. Although some authors belief that the toxicity in prion disease is explainable simply by a loss-of function of PrP^C (Nazor et al. 2007), we and others (Westergard et al. 2007) believe a gain of function is more likely, particularly because the phenotypes of PrP^C-deficient mice are very mild. However, a neuroprotective function that may be physiologically provided by PrP^C, which would protect neurons during prion infection, could be reduced following its conversion to PrP^{Sc}.

Our laboratory, and many others, has pursued the hypothesis that elucidating the physiological function of PrP^C might help researchers understand the mechanism involved in prion-induced neurodegeneration. The following discussion centers on the discovery that mice expressing deletion mutants of PrP^C develop severe neurotoxic syndromes and identifies the reasons why we believe that study of these syndromes may reveal the mechanisms operative in prion diseases.

PHYSIOLOGICAL FUNCTION OF THE CELLULAR PRION PROTEIN

The cellular prion protein PrP^C is a GPI-linked extracellular membrane protein with two N-linked complex glycosylation sites. PrP^C is highly abundant in the developing and mature nervous system, where it is expressed by neuronal and glial cells. This mature version originates from a precursor protein proteolytically processed in the endoplasmic reticulum and Golgi (Stahl et al. 1987). As revealed by its atomic structure, the mature PrP^C protein contains a well-defined carboxy-terminal globular domain comprising residues 127–231 (murine numbering), consisting of three alpha helices and two beta sheets (Hornemann et al. 1997; Riek et al. 1996) and a structurally less-defined amino proximal region containing a stretch of several octapeptide repeats, termed the OR, and framed by two positively charged charge clusters, CC$_1$ (aa 23–27) and CC$_2$ (aa 95–110). These domains are linked by a hydrophobic stretch of amino acids [aa 111–134, also termed hydrophobic core (HC)] (**Figure 3**).

HC: hydrophobic core

Prion Protein–Deficient Mice

An astonishing number of independent lines of mice lacking PrP^C have been generated by homologous recombination in embryonic stem cells in many laboratories. Mice with disruptive modifications restricted to the open reading frame are known as *Prnp$^{o/o}$* [Zürich I] (Bueler et al. 1992) or *Prnp$^{-/-}$* [Edinburgh] (Manson et al. 1994). They developed normally, and no severe pathologies were observed later in life. As predicted by the protein-only hypothesis, these mice were entirely resistant to prion infections (Bueler et al. 1993).

In contrast with these earliest lines, three lines generated afterwards: *Prnp$^{-/-}$* [Nagasaki],

Rcm0, and *Prnp*[−/−] [Zürich II] (Moore et al. 1999, Rossi et al. 2001, Sakaguchi et al. 1996) developed ataxia and Purkinje cell loss later in life. Because the phenotype was abolished by reintroduction of *Prnp* as a transgene, the originators of the Nagasaki mice concluded that it occurred because of the lack of PrP[C]. This, however, would run counter to the lack of pathology in *Prnp*[o/o] Zürich-I mice.

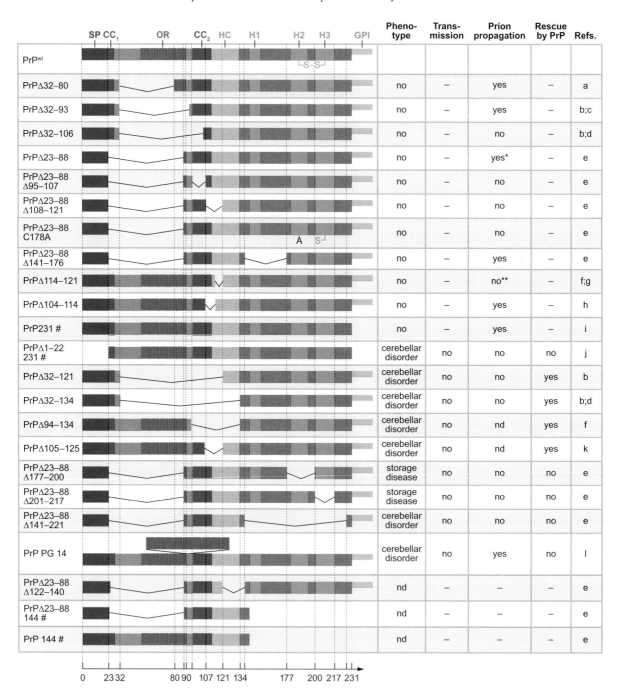

The discrepancy between the different lines of PrP knockout mice was not resolved until a novel gene (*Prnd*), encoding a protein called Doppel (Dpl), was discovered. *Prnd* is localized 16 kb downstream of *Prnp*. In all three lines of PrP^C-deficient mice developing ataxia and Purkinje cell loss, a splice acceptor site to the third exon of *Prnp* was deleted. This placed *Prnd* under transcriptional control of the *Prnp* promoter, resulting in the formation of chimeric transcripts and in overexpression of Dpl in the brain (Moore et al. 1999, Rossi et al. 2001, Sakaguchi et al. 1996). Precisely why the overexpression of Dpl is deleterious is still unclear. On the basis of the observation that Dpl expression induced heme oxygenase 1 (HO-1) and neuronal and inducible nitric oxide synthases (nNOS and iNOS), suggesting an increased oxidative stress in the brains of the Dpl-expressing *Prnp*$^{0/0}$ mice, Wong et al. (2001c) proposed that Dpl expression exacerbates oxidative damage by antagonizing wild-type PrP^C's antioxidative function.

The latency period before the various transgenic mice overexpressing Dpl develop pathological phenotypes is inversely correlated to the Dpl expression level in the brain, indicating a rather strict gene-dosage effect (Rossi et al. 2001). The Dpl-induced disease can be rescued by coexpression of wild-type PrP^C (Nishida et al. 1999, Rossi et al. 2001), indicating that toxicity of Dpl and the physiological function of wild-type PrP^C are not independent of each other, but rather are involved in a common pathway. Dpl-deficient mice suffer from steril-ity (Behrens et al. 2002), suggesting that the primary physiological function of Dpl is related to sperm maturation.

Functional Domains of the Prion Protein

For all the uncertainties surrounding the physiological and molecular functions of PrP^C, some knowledge was generated by expressing a series of partially deleted *Prnp* variants in cultured cells and transgenic mice. Some of these mutants were made to identify the essential domains necessary for restoring prion susceptibility. However, investigators found that domain expression provoked spontaneous neurodegenerative disease (**Figure 3**). In many instances, these syndromes were partially or fully counteracted by coexpression of wild-type PrP^C. Because the lack of PrP^C itself did not induce an obvious phenotype, the latter pathologies indicate pathways in which PrP^C is functionally active. Hence mice expressing PrP, which lacks defined domains, may allow for the identification of functionally relevant domains within PrP^C.

N-terminal deletion mutants of PrP. The OR has long been suspected to represent a major mediator of PrP^C's function, and insertion mutations affecting the OR are associated with hereditary human prion disease. However, transgenic studies indicate that the OR is not required for PrP^C to function or for its convertibility into PrP^{Sc} (Flechsig et al. 2000).

Figure 3

Murine PrP^C protein and transgenic mutant PrP. Schematic drawing of full-length murine PrP^C, including the signal peptide of the precursor protein (SP; *brown box*). Although amino acid numbering differs between human and mouse PrP, the organization of domains (including CC$_1$ and CC$_2$, OR, HC, and H1–H3) is similar to that of human PrP^C (see **Figure 2**). Mouse PrP also contains a disulphide bond (S–S) and a GPI-anchor. The left column denotes the individual mutants described in the text. The right columns indicate presence or absence of phenotypic abnormalities (Phenotype) in transgenic mice when expressed on a PrP-deficient genetic background, transmissibility of this phenotype to recipient mice (Transmission), and susceptibility of transgenic mice to prions after intracerebral inoculation with a mouse-adapted strain of scrapie prions. References: *a*, Fischer et al. (1996); *b*, Shmerling et al. (1998); *c*, Flechsig et al. (2000); *d*, E. Flechsig, I. Hegyi, A. Aguzzi, and C. Weissman (unpublished results); *e*, Muramoto et al. (1997); *f*, Baumann et al. (2007); *g*, Holscher et al. (1998); *h*, Hegde et al. (1998); *i*, Chesebro et al. (2005); *j*, Ma et al. (2002); *k*, Li et al. (2007); *l*, Chiesa et al. (1998).

The OR appears to have, at best, a modulating influence on PrP conversion. Mice expressing OR-deficient PrP^C mutants do not develop pathologies (Fischer et al. 1996, Muramoto et al. 1997, Shmerling et al. 1998). This was unexpected because a variety of in vitro data had identified the OR as being responsible for copper binding (Aronoff-Spencer et al. 2000, Chattopadhyay et al. 2005, Furlan et al. 2007, Leclerc et al. 2006, Qin et al. 2002, Stockel et al. 1998; reviewed in Vassallo & Herms 2003) and for conferring protection against oxidative stress (Brown et al. 1999, Fukuuchi et al. 2006, White et al. 1999, Wong et al. 2001a). On the other hand, transgenic mice expressing nine supernumerary octapeptide repeats, for a total of 14 proline and glycine-rich repeats (Chiesa et al. 1998)—which models a human familial CJD-linked mutation—develop ataxia and cerebellar atrophy, granule cell loss, gliosis, progressive myopathy, and PrP deposition. The latter phenotype resembles its human counterpart in some ways (Chiesa et al. 2000), yet transmission to wild-type mice failed (Chiesa et al. 2003).

In vitro studies indicate that the CC_1 region is involved in recycling and internalizing PrP^C from the cell surface (Sunyach et al. 2003, Taylor et al. 2005). Unfortunately, in vivo little evidence supports the latter contention. Lack of CC_1 in ($PrP_{\Delta 23-88}$) (Muramoto et al. 1997) did not induce pathologies in transgenic mice, and convertibility to PrP^{Sc} was retained. In $PrP_{\Delta 23-88}$ mice, a second charge cluster (CC_2) with several lysine residues around position 100 may replace the function of CC_1. However, mice bearing partial deletions of CC_2 ($PrP_{\Delta 23-88\,\Delta 95-107}$ and $PrP_{\Delta 23-88\,\Delta 108-121}$) are also healthy (Muramoto et al. 1997). The combination of amino-terminal deletion with the elimination of amino acids 141–176 ($PrP_{\Delta 23-88\,\Delta 141-176}$) was also innocuous and restored susceptibility to prion infection (Muramoto et al. 1996) despite a large deletion within the globular domain of PrP^C.

The function of PrP^C may depend on the HC region in concert with CC_2. With the exception of a small deletion between CC_2 and HC ($PrP_{\Delta 104-114}$) (Hegde et al. 1998), ablation of CC_2 in combination with a partial or complete deletion of HC elicits severe pathologies in mice. $PrP_{\Delta 32-121}$ and $PrP_{\Delta 32-134}$ transgenic mice suffer from ataxia and cerebellar granule cell loss in addition to widespread white matter disease (Radovanovic et al. 2005, Shmerling et al. 1998). The latter is also seen in mice expressing deletions encompassing all ($PrP_{\Delta 94-134}$) or part ($PrP_{\Delta 105-125}$) of the central domain (CD) (Baumann et al. 2007, Li et al. 2007). These pathologies are radically different from those seen in prion infections, and none of them goes along with pathological aggregation of PrP.

Each of these pathologies can be counteracted by coexpression of wild-type PrP^C (Baumann et al. 2007, Li et al. 2007, Shmerling et al. 1998), suggesting a competition of sorts between PrP^C and the toxic mutants. In one conceivable scenario, PrP^C and its variants may compete for a common ligand. Binding or complex assembly may represent the first step in a series of events that also involve the interaction of an effector domain located in or controlled by the central domain (CC_2 and HC), eventually resulting in signal transduction.

A partial deletion of HC ($PrP_{\Delta 114-121}$) (Baumann et al. 2007) is nontoxic, but its potential to counteract the toxicity of $PrP_{\Delta 32-134}$ is lower than that of wild-type PrP^C. Mice with deletion of CC_2 and HC ($PrP_{\Delta 32-121}$ and $PrP_{\Delta 32-134}$ as well as $PrP_{\Delta 104-114}$, $PrP_{\Delta 114-121}$) did not support prion propagation (Flechsig & Weissmann 2004; Hegde et al. 1998, Holscher et al. 1998), indicating an involvement of these regions in conversion.

Carboxy-proximal deletion mutants of PrP. Mice expressing PrP mutants with deletions affecting Helix 2 ($PrP_{\Delta 23-88\,\Delta 177-200}$), Helix 3 ($PrP_{\Delta 23-88\,\Delta 201-217}$), or both helices 2 and 3 ($PrP_{\Delta 23-88\,\Delta 141-221}$) suffer from ataxia and present with features of neuronal storage disease (Muramoto et al. 1997, Supattapone et al. 2001) but fail to replicate prions (Muramoto et al. 1996). Obviously at least Helix 2 and Helix 3 are indispensable for stabilizing the structure of PrP^C. None of these diseases proved to be transmissible to normal wild-type mice, and they all manifested themselves independently of the presence or absence of wild-type PrP. This stands in sharp contrast to the group of deletion mutants affecting CC$_2$ and HC.

Several attempts to generate mice expressing truncated carboxy-terminal mutants lacking membrane anchoring (PrP$_{\Delta 23-88\,144\#}$ and PrP$_{144\#}$) have failed (Fischer et al. 1996, Muramoto et al. 1997). As in the case of PrP$_{\Delta 23-88\,\Delta 122-140}$ (Muramoto et al. 1997), lines expressing high levels of mRNA were generated but protein was never detected. Essential signals may have been lost, thereby preventing correct sorting, processing, or folding of PrP and resulting in a short-lived polypeptide.

Mutations affecting the localization of PrP^C. The affinity of GPI-linked PrP^C for artificial membranes, as measured by surface plasmon resonance (Elfrink et al. 2007), suggests extremely strong interactions. One might therefore expect that most PrP^C is attached to cell membranes, with perhaps traces of PrP^C floating in body fluids. Thus it may come as a surprise that plasma contains conspicuous amounts of PrP^C (Volkel et al. 2001), and the concentration of PrP^C in cerebrospinal fluid is even higher (Castagna et al. 2002). However, it is unclear whether this soluble PrP^C is chemically identical to its membrane-bound isoform. Treatment of cultured cells with phosphatidylinositol phospholipase C efficiently releases PrP^C from cultured cell membranes (Stahl et al. 1987), and a similar mechanism may underlie the physiological shedding of PrP^C into body fluids.

Release of full-length secreted PrP^C was forced by deletion of its carboxy terminal hydrophobic domain (PrP$_{231\#}$), which is normally replaced by a GPI-anchor. This manipulation did not induce any pathological phenotype (Chesebro et al. 2005). In contrast, targeting PrP^C to the cytosol (cyPrP = PrP$_{\Delta 1-22\,231\#}$) by deleting its amino-terminal leader peptide (which targets PrP^C to the endoplasmic reticulum and to the secretory pathway) provoked

ataxia with cerebellar degeneration and gliosis (Ma et al. 2002). Coexpression of wild-type PrP^C did not influence the phenotypes of these mice. Whether cytoplasmic expression of PrP and its cytotoxicity represent realistic models of the events occurring during prion disease remains very hotly debated (Fioriti et al. 2005, Roucou et al. 2003).

Point Mutations within the Prion Protein

As previously described, a considerable set of point mutations within *PRNP* has been linked to various forms of human prion diseases. Some of these mutations have been expressed in mice. With the possible exception of some strains of mice expressing the P101L variant of PrP^C (Hsiao et al. 1994, Telling et al. 1996), none of these attempts succeeded in reproducing the infectiousness of bona fide prions. Point mutations affecting the two N-linked glycosylation sites of PrP^C proved, as expected, to alter its glycosylation (Kiachopoulos et al. 2005). Point mutations N182T, A198T, or N182T/A198T prevented glycosylation in transgenic mice without grossly affecting cellular sorting in cell culture. Mice developed normally and were readily susceptible to scrapie or BSE (Neuendorf et al. 2004). Knock-in mutants carrying either N180T or N196T, or both mutations, (Cancellotti et al. 2005) did not suffer from any constitutive phenotype, even if the complete blockade of glycosylation by the N180T/N196T double mutation led to a mainly intracellular localization of PrP^C. This finding is somewhat surprising because results from cultured cells had predicted that unglycosylated PrP^C would be prone to spontaneous aggregation (Korth et al. 2000, Priola & Lawson 2001).

Two sets of point mutations, PrP3AV (exchange of alanine to valine at positions 113, 115, and 118) (Prusiner & Scott 1997) and PrPKHII (exchange of lysine 109 and histidine 110 for isoleucine) (Hegde et al. 1998), generated PrP with altered topology, termed ^{Ctm}PrP in a cell free assay. ^{Ctm}PrP supposedly spans the membrane via the HC domain (Hegde et al. 1998). Transgenic mice expressing these proteins developed a fatal neurological disorder (Hegde et al. 1998). A similar phenotype was observed in transgenic mice with substitution of leucine 9 into arginine in addition to this 3AV mutation (Stewart et al. 2005, Stewart & Harris 2005). However, research never formally proved that ^{Ctm}PrP exists *in vivo*. It is still noteworthy that coexpression of wild-type PrP^C with mutants promoting the ^{Ctm}PrP topology aggravated their phenotype. Subtle changes, such as the removal of disulfide bridges (PrP$_{\Delta23-88\ C178A}$), are tolerated without inducing a spontaneous phenotype though reducing the susceptibility for conversion into PrP^{Sc} (Muramoto et al. 1997).

Evolution of the Prion Protein

PrP is present in a broad variety of species (**Figure 4**). Genes with similarities to *Prnp* exist in birds (Gabriel et al. 1992), reptiles (Simonic et al. 2000), amphibians (Strumbo et al. 2001), and possibly in fish (Favre-Krey et al. 2007, Oidtmann et al. 2003, Rivera-Milla et al. 2003, Suzuki et al. 2002) in addition to all mammals. However, more primitive organisms such as insects, cephalopods, and protozoa have not been reported to contain PrP homologs. All PrPs are glycosylated and membrane attached by a GPI anchor. The sequence identity among the known PrP homologs is limited, and protein length can vary between ~250 amino acids in tetrapods to ~600 amino acids in fish. Fish may have developed additional *Prnp*-like genes (Rivera-Milla et al. 2006). The putative fish PrP genes are thus far identified only on the basis of rather tenuous sequence similarities. The contention that these molecules indeed represent paralogs of PrP^C would be greatly strengthened if knockdown-induced phenotypes of zebrafish would be functionally corrected by mammalian PrP^C expression. Such experiments have not been reported.

Comparisons between the available structures and molecular models suggest that all PrPs share a common blueprint. A flexible amino-terminal tail, with a positively charged CC$_1$ at its far end and repetitive domains of variable numbers, is hooked to a globular carboxy-terminal domain. The fold of this domain is strongly conserved and stabilized by a disulfide bridge, although the primary sequence shows considerable diversity. These two domains are linked by a highly conserved hydrophobic linker having a second positive-charge cluster CC$_2$ at its amino-terminus. This linker region is by far the most conserved sequence motive of PrP in all species.

Cellular Processes Influenced by PrP^C Expression

Several cellular processes in the nervous system have been influenced by the *Prnp*-genotype, including neuronal survival; neurite outgrowth; synapse formation, maintenance, and function; and maintenance of myelinated fibers (**Figure 5**).

One of the most frequently suggested cellular functions of PrP^C is a survival-promoting effect on neuronal and nonneuronal cells, which has been observed *in vitro* as well as in *in vivo* studies.

This neuroprotective function, or cytoprotective function in general reviewed in Roucou & LeBlanc (2005), has been mediated by anti-apoptotic or antioxidative mechanisms.

Antiapoptotic function. Neurons derived from *Prnp*$^{-/-}$ mice were originally reported to be more susceptible to the induction of apoptosis by serum-deprivation than were cells expressing PrP^C (Kuwahara et al. 1999), but this effect may have been brought about by Dpl overexpression rather than by PrP^C ablation. However, several studies indicate that PrP^C has a cytoprotective function by decreasing the rate of apoptosis after particular apoptotic stimuli such as Bax overexpression or TNF-α. Bax overexpression induces apoptosis in human neuronal cells. Coexpression of wild-type PrP^C, but not of PrP lacking the octarepeats, reversed the Bax-mediated induction of apoptosis (Bounhar et al. 2001).

The presence of PrP in the cytosol, be it due to reverse translocation from the endoplasmic reticulum or through direct cytosolic expression, was virulently neurotoxic (Ma et al. 2002). However, other studies failed to confirm the toxicity of cytosolic PrP and claimed that it can instead protect against Bax-mediated apoptosis in human primary neurons (Roucou et al. 2003). In this context, PrP^C inhibited the proapoptotic conformational change of Bax and cytochrome c release from mitochondria (Roucou et al. 2005).

In a screening approach for proteins protecting cancer cells from apoptosis, researchers investigated the gene-expression profile in an established cell clone of MCF-7 breast cancer cell line resistant to TNFα-induced apoptosis. PrP^C was overexpressed 17-fold. Conversely, overexpression of PrP^C converted MCF-7 cells sensitive to TNFα-induced apoptosis into resistant cells (Diarra-Mehrpour et al. 2004).

The neuroprotective function of PrP^C in the postischemic rodent brain has been intensively studied. Levels of PrP^C after ischemia were increased compared with controls (Shyu et al. 2005, Weise et al. 2004). Moreover, adenovirus-mediated overexpression of PrP^C reduced infarct size in rat brain and improved neurological behavior after cerebral ischemia (Shyu et al. 2005). Conversely, in a mouse model of ischemic brain injury $Prnp^{o/o}$ mice displayed significantly increased infarct volumes when compared with wild-type mice (McLennan et al. 2004, Weise et al. 2006). Two groups of researchers showed that mice lacking PrP^C had enhanced postischemic caspase-3 activation (Spudich et al. 2005, Weise et al. 2006). An increase in Erk-1/-2, STAT-1, and JNK-1/-2 phosphorylation and activation was identified, suggesting PrP^C's possible involvement in cellular signaling (Spudich et al. 2005). Also, a reduced amount of phospho-Akt in the gray matter suggested that PrP^C deficiency brings about an impairment of the antiapoptotic phosphatidylinositol 3-kinase/Akt pathway (Weise et al. 2006). Finally, Mitteregger et al. (2007) claimed that the OR is required within PrP^C for the neuroprotection in the ischemic mouse

brain, although the genetic homogeneity of the mice tested in the latter experiment was not controlled for.

Protection against oxidative stress. Besides its possible antiapoptotic function, there are many reports about an antioxidative effect of PrP^C. These two effects are not necessarily mutually exclusive. Oxidative stress may be involved in TSE pathogenesis. However, one must remember that oxidative stress is very unspecific and is seen in different kinds of damage to the nervous system with impaired mitochondrial function such as defects in the ubiquitin-proteasome system, protein aggregation, and inflammation.

Many investigators believe that the main function of PrP^C consists of protecting against oxidative stress (see Milhavet & Lehmann 2002 for a review). First hints came from in vitro studies of rat pheochromocytoma cells. Those selected for resistance to copper toxicity or oxidative stress showed higher levels of PrP^C (Brown et al. 1997a). Primary neuronal cells lacking PrP^C were more susceptible to hydrogen peroxide (H_2O_2) than were wild-type cells. The increased peroxide toxicity went along with a significant decrease in glutathione reductase activity measured in PrP^C-deficient neurons (White et al. 1999). Also, PrP^C-deficient primary neurons were more susceptible to treatment with agents inducing oxidative stress compared with wild-type cells, a phenomenon that was explained by a reduced Cu/Zn superoxide dismutase (SOD) activity observed in vivo (Brown et al. 1997b, 2002). Higher levels of oxidative damage to proteins and lipids were identified in the brain lysates derived from $Prnp^{-/-}$ compared with wild-type mice (Klamt et al. 2001, Wong et al. 2001b).

PrP^C itself could have SOD activity and thereby mediate the antioxidative function (Brown et al. 1999). However, there is significant controversy about this alleged SOD activity. Others, ourselves included, failed to confirm this proposed SOD activity in vitro (Jones et al. 2005) and in vivo (Hutter et al. 2003). Furthermore, PrP^C expression level did

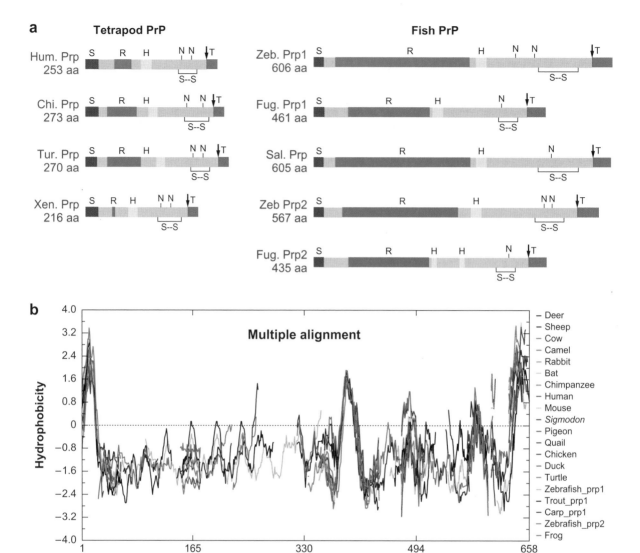

Figure 4

PrP structural diversity in vertebrates. (*a*) Schematic drawing of tetrapod PrPs and long (PrP1 and PrP2) fish PrPs. The species abbreviations refer to sequences from human (Hum), chicken (Chi), turtle (Tur), Xenopus (Xen), zebrafish (Zeb), salmon (Sal), and Fugu (Fug). The location and relative size of conserved structural features are indicated. However, these features were physically determined for the structure of human PrP^{C} and represent mere conjectures in the case of fish. Domains are indicated by different boxes and/or letters: S, signal peptide sequence; R, repetitive region; H, hydrophobic region; S—S, disulfide bridge; N, glycosylation site; arrow, GPI anchor residue; and T, hydrophobic tail. (*b*) Comparison of hydrophobicity plots. Sequences of indicated species were aligned using DNAMAN software (Lynnon BioSoft, Canada), and a hydrophobicity plot was generated using a window of nine amino-acid residues. Numbering of residues is according to alignment matrix. (*c*) 3-dimensional structures of human (hum based on 1QM2.pdb model) chicken (chi based on 1U3M.pdb) turtle (tur based on 1U5L.pdb), and frog (fro based on 1XOU.pdb); pdb files are from the protein database (Berman et al. 2000). Note the similarity of the carboxy-terminal globular domain. (*d*) Evolutionary relationships among vertebrate PrP sequences are based on distance methods (neighbor-joining). Bootstrap values are shown at relevant nodes using DNAMAN software (Lynnon BioSoft, Canada).

not significantly influence SOD activity in vivo (Hutter et al. 2003, Waggoner et al. 2000).

Mitochondria play an important role not only in oxidative stress but also in the induction of apoptosis. Morphological alterations in mitochondria have been described in scrapie-infected hamsters (Choi et al. 1998) and mice (Lee et al. 1999) as well as in mice lacking PrP, in which the number of mitochondria was reduced (Miele et al. 2002).

Role of PrP^C in synapses. Synapses have developed into a sort of hot spot in prion research. Several immuno-electron microscopy studies could show that PrP^C is localized in synaptic boutons, whereas it is mainly presynaptic (Fournier et al. 1995, Moya et al. 2000, Sales et al. 1998, Tateishi et al. 1996b). However, others described a much broader distribution of neuronal PrP^C (Laine et al. 2001, Mironov et al. 2003). Because PrP^C is processed and broken down into various fragments, not all of which

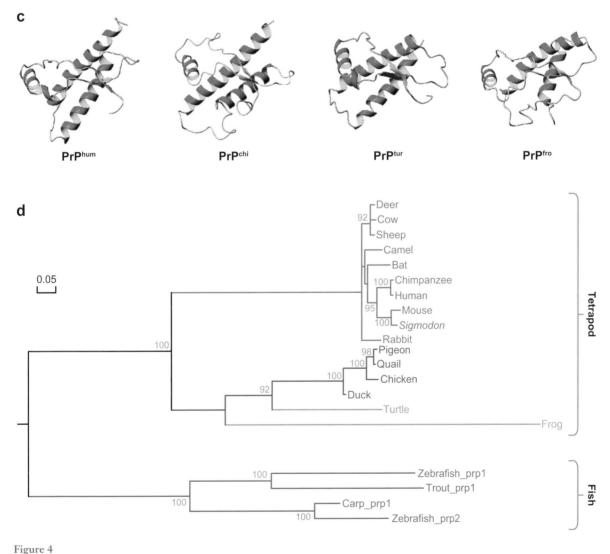

Figure 4

(*Continued*)

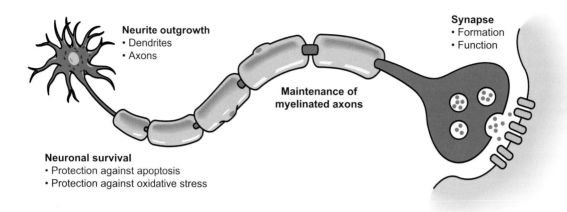

Neurite outgrowth
- Dendrites
- Axons

Synapse
- Formation
- Function

Maintenance of myelinated axons

Neuronal survival
- Protection against apoptosis
- Protection against oxidative stress

Figure 5

Physiological processes involving PrP^C. Several processes in the nervous system have been influenced by PrP^C. Neurite outgrowth, including growth of axons and dendrites, was observed to be reduced in neurons lacking PrP^C. PrP^C has often been reported to promote neuronal survival, in particular following apoptotic or oxidative stress. Cerebellar granule cell apoptosis was observed in mice expressing toxic N-terminal deletion mutants of PrP. In addition, the latter transgenic mice show an impaired maintenance of myelinated axons in the white matter. Another site of PrP^C action might be the synapse, which is often affected in the first stage of prion diseases and whose formation was found to be reduced in neuronal cultures devoid of PrP^C. Furthermore, electrophysiological studies indicate a role of PrP^C in synapse function, especially in neurotransmitter release.

are recognized by the antibodies used in these studies, one might speculate that some PrP^C degradation products acquire distinct subcellular topologies.

Early pathologic changes occurring in prion diseases involve synapse loss and PrP^{Sc} deposition in synaptic terminals (Grigoriev et al. 1999, Jeffrey et al. 2000, Kitamoto et al. 1992, Matsuda et al. 1999, Roikhel et al. 1983). Synaptic vesicle proteins associated with exosomes and neurotransmission are reduced in brains of patients with spongiform encephalopathy (Ferrer et al. 1999). Synaptic disorganization and loss are fundamental and constant features of prion disease, irrespective of the presence or absence of spongiform change, neuronal loss, and severe gliosis (Clinton et al. 1993). Abnormal electrophysiological recordings in scrapie-infected mouse and hamster hippocampal and cortical slices further support the synaptic dysfunction during the course of prion disease (Barrow et al. 1999, Johnston et al. 1998). In a terminal disease state, PrP^{Sc} accumulation in synaptosomes correlated with alterations in the GABAergic system (Bouzamondo-Bernstein et al. 2004). Despite the wealth of

the above evidence, however, it should not go undiscussed that synaptic changes can represent nonspecific phenomena that are seen in essentially all brain diseases at one stage or another.

The generally held view that PrP^C is an important protein in synapses is supported by electrophysiological studies of CA1 hippocampal neurons derived from PrP^C-deficient mice. Excitatory glutamatergic synaptic transmission, GABA$_A$ receptor–mediated fast inhibition, long-term potentiation, and late afterhyperpolarization were reduced or absent in mice lacking PrP^C (Carleton et al. 2001; Colling et al. 1994, 1996; Mallucci et al. 2002). Some of the findings could be explained by alterations in Ca-activated K$^+$ currents (Colling et al. 1996, Herms et al. 2001). However, the reader should note that alterations in synaptic transmission were not confirmed by others (Lledo et al. 1996), and glutamatergic synaptic transmission was even observed to be increased in PrP^C-deficient mice by yet another laboratory (Maglio et al. 2004, 2006). Another report indicates the impact of aging on these alterations describing a reduction in the level of posttetanic potentiation and long-term potentiation only in old PrP^C-deficient mice (Curtis et al. 2003). In summary, the impact of the loss of PrP^C on hippocampal electrophysiological parameters is still being hotly debated despite a full decade of research efforts. Some of the discrepancies may depend on additional genetic modifiers for which investigators have not rigorously controlled.

Other alterations in PrP^C-deficient mice might be related to synaptic dysfunction such as altered circadian rhythms and sleep (Tobler et al. 1996) and impaired hippocampal-dependent spatial learning (Criado et al. 2005). The neuromuscular junction is another site where PrP^C was concentrated, namely enriched in subsynaptic endosomes (Gohel et al. 1999). A potentiation of acetylcholine release from presynaptic axon terminals was observed after administration of recombinant PrP at nanomolar concentrations to mouse phrenic-diaphragm preparations (Re et al. 2006). The suggestion of an involvement of PrP^C in synapse formation originated from in vitro observations in hippocampal neurons, in which synaptic-like contacts were increased after addition of recombinant PrP (Kanaani et al. 2005).

It is unknown whether the role of PrP^C in synapses is related to its above-mentioned antiapoptotic or antioxidative effects or whether it is mainly the involvement of PrP^C in neurotransmitter release (e.g., via direct interaction with synapsin1 and synaptophysin). However, an as-yet-unidentified process could also play a key role, or several processes could work together.

Neurite outgrowth. Several lines of evidence indicate PrP^C's involvement in neuronal development, differentiation, and neurite outgrowth. Axon or dendrite outgrowth was associated with PrP^C-dependent activation of signal transduction pathways including p59Fyn kinase, cAMP/protein kinase A (PKA), protein kinase C (PKC), and MAP kinase activation (Chen et al. 2003, Kanaani et al. 2005, Lopes et al. 2005, Santuccione et al. 2005). P59Fyn kinase activation in this context was depen-

dent on the recruitment of neural cell adhesion molecule (N-CAM) to lipid rafts (Santuccione et al. 2005). Recent studies show that PrP^C positively regulates neural precursor proliferation during development and adult mammalian neurogenesis (Steele et al. 2006).

Maintenance of the white matter. Central nervous system white matter, composed mainly of myelinated axons, might be disrupted in prion diseases and prionopathies. In some cases of GSS, cerebellar and frontal white matter are affected (Itoh et al. 1994). In an experimental model of human TSEs in rodents, vacuolation in myelinated fibers with splitting of myelin lamellae was observed (Walis et al. 2003). PrP^C is present in purified myelin fractions derived from brain homogenates (Radovanovic et al. 2005). Several transgenic mice expressing deletion mutants of PrP^C (Baumann et al. 2007, Li et al. 2007, Shmerling et al. 1998) as well as $Prnp^{-/-}$ mice accidentally overexpressing Dpl (Nishida et al. 1999) show vacuolation and degeneration of myelinated fibers in the central nervous system.

PrP and the Immune System

The immune system plays a fundamental role in prion disease and PrP^C is expressed on cells of the immune and hematopoietic system, where it might have a physiological function. This topic is reviewed in depth in Isaacs et al. (2006). Also, Zhang et al. (2006) reported that PrP^C is involved in self-renewal of hematopoietic stem cells.

Molecular Mechanisms Mediating the Function of PrP^C

Despite the overwhelming number of reports about alterations in mice and cells lacking PrP^C summarized above, little is known about the molecular mechanisms involved in these cellular processes. **Figure 6** depicts some theoretical models of how PrP^C might influence cell signaling, endocytosis, and cell adhesion. Whether these events are mutually exclusive, or whether they occur only under specific circumstances in a diversity of tissues, or whether they can act in a combined way, remains speculative. In all cases, PrP^C is likely to mediate its function via one or more interaction partners.

One explanation for the diversity of the suggested physiological functions of this PrP^C is

a Endocytosis of PrP^c via Clathrin-coated pits or caveolae

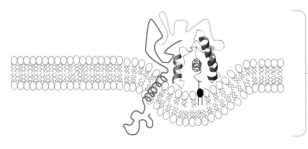

- Cointernalization of another cell component
- Modulation of signaling pathways
- Degradation of PrP^c cointernalized TM proteins

b Interaction with TM protein in *cis*

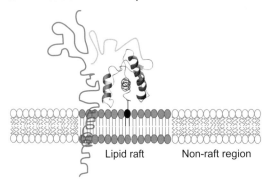

Lipid raft   Non-raft region

- Modulation of signal transduction pathways, e.g.activation of:
  - Fyn kinase
  - Erk1/2
  - cAMP
  - PKC

c Interaction with TM protein in *trans*

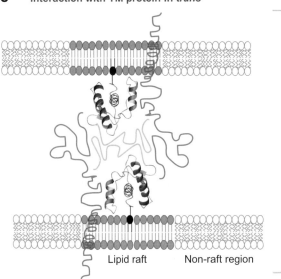

Lipid raft   Non-raft region

- Modulation of signaling pathways
- Modulation of cellular adhesion

that it may exert pleiotropic effects, thereby modulating the function of several cellular pathways. Examples for such a general cellular process would be stabilization of protein complexes and the targeting of cell components to certain cellular sites, such as rafts or endosomes.

Signaling. The attachment of PrP^C to the membrane by a GPI anchor, its localization in detergent-resistant membranes, also known as lipid rafts, in many cell types may suggest an involvement in cellular signaling (Shmerling et al. 1998) as is the case for other raft-associated proteins. Moreover, as we describe below, PrP^C could also influence cellular signaling events by its involvement in endocytotic pathways.

Several signaling pathways or signaling components, such as Akt, Fyn, cAMP, and Erk1/2, are modulated by PrP^C expression, its cross-linking, or its interaction with another protein. Antibody-mediated cross-linking of PrP^C induced activation of the p59Fyn kinase, a family member of nonreceptor Src-related kinases, in neuronal differentiated cells in a caveolin-1-dependent manner (Mouillet-Richard et al. 2000). As a downstream event, the same group claims to have identified Erk1/2 phosphorylation (Schneider et al. 2003). PrP^C cross-linking additionally modulates serotonergic receptor activity in differentiated neuronal cells (Mouillet-Richard et al. 2005). The finding that PrP^C cross-linking modulates activity of serotonergic receptors in differentiated neuronal cells await replication and *in vivo* confirmation. However, p59^{Fyn} activation and downstream activation of Erk1/2 were also seen in a hypothalamic cell line (Toni et al. 2006).

Several studies indicated PrP^C involvement in neurite outgrowth and neuronal survival. Chen et al. reported increased neuronal survival and neurite outgrowth from neurons when cultured on Chinese hamster ovary (CHO) cells transfected to express mouse PrP. Although p59Fyn kinase activity in this context was involved mainly in neurite outgrowth, the PI3 kinase/Akt pathway as well as regulation of Bcl-2 and Bax expression contributed to the survival effect elicited by PrP. Cyclic AMP/protein kinase A (PKA) and Erk signaling pathways contributed to both neurite outgrowth and neuronal survival (Chen et al. 2003, Santuccione et al. 2005).

Some investigators suggested that engagement of PrP^C with stress-inducible protein I (STI1) induces neuroprotective signals that rescue cells from apoptosis via cAMP/protein kinase A and the Erk signaling pathways (Chiarini et al. 2002, Zanata et al. 2002). Interaction with STI1 induced different signaling pathways, promoting neuroprotection by PKA activation and neuritogenesis by activation of the MAPK pathway (Lopes et al. 2005).

Endocytosis and internalization of PrP^C. PrP^C is rapidly internalized from the cell membrane. This internalization of PrP^C could be crucial for its function, e.g., by influencing signal transduction pathways. Endocytosis of membrane receptors does not necessarily downregulate receptor activity. While being internalized, both tyrosine kinase and G protein–coupled receptors may remain

---

Figure 6

Models of how PrP^C could exert its physiological function. (*a*) Endocytosis of PrP^C via clathrin-coated pits or caveolae may represent a mechanism for the downregulation of PrP^C on the cell surface. Alternatively, or additionally, endocytosis of PrP^C leads to cointernalization of another cell component, e.g., a proteinaceous interacting partner, thereby regulating the presence of the latter on the cell surface. This regulation could positively or negatively modulate the activity of signal transduction pathways, e.g., via inducing degradation of the cointernalized partner. (*b*) An interaction with a transmembrane (TM) protein in *cis* independent of an internalization process may lead to modulation of signal transduction pathways in the cell carrying PrP^C on its cell surface. (*c*) Similarly, an interaction with another protein in *trans* may lead to modulation of signal-transduction pathways or adhesion to adjacent cells.

activated and produce intracellular responses along the endosome-lysosome pathway (Prado et al. 2004). Internalization of tyrosine kinase receptors was functionally important in studies: TrkA receptors, for example, mediate nerve growth factor (NGF)-dependent cell survival while they are located at the cell membrane, whereas internalization is required for induction of neurite outgrowth (York et al. 2000, Zhang et al. 2000).

The mechanism of PrP^C internalization is still controversial because both raft/caveolae or caveolae-like (Kaneko et al. 1997, Marella et al. 2002, Peters et al. 2003, Vey et al. 1996) as well as clathrin-dependent endocytosis may be operative (Shyng et al. 1994, Taylor et al. 2005). These mechanisms might as well be equally important. In addition, internalization of the same ligand/receptor complex by distinct endocytotic pathways can result in different signaling outcomes. TGF-β receptor, for example, is degraded after endocytosis via caveolae, but internalization via clathrin-coated pits promotes its signaling (Di Guglielmo et al. 2003). However, in lymphocytes and neuronal cells that do not express caveolin, internalization can occur in a lipid raft–associated noncaveolar, clathrin-independent process (Kirkham & Parton 2005, Parton & Richards 2003). Therefore additional, less well-characterized endocytosis pathways including caveolae-like endocytosis might be involved in internalization of PrP^C.

For endocytosis by clathrin-coated pits, PrP^C would need to leave the lipid rafts prior to internalization because the rigid structure of raft lipids is unlikely to accommodate the tight curvature of coated pits. This phenomenon occurred after binding of copper to the OR (Sunyach et al. 2003, Taylor et al. 2005), but its physiological significance is unknown. Low-density lipoprotein receptor-related protein 1 (LRP1) was later shown to mediate PrP^C endocytosis (Taylor & Hooper 2007), and CC1 region was essential for its internalization in neuroblastoma cells (Sunyach et al. 2003, Taylor et al. 2005). Sunyach et al. (2003) suggested that heparin sulfate proteoglycans are part of the endocytotic complex involving PrP^C.

Glypican-1, a GPI-anchored heparan sulfate-containing cell-associated proteoglycan, interacts and cointernalizes with PrP^C in N2a after induction with copper ions. In cells expressing PrP, which lacks the OR, internalization of glypican-1 is reduced, suggesting a possible role for PrP^C in the cointernalization of other cellular components (Cheng et al. 2006).

PrP^C might participate in the correct localization of some other proteins in lipid rafts. Neuronal nitric oxide synthase (nNOS), for example, involved in various nervous system processes such as development, synaptic plasticity, regeneration, and regulation of transmitter release was associated with lipid rafts in wild-type mice. In contrast, in brains of PrP^C-deficient as well as scrapie-infected mice, nNOS was not associated with rafts, and activity of nNOS was reduced. Therefore PrP^C could be important for the proper cellular localization of other proteins (Keshet et al. 1999). Similarly N-CAM was recruited to lipid rafts by PrP^C (Santuccione et al. 2005).

However, PrP^C is involved in a number of cellular functions and how endocytosis influences them in vivo remains widely unknown; internalization of PrP^C could contribute to down-regulation of a signaling event but could also be necessary for signaling. A general involvement of PrP^C in vesicle formation could be a possible explanation for most of the suggested molecular functions of PrP^C because it could regulate signaling and influence synaptic transmission.

PrP and cell adhesion. Several reports are consistent with a possible function of PrP^C as a cell adhesion or recognition molecule. Some interaction partners of PrP^C identified so far have a role in adhesion, including laminin (Graner et al. 2000a,b), a structural component of basement membrane, laminin-receptor precursor (Gauczynski et al. 2001, Rieger et al. 1997), and N-CAM (Schmitt-Ulms et al. 2001). These three molecules are involved in adhesion in a diversity of signal transduction pathways, in differentiation, and in neurite outgrowth (Colognato & Yurchenco 2000, Maness & Schachner 2007). Interaction of laminin or N-CAM with PrP^C has been associated mainly with its suggested role in neuritogenesis and neurite outgrowth (Graner et al. 2000a,b; Santuccione et al. 2005). Sales et al. (1998) also proposed that PrP^C in its synaptic location might stabilize apposing synaptic membranes through adhesive mechanisms.

The evidence reported so far could indicate that PrP^C is involved in adhesive mechanisms, but this is likely not its sole function. Adhesion molecules that interact biochemically with PrP^C could also transduce a PrP^C-dependent signal. The neurite-outgrowth-promoting interaction between PrP^C and N-CAM, for example, was associated with activation of the p59Fyn kinase (Santuccione et al. 2005).

Interaction Partners of PrP^C

Investigators often hypothesized that PrP^C exerts its function via interaction with other cell-surface components. GPI-anchored proteins would need to interact with a transmembrane adaptor to influence intracellular signal-transduction pathways, thereby enabling the transduction of an extracellular signal. An example of such a protein is the urokinase-type plasminogen activator receptor uPAR, which is involved in cellular adhesion, differentiation, proliferation, and migration mediated by the interaction with transmembrane adaptors such as integrins, G protein–coupled receptors, and caveolins (Blasi & Carmeliet 2002). Analogously, PrP^C may bind to a transmembrane protein or to a protein complex that mediates functional association with intracellular pathways.

The toxic deletion mutants of PrP^C may destroy such a complex by competing for the binding of some complex components yet failing to interact with the signal-transducing components. Indeed, several models for the toxicity of PrP^C deletion mutants have proposed that PrP^C binds to a transmembrane receptor and that deletion mutants induce a toxic signal (Li et al. 2007) or prevent a survival signal (Baumann et al. 2007, Shmerling et al. 1998).

The commonly shared opinion that PrP^C binds to a receptor might explain the tremendous effort put into the identification of its interaction partners by different methods such as yeast-two hybrid, coimmunoprecipitations, and cross-linking experiments.

All PrP^C interaction partners identified thus far are summarized in **Table 1**. They include membrane proteins (receptors, enzymes, Caveolin-1, Na-K-ATPase, and a potassium channel), cytoplasmic proteins (components of the cytoskeleton, heat-shock proteins, and adaptor proteins involved in signaling), and even the nuclear protein CBP70. Tantalizingly, several of these interactions partners are known to play a role in synaptic vesicle function. Unfortunately, the physiological relevance of most of the proposed interaction partners remains unconfirmed. Proteins that are not even localized to the outer leaflet of the cell membrane, where PrP^C is located and believed to exert its function, would at least require an additional interacting cell component, meaning that their interaction with PrP^C must be indirect at best.

Another possible explanation for cytosolic interaction partners is the suggested presence of transmembranous variants of PrP^C, termed ^{Ntm}PrP and ^{Ctm}PrP. However, under normal conditions they have been described at best in minute amounts only (Hegde et al. 1998, 1999; Stewart & Harris 2001). Cytosolic PrP was later detected (Ma & Lindquist 2002, Ma et al. 2002). The significance of these findings still remains obscure.

One must consider methodical bottlenecks. Because PrP^C is exposed to the extracellular space, it is questionable whether a yeast two-hybrid screen that artificially exposes PrP^C to the cytosolic compartment with its different biochemical composition is the most appropriate method by which to study PrP^C interaction partners. A high number of false-positive results is to be expected, and therefore it is particularly important that additional methods are used to confirm interactions identified by yeast-two hybrid.

One item of paramount importance in immunoprecipitation experiments is the choice of the detergent conditions to allow for weak

Table 1 Molecules identified as interaction partners to PrP^C or PrP^{Sca}

Interaction partner	Subcellular localization	Method	Binding site	Function of interaction partner and references
Hsp60 and GoEL	Mitochondria	Y2H	180–210	Chaperone (Edenhofer et al. 1996)
Laminin receptor precursor	Plasma membrane	Y2H, Co-IP	144–179	Cell adhesion (Hundt et al. 2001, Rieger et al. 1997)
Laminin	Extracellular, basement membrane	Binding assay	Unknown	Signal transduction, cell adhesion, neuritogenesis (Graner et al. 2000a,b)
Synapsin Ib	Synaptic vesicles	Y2H, Co-IP	23–100 and 90–231	Synaptic vesicle formation, modulation of neurotransmitter release and in synaptogenesis (Spielhaupter & Schatzl 2001)
Grb2	Cytoplasm	Y2H, Co-IP	23–100, 90–231	Adaptor protein mediating signal transduction (Spielhaupter & Schatzl 2001)
Pint1	Unknown	Y2H, Co-IP	90–231	Unknown (Spielhaupter & Schatzl 2001)
N-CAM	Plasma membrane (transmembrane/ GPI-anchored)	Formaldehyde cross-linking	141–176	Cell adhesion, signaling, etc. (Maness & Schachner 2007, Schmitt-Ulms et al. 2001)
Stress-inducible protein 1 (STI 1)	Cytoplasm, plasma membrane?	Complementary hydropathy Co-IP	113–128	Heat shock protein (Chiarini et al. 2002, Zanata et al. 2002)
Caveolin-1	Caveolae, plasma membrane (transmembrane)	Co-IP	unknown	Caveolae formation and endocytosis; cross-linking of PrP induces Fyn activation (Mouillet-Richard et al. 2000)
Fyn kinase	Cytoplasm, associated with cytosolic side of plasma membrane	Co-IP	unknown	Signal transducer molecule (Mattei et al. 2004, Mouillet-Richard et al. 2000)
ZAP70	Cytoplasm	Co-IP	Unknown	Signal transduction during T cell activation (Mattei et al. 2004)
Synaptophysin	Synaptic vesicles (transmembrane)	Co-IP	Unknown	Presynaptic vesicle protein (Keshet et al. 2000)
Neuronal nitric oxide synthase (nNOS)	Intracellular, partly membrane bound	Co-IP	Unknown	Production of NO in neuronal tissue, involved in signaling, neurotransmission, etc. (Keshet et al. 2000)
β-dystroglycan	Plasma membrane (transmembrane, part of the dystroglycan complex)	Co-IP	Unknown	Transmembrane protein, binds extracellularly to α-dystroglycan (bound to laminin) and, intracellularly, to dystrophin (Keshet et al. 2000)
Dp71 (dystrophin)	Cytoplasmic face of plasma membrane (part of the dystroglycan complex)	Co-IP	Unknown	Cytoskeletal protein (Keshet et al. 2000)
α-syntrophin	Cytoplasmic face of membrane (part of the dystroglycan complex)	Co-IP	Unknown	Adaptor protein, in sarcolemma and the neuromuscular junction, binds dystrophins and nNOS (Keshet et al. 2000)
α-tubulin	Microtubules, cytoplasm	Cross-linking, Co-IP	Unknown	Microtubule subunit (Keshet et al. 2000, Nieznanski et al. 2005)
Glutamic acid decarboxylase (GAD)	Intracellular, vesicle membrane	Co-IP	Unknown	Enzyme that synthesizes the neurotransmitter GABA from glutamate (Keshet et al. 2000)

(Continued)

Table 1   (*Continued*)

Interaction partner	Subcellular localization	Method	Binding site	Function of interaction partner and references
β-actin	Intracellular cytoskeleton protein	Co-IP, affinity chromatography	Unknown	Subunit of microfilaments of the cytoskeleton (Keshet et al. 2000)
BACE-1	Plasma membrane (transmembrane)	Co-IP	Unknown	APP processing (Parkin et al. 2007)
TREK-1	Plasma membrane (transmembrane)	Bacterial two-hybrid, Co-IP	128–230	Two-pore potassium channel protein (Azzalin et al. 2006)
Bcl-2	Cytoplasmic face of mitochondrial outer membrane, nuclear envelop, and ER	Yeast two hybrid	72–254	Inhibition of apoptosis (Kurschner & Morgan 1995)
NRAGE (neurotrophin receptor–interacting MAGE homolog)	Cytoplasm, plasma membrane association when NGF is bound to p75NTR	Y2H, Co-IP	122–231	Binding to p75NTR, permits NGF/p75NTR-dependent apoptosis (Bragason & Palsdottir 2005, Salehi et al. 2000)
Mouse p45 NF-E2 related factor 2 (Nrf2)	Cytoplasmic or nuclear (regulated shuttling)	cDNA library screen, interaction with AP-tagged soluble PrP	Unknown	Transcription factor regulating expression of a set of detoxifying and antioxidant enzyme genes
Mouse amyloid precursor-like protein 1 (APLP1)	Plasma membrane (transmembrane)	cDNA library screen, interaction with AP tagged soluble PrP	Unknown	Neuronal survival and neurite outgrowth (Sakai & Hohjoh 2006, Yehiely et al. 1997)
αB-crystalline	Cytoplasm	Y2H	Unknown	Small heat-shock protein, chaperone
Ribosomal protein P0	Ribosomes	Affinity chromatography	Unknown	Important for ribosomal activity (Sun et al. 2005)
CNPase	Myelin, cytoplasmic face, microtubule-associated	Affinity chromatography	Unknown	Membrane anchor for microtubules, required for maintaining the integrity of paranodes and axons (Petrakis & Sklaviadis 2006)
Creatine kinase-B	Cytoplasm, intracellular enzyme	Affinity chromatography	Unknown	Catalyzes transfer of phosphate between ATP and various phosphogens like creatine phosphate (Petrakis & Sklaviadis 2006)
NSE (neuron-specific enolase)	Cytoplasm, intracellular enzyme	Affinity chromatography	Unknown	Enolase is a glycolytic enzyme; NSE contains γ-subunit, expressed primarily in neurons and in neuroendocrine cells (Petrakis & Sklaviadis 2006)
Clathrin heavy chain 1	Cytosolic face of coated vesicles and coated pits	Affinity chromatography	Unknown	Clathrin: major protein component of the cytoplasmic face of coated vesicles and coated pits, involved in intracellular trafficking of receptors and endocytosis (Petrakis & Sklaviadis 2006)
α-spectrin	Cytoskeletal protein, lines intracellular side of the plasma membrane	Affinity chromatography	Unknown	Maintenance of plasma membrane integrity and cytoskeletal structure (Petrakis & Sklaviadis 2006)

(*Continued*)

Table 1   (*Continued*)

Interaction partner	Subcellular localization	Method	Binding site	Function of interaction partner and references
Glial fibrillary acidic protein (GFAP)	Intracellular, intermediate filament in glial cells such as astrocytes	Affinity chromatography	Unknown	Different cellular processes, like cell structure and movement, cell communication (Petrakis & Sklaviadis 2006)
Na+/K+ ATPase α3	Plasma membrane	Affinity chromatography	Unknown	Catalytic subunit of P-type ATPase, active transport of cations across cell membranes, maintain ionic gradients (Petrakis & Sklaviadis 2006)
PLP (proteolipid protein)	Myelin (transmembrane)	Cross-linking	Unknown	Major component of central nervous system myelin (Petrakis & Sklaviadis 2006)
STXBP1 (Syntaxin binding protein 1)	Membrane-associated in presence of syntaxin	Cross-linking	Unknown	Participation in regulation of synaptic vesicle docking and fusion (Petrakis & Sklaviadis 2006)
ζ14–3–3	Intracellular	Cross-linking	Unknown	Phosphoserine or phosphothreonine binding protein binds diverse signaling proteins, including kinases, phosphatases, and transmembrane receptors, thereby regulating most cellular processes (Petrakis & Sklaviadis 2006)
BASP1	Bound to plasma membrane, calmodulin, and actin	Cross-linking	Unknown	Accumulation mainly in axon endings (growth cones and presynaptic area of synapses) (Petrakis & Sklaviadis 2006)
Vitronectin	Extracellular matrix glycoprotein	Binding and competition assay	105–119	Axonal growth in the peripheral nervous system (Hajj et al. 2007)
CBP70	Nuclear [and cytoplasmatic (Rousseau et al. 1997)]	Co-IP		Lectin (Rybner et al. 2002)
Glycosaminoglycans (GAG)	Connective tissues, covalently linked to proteins to form proteoglycans	ELISA, SPR, heparin-agarose binding assay	23–35, 23–52, 53–93, 110–128	Various functions (Caughey et al. 1994, Pan et al. 2002, Warner et al. 2002)
ApoE	Secreted	Co-IP, cross-linking	23–90 109–141	Main apolipoprotein of chylomicrons, involved in neurodegenerative diseases such as Alzheimer disease (Baumann et al. 2000, Gao et al. 2006, Schmitt-Ulms et al. 2004)
Plasminogen	Secreted	Immobilized serum proteins probed with PrP^C or PrP^{pres}	Unknown	Inactive zymogen form of plasmin, which participates in thrombolysis or extracellular matrix degradation (Ellis et al. 2002, Fischer et al. 2000)
Rdj2	Cytoplasmic face of membranes	Co-IP with GST fusion proteins	Unknown	Chaperone (Beck et al. 2006)
Casein kinase 2 α/α'	Cytoplasmic	Far-Western, SPR	105–242	Protein kinase (Meggio et al. 2000)

[a]Co-IP, coimmunoprecipitation; Y2H, yeast-two hybrid; SPR, surface plasmon resonance.

and transient protein-protein interactions but destroy artificial or unspecific interactions. The influence of the detergent used was exemplified in one study in which the interactions between PrP^C and the dystroglycan complex were studied as a function of the detergent used and on the integrity of lipid rafts (Keshet et al. 2000). Some of the interactions identified by immunoprecipitation may be artifactual—a result that could be avoided by using stringent controls including knockout tissues and specific antibody competition experiments.

CONCLUSIONS

Despite the progress discussed above, several important issues in the prion field remain unresolved. Most conspicuously, both the physiological function of PrP^C and the molecular pathways leading to the fatal neurodegeneration in prion diseases remain unknown. These two issues may be linked, and elucidation of the physiological function of PrP^C has the potential to help researchers understand the mechanisms involved in prion-induced neurodegeneration.

Studies of mice carrying targeted disruptions of any given gene have often provided researchers with useful tools to identify the respective gene product's function. However, the many lines of mice lacking PrP^C that have been generated independently by homologous recombination have failed to uncover a clear molecular physiological function of PrP^C. The most obvious phenotype was their resistance to prion infection. Nevertheless, an overwhelming number of molecular, structural, or functional alterations have been reported in *Prnp*$^{-/-}$ mice.

Cellular processes in the nervous system that have been influenced by the *Prnp* genotype include neuronal survival, neurite outgrowth, synapse formation, maintenance, and function, as well as maintenance of myelinated fibers.

One of the most frequently cited cellular functions of PrP^C is a survival-promoting effect on neuronal and nonneuronal cells. This cytoprotective function has been mediated by antiapoptotic or antioxidative mechanisms, but nothing is known about the proximal mechanisms mediating such effects.

The frequently voiced opinion that PrP^C binds to a receptor has driven a large effort toward the identification of its interaction partners by different methods such as yeast-two hybrid, coimmunoprecipitations, and cross-linking experiments. A rather large number of interaction partners have been identified, yet a functional interaction was unambiguously discovered for none of them.

Some of the latter results may need to be reconsidered critically. Several reports suffer from intrinsic methodological shortcomings, some of which were unavoidable at the time of publication. Furthermore, the likelihood of interactions between membrane-attached extracellular PrP^C with cytosolic or mitochondrial molecule is counterintuitive and may point to artifactual effects.

Although it is highly plausible in our opinion, the connection between the normal function of PrP^C and the neurotoxicity of prions remains admittedly hypothetical and lacks experimental confirmation. Alternative scenarios are thinkable, and it is not impossible that the cascade of events outlined above will prove incorrect. However, the depth of the knowledge gaps in prion biology, along with the medical importance of the neurodegeneration problem, indicates that many exciting discoveries still lie ahead. Therefore, despite the disappearance of "mad cow disease" from the media and from public awareness, the field of prion science remains an excellent choice for ambitious PhD students and postdocs willing to make an impact in current experimental biology.

DISCLOSURE STATEMENT

The authors are not aware of any biases that might be perceived as affecting the objectivity of this review.

LITERATURE CITED

Aguzzi A. 1996. Between cows and monkeys. *Nature* 381:734–55

Aguzzi A. 2007. Prion biology: the quest for the test. *Nat. Methods* 4:614–16

Aguzzi A, Glatzel M. 2006. Prion infections, blood and transfusions. *Nat. Clin. Pract. Neurol.* 2:321–29

Aguzzi A, Weissmann C. 1996. Spongiform encephalopathies: a suspicious signature. *Nature* 383:666–67

Andrews NJ, Farrington CP, Ward HJ, Cousens SN, Smith PG, et al. 2003. Deaths from variant Creutzfeldt-Jakob disease in the UK. *Lancet* 361:751–52

Aronoff-Spencer E, Burns CS, Avdievich NI, Gerfen GJ, Peisach J, et al. 2000. Identification of the Cu^{2+} binding sites in the N-terminal domain of the prion protein by EPR and CD spectroscopy. *Biochemistry* 39:13760–71

Atarashi R, Moore RA, Sim VL, Hughson AG, Dorward DW, et al. 2007. Ultrasensitive detection of scrapie prion protein using seeded conversion of recombinant prion protein. *Nat. Methods* 4:645–50

Azzalin A, Ferrara V, Arias A, Cerri S, Avella D, et al. 2006. Interaction between the cellular prion (PrP^C) and the 2P domain K^+ channel TREK-1 protein. *Biochem. Biophys. Res. Commun.* 346:108–15

Barrow PA, Holmgren CD, Tapper AJ, Jefferys JG. 1999. Intrinsic physiological and morphological properties of principal cells of the hippocampus and neocortex in hamsters infected with scrapie. *Neurobiol. Dis.* 6:406–23

Baumann F, Tolnay M, Brabeck C, Pahnke J, Kloz U, et al. 2007. Lethal recessive myelin toxicity of prion protein lacking its central domain. *EMBO J.* 26:538–47

Baumann MH, Kallijarvi J, Lankinen H, Soto C, Haltia M. 2000. Apolipoprotein E includes a binding site which is recognized by several amyloidogenic polypeptides. *Biochem. J.* 349:77–84

Beck KE, Kay JG, Braun JE. 2006. Rdj2, a J protein family member, interacts with cellular prion PrP(C). *Biochem. Biophys. Res. Commun.* 346:866–71

Behrens A, Genoud N, Naumann H, Rulicke T, Janett F, et al. 2002. Absence of the prion protein homologue Doppel causes male sterility. *EMBO J.* 21:3652–58

Berman HM, Westbrook J, Feng Z, Gilliland G, Bhat TN, et al. 2000. The Protein Data Bank. *Nucleic Acids Res.* 28:235–42

Bessen RA, Kocisko DA, Raymond GJ, Nandan S, Lansbury PT, Caughey B. 1995. Non-genetic propagation of strain-specific properties of scrapie prion protein. *Nature* 375:698–700

Blasi F, Carmeliet P. 2002. uPAR: a versatile signalling orchestrator. *Nat. Rev. Mol. Cell Biol.* 3:932–43

Bounhar Y, Zhang Y, Goodyer CG, LeBlanc A. 2001. Prion protein protects human neurons against Bax-mediated apoptosis. *J. Biol. Chem.* 276:39145–49

Bouzamondo-Bernstein E, Hopkins SD, Spilman P, Uyehara-Lock J, Deering C, et al. 2004. The neurodegeneration sequence in prion diseases: evidence from functional, morphological and ultrastructural studies of the GABAergic system. *J. Neuropathol. Exp. Neurol.* 63:882–99

Bragason BT, Palsdottir A. 2005. Interaction of PrP with NRAGE, a protein involved in neuronal apoptosis. *Mol. Cell Neurosci.* 29:232–44

Brandner S, Isenmann S, Raeber A, Fischer M, Sailer A, et al. 1996. Normal host prion protein necessary for scrapie-induced neurotoxicity. *Nature* 379:339–43

Brown DR, Nicholas RS, Canevari L. 2002. Lack of prion protein expression results in a neuronal phenotype sensitive to stress. *J. Neurosci. Res.* 67:211–24

Brown DR, Schmidt B, Kretzschmar HA. 1997a. Effects of oxidative stress on prion protein expression in PC12 cells. *Int. J. Dev. Neurosci.* 15:961–72

Brown DR, Schulz-Schaeffer WJ, Schmidt B, Kretzschmar HA. 1997b. Prion protein-deficient cells show altered response to oxidative stress due to decreased SOD-1 activity. *Exp. Neurol.* 146:104–12

Brown DR, Wong BS, Hafiz F, Clive C, Haswell SJ, Jones IM. 1999. Normal prion protein has an activity like that of superoxide dismutase. *Biochem. J.* 344(Pt.1):1–5

Brown P, Gibbs CJ Jr, Rodgers-Johnson P, Asher DM, Sulima MP, et al. 1994. Human spongiform encephalopathy: the National Institutes of Health series of 300 cases of experimentally transmitted disease. *Ann. Neurol.* 35:513–29

Bruce ME, Will RG, Ironside JW, McConnell I, Drummond D, et al. 1997. Transmissions to mice indicate that 'new variant' CJD is caused by the BSE agent. *Nature* 389:498–501

Bueler H, Aguzzi A, Sailer A, Greiner RA, Autenried P, et al. 1993. Mice devoid of PrP are resistant to scrapie. *Cell* 73:1339–47

Bueler H, Fischer M, Lang Y, Bluethmann H, Lipp HP, et al. 1992. Normal development and behaviour of mice lacking the neuronal cell-surface PrP protein. *Nature* 356:577–82

Cancellotti E, Wiseman F, Tuzi NL, Baybutt H, Monaghan P, et al. 2005. Altered glycosylated PrP proteins can have different neuronal trafficking in brain but do not acquire scrapie-like properties. *J. Biol. Chem.* 280:42909–18

Carleton A, Tremblay P, Vincent JD, Lledo PM. 2001. Dose-dependent, prion protein (PrP)-mediated facilitation of excitatory synaptic transmission in the mouse hippocampus. *Pflugers Arch.* 442:223–29

Castagna A, Campostrini N, Farinazzo A, Zanusso G, Monaco S, Righetti PG. 2002. Comparative two-dimensional mapping of prion protein isoforms in human cerebrospinal fluid and central nervous system. *Electrophoresis* 23:339–46

Castilla J, Saa P, Hetz C, Soto C. 2005a. In vitro generation of infectious scrapie prions. *Cell* 121:195–206

Castilla J, Saa P, Soto C. 2005b. Detection of prions in blood. *Nat. Med.* 11:982–85

Caughey B, Brown K, Raymond GJ, Katzenstein GE, Thresher W. 1994. Binding of the protease-sensitive form of PrP (prion protein) to sulfated glycosaminoglycan and congo red. *J. Virol.* 68:2135–41

Chattopadhyay M, Walter ED, Newell DJ, Jackson PJ, Aronoff-Spencer E, et al. 2005. The octarepeat domain of the prion protein binds Cu(II) with three distinct coordination modes at pH 7.4. *J. Am. Chem. Soc.* 127:12647–56

Chen S, Mangé A, Dong L, Lehmann S, Schachner M. 2003. Prion protein as trans-interacting partner for neurons is involved in neurite outgrowth and neuronal survival. *Mol. Cell Neurosci.* 22:227–33

Cheng F, Lindqvist J, Haigh CL, Brown DR, Mani K. 2006. Copper-dependent cointernalization of the prion protein and glypican-1. *J. Neurochem.* 98:1445–57

Chesebro B, Trifilo M, Race R, Meade-White K, Teng C, et al. 2005. Anchorless prion protein results in infectious amyloid disease without clinical scrapie. *Science* 308:1435–39

Chiarini LB, Freitas AR, Zanata SM, Brentani RR, Martins VR, Linden R. 2002. Cellular prion protein transduces neuroprotective signals. *EMBO J.* 21:3317–26

Chiesa R, Drisaldi B, Quaglio E, Migheli A, Piccardo P, et al. 2000. Accumulation of protease-resistant prion protein (PrP) and apoptosis of cerebellar granule cells in transgenic mice expressing a PrP insertional mutation. *Proc. Natl. Acad. Sci. USA* 97:5574–79

Chiesa R, Piccardo P, Ghetti B, Harris DA. 1998. Neurological illness in transgenic mice expressing a prion protein with an insertional mutation. *Neuron* 21:1339–51

Chiesa R, Piccardo P, Quaglio E, Drisaldi B, Si-Hoe SL, et al. 2003. Molecular distinction between pathogenic and infectious properties of the prion protein. *J. Virol.* 77:7611–22

Choi SI, Ju WK, Choi EK, Kim J, Lea HZ, et al. 1998. Mitochondrial dysfunction induced by oxidative stress in the brains of hamsters infected with the 263 K scrapie agent. *Acta Neuropathol.* 96:279–86

Clinton J, Forsyth C, Royston MC, Roberts GW. 1993. Synaptic degeneration is the primary neuropathological feature in prion disease: a preliminary study. *NeuroReport* 4:65–68

Colling SB, Collinge J, Jefferys JG. 1996. Hippocampal slices from prion protein null mice: disrupted $Ca^{(2+)}$-activated K^+ currents. *Neurosci. Lett.* 209:49–52

Collinge J, Whitfield J, McKintosh E, Beck J, Mead S, et al. 2006. Kuru in the 21st century—an acquired human prion disease with very long incubation periods. *Lancet* 367:2068–74

Collinge J, Whittington MA, Sidle KC, Smith CJ, Palmer MS, et al. 1994. Prion protein is necessary for normal synaptic function. *Nature* 370:295–97

Collins S, McLean CA, Masters CL. 2001. Gerstmann-Sträussler-Scheinker syndrome, fatal familial insomnia, and kuru: a review of these less common human transmissible spongiform encephalopathies. *J. Clin. Neurosci.* 8:387–97

Colognato H, Yurchenco PD. 2000. Form and function: the laminin family of heterotrimers. *Dev. Dyn.* 218:213–34

Creutzfeldt HG. 1920. Über eine eigenartige herdförmige erkrankung des zentralnervensystems. *Z. Ges. Neurol. Psychiatr.* 57:1–19

Criado JR, Sánchez-Alavez M, Conti B, Giacchino JL, Wills DN, et al. 2005. Mice devoid of prion protein have cognitive deficits that are rescued by reconstitution of PrP in neurons. *Neurobiol. Dis.* 19:255–65

Cuille J, Chellè PL. 1939. Experimental transmission of trembling to the goat. *C. R. Acad. Sci.* 208:1058–160

Curtis J, Errington M, Bliss T, Voss K, MacLeod N. 2003. Age-dependent loss of PTP and LTP in the hippocampus of PrP-null mice. *Neurobiol. Dis.* 13:55–62

Deleault NR, Geoghegan JC, Nishina K, Kascsak R, Williamson RA, Supattapone S. 2005. Protease-resistant prion protein amplification reconstituted with partially purified substrates and synthetic polyanions. *J. Biol. Chem.* 280:26873–79

Deleault NR, Harris BT, Rees JR, Supattapone S. 2007. Formation of native prions from minimal components in vitro. *Proc. Natl. Acad. Sci. USA* 104:9741–46

Diarra-Mehrpour M, Arrabal S, Jalil A, Pinson X, Gaudin C, et al. 2004. Prion protein prevents human breast carcinoma cell line from tumor necrosis factor α-induced cell death. *Cancer Res.* 64:719–27

Di Guglielmo GM, Le Roy C, Goodfellow AF, Wrana JL. 2003. Distinct endocytic pathways regulate TGF-β receptor signalling and turnover. *Nat. Cell Biol.* 5:410–21

Duffy P, Wolf J, Collins G, DeVoe AG, Streeten B, Cowen D. 1974. Possible person-to-person transmission of Creutzfeldt-Jakob disease. *N. Engl. J. Med.* 290:692–93

Edenhofer F, Rieger R, Famulok M, Wendler W, Weiss S, Winnacker EL. 1996. Prion protein PrPc interacts with molecular chaperones of the Hsp60 family. *J. Virol.* 70:4724–28

Elfrink K, Nagel-Steger L, Riesner D. 2007. Interaction of the cellular prion protein with raft-like lipid membranes. *Biol. Chem.* 388:79–89

Ellis V, Daniels M, Misra R, Brown DR. 2002. Plasminogen activation is stimulated by prion protein and regulated in a copper-dependent manner. *Biochemistry* 41:6891–96

Favre-Krey L, Theodoridou M, Boukouvala E, Panagiotidis CH, Papadopoulos AI, et al. 2007. Molecular characterization of a cDNA from the gilthead sea bream (*Sparus aurata*) encoding a fish prion protein. *Comp. Biochem. Physiol. B* 147:566–73

Ferrer I, Rivera R, Blanco R, Martí E. 1999. Expression of proteins linked to exocytosis and neurotransmission in patients with Creutzfeldt-Jakob disease. *Neurobiol. Dis.* 6:92–100

Fioriti L, Dossena S, Stewart LR, Stewart RS, Harris DA, et al. 2005. Cytosolic prion protein (PrP) is not toxic in N2a cells and primary neurons expressing pathogenic PrP mutations. *J. Biol. Chem.* 280:11320–28

Fischer M, Rülicke T, Raeber A, Sailer A, Moser M, et al. 1996. Prion protein (PrP) with amino-proximal deletions restoring susceptibility of PrP knockout mice to scrapie. *EMBO J.* 15:1255–64

Fischer MB, Roeckl C, Parizek P, Schwarz HP, Aguzzi A. 2000. Binding of disease-associated prion protein to plasminogen. *Nature* 408:479–83

Flechsig E, Shmerling D, Hegyi I, Raeber AJ, Fischer M, et al. 2000. Prion protein devoid of the octapeptide repeat region restores susceptibility to scrapie in PrP knockout mice. *Neuron* 27:399–408

Flechsig E, Weissmann C. 2004. The role of PrP in health and disease. *Curr. Mol. Med.* 4:337–53

Fournier JG, Escaig-Haye F, Billette de Villemeur T, Robain O. 1995. Ultrastructural localization of cellular prion protein (PrPc) in synaptic boutons of normal hamster hippocampus. *C. R. Acad. Sci. III* 318:339–44

Fukuuchi T, Doh-Ura K, Yoshihara S, Ohta S. 2006. Metal complexes with superoxide dismutase-like activity as candidates for antiprion drug. *Bioorg. Med. Chem. Lett.* 16:5982–87

Furlan S, La Penna G, Guerrieri F, Morante S, Rossi GC. 2007. Ab initio simulations of Cu binding sites on the N-terminal region of prion protein. *J. Biol. Inorg. Chem.* 12:571–83

Gabriel JM, Oesch B, Kretzschmar H, Scott M, Prusiner SB. 1992. Molecular cloning of a candidate chicken prion protein. *Proc. Natl. Acad. Sci. USA* 89:9097–101

Gajdusek DC, Gibbs CJ Jr, Alpers M. 1966. Experimental transmission of a Kuru-like syndrome to chimpanzees. *Nature* 209:794–96

Gajdusek DC, Gibbs CJ Jr, Alpers M. 1967. Transmission and passage of experimenal "kuru" to chimpanzees. *Science* 155:212–14

Gajdusek DC, Gibbs CJ Jr, Asher DM, David E. 1968. Transmission of experimental kuru to the spider monkey (*Ateles geoffreyi*). *Science* 162:693–94

Gajdusek DC, Reid LH. 1961. Studies on kuru. IV. The kuru pattern in Moke, a representative Fore village. *Am. J. Trop. Med. Hyg.* 10:628–38

Gajdusek DC, Zigas V. 1957. Degenerative disease of the central nervous system in New Guinea; the endemic occurrence of kuru in the native population. *N. Engl. J. Med.* 257:974–78

Gao C, Lei YJ, Han J, Shi Q, Chen L, et al. 2006. Recombinant neural protein PrP can bind with both recombinant and native apolipoprotein E in vitro. *Acta Biochim. Biophys. Sin.* 38:593–601

Gauczynski S, Peyrin JM, Haik S, Leucht C, Hundt C, et al. 2001. The 37-kDa/67-kDa laminin receptor acts as the cell-surface receptor for the cellular prion protein. *EMBO J.* 20:5863–75

Gerstmann J. 1928. Über ein noch nicht beschriebenes Reflexphanomen bei einer Erkrankung des zerebellaren Systems. *Wien. Medizin Wochenschr.* 78:906–8

Gerstmann J, Sträussler E, Scheinker I. 1936. Über eine eigenartige hereditar-familiäre Erkrankung des Zentralnervensystems. Zugleich ein Beitrag zur Frage des vorzeitigen lokalen Alterns. *Z. Neurol.* 154:736–62

Gohel C, Grigoriev V, Escaig-Haye F, Lasmezas CI, Deslys JP, et al. 1999. Ultrastructural localization of cellular prion protein (PrPc) at the neuromuscular junction. *J. Neurosci. Res.* 55:261–67

Goldfarb LG, Petersen RB, Tabaton M, Brown P, LeBlanc AC, et al. 1992. Fatal familial insomnia and familial Creutzfeldt-Jakob disease: disease phenotype determined by a DNA polymorphism. *Science* 258:806–8

Graner E, Mercadante AF, Zanata SM, Forlenza OV, Cabral AL, et al. 2000a. Cellular prion protein binds laminin and mediates neuritogenesis. *Brain Res. Mol. Brain Res.* 76:85–92

Graner E, Mercadante AF, Zanata SM, Martins VR, Jay DG, Brentani RR. 2000b. Laminin-induced PC-12 cell differentiation is inhibited following laser inactivation of cellular prion protein. *FEBS Lett.* 482:257–60

Grigoriev V, Escaig-Haye F, Streichenberger N, Kopp N, Langeveld J, et al. 1999. Submicroscopic immunodetection of PrP in the brain of a patient with a new-variant of Creutzfeldt-Jakob disease. *Neurosci. Lett.* 264:57–60

Hadlow WJ. 1959. Scrapie and kuru. *Lancet* 2:289–90

Hajj GN, Lopes MH, Mercadante AF, Veiga SS, da Silveira RB, et al. 2007. Cellular prion protein interaction with vitronectin supports axonal growth and is compensated by integrins. *J. Cell Sci.* 120:1915–26

Hegde RS, Mastrianni JA, Scott MR, DeFea KA, Tremblay P, et al. 1998. A transmembrane form of the prion protein in neurodegenerative disease. *Science* 279:827–34

Hegde RS, Tremblay P, Groth D, DeArmond SJ, Prusiner SB, Lingappa VR. 1999. Transmissible and genetic prion diseases share a common pathway of neurodegeneration. *Nature* 402:822–26

Herms JW, Tings T, Dunker S, Kretzschmar HA. 2001. Prion protein affects Ca^{2+}-activated K^+ currents in cerebellar purkinje cells. *Neurobiol. Dis.* 8:324–30

Hill AF, Desbruslais M, Joiner S, Sidle KC, Gowland I, et al. 1997. The same prion strain causes vCJD and BSE. *Nature* 389:448–50, 526

Holscher C, Delius H, Burkle A. 1998. Overexpression of nonconvertible PrPc delta114–121 in scrapie-infected mouse neuroblastoma cells leads to trans-dominant inhibition of wild-type PrP(Sc) accumulation. *J. Virol.* 72:1153–59

Hornemann S, Korth C, Oesch B, Riek R, Wider G, et al. 1997. Recombinant full-length murine prion protein, mPrP(23–231): purification and spectroscopic characterization. *FEBS Lett.* 413:277–81

Hsiao K, Baker HF, Crow TJ, Poulter M, Owen F, et al. 1989. Linkage of a prion protein missense variant to Gerstmann-Sträussler syndrome. *Nature* 338:342–45

Hsiao KK, Groth D, Scott M, Yang SL, Serban H, et al. 1994. Serial transmission in rodents of neurodegeneration from transgenic mice expressing mutant prion protein. *Proc. Natl. Acad. Sci. USA* 91:9126–30

Hundt C, Peyrin JM, Haik S, Gauczynski S, Leucht C, et al. 2001. Identification of interaction domains of the prion protein with its 37-kDa/67-kDa laminin receptor. *EMBO J.* 20:5876–86

Hutter G, Heppner FL, Aguzzi A. 2003. No superoxide dismutase activity of cellular prion protein in vivo. *Biol. Chem.* 384:1279–85

Isaacs JD, Jackson GS, Altmann DM. 2006. The role of the cellular prion protein in the immune system. *Clin. Exp. Immunol.* 146:1–8

Itoh Y, Yamada M, Hayakawa M, Shozawa T, Tanaka J, et al. 1994. A variant of Gerstmann-Sträussler-Scheinker disease carrying codon 105 mutation with codon 129 polymorphism of the prion protein gene: a clinicopathological study. *J. Neurol. Sci.* 127:77–86

Jakob A. 1921. Über eigenartige erkrankungen des zentralnervensystems mit bemerkenswertem anatomischem befunde (spastische pseudosklerose-encephalomyelopathie mit disseminierten degenerationsherden). *Z. Ges. Neurol. Psychiatr.* 64:147–228

Jeffrey M, Halliday WG, Bell J, Johnston AR, MacLeod NK, et al. 2000. Synapse loss associated with abnormal PrP precedes neuronal degeneration in the scrapie-infected murine hippocampus. *Neuropathol. Appl. Neurobiol.* 26:41–54

Johnston AR, Fraser JR, Jeffrey M, MacLeod N. 1998. Synaptic plasticity in the CA1 area of the hippocampus of scrapie-infected mice. *Neurobiol. Dis.* 5:188–95

Jones S, Batchelor M, Bhelt D, Clarke AR, Collinge J, Jackson GS. 2005. Recombinant prion protein does not possess SOD-1 activity. *Biochem. J.* 392:309–12

Kanaani J, Prusiner SB, Diacovo J, Baekkeskov S, Legname G. 2005. Recombinant prion protein induces rapid polarization and development of synapses in embryonic rat hippocampal neurons in vitro. *J. Neurochem.* 95:1373–86

Kaneko K, Vey M, Scott M, Pilkuhn S, Cohen FE, Prusiner SB. 1997. COOH-terminal sequence of the cellular prion protein directs subcellular trafficking and controls conversion into the scrapie isoform. *Proc. Natl. Acad. Sci. USA* 94:2333–38

Keshet GI, Bar-Peled O, Yaffe D, Nudel U, Gabizon R. 2000. The cellular prion protein colocalizes with the dystroglycan complex in the brain. *J. Neurochem.* 75:1889–97

Keshet GI, Ovadia H, Taraboulos A, Gabizon R. 1999. Scrapie-infected mice and PrP knockout mice share abnormal localization and activity of neuronal nitric oxide synthase. *J. Neurochem.* 72:1224–31

Kiachopoulos S, Bracher A, Winklhofer KF, Tatzelt J. 2005. Pathogenic mutations located in the hydrophobic core of the prion protein interfere with folding and attachment of the glycosylphosphatidylinositol anchor. *J. Biol. Chem.* 280:9320–29

Kirkham M, Parton RG. 2005. Clathrin-independent endocytosis: new insights into caveolae and noncaveolar lipid raft carriers. *Biochim. Biophys. Acta* 1746:349–63

Kitamoto T, Shin RW, Doh-ura K, Tomokane N, Miyazono M, et al. 1992. Abnormal isoform of prion proteins accumulates in the synaptic structures of the central nervous system in patients with Creutzfeldt-Jakob disease. *Am. J. Pathol.* 140:1285–94

Klamt F, Dal-Pizzol F, Conte da Frota MJ, Walz R, Andrades ME, et al. 2001. Imbalance of antioxidant defense in mice lacking cellular prion protein. *Free Radic. Biol. Med.* 30:1137–44

Kocisko DA, Priola SA, Raymond GJ, Chesebro B, Lansbury PT Jr, Caughey B. 1995. Species specificity in the cell-free conversion of prion protein to protease-resistant forms: a model for the scrapie species barrier. *Proc. Natl. Acad. Sci. USA* 92:3923–27

Korth C, Kaneko K, Prusiner SB. 2000. Expression of unglycosylated mutated prion protein facilitates PrP(Sc) formation in neuroblastoma cells infected with different prion strains. *J. Gen. Virol.* 81:2555–63

Kovacs GG, Trabattoni G, Hainfellner JA, Ironside JW, Knight RS, Budka H. 2002. Mutations of the prion protein gene phenotypic spectrum. *J. Neurol.* 249:1567–82

Kurschner C, Morgan JI. 1995. The cellular prion protein (PrP) selectively binds to Bcl-2 in the yeast two-hybrid system. *Brain Res. Mol. Brain Res.* 30:165–68

Kuwahara C, Takeuchi AM, Nishimura T, Haraguchi K, Kubosaki A, et al. 1999. Prions prevent neuronal cell-line death. *Nature* 400:225–26

Laine J, Marc ME, Sy MS, Axelrad H. 2001. Cellular and subcellular morphological localization of normal prion protein in rodent cerebellum. *Eur. J. Neurosci.* 14:47–56

Leclerc E, Serban H, Prusiner SB, Burton DR, Williamson RA. 2006. Copper induces conformational changes in the N-terminal part of cell-surface PrPC. *Arch. Virol.* 151:2103–9

Lee DW, Sohn HO, Lim HB, Lee YG, Kim YS, et al. 1999. Alteration of free radical metabolism in the brain of mice infected with scrapie agent. *Free Radic. Res.* 30:499–507

Legname G, Baskakov IV, Nguyen HO, Riesner D, Cohen FE, et al. 2004. Synthetic mammalian prions. *Science* 305:673–76

Legname G, Nguyen HO, Baskakov IV, Cohen FE, Dearmond SJ, Prusiner SB. 2005. Strain-specified characteristics of mouse synthetic prions. *Proc. Natl. Acad. Sci. USA* 102:2168–73

Li A, Christensen HM, Stewart LR, Roth KA, Chiesa R, Harris DA. 2007. Neonatal lethality in transgenic mice expressing prion protein with a deletion of residues 105–125. *EMBO J.* 26:548–58

Lledo PM, Tremblay P, DeArmond SJ, Prusiner SB, Nicoll RA. 1996. Mice deficient for prion protein exhibit normal neuronal excitability and synaptic transmission in the hippocampus. *Proc. Natl. Acad. Sci. USA* 93:2403–7

Llewelyn CA, Hewitt PE, Knight RS, Amar K, Cousens S, et al. 2004. Possible transmission of variant Creutzfeldt-Jakob disease by blood transfusion. *Lancet* 363:417–21

Lopes MH, Hajj GN, Muras AG, Mancini GL, Castro RM, et al. 2005. Interaction of cellular prion and stress-inducible protein 1 promotes neuritogenesis and neuroprotection by distinct signaling pathways. *J. Neurosci.* 25:11330–39

Lugaresi E, Medori R, Montagna P, Baruzzi A, Cortelli P, et al. 1986. Fatal familial insomnia and dysautonomia with selective degeneration of thalamic nuclei. *N. Engl. J. Med.* 315:997–1003

Ma J, Lindquist S. 2002. Conversion of PrP to a self-perpetuating PrPSc-like conformation in the cytosol. *Science* 298:1785–88

Ma J, Wollmann R, Lindquist S. 2002. Neurotoxicity and neurodegeneration when PrP accumulates in the cytosol. *Science* 298:1781–85

Maglio LE, Martins VR, Izquierdo I, Ramirez OA. 2006. Role of cellular prion protein on LTP expression in aged mice. *Brain Res.* 1097:11–18

Maglio LE, Perez MF, Martins VR, Brentani RR, Ramirez OA. 2004. Hippocampal synaptic plasticity in mice devoid of cellular prion protein. *Brain Res. Mol. Brain Res.* 131:58–64

Mallucci G, Dickinson A, Linehan J, Klöhn PC, Brandner S, Collinge J. 2003. Depleting neuronal PrP in prion infection prevents disease and reverses spongiosis. *Science* 302:871–74

Mallucci GR, Ratte S, Asante EA, Linehan J, Gowland I, et al. 2002. Post-natal knockout of prion protein alters hippocampal CA1 properties, but does not result in neurodegeneration. *EMBO J.* 21:202–10

Maness PF, Schachner M. 2007. Neural recognition molecules of the immunoglobulin superfamily: signaling transducers of axon guidance and neuronal migration. *Nat. Neurosci.* 10:19–26

Manson JC, Clarke AR, Hooper ML, Aitchison L, McConnell I, Hope J. 1994. 129/Ola mice carrying a null mutation in PrP that abolishes mRNA production are developmentally normal. *Mol. Neurobiol.* 8:121–27

Marella M, Lehmann S, Grassi J, Chabry J. 2002. Filipin prevents pathological prion protein accumulation by reducing endocytosis and inducing cellular PrP release. *J. Biol. Chem.* 277:25457–64

Masters CL, Gajdusek DC, Gibbs CJ Jr. 1981. Creutzfeldt-Jakob disease virus isolations from the Gerstmann-Straussler syndrome with an analysis of the various forms of amyloid plaque deposition in the virus-induced spongiform encephalopathies. *Brain* 104:559–88

Mastrianni JA, Nixon R, Layzer R, Telling GC, Han D, et al. 1999. Prion protein conformation in a patient with sporadic fatal insomnia. *N. Engl. J. Med.* 340:1630–38

Matsuda H, Mitsuda H, Nakamura N, Furusawa S, Mohri S, Kitamoto T. 1999. A chicken monoclonal antibody with specificity for the N-terminal of human prion protein. *FEMS Immunol. Med. Microbiol.* 23:189–94

Mattei V, Garofalo T, Misasi R, Circella A, Manganelli V, et al. 2004. Prion protein is a component of the multimolecular signaling complex involved in T cell activation. *FEBS Lett.* 560:14–18

McLennan NF, Brennan PM, McNeill A, Davies I, Fotheringham A, et al. 2004. Prion protein accumulation and neuroprotection in hypoxic brain damage. *Am. J. Pathol.* 165:227–35

Medori R, Tritschler HJ, LeBlanc A, Villare F, Manetto V, et al. 1992. Fatal familial insomnia, a prion disease with a mutation at codon 178 of the prion protein gene. *N. Engl. J. Med.* 326:444–49

Meggio F, Negro A, Sarno S, Ruzzene M, Bertoli A, et al. 2000. Bovine prion protein as a modulator of protein kinase CK2. *Biochem. J.* 352(Pt. 1):191–96

Miele G, Jeffrey M, Turnbull D, Manson J, Clinton M. 2002. Ablation of cellular prion protein expression affects mitochondrial numbers and morphology. *Biochem. Biophys. Res. Commun.* 291:372–77

Milhavet O, Lehmann S. 2002. Oxidative stress and the prion protein in transmissible spongiform encephalopathies. *Brain Res. Brain Res. Rev.* 38:328–39

Mironov A Jr, Latawiec D, Wille H, Bouzamondo-Bernstein E, Legname G, et al. 2003. Cytosolic prion protein in neurons. *J. Neurosci.* 23:7183–93

Mitteregger G, Vosko M, Krebs B, Xiang W, Kohlmannsperger V, et al. 2007. The role of the octarepeat region in neuroprotective function of the cellular prion protein. *Brain Pathol.* 17:174–83

Moore RC, Lee IY, Silverman GL, Harrison PM, Strome R, et al. 1999. Ataxia in prion protein (PrP)-deficient mice is associated with upregulation of the novel PrP-like protein Doppel. *J. Mol. Biol.* 292:797–817

Mouillet-Richard S, Ermonval M, Chebassier C, Laplanche JL, Lehmann S, et al. 2000. Signal transduction through prion protein. *Science* 289:1925–28

Mouillet-Richard S, Pietri M, Schneider B, Vidal C, Mutel V, et al. 2005. Modulation of serotonergic receptor signaling and cross-talk by prion protein. *J. Biol. Chem.* 280:4592–601

Moya KL, Sales N, Hassig R, Creminon C, Grassi J, Di Giamberardino L. 2000. Immunolocalization of the cellular prion protein in normal brain. *Microsc. Res. Tech.* 50:58–65

Muramoto T, DeArmond SJ, Scott M, Telling GC, Cohen FE, Prusiner SB. 1997. Heritable disorder resembling neuronal storage disease in mice expressing prion protein with deletion of an α-helix. *Nat. Med.* 3:750–55

Muramoto T, Scott M, Cohen FE, Prusiner SB. 1996. Recombinant scrapie-like prion protein of 106 amino acids is soluble. *Proc. Natl. Acad. Sci. USA* 93:15457–62

Nazor KE, Kuhn F, Seward T, Green M, Zwald D, et al. 2005. Immunodetection of disease-associated mutant PrP, which accelerates disease in GSS transgenic mice. *EMBO J.* 24:2472–80

Nazor KE, Seward T, Telling GC. 2007. Motor behavioral and neuropathological deficits in mice deficient for normal prion protein expression. *Biochim. Biophys. Acta* 1772:645–53

Neuendorf E, Weber A, Saalmueller A, Schatzl H, Reifenberg K, et al. 2004. Glycosylation deficiency at either one of the two glycan attachment sites of cellular prion protein preserves susceptibility to bovine spongiform encephalopathy and scrapie infections. *J. Biol. Chem.* 279:53306–16

Nieznanski K, Nieznanska H, Skowronek KJ, Osiecka KM, Stepkowski D. 2005. Direct interaction between prion protein and tubulin. *Biochem. Biophys. Res. Commun.* 334:403–11

Nishida N, Tremblay P, Sugimoto T, Shigematsu K, Shirabe S, et al. 1999. A mouse prion protein transgene rescues mice deficient for the prion protein gene from Purkinje cell degeneration and demyelination. *Lab. Invest.* 79:689–97

Oidtmann B, Simon D, Holtkamp N, Hoffmann R, Baier M. 2003. Identification of cDNAs from Japanese pufferfish (*Fugu rubripes*) and Atlantic salmon (*Salmo salar*) coding for homologues to tetrapod prion proteins. *FEBS Lett.* 538:96–100

Pan T, Wong BS, Liu T, Li R, Petersen RB, Sy MS. 2002. Cell-surface prion protein interacts with glycosaminoglycans. *Biochem. J.* 368:81–90

Parchi P, Capellari S, Chin S, Schwarz HB, Schecter NP, et al. 1999. A subtype of sporadic prion disease mimicking fatal familial insomnia. *Neurology* 52:1757–63

Parkin ET, Watt NT, Hussain I, Eckman EA, Eckman CB, et al. 2007. Cellular prion protein regulates beta-secretase cleavage of the Alzheimer's amyloid precursor protein. *Proc. Natl. Acad. Sci. USA* 104:11062–67

Parton RG, Richards AA. 2003. Lipid rafts and caveolae as portals for endocytosis: new insights and common mechanisms. *Traffic* 4:724–38

Peden AH, Head MW, Ritchie DL, Bell JE, Ironside JW. 2004. Preclinical vCJD after blood transfusion in a PRNP codon 129 heterozygous patient. *Lancet* 364:527–29

Peters PJ, Mironov A Jr, Peretz D, van Donselaar E, Leclerc E, et al. 2003. Trafficking of prion proteins through a caveolae-mediated endosomal pathway. *J. Cell Biol.* 162:703–17

Petrakis S, Sklaviadis T. 2006. Identification of proteins with high affinity for refolded and native PrPC. *Proteomics* 6:6476–84

Prado MA, Alves-Silva J, Magalhaes AC, Prado VF, Linden R, et al. 2004. PrPc on the road: trafficking of the cellular prion protein. *J. Neurochem.* 88:769–81

Priola SA, Lawson VA. 2001. Glycosylation influences cross-species formation of protease-resistant prion protein. *EMBO J.* 20:6692–99

Prusiner SB, Scott MR. 1997. Genetics of prions. *Annu. Rev. Genet.* 31:139–75

Qin K, Yang Y, Mastrangelo P, Westaway D. 2002. Mapping Cu(II) binding sites in prion proteins by diethyl pyrocarbonate modification and matrix-assisted laser desorption ionization-time of flight (MALDI-TOF) mass spectrometric footprinting. *J. Biol. Chem.* 277:1981–90

Radovanovic I, Braun N, Giger OT, Mertz K, Miele G, et al. 2005. Truncated prion protein and Doppel are myelinotoxic in the absence of oligodendrocytic PrPC. *J. Neurosci.* 25:4879–88

Re L, Rossini F, Re F, Bordicchia M, Mercanti A, et al. 2006. Prion protein potentiates acetylcholine release at the neuromuscular junction. *Pharmacol. Res.* 53:62–68

Rieger R, Edenhofer F, Lasmezas CI, Weiss S. 1997. The human 37-kDa laminin receptor precursor interacts with the prion protein in eukaryotic cells. *Nat. Med.* 3:1383–88

Riek R, Hornemann S, Wider G, Billeter M, Glockshuber R, Wuthrich K. 1996. NMR structure of the mouse prion protein domain PrP(121–321). *Nature* 382:180–82

Rivera-Milla E, Oidtmann B, Panagiotidis CH, Baier M, Sklaviadis T, et al. 2006. Disparate evolution of prion protein domains and the distinct origin of Doppel- and prion-related loci revealed by fish-to-mammal comparisons. *FASEB J.* 20:317–19

Rivera-Milla E, Stuermer CA, Malaga-Trillo E. 2003. An evolutionary basis for scrapie disease: identification of a fish prion mRNA. *Trends Genet.* 19:72–75

Roikhel VM, Fokina GI, Sobolev SG, Korolev MB, Ravkina LI, Pogodina VV. 1983. Study of early stages of the pathogenesis of scrapie in experimentally infected mice. *Acta Virol.* 27:147–53

Rossi D, Cozzio A, Flechsig E, Klein MA, Rulicke T, et al. 2001. Onset of ataxia and Purkinje cell loss in PrP null mice inversely correlated with Dpl level in brain. *EMBO J.* 20:694–702

Roucou X, Giannopoulos PN, Zhang Y, Jodoin J, Goodyer CG, LeBlanc A. 2005. Cellular prion protein inhibits proapoptotic Bax conformational change in human neurons and in breast carcinoma MCF-7 cells. *Cell Death Differ.* 12:783–95

Roucou X, Guo Q, Zhang Y, Goodyer CG, LeBlanc AC. 2003. Cytosolic prion protein is not toxic and protects against Bax-mediated cell death in human primary neurons. *J. Biol. Chem.* 278:40877–81

Roucou X, LeBlanc AC. 2005. Cellular prion protein neuroprotective function: implications in prion diseases. *J. Mol. Med.* 83:3–11

Rousseau C, Felin M, Doyennette-Moyne MA, Seve AP. 1997. CBP70, a glycosylated nuclear lectin. *J. Cell. Biochem.* 66:370–85

Rybner C, Finel-Szermanski S, Felin M, Sahraoui T, Rousseau C, et al. 2002. The cellular prion protein: a new partner of the lectin CBP70 in the nucleus of NB4 human promyelocytic leukemia cells. *J. Cell. Biochem.* 84:408–19

Saa P, Castilla J, Soto C. 2006. Presymptomatic detection of prions in blood. *Science* 313:92–94

Sakaguchi S, Katamine S, Nishida N, Moriuchi R, Shigematsu K, et al. 1996. Loss of cerebellar Purkinje cells in aged mice homozygous for a disrupted PrP gene. *Nature* 380:528–31

Sakai T, Hohjoh H. 2006. Gene silencing analyses against amyloid precursor protein (APP) gene family by RNA interference. *Cell Biol. Int.* 30:952–56

Salehi AH, Roux PP, Kubu CJ, Zeindler C, Bhakar A, et al. 2000. NRAGE, a novel MAGE protein, interacts with the p75 neurotrophin receptor and facilitates nerve growth factor-dependent apoptosis. *Neuron* 27:279–88

Sales N, Rodolfo K, Hassig R, Faucheux B, Di Giamberardino L, Moya KL. 1998. Cellular prion protein localization in rodent and primate brain. *Eur. J. Neurosci.* 10:2464–71

Santuccione A, Sytnyk V, Leshchyns'ka I, Schachner M. 2005. Prion protein recruits its neuronal receptor NCAM to lipid rafts to activate p59fyn and to enhance neurite outgrowth. *J. Cell Biol.* 169:341–54

Schmitt-Ulms G, Hansen K, Liu J, Cowdrey C, Yang J, et al. 2004. Time-controlled transcardiac perfusion cross-linking for the study of protein interactions in complex tissues. *Nat. Biotechnol.* 22:724–31

Schmitt-Ulms G, Legname G, Baldwin MA, Ball HL, Bradon N, et al. 2001. Binding of neural cell adhesion molecules (N-CAMs) to the cellular prion protein. *J. Mol. Biol.* 314:1209–25

Schneider B, Mutel V, Pietri M, Ermonval M, Mouillet-Richard S, Kellermann O. 2003. NADPH oxidase and extracellular regulated kinases 1/2 are targets of prion protein signaling in neuronal and nonneuronal cells. *Proc. Natl. Acad. Sci. USA* 100:13326–31

Shmerling D, Hegyi I, Fischer M, Blattler T, Brandner S, et al. 1998. Expression of amino-terminally truncated PrP in the mouse leading to ataxia and specific cerebellar lesions. *Cell* 93:203–14

Shyng SL, Heuser JE, Harris DA. 1994. A glycolipid-anchored prion protein is endocytosed via clathrin-coated pits. *J. Cell. Biol.* 125:1239–50

Shyu WC, Lin SZ, Chiang MF, Ding DC, Li KW, et al. 2005. Overexpression of PrPC by adenovirus-mediated gene targeting reduces ischemic injury in a stroke rat model. *J. Neurosci.* 25:8967–77

Simonic T, Duga S, Strumbo B, Asselta R, Ceciliani F, Ronchi S. 2000. cDNA cloning of turtle prion protein. *FEBS Lett.* 469:33–38

Spielhaupter C, Schatzl HM. 2001. PrPC directly interacts with proteins involved in signaling pathways. *J. Biol. Chem.* 276:44604–12

Spudich A, Frigg R, Kilic E, Kilic U, Oesch B, et al. 2005. Aggravation of ischemic brain injury by prion protein deficiency: role of ERK-1/-2 and STAT-1. *Neurobiol. Dis.* 20:442–49

Stahl N, Borchelt DR, Hsiao K, Prusiner SB. 1987. Scrapie prion protein contains a phosphatidylinositol glycolipid. *Cell* 51:229–40

Steele AD, Emsley JG, Ozdinler PH, Lindquist S, Macklis JD. 2006. Prion protein (PrPc) positively regulates neural precursor proliferation during developmental and adult mammalian neurogenesis. *Proc. Natl. Acad. Sci. USA* 103:3416–21

Stewart RS, Harris DA. 2001. Most pathogenic mutations do not alter the membrane topology of the prion protein. *J. Biol. Chem.* 276:2212–20

Stewart RS, Harris DA. 2005. A transmembrane form of the prion protein is localized in the Golgi apparatus of neurons. *J. Biol. Chem.* 280:15855–64

Stewart RS, Piccardo P, Ghetti B, Harris DA. 2005. Neurodegenerative illness in transgenic mice expressing a transmembrane form of the prion protein. *J. Neurosci.* 25:3469–77

Stockel J, Safar J, Wallace AC, Cohen FE, Prusiner SB. 1998. Prion protein selectively binds copper(II) ions. *Biochemistry* 37:7185–93

Strumbo B, Ronchi S, Bolis LC, Simonic T. 2001. Molecular cloning of the cDNA coding for *Xenopus laevis* prion protein. *FEBS Lett.* 508:170–74

Sun G, Guo M, Shen A, Mei F, Peng X, et al. 2005. Bovine PrPC directly interacts with αB-crystalline. *FEBS Lett.* 579:5419–24

Sunyach C, Jen A, Deng J, Fitzgerald KT, Frobert Y, et al. 2003. The mechanism of internalization of glycosylphosphatidylinositol-anchored prion protein. *EMBO J.* 22:3591–601

Supattapone S, Bouzamondo E, Ball HL, Wille H, Nguyen HO, et al. 2001. A protease-resistant 61-residue prion peptide causes neurodegeneration in transgenic mice. *Mol. Cell Biol.* 21:2608–16

Suzuki T, Kurokawa T, Hashimoto H, Sugiyama M. 2002. cDNA sequence and tissue expression of Fugu rubripes prion protein-like: a candidate for the teleost orthologue of tetrapod PrPs. *Biochem. Biophys. Res. Commun.* 294:912–17

Tateishi J, Brown P, Kitamoto T, Hoque ZM, Roos R, et al. 1995. First experimental transmission of fatal familial insomnia. *Nature* 376:434–35

Tateishi J, Kitamoto T, Hashiguchi H, Shii H. 1988. Gerstmann-Sträussler-Scheinker disease: immunohistological and experimental studies. *Ann. Neurol.* 24:35–40

Tateishi J, Kitamoto T, Hoque MZ, Furukawa H. 1996a. Experimental transmission of Creutzfeldt-Jakob disease and related diseases to rodents. *Neurology* 46:532–37

Tateishi J, Kitamoto T, Kretzschmar H, Mehraein P. 1996b. Immunhistological evaluation of Creutzfeldt-Jakob disease with reference to the type PrPres deposition. *Clin. Neuropathol.* 15:358–60

Taylor DR, Hooper NM. 2007. The low-density lipoprotein receptor-related protein 1 (LRP1) mediates the endocytosis of the cellular prion protein. *Biochem. J.* 402:17–23

Taylor DR, Watt NT, Perera WS, Hooper NM. 2005. Assigning functions to distinct regions of the N-terminus of the prion protein that are involved in its copper-stimulated, clathrin-dependent endocytosis. *J. Cell Sci.* 118:5141–53

Telling GC, Haga T, Torchia M, Tremblay P, DeArmond SJ, Prusiner SB. 1996. Interactions between wild-type and mutant prion proteins modulate neurodegeneration in transgenic mice. *Genes Dev.* 10:1736–50

Tobler I, Gaus SE, Deboer T, Achermann P, Fischer M, et al. 1996. Altered circadian activity rhythms and sleep in mice devoid of prion protein. *Nature* 380:639–42

Toni M, Spisni E, Griffoni C, Santi S, Riccio M, et al. 2006. Cellular prion protein and caveolin-1 interaction in a neuronal cell line precedes fyn/erk 1/2 signal transduction. *J. Biomed. Biotechnol.* 2006:69469

Vassallo N, Herms J. 2003. Cellular prion protein function in copper homeostasis and redox signalling at the synapse. *J. Neurochem.* 86:538–44

Vey M, Pilkuhn S, Wille H, Nixon R, DeArmond SJ, et al. 1996. Subcellular colocalization of the cellular and scrapie prion proteins in caveolae-like membranous domains. *Proc. Natl. Acad. Sci. USA* 93:14945–49

Volkel D, Zimmermann K, Zerr I, Bodemer M, Lindner T, et al. 2001. Immunochemical determination of cellular prion protein in plasma from healthy subjects and patients with sporadic CJD or other neurologic diseases. *Transfusion* 41:441–48

Waggoner DJ, Drisaldi B, Bartnikas TB, Casareno RL, Prohaska JR, et al. 2000. Brain copper content and cuproenzyme activity do not vary with prion protein expression level. *J. Biol. Chem.* 275:7455–58

Walis A, Bratosiewicz J, Sikorska B, Brown P, Gajdusek DC, Liberski PP. 2003. Ultrastructural changes in the optic nerves of rodents with experimental Creutzfeldt-Jakob Disease (CJD), Gerstmann-Sträussler-Scheinker disease (GSS) or scrapie. *J. Comp. Pathol.* 129:213–25

Walker L, Levine H, Jucker M. 2006. Koch's postulates and infectious proteins. *Acta Neuropathol.* 112:1–4

Warner RG, Hundt C, Weiss S, Turnbull JE. 2002. Identification of the heparan sulfate binding sites in the cellular prion protein. *J. Biol. Chem.* 277:18421–30

Weise J, Crome O, Sandau R, Schulz-Schaeffer W, Bahr M, Zerr I. 2004. Upregulation of cellular prion protein (PrPc) after focal cerebral ischemia and influence of lesion severity. *Neurosci. Lett.* 372:146–50

Weise J, Sandau R, Schwarting S, Crome O, Wrede A, et al. 2006. Deletion of cellular prion protein results in reduced Akt activation, enhanced postischemic caspase-3 activation, and exacerbation of ischemic brain injury. *Stroke* 37:1296–300

Weissmann C, Flechsig E. 2003. PrP knock-out and PrP transgenic mice in prion research. *Br. Med. Bull.* 66:43–60

Wells GA, Scott AC, Johnson CT, Gunning RF, Hancock RD, et al. 1987. A novel progressive spongiform encephalopathy in cattle. *Vet. Rec.* 121:419–20

Westaway D, DeArmond SJ, Cayetano-Canlas J, Groth D, Foster D, et al. 1994. Degeneration of skeletal muscle, peripheral nerves, and the central nervous system in transgenic mice overexpressing wild-type prion proteins. *Cell* 76:117–29

Westergard L, Christensen HM, Harris DA. 2007. The cellular prion protein (PrP(C)): its physiological function and role in disease. *Biochim. Biophys. Acta* 1772:629–44

White AR, Collins SJ, Maher F, Jobling MF, Stewart LR, et al. 1999. Prion protein-deficient neurons reveal lower glutathione reductase activity and increased susceptibility to hydrogen peroxide toxicity. *Am. J. Pathol.* 155:1723–30

Will RG. 2003. Acquired prion disease: iatrogenic CJD, variant CJD, kuru. *Br. Med. Bull.* 66:255–65

Wong BS, Brown DR, Pan T, Whiteman M, Liu T, et al. 2001a. Oxidative impairment in scrapie-infected mice is associated with brain metals perturbations and altered antioxidant activities. *J. Neurochem.* 79:689–98

Wong BS, Liu T, Li R, Pan T, Petersen RB, et al. 2001b. Increased levels of oxidative stress markers detected in the brains of mice devoid of prion protein. *J. Neurochem.* 76:565–72

Wong BS, Liu T, Paisley D, Li R, Pan T, et al. 2001c. Induction of HO-1 and NOS in Doppel-expressing mice devoid of PrP: implications for Doppel function. *Mol. Cell Neurosci.* 17:768–75

Wroe SJ, Pal S, Siddique D, Hyare H, Macfarlane R, et al. 2006. Clinical presentation and premortem diagnosis of variant Creutzfeldt-Jakob disease associated with blood transfusion: a case report. *Lancet* 368:2061–67

Yehiely F, Bamborough P, Da Costa M, Perry BJ, Thinakaran G, et al. 1997. Identification of candidate proteins binding to prion protein. *Neurobiol. Dis.* 3:339–55

York RD, Molliver DC, Grewal SS, Stenberg PE, McCleskey EW, Stork PJ. 2000. Role of phosphoinositide 3-kinase and endocytosis in nerve growth factor-induced extracellular signal-regulated kinase activation via Ras and Rap1. *Mol. Cell Biol.* 20:8069–83

Zanata SM, Lopes MH, Mercadante AF, Hajj GN, Chiarini LB, et al. 2002. Stress-inducible protein 1 is a cell surface ligand for cellular prion that triggers neuroprotection. *EMBO J.* 21:3307–16

Zarranz JJ, Digon A, Atares B, Rodriguez-Martinez AB, Arce A, et al. 2005. Phenotypic variability in familial prion diseases due to the D178N mutation. *J. Neurol. Neurosurg. Psychiatry* 76:1491–96

Zhang CC, Steele AD, Lindquist S, Lodish HF. 2006. Prion protein is expressed on long-term repopulating hematopoietic stem cells and is important for their self-renewal. *Proc. Natl. Acad. Sci. USA* 103:2184–89

Zhang Y, Moheban DB, Conway BR, Bhattacharyya A, Segal RA. 2000. Cell surface Trk receptors mediate NGF-induced survival while internalized receptors regulate NGF-induced differentiation. *J. Neurosci.* 20:5671–78

Mechanisms Underlying Development of Visual Maps and Receptive Fields

Andrew D. Huberman,[1] Marla B. Feller,[2] and Barbara Chapman[3]

[1] Department of Neurobiology, Stanford University School of Medicine, Palo Alto, California 94305; email: adh1@stanford.edu

[2] Department of Molecular and Cell Biology and Helen Wills Neuroscience Institute, University of California, Berkeley, Berkeley, California 94720; email: mfeller@berkeley.edu

[3] Center for Neuroscience, University of California, Davis, California 95616; email: bxchapman@ucdavis.edu

Annu. Rev. Neurosci. 2008. 31:479–509

The *Annual Review of Neuroscience* is online at neuro.annualreviews.org

This article's doi: 10.1146/annurev.neuro.31.060407.125533

Key Words

synaptogenesis, activity-dependent, visual plasticity, axonal refinement, retinal waves, axon guidance

Abstract

Patterns of synaptic connections in the visual system are remarkably precise. These connections dictate the receptive field properties of individual visual neurons and ultimately determine the quality of visual perception. Spontaneous neural activity is necessary for the development of various receptive field properties and visual feature maps. In recent years, attention has shifted to understanding the mechanisms by which spontaneous activity in the developing retina, lateral geniculate nucleus, and visual cortex instruct the axonal and dendritic refinements that give rise to orderly connections in the visual system. Axon guidance cues and a growing list of other molecules, including immune system factors, have also recently been implicated in visual circuit wiring. A major goal now is to determine how these molecules cooperate with spontaneous and visually evoked activity to give rise to the circuits underlying precise receptive field tuning and orderly visual maps.

Contents

INTRODUCTION

The wiring diagram of the vertebrate visual system is remarkably precise. Throughout the visual pathway, neurons are tuned to respond to specific features of the visual scene by virtue of the types and patterns of synaptic connections they receive. Moreover, visual connections are arranged into regular feature maps such as retinotopic maps, eye-specific layers, ocular dominance columns, orientation maps, and direction preference maps. Visual receptive fields and feature maps develop through a refinement process in which imprecise connections are weakened and eliminated and correctly targeted connections are strengthened and maintained (Katz & Shatz 1996). Much of this refinement is well underway or even completed before vision begins (Chapman et al. 1996, Chapman & Stryker 1993, Godement et al. 1984, Horton & Hocking 1996, Linden et al. 1981, McLaughlin & O'Leary 2005, Rakic 1976). Some visual circuit connections require vision to be maintained (White et al. 2001), whereas others are refractory to disruptions in visual experience (Wiesel & Hubel 1963a). Still others form only after eye-opening and require normal visual experience in order to develop in the first place (Li et al. 2006). Here, we review recent findings on mammalian visual circuit development. As a general framework for thinking about this process, we first review the sorts of activity-based and molecular-based cues that, in theory, are sufficient to instruct development of precise visual circuits. Second, we review what is currently known about the role of these cues in the development of particular types of receptive field properties and visual feature maps.

MECHANISMS TO INSTRUCT VISUAL CIRCUIT DEVELOPMENT

How do precise visual connections form? More than 40 years ago, Roger Sperry (1963) put forth the chemoaffinity hypothesis, the idea that molecular cues guide the formation of orderly connections between the eye and the brain (Sperry 1963). Around the same time, David Hubel and Torsten Wiesel discovered the importance of early visual experience for plasticity of visual circuits (Wiesel & Hubel 1963a, 1965a,b). Sperry, Hubel, and Wiesel all noted the remarkable degree of anatomical and functional precision that is present in the visual pathway before the onset of vision, and they concluded that the wiring of the visual system relies on "innate cues" (Sperry 1963, Wiesel & Hubel 1963b). Modern methods have further demonstrated the high degree of functional and anatomical precision that is attained in mammalian visual circuits, even before vision is possible. Retinotopic maps in the superior colliculus (SC) (Chalupa & Snider 1998, King et al. 1998, McLaughlin & O'Leary 2005), lateral geniculate nucleus (LGN) (Jeffery 1985, Pfeiffenberger et al. 2006), and primary visual cortex (V1) (Cang et al. 2005b), as well as

Superior colliculus (SC): a major subcortical target of RGC axons important for visual orienting. The mammalian homolog to the optic tectum

Lateral geniculate nucleus (LGN): also often referred to as the dorsal LGN. Receives direct input from RGCs and relays visual information to the visual cortex

V1: primary visual cortex

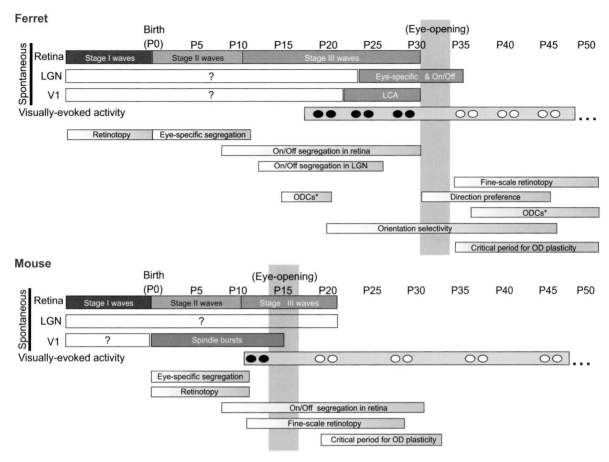

Figure 1

Timing of the development of each of the major visual circuit properties in ferret and mouse (*shown as bars with light blue gradients*). Orange bars indicate ionotropic glutamate receptor-mediated retinal waves. Turquoise bars indicate nicotinic acetylcholine receptor–mediated retinal waves. *Blue* bars (stage I waves) indicate cholinergic and gap-junction-mediated activity. *Gray* column indicates the approximate age for eye opening in each species. *Black* icons within the "visually-evoked activity" *yellow* bar indicates the period in which vision occurs through naturally closed eye lids. *White* icons indicate vision through open eyelids. Asterisks in the two bars labeled "ODCs" indicate that two different time frames have been reported for ODC development in ferrets (early development of ODCs, Crowley & Katz 2000; later development of ODCs, Ruthazer et al. 1999) and which report is accurate remains controversial owing to technical limitations of the tracing techniques used in both studies (see main text). Note that ODCs, orientation maps, and direction maps are not listed for the mouse because mice lack these anatomical features, although single cells in mouse V1 do exhibit responses tuned according to these stimulus features. The timing of the emergence of these physiological features in the mouse has not yet been reported.

eye-specific inputs to the LGN (Godement et al. 1984, Linden et al. 1981, Rakic 1976, Shatz 1983) and ocular dominance columns in V1 (Crowley & Katz 2002, Horton & Hocking 1996), all develop before photoreceptors can transduce light (**Figure 1**). Orientation-selective circuits in V1 (Chapman et al. 1996, Chapman & Stryker 1993, Issa et al. 1999,

White et al. 2001) also begin to form before visual experience begins (**Figure 1**). The fact that such a high degree of wiring precision is achieved prior to the onset of visual experience implies that activity-independent cues such as axon guidance molecules establish the basic blueprint of visual circuit connections, and then visual experience subsequently modifies these connections during the "critical period" (reviewed in Hensch 2004). However, studies over the past two decades reveal that neurons in the developing visual system are spontaneously active long before the critical period and, indeed, even before photoreceptors become capable of responding to light (Galli & Maffei 1988, Maffei & Galli-Resta 1990). The spontaneous activity patterns present in the developing visual pathway are highly structured; retinotopic information, on-center vs. off-center receptive field information, and eye-specific information can all be found encoded in the patterns of spontaneous firing of retinal, LGN, and cortical neurons (Chiu & Weliky 2001, 2002; Feller 1999; Weliky & Katz 1999; Wong 1999). Once the visual system becomes capable of responding to light, sensory-evoked activity then stabilizes some of these nascent visual connections (Chapman et al. 1996, Chapman & Stryker 1993, Issa et al. 1999, White et al. 2001), refines them further (Tian & Copenhagen 2003), or induces additional circuit properties (Li et al. 2006). Some species also progress through an intermediate stage before eye-opening during which both spontaneous activity and vision through naturally closed eyelids coexist (Demas et al. 2003, Krug et al. 2001) and together drive refinement of visual connections (Akerman et al. 2002, Tian & Copenhagen 2003, White et al. 2001).

Here we review the types and patterns of spontaneous and visually evoked activity that can impact developing visual connections. We then discuss the role of molecules, such as axon guidance cues and cell adhesion molecules, in visual circuit development. For clarity, we present activity and molecular cues separately, but the reader should note that whether activity and guidance molecules are in fact independent of one another must be established on a case-by-case basis, especially in light of recent findings that guidance molecules can be modulated by altering patterns of spontaneous neural activity (Hanson & Landmesser 2004, Ming et al. 2001) and that guidance molecules can impact neural activity patterns as well (Bouzioukh et al. 2006, Sahay et al. 2005). In the second part of this re-

view we consider how neural activity and axon guidance molecules contribute to the development of specific types of visual receptive fields and orderly feature maps.

Neural Activity

Spontaneous retinal waves. Prior to the onset of visual experience, the developing vertebrate retina spontaneously generates retinal waves. During a wave, retinal ganglion cells (RGCs) spontaneously fire correlated bursts of action potentials that propagate across the retina. Waves initiate from random locations in the retina and propagate over spatially restricted domains such that over time the whole retina is tiled by locally correlated activity (Feller et al. 1996, 1997). Retinal waves have been found in chickens, turtles, mice, ferrets, and monkeys (Warland et al. 2006, Wong 1999). In some species, such as primates, waves occur only prenatally, whereas in other species, such as mice and ferrets, they occur both prenatally and postnatally. In mammals, retinal waves begin prior to when photoreceptors are capable of transducing light, and they disappear around the time of eye-opening, regardless of whether visual experience is prevented or not (Demas et al. 2003). The circuits that mediate retinal waves have been divided into three stages that are remarkably similar across species (Firth et al. 2005, Sernagor et al. 2001, Wong 1999). The waves that occur at each stage have somewhat different properties such as wave front velocities, domain sizes, and frequencies; thus the waves particular to each stage are positioned to instruct the development of particular types or aspects of visual circuit development.

Stage I waves emerge before birth (Bansal et al. 2000, Syed et al. 2004) (**Figure 1**). Gap junction blockers can inhibit stage I retinal waves in rabbit (Syed et al. 2004), whereas blocking nicotinic acetylcholine receptor (nAChR) inhibits some but not all of these early waves in mice (Bansal et al. 2000). Stage I waves are relatively infrequent, occurring every ~90–120 s.

Stage II waves emerge around the time of birth in mice and ferrets (Bansal et al. 2000, Syed et al. 2004). Stage II waves then last the first 1–2 postnatal weeks, coincident with retinotopic and eye-specific refinement (**Figure 1**). Stage II waves are driven by ACh released from starburst amacrine cells (Feller et al. 1996, Zheng et al. 2004). Acute application of nAChR antagonists blocks Stage II retinal waves (for reviews, see Firth et al. 2005, Zhou 1998). Stage II waves occur relatively infrequently (every 1–2 min) and propagate with an average wave-front velocity ranging from 100 to 300 microns/s, varying with species (Bansal et al. 2000, Feller et al. 1996). The propagating and restricted domain size of stage II waves allows them to relay information about the retinotopic relationship of RGCs to brain targets in an activity-dependent manner (Butts 2002). Also, because stage II waves are relatively infrequent, there is a low probability that RGCs residing at retinotopically similar positions in the two eyes will fire simultaneously. Thus, stage II waves also relay information identifying from which eye a given RGC axon originates to retinorecipient targets in the brain (Eglen 1999, Butts et al. 2007).

Mice lacking β2-subunits of neuronal nAChRs ($\beta 2nAChR^{-/-}$) do not exhibit stage II retinal waves (Bansal et al. 2000, Muir-Robinson et al. 2002) RGCs remain active in the $\beta 2nAChR^{-/-}$ mouse, but rather than firing in correlated fashion, the RGCs in these mutants exhibit spontaneous spiking activity that is not correlated among neighboring RGCs and there is no propagating activity (Bansal et al. 2000, Muir-Robinson et al. 2002). Remarkably, waves at other developmental stages (I and III) are unaffected in the $\beta 2nAChR^{-/-}$ mouse. The $\beta 2nAChR^{-/-}$ mouse is thus an important animal model for studying the role of stage II waves in visual circuit development. Similarly, the cholinergic agonist epibatidine blocks stage II waves by desensitizing nAChRs (Penn et al. 1998). Chronic intraocular application of epibatidine to early postnatal mice or ferrets is a paradigm also commonly used to disrupt stage II waves

and thereby study their role in visual circuit development.

Stage III waves begin at approximately P10–P12 in mice and ferrets and are mediated by glutamate released from retinal bipolar cells (Bansal et al. 2000, Wong et al. 2000, Zhou & Zhao 2000) (**Figure 1**). Stage III waves persist until around the time of eye-opening (Bansal et al. 2000, Demas et al. 2003, Syed et al. 2004, Wong et al. 1993). During stage III waves, activity still propagtes across the retina, but as it does so, it causes RGC firing to become more correlated among RGCs of the same sign (On-On or Off-Off) than it does between RGCs of the opposite sign (On vs. Off), in part because Off-RGCs manifest a higher frequency of burst firing than On-RGCs (Myhr et al. 2001, Wong & Oakley 1996). Thus, On vs. Off information is encoded in stage III waves and is relayed to central brain targets as differences in the temporal pattern of spiking and mean firing rate across these two RGC populations. Other types of functionally distinct RGCs, such as alpha, beta, and gamma RGCs, also exhibit differences in their spontaneous firing patterns during stage III waves (Liets et al. 2003), and these differences may contribute to the wiring of their distinct patterns of connectivity.

Spontaneous activity in the lateral geniculate nucleus. A few landmark studies used in vitro explants to confirm that the RGC spiking evoked during waves induces LGN neuron spiking (Mooney et al. 1996) and that such synaptic transmission can influence plasticity at developing retinogeniculate synapses (Mooney et al. 1996), two requisite features for wave-based activity-dependent models of visual circuit refinement (Butts 2002, Butts et al. 2007, Katz & Shatz 1996). Aside from these studies, however, virtually all our knowledge about spontaneous activity patterns in the developing LGN comes from experiments in which LGN activity was recorded with chronically implanted multi-electrodes in nonanesthetized ferret pups (Weliky & Katz 1999). These studies revealed that spontaneous activity is more correlated among neurons in same-eye

Epibatidine: a nicotinic cholinergic agonist that can block retinal waves (contralateral or ipsilateral) layers of the LGN than among neurons in opposite eye (contralateral vs. ipsilateral) layers. They also revealed higher correlations among neurons in same-sign (On or Off) sublaminae of the LGN than among neurons situated in opposite sign (On vs. Off) sublaminae. The youngest animals that Weliky and Katz were able to record from were ~P24, which is about one week before eye-opening in this species and coincides with stage III glutamate-mediated retinal waves (**Figure 1**). Indeed, spontaneous correlated neuron firing in the developing LGN was strongly enhanced by input from the contralateral retina and the ipsilateral visual cortex. Thus, the developing LGN is capable of intrinsically generating spontaneous correlated activity that is eye specific and On- vs. Off-center specific. Activity from the retina and the cortex further strengthens these eye-specific and On/Off correlations.

Spontaneous activity in visual cortex. A few research groups have recorded spontaneous activity in the developing visual cortex of intact, nonanesthetized animals at stages prior to the onset of visual experience. Fiber optic recordings of calcium-sensitive dyes revealed the presence of propagating calcium waves in the cortex of newborn mice (Adelsberger et al. 2005). Other groups have recorded spindle burst oscillations in V1 of unanesthetized newborn rodents. From birth until ~P10, these spindle bursts are triggered by stage II retinal waves in the contralateral eye. After P10 they are triggered by light stimulation of the contralateral eye (Hanganu et al. 2006). The specific role of spindle bursts in visual circuit wiring remains unknown.

Does spontaneous activity in V1 contribute to visual circuit refinement? Multisite electrode recordings from the visual cortex of P22–P28 ferrets revealed patches of V1 neurons that spontaneously exhibit synchronous firing. These patches were separated by ~1 mm and resembled ferret ocular dominance columns (ODCs). This long-range correlated activity (LCA) (Chiu & Weliky 2001) is intrinsic to the

cortex because it persists despite LGN activity blockade, but it is modulated by subcortical inputs. A subsequent study combined light stimulation of the retina and multisite recordings from V1 to show that LCA domains are associated with ODCs (Chiu & Weliky 2002). However, LCA was also observed in large monocular portions of V1 that do not have ODCs. Thus LCA in developing V1 is generated by circuits intrinsic to V1 and is modified by eye-specific connections from the retina via the LGN.

Collectively, the above-described studies demonstrate that patterns of spontaneous activity generated within the developing retina, LGN, and V1 are highly structured and that, by virtue of nascent connections within and between these visual areas, the overall strength of eye-specific and On/Off correlations is enhanced. This arrangement suggests a hierarchical organization whereby increasingly specific information is encoded in the patterns of spontaneous activity as the visual pathway matures and visual experience begins.

Early patterns of visually driven activity. Vision begins at birth in species born with their eyes open, and it begins several days to weeks after birth in species born with their eyes closed such as mice, ferrets, and cats. It is tempting to assume that the statistical properties of the visual scene reliably predict the activity patterns in the developing visual pathway from the onset of vision, and that those activity patterns contribute to visual receptive field development. However, very few studies have examined what sorts of visually driven activity patterns are actually present in the visual pathway at the earliest stages of vision. Moreover, in mice and ferrets, visual experience begins before eye-opening through naturally closed eye lids (Krug et al. 2001) at a time when stage III spontaneous retinal waves and correlated LGN and V1 activity are also occuring (**Figure 1**). Although dark-rearing before eye-opening disrupts some aspects of visual circuit refinement (Akerman et al. 2002, White et al. 2001), determining what activity patterns are present in the visual pathway at these stages requires that the investigators record activity from awake unanaesthetized animals (so as not to eliminate spontaneous activity) while the animals view natural scenes through closed eyelids (to recapitulate normal experience). LGN recordings under these conditions have recently been carried out in postnatal ferrets (Ohshiro & Weliky 2006). These experiments revealed that the correlations among like sign (On- or Off-center) LGN neurons generated by spontaneous activity closely match the correlations generated by the animal viewing natural scenes. Thus, spontaneous and visually generated neural activity patterns could sharpen or induce receptive fields in the visual pathway, even before eye-opening.

Molecular Guidance Cues

Molecules such as axon guidance cues (Goodman & Shatz 1993) and cell adhesion molecules (Sanes & Yamagata 1999, Yamagata et al. 2002) that are traditionally thought of as activity independent can induce a considerable degree of wiring precision in the developing visual pathway. Indeed, the development of virtually all the visual circuit properties we discuss below could, in theory, involve molecular guidance cues. Although we do not focus extensively on the mechanisms by which these guidance cues function, several excellent reviews on this topic are cited as further reading. We now consider the types of visual circuit wiring that molecular guidance cues can induce.

Graded expression of axon guidance cues. Sperry (1963) first hypothesized that graded expression of molecules in inputs and targets regulates visual map development. Considerable evidence now demonstrates that gradients of the ephrin/Eph family of axon guidance cues are key to the development of topographic maps not only in the visual system, but also in many other sensory systems (reviewed in Flanagan 2006, McLaughlin & O'Leary 2005) (**Figure 2a**). Quantitative models demonstrate how graded expression of Ephs in the retina and ephrin ligands in the target instructs mapping

of the cardinal retinal axes [(nasal-temporal (N-T), dorsal-ventral (D-V)], although, as we discuss below, such models also need to incorporate activity-dependent synaptic refinements to fully explain map development as it occurs in vivo.

Graded expression of ephrins can also induce binary axon guidance decisions (Dufour et al. 2003, Huberman et al. 2005b). This sort of pathfinding is well-suited to segregate many of the highly specialized functionally visual processing streams (motion, form, color, etc.) from one another along fiber pathways and within target structures (Callaway 2005).

Layer specific adhesion molecules. One hallmark feature of the visual system is layer-specific connectivity. In the retina, SC, LGN, and visual cortex, functionally distinct visual pathways arise through layer-specific axonal and dendritic connections. A few molecules have been identified that regulate layer-specific connectivity in the retina, such as sidekicks and DSCAMS (Yamagata et al. 2002, Yamagata & Sanes 2008) (**Figure 2b**). Also N-cadherin, in collaboration with neural activity, regulates layer specificity of geniculo-cortical projections (Poskanzer et al. 2003, Uesaka et al. 2007, Yamamoto et al. 2000) (**Figure 2b**). No one has yet reported regarding the molecules that mediate layer-specific targeting of functionally defined visual processing streams, although such cues have been proposed to exist (Meissirel et al. 1997; Kawasaki et al. 2004).

Axon-axon recognition molecules. One way to ensure selective wiring of axons or dendrites arising from one functional subclass of visual neurons (say, for example, On-center or motion-sensitive RGCs) is to have those cells express adhesion molecules that keep them bound together along the fiber tracts and/or within their targets (**Figure 2c**). Indeed, axons arising from RGCs in retinotopically similar locations in the retina establish retinotopic order within the optic tract prior to target innervation (reviewed in Plas et al. 2005), and the axons of functionally distinct RGCs segregate

a Graded axon guidance cues

b Homophilic adhesion cues

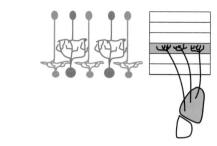

c Axon-axon recognition cues

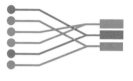

d Sub-cellular adhesion cues

Figure 2
_____

Schematic diagrams of the types of molecular cues that direct wiring specificity in developing visual circuits.
(*a*) Graded expression of guidance cues in axons and in their targets can guide specific patterns of visual
connections according to matching of ligand and receptor levels (see reviews by McLaughlin & O'Leary
2005, Flanagan 2006) and see **Figure 3**. (*b*) Homophilic adhesion cues expressed in the axons and dendrites
of pre- and postsynaptic neurons can lead to highly specific patterns of connectivity (red cells connect to red
cells, green cells to green cells, etc.) (e.g., Yamagata et al. 2002). Similarly, the expression of adhesion
molecules in neurons within one structure (*shown in blue*; schematic of the mouse LGN) and within the
specific layer of their targets to which they project (also shown in blue; schematic of cortex) can induce highly
specific patterns of connectivity (e.g., Poskanzer et al. 2003). (*c*) Adhesive cues expressed among axons arising
from a common cell type (schematized here as *red, green or yellow*) can segregate these axons into distinct
fiber tracts and/or portions of fiber tracts, which can then lead to segregation of their axons within the final
target (reviewed in Chen & Flanagan 2006, Mombaerts 2006). (*d*) Different adhesion cues expressed at
different sites along the dendritic arbor of an individual postsynaptic neuron can segregate synaptic inputs
arising from different cells/sources at the subcellular level (e.g., Ango et al. 2004, Di Cristo et al. 2004) and
thereby impact the receptive field properties of the postsynaptic neuron (e.g., Sherman 2004).

from one another along fiber pathways (Reese
1987, Reese & Cowey 1988) and segregate
into different layers within their targets (Sur
& Sherman 1982). The molecules that regulate
axon-axon recognition and segregation in the
mammalian visual system have not been identified, but one such molecule—heparin sulfate—
can control axon-axon segregation of retinotopically distinct RGCs in the optic tract of
zebrafish (Lee et al. 2004). Neural activity may
also regulate axon-axon recognition and targeting, but purely activity-dependent models cannot explain why, for instance eye-specific lay-

ers or On-center versus Off-center sublaminae
always form in the same stereotyped locations
in their target (Stryker & Zahs 1983). Therefore, a molecular or other activity-independent
mechanism, such as variation in ingrowth
timing (Walsh & Guillery 1985), must ensure this feature. In this context, axon sorting on the basis of molecular guidance cues is
a well-established phenomenon in developing
olfactory circuits, but neural activity also contributes to axon sorting in that system (Chen
& Flanagan 2006, Mombaerts 2006, Yu et al.
2004).

Subcellular recognition molecules. The specific receptive field properties of a given visual neuron arise by virtue of the type, number, strength, and location of synaptic inputs onto that cell (Chapman et al. 1991, Nelson et al. 1978, Reid & Alonso 1995, Sherman 2004, Tavazoie & Reid 2000). In recent years, studies have begun to explore the factors that regulate where particular inputs form synapses on a postsynaptic cell. Huang and coworkers have shown that the perisomatic innervation of pyramidal neurons by GABAergic parvalbumin neurons in the visual cortex occurs even in the complete absence of sensory experience (Di Cristo et al. 2004). The same group has also shown, however, that GABAergic signaling plays an important role in the emergence of perisomatic inhibition (Chattopadhyaya et al. 2007). Thus both restricted expression of cell adhesion molecules (Ango et al. 2004) (**Figure 2d**) and neural activity likely control the targeting of highly specific patterns of subcellular connectivity observed in the mature visual system. We now consider the role of neural activity and molecular cues in the development of connectivity underlying specific types of receptive field properties and visual feature maps.

DEVELOPMENT OF VISUAL MAPS AND RECEPTIVE FIELD PROPERTIES

Retinotopic Mapping

Visual connections are organized to convey information about the position of stimuli in the visual field. Neighboring RGCs in the retina project to neighboring cells in the brain, thereby forming retinotopic maps of the visual field in the LGN and SC. Retinotopic mapping is also present in the pattern of projections from the LGN to V1. Numerous studies have shown that retinotopic maps develop through a process of refinement; when RGC axons initially innervate the SC they overshoot their correct termination zone. A fine precision retinotopic map gradually emerges as RGCs arborize interstitial branches in the appropriate target loca-

tion and the overshooting portion of the axon degenerates (**Figure 3a** and **b**). Retinotopic refinement is completed prior to eye-opening and the onset of vision (**Figure 1**). Two major forces instruct refinement of retinotopic maps: gradients of molecular cues and correlated neural activity (reviewed in McLaughlin & O'Leary 2005).

The ephrin family of axon guidance molecules organizes the overall layout of retinotopic maps. Gradients of ephrins and their receptors, Ephs, are expressed in the retina and its central visual targets (Flanagan 2006, McLaughlin & O'Leary 2005). In mammals, EphA5 is expressed in a low nasal to high temporal gradient in RGCs, and the repellant ligands, ephrin-A2/A5, are expressed in low anterior to high posterior gradients in the SC (Feldheim et al. 1998, McLaughlin & O'Leary 2005). Mutant mice lacking ephrin-A2/5 (*ephrin-A2/5$^{-/-}$*) or EphA5 (*EphA5$^{-/-}$*) show aberrant retinotopic mapping of the N-T retinal axis (visual field azimuth representation) in the SC and in the LGN (Feldheim et al. 2000, 2004). If an anterograde tracer is focally injected at one location in the retina of a wildtype (*wt*) mouse, a single dense focus of RGC terminal arbors is seen in the retinotopically appropriate location in the SC and in the LGN. When such injections are made in mice lacking ephrin-A2 and ephrin-A5, however, multiple topographically incorrect RGC termination zones are observed, consistent with the loss of the repellent cues in the targets (Feldheim et al. 1998, 2000; Frisen et al. 1998; Pfeiffenberger et al. 2006) (**Figure 3c**). Similarly, ectopic expression studies indicate that a gradient of EphA5 expression in the LGN and ephrin-A5 in the visual cortex controls retinotopic mapping of the N-T retina axis (visual field azimuth) in the geniculocortical projection (Cang et al. 2005a).

Retinotopic targeting of RGCs along the D-V axis of the retina (corresponding to visual field elevation) relies on ephrin-/EphBs and wnt/ryk signaling. Wnt3 and EphrinB ligands are expressed in high-to-low gradients along the medial-lateral axis of the SC/tectum, and their receptors, ryk and EphBs respectively, are

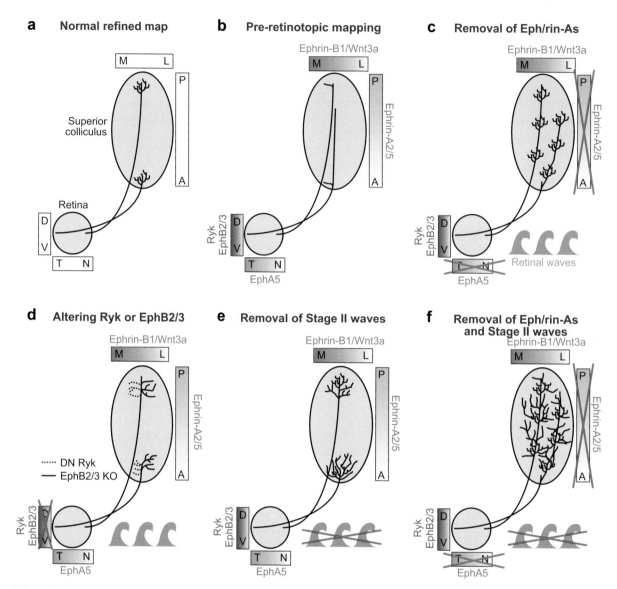

a Normal refined map

M L

P

Superior
colliculus

A

Retina

D

V

T N

b Pre-retinotopic mapping

Ephrin-B1/Wnt3a

M L

P

Ephrin-A2/5

A

Ryk
EphB2/3

D

V

T N

EphA5

c Removal of Eph/rin-As

Ephrin-B1/Wnt3a

M L

P

Ephrin-A2/5

A

Ryk
EphB2/3

D

V

T N

EphA5

Retinal waves

d Altering Ryk or EphB2/3

Ephrin-B1/Wnt3a

M L

P

Ephrin-A2/5

A

······ DN Ryk
——— EphB2/3 KO

Ryk
EphB2/3

D

V

T N

EphA5

e Removal of Stage II waves

Ephrin-B1/Wnt3a

M L

P

Ephrin-A2/5

A

Ryk
EphB2/3

D

V

T N

EphA5

**f Removal of Eph/rin-As
and Stage II waves**

Ephrin-B1/Wnt3a

M L

P

Ephrin-A2/5

A

Ryk
EphB2/3

D

V

T N

EphA5

Figure 3

Schematic diagrams of (*a*) the mature retinotopic map in the SC, (*b*) the immature unrefined retinotopic map in the SC (and the expression of molecular cues that guide formation of retinotopic maps) (see McLaughlin & O'Leary 2005). (*c*) Disruptions in retinotopic mapping in *ephrin-A2/5*$^{-/-}$ or *EphA5*$^{-/-}$ mice. Multiple dense termination zones are observed along the N-T axis of the target (Feldheim et al. 2000, 2004). Waves denote that retinal waves are intact in these animals. (*d*) Disruptions in retinotopic mapping in *EphB2/3*$^{-/-}$ mice or in response to disrupting Wnt/ryk signaling; RGC terminals shift more medially (ryk disruption; *dashed lines in target*) (Schmitt et al. 2006) or shift more laterally (EphB2/3 knockout; *solid lines*) (Hindges et al. 2002). Similar results are observed after disruption of BMPs (Chandrasekaran et al. 2005). (*e*) The overall retinotopic map forms when stage II retinal waves are eliminated (because ephrin-A2/5 signaling is still intact), but RGC axons fail to form dense terminal arbors in their correct topopographic locations and are abnormally broad (McLaughlin et al. 2003, Grubb et al. 2003). (*f*) When stage II waves are prevented in *ephrin-A2/5*$^{-/-}$ mice, N-T mapping of RGC projections is abolished (Pfeiffenberger et al. 2006). For all the manipulations shown here, the retino-SC projection is depicted. The same general defect pattern is observed in the LGN and V1, where ephrin-As and Bs and retinal waves regulate topographic map formation (see main text for details).

expressed in high-to-low gradients along the ventral-to-dorsal axis in the retina (Hindges et al. 2002) and optic tectum (Braisted et al. 1997) (**Figure 3***d*). *EphB2*$^{-/-}$ and *EphB3*$^{-/-}$ or kinase-inactive EphB2 mice exhibit D-V mapping errors in the SC (Hindges et al. 2002), and disruption of Wnt signaling by dominant-negative ryk expression in the retina also alters D-V mapping but in the opposite direction of *EphB2/3*$^{-/-}$ mice (Schmitt et al. 2006) (**Figure 3***d*). Thus the balance between Ephrin-B and Wnt/ryk signaling guides RGC axons to their correct medial-lateral termination zone in the SC/tectum (see review by Salinas & Zou 2008, in this volume).

Computational models suggest that molecular cues alone cannot specify the precision of retinotopic mapping observed in vivo (McLaughlin & O'Leary 2005, O'Leary & McLaughlin 2005, Yates et al. 2004) and that the correlated firing of neighboring RGCs is necessary to enhance the precision of retinotopic mapping through Hebbian "fire together wire together" type mechanisms (Butts et al. 2007, Yates et al. 2004). Such models predict that, in the absence of correlated RGC firing, ephrins will guide RGC axons to the correct locations in their targets, but the axons will fail to form dense terminations in those locations, effectively blurring the local visual field representation. Experimental evidence now supports this idea. Retinotopic refinement of RGC and LGN projections occurs during the same developmental period as stage II retinal waves (**Figure 1**). In the *β2nAChR*$^{-/-}$ mouse or in mice that receive intraocular injections of epibatidine, stage II waves are abolished, and RGC axons project to the correct retinotopic location in the SC and LGN but therein form abnormally large, diffuse axonal arborizations (Grubb et al. 2003, McLaughlin et al. 2003, Pfeiffenberger et al. 2006) (**Figure 3***e*). As a result of these diffuse arbors, SC and LGN neurons exhibit larger-than-normal receptive fields in these animals (Chandrasekaran et al. 2005, Grubb et al. 2003, Mrsic-Flogel et al. 2005). Likewise, epibatidine-treated and *β2nAChR*$^{-/-}$ mice also exhibit altered retinotopy in the

geniculocortical pathway (Cang et al. 2005b). Poorly refined retinotopic maps are also observed if the readout of correlated activity is prevented in target neurons by N-methyl-D-aspartate (NMDA) receptor blockers (Huang & Pallas 2001, Simon et al. 1992). These findings strongly reinforce Hebbian activity–dependent models of retinotopic map refinement (Butts 2002).

Rather than regulate each other, activity and ephrin-As act in parallel to control mapping of the nasal temporal retinal axis in visual targets. When *β2nAChR*$^{-/-}$ mice are crossed to *ephrin-A2/5*$^{-/-}$ mice, mapping of the N-T axis is abolished (Cang et al. 2008, Pfeiffenberger et al. 2006) (**Figure 3***f*). Correlated activity refines the retinotopic map predominantly along the N-T axis; mapping of the D-V axis is unaffected by loss of waves (Grubb et al. 2003; Pfeiffenberger et al. 2006; Cang et al. 2008). Why waves affect only N-T mapping remains unknown. Waves are not thought to exhibit locational differences along the N-T or D-V axes or according to eccentricity (distance from the center of the retina), but it should be noted that regional differences in wave properties have never been examined systematically. Mapping along the D-V axis may be refractory to altering stage II waves because D-V mapping is established before target innervation within the optic tract (Plas et al. 2005). D-V mapping in the optic tract may be controlled by as-yet-undiscovered axon-axon recognition cues (**Figure 1***c*) and/or by stage I waves.

Fine-Scale Retinotopic Refinement

After RGC axons map to their appropriate retinotopic locations in their targets and form dense axonal arborizations at those locations, they continue to undergo fine-scale retinotopic refinement. In vitro experiments that combine optic tract stimulation with whole-cell recordings from LGN neurons indicate that on P10, approximately 12-30 RGCs provide weak synaptic input onto each LGN neuron. Significant elimination of retinogeniculate synapses occurs within the following week so that by

P16 each LGN neuron receives strong input from only 1–3 RGCs (Chen & Regehr 2000, Hooks & Chen 2006, Jaubert-Miazza et al. 2005) (reviewed in Huberman 2007). This fine-scale retinotopic refinement has direct implications for sharpening receptive fields in the LGN. Indeed, in ferrets around the time of eye-opening, LGN receptive fields are significantly larger and more elongated than they are three weeks later when the loss of all but one to two RGC inputs onto each LGN neuron has occurred (Tavazoie & Reid 2000). Stage III retinal waves coincide with fine-scale retinotopic refinement (**Figure 1**) and drive this process. Hooks & Chen (2006) used chronic intravitreal application of tetrodotoxin (TTX) from P11 to P15 to block stage III wave-induced RGC spiking. TTX prevented both the normal elimination of weak RGC inputs onto single LGN neurons and the strengthening of any of the supernumerary weak inputs. Vision in mice begins at ~P11, which is about four days before eye-opening (**Figure 1**). Hooks & Chen (2006) therefore tested the role of visual experience in fine-scale retinotopic refinement. Visual deprivation from birth until P28 had no impact on this process, but surprisingly, dark-rearing from P25 to P30 caused the retinogeniculate projections to revert to a multiply innervated state of ~10–12 inputs. These findings indicate that spontaneous activity is critical for fine-scale retinotopic refinement in the LGN and suggest that light-induced maturation of inhibitory circuits and/or cortical feedback may be involved in maintenance of newly refined circuits. TTX likely eliminates all RGC spiking induced by stage III waves. Thus the issue of whether the specific patterns of RGC spiking generated during stage III waves are instructive for fine-scale refinement still needs to be determined using manipulations that alter the structure of RGC activity during stage III waves but do not eliminate RGC spiking altogether. Tools to accomplish this have not yet been identified.

Some of the molecular cues that act downstream of activity to mediate fine-scale retinotopic refinement have been identified. Components of the classical complement cascade of immune proteins are expressed at developing retinogeniculate synapses, and in their absence, much of the normal synaptic elimination that occurs during fine-scale retinotopic refinement is prevented (Stevens et al. 2007). The expression of other immune molecules, such as MHC I (Corriveau et al. 1998, Huh et al. 2000) and neuronal pentraxins (NPs) (Bjartmar et al. 2006), is regulated by activity. MHC I is essential for plasticity events leading to synaptic weakening (Huh et al. 2000), and NPs are critical for glutamate receptor insertion at developing retinogeniculate synapses (S. Koch and E. Ullian submitted). Proteins in the complement pathway as well as MHC I and NPs are all also required for anatomical eye-specific refinement (Bjartmar et al. 2006, Huh et al. 2000, S. Koch and E. Ullian submitted) (**Figure 4t**). Immune genes are thus emerging as important mediators of activity-dependent plasticity in visual circuit refinement.

Eye-Specific Segregation in the LGN

In mammals, RGC axons from the two eyes project to both sides of the brain and are segregated into eye-specific domains within their targets, including the LGN, the SC, vLGN, pretectum, and suprachiasmatic nucleus (Muscat et al. 2003). Whether eye-specific segregation is important for visual processing remains unknown. Nevertheless, the development of segregated eye-specific projections is a prominent model system for exploring how precise patterns of synaptic connections emerge during development and, in particular, for exploring the role of neural activity in axonal refinement. The development of eye-specific segregation has been studied mostly in the LGN because the patterns of eye-specific projections are so stereotyped in this target. Within a given species, axons from the contralateral and ipsilateral eye terminate in domains of stereotyped shape, position, and size (**Figure 4**). In carnivores and primates, where much of the work on the development of eye-specific retinogeniculate projections has been carried out, eye-specific inputs are also mirrored by

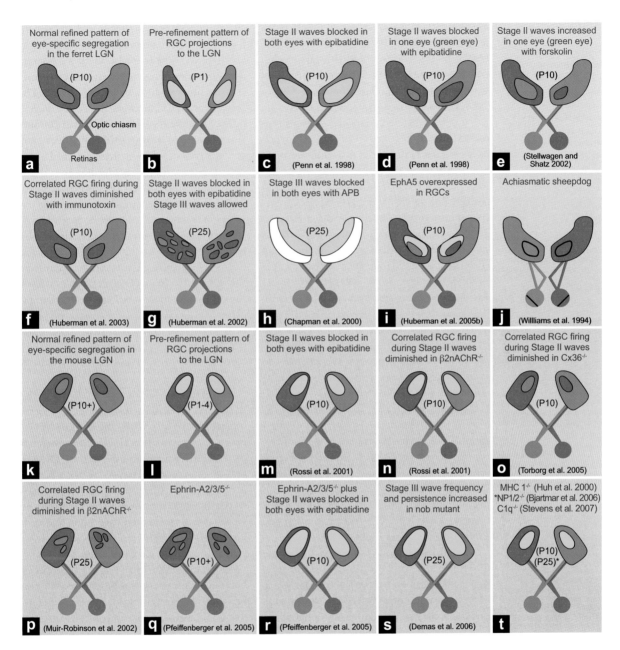

Figure 4

Schematic representations of the eye-specific projection patterns to the LGN of the ferret (panels *a–j*) and the mouse (panels *k–t*) during normal development and the results of experiments examining the role of spontaneous retinal activity, ephrin-As, and immune genes in eye-specific retino-LGN segregation. (*b, l*) The early prerefined pattern of RGC inputs to the LGN in the newborn ferret (*b*) and mouse (*l*). Red areas of the LGN correspond to territory occupied by RGC axons arising from the right (*red*) eye, and *green* areas correspond the territory occupied by RGC axons from the left (*green*) eye. *Yellow* corresponds to the LGN territory where *red* and *green* axons from the two eyes overlap. The ages depicted are shown in parentheses. The manipulations leading to each phenotype are described in each box, as well as in the main text. The asterisk above *NP1/2 and *P25 in panel *t* refers to the fact that the lack of eye-specific segregation observed in the P10 *NP1/2*−/− mouse changes to a pattern similar to panel *p* by P25. By contrast, *C1q*−/− mice and *MHCI*−/− mice exhibit defects in eye-specific segregation until at least P25.

eye-specific cellular layers in the LGN (Linden et al. 1981, Rakic 1976, Shatz 1983). This feature allows one to study the relationship between axon targeting and cytoarchitectural development as well (Casagrande & Condo 1988, Huberman et al. 2002).

Development of eye-specific connections occurs before the onset of visual experience (**Figure 1**) through a refinement process whereby RGC axons from the two eyes initially overlap (**Figure 4a,b**). In macaques and humans, eye-specific retinogeniculate segregation occurs in the second trimester (Hevner 2000, Huberman et al. 2005a, Rakic 1976). In cats this process occurs shortly before birth (E49–60) (Shatz 1983). In mice and ferrets, eye-specific retinogeniculate segregation occurs during the first ten days of postnatal life (P1–10) (Godement et al. 1984, Jaubert-Miazza et al. 2005) (**Figure 1**). Individual-axon labeling studies indicate that eye-specific segregation reflects two general processes: (*a*) synapse and axon arbor elaboration and (*b*) synapse and axon arbor elimination (Campbell & Shatz 1992; Sretavan & Shatz 1984, 1987). Electrophysiological recordings indicate that before segregation, RGC axons from both eyes can drive activity in individual LGN neurons. Then as segregation proceeds, inputs from one eye are weakened and removed and inputs from the remaining eye are strengthened and maintained (Jaubert-Miazza et al. 2005, Shatz & Kirkwood 1984).

Competitive interactions between RGC axons from the two eyes are critical for eye-specific segregation because if one eye is removed at the overlap stage, RGC axons from the intact eye maintain arbors throughout the LGN (Morgan & Thompson 1993, Sretavan & Shatz 1986). Early studies used intracranial infusions of TTX to demonstrate that spiking of LGN neurons is necessary for eye-specific segregation (Sretavan et al. 1988).

Is spontaneous retinal activity necessary for eye-specific segregation? Stage II retinal waves coincide with the period of eye-specific segregation in ferrets (Penn et al. 1998) and in mice (Bansal et al. 2000) (**Figure 1**). As described above, stage II waves encode eye-specific information. In a landmark study, Penn et al. (1998) tested whether stage II waves are necessary for eye-specific retinogeniculate segregation. To block stage II waves, they applied epibatidine at high concentrations to one or both retinas of ferrets from P1 to P10 (the normal period of eye-specific segregation in this species). In P10 ferrets that had waves blocked in both eyes, eye-specific segregation was completely prevented; RGC axons from the two eyes remained intermingled to the same extent observed in normal P1 ferrets (before segregation) (**Figure 4c**). When wave activity was blocked in just one eye, however, retinogeniculate projections still segregated: The territory arising from the activity-blocked eye shrank dramatically at the expense of the normally active eye, which expanded its axonal territory (**Figure 4d**). Penn et al. (1998) was thus the first to demonstrate that spontaneous retinal activity mediates the binocular competition underlying eye-specific segregation.

Subsequent studies used intraocular injections of cAMP analogs to increase the size and frequency of retinal waves in one or both eyes of ferrets (Stellwagen et al. 1999, Stellwagen & Shatz 2002). Results of these studies indicated that the more active eye always wins more territory in the LGN (**Figure 4e**). Thus the relative levels of spontaneous activity in the two retinas, rather than a requirement for some threshold level of retinal activity, dictate how much LGN territory axons from a given retina will occupy.

Do the specific patterns of RGC spiking induced by retinal waves instruct eye-specific segregation? Answering this question requires that one abolish the correlated firing pattern of neighboring RGCs without eliminating RGC spiking or changing overall retinal activity levels. The first study to accomplish this task used immunotoxin ablation of starburst amacrine cells in newborn ferrets to significantly reduce the correlated firing of neighboring RGCs. Eye-specific retinogeniculate segregation was normal in these animals (Huberman et al. 2003) (**Figure 4f**). It was therefore concluded that spontaneous retinal activity is permissive rather than instructive for eye-specific segregation. By contrast, in the $\beta 2nAChR^{-/-}$ mouse, which has altered stage-II waves, eye-specific segregation fails to occur in the normal time frame (Muir-Robinson et al. 2002, Rossi et al. 2001) (**Figure 4n**).

Pharmacologic studies confirm that the lack of segregation observed in the $\beta 2nAChR^{-/-}$ mouse is not caused by a lack of $\beta 2nAChRs$ in the LGN or other brain regions but rather is due to defects in stage II spontaneous retinal activity (Cang et al. 2005b, Chandrasekaran et al. 2005, Rossi et al. 2001) (**Figure 4m**). Therefore, how can researchers explain the discrepancy between the results seen in the $\beta 2nAChR^{-/-}$ mouse and immunotoxin-treated ferrets? Several possibilities exist. First, ferrets have eye-specific cellular layers in the LGN, whereas mice do not. Cues besides patterned activity—such as graded (Huberman et al. 2005b) or layer-specific adhesion molecules (**Figure 2**)—might cooperate with activity to induce eye-specific segregation in species that have eye-specific cellular laminae in the LGN. Second, eye-specific segregation may fail to occur in $\beta 2nAChR^{-/-}$ mice, not because correlated RGC firing is abolished in this animal but because a fraction of RGCs are rendered silent in these animals (McLaughlin et al. 2003). However, some RGCs are rendered silent in immunotoxin treated retinas too, yet this does not impact eye-specific segregation (Huberman et al. 2002, 2003). Third, in the immunotoxin studies, early calcium waves were spared, and whether RGCs were correlated over a larger distance in the retina was not investigated (Huberman et al. 2003) and might explain why eye-specific segregation persisted in these animals.

To define which spatiotemporal features of retinal activity are critical for eye-specific targeting, a recent study examined a transgenic mouse that lacks the gap junction protein, Cx36 ($Cx36^{-/-}$). $Cx36^{-/-}$ mice continue to exhibit retinal waves but show significant increases in the number of asynchronous RGC action potentials, which significantly reduces the nearest neighbor correlation observed in normal mice (Torborg et al. 2005). In contrast with $\beta 2nAChR^{-/-}$ mice, $Cx36^{-/-}$ mice exhibit normal eye-specific segregation in the LGN (**Figure 4o**). By comparing the firing patterns of normal mice and $Cx36^{-/-}$ mice to $\beta 2nAChR^{-/-}$, investigators concluded that high-frequency bursts synchronized across nearby RGCs are correlated with eye-specific segregation, whereas additional asynchronous spikes do not inhibit segregation (Torborg et al. 2005).

The debate about whether waves are instructive or permissive for eye-specific segregation remains unresolved. Binocular TTX injections from birth to P10 delay but do not prevent eye-specific segregation (Cook et al. 1999). But whether TTX blocked all RGC spiking in those animals remains unclear. A recent study concluded that waves are permissive for the proper readout of ephrin-induced repulsion during RGC axonal targeting (Nicol et al. 2007), but these results have not yet been confirmed in vivo. Experiments that alter or, ideally, control RGC spiking patterns in reliable ways are needed to resolve the controversy about whether correlated RGC firing induced by waves is permissive or instructive for eye-specific segregation.

In the $\beta 2nAChR^{-/-}$ mouse and in mice or ferrets treated with epibatidine in both eyes from P1 to P10, eye-specific segregation fails to occur (Penn et al. 1998, Rossi et al. 2001) (**Figure 4c,m,n**). But if these animals are allowed to survive until P25, stage III waves return and rescue eye-specific segregation (Huberman et al. 2002, Muir-Robinson et al. 2002). The overall organization of eye-specific zones that form, however, is highly perturbed; whereas normally eye-specific territories are highly stereotyped, stage II wave blockade followed by a recovery period induces a patchy, random pattern of eye-specific zones in the LGN, and the layout of these zones is highly variable from one animal to the next (Huberman et al. 2002, Muir-Robinson et al. 2002) (**Figure 4g,p**). These findings indicate that retinal activity is necessary for eye-specific segregation but that factors other than activity that are available only during the normal period of eye-specific segregation (P1-P10 in the ferrets and mice) control the location in the target in which the projections from one or the other eye arborize.

Ephrin-As control the spatial patterning of eye-specific layers in the LGN. This is perhaps not surprising given that "eye-specific" projections arise from nasal-retina (contralateral) versus temporal-retina (ipsilateral) RGC axons, and EphA5 is differentially expressed across the N-T retinal axis. In *ephrin-A2/3/5$^{-/-}$* mice (ephrin-A2/3/5 are the ephrin ligands normally expressed in the LGN during eye-specific segregation), eye-specific segregation occurs but in a patchy, random pattern rather than in stereotyped contralateral and ipsilateral eye territories (Pfeiffenberger et al. 2005) (**Figure 4q**). Likewise, EphAs and ephrin-As are expressed in the ferret retina and LGN during eye-specific segregation in a pattern suggestive of an eye-specific pathfinding role. If EphA3 or EphA5 are ectopically expressed in RGCs of newborn ferrets, eye-specific patterning and segregation are perturbed (Huberman et al. 2005b) (**Figure 4i**). Together these findings indicate that ephrin-A/EphA interactions are critical for controlling the position, shape, and size of eye-specific territories in the LGN.

Both the timing and pattern of ephrin-A expression in the LGN are critical for patterning of eye-specific territories. Ephrin-As are downregulated in the LGN after the normal period of eye-specific segregation. Misexpression of EphA receptors in RGCs after the normal period of eye-specific segregation has no impact on patterning of RGC projections to the LGN (Huberman et al. 2005b). Thus, timed ephrin-A expression in the LGN regulates the early critical period for eye-specific LGN layer formation observed in other studies (Muir-Robinson et al. 2002, Huberman et al. 2002). Also, in experiments where RGC axons are experimentally rewired to novel, non-visual targets, ipsilateral RGC axons avoid regions of high ephrin-A expression and segregate from contralateral eye RGC axons in those novel targets (Ellsworth et al. 2005), consistent with the idea that the expression pattern of ephrin-As unique to each subcortical RGC target dictates the characteristic spatial organization of eye-specific territories that will form in that target.

Collectively, the above-described studies indicate that spontaneous RGC spiking that occurs during stage II waves is critical for eye-specific segregation and that ephrin-As dictate where eye-specific domains form within the LGN. Indeed, if *ephrin-A2/3/5$^{-/-}$* mice receive binocular injections of epibatidine from P1 to P10, to eliminate stage II waves, RGC axons from the two eyes remain permanently intermingled in a salt and pepper fashion throughout the entire LGN (Pfeiffenberger et al. 2005; D. Feldheim, personal communication) (**Figure 4r**). As with retinotopic mapping, ephrin-As and activity act in parallel to induce eye-specific pathfinding; removal or overexpression of Eph/ephrin-As does not effect neural activity patterns, nor does altering activity impact ephrin expression in the retina or LGN (Huberman et al. 2005b, Pfeiffenberger et al. 2005). Whether specific patterns of RGC spiking are critical to instruct segregation remains controversial and demands further investigation.

Maintenance of Eye-Specific Projections

After eye-specific segregation is complete, retinal waves help maintain segregation of RGC axons from the two eyes. In young ferrets (P10–P30), the drug APB completely blocks stage III waves (Chapman 2000). If binocular injections of APB are applied to both eyes of ferrets from P10 to P25, axons from the two eyes revert to an overlapping state in the LGN (Chapman 2000). The pattern of overlap caused by a stage III wave blockade noticeably differs from that observed by stage II wave blockade. Whereas blocking stage II waves causes contralateral and ipsilateral eye axons to overlap throughout most of the LGN (**Figure 4c,m,n**), blocking stage III waves causes RGC axons from the ipsilateral eye to collapse from layer A1 into the inner segment of the LGN (termed layer A in ferrets and cats) where they overlap with axons from the contralateral eye, leaving the territory normally occupied by axons from the ipsilateral eye (A1) is left devoid of retinal input (Chapman 2000) (**Figure 4b**). These findings suggest that stage III retinal waves maintain eye-specific segregation by preventing collapse of ipsilateral RGC axons into the more attractive (or less repulsive) contralateral eye territory. Interestingly, expression of the axon-repellant ligands ephrin-A2/5 is lower in contralateral eye territory (layer A) than it is in ipsilateral eye (A1) territory (Huberman et al. 2005b).

Do the specific activity patterns that occur during stage III waves instruct maintenance of eye-specific projections, or is RGC spiking simply permissive to maintain segregation? Demas et al. (2006) discovered that in *nob* mice (no b-wave mice) stage II waves and eye-specific segregation are normal, but stage III retinal waves are excessively frequent and persist abnormally past eye-opening. These aberrantly patterned and persistent stage III waves cause desegregation of eye-specific retinogeniculate inputs, indicating that normally stage III spontaneous retinal activity patterns instruct stabilization of newly refined eye-specific circuits (Demas et al. 2006) (reviewed in Huberman 2006) (**Figure 4s**). As in the earlier APB studies (Chapman 2000), Demas et al. (2006) found that ipsilateral eye axons were preferentially disrupted in nob mice. Why might ipsilateral axons be affected differently by activity manipulations than would other RGC axons? The ipsilateral visual pathway is less efficient in driving postsynaptic activity than the contralateral pathway early in development (Crair et al. 1998, 2001). Also, the RGCs that project ipsilaterally are molecularly distinct in terms of their expression of certain EphAs (Huberman et al. 2005b) and EphBs (Williams et al. 2003, 2006) and zinc transporters (Land & Shamalla-Hannah 2001). Indeed, even in animals that lack an optic chiasm and in which all RGC axons project to the same side of the brain, axons from RGCs in the temporal retina (that normally do not cross at the chiasm) segregate from the axons arising

from other RGCs in the retina (that normally cross at the chiasm) in the LGN (Williams et al. 1994) (**Figure 4j**), indicating that these two axon populations are molecularly distinct. Microarray gene profiling of contralaterally vs. ipsilaterally projecting RGCs would be useful for discovering the genes that regulate the differential targeting of these two cell populations.

Ocular Dominance Column Development

In many carnivore and primate species, including humans, segregation of eye-specific pathways is evident not only in the pattern of eye-specific inputs to the LGN, but also in the pattern of LGN projections to V1 (Hubel & Wiesel 1962). In these species, the axons of LGN neurons located in one or the other eye-specific layer project to V1 in a series of nonoverlapping, alternating stripes called ocular dominance columns (ODCs) (Wiesel et al. 1974). ODCs are a classic model system for exploring the role of early visual experience in the plasticity of cortical circuits (Katz & Crowley 2002). Patterns of visually evoked activity during the critical period instruct the layout of ODCs through binocular competition and Hebbian-type plasticity (reviewed in Hensch 2004). ODCs emerge before the critical period, however, and do not require vision to form (Crair et al. 1998, Horton & Hocking 1996, Wiesel & Hubel 1974).

The anatomical events that underlie ODC development still remain unclear. Early studies that used monocular injection of transneuronal tracers (retino>LGN>V1) to label geniculocortical axons found that in young animals the pattern of proline label in V1 was continuous, whereas in more mature animals, an alternating pattern of proline label (ODCs) was seen (LeVay et al. 1978, Rakic 1976, Ruthazer et al. 1999). Thus investigators concluded that LGN axons representing the two eyes intermingle extensively before segregating into ODCs. One potential problem with this conclusion, however, is that in young animals transneuronal tracers spill over into neighboring eye-specific layers in the LGN (LeVay et al. 1978, Ruthazer et al. 1999, Stryker & Harris 1986), which can lead to incorrect conclusions regarding the absence of ODCs in V1. Indeed, subsequent optical imaging studies (Crair et al. 1998, 2001) and anatomical studies that used focal injections of anterograde tracers into the LGN (Crowley & Katz 2000, 2002) concluded that ODCs are present much earlier than can be observed by transneuronal labeling from the eye (LeVay et al. 1978, Ruthazer et al. 1999, Stryker & Harris 1986). Neither set of studies resolved, however, whether ODCs form precisely from the outset or through refinement. Researchers in this field eagerly await the discovery of anterograde trans-synaptic tracers that are not prone to spillover to firmly resolve this controversial issue.

Stage II and III retinal waves and spontaneous LGN activity are present throughout the ODC development period (Wong et al. 1993, Wong & Oakley 1996) (**Figure 1**), and as discussed above, all these sources of activity are structured to relay eye-specific information to V1. Classic experiments concluded that blocking retinal activity with TTX disrupts the formation of ODCs (Stryker & Harris 1986) and that stimulating the two optic nerves simultaneously prevents ODC segregation (Stryker & Strickland, 1984). However, optical imaging studies (Crair et al. 1998, 2001) later showed that these experiments were carried out after ODCs had already formed and thus could implicate activity only in the maintenance of ODCs, not in their initial formation.

The idea that activity-independent cues drive the initial ODC segregation was actually first proposed by Hubel & Wiesel (1963). Crowley & Katz (1999) tested this idea by removing one or both eyes from newborn ferrets then allowed the animals to mature into adulthood, when they labeled thalamocortical axons with anterograde tracers injected directly into the inner or outer half of the LGN. When the eyes are removed early in development, eye-specific cellular layers fail to form in the LGN (Guillery et al. 1985, Morgan & Thompson 1993)]. In the ferrets lacking eyes from birth,

patches of label with a periodicity similar to that of ODCs were observed in V1 (Crowley & Katz 1999). Although these eye-removal experiments did not completely rule out a role for neural activity in ODC formation, they did challenge the traditional model in which retinal activity patterns control the process (reviewed in Crowley & Katz 2002). Indeed, prior studies from the same lab showed that eye removal significantly reduces (but does not eliminate) the strength of eye-specific correlations in the developing LGN, at least for ages P24 and older (Weliky & Katz 1999) (**Figure 1**). Crowley & Katz (1999), therefore, concluded that ODCs form according to eye-specific molecular cues in the LGN and/or V1 and not on the basis of Hebbian-type plasticity. Subsequent studies showed that ODCs appeared earlier in development than previously thought and were well segregated from the outset without undergoing significant refinement (Crowley & Katz 2000). These findings reinforced the hypothesis that innate (molecular) cues induce ODC development, though others have argued that the patches of label observed in V1 after focal LGN injections are not actually ODCs, but instead reflect some other patchy organization of the thalamocortical pathway. Given the potential implications of these findings for models of ODC development, and thalamocortical patterning in general, this issue is crucial to resolve. One way to determine if the patches observed in V1 after focal injections into the LGN are indeed ODCs would be to focally inject an anterograde tracer into the monocular portion of the LGN. This region of the LGN is innervated by the nasal portion of the contralateral eye and never projects to V1 in ODCs. Thus, such an injection should lead to a uniform label in V1, not patches. Molecularly distinct portions of the LGN that corresponded to functionally specific, but not eye-specific, visual pathways have been identified (Kawasaki et al. 2004). Still, molecules that mediate ODC formation may exist. Considering the established role of ephrin-As in eye-specific pathfinding in the LGN (Huberman et al. 2005b, Pfeiffenberger et al. 2005), it would be interesting to examine ephrin-A expression in V1 of species such as ferrets, cats, or macaques, which have ODCs. Other cues that might regulate ODC formation are the molecular cues that are differentially expressed in RGCs that take different courses at the optic chiasm, such as zinc transporters (Land & Shamalla-Hannah 2001), EphBs, and zic transcription factors (see review by Petros et al. 2008, in this volume). The expression of these molecules in V1 of species that have ODCs has not been reported.

Given the controversy over whether spontaneous retinal activity mediates ODC development, a recent study used binocular epibatidine injections to prevent stage II retinal waves in ferrets and then transneuronally labeled thalamo-cortical projections in those animals at adulthood, when spillover is not an issue. A key feature of this study's design is that epibatidine treatment was carried out before ODCs normally form. Results showed that preventing stage II waves severely disrupts ODC segregation and patterning (Huberman et al. 2006). Moreover, eliminating stage II waves caused the size of binocular receptive fields in V1 to increase ~30-fold, whereas monocular receptive fields were unaffected. These findings indicate that stage II waves mediate ODC formation and binocular competition leading to binocular receptive field patterning in V1 and generally support the activity-dependent model of ODC formation.

Epibatidine treatment only alters spontaneous retinal activity patterns in ferrets from P1 until ~P10–P11 (Huberman et al. 2002, Penn et al. 1998). At those ages, LGN axons reside in the subplate (Herrmann et al. 1994), a transient structure that sends massive synaptic input to the developing cortex and is critical for LGN axon pathfinding to V1 (Ghosh et al. 1990, Ghosh & Shatz 1992b), as well as for ODC and ODC plasticity formation (Kanold et al. 2003) and maintenance (Ghosh & Shatz 1992a) during the critical period (Kanold & Shatz 2006). Thus stage II waves may cause LGN axons to segregate into ODCs while they still reside in the subplate, which might explain why Crowley & Katz (2000) observed that ODCs are seg-

regated as they grow out of the subplate and into cortical layer four. Given its importance in so many aspects of thalamo-cortical development, the subplate is a terrific source for candidate cues (activity-dependent or otherwise) that regulate ODC development and plasticity. Microarray gene profiling of cells in the developing subplate under normal conditions and under conditions of binocular epibatidine-induced wave blockade (which prevents ODC segregation) ought to be very informative.

Development of Orientation Selectivity

Neurons in the primary visual cortex of adult mammals are tuned for the orientation of visual stimuli. In carnivores and primates, orientation selectivity is organized in a columnar fashion (Hubel & Wiesel 1962, Thompson et al. 1983) and mapped smoothly across cortex (Blasdel & Salama 1986, Grinvald et al. 1986). The development of orientation selectivity was first studied in traditional animal models of the visual system, monkeys and cats. In rhesus monkeys, at least some cells demonstrate orientation-specific responses at birth (Wiesel & Hubel 1974). In kittens studied at the end of the first postnatal week, before natural eye opening, the degree of orientation tuning reported in the literature varied widely, from 0% of cells exhibiting orientation selectivity (Barlow & Pettigrew 1971, Pettigrew 1974), to 25%–30% (Blakemore & Van Sluyters 1975, Buisseret & Imbert 1976, Fregnac & Imbert 1978), to 100% (Hubel & Wiesel 1963). These discrepancies may be due to the difficulty of performing electrophysiological recording in young kittens where cells respond sluggishly and habituate rapidly (Hubel & Wiesel 1963), and where small changes in blood pressure or expired carbon dioxide levels can change an orientation-selective cell into a nonselective or unresponsive cell (Blakemore & Van Sluyters 1975). For this reason, more recent studies of orientation selectivity development have used the ferret as an animal model. The ferret has a visual system quite similar to that of the cat (Law et al. 1988) but is born about three weeks earlier in development (Linden et al. 1981). The ferret therefore provides a terrific model system for exploring the mechanisms underlying development of orientation selectivity (Chapman & Stryker 1993, Krug et al. 2001).

Studies of ferret primary visual cortex have shown that some degree of orientation tuning is present as early as visual responses can be elicited, two weeks before the time of natural eye opening, but that adult-like tuning levels are not reached until about a week after eye opening (Chapman & Stryker 1993, Krug et al. 2001) (**Figure 1**). Studies of the development of orientation maps in V1 have shown that orientation maps are observed prior to eye opening and that the overall map layout is very stable throughout development (Chapman et al. 1996).

The role of visual experience in the development of orientation selectivity has been widely studied. The fact that orientation-selective responses are already present at or before the time of eye opening suggested that vision is not necessary for the initial emergence of cortical cell orientation tuning. However, cortical neurons can be driven in an orientation-selective manner, even before natural eye-opening, through closed eyelids (Eysel 1979, Krug et al. 2001, Spear et al. 1978), so dark-rearing beginning at stages prior to natural eye-opening were carried out. Experiments comparing the orientation maps in normal and dark-reared animals showed that orientation selectivity can in fact develop in the absence of visual experience but that normal mature levels of selectivity are never achieved (White et al. 2001). In contrast, binocular deprivation by lid suture after the normal period of eye-opening produced much more devastating effects on orientation selectivity, allowing little orientation selectivity maturation beyond that seen in very young normal ferrets (Chapman & Stryker 1993, White et al. 2001). These results suggest that visual experience is not necessary for the initial phase of orientation development but that vision is needed for its maturation. The impact of early dark-rearing furthermore shows that abnormal visual experience through closed eyelids disrupts the initial emergence of circuits underlying orientation selectivity. A somewhat different picture emerges from studies of orientation selectivity development in the cat visual cortex: In these studies, binocular visual deprivation did not prevent the initial development of orientation selectivity, although it did prevent map maintenance (Crair et al. 1998). It is not clear whether this difference is due to a difference in the amount of vision possible through a closed eyelid in the two species (i.e., that a lid-sutured cat may experience very little vision and therefore respond like a dark-reared ferret) or whether it is due to a real difference in the role of visual experience in the initial development of orientation selectivity between the two species.

Although visual experience is not necessary for the initial development of orientation selectivity, spontaneous neuronal activity is necessary. Silencing all spontaneous action potential activity in the visual cortex with TTX completely abolishes the maturation of orientation selectivity (Chapman & Stryker 1993). Several experiments have addressed whether activity is playing an instructive role in orientation selectivity development, that is, whether altering the spatio-temporal properties of neuronal activity produces an effect. Computational models have suggested that the patterns of spontaneous activity in On- and Off-center RGCs may instruct the development of orientation tuning in cortical cells. Correlations in neighboring RGC firing of like center type (On or Off) and anticorrelations in the firing of cells with opposite center type result in the development of oriented cortical cell receptive fields in these models (Miller 1992, 1994). Pharmacologically silencing On-RGC activity does indeed prevent the maturation of orientation selectivity, supporting this class of computational model and suggesting that activity is, in fact, playing an instructive role (Chapman & Godecke 2000). An alternative method of altering activity patterns during development, artificially increasing correlations of all RGCs by electrically stimulating the optic nerve, likewise weakened orientation selectivity maturation (Weliky & Katz 1997). Thus correlation among On- versus Off-center

LGN inputs to V1 appears to instruct development of orientation tuning. The molecular basis for detecting the correlational structure of On- and Off-center inputs to cortical cells is likely to be NMDA receptor activation; blocking NMDA receptor function prevents orientation selectivity maturation (Ramoa et al. 2001).

The exact mechanisms underlying the development of cortical cell orientation selectivity remain to be elucidated. Patterns of spontaneous activity before eye-opening are clearly involved in the initial development of orientation tuning, and visual experience is involved in later maturation and maintenance of the circuits underlying this receptive field feature. A role for molecular cues in orientation selectivity development has not been tested. However, the positional stability of the overall layout of orientation maps as they emerge (Chapman et al. 1996) suggests that factors other than activity are involved.

Development of Direction Preference

Neurons at various stages of the visual pathway are tuned to the movement direction of stimuli in the visual field. In some species, such as rabbits and mice, RGCs exhibit direction selectivity, whereas in other species such as ferrets, cats, and monkeys, this property emerges first in V1. The circuit that underlies retinal direction selectivity has been extensively studied (Demb 2007, Vaney & Taylor 2002), but the mechanisms that control retinal direction selectivity development remain a mystery.

Some neurons in the SC exhibit direction-selective responses (Hoffmann & Sherman 1974). Manipulations that alter early activity patterns, such as strobe rearing, disrupt direction-selective cell tuning in the SC (Chalupa & Rhoades 1978, Flandrin & Jeannerod 1975); however, dark-rearing animals during the same period does not disrupt direction tuning of SC cells (Chalupa & Rhoades 1978), which indicates that direction-selective circuits in the SC do not require vision to form.

In some species, neurons in V1 exhibit direction preference, and optical imaging experiments have shown that V1 cells are organized into regular maps for preferred direction (Weliky et al. 1996). Direction selectivity in the cortex is absent at eye opening in ferrets and then develops during the subsequent two weeks from P30 to P45, making direction one of the last basic visual circuit properties to form. Patterned visual experience is the cue that instructs development of cortical direction selectivity. If ferrets are dark reared after eye-opening, direction maps fail to form and single V1 neurons lack direction tuning (Li et al. 2006). However, a brief exposure to moving visual stimuli in a single direction will allow the development of cells tuned to that direction, whereas other cells remain untuned for direction. This finding indicates that visual experience is instructive, not merely permissive, for the emergence of cortical direction selectivity.

Allowing vision to return at P45 rescues many response properties of V1 neurons such as contrast sensitivity and orientation, but direction preference is permanently abolished. This result indicates that, even after eye opening, the initial development—not just the plasticity—of direction-selective circuits in V1 encounters an early critical period. The critical period for ocular dominance plasticity begins in the ferret at ~P35 and peaks at P42 (Issa et al. 1999) (**Figure 1**). Thus even though the cortical circuits underlying eye specificity form before eye-opening and then are subject to plastic changes at this time, some circuit properties such as direction selectivity are still in the process of developing. Why direction-selective circuits in retina and SC do not require vision to form, whereas direction-selective circuits in V1 do, remains unclear. A better understanding of the neural circuits underlying direction selectivity in each visual area (retina, SC, V1) will shed light on this issue.

SYNTHESIS, CONCLUSIONS, FUTURE DIRECTIONS

Spontaneous and early visually evoked neural activity is necessary for anatomical and functional refinement of developing visual circuits.

In the past decade, investigators have debated whether specific activity patterns matter for this refinement or whether activity is permissive for readout of guidance cues. The emerging model posits that activity patterns do matter, but which specific patterns matter and which plasticity rules they engage at nascent synapses to refine connections and induce each circuit property remain a mystery. With the discovery of molecules that translate activity into structural changes in the developing visual system (Huh et al. 2000, Bjartmar et al. 2006, Stevens et al. 2007), it will be interesting to see if activity-dependent synaptic refinement employs general or specialized mechanisms depending on the circuit property in question. At the same time, growing evidence indicates that activity-independent factors are responsible for inducing the highly stereotyped features of visual circuitry such as layer position and overall map layout. Again, the question arises as to whether general rules and principles will emerge for this category of cues and if they are truly activity independent (as appears to be the case for ephrin-As in retinotopic and eye-specific map formation). Much remains to be learned about the development of visual maps and receptive field properties, both for the sake of understanding visual system development and for the sake of understanding the mechanisms of neural circuit refinement throughout the central nervous system.

DISCLOSURE STATEMENT

The authors are not aware of any biases that might be perceived as affecting the objectivity of this review.

ACKNOWLEDGMENTS

Work in the laboratory of M.B. Feller related to the topic of this review was supported by Klingenstein Foundation, Whitehall Foundation, March of Dimes, McKnight Scholars Fund and NIH RO1 EY13528. Work in the laboratory of B. Chapman related to the topic of this review was supported by NIH EY11369. A. D. Huberman was supported by a Helen Hay Whitney Postdoctoral Fellowship. We are grateful to David Feldheim, Michael Susman, and Jocelyn Krey for critical reading and helpful comments.

LITERATURE CITED

Adelsberger H, Garaschuk O, Konnerth A. 2005. Cortical calcium waves in resting newborn mice. *Nat. Neurosci.* 8:988–90

Akerman CJ, Smyth D, Thompson ID. 2002. Visual experience before eye-opening and the development of the retinogeniculate pathway. *Neuron* 36:869–79

Ango F, Di Cristo G, Higashiyama H, Bennett V, Wu P, Huang ZJ. 2004. Ankyrin-based subcellular gradient of neurofascin, an immunoglobulin family protein, directs GABAergic innervation at Purkinje axon initial segment. *Cell* 119:257–72

Bansal A, Singer JH, Hwang BJ, Xu W, Beaudet A, Feller MB. 2000. Mice lacking specific nicotinic acetylcholine receptor subunits exhibit dramatically altered spontaneous activity patterns and reveal a limited role for retinal waves in forming ON and OFF circuits in the inner retina. *J. Neurosci.* 20:7672–81

Barlow HB, Pettigrew JD. 1971. Lack of specificity of neurones in the visual cortex of young kittens. *J. Physiol.* 218:P98–100

Bjartmar L, Huberman AD, Ullian EM, Renteria RC, Liu X, et al. 2006. Neuronal pentraxins mediate synaptic refinement in the developing visual system. *J. Neurosci.* 26:6269–81

Blakemore C, Van Sluyters RC. 1975. Innate and environmental factors in the development of the kitten's visual cortex. *J. Physiol.* 248:663–716

Blasdel GG, Salama G. 1986. Voltage-sensitive dyes reveal a modular organization in monkey striate cortex. *Nature* 321:579–85

Bouzioukh F, Daoudal G, Falk J, Debanne D, Rougon G, Castellani V. 2006. Semaphorin3A regulates synaptic function of differentiated hippocampal neurons. *Eur. J. Neurosci.* 23:2247–54

Braisted JE, McLaughlin T, Wang HU, Friedman GC, Anderson DJ, O'Leary DD. 1997. Graded and lamina-specific distributions of ligands of EphB receptor tyrosine kinases in the developing retinotectal system. *Dev. Biol.* 191:14–28

Buisseret P, Imbert M. 1976. Visual cortical cells: their developmental properties in normal and dark reared kittens. *J. Physiol.* 255:511–25

Butts DA. 2002. Retinal waves: implications for synaptic learning rules during development. *Neuroscientist* 8:243–53

Butts DA, Kanold PO, Shatz CJ. 2007. A burst-based "Hebbian" learning rule at retinogeniculate synapses links retinal waves to activity-dependent refinement. *PLoS Biol.* 5:e61

Callaway EM. 2005. Structure and function of parallel pathways in the primate early visual system. *J. Physiol.* 566:13–19

Campbell G, Shatz CJ. 1992. Synapses formed by identified retinogeniculate axons during the segregation of eye input. *J. Neurosci.* 12:1847–58

Cang J, Kaneko M, Yamada J, Woods G, Stryker MP, Feldheim DA. 2005a. Ephrin-as guide the formation of functional maps in the visual cortex. *Neuron* 48:577–89

Cang J, Renteria RC, Kaneko M, Liu X, Copenhagen DR, Stryker MP. 2005b. Development of precise maps in visual cortex requires patterned spontaneous activity in the retina. *Neuron* 48:797–809

Cang J, Niell CM, Liu X, Pfeiffenberger C, Feldheim DA, Stryker MP. 2008. Selective disruption of one cartesian axis of cortical maps and receptive fields by deficiency in ephrin-As and structured activity. *Neuron* 57:511–23

Casagrande VA, Condo GJ. 1988. The effect of altered neuronal activity on the development of layers in the lateral geniculate nucleus. *J. Neurosci.* 8:395–416

Chalupa LM, Rhoades RW. 1978. Directional selectivity in hamster superior colliculus is modified by strobe-rearing but not by dark-rearing. *Science* 199:998–1001

Chalupa LM, Snider CJ. 1998. Topographic specificity in the retinocollicular projection of the developing ferret: an anterograde tracing study. *J. Comp. Neurol.* 392:35–47

Chandrasekaran AR, Plas DT, Gonzalez E, Crair MC. 2005. Evidence for an instructive role of retinal activity in retinotopic map refinement in the superior colliculus of the mouse. *J. Neurosci.* 25:6929–38

Chapman B. 2000. Necessity for afferent activity to maintain eye-specific segregation in ferret lateral geniculate nucleus. *Science* 287:2479–82

Chapman B, Godecke I. 2000. Cortical cell orientation selectivity fails to develop in the absence of ON-center retinal ganglion cell activity. *J. Neurosci.* 20:1922–30

Chapman B, Stryker MP. 1993. Development of orientation selectivity in ferret visual cortex and effects of deprivation. *J. Neurosci.* 13:5251–62

Chapman B, Stryker MP, Bonhoeffer T. 1996. Development of orientation preference maps in ferret primary visual cortex. *J. Neurosci.* 16:6443–53

Chapman B, Zahs KR, Stryker MP. 1991. Relation of cortical cell orientation selectivity to alignment of receptive fields of the geniculocortical afferents that arborize within a single orientation column in ferret visual cortex. *J. Neurosci.* 11:1347–58

Chattopadhyaya B, Di Cristo G, Wu CZ, Knott G, Kuhlman S, et al. 2007. GAD67-mediated GABA synthesis and signaling regulate inhibitory synaptic innervation in the visual cortex. *Neuron* 54:889–903

Chen C, Regehr WG. 2000. Developmental remodeling of the retinogeniculate synapse. *Neuron* 28:955–66

Chen Y, Flanagan JG. 2006. Follow your nose: axon pathfinding in olfactory map formation. *Cell* 127:881–84

Chiu C, Weliky M. 2001. Spontaneous activity in developing ferret visual cortex in vivo. *J. Neurosci.* 21:8906–14

Chiu C, Weliky M. 2002. Relationship of correlated spontaneous activity to functional ocular dominance columns in the developing visual cortex. *Neuron* 35:1123–34

Cook PM, Prusky G, Ramoa AS. 1999. The role of spontaneous retinal activity before eye opening in the maturation of form and function in the retinogeniculate pathway of the ferret. *Vis. Neurosci.* 16:491–501

Corriveau RA, Huh GS, Shatz CJ. 1998. Regulation of class I MHC gene expression in the developing and mature CNS by neural activity. *Neuron* 21:505–20

Crair MC, Gillespie DC, Stryker MP. 1998. The role of visual experience in the development of columns in cat visual cortex. *Science* 279:566–70

Crair MC, Horton JC, Antonini A, Stryker MP. 2001. Emergence of ocular dominance columns in cat visual cortex by 2 weeks of age. *J. Comp. Neurol.* 430:235–49

Crowley JC, Katz LC. 1999. Development of ocular dominance columns in the absence of retinal input. *Nat. Neurosci.* 2:1125–30

Crowley JC, Katz LC. 2000. Early development of ocular dominance columns. *Science* 290:1321–24

Crowley JC, Katz LC. 2002. Ocular dominance development revisited. *Curr. Opin. Neurobiol.* 12:104–9

Demas J, Eglen SJ, Wong RO. 2003. Developmental loss of synchronous spontaneous activity in the mouse retina is independent of visual experience. *J. Neurosci.* 23:2851–60

Demas J, Sagdullaev BT, Green E, Jaubert-Miazza L, McCall MA, et al. 2006. Failure to maintain eye-specific segregation in nob, a mutant with abnormally patterned retinal activity. *Neuron* 50:247–59

Demb JB. 2007. Cellular mechanisms for direction selectivity in the retina. *Neuron* 55:179–86

Di Cristo G, Wu C, Chattopadhyaya B, Ango F, Knott G, et al. 2004. Subcellular domain-restricted GABAergic innervation in primary visual cortex in the absence of sensory and thalamic inputs. *Nat. Neurosci.* 7:1184–86

Dufour A, Seibt J, Passante L, Depaepe V, Ciossek T, et al. 2003. Area specificity and topography of thalamocortical projections are controlled by ephrin/Eph genes. *Neuron* 39:453–65

Eglen SJ. 1999. The role of retinal waves and synaptic normalization in retinogeniculate development. *Philos. Trans. R. Soc. London Ser. B* 354:497–506

Ellsworth CA, Lyckman AW, Feldheim DA, Flanagan JG, Sur M. 2005. Ephrin-A2 and -A5 influence patterning of normal and novel retinal projections to the thalamus: conserved mapping mechanisms in visual and auditory thalamic targets. *J. Comp. Neurol.* 488:140–51

Eysel UT. 1979. Maintained activity, excitation and inhibition of lateral geniculate neurons after monocular deafferentation in the adult cat. *Brain Res.* 166:259–71

Feldheim DA, Kim YI, Bergemann AD, Frisen J, Barbacid M, Flanagan JG. 2000. Genetic analysis of ephrin-A2 and ephrin-A5 shows their requirement in multiple aspects of retinocollicular mapping. *Neuron* 25:563–74

Feldheim DA, Nakamoto M, Osterfield M, Gale NW, DeChiara TM, et al. 2004. Loss-of-function analysis of EphA receptors in retinotectal mapping. *J. Neurosci.* 24:2542–50

Feldheim DA, Vanderhaeghen P, Hansen MJ, Frisen J, Lu Q, et al. 1998. Topographic guidance labels in a sensory projection to the forebrain. *Neuron* 21:1303–13

Feller MB. 1999. Spontaneous correlated activity in developing neural circuits. *Neuron* 22:653–56

Feller MB, Butts DA, Aaron HL, Rokhsar DS, Shatz CJ. 1997. Dynamic processes shape spatiotemporal properties of retinal waves. *Neuron* 19:293–306

Feller MB, Wellis DP, Stellwagen D, Werblin FS, Shatz CJ. 1996. Requirement for cholinergic synaptic transmission in the propagation of spontaneous retinal waves. *Science* 272:1182–87

Firth SI, Wang CT, Feller MB. 2005. Retinal waves: mechanisms and function in visual system development. *Cell Calcium* 37:425–32

Flanagan JG. 2006. Neural map specification by gradients. *Curr. Opin. Neurobiol.* 16:59–66

Flandrin JM, Jeannerod M. 1975. Superior colliculus: environmental influences on the development of directional responses in the kitten. *Brain Res.* 89:348–52

Fregnac Y, Imbert M. 1978. Early development of visual cortical cells in normal and dark-reared kittens: relationship between orientation selectivity and ocular dominance. *J. Physiol.* 278:27–44

Frisen J, Yates PA, McLaughlin T, Friedman GC, O'Leary DD, Barbacid M. 1998. Ephrin-A5 (AL-1/RAGS) is essential for proper retinal axon guidance and topographic mapping in the mammalian visual system. *Neuron* 20:235–43

Galli L, Maffei L. 1988. Spontaneous impulse activity of rat retinal ganglion cells in prenatal life. *Science* 242:90–91

Ghosh A, Antonini A, McConnell SK, Shatz CJ. 1990. Requirement for subplate neurons in the formation of thalamocortical connections. *Nature* 347:179–81

Ghosh A, Shatz CJ. 1992a. Involvement of subplate neurons in the formation of ocular dominance columns. *Science* 255:1441–43

Ghosh A, Shatz CJ. 1992b. Pathfinding and target selection by developing geniculocortical axons. *J. Neurosci.* 12:39–55

Godement P, Salaun J, Imbert M. 1984. Prenatal and postnatal development of retinogeniculate and retinocollicular projections in the mouse. *J. Comp. Neurol.* 230:552–75

Goodman CS, Shatz CJ. 1993. Developmental mechanisms that generate precise patterns of neuronal connectivity. *Cell* 72(Suppl.):77–98

Grinvald A, Lieke E, Frostig RD, Gilbert CD, Wiesel TN. 1986. Functional architecture of cortex revealed by optical imaging of intrinsic signals. *Nature* 324:361–64

Grubb MS, Rossi FM, Changeux JP, Thompson ID. 2003. Abnormal functional organization in the dorsal lateral geniculate nucleus of mice lacking the ß2 subunit of the nicotinic acetylcholine receptor. *Neuron* 40:1161–72

Guillery RW, LaMantia AS, Robson JA, Huang K. 1985. The influence of retinal afferents upon the development of layers in the dorsal lateral geniculate nucleus of mustelids. *J. Neurosci.* 5:1370–79

Hanganu IL, Ben-Ari Y, Khazipov R. 2006. Retinal waves trigger spindle bursts in the neonatal rat visual cortex. *J. Neurosci.* 26:6728–36

Hanson MG, Landmesser LT. 2004. Normal patterns of spontaneous activity are required for correct motor axon guidance and the expression of specific guidance molecules. *Neuron* 43:687–701

Hensch TK. 2004. Critical period regulation. *Annu. Rev. Neurosci.* 27:549–79

Herrmann K, Antonini A, Shatz CJ. 1994. Ultrastructural evidence for synaptic interactions between thalamocortical axons and subplate neurons. *Eur. J. Neurosci.* 6:1729–42

Hevner RF. 2000. Development of connections in the human visual system during fetal midgestation: a DiI-tracing study. *J. Neuropathol. Exp. Neurol.* 59:385–92

Hindges R, McLaughlin T, Genoud N, Henkemeyer M, O'Leary DD. 2002. EphB forward signaling controls directional branch extension and arborization required for dorsal-ventral retinotopic mapping. *Neuron* 35:475–87

Hoffmann KP, Sherman SM. 1974. Effects of early monocular deprivation on visual input to cat superior colliculus. *J. Neurophysiol.* 37:1276–86

Hooks BM, Chen C. 2006. Distinct roles for spontaneous and visual activity in remodeling of the retinogeniculate synapse. *Neuron* 52:281–91

Horton JC, Hocking DR. 1996. An adult-like pattern of ocular dominance columns in striate cortex of newborn monkeys prior to visual experience. *J. Neurosci.* 16:1791–807

Huang L, Pallas SL. 2001. NMDA antagonists in the superior colliculus prevent developmental plasticity but not visual transmission or map compression. *J. Neurophysiol.* 86:1179–94

Hubel DH, Wiesel TN. 1962. Receptive fields, binocular interaction and functional architecture in the cat's visual cortex. *J. Physiol.* 160:106–54

Hubel DH, Wiesel TN. 1963. Receptive fields of cells in striate cortex of very young, visually inexperienced kittens. *J. Neurophysiol.* 26:994–1002

Huberman AD. 2006. Nob mice wave goodbye to eye-specific segregation. *Neuron* 50:175–77

Huberman AD. 2007. Mechanisms of eye-specific visual circuit development. *Curr. Opin. Neurobiol.* 17:73–80

Huberman AD, Dehay C, Berland M, Chalupa LM, Kennedy H. 2005a. Early and rapid targeting of eye-specific axonal projections to the dorsal lateral geniculate nucleus in the fetal macaque. *J. Neurosci.* 25:4014–23

Huberman AD, Murray KD, Warland DK, Feldheim DA, Chapman B. 2005b. Ephrin-As mediate targeting of eye-specific projections to the lateral geniculate nucleus. *Nat. Neurosci.* 8:1013–21

Huberman AD, Speer CM, Chapman B. 2006. Spontaneous retinal activity mediates development of ocular dominance columns and binocular receptive fields in v1. *Neuron* 52:247–54

Huberman AD, Stellwagen D, Chapman B. 2002. Decoupling eye-specific segregation from lamination in the lateral geniculate nucleus. *J. Neurosci.* 22:9419–29

Huberman AD, Wang GY, Liets LC, Collins OA, Chapman B, Chalupa LM. 2003. Eye-specific retinogeniculate segregation independent of normal neuronal activity. *Science* 300:994–98

Huh GS, Boulanger LM, Du H, Riquelme PA, Brotz TM, Shatz CJ. 2000. Functional requirement for class I MHC in CNS development and plasticity. *Science* 290:2155–59

Issa NP, Trachtenberg JT, Chapman B, Zahs KR, Stryker MP. 1999. The critical period for ocular dominance plasticity in the ferret's visual cortex. *J. Neurosci.* 19:6965–78

Jaubert-Miazza L, Green E, Lo FS, Bui K, Mills J, Guido W. 2005. Structural and functional composition of the developing retinogeniculate pathway in the mouse. *Vis. Neurosci.* 22:661–76

Jeffery G. 1985. Retinotopic order appears before ocular separation in developing visual pathways. *Nature* 313:575–76

Kanold PO, Kara P, Reid RC, Shatz CJ. 2003. Role of subplate neurons in functional maturation of visual cortical columns. *Science* 301:521–25

Kanold PO, Shatz CJ. 2006. Subplate neurons regulate maturation of cortical inhibition and outcome of ocular dominance plasticity. *Neuron* 51:627–38

Katz LC, Crowley JC. 2002. Development of cortical circuits: lessons from ocular dominance columns. *Nat. Rev. Neurosci.* 3:34–42

Katz LC, Shatz CJ. 1996. Synaptic activity and the construction of cortical circuits. *Science* 274:1133–38

Kawasaki H, Crowley JC, Livesey FJ, Katz LC. 2004. Molecular organization of the ferret visual thalamus. *J. Neurosci.* 24:9962–70

King AJ, Schnupp JW, Thompson ID. 1998. Signals from the superficial layers of the superior colliculus enable the development of the auditory space map in the deeper layers. *J. Neurosci.* 18:9394–408

Krug K, Akerman CJ, Thompson ID. 2001. Responses of neurons in neonatal cortex and thalamus to patterned visual stimulation through the naturally closed lids. *J. Neurophysiol.* 85:1436–43

Land PW, Shamalla-Hannah L. 2001. Transient expression of synaptic zinc during development of uncrossed retinogeniculate projections. *J. Comp. Neurol.* 433:515–25

Law MI, Zahs KR, Stryker MP. 1988. Organization of primary visual cortex (area 17) in the ferret. *J. Comp. Neurol.* 278:157–80

Lee JS, von der Hardt S, Rusch MA, Stringer SE, Stickney HL, et al. 2004. Axon sorting in the optic tract requires HSPG synthesis by ext2 (dackel) and extl3 (boxer). *Neuron* 44:947–60

LeVay S, Stryker MP, Shatz CJ. 1978. Ocular dominance columns and their development in layer IV of the cat's visual cortex: a quantitative study. *J. Comp. Neurol.* 179:223–44

Li Y, Fitzpatrick D, White LE. 2006. The development of direction selectivity in ferret visual cortex requires early visual experience. *Nat. Neurosci.* 9:676–81

Liets LC, Olshausen BA, Wang GY, Chalupa LM. 2003. Spontaneous activity of morphologically identified ganglion cells in the developing ferret retina. *J. Neurosci.* 23:7343–50

Linden DC, Guillery RW, Cucchiaro J. 1981. The dorsal lateral geniculate nucleus of the normal ferret and its postnatal development. *J. Comp. Neurol.* 203:189–211

Maffei L, Galli-Resta L. 1990. Correlation in the discharges of neighboring rat retinal ganglion cells during prenatal life. *Proc. Natl. Acad. Sci. USA* 87:2861–64

McLaughlin T, O'Leary DD. 2005. Molecular gradients and development of retinotopic maps. *Annu. Rev. Neurosci.* 28:327–55

McLaughlin T, Torborg CL, Feller MB, O'Leary DD. 2003. Retinotopic map refinement requires spontaneous retinal waves during a brief critical period of development. *Neuron* 40:1147–60

Meissirel C, Wikler KC, Chalupa LM, Rakic P. 1997. Early divergence of magnocellular and parvocellular functional subsystems in the embryonic primate visual system. *Proc. Natl. Acad. Sci. USA* 94:5900–5

Miller KD. 1992. Development of orientation columns via competition between ON- and OFF-center inputs. *NeuroReport* 3:73–76

Miller KD. 1994. Models of activity-dependent neural development. *Prog. Brain Res.* 102:303–18

Ming G, Henley J, Tessier-Lavigne M, Song H, Poo M. 2001. Electrical activity modulates growth cone guidance by diffusible factors. *Neuron* 29:441–52

Mombaerts P. 2006. Axonal wiring in the mouse olfactory system. *Annu. Rev. Cell Dev. Biol.* 22:713–37

Mooney R, Penn AA, Gallego R, Shatz CJ. 1996. Thalamic relay of spontaneous retinal activity prior to vision. *Neuron* 17:863–74

Morgan J, Thompson ID. 1993. The segregation of ON- and OFF-center responses in the lateral geniculate nucleus of normal and monocularly enucleated ferrets. *Vis. Neurosci.* 10:303–11

Mrsic-Flogel TD, Hofer SB, Creutzfeldt C, Cloëz-Tayarani I, Changeux JP, et al. 2005. Altered map of visual space in the superior colliculus of mice lacking early retinal waves. *J. Neurosci.* 25:6921–28

Muir-Robinson G, Hwang BJ, Feller MB. 2002. Retinogeniculate axons undergo eye-specific segregation in the absence of eye-specific layers. *J. Neurosci.* 22:5259–64

Muscat L, Huberman AD, Jordan CL, Morin LP. 2003. Crossed and uncrossed retinal projections to the hamster circadian system. *J. Comp. Neurol.* 466:513–24

Myhr KL, Lukasiewicz PD, Wong RO. 2001. Mechanisms underlying developmental changes in the firing patterns of ON and OFF retinal ganglion cells during refinement of their central projections. *J. Neurosci.* 21:8664–71

Nelson R, Famiglietti EV Jr, Kolb H. 1978. Intracellular staining reveals different levels of stratification for on- and off-center ganglion cells in cat retina. *J. Neurophysiol.* 41:472–83

Nicol X, Voyatzis S, Muzerelle A, Narboux-Nême N, Südhof TC, et al. 2007. cAMP oscillations and retinal activity are permissive for ephrin signaling during the establishment of the retinotopic map. *Nat. Neurosci.* 10:340–47

Ohshiro T, Weliky M. 2006. Simple fall-off pattern of correlated neural activity in the developing lateral geniculate nucleus. *Nat. Neurosci.* 9:1541–48

O'Leary DD, McLaughlin T. 2005. Mechanisms of retinotopic map development: Ephs, ephrins, and spontaneous correlated retinal activity. *Prog. Brain Res.* 147:43–65

Penn AA, Riquelme PA, Feller MB, Shatz CJ. 1998. Competition in retinogeniculate patterning driven by spontaneous activity. *Science* 279:2108–12

Petros TJ, Rebsam A, Mason C. 2008. Retinal axon growth at the optic chasm: to cross or not to cross. *Annu. Rev. Neurosci.* 31:295–315

Pettigrew JD. 1974. The effect of visual experience on the development of stimulus specificity by kitten cortical neurones. *J. Physiol.* 237:49–74

Pfeiffenberger C, Cutforth T, Woods G, Yamada J, Renteria RC, et al. 2005. Ephrin-As and neural activity are required for eye-specific patterning during retinogeniculate mapping. *Nat. Neurosci.* 8:1022–27

Pfeiffenberger C, Yamada J, Feldheim DA. 2006. Ephrin-As and patterned retinal activity act together in the development of topographic maps in the primary visual system. *J. Neurosci.* 26:12873–84

Plas DT, Lopez JE, Crair MC. 2005. Pretarget sorting of retinocollicular axons in the mouse. *J. Comp. Neurol.* 491:305–19

Poskanzer K, Needleman LA, Bozdagi O, Huntley GW. 2003. N-cadherin regulates ingrowth and laminar targeting of thalamocortical axons. *J. Neurosci.* 23:2294–305

Rakic P. 1976. Prenatal genesis of connections subserving ocular dominance in the rhesus monkey. *Nature* 261:467–71

Ramoa AS, Mower AF, Liao D, Jafri SI. 2001. Suppression of cortical NMDA receptor function prevents development of orientation selectivity in the primary visual cortex. *J. Neurosci.* 21:4299–309

Reese BE. 1987. The position of the crossed and uncrossed optic axons, and the nonoptic axons, in the optic tract of the rat. *Neuroscience* 22:1025–39

Reese BE, Cowey A. 1988. Segregation of functionally distinct axons in the monkey's optic tract. *Nature* 331:350–51

Reid RC, Alonso JM. 1995. Specificity of monosynaptic connections from thalamus to visual cortex. *Nature* 378:281–84

Rossi FM, Pizzorusso T, Porciatti V, Marubio LM, Maffei L, Changeux JP. 2001. Requirement of the nicotinic acetylcholine receptor beta 2 subunit for the anatomical and functional development of the visual system. *Proc. Natl. Acad. Sci. USA* 98:6453–58

Ruthazer ES, Baker GE, Stryker MP. 1999. Development and organization of ocular dominance bands in primary visual cortex of the sable ferret. *J. Comp. Neurol.* 407:151–65

Sahay A, Kim CH, Sepkuty JP, Cho E, Huganir RL, et al. 2005. Secreted semaphorins modulate synaptic transmission in the adult hippocampus. *J. Neurosci.* 25:3613–20

Salinas PC, Zou Y. 2008. Wnt signaling and neuronal circuit assembly. *Annu. Rev. Neurosci.* 31:339–58

Sanes JR, Yamagata M. 1999. Formation of lamina-specific synaptic connections. *Curr. Opin. Neurobiol.* 9:79–87

Schmitt AM, Shi J, Wolf AM, Lu CC, King LA, Zou Y. 2006. Wnt-Ryk signalling mediates medial-lateral retinotectal topographic mapping. *Nature* 439:31–37

Sernagor E, Eglen SJ, Wong RO. 2001. Development of retinal ganglion cell structure and function. *Prog. Retin. Eye Res.* 20:139–74

Shatz CJ. 1983. The prenatal development of the cat's retinogeniculate pathway. *J. Neurosci.* 3:482–99

Shatz CJ, Kirkwood PA. 1984. Prenatal development of functional connections in the cat's retinogeniculate pathway. *J. Neurosci.* 4:1378–97

Sherman SM. 2004. Interneurons and triadic circuitry of the thalamus. *Trends Neurosci.* 27:670–75

Simon DK, Prusky GT, O'Leary DD, Constantine-Paton M. 1992. N-methyl-D-aspartate receptor antagonists disrupt the formation of a mammalian neural map. *Proc. Natl. Acad. Sci. USA* 89:10593–97

Spear PD, Tong L, Langsetmo A. 1978. Striate cortex neurons of binocularly deprived kittens respond to visual stimuli through the closed eyelids. *Brain Res.* 155:141–46

Sperry RW. 1963. Chemoaffinity in the orderly growth of nerve fiber patterns and connections. *Proc. Natl. Acad. Sci. USA* 50:703–10

Sretavan D, Shatz CJ. 1984. Prenatal development of individual retinogeniculate axons during the period of segregation. *Nature* 308:845–48

Sretavan DW, Shatz CJ. 1986. Prenatal development of cat retinogeniculate axon arbors in the absence of binocular interactions. *J. Neurosci.* 6:990–1003

Sretavan DW, Shatz CJ. 1987. Axon trajectories and pattern of terminal arborization during the prenatal development of the cat's retinogeniculate pathway. *J. Comp. Neurol.* 255:386–400

Sretavan DW, Shatz CJ, Stryker MP. 1988. Modification of retinal ganglion cell axon morphology by prenatal infusion of tetrodotoxin. *Nature* 336:468–71

Stellwagen D, Shatz CJ. 2002. An instructive role for retinal waves in the development of retinogeniculate connectivity. *Neuron* 33:357–67

Stellwagen D, Shatz CJ, Feller MB. 1999. Dynamics of retinal waves are controlled by cyclic AMP. *Neuron* 24:673–85

Stevens B, Allen NJ, Vasquez LE, Christopherson KS, Nouri N, et al. 2007. The classical complement cascade mediates developmental synapse elimination. *Cell* 131:1164–78

Stryker MP, Harris WA. 1986. Binocular impulse blockade prevents the formation of ocular dominance columns in cat visual cortex. *J. Neurosci.* 6:2117–33

Stryker MP, Strickland S. 1984. Physiological segregation of ocular dominance columns depends on the pattern of afferent electrical activity. *Invest. Opthal. Vis. Sci. (Suppl).* 25:278

Stryker MP, Zahs KR. 1983. On and off sublaminae in the lateral geniculate nucleus of the ferret. *J. Neurosci.* 3:1943–51

Sur M, Sherman SM. 1982. Retinogeniculate terminations in cats: morphological differences between X and Y cell axons. *Science* 218:389

Syed MM, Lee S, Zheng J, Zhou ZJ. 2004. Stage-dependent dynamics and modulation of spontaneous waves in the developing rabbit retina. *J. Physiol.* 560:533–49

Tavazoie SF, Reid RC. 2000. Diverse receptive fields in the lateral geniculate nucleus during thalamocortical development. *Nat. Neurosci.* 3:608–16

Thompson ID, Kossut M, Blakemore C. 1983. Development of orientation columns in cat striate cortex revealed by 2-deoxyglucose autoradiography. *Nature* 301:712–15

Tian N, Copenhagen DR. 2003. Visual stimulation is required for refinement of ON and OFF pathways in postnatal retina. *Neuron* 39:85–96

Torborg CL, Hansen KA, Feller MB. 2005. High frequency, synchronized bursting drives eye-specific segregation of retinogeniculate projections. *Nat. Neurosci.* 8:72–78

Uesaka N, Hayano Y, Yamada A, Yamamoto N. 2007. Interplay between laminar specificity and activity-dependent mechanisms of thalamocortical axon branching. *J. Neurosci.* 27:5215–23

Vaney DI, Taylor WR. 2002. Direction selectivity in the retina. *Curr. Opin. Neurobiol.* 12:405–10

Walsh C, Guillery RW. 1985. Age-related fiber order in the optic tract of the ferret. *J. Neurosci.* 5:3061–69

Warland DK, Huberman AD, Chalupa LM. 2006. Dynamics of spontaneous activity in the fetal macaque retina during development of retinogeniculate pathways. *J. Neurosci.* 26:5190–97

Weliky M, Bosking WH, Fitzpatrick D. 1996. A systematic map of direction preference in primary visual cortex. *Nature* 379:725–28

Weliky M, Katz LC. 1997. Disruption of orientation tuning in visual cortex by artificially correlated neuronal activity. *Nature* 386:680–85

Weliky M, Katz LC. 1999. Correlational structure of spontaneous neuronal activity in the developing lateral geniculate nucleus in vivo. *Science* 285:599–604

White LE, Coppola DM, Fitzpatrick D. 2001. The contribution of sensory experience to the maturation of orientation selectivity in ferret visual cortex. *Nature* 411:1049–52

Wiesel TN, Hubel DH. 1963a. Effects of visual deprivation on morphology and physiology of cells in the cats lateral geniculate body. *J. Neurophysiol.* 26:978–93

Wiesel TN, Hubel DH. 1963b. Single-cell responses in striate cortex of kittens deprived of vision in one eye. *J. Neurophysiol.* 26:1003–17

Wiesel TN, Hubel DH. 1965a. Comparison of the effects of unilateral and bilateral eye closure on cortical unit responses in kittens. *J. Neurophysiol.* 28:1029–40

Wiesel TN, Hubel DH. 1965b. Extent of recovery from the effects of visual deprivation in kittens. *J. Neurophysiol.* 28:1060–72

Wiesel TN, Hubel DH. 1974. Ordered arrangement of orientation columns in monkeys lacking visual experience. *J. Comp. Neurol.* 158:307–18

Wiesel TN, Hubel DH, Lam DM. 1974. Autoradiographic demonstration of ocular-dominance columns in the monkey striate cortex by means of transneuronal transport. *Brain Res.* 79:273–79

Williams RW, Hogan D, Garraghty PE. 1994. Target recognition and visual maps in the thalamus of achiasmatic dogs. *Nature* 367:637–39

Williams SE, Grumet M, Colman DR, Henkemeyer M, Mason CA, Sakurai T. 2006. A role for Nr-CAM in the patterning of binocular visual pathways. *Neuron* 50:535–47

Williams SE, Mann F, Erskine L, Sakurai T, Wei S, et al. 2003. Ephrin-B2 and EphB1 mediate retinal axon divergence at the optic chiasm. *Neuron* 39:919–35

Wong RO. 1999. Retinal waves and visual system development. *Annu. Rev. Neurosci.* 22:29–47

Wong RO, Meister M, Shatz CJ. 1993. Transient period of correlated bursting activity during development of the mammalian retina. *Neuron* 11:923–38

Wong RO, Oakley DM. 1996. Changing patterns of spontaneous bursting activity of on and off retinal ganglion cells during development. *Neuron* 16:1087–95

Wong WT, Myhr KL, Miller ED, Wong RO. 2000. Developmental changes in the neurotransmitter regulation of correlated spontaneous retinal activity. *J. Neurosci.* 20:351–60

Yamagata M, Sanes JR. 2008. Dscam and Sidekick proteins direct lamina-specific synaptic connections in vertebrate retina. *Nature* 451:465–69

Yamagata M, Weiner JA, Sanes JR. 2002. Sidekicks: synaptic adhesion molecules that promote lamina-specific connectivity in the retina. *Cell* 110:649–60

Yamamoto N, Matsuyama Y, Harada A, Inui K, Murakami F, Hanamura K. 2000. Characterization of factors regulating lamina-specific growth of thalamocortical axons. *J. Neurobiol.* 42:56–68

Yates PA, Holub AD, McLaughlin T, Sejnowski TJ, O'Leary DD. 2004. Computational modeling of retinotopic map development to define contributions of EphA-ephrinA gradients, axon-axon interactions, and patterned activity. *J. Neurobiol.* 59:95–113

Yu CR, Power J, Barnea G, O'Donnell S, Brown HE, et al. 2004. Spontaneous neural activity is required for the establishment and maintenance of the olfactory sensory map. *Neuron* 42:553–66

Zheng JJ, Lee S, Zhou ZJ. 2004. A developmental switch in the excitability and function of the starburst network in the mammalian retina. *Neuron* 44:851–64

Zhou ZJ. 1998. Direct participation of starburst amacrine cells in spontaneous rhythmic activities in the developing mammalian retina. *J. Neurosci.* 18:4155–65

Zhou ZJ, Zhao D. 2000. Coordinated transitions in neurotransmitter systems for the initiation and propagation of spontaneous retinal waves. *J. Neurosci.* 20:6570–77

Neural Substrates
of Language Acquisition

Patricia Kuhl and Maritza Rivera-Gaxiola

Institute for Learning and Brain Sciences, University of Washington, Seattle, Washington 98195; email: pkkuhl@u.washington.edu

Annu. Rev. Neurosci. 2008. 31:511–34

The *Annual Review of Neuroscience* is online at neuro.annualreviews.org

This article's doi:
10.1146/annurev.neuro.30.051606.094321

Key Words

brain measures, language development, speech perception, infants

Abstract

Infants learn language(s) with apparent ease, and the tools of modern neuroscience are providing valuable information about the mechanisms that underlie this capacity. Noninvasive, safe brain technologies have now been proven feasible for use with children starting at birth. The past decade has produced an explosion in neuroscience research examining young children's processing of language at the phonetic, word, and sentence levels. At all levels of language, the neural signatures of learning can be documented at remarkably early points in development. Individual continuity in linguistic development from infants' earliest responses to phonemes is reflected in infants' language abilities in the second and third year of life, a finding with theoretical and clinical implications. Developmental neuroscience studies using language are beginning to answer questions about the origins of humans' language faculty.

Contents

INTRODUCTION

Language and neuroscience have deep historical connections. Neuroscientists and psycholinguists have mined the link between language and the brain since the neuropsychologist Paul Broca described his patient, nicknamed "Tan" for the only word he could utter after his devastating accident. "Tan" was found to have a brain lesion in what is now known as Broca's area. This finding set the stage for "mapping" language functions in the brain.

Humans' linguistic capacities have intrigued philosophers and empiricists for centuries. Young children learn language rapidly and effortlessly, transitioning from babbling at 6 months of age to full sentences by the age of 3, following a consistent developmental path regardless of culture. Linguists, psychologists, and neuroscientists have struggled to explain how children acquire language and ponder how such regularity is achieved if the acquisition mechanism depends on learning and environmental input. Studies of the infant brain using modern neuroscience techniques are providing the first glimpses of the mechanisms underlying the human capacity for language and are examining whether these mechanisms are specific to speech and to the human species.

Infants begin life with the ability to detect differences among all the phonetic distinctions used in the world's languages (Eimas et al. 1971, Streeter 1976). Before the end of the first year they learn from experience and develop a language-specific phonetic capacity and detect native-language word forms, learning implicitly and informally (Kuhl et al. 1992, Saffran et al. 1996, Maye et al. 2002, Newport & Aslin 2004a). The goal of experiments on infants has been to determine whether the initial state and the learning mechanisms are speech specific and species specific.

Tests on nonhuman animals and in humans using nonspeech signals have examined the species specificity and domain specificity of the initial state. Animal tests on phonetic perception (Kuhl & Miller 1975) and on the learning strategies for words (Hauser et al. 2002, Newport & Aslin 2004b), as well as tests on human infants' perception of nonspeech signals resembling speech (Jusczyk et al. 1977), suggested that general perceptual and cognitive abilities may account for infants' initial speech perception abilities (Aslin et al. 1998, Saffran et al. 1999). These data promoted a shift in theoretical positions regarding language, one that advanced the domain-general hypothesis to describe the initial state more strongly than had been done previously (Jusczyk 1997, Kuhl 2000, Saffran 2003, Newport & Aslin 2004a,b).

Theoretical attention is increasingly focused on the mechanisms that underlie language learning. Experiments showed that infants' initial learning from speech experience involves computational skills that are not restricted to speech or to humans (Saffran 2003, Newport & Aslin 2004a). However, there is new evidence to suggest that social interaction (Kuhl et al. 2003) and an interest in speech (Kuhl et al 2005a) may be essential in this process. The combination of computational and social abilities may be exclusive to humans (Kuhl 2007).

Ultimately, determining the specificity of humans' language capacity will require functional measures of the infant brain. Brain measures are ideal for language because infants simply have to listen—there is no need for them to make an overt response—as new brain technologies record the infant brain's language processing. The promise that neural measures will provide the answer to the initial state query as well as the query regarding the specificity of the learning mechanisms has motivated intense work on the development of brain technologies that are safe and feasible with infants and young children.

The goal of this review is to examine the neuroscience techniques now in use with infants and young children and the data these techniques have produced. Neural substrates uncovered using these techniques now extend from the smallest building blocks of language—phonemes—to children's encoding of early words and to studies examining children's processing of the semantic and syntactic information in sentences. We review the neural signatures of learning at each of these levels of language. Using brain measures, we can now link infants' processing at these various levels of language: Brain measures of perception of the elemental phonetic units in the first year are strongly associated with infants' processing of words and syntax in the second and third year of life, showing continuity in the development of language. The studies suggest that exposure to language in the first year of life begins to set the neural architecture in a way that vaults the infant forward in the acquisition of language. Our goal is to explore the neural mechanisms that underlie this capacity.

WINDOWS TO THE YOUNG BRAIN

Noninvasive techniques that examine language processing in infants and young children have advanced rapidly (**Figure 1**) and include electroencephalography (EEG)/event-related potentials (ERPs), magnetoencephalography (MEG), functional magnetic resonance imaging (fMRI), and near-infrared spectroscopy (NIRS).

ERPs have been widely used to study speech and language processing in infants and young children (for reviews, see Kuhl 2004, Friederici 2005, Conboy et al. 2008). ERPs, part of the EEG, reflect electrical activity that is time-locked to the presentation of a specific sensory stimulus (for example, syllables, words) or a cognitive process (recognition of a semantic violation within a sentence or phrase). By placing sensors on a child's scalp, the activity of neural networks firing in a coordinated and synchronous fashion in open field configurations can be measured, and voltage changes occurring as a function of cortical neural activity can be detected. ERPs provide precise time resolution (milliseconds), making them well suited for studying the high-speed and temporally ordered structure of human speech. ERP experiments can also be carried out in populations who, because of age or cognitive impairment, cannot provide overt responses. Spatial resolution of the source of brain activation is limited, however.

MEG is another brain-imaging technique that tracks activity in the brain with exquisite temporal resolution. The SQUID (superconducting quantum interference device) sensors located within the MEG helmet measure the minute magnetic fields associated with electrical currents produced by the brain when it is performing sensory, motor, or cognitive tasks. MEG allows precise spatial localization of the neural currents responsible for the sources of the magnetic fields. Cheour et al. (2004), Kujala et al. (2004), and Imada et al. (2006) have shown phonetic discrimination using MEG in newborns and infants in the first year of life. The newest studies employ sophisticated head-tracking software and hardware that allow correction for infants' head movements and examine multiple brain areas as infants listen to speech (Imada et al. 2006). MEG (as well as EEG) techniques are completely safe and noiseless.

Magnetic resonance imaging (MRI) can be combined with MEG and/or EEG to provide static structural/anatomical pictures of

---

Phonemes: the contrasting element in word pairs that signals a difference in meaning for a specific language (e.g., rake-lake; far-fall)

Phonetic units: sets of articulatory gestures that constitute phoneme categories in a language

EEG: electroencephalography

Event-related potentials (ERPs): scalp recordings that measure electrical activity in the brain that can be time-locked to specific sensory stimuli or cognitive processes

Magnetoencephalography (MEG): measures magnetic fields from electrical currents produced by the brain during sensory, motor, or cognitive tasks

Functional magnetic resonance imaging (fMRI): measures changes in blood oxygenation levels that occur in response to neural firing, allowing precise localization of brain activity

Near-infrared spectroscopy (NIRS): uses infrared light to measure changes in blood concentration as an indicator of neural activity throughout the cortex

Neuroscience techniques used with infants

Inexpensive

EEG/ERP: Electrical potential changes
- Excellent temporal resolution
- Studies cover the life span
- Sensitive to movement
- Noiseless

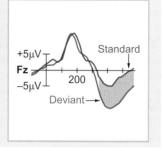

Expensive

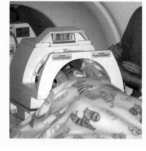

MEG: Magnetic field changes
- Excellent temporal and spatial resolution
- Studies on adults and young children
- Head tracking for movement calibration
- Noiseless

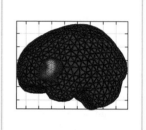

Expensive

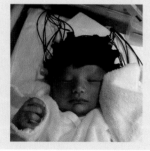

fMRI: Hemodynamic changes
- Excellent spatial resolution
- Studies on adults and a few on infants
- Extremely sensitive to movement
- Noise protectors needed

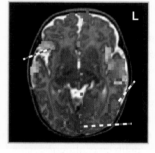

Moderate

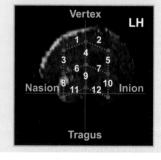

NIRS: Hemodynamic changes
- Good spatial resolution
- Studies on infants in the first 2 years
- Sensitive to movement
- Noiseless

Figure 1

Four neuroscience techniques now used with infants and young children to examine their brain responses to linguistic signals.

the brain. Structural MRIs show maturational anatomical differences in brain regions across the life span. Individual or averaged MRIs can be used to superimpose the physiological activity detected by MEG or EEG to refine the spatial localization of brain activities.

fMRI is a popular method of neuroimaging in adults because it provides high spatial-resolution maps of neural activity across the entire brain (e.g., Gernsbacher & Kaschak 2003). Unlike EEG and MEG, fMRI does not directly detect neural activity, but rather detects the changes in blood oxygenation that occur in response to neural activation/firing. Neural events happen in milliseconds; however, the blood-oxygenation changes that they induce are spread out over several seconds, thereby severely limiting fMRI's temporal resolution. Few studies have attempted fMRI with children because the technique requires infants to be perfectly still and because the MRI device produces loud sounds making it necessary to shield infants' ears while delivering the sound stimuli (Dehaene-Lambertz et al. 2002, 2006).

Near-infrared spectroscopy (NIRS) also measures cerebral hemodynamic responses in relation to neural activity (Aslin & Mehler 2005). NIRS utilizes near-infrared light to measure changes in blood oxy- and deoxy-hemoglobin concentrations in the brain as well as total blood volume changes in various regions of the cerebral cortex. The NIRS system can determine the activity in specific regions of the brain by continuously monitoring blood hemoglobin level. Reports have begun to appear on infants in the first two years of life, testing infants' responses to phonemes as well as to longer stretches of speech such as sentences (Peña et al. 2002, Homae et al. 2006, Bortfeld et al. 2007, Taga & Asakawa 2007). As with other hemodynamic techniques such as fMRI, NIRS does not provide good temporal resolution. One of the strengths of this technique is that coregistration with other testing techniques such as EEG and MEG may be possible.

Each of these techniques is being applied to infants and young children as they listen to speech, from phonemes to words to sentences.

In the following sections, we review the findings at each level of language, noting the advances made using neural measures.

NEURAL SIGNATURES OF PHONETIC LEARNING

Perception of the phonetic units of speech—the vowels and consonants that constitute words—is one of the most widely studied behaviors in infancy and adulthood. Phonetic perception can be studied in children at birth and during development as they are bathed in a particular language, in adults from different cultures, in children with developmental disabilities, and in nonhuman animals. Phonetic perception studies conducted critical tests of theories of language development and its evolution. An extensive literature on developmental speech perception exists, and brain measures are adding substantially to our knowledge of phonetic development and learning (see Kuhl 2004, Werker & Curtin 2005).

Behavioral studies demonstrated that at birth young infants exhibit a universal capacity to detect differences between phonetic contrasts used in the world's languages (Eimas et al. 1971, Streeter 1976). This capacity is dramatically altered by language experience starting as early as 6 months for vowels and by 10 months for consonants. Two important changes occur at the transition in phonetic perception: Native language phonetic abilities significantly increase (Cheour et al. 1998; Kuhl et al. 1992, 2006; Rivera-Gaxiola et al. 2005b, Sundara et al. 2006) while the ability to discriminate phonetic contrasts not relevant to the language of the culture declines (Werker & Tees 1984, Cheour et al. 1998, Best & McRoberts 2003, Rivera-Gaxiola et al. 2005b, Kuhl et al. 2006).

By the end of the first year, the infant brain is no longer universally prepared for all languages, but instead primed to acquire the language(s) to which the infant has been exposed. What was once a universal phonetic capacity—phase 1 of development—narrows as learning proceeds in phase 2. This perceptual narrowing is widespread, affecting the perception of

SQUID: superconducting quantum interference device, a mechanism used to measure extremely weak signals, such as subtle changes in the human body's electromagnetic energy field

Magnetic resonance imaging (MRI): provides static structural and anatomical information, such as grey matter and white matter, in various regions of the brain

signs from American Sign Language (Krentz & Corina 2008, Klarman et al. 2007), as well as the perception of visual speech information from talking faces in monolingual (Lewkowicz & Ghazanfar 2006) and bilingual infants (Weikum et al. 2007). In all cases, perception changes from a more universal ability to one that is more restricted and focused on the properties important to the language(s) to which the infant is exposed. How learning produces this narrowing of perceptual abilities has become the focus of intense study because it demonstrates the brain's shaping by experience during a critical period in language development. Scientific understanding of the critical period is advancing (see Bongaerts et al. 1995 and Kuhl et al. 2005b for review). Data now suggest that an adult's difficulty in learning a second language is affected not only by maturation (Johnson & Newport 1989, Newport 1990, Neville et al. 1997, Mayberry & Lock 2003), but also by learning itself (Kuhl et al. 2005b).

Kuhl et al. (2005b, Kuhl et al. 2008) explored the relationship between the transition in phonetic perception and later language. This work examined a critical test stemming from the native language neural commitment (NLNC) hypothesis (Kuhl 2004). According to NLNC, initial native language experience produces changes in the neural architecture and connections that reflect the patterned regularities contained in ambient speech. The brain's "neural commitment" to native-language patterns has bidirectional effects: Neural coding facilitates the detection of more complex language units (words) that build on initial learning, while simultaneously reducing attention to alternate patterns, such as those of a nonnative language. This formulation suggests that infants with excellent native phonetic skills should advance more quickly toward language. In contrast, excellent nonnative phonetic abilities would not promote native-language learning. Nonnative skills reflect the degree to which the brain remains uncommitted to native-language patterns—phase 1 of development—when perception is universally good for all phonetic contrasts. In phase 1, infants are open to all possible languages, and the brain's language areas are uncommitted. On this reasoning, infants who remain in phase 1 for a longer period of time, showing excellent discrimination of nonnative phonetic units, would be expected to show a slower progression toward language.

The NLNC hypothesis received support from both behavioral (Conboy et al. 2005, Kuhl et al. 2005b) and ERP tests on infants (Rivera-Gaxiola et al. 2005a,b; Silven et al. 2006; Kuhl et al. 2008). ERP studies conducted on infants (**Figure 2a**) used both a nonnative contrast (Mandarin /ɕ-tɕʰ/ or Spanish /t-d/) and a native contrast /p-t/. The results revealed that individual variation in both native and nonnative discrimination, measured neurally at 7.5 months of age, were significantly correlated with later language abilities. As predicted by the NLNC hypothesis, the patterns of prediction were in opposite directions (Kuhl et al. 2008). The measure used was mismatch negativity, which has been shown in adults to be a neural correlate of phonetic discrimination (Näätänen et al. 1997). The MMN-like ERP component was elicited in infants between 250–400 ms for both contrasts, tested in counterbalanced order (**Figure 2b**). Better neural discrimination of the native phonetic contrast at 7.5 months predicted advanced language acquisition at all levels—word production at 24 months, sentence complexity at 24 months, and mean length of utterance at 30 months of age. In contrast, greater neural discrimination of the nonnative contrast at the same age predicted less advanced language development at the same future points in time—lower word production, less complex sentences, and shorter mean utterance length. The behavioral (Kuhl et al. 2005b) and brain (Kuhl et al. 2008) measures, collected on the same infants, were highly correlated.

The growth of vocabulary clearly shows this relationship: Better discrimination of the native contrast resulted in faster vocabulary growth (**Figure 2c**, *left column*), whereas better discrimination of the nonnative contrast resulted in slower vocabulary growth (**Figure 2c**, *right column*). Children's vocabulary growth was measured from 14 to 30 months and was

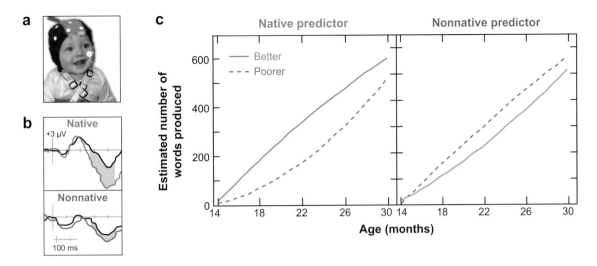

Figure 2

(*a*) A 7.5-month-old infant wearing an ERP electrocap. (*b*) Infant ERP waveforms at one sensor location (CZ) for one infant are shown in response to a native (English) and nonnative (Mandarin) phonetic contrast at 7.5 months. The mismatch negativity (MMN) is obtained by subtracting the standard waveform (*black*) from the deviant waveform (English = *red*; Mandarin = *blue*). This infant's response suggests that native-language learning has begun because the MMN negativity in response to the native English contrast is considerably stronger than that to the nonnative contrast. (*c*) Hierarchical linear growth modeling of vocabulary growth between 14 and 30 months for MMN values of +1SD and −1SD on the native contrast at 7.5 months (*c, left*) and vocabulary growth for MMN values of +1SD and −1SD on the nonnative contrast at 7.5 months (*c, right*). Analyses show that both contrasts predict vocabulary growth but that the effects of better discrimination are reversed for the native and nonnative contrasts. (From Kuhl et al. 2008)

examined using hierarchical linear growth curve modeling (Raudenbush et al. 2005). Analyses show that both native and nonnative discrimination at 7.5 months are significant predictors of vocabulary growth, but also that the effects of good phonetic discrimination are reversed for the native and nonnative predictors. A study of Finnish 11-month-old infants replicated this pattern using Finnish and Russian contrasts (Silven et al. 2006).

Rivera-Gaxiola and colleagues (2005a) demonstrated a similar pattern of prediction using a different nonnative contrast. They recorded auditory ERP complexes in 7- and 11-month-old American infants in response to both Spanish and English voicing contrasts. They found that infants' responses to the nonnative contrast predicted the number of words produced at 18, 22, 25, 27, and 30 months of age (Rivera-Gaxiola et al. 2005a). Infants showing better neural discrimination (larger negativities) to the nonnative contrast at 11 months of age produced significantly fewer words at

all ages when compared with infants showing poorer neural discrimination of the nonnative contrast. Thus in both Kuhl et al. (2005b, 2008) and Rivera-Gaxiola et al. (2005a), better discrimination of a nonnative phonetic contrast was associated with slower vocabulary development. Predictive measures now demonstrate continuity in development from a variety of early measures of pre-language skills to measures of later language abilities. Infants' early phonetic perception abilities (Tsao et al. 2004; Kuhl et al. 2005b, 2008; Rivera-Gaxiola et al. 2005a), their pattern-detection skills for speech (Newman et al. 2006), and their processing efficiency for words (Fernald et al. 2006) have all been linked to advanced later language abilities. These studies form bridges between the early precursors to language in infancy and measures of language competencies in early childhood that are important to theory building as well as to understanding clinical populations with developmental disabilities involving language.

THE ROLE OF SOCIAL FACTORS IN EARLY LANGUAGE LEARNING

Although studies have shown that young infants' computational skills assist language acquisition at the phonological and word level (Saffran et al. 1996, Maye et al. 2002), recent data suggest that another component, social interaction, plays a more significant role in early language learning than previously thought, at least in natural language-learning situations (Kuhl et al. 2003, Kuhl 2007, Conboy et al. 2008). Neuroscience research that reveals the source of this interaction between social and linguistic processing will be of future interest.

The impact of social interaction on speech learning was demonstrated in a study investigating whether infants are capable of phonetic and word learning at nine months of age from natural first-time exposure to a foreign language. Mandarin Chinese was used in the first foreign-language intervention experiment (Kuhl et al. 2003). Infants heard 4 native speakers of Mandarin (male and female) during 12 25-min sessions of book reading and play across a 4–6 week period. A control group of infants also came into the laboratory for the same number and variety of reading and play sessions but heard only English. Two additional groups were exposed to the identical Mandarin material over the same number of sessions via either standard television or audio-only presentation. After exposure, Mandarin syllables that are not phonemic in English were used to test infant learning using both behavioral (Kuhl et al. 2003) and brain (Kuhl et al. 2008) tests.

Infants learned from live exposure to Mandarin tutors, as shown by comparisons with the English control group, indicating that phonetic learning from first-time exposure could occur at nine months of age. However, infants' Mandarin discrimination scores after exposure to television or audio-only tutors were no greater than those of the control infants who had not experienced Mandarin at all (Kuhl et al. 2003) (**Figure 3**). Learning in the live condition was robust and durable. Behavioral tests of infant learning were conducted 2–12 days (median = 6 days) after the final language-exposure session, and the ERP tests were conducted between 12 and 30 days (median = 15 days) after the final exposure session, with no observable differences in infant performance as a function of the delay.

In further experiments, 9-month-old English-learning infants were exposed to Spanish, and ERPs were used to test learning of both Spanish phonemes and Spanish words after 12 exposure sessions (Conboy & Kuhl 2007). The results showed learning of both Spanish phonemes and words. Moreover, the study was designed to test the hypothesis that social interaction during the exposure sessions would predict the degree to which individual infants learned, and this hypothesis was confirmed. Social factors, such as overall attention during the exposure sessions, and specific measures of shared visual attention between the infant and tutor predicted the degree to which individual infants learned Spanish phonemes and words (Conboy et al. 2008).

Social interaction may be essential for learning in complex natural language-learning situations—the neurobiological mechanisms underlying the evolution of language likely utilized the kinds of interactional cues available only in a social setting (Kuhl 2007). Humans are not the only species in which social interaction plays a significant role in communication learning. In other species, such as songbirds, social contact can be essential for communicative learning (see e.g., Immelmann 1969, Baptista & Petrinovich 1986, Brainard & Knudsen 1998, Goldstein et al. 2003).

The importance of social interaction in human language learning is illustrated by its impact on children with autism. A lack of interest in listening to speech signals is correlated with aberrant neural responses to speech and with the severity of autism symptoms, indicating the tight coupling between language and social interaction (Kuhl et al. 2005a). Typically developing infants prefer speech even from birth (Vouloumanos & Werker 2004), and particularly infant-directed (ID) speech (see sidebar on

a Foreign-language exposure

Live exposure

Television exposure

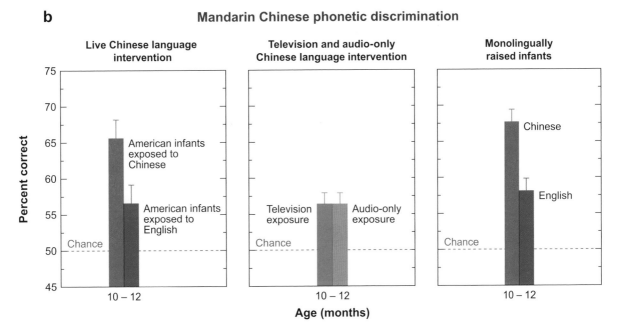

b Mandarin Chinese phonetic discrimination

Figure 3

(*a*) Effects of natural foreign-language intervention using Mandarin Chinese at 9 months of age. Infants experience either live exposure or a television presentation. (*b*) Live exposure (*left panel*) produced significant learning of a Mandarin Chinese phonetic contrast when compared with a control group exposed only to English, whereas television exposure or audio-alone exposure did not produce learning (*middle panel*). Performance after 12 live exposure sessions beginning at 9 months produced performance in American infants (*left panel*) that equaled that of monolingual Taiwanese infants who had listened to Mandarin for 11 months (*right panel*) (from Kuhl et al. 2003).

Motherese) (Fernald & Kuhl 1987, Grieser & Kuhl 1988).

The data on phonetic and word learning from a foreign language indicate that at the earliest stages of language learning, social factors play a significant role, perhaps by gating the computational mechanisms underlying language learning (Kuhl 2007). Interaction between the brain mechanisms underlying linguistic and social processing will be of strong interest in the future. Neuroscience research focused on shared neural systems for perception and action has a long tradition in speech, and interest in "mirror systems" for social

MOTHERESE

Adults use a special speech "register," called "motherese,"when addressing infants and children. Motherese has a unique acoustic signature: It is slower, has a higher average pitch, and contains exaggerated pitch contours (Fernald & Simon 1984). Typically developing infants and children (Fernald & Kuhl 1987), though not children with autism (Kuhl et al. 2005a), prefer Motherese over adult-directed speech or a nonspeech analog signal when given a choice. Motherese exaggerates the phonetic differences in speech, both for vowels (Kuhl et al. 1997, Burnham et al. 2002) and for consonants (Liu et al. 2007), making the words contained in Motherese easier to discriminate for infants. Some evidence indicates that Motherese may aid language learners: Mothers who stretch the acoustic cues in phonemes to a greater extent in infant-directed speech early in life have infants who are better able to hear subtle speech distinctions when tested in the laboratory months later (Liu et al. 2003). Motherese also facilitates word recognition (Thiessen & Saffran 2003). Brain measures of infants' responses to ID speech are enhanced, as shown by NIRS—more activation was seen in left temporal areas when infants were presented with ID speech as opposed to backward speech or silence (Peña et al. 2002), and also when six- to nine-month-old infants were presented with audio-visual ID speech (Bortfeld et al. 2007).

cognition (Kuhl & Meltzoff 1996, Meltzoff & Decety 2003, Rizzolatti & Craighero 2004, Pulvermüller 2005, Rizzolatti 2005, Kuhl 2007) has reinvigorated this tradition (see sidebar on Mirror Systems and Speech). Neuroscience studies using speech and imaging techniques have the capacity to examine this question in infants from birth (e.g., Imada et al. 2006).

NEURAL SIGNATURES OF WORD LEARNING

A sudden increase in vocabulary typically occurs between 18 and 24 months of age—a "vocabulary explosion" (Ganger & Brent 2004, Fernald et al. 2006)—but word learning begins much earlier. Infants show recognition of their own name at four and a half months (Mandel et al. 1995). At six months, infants recognize their own names or the word Mommy as a word seg-

mentation cue (Bortfeld et al. 2005) and look appropriately to pictures of their mothers or fathers when hearing "Mommy" or "Daddy" (Tincoff & Jusczyk 1999). By 7 months, infants listen longer to passages containing words they previously heard rather than passages containing words they have not heard (Jusczyk & Hohne 1997), and by 11 months infants prefer to listen to words that are highly frequent in language input over infrequent words (Halle & de Boysson-Bardies 1994).

One question has been how infants recognize potential words in running speech given that speech is continuous and has no acoustic breaks between words (unlike the spaces between words in written text). Behavioral studies indicate that by 8 months, infants use various strategies to identify potential words in speech. For example, infants treated adjacent syllables with higher transitional probabilities as word-like units in strings of nonsense syllables (Saffran et al. 1996, Saffran 2003, Newport & Aslin 2004a); real words contain higher transitional probabilities between syllables and this strategy would provide infants with a reliable cue. Infants also use the typical pattern of stress in their language to detect potential words. English, for example, puts emphasis on the first syllable, as in the word BASEball, whereas other languages, such as Polish, emphasize the second. English-learning infants treated syllables containing stress as the beginning of a potential word (Cutler & Norris 1988, Johnson & Jusczyk 2001, Nazzi et al. 2006, Hohle et al. 2008). Both transitional probabilities between adjacent syllables and stress cues thus provide infants with clues that allow them to identify potential words in speech.

How is early word recognition evidenced in the brain? ERPs in response to words index word familiarity as early as 9 months of age and word meaning by 13–17 months of age: ERP studies have shown differences in amplitude and scalp distributions for components related to words that are known to the child vs. those that are unknown to the child (Molfese 1990; Molfese et al. 1990, 1993; Mills et al. 1993, 1997, 2005; Thierry et al. 2003).

Toddlers with larger vocabularies showed a larger N200 for known words vs. unknown words at left temporal and parietal electrode sites. In contrast, children with smaller vocabularies showed brain activation that was more broadly distributed (Mills et al. 1993). This distributional difference, with more focalized brain activation linked to greater vocabulary skill, also distinguished typically developing preschool children from preschool children with autism (Coffey-Corina et al. 2007). Increased focalization of brain activation has also been seen in adults for native- as opposed to nonnative phonemes and words in MEG studies (Zhang et al. 2005). This indicates that focal activation has the potential to index language experience and proficiency not just in childhood, but over the life span.

Mills et al. (2005) used ERPs in 20-month-old toddlers to examine new word learning. The children listened to known and unknown words and to nonwords that were phonotactically legal in English. ERPs were recorded as the children were presented with novel objects paired with the nonwords. After the learning period, ERPs to the nonwords that had been paired with novel objects were shown to be similar to those of previously known words, suggesting that new words may be encoded in the same neural regions as previously learned words.

ERP studies on German infants reveal the development of word-segmentation strategies based on the typical stress patterns of German words. When presented with bisyllabic strings with either stress on the first syllable (typical in German) or the second syllable, infants who heard first-syllable stress patterns embedded in a string of syllables showed the N200 ERP component similar to that elicited in response to a known word, whereas infants presented with the nonnative stress pattern showed no response (Weber et al. 2004). The data suggest that German infants at this age are applying learned stress rules about their native language to segment nonsense speech strings into word-like units, in agreement with behavioral data (Hohle et al. In press).

MIRROR SYSTEMS AND SPEECH

The field has a long tradition of theoretical linkage between perception and action in speech. The Motor Theory (Liberman et al. 1967) and Direct Realism (Fowler 1986) posited close interaction between speech perception and production. The discovery of mirror neurons in monkeys that react both to the sight of others' actions and to the same actions they themselves produced (Rizzolatti et al. 1996, 2002; Gallese 2003) has rekindled interest in a potential mirror system for speech, as has work on the origins of infant imitation (Meltzoff & Moore 1997). Liberman & Mattingly (1985) view the perception-action link for speech as potentially innate, whereas Kuhl & Meltzoff (1982, 1996) view it as forged early in development through experience. Two new infant studies shed light on the developmental issue. Imada et al. (2006) used magnetoencephalography (MEG) to study newborns, 6-month-old infants, and 12-month-old infants while they listened to nonspeech, harmonics, and syllables (**Figure 4**). Dehaene-Lambertz et al. (2006) used fMRI to scan three-month-old infants while they listened to sentences. Both studies showed activation in brain areas responsible for speech production (the inferior frontal, Broca's area) in response to auditorily presented speech. Imada et al. reported synchronized activation in response to speech in auditory and motor areas at 6 and 12 months, and Dehaene-Lambertz et al. reported activation in motor speech areas in response to sentences in three-month-olds. Is activation of Broca's area to the pure perception of speech present at birth? Newborns tested by Imada et al. showed no activation in motor speech areas for any signals, whereas auditory areas responded robustly to all signals, suggesting that perception-action linkages for speech develop by three months of age as infants produce vowel-like sounds. Further work must be done to determine whether binding of perception and action requires experience. Using the tools of modern neuroscience, we can now ask how the brain systems responsible for speech perception and production forge links in early development, a necessary step for communication development.

INFANTS' EARLY LEXICONS

Young children's growing lexicons must code words in a way that distinguishes them from one another. Words—produced by different speakers, at different rates, and in different contexts—have variations that are not relevant to word meaning. Infants have to code the

Auditory (superior temporal) Wernicke's area

Newborns	6-month-olds	12-month-olds

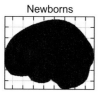 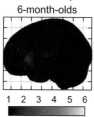 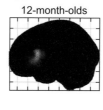

Motor (inferior frontal) Broca's area

Newborns	6-month-olds	12-month-olds

1 2 3 4 5 6

Z score relative to 100 to 0 ms baseline

Figure 4

(*top*) Neuromagnetic signals were recorded in newborns, 6-month-old (shown) and 12-month-old infants in the MEG machine while listening to speech (shown) and nonspeech auditory signals. (*bottom*) Brain activation in response to speech recorded in auditory (*top row*) and motor (*bottom row*) brain regions showed no activation in the motor speech areas in the newborn in response to auditory speech, but increasing activity that was temporally synchronized between the auditory and motor brain regions was seen in 6- and 12-month-old infants when listening to speech (from Imada et al. 2006).

critical features of words in sufficient detail to allow them to be distinguished.

One approach to this question is to test how children of different ages react to mispronounced words. Reactions to mispronunciations—examining whether children accept "tup" for cup or "bog" for dog—provide information about the level of phonological detail in their mental representations of words.

Studies across languages showed that by one year of age infants do not accept mispronunciations of common words (Jusczyk & Aslin 1995, Fennell & Werker 2003), words in stressed syllables (Vihman et al. 2004), or monosyllabic words (Swingley 2005), indicating that their representations of these words are well-specified by that age. Other studies using visual fixation of two targets (e.g., apple and ball) while one is named ("Where's the ball?") indicated that between 14 and 25 months of age children's tendencies to fixate the target item when it is mispronounced diminishes over time (Swingley & Aslin 2000, 2002; Bailey & Plunkett 2002, Ballem & Plunkett 2005).

Studies also indicate that when learning new words, 14-month-old children's phonological skills are taxed. Stager & Werker (1997) demonstrated that 14-month-old infants failed to learn new words when similar-sounding phonetic units were used to distinguish those words ("bih" and "dih") but did learn if the two new words were distinct phonologically ("leef" and "neem"). By 17 months of age, infants learned to associate similar-sounding nonsense words with novel objects (Bailey & Plunkett 2002, Werker et al. 2002). Infants with larger vocabularies succeeded on this task even at the younger age, suggesting that infants with greater phonetic learning skills acquire new words more rapidly. This is consistent with studies showing that children with better native-language phonetic learning skills showed advanced vocabulary development (Tsao et al. 2004; Kuhl et al. 2005b, 2008; Rivera-Gaxiola et al. 2005a).

ERP methods corroborated these results. Mills et al. (2004) compared children's ERP responses when responding to familiar words that were either correctly pronounced or mispronounced, as well as some nonwords (**Figure 5**). At the earliest age tested, 14 months, a negative ERP component (N200–400) distinguished known vs. dissimilar nonsense words ("bear" vs. "kobe") but not known vs. phonetically similar nonsense words ("bear" vs. "gare"). By 20 months, this same ERP component distinguished correct pronunciations, mispronunciations, and nonwords, supporting the idea that

between 14 and 20 months, children's phonological representations of early words become
increasingly detailed.

Other evidence of early processing limitations stems from infants' failure to learn a novel
word when its auditory label closely resembles
a word they already know ("gall," which closely
resembles "ball"), suggesting lexical competition effects (Swingley & Aslin 2007). How phonetic and word learning interact—and whether
the progression is from phonemes to words,
words to phonemes, or bi-directional—is a
topic of strong interest that will be aided by
using neuroscientific methods.

Two recent models/frameworks of early language acquisition, the native language magnet theory, expanded (NLM-e) (Kuhl et al.
2008) and the processing rich information
from multidimensional interactive representations (PRIMIR) framework (Werker & Curtin
2005), suggest that phonological and word
learning bidirectionally influence one another.
According to NLM-e, infants with better phonetic learning skills advance more quickly toward language because phonetic skills assist
phonotactic pattern and word learning (Kuhl
et al. 2005b, 2008). At the same time, the more
words children learn, the more crowded lexical
space becomes, pressuring children to attend
to the phonetic units that distinguish words
from one another (see Swingley & Aslin 2007
for discussion). Further studies examining both
phoneme and word learning in the same children will help address this issue.

NEURAL SIGNATURES OF EARLY SENTENCE PROCESSING

To understand sentences, a child must have
exquisite phonological abilities that allow segmentation of the speech signal into words, and
the ability to extract word meaning. In addition, the relationship among words composing
the sentence—between a subject, its verb, and
its accompanying object—must be deciphered
to arrive at a full understanding of the sentence. Human language is based on the ability

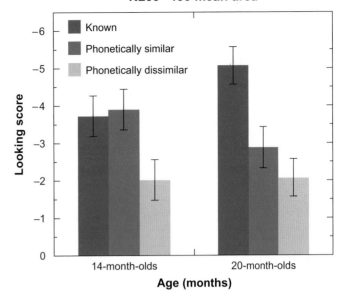

N200–400 mean area

Figure 5

ERP responses to known words, phonetically similar nonsense words, and
phonetically distinct nonsense words at 14- and 20-months of age. At
14 months, infants' brain responses are indistinguishable for known and
similar-sounding nonsense words, although they were distinct for dissimilar
nonsense words. By 20 months, infants' brain responses are distinct for all three
types of words (from Mills et al. 2005).

to process hierarchically structured sequences
(Friederici et al. 2006).

Electrophysiological components of sentence processing have been recorded in children and contribute to our knowledge of when
and how the young brain decodes syntactic and
semantic information in sentences. In adults,
specific neural systems process semantic vs. syntactic information within sentences, and the
ERP components elicited in response to syntactic and semantic anomalies are well established (**Figure 6**). For example, a negative ERP
wave occurring between 250 and 500 ms that
peaks around 400 ms, referred to as the N400,
is elicited to semantically anomalous words in
sentences (Kutas 1997). A late positive wave
peaking at ∼600 ms and that is largest at parietal
sites, known as the P600, is elicited in response
to syntactically anomalous words in sentences
(Friederici 2002). And a negative wave over
frontal sites between 300 and 500 ms, known

Native language magnet model, expanded (NLM-e): proposes that early language experience shapes neural architecture, affecting later language learning, and both computational and social abilities affect learning

Processing rich information from multidimensional interactive representations (PRIMIR): a framework for early language that links levels of processing and describes their interaction

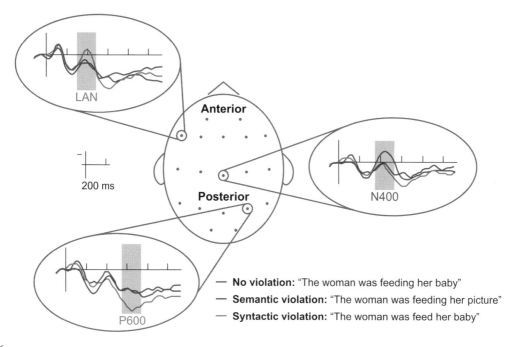

No violation: "The woman was feeding her baby"
Semantic violation: "The woman was feeding her picture"
Syntactic violation: "The woman was feed her baby"

Figure 6

Distribution and polarity of the ERP responses to normal sentences and sentences with either semantic or syntactic anomalies in adults (from Kuhl & Damasio 2008).

as the late anterior negativity (LAN), is elicited in response to syntactic and morphological violations (Friederici 2002).

Beginning in a child's second year of life, ERP data on sentence processing in children suggest that adult-like components can be elicited in response to violations in semantic and syntactic components, but that differences exist in the latencies and scalp distributions of these components in children vs. those in adults (Harris 2001; Oberecker et al. 2005; Friedrich & Friederici 2005, 2006; Silva-Pereyra et al. 2005a,b, 2007; Oberecker & Friederici 2006). Holcomb et al. (1992) reported the N400 in response to semantic anomaly in children from five years of age to adolescence; the latency of the response declined systematically with age (see also Neville et al. 1993, Hahne et al. 2004). Studies also show that syntactically anomalous sentences elicit the P600 in children between 7 and 13 years of age (Hahne et al. 2004).

Recent studies have examined these ERP components in preschool children. Harris

(2001) reported an N400-like effect in 36–38-month-old children; the N400 was largest over posterior regions of both hemispheres. Friedrich & Friederici (2005) observed an N400-like wave to semantic anomalies in 19- and 24-month-old German-speaking children.

Silva-Pereyra et al. (2005b) recorded ERPs in children between 36 and 48 months of age in response to semantic ("My uncle will blow the movie") and syntactic anomalies ("My uncle will watching the movie") when compared with control sentences. In both cases, the ERP effects in children were more broadly distributed and elicited at later latencies than in adults. In work with even younger infants (30-month-olds), Silva-Pereyra et al. (2005a) used the same stimuli and observed late positivities distributed broadly posteriorly in response to syntactic anomalies and anterior negativities in response to semantically anomalous sentences. In both cases, children's latencies were longer than those seen in older children and adults (**Figure 7**), a pattern seen repeatedly and

Phonotactic patterns: sequential constraints governing permissible strings of phonemes in a given language. Each language allows different sequences

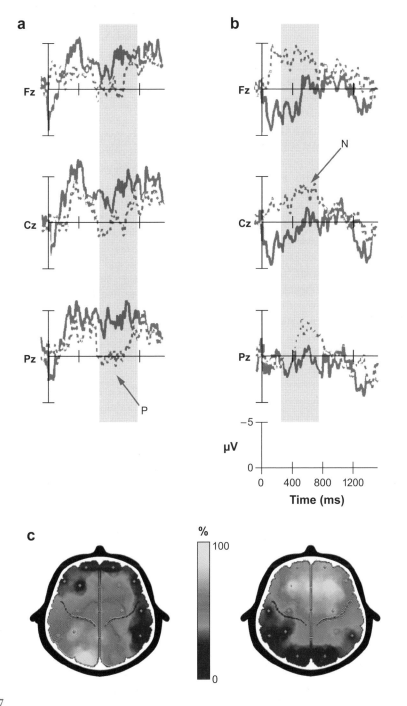

Figure 7

ERP waveforms elicited from 36- and 48-month-old children in response to sentences with syntactic (*a*) or semantic (*b*) violations. Children's ERP responses resemble those of adults (see **Figure 6**) but have longer latencies and are more broadly distributed (*c*) (from Silva-Pereyra et al. 2005b).

attributed to the immaturities of the developing neural mechanisms.

Syntactic processing of sentences with semantic content information removed—"jabberwocky sentences"—has also been tested using ERP measures with children. Silva-Pereyra and colleagues (2007) recorded ERPs to phrase structure violations in 36-month-old children using sentences in which the content words were replaced with pseudowords while grammatical function words were left intact. The pseudowords differed from real words by only a few phonemes: "My uncle watched a movie about my family" (control sentence) became "My macle platched a flovie about my garily" (jabberwocky sentence without violation) and "My macle platched about a flovie my garily" (jabberwocky sentence with phrase structure violation). The ERP components elicited to the jabberwocky phrase structure violations differed as compared with the same violations in real sentences. Two negative components were observed: one from 750–900 ms and the other from 950–1050 ms, rather than the positivities seen in response to phrase structure violations in real sentences in the same children. Jabberwocky studies with adults (Munte et al. 1997, Canseco-Gonzalez 2000, Hahne & Jescheniak 2001) have also reported negative-going waves for jabberwocky sentences, though at much shorter latencies.

Recent studies on language and pre-reading skills in 5-year old children prior to their entry into school (Raizada et al. 2008), as well as in first- through third-grade children (Noble et al. 2006, 2007), indicate that in healthy, socio-economically diverse populations assessed with various linguistic, cognitive, and environmental measures, significant correlations exist between low socio-economic status and behavioral and brain (fMRI) measures of language skills. Future work on the complex socio-economic factors that potentially mediate language and reading skills in children could potentially lead to interventions that would improve the skills needed for reading prior to the time children enter school.

BILINGUAL INFANTS: TWO LANGUAGES, ONE BRAIN

One of the most interesting questions is how infants map two distinct languages in the brain. From phonemes to words, and then to sentences, how do infants simultaneously bathed in two languages develop the neural networks necessary to respond in a native-like manner to two different codes?

Word development in bilingual children has just begun to be studied using ERP techniques. Conboy & Mills (2006) recorded ERPs to known and unknown English and Spanish words in bilingual children at 19–22 months. Expressive vocabulary sizes were obtained in both English and Spanish and used to determine language dominance for each child. A conceptual vocabulary score was calculated by summing the total number of words in both languages and then subtracting the number of times a pair of conceptually equivalent words (e.g., "water" and "agua") occurred in the two languages.

ERP differences to known and unknown words in the dominant language occurred as early as 200–400 and 400–600 ms in these 19- to 22-month-old infants and were broadly distributed over the left and right hemispheres, resembling patterns observed in younger (13- to 17-month-old) monolingual children (Mills et al. 1997). In the nondominant language of the same children, these differences were not apparent until late in the waveform, from 600 to 900 ms. Moreover, children with high vs. low conceptual vocabulary scores produced a greater number of responses to known words in the left hemisphere, particularly for the dominant language (Conboy & Mills 2006).

Using ERPs, Neville and colleagues have shown that longer latencies and/or durations of semantic ERP effects are noted for later- vs. early-acquired second languages across individuals (Neville et al. 1992, Weber-Fox & Neville 1996, Neville et al. 1997). In the syntactic domain, ERP effects typically elicited by syntactically anomalous sentences and closed- vs. open-class words have been absent or attenuated for second languages acquired after infancy (Neville et al. 1992, 1997; Weber-Fox & Neville 1996).

A second question about bilingual language development is whether it takes longer to acquire two languages as compared with acquiring one. Bilingual language experience could impact the rate of language development, not because the learning process is different but because learning might require more time for sufficient data from both languages to be experienced. Infants learning two languages simultaneously might therefore reach the transition from universal phonetic listening (phase 1 of development) to language-specific listening (phase 2 in development) at a later point in development than do infants learning either language monolingually. This delay could depend on factors such as the number of people in the infants' environment producing the two languages in speech directed toward the child and the amount of input these people provide. Such factors could change the development rate in bilingual infants.

Very little data address this question thus far, and what data do exist are mixed. Some studies suggest that infants exposed to two languages show a different pattern of phonetic perception development when compared with monolingual infants (Bosch & Sebastián-Gallés 2003a,b). Other studies report that phonetic perception in bilingual infants is identical to that occuring in monolingual infants (Burns et al. 2007). The data are at present not sufficient to allow us to answer how bilingual vs. monolingual language exposure affects phonetic perception development and the timing of the transition in phonetic perception from one of universal phonetic perception to one reflecting the language(s) to which infants have been exposed.

CONCLUSIONS

Knowledge of infant language acquisition is now beginning to reap benefits from information obtained by experiments that directly examine the human brain's response to linguis-

tic material as a function of experience. EEG, MEG, fMRI, and NIRS technologies—all safe, noninvasive, and proven feasible—are now being used in studies with very young infants, including newborns, as they listen to the phonetic units, words, and sentences of a specific language. Brain measures now document the neural signatures of learning as early as 7 months for native-language phonemes, 9 months for familiar words, and 30 months for semantic and syntactic anomalies in sentences.

Studies now show continuity from the earliest phases of language learning in infancy to the complex processing evidenced at the age of three when all typically developing children show the ability to carry on a sophisticated conversation. Individual variation in language-specific processing at the phonetic level—at the cusp of the transition from phase 1, in which all phonetic contrasts are discriminated, to phase 2, in which infants focus on the distinctions relevant to their native language—is strongly linked to infants' abilities to process words and sentences. This is important theoretically but is also vital to the eventual use of these early precursors to speech to diagnose children with developmental disabilities that involve language. Furthermore, the fact that the earliest stages of learning affect brain processing of both the signals being learned (native patterns) and the signals to which the infant is not exposed (nonnative patterns) may play a role in our understanding of the mechanisms underlying the critical period, at least at the phonetic level, showing that learning itself, not merely time, affects one's future ability to learn.

Whole-brain imaging now allows us to examine multiple brain areas that are responsive to speech, including those responsible for perception as well as production, revealing the possible existence of a mirror system for speech. Research has begun to use these measures to understand how the bilingual brain maps two distinct languages. Brain studies are beginning to show, in children with developmental disabilities, how particular disabilities affect the brain structures underlying language learning. Answers to the classic questions about the unique human capacity to acquire language will continue to be enriched by studies that utilize modern neuroscience tools to examine the infant brain.

FUTURE DIRECTIONS

1. Are the brain structures activated in infants in response to language the same as those activated in adults, and in both cases, are these brain systems speech specific?

2. Why do infants fail to learn language from television presentations—how does social interaction during language exposure affect the brain's ability to learn?

3. Is the neural network connecting speech perception and speech production innate, and if so, is this network activated exclusively in response to language?

4. How is language mapped in the bilingual brain? Does experience with two or more languages early in development affect the brain systems underlying social and/or cognitive processing?

5. How do developmental disabilities such as autism, dyslexia, and specific language impairment affect the brain's processing of speech?

6. Which causal mechanisms underlie the critical period for second language acquisition—why are adults, with their superior cognitive skills, unable to learn as well as young infants? Can techniques be developed to help adults learn a second language?

DISCLOSURE STATEMENT

The authors are not aware of any biases that might be perceived as affecting the objectivity of this review.

ACKNOWLEDGMENTS

The authors are supported by a grant from the National Science Foundation Science of Learning Center Program to the University of Washington LIFE Center, by grants to P.K. from the National Institutes of Health (HD 37954, HD 34565, HD 55782).

LITERATURE CITED

Aslin RN, Mehler J. 2005. Near-infrared spectroscopy for functional studies of brain activity in human infants: promise, prospects, and challenges. *J. Biomed. Opt.* 10:011009

Aslin RN, Saffran JR, Newport EL. 1998. Computation of conditional probability statistics by 8-month-old infants. *Psychol. Sci.* 9:321–24

Bailey TM, Plunkett K. 2002. Phonological specificity in early words. *Cogn. Dev.* 17:1265–82

Ballem KD, Plunkett K. 2005. Phonological specificity in children at 1;2. *J. Child Lang.* 32:159–73

Baptista LF, Petrinovich L. 1986. Song development in the white-crowned sparrow: Social factors and sex differences. *Anim. Behav.* 34:1359–71

Best C, McRoberts GW. 2003. Infant perception of non-native consonant contrasts that adults assimilate in different ways. *Lang. Speech.* 46:183–216

Bongaerts T, Planken B, Schils E. 1995. Can late starters attain a native accent in a foreign language: a test of the critical period hypothesis. In *The Age Factor in Second Language Acquisition*, ed. D Singleton, Z Lengyel, pp. 30–50. Clevedon, UK: Multiling. Matters

Bortfeld H, Wruck E, Boas DA. 2007. Assessing infants' cortical response to speech using near-infrared spectroscopy. *Neuroimage.* 34:407–15

Bortfeld HM, Morgan JL, Golinkoff RM, Rathbun K. 2005. Mommy and me: familiar names help launch babies into speech-stream segmentation. *Psychol. Sci.* 16:298–304

Bosch L, Sebastián-Gallés N. 2003a. Language experience and the perception of a voicing contrast in fricatives: infant and adult data. *Proc. Int. Congr. Phon. Sci., 15th, Barcelona*, ed. MJ Solé, D Recasens, pp. 1987–90. Adelaide, Aust.: Casual Prod.

Bosch L, Sebastián-Gallés N. 2003b. Simultaneous bilingualism and the perception of a language-specific vowel contrast in the first year of life. *Lang. Speech.* 46:217–43

Brainard MS, Knudsen EI. 1998. Sensitive periods for visual calibration of the auditory space map in the barn owl optic tectum. *J. Neurosci.* 18:3929–42

Burnham D, Kitamura C, Vollmer-Conna U. 2002. What's new, pussycat? On talking to babies and animals. *Science.* 296:1435

Burns TC, Yoshida KA, Hill K, Werker JF. 2007. The development of phonetic representation in bilingual and monolingual infants. *Appl. Psycholinguist.* 28:455–74

Canseco-Gonzalez E. 2000. Using the recording of event-related brain potentials in the study of sentence processing. In *Language and the Brain: Representation and Processing*, pp. 229–66. San Diego: Academic

Cheour M, Ceponiene R, Lehtokoski A, Luuk A, Allik J, et al. 1998. Development of language-specific phoneme representations in the infant brain. *Nat. Neurosci.* 1:351–53

Cheour M, Imada T, Taulu S, Ahonen A, Salonen J, et al. 2004. Magnetoencephalography (MEG) is feasible for infant assessment of auditory discrimination. *Exp. Neurol.* 190:S44–51

Coffey-Corina S, Padden D, Kuhl PK, Dawson G. 2007. *Electrophysiological processing of single words in toddlers and school-age children with Autism Spectrum Disorder.* Presented at Annu. Meet. Cogn. Neurosci. Soc., New York

Conboy BT, Brooks R, Meltzoff AN, Kuhl PK. 2008. *Joint engagement with language tutors predicts brain and behavioral responses to second-language phonetic stimuli.* Presented at ICIS, 16th, Vancouver

Conboy BT, Kuhl PK. 2007. *ERP mismatch negativity effects in 11-month-old infants after exposure to Spanish.* Presented at SRCD Bien Meet., Boston

Conboy BT, Mills DL. 2006. Two languages, one developing brain: event-related potentials to words in bilingual toddlers. *Dev. Sci.* 9:F1–12

Conboy BT, Rivera-Gaxiola M, Klarman L, Aksoylu E, Kuhl PK. 2005. Associations between native and nonnative speech sound discrimination and language development at the end of the first year. *Proc. Suppl. Annu. Boston Univ. Conf. Lang. Dev., 29th, Boston* **http://www.bu.edu/linguistics/APPLIED/BUCLD/supp29.html**

Cutler A, Norris D. 1988. The role of strong syllables in segmentation for lexical access. *J. Exp. Psychol. Hum. Percept. Perform.* 14:113–21

Dehaene-Lambertz G, Dehaene S, Hertz-Pannier L. 2002. Functional neuroimaging of speech perception in infants. *Science.* 298:2013–15

Dehaene-Lambertz G, Hertz-Pannier L, Dubois J, Meriaux S, Roche A, et al. 2006. Functional organization of perisylvian activation during presentation of sentences in preverbal infants. *Proc. Natl. Acad. Sci. USA* 103:14240–45

Eimas PD, Siqueland ER, Jusczyk P, Vigorito J. 1971. Speech perception in infants. *Science.* 171:303–6

Fennell CT, Werker JF. 2003. Early word learners' ability to access phonetic detail in well-known words. *Lang. Speech.* 3:245–64

Fernald A, Kuhl P. 1987. Acoustic determinants of infant preference for Motherese speech. *Infant Behav. Dev.* 10:279–93

Fernald A, Perfors A, Marchman VA. 2006. Picking up speed in understanding: speech processing efficiency and vocabulary growth across the 2nd year. *Dev. Psychol.* 42:98–116

Fernald A, Simon T. 1984. Expanded intonation contours in mothers' speech to newborns. *Dev. Psychol.* 20:104–13

Fowler CA. 1986. An event approach to the study of speech perception from a direct-realist perspective. *J. Phon.* 14:3–28

Friederici AD. 2002. Towards a neural basis of auditory sentence processing. *Trends Cogn. Sci.* 6:78–84

Friederici AD. 2005. Neurophysiological markers of early language acquisition: from syllables to sentences. *Trends Cogn. Sci.* 9:481–88

Friederici AD, Fiebach CJ, Schlesewsky M, Bornkessel ID, von Cramon DY. 2006. Processing linguistic complexity and grammaticality in the left frontal cortex. *Cereb. Cortex.* 16:1709–17

Friedrich M, Friederici AD. 2005. Lexical priming and semantic integration reflected in the event-related potential of 14-month-olds. *NeuroReport.* 16:653–56

Friedrich M, Friederici AD. 2006. Early N400 development and later language acquisition. *Psychophysiology.* 43:1–12

Gallese V. 2003. The manifold nature of interpersonal relations: the quest for a common mechanism. *Philos. Trans. R. Soc. B* 358:517–28

Ganger J, Brent MR. 2004. Reexamining the vocabulary spurt. *Dev. Psychol.* 40:621–32

Gernsbacher MA, Kaschak MP. 2003. Neuroimaging studies of language production and comprehension. *Annu. Rev. Psychol.* 54:91–114

Goldstein M, King A, West M. 2003. Social interaction shapes babbling: testing parallels between birdsong and speech. *Proc. Natl. Acad. Sci. USA* 100:8030–35

Grieser DL, Kuhl PK. 1988. Maternal speech to infants in a tonal language: support for universal prosodic features in motherese. *Dev. Psychol.* 24:14–20

Hahne A, Eckstein K, Friederici AD. 2004. Brain signatures of syntactic and semantic processes during children's language development. *J. Cogn. Neurosci.* 16:1302–18

Hahne A, Jescheniak JD. 2001. What's left if the jabberwock gets the semantics? An ERP investigation into semantic and syntactic processes during auditory sentence comprehension. *Cogn. Brain Res.* 11:199–212

Halle P, de Boysson-Bardies B. 1994. Emergence of an early receptive lexicon: infants' recognition of words. *Infant Behav. Dev.* 17:119–29

Harris A. 2001. Processing semantic and grammatical information in auditory sentences: electrophysiological evidence from children and adults. *Diss. Abstr. Int. Sect. B. Sci. Eng.* 61:6729

Hauser MD, Weiss D, Marcus G. 2002. Rule learning by cotton-top tamarins. *Cognition.* 86:B15–22

Hohle B, Bijeljac-Babic R, Nazzi T, Herold B, Weissenborn J. 2008. The emergence of language specific prosodic preferences during the first half year of life: evidence from German and French infants. *Cogn. Psychol.* In press

Holcomb PJ, Coffey SA, Neville HJ. 1992. Visual and auditory sentence processing: a developmental analysis using event-related brain potentials. *Dev. Neuropsychol.* 8:203–41

Homae F, Watanabe H, Nakano T, Asakawa K, Taga G. 2006. The right hemisphere of sleeping infant perceives sentential prosody. *Neurosci. Res.* 54:276–80

Imada T, Zhang Y, Cheour M, Taulu S, Ahonen A, et al. 2006. Infant speech perception activates Broca's area: a developmental magnetoencephalography study. *NeuroReport* 17:957–62

Immelmann K. 1969. Song development in the zebra finch and other estrildid finches. In *Bird Vocalizations*, ed. R Hinde, pp. 61–74. London: Cambridge Univ. Press

Johnson EK, Jusczyk PW. 2001. Word segmentation by 8-month-olds: when speech cues count more than statistics. *J. Mem. Lang.* 44:548–67

Johnson J, Newport E. 1989. Critical period effects in sound language learning: the influence of maturation state on the acquisition of English as a second language. *Cogn. Psychol.* 21:60–99

Jusczyk PW. 1997. *The Discovery of Spoken Language.* Cambridge, MA: MIT Press

Jusczyk PW, Aslin R. 1995. Infants' detection of the sound patterns of words in fluent speech. *Cogn. Psychol.* 29:1–23

Jusczyk PW, Hohne EA. 1997. Infants' memory for spoken words. *Science.* 277:1984–86

Jusczyk PW, Rosner BS, Cutting JE, Foard CF, Smith LB. 1977. Categorical perception of nonspeech sounds by 2-month-old infants. *Percep. Psychophys.* 21:50–54

Klarman L, Krentz U, Brinkley B, Corina D, Kuhl PK. 2007. *Deaf and hearing infants preference for American Sign Language.* Poster presented at Annu. Meet. Am. Psychol. Assoc., 115th, San Francisco

Krentz UC, Corina DP. 2008. Preference for language in early infancy: the human language bias is not speech specific. *Dev. Sci.* 11:1–9

Kuhl PK. 2000. A new view of language acquisition. *Proc. Natl. Acad. Sci. USA* 97:11850–57

Kuhl PK. 2004. Early language acquisition: cracking the speech code. *Nat. Rev. Neurosci.* 5:831–43

Kuhl PK. 2007. Is speech learning 'gated' by the social brain? *Dev. Sci.* 10:110–20

Kuhl PK, Andruski JE, Chistovich IA, Chistovich LA, Kozhevnikova EV, et al. 1997. Cross-language analysis of phonetic units in language addressed to infants. *Science.* 277:684–86

Kuhl PK, Coffey-Corina S, Padden D, Dawson G. 2005a. Links between social and linguistic processing of speech in preschool children with autism: Behavioral and electrophysiological evidence. *Dev. Sci.* 8:1–12

Kuhl PK, Conboy BT, Coffey-Corina S, Padden D, Nelson T, et al. 2005b. Early speech perception and later language development: implications for the "critical period." *Lang. Learn. Dev.* 1:237–64

Kuhl PK, Conboy BT, Padden D, Rivera-Gaxiola M, Nelson T. 2008. Phonetic learning as a pathway to language: new data and native language magnet theory expanded (NLM-e). *Philos. Trans. R. Soc. B.* 363:979–1000

Kuhl PK, Damasio A. 2008. Language. In *Principles of Neural Science*, ed. ER Kandel, JH Schwartz, TM Jessell, S Siegelbaum, J Hudspeth. New York: McGraw Hill. 5th ed. In press

Kuhl PK, Meltzoff AN. 1982. The bimodal perception of speech in infancy. *Science.* 218:1138–41

Kuhl PK, Meltzoff AN. 1996. Infant vocalizations in response to speech: vocal imitation and developmental change. *J. Acoust. Soc. Am.* 100:2425–38

Kuhl PK, Miller JD. 1975. Speech perception by the chinchilla: voiced-voiceless distinction in alveolar plosive consonants. *Science.* 190:69–72

Kuhl PK, Stevens E, Hayashi A, Deguchi T, Kiritani S, et al. 2006. Infants show a facilitation effect for native language phonetic perception between 6 and 12 months. *Dev. Sci.* 9:F13–21

Kuhl PK, Tsao F-M, Liu H-M. 2003. Foreign-language experience in infancy: effects of short-term exposure and social interaction on phonetic learning. *Proc. Natl. Acad. Sci. USA* 100:9096–101

Kuhl PK, Williams KA, Lacerda F, Stevens KN, Lindblom B. 1992. Linguistic experience alters phonetic perception in infants by 6 months of age. *Science.* 255:606–8

Kujala A, Alho K, Service E, Ilmoniemi RJ, Connolly JF. 2004. Activation in the anterior left auditory cortex associated with phonological analysis of speech input: localization of the phonological mismatch negativity response with MEG. *Cogn. Brain Res.* 21:106–13

Kutas M. 1997. Views on how the electrical activity that the brain generates reflects the functions of different language structures. *Psychophysiology.* 34:383–98

Lewkowicz DJ, Ghazanfar AA. 2006. The decline of cross-species intersensory perception in human infants. *Proc. Natl. Acad. Sci. USA* 103:6771–74

Liberman AM, Cooper FS, Shankweiler DP, Studdert-Kennedy M. 1967. Perception of the speech code. *Psychol. Rev.* 74:431–61

Liberman AM, Mattingly IG. 1985. The motor theory of speech perception revised. *Cognition.* 21:1–36

Liu H-M, Kuhl PK, Tsao F-M. 2003. An association between mothers' speech clarity and infants' speech discrimination skills. *Dev. Sci.* 6:F1–10

Liu H-M, Tsao F-M, Kuhl PK. 2007. Acoustic analysis of lexical tone in Mandarin infant-directed speech. *Dev. Psychol.* 43:912–17

Mandel DR, Jusczyk PW, Pisoni D. 1995. Infants' recognition of the sound patterns of their own names. *Psychol. Sci.* 6:314–17

Mayberry RI, Lock E. 2003. Age constraints on first versus second language acquisition: evidence for linguistic plasticity and epigenesis. *Brain Lang.* 87:369–84

Maye J, Werker JF, Gerken L. 2002. Infant sensitivity to distributional information can affect phonetic discrimination. *Cognition.* 82:B101–11

Meltzoff AN, Decety J. 2003. What imitation tells us about social cognition: a rapprochement between developmental psychology and cognitive neuroscience. *Philos. Trans R. Soc. B.* 358:491–500

Meltzoff AN, Moore MK. 1997. Explaining facial imitation: a theoretical model. *Early Dev. Parent.* 6:179–92

Mills DL, Coffey-Corina SA, Neville HJ. 1993. Language acquisition and cerebral specialization in 20-month-old infants. *J. Cogn. Neurosci.* 5:317–34

Mills DL, Coffey-Corina S, Neville HJ. 1997. Language comprehension and cerebral specialization from 13–20 months. *Dev. Neuropsychol.* 13:233–37

Mills DL, Plunkett K, Prat C, Schafer G. 2005. Watching the infant brain learn words: effects of vocabulary size and experience. *Cogn. Dev.* 20:19–31

Mills DL, Prat C, Zangl R, Stager CL, Neville HJ, et al. 2004. Language experience and the organization of brain activity to phonetically similar words: ERP evidence from 14- and 20-month-olds. *J. Cogn. Neurosci.* 16:1452–64

Molfese DL. 1990. Auditory evoked responses recorded from 16-month-old human infants to words they did and did not know. *Brain Lang.* 38:345–63

Molfese DL, Morse PA, Peters CJ. 1990. Auditory evoked responses to names for different objects: cross-modal processing as a basis for infant language acquisition. *Dev. Psychol.* 26:780–95

Molfese DL, Wetzel WF, Gill LA. 1993. Known versus unknown word discriminations in 12-month-old human infants: electrophysiological correlates. *Dev. Neuropsychol.* 9:241–58

Munte TF, Matzke M, Johanes S. 1997. Brain activity associated with syntactic incongruencies in words and psuedowords. *J. Cogn. Neurosci.* 9:318–29

Näätänen R, Lehtokoski A, Lennes M, Cheour M, Huotilainen M, et al. 1997. Language-specific phoneme representations revealed by electric and magnetic brain responses. *Nature.* 385:432–34

Nazzi TL, Iakimova G, Bertoncini J, Fredonie S, Alcantara C. 2006. Early segmentation of fluent speech by infants acquiring French: emerging evidence for crosslinguistic differences. *J. Mem. Lang.* 54:283–99

Neville HJ, Coffey SA, Holcomb PJ, Tallal P. 1993. The neurobiology of sensory and language processing in language-impaired children. *Cogn. Neurosci.* 5:235–53

Neville HJ, Coffey SA, Lawson DS, Fischer A, Emmorey K, et al. 1997. Neural systems mediating American Sign Language: effects of sensory experience and age of acquisition. *Brain Lang.* 57:285–308

Neville HJ, Mills DL, Lawson DS. 1992. Fractionating language: different neural subsystems with different sensitive periods. *Cereb. Cortex* 2:244–58

Newman R, Ratner NB, Jusczyk AM, Jusczyk PW, Dow KA. 2006. Infants' early ability to segment the conversational speech signal predicts later language development: a retrospective analysis. *Dev. Psychol.* 42:643–55

Newport E. 1990. Maturational constraints on language learning. *Cogn. Sci.* 14:11–28

Newport EL, Aslin RN. 2004a. Learning at a distance I. Statistical learning of nonadjacent dependencies. *Cogn. Psychol.* 48:127–62

Newport EL, Aslin RN. 2004b. Learning at a distance II. Statistical learning of nonadjacent dependencies in a nonhuman primate. *Cogn. Psychol.* 49:85–117

Noble KG, McCandliss BD, Farah MJ. 2007. Socioeconomic gradients predict individual differences in neurocognitive abilities. *Dev. Sci.* 10:464–80

Noble KG, Wolmetz ME, Ochs LG, Farah MJ, McCandliss BD. 2006. Brain-behavior relationships in reading acquisition are modulated by socioeconomic factors. *Dev. Sci.* 9:642–54

Oberecker R, Friederici AD. 2006. Syntactic event-related potential components in 24-month-olds' sentence comprehension. *NeuroReport* 17:1017–21

Oberecker R, Friedrich M, Friederici AD. 2005. Neural correlates of syntactic processing in two-year-olds. *J. Cogn. Neurosci.* 17:1667–78

Peña M, Bonatti L, Nespor M, Mehler J. 2002. Signal-driven computations in speech processing. *Science* 298:604–7

Pulvermüller F. 2005. The neuroscience of language: on brain circuits of words and serial order. Cambridge, UK: Med. Res. Counc., Cambridge Univ. Press

Raudenbush SW, Bryk AS, Cheong YF, Congdon R. 2005. *HLM-6: Hierarchical Linear and Nonlinear Modeling*. Lincolnwood, IL: Sci. Softw. Int.

Raizada RD, Richards TL, Meltzoff AN, Kuhl PK. 2008. Socioeconomic status predicts hemispheric specialization of the left inferior frontal gyrus in young children. *NeuroImage.* 40:1392–401

Rivera-Gaxiola M, Klarman L, Garcia-Sierra A, Kuhl PK. 2005a. Neural patterns to speech and vocabulary growth in American infants. *NeuroReport* 16:495–98

Rivera-Gaxiola M, Silva-Pereyra J, Kuhl PK. 2005b. Brain potentials to native and non-native speech contrasts in 7- and 11-month-old American infants. *Dev. Sci.* 8:162–72

Rizzolatti G. 2005. The mirror neuron system and its function in humans. *Anat. Embryol.* 210:419–21

Rizzolatti G, Craighero L. 2004. The mirror neuron system. *Annu. Rev. Neurosci.* 27:169–92

Rizzolatti G, Fadiga L, Fogassi L, Gallese V. 2002. From mirror neurons to imitation, facts, and speculations. In *The Imitative Mind: Development, Evolution, and Brain Bases*, ed. AN Meltzoff, W Prinz, pp. 247–66. Cambridge, UK: Cambridge Univ. Press

Rizzolatti G, Fadiga L, Gallese V, Fogassi L. 1996. Premotor cortex and the recognition of motor actions. *Cogn. Brain Res.* 3:131–41

Saffran JR. 2003. Statistical language learning: mechanisms and constraints. *Curr. Dir. Psychol. Sci.* 12:110–14

Saffran JR, Aslin RN, Newport EL. 1996. Statistical learning by 8-month old infants. *Science.* 274:1926–28

Saffran JR, Johnson EK, Aslin RN, Newport EL. 1999. Statistical learning of tone sequences by human infants and adults. *Cognition.* 70:27–52

Silva-Pereyra J, Conboy BT, Klarman L, Kuhl PK. 2007. Grammatical processing without semantics? An event-related brain potential study of preschoolers using jabberwocky sentences. *J. Cogn. Neurosci.* 19:1050–65

Silva-Pereyra J, Klarman L, Lin Jo-Fu L, Kuhl PK. 2005a. Sentence processing in 30-month-old children: an ERP study. *NeuroReport.* 16:645–48

Silva-Pereyra J, Rivera-Gaxiola M, Kuhl PK. 2005b. An event-related brain potential study of sentence comprehension in preschoolers: semantic and morphosyntatic processing. *Cogn. Brain Res.* 23:247–85

Silven M, Kouvo A, Haapakoski M, Lähteenmäki V, Voeten M, et al. 2006. *Early speech perception and vocabulary growth in bilingual infants exposed to Finnish and Russian.* Poster presented at Lang. Acquis. Biling. Conf., Toronto

Stager CL, Werker JF. 1997. Infants listen for more phonetic detail in speech perception than in word-learning tasks. *Nature.* 388:381–82

Streeter LA. 1976. Language perception of 2-month-old infants shows effects of both innate mechanisms and experience. *Nature.* 259:39–41

Sundara M, Polka L, Genesee F. 2006. Language-experience facilitates discrimination of /d-ð/ in monolingual and bilingual acquisition of English. *Cognition* 100:369–88

Swingley D. 2005. 11-month-olds' knowledge of how familiar words sound. *Dev. Sci.* 8:432–43

Swingley D, Aslin RN. 2000. Spoken word recognition and lexical representation in very young children. *Cognition.* 76:147–66

Swingley D, Aslin RN. 2002. Lexical neighborhoods and the word-form representations of 14-month-olds. *Psychol. Sci.* 13:480–84

Swingley D, Aslin RN. 2007. Lexical competition in young children's word learning. *Cogn. Psychol.* 54:99–132

Taga G, Asakawa K. 2007. Selectivity and localization of cortical response to auditory and visual stimulation in awake infants aged 2 to 4 months. *Neuroimage* 36:1246–52

Thierry G, Vihman M, Roberts M. 2003. Familiar words capture the attention of 11-month-olds in less than 250 ms. *NeuroReport.* 14:2307–10

Thiessen ED, Saffran JR. 2003. When cues collide: use of stress and statistical cues to word boundaries by 7- to 9-month-old infants. *Dev. Psychol.* 39:706–16

Tincoff R, Jusczyk PW. 1999. Some beginnings of word comprehension in 6-month-olds. *Psychol. Sci.* 10:172–75

Tsao F-M, Liu H-M, Kuhl PK. 2004. Speech perception in infancy predicts language development in the second year of life: a longitudinal study. *Child Dev.* 75:1067–84

Vihman MM, Nakai S, DePaolis RA, Halle P. 2004. The role of accentual pattern in early lexical representation. *J. Mem. Lang.* 50:336–53

Vouloumanos A, Werker JF. 2004. Tuned to the signal: the privileged status of speech for young infants. *Dev. Sci.* 7:270–76

Weber C, Hahne A, Friedrich M, Friederici AD. 2004. Discrimination of word stress in early infant perception: electrophysiological evidence. *Cogn. Brain Res.* 18:149–61

Weber-Fox C, Neville HJ. 1996. Maturational constraints on functional specializations for language processing: ERP and behavioral evidence in bilingual speakers. *J. Cogn. Neurosci.* 8:231–56

Weikum WM, Vouloumanos A, Navarra J, Soto-Faraco S, Sebastian-Galles N, et al. 2007. Visual language discrimination in infancy. *Science* 316:1159

Werker JF, Curtin S. 2005. PRIMIR: a developmental framework of infant speech processing. *Lang. Learn. Dev.* 1:197–234

Werker JF, Fennell CT, Corcoran KM, Stager CL. 2002. Infants' ability to learn phonetically similar words: effects of age and vocabulary size. *Infancy.* 3:1–30

Werker JF, Tees RC. 1984. Cross-language speech perception: evidence for perceptual reorganization during the first year of life. *Infant Behav. Dev.* 7:49–63

Zhang Y, Kuhl PK, Imada T, Kotani M, Tohkura Y. 2005. Effects of language experience: neural commitment to language-specific auditory patterns. *NeuroImage* 26:703–20

Axon-Glial Signaling and the Glial Support of Axon Function

Klaus-Armin Nave[1] and Bruce D. Trapp[2]

[1] Department of Neurogenetics, Max Planck Institute of Experimental Medicine, D-37075 Göttingen, Germany; email: nave@em.mpg.de

[2] Department of Neurosciences, Lerner Research Institute, Cleveland Clinic, Cleveland, Ohio 44195; email: trappb@ccf.org

Annu. Rev. Neurosci. 2008. 31:535–61

The *Annual Review of Neuroscience* is online at neuro.annualreviews.org

This article's doi: 10.1146/annurev.neuro.30.051606.094309

Key Words

oligodendrocytes, Schwann cells, myelination, axonal transport, growth factors, energy metabolism, neurodegenerative diseases

Abstract

Oligodendrocytes and Schwann cells are highly specialized glial cells that wrap axons with a multilayered myelin membrane for rapid impulse conduction. Investigators have recently identified axonal signals that recruit myelin-forming Schwann cells from an alternate fate of simple axonal engulfment. This is the evolutionary oldest form of axon-glia interaction, and its function is unknown. Recent observations suggest that oligodendrocytes and Schwann cells not only myelinate axons but also maintain their long-term functional integrity. Mutations in the mouse reveal that axonal support by oligodendrocytes is independent of myelin assembly. The underlying mechanisms are still poorly understood; we do know that to maintain axonal integrity, mammalian myelin-forming cells require the expression of some glia-specific proteins, including CNP, PLP, and MAG, as well as intact peroxisomes, none of which is necessary for myelin assembly. Loss of glial support causes progressive axon degeneration and possibly local inflammation, both of which are likely to contribute to a variety of neuronal diseases in the central and peripheral nervous systems.

Contents

INTRODUCTION

In all complex nervous systems (except for those of the coelenterates) neuronal cells coexist with glial cells. This finding suggests that neuron-glia interactions are principal features of neural function. For example, in the *Drosophila* nervous system, glial cells form diffusion barriers, cover neuronal cell bodies, and align with axons in the peripheral nervous system (PNS). Indeed, the glial cells that wrap anterior and posterior commissural fibers in *Drosophila* (**Figure 1**) resemble the nonmyelin-forming Schwann cells of vertebrates because they associate with multiple axons without myelinating them (Klämbt et al. 2001). Although the numerical glia-to-neuron ratio is low in the ganglia and fiber tracts of invertebrates, this ratio has increased as the vertebrate nervous system has grown larger. Although glial cell numbers may have been over-estimated in the past (Herculano-Houzel & Lent 2005), glia clearly outnumber neurons in the primate brain (Sherwood et al. 2006).

PNS: peripheral nervous system

CNS: central nervous system

Researchers generally assume that glial cells, which lack electrical excitability, support and modulate neuronal function, but what exactly is meant by the term support is still poorly understood. Moreover, experimental in vivo evidence for the known functions of glia, except for myelination (see below), is remarkably scarce. Research has identified few clearly microglia- or astrocyte-specific human diseases. Astrocytes have been implicated in forming the blood-brain barrier (Hawkins & Davis 2005), neurotransmitter reuptake (Hertz & Zielke 2004), metabolic coupling of synaptic activity (Magistretti 2006), and injury response. Whereas astrocytes are the predominant glial cell type in the mammalian cortex, the largest proportion of all central nervous system (CNS) glia is oligodendrocytes, which are most abundant in (but not restricted to) white-matter tracts.

In this review, we first compare the development and morphology of oligodendrocytes with Schwann cells in the peripheral nervous system (PNS). This will lead to a discussion of the first functional aspect of axon-glia interactions, the relevant axonal signals that initiate and regulate glial cell differentiation and myelination, processes with clear differences between the PNS and CNS. Here, key signaling molecules have recently been identified, including neuregulins and neurotrophins. Axonal electrical activity is also a likely regulator of glial ensheathment. In the second part of this review of axon-glia interactions, we review unexpected findings revealing that long-term axonal function and survival depend on the association of axons with ensheathing glial cells, not necessarily on the myelin sheath. Although it is mechanistically least understood, the view emerges that glial cell ability to support axons in the brain white matter could have a major impact on the course of human neurological and psychiatric diseases.

MYELINATING GLIAL CELLS

Oligodendrocytes are known for their role in axon myelination, which enables rapid saltatory impulse propagation. In fact, the

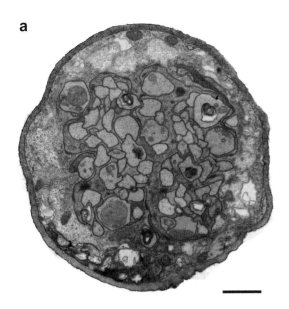

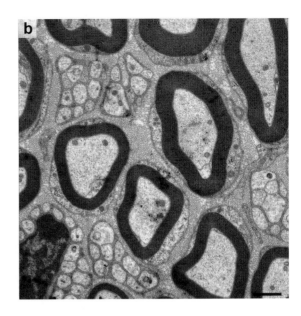

Figure 1

Electron micrograph of a peripheral nerve in *Drosophila* (panel *a* provided by C. Klämbt), showing glial cell processes that interdigitate and engulf single axons. The relationship is strikingly similar to that of nonmyelin-forming Schwann cells that engulf small-caliber C-fiber axons in the mouse sciatic nerve (*b*). Scale bar, 1 μm.

electrophysiology of myelinated nerves is one of the best understood concepts of nervous system function. Moreover, some human neurological diseases are caused primarily by myelinating glial cells, such as leukodystrophies, in the CNS. In the PNS, Schwann cells outnumber the axons that they ensheath because each cell is associated with a short axonal segment. Schwann cell dysfunctions lead to demyelinating neuropathies, which include some frequently inherited neurological disorders. One emerging concept indicates that primary myelin diseases can affect in many ways the functional integrity and survival of the axons ensheathed by defective glial cells. As visualized in different myelin mutant mice, glia's failure to support axon function and survival is surprisingly not proportional to the more obvious structural defects of the ensheathment.

Myelin extends from the glial plasma membrane, which spirally enwraps an axonal segment and is condensed into a multilamellar-compacted sheath, typically depicted by electron microscopy in cross-sections (**Figure 1***b*).

Although cell biologists have understood for many years the ultrastructure of mature myelin and the relationship between axons and Schwann cells or oligodendrocytes, knowledge about the spatiotemporal development of myelin as an extracellular organelle is still elusive. Although axon-glial contact, axonal engulfment, and the first rounds of spiral ensheathment can be documented, all later stages of myelination are difficult to capture by light or electron microscopy, including the dynamics of myelin membrane deposition and membrane compaction, as well as the formation of nodal specializations and paranodal junctions.

Whereas glial engulfment of axons is, in evolutionary terms, an early feature of complex nervous systems, myelination is a late invention. In fact, spiral membrane ensheathment of axonal segments with ion channel clustering at node-like structures has been independently developed in vertebrates and several invertebrate clades (Davis et al. 1999, Schweigreiter et al. 2006, Hartline & Colman 2007). This finding suggests that glials have been repeatedly recruited from a more ancestral, poorly defined function within axon bundles into a new role of providing a multilayered membrane ensheathment for saltatory impulse propagation.

Although the morphological design and ultrastructure of multilayered glial ensheathments vary between different (invertebrate and vertebrate) species, some basic features are conserved. For example, large axons are the first to recruit glial wrappings, and these receive more layers than do smaller axons. In vertebrates, this well-preserved ratio between axonal diameter and myelinated fiber diameter was recognized many years ago (Donaldson & Hoke 1905, Friede 1972) and is now frequently used as a measure (termed the g-ratio) to quantify myelination in development and disease states. These comparisons across species strongly suggest that axon-derived signals recruit glial cells to differentiate into myelinating glia and that this axonal signal is quantitatively related to axon size. Such a recruitment must have occurred independently several times during evolution.

In vertebrates, all myelinating glial cells share key subcellular and ultrastructural features, including the spiral ensheathment of axons, nodes, and paranodal specializations, and tight membrane compaction, suggesting a single developmental origin of myelin in the vertebrate lineage. Only the most ancestral surviving species of fish (e.g., cyclostomes, such as lamprey or hagfish) have unmyelinated nerves (Bullock et al. 1984), whereas beginning with cartilaginous fish, all present-day vertebrates exhibit myelinated central and peripheral axons. In the CNS, multipolar oligodendrocytes generally interact with multiple axons. In white-matter tracts, only larger axons are myelinated, whereas in the optic nerve or within the cortex, for example, oligodendrocytes can also ensheath very small axons. In the PNS, myelination begins when Schwann cells sort out single axons from the bundle of multiple axons that they are engulfing (Jessen & Mirsky 2005). Typically, only axons larger than 1 μm in diameter are sorted, and myelination itself increases axonal diameter (de Waegh et al. 1992).

Smaller axons remain engulfed by nonmyelinating Schwann cells and stay grouped together in Remak bundles, with thin glial processes interdigitating each axon, a process remarkably similar to the glial engulfment of axons in invertebrates such as *Drosophila* (**Figure 1a**).

Vertebrate myelin evolved ~600 million years ago and has allowed fish to develop fast escape reflexes, leading to the genetic selection of both myelinated prey and myelinated predators. Thus, myelination must have been an important driving force in early vertebrate evolution and has become essential for all freely moving terrestrial vertebrates. Its vital function becomes obvious in animals that fail to make myelin and in myelin disease in humans. Natural mouse mutants with central or peripheral dysmyelination suffer from severe motor defects, ataxia, and seizures and die prematurely, often just days or weeks after birth. In humans, the absence of myelination can be observed in leukodystrophies and neuropathies, two heterogenous groups of neurological disorders. In the primate nervous system, the integrity of myelin is subject to clear degenerative changes with aging (Sandell & Peters 2003).

Novel mouse mutants have recently been generated that harbor oligodendrocyte-specific gene defects and are fully myelinated but that later develop progressive axonal loss and die prematurely. These models provide in vivo proof that myelinating glial cells preserve axon function and survival, independent of myelination. Moreover, human patients with peripheral neuropathies have recently been identified who suffer, by genetic criteria, from a primary Schwann cell disease, but exhibit mostly clinical features of axonal degeneration in apparently well-myelinated nerves. As detailed below, novel disease models may help better define the supportive function of axon-ensheathing glia that is independent of myelination. Nonmyelinating (Remak) Schwann cells may represent an ancestral type of glia that has been preserved in the mammalian PNS. Exploring Remak cell function will lead to a better understanding of fundamental axon-glia interactions. The evolutionary steps leading to myelination are difficult to study. However, a related question is how the two alternate fates of present-day Schwann cells are molecularly controlled (i.e., the fate of a myelinating versus nonmyelinating cell). Recent experiments involving transgenic and mutant mice revealed that the axonal growth factor neuregulin-1 (NRG1) plays an important role not only in this developmental decision, but also in the subsequent regulation of myelin membrane growth.

Through axonal growth factor expression, vertebrate neurons recruit immature glial cells to multiply and to follow and ensheath the growing axonal process to secure long-term glial support. This vital function is still required when the ensheathing cells begin to wrap the axonal segments with a multilayered myelin membrane for rapid impulse propagation.

AXONAL SIGNALS THAT RECRUIT GLIAL CELLS

In the PNS, the developmental program of glial cells is controlled entirely by axonal signals, from the proliferation of the neural crest–derived precursor cells to the differentiation of mature myelin-forming Schwann cells (for a detailed review, see Jessen & Mirsky 2005). The dependency on axonal influences is also a feature of developing oligodendrocytes but is much less pronounced compared with Schwann cells in corresponding in vivo (axonal lesion) and in vitro (neuron-glia coculture) studies. This difference and the distinct response of cultured oligodendrocytes and Schwann cells to neuronal growth factors (see below) suggest that CNS glial cells have acquired additional mechanisms to control myelin. Several observations support the idea that neurons have recruited ensheathing glial cells to obtain glial support and myelination (see Colello & Pott 1997).

That axons control myelination, and thus exert visible control over Schwann cell behavior, was first suggested by nerve-grafting experiments. Classical studies by Aguayo and colleagues showed that Schwann cells that cannot make myelin in unmyelinated nerves can nev-ertheless do so in a remyelination experiment (i.e., when confronted with regenerating axons from a myelinated nerve as a tissue graft). Thus axonal signals rather than Schwann cell lineage must be responsible for myelination control (Aguayo et al. 1976, Weinberg & Spencer 1976).

Peripheral axons more than 1 μm in diameter are typically myelinated (Peters et al. 1991), whereas smaller C-fiber axons remain unmyelinated and grouped in Remak bundles, suggesting that myelination is a function of axon size (Murray 1968, Smith et al. 1982). Voyvodic (1989) experimentally showed a direct correlation between axon size and myelination, which strongly suggested a causal relationship. When unmyelinated postganglionic nerves of the salivary gland were surgically hemisected, the surviving axons were left with approximately twice the normal target size and neurotrophic support. This resulted in neurotrophin-induced axon growth, and the increase of axonal diameter was sufficient to recruit resident Remak Schwann cells for myelination. The most likely explanation was that the number of signaling molecules from the enlarged axonal surface increased, reaching a critical threshold level for myelination, but these studies could not identify the responsible signals.

Numerous in vitro studies have demonstrated that cultured Schwann cells respond to axonal signals, such as contact-dependent mitogens (Salzer et al. 1980). Pure glial cultures that myelinate axons in vitro can be experimentally manipulated and have led to the identification of axon-derived regulatory factors (see below) and second messenger pathways (Jessen & Mirsky 2005). Oligodendrocytes also respond to axonal signals (Goto et al. 1990, Kidd et al. 1990, McPhilemy et al. 1991, Scherer et al. 1992), but their differentiation in culture also proceeds in the absence of neuronal signals (Dubois-Dalcq et al. 1986, Ueda et al. 1999). Moreover, myelination by oligodendrocytes is regulated by a balance of promoting and inhibiting factors (Coman et al. 2005, Rosenberg et al. 2006). For example, axonal ensheathment by oligodendrocyte processes is inhibited by the

NRG1: neuregulin-1

presence of PSA-NCAM and L1 on the axon (Charles et al. 2000). The extent to which the downregulation of these inhibitors is a physiological switch and rate-limiting step of myelination is not known. Neurons may utilize a battery of signals to control the mitotic division and differentiation of associated glial cells that provide long-term axonal support. Later in development, axonal signals specifiy which axons are myelinated and help match myelin sheath thickness to axon caliber.

Neuregulins

NRG1 is part of a family of neuronal growth factors that stimulate myelinating glial cells in vitro and in vivo. NRG1 is also found outside the nervous system and is essential for normal mammary and cardiac development. With different start sites for transcription and alternative mRNA splicing, the *NRG1* gene codes for at least 15 different proteins. These proteins share an epidermal growth factor–like signaling domain that is necessary and sufficient for the activation of ErbB receptor tyrosine kinases, which are expressed by oligodendrocytes and Schwann cells (for details, see Adlkofer & Lai 2000, Falls 2003, Esper et al. 2006, Nave & Salzer 2006).

NRG1 binds to ErbB3, a membrane protein lacking a kinase domain, or to ErbB4 but not directly to ErbB2 receptor tyrosin kinase (which must heterodimerize with either ErbB3 or ErbB4). Three NRG1 subgroups have been defined on the basis of their amino termini; NRG1 type I (also termed heregulin, neu differentiation factor, or acetylcholine receptor–inducing activity) and NRG1 type II (glial growth factor) were either secreted or shed from the axon following proteolytic processing. These factors have an immunoglobulin-like domain, bind to heparan sulfate proteoglycans in the extracellular matrix, and thus act in a paracrine fashion. In contrast, NRG1 type III (also termed SMDF) is defined by a second transmembrane (cysteine-rich) domain, remains associated with the membrane after proteolytic cleavage, and serves as a juxtacrine axonal signal

(Schroering & Carey 1998). NRG1 processing enzymes include the tumor-necrosis factor-alpha-converting enzyme for NRG1 type I shedding and the beta-amyloid converting enzyme, which likely activates NRG1 type III (Horiuchi et al. 2005, Hu et al. 2006, Willem et al. 2006).

Although NRG1 has been implicated in numerous neural functions (such as neuronal migration, synaptogenesis, and glutamatergic neurotransmission), its best understood function is the neuronal and axonal regulation of Schwann cell development (reviewed in Garratt et al. 2000a, Corfas et al. 2004, Britsch 2007). Beginning with the specification of glial precursors in the embryonic neural crest (Shah et al. 1994), the entire Schwann cell lineage is controlled, at least in part, by NRG1. The classical finding that axonal membranes stimulate Schwann cell proliferation in vitro (Salzer et al. 1980) can be explained largely by the activity of membrane-associated NRG1 as a Schwann cell mitogen. Its expression by axons secures the necessary number of glial cells for normal ensheathment (Morrissey et al. 1995, Jessen & Mirsky 2005), provided that Schwann cells express ErbB2 and ErbB3 receptors (Riethmacher et al. 1997, Garratt et al. 2000b). NRG1/ErbB signals are amplified by the PI3 kinase pathway (Maurel & Salzer 2000, Ogata et al. 2004). Repopulation and remyelination of a crush-injured nerve by dedifferentiated Schwann cells do not require ErbB2 expression, as suggested by inducible gene targeting in adult mice (Atanasoski et al. 2006).

Following expansion of the Schwann cell pool in the developing peripheral nerve, axon-bound NRG1 type III is required for the differentiation of the myelinating Schwann cell phenotype (Leimeroth et al. 2002, Taveggia et al. 2005). This function includes the quantitative control of myelin membrane growth because myelin sheath thickness (as determined by g-ratio measurements) is a function of total axonal NRG1 that is presented to the ensheathing Schwann cell. Hence, in mice with reduced *NRG1* gene expression, peripheral myelin sheaths are thinner than in wild type, and in transgenic mice that overexpress NRG1 type III [in dorsal root ganglia (DRG) and motoneurons], peripheral myelin is thicker than normal (Michailov et al. 2004). The proper axonal presentation of NRG1 is essential for myelination control, possibly in the context of axonal laminin (Colognato et al. 2002), because it cannot be replaced by paracrine NRG1 signaling (Zanazzi et al. 2001). If the level of axonal NRG1 type III stays below the threshold, the associated axons are not sorted and myelinated but remain grouped together as a Remak bundle with a single nonmyelinating Schwann cell. Experimental NRG1 type III axonal overexpression is sufficient, however, to trigger axonal sorting and myelination in vitro (Taveggia et al. 2005). Thus, the expression level of NRG1 in neurons and on the axonal surface is responsible for a lineage decision made by mammalian Schwann cells. A similar genetic switch in early vertebrate evolution may have led to the recruitment of ensheathing glia and the invention of myelin.

Many axons are myelinated first by Schwann cells and then by oligodendrocytes as they enter the spinal cord and vice versa. Moreover, Schwann cells can invade the injured spinal cord and ensheath central axons. These observations suggest that the axonal signals for myelination are conserved in the CNS and PNS (Colello & Pott 1997). Indeed, oligodendrocytes respond to NRG1 in vitro and ex vivo (Canoll et al. 1996; Vartanian et al. 1997, 1999; Fernandez et al. 2000; Flores et al. 2000; Calaora et al. 2001; Sussman et al. 2005), and transgenic expression of a dominant-negative ErbB4 construct in oligodendrocytes leads to hypomyelination (Roy et al. 2007). The analysis of conditional mouse mutants that completely lack NRG1 in cortical projection neurons (as early as E11.5) unexpectedly failed to show any reduction of myelin assembly in the subcortical white matter or the spinal cord (B. Brinkmann et al., under review). Nevertheless, transgenic NRG1 overexpression in cortical neurons induced a significant hypermyelination that was not restricted to the NRG1 type III isoform (B. Brinkmann et al., under review). These observations suggest that central axons regulate oligodendrocytes using distinct mechanisms from Schwann cells, no longer requiring NRG1 as an instructive myelination signal. Perhaps a simple system of axon-glia interactions (represented by NRG1 type III/ErbB signaling to Schwann cells) has been superseded in CNS evolution by a more complex regulation involving other growth factors and signaling systems still to be identified. This hypothesis is consistent with distinct responses from oligodendrocytes and Schwann cells to neurotrophins.

Neurotrophins

Neurotrophins compose a family of target-derived growth factors [nerve growth factor (NGF), BDNF, NT3, and NT4/5] well-known for their effect on neuronal survival. They also play a role in dendritic pruning and neurotransmitter release and have been implicated in neurodegenerative diseases (reviewed in Chao et al. 2006). Growth factors that regulate neuron survival may later stimulate axon myelination. Voyvodic (1989) suggested that small-caliber axons with elevated access to target-derived neurotrophins can grow in diameter and trigger Schwann cell myelination. In vitro, specific candidate factors can be tested by coculturing myelinating glial cells with DRG axons. Chan et al. (2004) demonstrated that NGF stimulates the myelination of such DRG axons by Schwann cells. NGF's effect was restricted to TrkA-expressing neurons, however, which suggests an indirect mechanism mediated by another axonal signal rather than direct glial stimulation (Rosenberg et al. 2006). This factor could be NRG1 because (Schwann cell–derived) NGF causes a rapid release of soluble NRG1 isoforms from the axon (Esper & Loeb 2004). Myelination in similar DRG-oligodendrocyte cocultures was unexpectedly inhibited by NGF, which suggests that oligodendrocytes and Schwann cells respond differently to the same axonal signals (Chan et al. 2004). Unfortunately, it is difficult to exclude the notion that NGF perturbs oligodendrocyte differentiation

MS: multiple sclerosis

CMT: Charcot-Marie-Tooth disease

SPG2: spastic paraplegia type 2

in vitro (Casaccia-Bonnefil et al. 1996), thereby overriding axonal myelination signals (Rosenberg et al. 2006). In Schwann cells (but not in oligodendrocytes), the low-affinity neurotrophin receptor p75^{NTR} is required for efficient myelination (as shown by blocking antibodies) and mediates the stimulatory effects of BDNF. In contrast, the neurotrophin NT3 is a Schwann cell mitogenic signal, acting via TrkC receptors, and with proliferation and differentiation being incompatible, NT3 is an inhibitor of myelination (Cosgaya et al. 2002).

Electrical Activity

Oligodendrocytes proliferate poorly in the postnatal optic nerve of rodents injected with tetrodotoxin to block sodium-dependent action potentials (Barres & Raff 1993). Demerens et al. (1996) showed that tetrodotoxin blocks myelination in vivo and in vitro. Indeed, axon-glial signaling includes axonal spiking activity because ATP release and adenosin generation are monitored by glial cells (for details, see Fields & Burnstock 2006). In the PNS, the axonal release of ATP (Stevens & Fields 2000) inhibits Schwann cell differentiation and myelination via purinergic P2 receptor signaling. In contrast, oligodendrocytes expressing purinergic P1 receptors are stimulated by the axonal release of adenosin (Stevens et al. 2002). Recent studies demonstrated that the activity-dependent release of ATP from axons in the CNS stimulates nearby astrocytes to release the cytokine leukemia inhibiting factor, which in turn stimulates oligodendrocyte myelination (Ishibashi et al. 2006), and defects in this tripartite pathway may cause dysmyelination. For example, in Alexander disease (a rare human leukodystrophy), the primary defect resides in astrocytes that lack the expression of the glial fibrillary acidic protein. Given that myelinated axons remain myelinated as they enter or exit the spinal cord, the identification of the many different mechanisms by which axonal signals regulate oligodendrocytes and Schwann cells was quite unexpected.

RECRUITED GLIAL CELLS PROTECT AXON FUNCTION AND SURVIVAL

Axon-ensheathing cells have been recruited throughout evolution specifically for myelination and rapid impulse propagation, yet genetic evidence suggests the existence of a more ancestral function of these glia in axonal support. This hypothesis is strongly supported by the positive influence of myelinating glial on axon caliber (Windebank et al. 1985, Colello et al. 1994, Kirkpatrick et al. 2001) and progressive axonal degeneration found in human neurological diseases that affect oligodendrocytes, such as multiple sclerosis (MS) (Trapp & Nave 2008) and leukodystrophies (see below). Inherited peripheral neuropathies, when caused by Schwann cell dysfunction [Charcot-Marie-Tooth (CMT) disease type 1], also present with progressive axon loss that marks a clinically relevant final common pathway for all CMT diseases (Nave et al. 2007). For oligodendrocytes, animal models with spontaneous or induced mutations have provided experimental evidence supporting an oligodendrocytic role in endogenous neuroprotection, which must be distinct from myelin's role.

MYELIN PROTEINS IN NEURODEGENERATIVE DISEASE

Developmental defects of oligodendrocytes and CNS myelination cause leukodystrophies. There has been considerable research on Pelizaeus-Merzbacher disease (PMD)/spastic paraplegia type-2 (SPG2), a prototype of leukodystrophy with early onset dysmyelination and demyelination (Pelizaeus 1885, Merzbacher 1909, Johnston & McKusick 1962; reviewed in Nave & Boespflug-Tanguy 1996, Garbern 2007). With the discovery of *Plp1* mutations in corresponding mouse models (Nave et al. 1986, Hudson et al. 1987, Schneider et al. 1992) and human PMD and SPG2 patients (Hudson et al. 1989, Saugier-Veber et al. 1994), PMD/SPG2 is now defined as a genetic defect of the X-linked proteolipid protein (*PLP1*) gene (for details, see Inoue 2005).

PLP (30 kDa) and its smaller splice isoform (DM20) are abundant tetraspan proteins found in CNS myelin of higher vertebrates. One cellular function of PLP/DM20, which is not essential for myelination itself, is the stabilization of compacted myelin membranes by serving as molecular struts (Klugmann et al. 1997). However, some subtle developmental functions of myelin in PLP-deficient mice are now well understood (Boison et al. 1995, Yool et al. 2001, Rosenbluth et al. 2006). PLP is a cholesterol-binding protein in lipid rafts (Simons et al. 2000), and mice doubly deficient in PLP and the closely related proteolipid M6B (also binding to cholesterol) exhibit a severe CNS dysmyelination with an altered myelin cholesterol content (H.B. Werner & K.-A. Nave, manuscript in preparation). Thus, proteolipids may be required to enrich cholesterol as an essential lipid of myelination (Saher et al. 2005) and to transport other proteins into the myelin compartment efficiently (Werner et al. 2007).

In mice and humans, most *Plp1* point mutations and even duplications of the human *PLP1* gene cause severe dysmyelination, triggered by PLP/DM20 misfolding (Jung et al. 1996), abnormal cysteine-cross-links (Dhaunchak & Nave 2007), endoplasmic reticulum retention (Gow & Lazzarini 1996), the unfolded protein response (Southwood et al. 2002), and finally oligodendrocyte death. The short life span of mutant mice therefore masks the functions of oligodendrocytes in adult mice for long-term axonal function and integrity. For example, *Plp1* mutant *jimpy* mice die at four weeks of age at a time when most of their axons are intact (Meier & Bischoff 1975). Moreover, all axonal abnormalities seen in *jimpy* mice (or *Plp1* mutant *md* rats) were originally thought to result from the absence of myelin (Rosenfeld & Friedrich 1983, Barron et al. 1987).

The first insights into oligodendrocytes supporting axonal survival independently of myelination came with the analysis of genetic null

mutants in the *Plp1* gene. These mice develop normally and are long-lived, which suggests that PLP is dispensable for myelination and that oligodendrocyte death in *jimpy* mice follows the expression of truncated PLP (Nave et al. 1986, Klugmann et al. 1997). *Plp1 null* mice, however, develop a late-onset (>12 months) neurodegenerative disease caused by progressive axonal loss throughout the CNS, preferentially small-caliber axons in long spinal tracts, and eventually premature death (Griffiths et al. 1998). Many months before the onset of clinical symptoms, this type of Wallerian degeneration is preceded by axonal swelling. Swellings occur in fully myelinated axons and are either organelle rich or filled with nonphosphorylated neurofilaments. They begin at the paranodal region, which may possibly be a bottleneck of axonal transport. Swellings and transection bulbs can be readily visualized by immunohistochemical staining of the amyloid precursor protein, when locally trapped, or by electron microscopy (**Figure 2**). Abnormal swellings likely reflect the complete breakdown of the fast axonal transport because they are developmentally preceded by a significant reduction of the axonal transport rate (initially the retrograde transport). This has been demonstrated for the optic nerve of 60-day-old *Plp1 null* mice that are clinically and histopathologically still unaffected (Edgar et al. 2004b).

With this late onset of ataxia (but no tremor or seizures), *Plp1 null* mice are genetically and clinically bona fide models for SPG2, a milder allelic form of PMD (Johnston & McKusick 1962, Saugier-Veber et al. 1994). Likewise, human patients with a null mutation of *PLP1* (Raskind et al. 1991) have a milder course of disease, dominated by slowly progressive degeneration of long spinal cord axons (Garbern et al. 2002) rather than dysmyelination at infant age. Reduced N-acetyl aspartate levels in the brains of PMD patients (Bonavita et al. 2001), however, suggest that some axonal involvement is a general feature of PLP-related diseases, provided there is a long enough survival time (see below).

PLP: proteolipid protein

DM20: smaller splice isoform of PLP

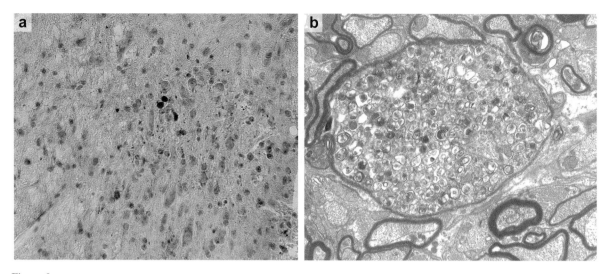

Figure 2

(*a*) Immunostaining and (*b*) electron microscopy of axonal swelling in the white matter of adult mice with a disruption of the oligodendroglial *Cnp1* gene. Note the presence of normally myelinated axons. Figure adapted from Lappe-Siefke et al. (2003).

The exact role of PLP or its alternative spliced isoform DM20 in axonal preservation is still unknown. Mice with the *rumpshaker* mutation in *Plp1* (Schneider et al. 1992) are long-lived and hypomyelinated and have reduced amounts of mutant DM20 and trace amounts of PLP incorporated into myelin. These mice also exhibit late-onset axonal degeneration (Edgar et al. 2004a), suggesting that *rumpshaker* DM20 may be sufficiently folded to reach the myelin compartment but unable to support axonal integrity fully. One study found axonal swellings in myelinated mice that lack selective expression of the PLP isoform (Stecca et al. 2000), although this phenotype was not as early as in the absence of both PLP and DM20 (Spörkel et al. 2002).

A second gene specifically expressed in myelinating glial cells and recently associated with the oligodendroglial support of axons is *Cnp1*. The encoded protein 2′,3′-cyclic nucleotide 3′-phosphodiesterase (CNP or RIP) is widely used as a marker protein for myelin-forming glial cells (Vogel & Thompson 1988, Watanabe et al. 2006). CNP is specifically associated with noncompacted myelin regions (i.e., inner mesaxon, paranodal loops, and Schmidt-

Lantermann incisures) but is absent from the compacted sheath (Braun et al. 1988, Trapp et al. 1988). The enzymatic activity of CNP in vivo is unclear because 2′,3′-cyclic nucleotides are not found in the brain and only known from RNA metabolism. CNP, when experimentally overexpressed, induces process outgrowth in cultured cells (Lee et al. 2005) and premature abnormal myelination in transgenic mice (Gravel et al. 1996, Yin et al. 1997). CNP is expressed outside the nervous system but at lower levels. A CNP-related protein in fish, termed RICH (regeneration induced CNP homolog), is expressed by retinal ganglion cells during axonal outgrowth, which suggests that CNP/RICH performs certain functions during active membrane growth. Thus, similar to other myelin proteins, CNP may have been recruited in evolution by myelin-forming glia from a more general cellular function. The *Cnp1* gene encodes two CNP isoforms; the larger one (CNP2) harbors an amino-terminal mitochondrial targeting sequence (Lee et al. 2006). Both forms are acylated and isoprenylated at their carboxyl terminus (Gravel et al. 1994), which explains their efficient association with cellular membranes. CNP also interacts with the actin

CNP: 2′,3′-cyclic nucleotide 3′-phosphodiesterase

skeleton and microtubules (DeAngelis & Braun 1996, Bifulco et al. 2002, Lee et al. 2005), as well as with mitochondria, in which the longer CNP isoform can be imported, at least in nonglial cells (McFerran & Burgoyne 1997, Lee et al. 2006). Whether CNP functions by associating mitochondria and/or RNA with the oligodendroglial cytoskeleton, as speculated, requires experimental support.

Mice with targeted disruption of the *Cnp1* gene develop on schedule and are fully myelinated, but they develop widespread and progressive axonal swellings (Lappe-Siefke et al. 2003). The phenotype is clinically more severe than that of *Plp1 null* mice, with a much earlier onset and premature death by ~12 months of age. Myelin sheaths are normally compacted in the absence of CNP, which emphasizes that axonal problems are not caused by a thin or physically unstable myelin sheath. Many oligodendroglial paranodes become disorganized before the onset of clinical symptoms (Rasband et al. 2005), which suggests that the absence of CNP alters the normal communication between axons and oligodendrocytes. That the paranodal changes are the major cause of axonal dysfunction and degeneration is unlikely because the first swellings (as early as postnatal day 5 in the optic nerve) and enlargement of the inner tongue precede the corresponding paranodal changes (J.M. Edgar et al., submitted). The axonal swellings also ultimately indicate an energy-related metabolic problem. The exact molecular mechanisms remain to be defined.

PLP and CNP are myelin-associated proteins expressed by oligodendrocytes and Schwann cells, but the axonal degeneration phenotype in null mutant mice is CNS specific. Yin et al. (1998) reported perturbed axonal integrity in the PNS for the myelin-associated glycoprotein (MAG). MAG is a member of the immunoglobulin superfamily (Lai et al. 1987) and has properties of a cell-adhesion protein and signaling molecule localized to the periaxonal membranes of all myelinating glia, as well as to paranodal loops and Schmidt-Lantermann incisures in Schwann

cells (Martini & Schachner 1986, Trapp et al. 1989). MAG is not present on axons. MAG's function was expected to involve specific axon-glia recognition, adhesion, and signaling, but the phenotype of MAG-deficient mice (Li et al. 1994, Montag et al. 1994) was not informative, most likely because MAG's function was masked by the presence of other adhesion molecules such as N-CAM and L1 (for details, see Bartsch 2003) or the recently identified Necl4 protein (Maurel et al. 2007, Spiegel et al. 2007). Some minor developmental abnormalities were documented, including multiply ensheathed axons or the delay of optic nerve myelination (Bartsch et al. 1995, 1997). Moreover, oligodendrocytes in aged mice exhibited degenerative changes (dying back oligodendrogliopathy), a pathological feature previously described in some MS lesions (Lassmann et al. 1997, Weiss et al. 2000). These phenotypes, however, fail to identify the normal function of this protein. Instead, MAG's role as one of several myelin-associated and Nogo receptor-dependent inhibitors of axonal regeneration has gained much attention (e.g., Domeniconi et al. 2002). Studies suggest that myelin inhibitors prevent inappropriate axonal sprouting. Resident microglial cells are another likely target of repulsive MAG signaling (F. Orfaniotou & K.-A. Nave, manuscript in preparation).

The inhibition of axonal growth cones by MAG demonstrates that this myelin protein possesses signaling domains that communicate with neuronal/axonal receptors. An important finding in this respect is that MAG modulates the physical caliber of the myelinated axons on Schwann cells. Despite the morphologically normal myelination of *Mag* mutant mice, peripheral axons remain significantly smaller in diameter when compared with wild-type mice (Yin et al. 1998), presumably because their neurofilaments are hypophosphorylated and thus more densely packed. The reduction in axonal diameter was strikingly more pronounced in paranodal regions of the myelin internode. Because much of the paranodal atrophy occurred after myelin was

MAG: myelin-associated glycoprotein

formed, and because myelin collapsed on the shrunken axon, axonal segments were hypermyelinated. The occurrence of hypermyelinated focal segments or tomacula, as found in certain demyelinating neuropathies (e.g., CMT4B), suggests that MAG may be one of the hypothesized glia-to-axon signals that may explain why normal myelination must maintain axonal calibers (de Waegh et al. 1992). This dependence of axon size on glial ensheathment was originally discovered in the myelin-deficient *Trembler* mouse, a natural point mutation of the *Pmp22* gene, and was later confirmed in the CNS of dysmyelinated *shiverer* mice, a natural deletion mutant of the *Mbp* gene (Roach et al. 1985, Brady et al. 1999).

In humans, inherited demyelinating PNS neuropathies (CMT disease) can be caused by a plethora of primary glial defects, comprising mutations in Schwann cell–specific or ubiquitously expressed genes (for details, see Suter & Scherer 2003, Berger et al. 2006, Nave et al. 2007). Most detailed clinical studies and nerve biopsy analyses have been performed using patients with defects of the myelin-associated proteins PMP22 (in CMT1A), MPZ (in CMT1B), and connexin-32 (in CMT1X). Transgenic and mutant animal models of the corresponding *Pmp22*, *Mpz*, and *Gjp1* genes provided formal proof of concept for the genetic cause of each disease and also helped further dissect disease mechanisms. Demyelinating CMT diseases present with progressive muscle weakness and sensory loss, in addition to the strong reduction of nerve conduction velocity (NCV), which demonstrates that a length-dependent loss of motor and sensory axons must be caused by primary Schwann cell dysfunction. In *Mpz* mutant mice, peripheral axon loss is also preceded by focal swellings (Ey et al. 2007). In a transgenic model of demyelinating neuropathy, caused by overexpression of the wild-type *Mpz* gene in Schwann cells (Yin et al. 2004), the authors observed synaptic retraction and loss of the neuromuscular end plate that preceded the visible axonal degeneration of lower motoneurons. Although the molecular details are not understood, presynaptic dysfunctions may

constitute an independent mechanism of target denervation that is clinically relevant to myelin disease.

By definition, all demyelinating neuropathies exhibit morphological signs of myelin pathology as the underlying cause of slowed NCV, which suggests that myelin per se could be required to maintain axonal integrity. It was therefore a major advance when specific mutations in the Schwann cell–specific *MPZ* gene were identified that caused CMT type 2 with normal NCV (i.e., the axonal form of CMT disease) rather than CMT1B (Marrosu et al. 1998, Senderek et al. 2000, Boerkoel et al. 2002). Although they are mechanistically not well understood, separate functions of Schwann cells in myelination (preserved in CMT2) and in axonal support (lost in all CMT forms) may have been uncoupled, reminiscent of *Cnp1* and *Plp null* mutations in myelinating oligodendrocytes. Nonmyelinating Schwann cells also support axonal integrity, as shown for L1-mutant mice, in which Remak Schwann cells fail to properly engulf C-fiber axons that degenerate (Haney et al. 1999). Transgenic overexpression of nonfunctional ErbB receptors in these Schwann cells also causes dysfunction and loss of C-fiber axons (Chen et al. 2003), which supports the hypothesis that axonal support is an ancestral glial function that precedes myelination.

Thus, when comparing myelin disorders in the CNS and PNS, all ensheathing glial cells are essential to maintain long-term axonal integrity. However, the consequences of demyelination and axon loss are regionally different. The enormous plasticity of the brain may be able to mask a slowly progressive degeneration of the subcortical axons for a long time, and a substantial axonal loss may remain clinically silent in MS patients (Trapp & Nave 2008). However, length-dependent axon loss in the spinal cord (e.g., in patients with SPG2) is more difficult to mask. Finally, the progressive loss of peripheral axons (e.g., in patients with CMT disease) causes invariable muscle denervation and sensory losses that cannot be hidden by neural plasticity.

POSSIBLE MECHANISMS OF GLIAL SUPPORT

Long-term demyelination is associated with axon loss (Raine & Cross 1989, Trapp et al. 1998). However, in comparing axon loss in different myelin diseases in mice and humans, there is no simple correlation between the degree of demyelination and axonal involvement. This discrepancy suggests that the absence of myelin alone is not the only cause of neurodegeneration. Nevertheless, demyelination is likely to increase dramatically the energy consumption of fast spiking axons in white-matter tracts, which may perturb axon function. One hypothetical model of how reduced axonal energy balance can trigger calcium-dependent proteolysis of a demyelinated axon is discussed elsewhere in this volume (Trapp & Nave 2008). Glial support, as discussed here, is a feature of fully myelinated axons that requires an alternative model.

Myelin-Associated Toxicity

One must critically discuss whether axonal perturbation as observed in myelinated *Plp* and *Cnp1 null* mice results from the loss of support, or alternatively from a possible gain of toxicity of the PLP-deficient and CNP-deficient myelin wraps. Although no toxic mechanisms are presently known, this alternate hypothesis is difficult to disprove. In a natural mosaic situation (e.g., random inactivation of the X chromosome–linked *Plp* gene in heterozygous females), axonal swellings are clearly detectable, although at reduced frequency (Griffiths et al. 1998). Likewise, after transplantation of *Plp* mutant oligodendrocytes into *shiverer* white-matter tracts, axonal swellings can be locally induced (Edgar et al. 2004b). Although they are compatible with some glial toxicity, these observations are equally well explained by a short range of oligodendroglial support. Only myelinated axons are dependent on support (Griffiths et al. 1998). One might thus speculate that myelination per se could be toxic for axons unless the ensheathing glial cells can support the axonal segment that they have myelinated. Indeed, compacted myelin membranes may act as a destructive shield to impair the axon's otherwise ready access to outside nutrients, metabolites, oxygen, and other important molecules. Thus a specific glial transport apparatus might be required to compensate for these restrictions and to prevent toxic myelination effects. These hypothetical functions would collectively qualify as glial support.

Myelin-Independent Mechanisms

The prodromal phenotype of *Plp null* mice includes axonal transport defects reminiscent of mitochondrial disorders in the CNS, and it suggests that in the presence of myelin, an underlying low energy balance leads to axon loss. We suggest that myelin-independent glial support requires axonal engulfment and a molecular apparatus that is part of, but not identical to, the myelin sheath. In addition to the genes mentioned above, other oligodendrocyte defects have been associated with a similar loss of axonal support, for example, in mice lacking the synthesis of GalC and sulfatide, two myelin-specific glycosphingolipids that are essential for paranodes forming axo-glial junctions (Garcia-Fresco et al. 2006, Marcus et al. 2006). However, not all myelin defects are associated with axonal loss. For example, *Mbp null* (i.e., *shiverer*) mice lack the expression of myelin basic protein (MBP), are severely dysmyelinated in the CNS, and die prematurely. However, they exhibit no obvious signs of axonal swelling or degeneration (Rosenbluth 1980, Inoue et al. 1981, Nixon 1982, Griffiths et al. 1998). In *shiverer* mice, electrophysiological signs of conduction block are caused by an abnormal distribution of potassium channels (Sinha et al. 2006) not by Wallerian degeneration. Increased energy needs in *shiverer* cause a twofold-higher density of mitochondria (Andrews et al. 2006). Thus, *Mbp null* mice are dysmyelinated with functional axons, whereas *Plp null* mice are fully myelinated with widespread axonal degeneration. Clearly, PLP must serve a unique function in axonal support that goes beyond its

CNTF: ciliary
neurotrophic factor

structural role of stabilizing compacted myelin membranes (Boison et al. 1995, Klugmann et al. 1997). This notion has been independently proven in transgenic knockout mice, in which PLP oligodendroglial expression was swapped with MPZ expression, the major cell adhesion molecule from PNS myelin (Yin et al. 2006). Although MPZ could form an adhesive strut between adjacent myelin membranes (similar to PLP), it did not allow oligodendrocytes to support axonal survival.

PLP is a tetraspan myelin protein and, by itself, is unlikely to support axonal survival. Recently, the biochemical analysis of myelin purified from *Plp null* mice revealed secondary abnormalities of the myelin proteome, which suggested that PLP is required for the transport of other proteins that may contribute to oligodendroglial support into the growing myelin compartment. One protein that is nearly absent from PLP-deficient myelin is Sirt2, an NAD^+-dependent deacetylase expressed most strongly in oligodendrocytes (Werner et al. 2007).

Our understanding of the glial support of axons is still in its infancy, and the mechanisms involved are likely complex. Comparisons at the ultrastructural level of just *Plp* and *Cnp1 null* mice revealed significant differences in the onset and distribution of prodromal axonal changes (J.M. Edgar et al., manuscript in preparation). The observation that *Plp*Cnp1* double mutants exhibit an earlier and much more severe course of disease than does either single mutant indicates additive effects that are compatible with the involvement of different protective mechanisms (H.A. Werner & K.A. Nave, unpublished observations).

Neurotrophic Factors

Myelinating glia may provide trophic support for axons, but the extent to which this support involves neurotrophins and related forms of glia-to-neuron growth factor signaling is not known. The function of neurotrophins has been reviewed in detail (Huang & Reichardt 2001), also with respect to myelinating glial cells (Rosenberg et al. 2006). Target-derived neurotrophins stimulate axon-bearing neurons, which likely alters the expression of axonal cues for myelination (Chan et al. 2004). In addition, ample evidence demonstrates that neurotrophins can stimulate the proliferation and differentiation of myelinating glial cells directly (Barres et al. 1994, Kumar et al. 1998, McTigue et al. 1998, Du et al. 2003), but they may also have detrimental effects (Casaccia-Bonnefil et al. 1996). Although all these functions are important for oligodendrocyte development and remyelination (and thus indirectly for axonal support), there is scarce experimental evidence that myelinating glia provide continuous axonal support by releasing trophic factors. Neurotrophins are expressed by Schwann cells (Chan et al. 2001) and in satellite glia of cultured DRG neurons. Here, the expression of NT3 and NGF can be stimulated by axotomy, causing the release of nitric oxide, and nitric oxide–induced neurotrophins are likely neuroprotective (Thippeswamy et al. 2005). There is obviously great interest in using neurotrophins in the prevention of axon loss in CMT disease. The systemic application of NT3 has even been tested clinically in a small number of patients with demyelinating neuropathy CMT1A with encouraging results (Sahenk et al. 2005).

Two other factors with reported survival functions in the nervous system and some expression in myelinating glia must be considered as candidates for glial trophic support. The ciliary neurotrophic factor (CNTF) is expressed by Schwann cells (Sendtner et al. 1994), but the mechanisms of release are unclear because this cytoplasmic cytokine is not exocytosed like growth factors are. In mice, CNTF is required for motoneuron survival and the long-term integrity of myelinated axons in the PNS (Gatzinsky et al. 2003). Also, in mice with experimental allergic encephalomyelitis, an immunological model of human MS, the absence of CNTF causes a more severe disease (Linker et al. 2002). A small but significant percentage (2.3%) of the Japanese population is homozygous for a null mutation of this gene (Takahashi et al. 1994), and this genetic polymorphism is not obviously linked to a neurological disease such as amyotrophic lateral sclerosis or CMT (Van Vught et al. 2007). This observation raises the question, at least in humans, of the relevance of CNTF in endogenous neuroprotection by myelinating glial cells.

The glia cell line–derived growth factor (GDNF) is a ligand of the ret receptor tyrosine kinase (Durbec et al. 1996) and signals independently by N-CAM and the fyn serine/threonine kinase (Paratcha et al. 2003). GDNF, along with several related proteins, is expressed inside and outside the nervous system (Suter-Crazzolara & Unsicker 1994), including Schwann cells (Springer et al. 1995, Trupp et al. 1995) in which GDNF is upregulated upon axotomy (Hammaberg et al. 1996). Mouse mutants of this gene model Hirschsprungs disease (Moore et al. 1996, Pichel et al. 1996, Sanchez et al. 1996) with the loss of enteric ganglia, which resembles early glial and neural crest defects (Inoue et al. 1999, Paratore et al. 2001). GDNF has been intensively studied for its survival effect on dopaminergic neurons (Beck et al. 1995), but it is also a survival factor for embryonic motoneurons (Henderson et al. 1994). The beneficial role of GDNF in the treatment of neuropathic pain (Boucher et al. 2000) could relate to neuroprotective effects normally exerted by Schwann cells. However, direct evidence showing that GDNF would be required to maintain axonal integrity in myelinated fibers is lacking. Similar to neurotrophins, promoting the differentiation of myelinating glia could provide a neuroprotective effect (Wilkins et al. 2003). When GDNF was injected into intact peripheral nerves, it caused Schwann cells to proliferate and to sort and myelinate small-caliber C-fiber axons (Hoke et al. 2003). This study also demonstrates that trophic factors can, in principle, overcome developmental thresholds and change the glial phenotype from non-myelinating Schwann cells to myelin-forming Schwann cells, similar to axonal NRG1.

An endogenous neuroprotective molecule, synthesized by oligodendrocytes, may be prostaglandin D. In the dysmyelinated mouse mutant *twitcher*, upregulated expression of the oligodendroglial prostaglandin D synthase correlated with neuronal survival, and *Pgds* *twitcher* double mutants had a more severe phenotype (Taniike et al. 2002). This enzyme is also induced in MS lesions (Kagitani-Shimono et al. 2006), but the mechanisms of action are not known.

Metabolic Support

Both oligodendrocyte and Schwann cell defects have been associated with a length-dependent axon loss. The most plausible, but least understood, mechanism of axonal preservation by ensheathing glial cells may be metabolic support (Spencer et al. 1979). Rapidly conducting white-matter axons consume a large fraction of the brain's energy supply, and oligodendrocytes are vulnerable to hypoxia and glucose deprivation (Fern et al. 1998). In dysmyelinated and demyelinated axons, the number of mitochondria is increased (Mutsaers & Carrol 1998, Andrews et al. 2006). Most ATP is required for membrane repolarization and for fast axonal transport of vesicles and organelles, occurring at a rate of micrometers per second (Hollenbeck & Saxton 2006). The latter includes axonal mitochondria, which travel with frequent stops and restarts (Misgeld et al. 2007). Mitochondria often pause at the node of Ranvier, which is thought to indicate the site of highest metabolic activity (Fabricius et al. 1993), an observation more obvious in the PNS (I. Griffiths, personal communication). Although the nodal region harbors Na^+/K^+ channels and is the site of most transmembrane ionic flow, most ATP is consumed by membrane repolarization, and Na^+/K^+ ATPases are dispersed along the entire internodal membrane of myelinated PNS and CNS axons (Alberti et al. 2007; E. Young, J.H. Fowler, G.J. Kidd, A. Chang, R. Rudick, E. Fisher & B. Trapp, submitted). We also note that mitochondria, when pausing at the nodal region, are in closest proximity to the glial paranodal loops, which form a highly specialized axon-glial junction. In the PNS of Caspr-deficient mice, which lack the normal paranodal ultrastructure, even more

GDNF: glial cell line–derived factor

mitochondria are abnormally retained at the intra-axonal membrane surface beneath the disrupted paranode, and many of these mitochondria have a swollen morphology (Einheber et al. 2006). This finding suggests that axonal mitochondrial transport is regulated, at least in part, by glial contact and that myelinating cells contribute to normal mitochondrial functions within axons.

In healthy axons, ATP is present in micromolar concentrations, but we predict that an impairment of (ATP-dependent) axonal transport ultimately affects the normal turnover rate and integrity of mitochondria themselves. This indicates a potentially vicious cycle for even minor or highly localized metabolic problems. Mitochondrial involvement can lead to local entrapment and axonal swellings as visible signs of system collapse, which leads to Wallerian degeneration. This hypothesis is supported by the axonal pathology of specific myelin mutants with reduced axonal transport and axonal swellings that closely resemble those in mitochondrial disorders (Ferreirinha et al. 2004), constituting the final common pathway of hereditary spastic paraplegia. Moreover, reduced axonal ATP decreases Na^+/K^+ ATPase (thus elevating axoplasmic Na^+) and ATP-dependent Na^+/Ca^{2+} exchanger activity. The latter can operate in reverse and exchange axoplasmic Na^+ for extracellular Ca^{2+}, at least under pathological conditions (low ATP, high Na^+). Elevated axoplasmic Ca^{2+} causes further damage to axonal and mitochondrial proteins, which introduces a second vicious pathological cycle (Stys et al. 1992, Li et al. 2000).

In the CNS, astrocytes provide free lactate to neurons for ATP generation (hypothesized as the "lactate shuttle"; for details, see Magistretti 2006). In white-matter tracts, this action can involve the mobilization of stored glycogen (Brown et al. 2003). Whether myelinating glial cells also provide energy-rich metabolites to the axons they ensheath is not well analyzed in vivo. In explant systems, non-myelin-forming Schwann cells take up glucose, which they transfer to axons, possibly as lactate equivalents via gap junctions (Vega et al. 2003). The generation of PEX5-deficient mutants lacking functional peroxisomes selectively in oligodendrocytes in vivo recently provided a link between the loss of specific peroxisomal pathways in myelinating glia and the progressive degeneration of axonal integrity in adult mice (Kassmann et al. 2007). We anticipate that a systematic genetic dissection of metabolic pathways in oligodendrocytes and Schwann cells of mutant mice will identify those pathways required for axonal support.

Mouse mutants lacking functional peroxisomes in glia myelinate normally, but similar to dysmyelinated mice that overexpress PLP in oligodendrocytes (Readhead et al. 1994), they exhibit axon loss, late-onset demyelination, and neuroinflammation in areas affected by degenerative changes. Unlike other myelin mutants, inflammation includes the infiltration of activated CD8(+) T cells, and, in peroxisomal mutants, perivascular B-cell infiltrates as well (Ip et al. 2006, Kassmann et al. 2007). Such observations demonstrate that specific oligodendrocyte defects contribute to, if not trigger, inflammatory demyelinating diseases, which is relevant to the unknown etiology of human MS (Trapp & Nave 2008).

CONCLUSION

Research on glial cells in the long-term support of axonal function and in endogenous neuroprotection has just begun. One conceptual link between axonal myelination in higher vertebrates and myelin-independent axonal support is that both mechanisms have evolved to meet the energy demands of rapidly conducting long axonal tracts. Loss of these functions can trigger a vicious cycle of pathological changes, including reduced axonal transport and retention of mitochondria, leading to axonal swelling and calcium-dependent Wallerian degeneration. Genetic data have now provided strong in vivo evidence that neurodegenerative diseases of the CNS and PNS can result from primary defects in oligodendrocytes and Schwann cells, respectively, although the affected glial cells reveal no obvious developmental defects. The view emerges that, except for clearly dysmyelinating and demyelinating disorders, some glial diseases may share features of neurodegenerative disorders. Because axonal degeneration is seen in many neurological diseases, dysfunction of glia and glial cell aging should be considered disease-modifying factors.

DISCLOSURE STATEMENT

The authors are not aware of any biases that might be perceived as affecting the objectivity of this review.

ACKNOWLEDGMENTS

The authors apologize to colleagues whose relevant work could not be cited because of space restrictions. We thank C. Klämbt and G. Saher for providing electron micrographs. We also thank J. Edgar, I. Griffiths, C. Kassmann, G. Saher, and H. Werner for helpful discussion and critical reading of the manuscript. K.-A.N. is supported by grants from the Deutsche Forschungsgemeinschaft, the European Union (FP6), the BMBF (Leukonet), the Hertie Institute of MS Research, the National MS Society, the Del Marmol Foundation, and Olivers Army. B.D.T. is supported by NIH grants NS038186, NS038667, and NS029818.

LITERATURE CITED

Adlkofer K, Lai C. 2000. Role of neuregulins in glial cell development. *Glia* 29:104–11

Aguayo AJ, Charron L, Bray GM. 1976. Potential of Schwann cells from unmyelinated nerves to produce myelin: a quantitative ultrastructural and radiographic study. *J. Neurocytol.* 5:565–73

Alberti S, Gregorio EA, Spadella CT, Cojocel C. 2007. Localization and irregular distribution of Na,K-ATPase in myelin sheath from rat sciatic nerve. *Tissue Cell* 39:195–201

Andrews H, White K, Thomson C, Edgar J, Bates D, et al. 2006. Increased axonal mitochondrial activity as an adaptation to myelin deficiency in the Shiverer mouse. *J. Neurosci. Res.* 83:1533–39

Atanasoski S, Scherer SS, Sirkowski E, Leone D, Garratt AN, et al. 2006. ErbB2 signaling in Schwann cells is mostly dispensable for maintenance of myelinated peripheral nerves and proliferation of adult Schwann cells after injury. *J. Neurosci.* 26:2124–31

Barres BA, Raff MC. 1993. Proliferation of oligodendrocyte precursor cells depends on electrical activity in axons. *Nature* 361:258–60

Barres BA, Raff MC, Gaese F, Bartke I, Dechant G, Barde YA. 1994. A crucial role for neurotrophin-3 in oligodendrocyte development. *Nature* 367:371–75

Barron KD, Dentinger MP, Csiza CK, Keegan SM, Mankes R. 1987. Abnormalities of central axons in a dysmyelinative rat mutant. *Exp. Mol. Pathol.* 47:125–42

Bartsch U. 2003. Neural CAMS and their role in the development and organization of myelin sheaths. *Front Biosci.* 8:477–90

Bartsch U, Montag D, Bartsch S, Schachner M. 1995. Multiply myelinated axons in the optic nerve of mice deficient for the myelin-associated glycoprotein. *Glia* 14:115–22

Bartsch S, Montag D, Schachner M, Bartsch U. 1997. Increased number of unmyelinated axons in optic nerves of adult mice deficient in the myelin-associated glycoprotein (MAG). *Brain Res.* 762:231–34

Beck KD, Valverde J, Alexi T, Poulsen K, Moffat B, et al. 1995. Mesencephalic dopaminergic neurons protected by GDNF from axotomy-induced degeneration in the adult brain. *Nature* 373:339–41

Berger P, Niemann A, Suter U. 2006. Schwann cells and the pathogenesis of inherited motor and sensory neuropathies (Charcot-Marie-Tooth disease). *Glia* 54:243–57

Bifulco M, Laezza C, Stingo S, Wolff J. 2002. 2′,3′-cyclic nucleotide 3′-phosphodiesterase: a membrane-bound, microtubule-associated protein and membrane anchor for tubulin. *Proc. Natl. Acad. Sci. USA* 99:1807–12

Boerkoel CF, Takashima H, Garcia CA, Olney RK, Johnson J, et al. 2002. Charcot-Marie-Tooth disease and related neuropathies: mutation distribution and genotype-phenotype correlation. *Ann. Neurol.* 51:190–201

Boison D, Bussow H, D'Urso D, Muller HW, Stoffel W. 1995. Adhesive properties of proteolipid protein are responsible for the compaction of CNS myelin sheaths. *J. Neurosci.* 15:5502–13

Bonavita S, Schiffmann R, Moore DF, Frei K, Choi B, et al. 2001. Evidence for neuroaxonal injury in patients with proteolipid protein gene mutations. *Neurology* 56:785–88

Boucher TJ, Okuse K, Bennett DLH, Munson JB, Wood JN, McMahon SB. 2000. Potent analgesic effects of GDNF in neuropathic pain states. *Science* 290:124–27

Brady ST, Witt AS, Kirkpatrick LL, de Waegh SM, Readhead C, et al. 1999. Formation of compact myelin is required for maturation of the axonal cytoskeleton. *J. Neurosci.* 19:7278–88

Braun PE, Sandillon F, Edwards A, Matthieu JM, Privat A. 1988. Immunocytochemical localization by electron microscopy of 2′3′-cyclic nucleotide 3′-phosphodiesterase in developing oligodendrocytes of normal and mutant brain. *J. Neurosci.* 8:3057–66

Britsch S. 2007. The neuregulin-I/ErbB signaling system in development and disease. *Adv. Anat. Embryol. Cell Biol.* 190:1–65

Brown AM, Tekkök SB, Ransom BR. 2003. Glycogen regulation and functional role in mouse white matter. *J. Physiol.* 549:501–12

Bullock TH, Moore JK, Fields RD. 1984. Evolution of myelin sheaths: both lamprey and hagfish lack myelin. *Neurosci. Lett.* 48:145–48

Calaora V, Rogister B, Bismuth K, Murray K, Brandt H, et al. 2001. Neuregulin signaling regulates neural precursor growth and the generation of oligodendrocytes in vitro. *J. Neurosci.* 21:4740–51

Canoll PD, Musacchio JM, Hardy R, Reynolds R, Marchionni MA, Salzer JL. 1996. GGF/neuregulin is a neuronal signal that promotes the proliferation and survival and inhibits the differentiation of oligodendrocyte progenitors. *Neuron* 17:229–43

Casaccia-Bonnefil P, Carter BD, Dobrowsky RT, Chao MV. 1996. Death of oligodendrocytes mediated by the interaction of nerve growth factor with its receptor p75. *Nature* 383:716–19

Chan JR, Cosgaya JM, Wu YJ, Shooter EM. 2001. Neurotrophins are key mediators of the myelination program in the peripheral nervous system. *Proc. Natl. Acad. Sci. USA* 98:14661–68

Chan JR, Watkins TA, Cosgaya JM, Zhang C, Chen L, et al. 2004. NGF controls axonal receptivity to myelination by Schwann Cells or oligodendrocytes. *Neuron* 43:183–91

Chao MV, Rajagopal R, Lee FS. 2006. Neurotrophin signalling in health and disease. *Clin. Sci.* 110:167–73

Charles P, Hernandez MP, Stankoff B, Aigrot MS, Colin C, et al. 2000. Negative regulation of central nervous system myelination by polysialylated-neural cell adhesion molecule. *Proc. Natl. Acad. Sci. USA* 97:7585–90

Chen S, Rio C, Ji RR, Dikkes P, Coggeshall RE, et al. 2003. Disruption of ErbB receptor signaling in adult nonmyelinating Schwann cells causes progressive sensory loss. *Nat. Neurosci.* 6:1186–93

Colello RJ, Pott U. 1997. Signals that initiate myelination in the developing mammalian nervous system. *Mol. Neurobiol.* 15:83–100

Colello RJ, Pott U, Schwab ME. 1994. The role of oligodendrocytes and myelin on axon maturation in the developing rat retinofugal pathway. *J. Neurosci.* 14:2594–605

Colognato H, Baron W, Avellana-Adalid V, Relvas JB, Baron-Van Evercooren A, et al. 2002. CNS integrins switch growth factor signalling to promote target-dependent survival. *Nat. Cell Biol.* 4:833–41

Coman I, Barbin G, Charles P, Zalc B, Lubetzki C. 2005. Axonal signals in central nervous system myelination, demyelination and remyelination. *J. Neurol. Sci.* 233:67–71

Corfas G, Velardez MO, Ko CP, Ratner N, Peles E. 2004. Mechanisms and roles of axon-Schwann cell interactions. *J. Neurosci.* 24:9250–60

Cosgaya JM, Chan JR, Shooter EM. 2002. The neurotrophin receptor p75NTR as a positive modulator of myelination. *Science* 298:1245–48

Davis AD, Weatherby TM, Hartline DK, Lenz PH. 1999. Myelin-like sheaths in copepod axons. *Nature* 398:571

De Angelis DA, Braun PE. 1996. 2′,3′-cyclic nucleotide 3′-phosphodiesterase binds to actin-based cytoskeletal elements in an isoprenylation-dependent manner. *J. Neurochem.* 67:943–51

Demerens C, Stankoff B, Logak M, Anglade P, Allinquant B, et al. 1996. Induction of myelination in the central nervous system by electrical activity. *Proc. Natl. Acad. Sci. USA* 93:9887–92

de Waegh SM, Lee VM, Brady ST. 1992. Local modulation of neurofilament phosphorylation, axonal caliber, and slow axonal transport by myelinating Schwann cells. *Cell* 68:451–63

Dhaunchak A, Nave KA. 2007. A common mechanism of PLP/DM20 misfolding leading to cysteine-mediated ER retention in oligodendrocytes and Pelizaeus-Merzbacher disease. *Proc. Natl. Acad. Sci. USA* 104:17815–18

Domeniconi M, Cao Z, Spencer T, Sivasankaran R, Wang KC, et al. 2002. Myelin-associated glycoprotein interacts with the Nogo66 receptor to inhibit neurite outgrowth. *Neuron* 35:283–90

Donaldson HH, Hoke GW. 1905. The areas of the axis cylinder and medullary sheath as seen in cross sections of the spinal nerves of vertebrates. *J. Comp. Neurol.* 15:1–16

Du Y, Fischer TZ, Lee LN, Lercher LD, Dreyfus CF. 2003. Regionally specific effects of BDNF on oligodendrocytes. *Dev. Neurosci.* 25:116–26

Dubois-Dalcq M, Behar T, Hudson L, Lazzarini RA. 1986. Emergence of three myelin proteins in oligodendrocytes cultured without neurons. *J. Cell Biol.* 102:384–92

Durbec P, Marcos-Gutierrez CV, Kilkenny C, Grigoriou M, Wartiowaara K, et al. 1996. GDNF signalling through the Ret receptor tyrosine kinase. *Nature* 381:789–93

Edgar JM, McLaughlin M, Barrie JA, McCulloch MC, Garbern J, Griffiths IR. 2004a. Age-related axonal and myelin changes in the rumpshaker mutation of the Plp gene. *Acta Neuropathol. (Berl.)* 107:331–35

Edgar JM, McLaughlin M, Yool D, Zhang SC, Fowler JH, et al. 2004b. Oligodendroglial modulation of fast axonal transport in a mouse model of hereditary spastic paraplegia. *J. Cell Biol.* 166:121–31

Einheber S, Bhat MA, Salzer JL. 2006. Disrupted axo-glial junctions result in accumulation of abnormal mitochondria at nodes of Ranvier. *Neuron Glia Biol.* 2:165–74

Esper RM, Loeb JA. 2004. Rapid axoglial signaling mediated by neuregulin and neurotrophic factors. *J. Neurosci.* 24:6218–27

Esper RM, Pankonin MS, Loeb JA. 2006. Neuregulins: versatile growth and differentiation factors in nervous system development and human disease. *Brain Res. Rev.* 51:161–75

Ey B, Kobsar I, Blazyca H, Kroner A, Martini R. 2007. Visualization of degenerating axons in a dysmyelinating mouse mutant with axonal loss. *Mol. Cell Neurosci.* 35:153–60

Fabricius C, Berthold CH, Rydmark M. 2003. Axoplasmic organelles at nodes of Ranvier. II. Occurrence and distribution in large myelinated spinal cord axons of the adult cat. *J. Neurocytol.* 22:941–54

Falls DL. 2003. Neuregulins: functions, forms, and signaling strategies. *Exp. Cell Res.* 284:14–30

Fern R, Davis P, Waxman SG, Ransom BR. 1998. Axon conduction and survival in CNS white matter during energy deprivation: a developmental study. *J. Neurophysiol.* 79:95–105

Fernandez PA, Tang DG, Cheng L, Prochiantz A, Mudge AW, Raff MC. 2000. Evidence that axon-derived neuregulin promotes oligodendrocyte survival in the developing rat optic nerve. *Neuron* 28:81–90

Ferreirinha F, Quattrini A, Pirozzi M, Valsecchi V, Dina G, et al. 2004. Axonal degeneration in paraplegin-deficient mice is associated with abnormal mitochondria and impairment of axonal transport. *J. Clin. Invest.* 113:231–42

Fields RD, Burnstock G. 2006. Purinergic signalling in neuron-glia interactions. *Nat. Rev. Neurosci.* 7:423–36

Flores AI, Mallon BS, Matsui T, Ogawa W, Rosenzweig A, et al. 2000. Akt-mediated survival of oligodendrocytes induced by neuregulins. *J. Neurosci.* 20:7622–30

Friede RL. 1972. Control of myelin formation by axon caliber (with a model of the control mechanism). *J. Comp. Neurol.* 144:233–52

Garbern JY. 2007. Pelizaeus-Merzbacher disease: genetic and cellular pathogenesis. *Cell. Mol. Life Sci.* 64:50–65

Garbern JY, Yool DA, Moore GJ, Wilds IB, Faulk MW, et al. 2002. Patients lacking the major CNS myelin protein, proteolipid protein 1, develop length-dependent axonal degeneration in the absence of demyelination and inflammation. *Brain* 125:551–61

Garcia-Fresco GP, Sousa AD, Pillai AM, Moy SS, Crawley JN, et al. 2006. Disruption of axoglial junctions causes cytoskeletal disorganization and degeneration of Purkinje neuron axons. *Proc. Natl. Acad. Sci. USA* 103:5137–42

Garratt AN, Britsch S, Birchmeier C. 2000a. Neuregulin, a factor with many functions in the life of a Schwann cell. *Bioessays* 22:987–96

Garratt AN, Voiculescu O, Topilko P, Charnay P, Birchmeier C. 2000b. A dual role of erbB2 in myelination and in expansion of the Schwann cell precursor pool. *J. Cell Biol.* 148:1035–46

Gatzinsky KP, Holtmann B, Daraie B, Berthold CH, Sendtner M. 2003. Early onset of degenerative changes at nodes of Ranvier in alpha-motor axons of Cntf null (−/−) mutant mice. *Glia* 42:340–49

Goto K, Kurihara T, Takahashi Y, Kondo H. 1990. Expression of genes for the myelin-specific proteins in oligodendrocytes in vivo demands the presence of axons. *Neurosci. Lett.* 117:269–74

Gow A, Lazzarini RA. 1996. A cellular mechanism governing the severity of Pelizaeus-Merzbacher disease. *Nat. Genet.* 13:422–28

Gravel M, DeAngelis D, Braun PE. 1994. Molecular cloning and characterization of rat brain 2′,3′-cyclic nucleotide 3′-phosphodiesterase isoform 2. *J. Neurosci. Res.* 38:243–47

Gravel M, Peterson J, Yong VW, Kottis V, Trapp B, Braun PE. 1996. Overexpression of 2′,3′-cyclic nucleotide 3′-phosphodiesterase in transgenic mice alters oligodendrocyte development and produces aberrant myelination. *Mol. Cell Neurosci.* 7:453–66

Griffiths I, Klugmann M, Anderson T, Yool D, Thomson C, et al. 1998. Axonal swellings and degeneration in mice lacking the major proteolipid of myelin. *Science* 280:1610–13

Haney CA, Sahenk Z, Li C, Lemmon VP, Roder J, Trapp BD. 1999. Heterophilic binding of L1 on unmyelinated sensory axons mediates Schwann cell adhesion and is required for axonal survival. *J. Cell Biol.* 146:1173–84

Hartline DK, Colman DR. 2007. Rapid conduction and the evolution of giant axons and myelinated fibers. *Curr. Biol.* 17:R29–35

Hawkins BT, Davis TP. 2005. The blood-brain barrier/neurovascular unit in health and disease. *Pharmacol. Rev.* 57:173–85

Henderson CE, Phillips HS, Pollock RA, Davies AM, Lemeulle C, et al. 1994. GDNF: a potent survival factor for motoneurons present in peripheral nerve and muscle. *Science* 266:1062–64

Herculano-Houzel S, Lent R. 2005. Isotropic fractionator: a simple, rapid method for the quantification of total cell and neuron numbers in the brain. *J. Neurosci.* 25:2518–21

Hertz L, Zielke HR. 2004. Astrocytic control of glutamatergic activity: astrocytes as stars of the show. *Trends Neurosci.* 27:735–43

Hoke A, Ho T, Crawford TO, LeBel C, Hilt D, Griffin JW. 2003. Glial cell line derived neurotrophic factor alters axon Schwann cell units and promotes myelination in unmyelinated nerve fibers. *J. Neurosci.* 23:561–67

Hollenbeck PJ, Saxton WM. 2005. The axonal transport of mitochondria. *J. Cell Sci.* 118:5411–19

Horiuchi K, Zhou HM, Kelly K, Manova K, Blobel CP. 2005. Evaluation of the contributions of ADAMs 9, 12, 15, 17, and 19 to heart development and ectodomain shedding of neuregulins beta1 and beta2. *Dev. Biol.* 283:459–71

Hu X, Hicks CW, He W, Wong P, Macklin WB, et al. 2006. Bace1 modulates myelination in the central and peripheral nervous system. *Nat. Neurosci.* 9:1520–25

Huang EJ, Reichardt LF. 2001. Neurotrophins: roles in neuronal development and function. *Annu. Rev. Neurosci.* 24:677–736

Hudson LD, Berndt JA, Puckett C, Kozak CA, Lazzarini RA. 1987. Aberrant splicing of proteolipid protein mRNA in the dysmyelinating jimpy mutant mouse. *Proc. Natl. Acad. Sci. USA* 84:1454–58

Hudson LD, Puckett C, Berndt J, Chan J, Gencic S. 1989. Mutation of the proteolipid protein gene PLP in a human X chromosome-linked myelin disorder. *Proc. Natl. Acad. Sci. USA* 86:8128–31

Inoue Y, Nakamura R, Mikoshiba K, Tsukada Y. 1981. Fine structure of the central myelin sheath in the myelin deficient mutant Shiverer mouse, with special reference to the pattern of myelin formation by oligodendroglia. *Brain Res.* 219:85–94

Inoue K. 2005. PLP1-related inherited dysmyelinating disorders: Pelizaeus-Merzbacher disease and spastic paraplegia type 2. *Neurogenetics* 6:1–16

Inoue K, Tanabe Y, Lupski JR. 1999. Myelin deficiencies in both the central and the peripheral nervous systems associated with a SOX10 mutation. *Ann. Neurol.* 46:313–38

Ip CW, Kroner A, Bendszus M, Leder C, Kobsar I, et al. 2006. Immune cells contribute to myelin degeneration and axonopathic changes in mice overexpressing proteolipid protein in oligodendrocytes. *J. Neurosci.* 26:8206–16

Ishibashi T, Dakin KA, Stevens B, Lee PR, Kozlov SV, et al. 2006. Astrocytes promote myelination in response to electrical impulses. *Neuron* 49:823–32

Jessen KR, Mirsky R. 2005. The origin and development of glial cells in peripheral nerves. *Nat. Rev. Neurosci.* 6:671–82

Johnston AW, McKusick VA. 1962. A sex-linked recessive form of spastic paraplegia. *Am. J. Hum. Genet.* 14:83–94

Jung M, Sommer I, Schachner M, Nave KA. 1996. Monoclonal antibody O10 defines a conformationally sensitive cell-surface epitope of proteolipid protein (PLP): evidence that PLP misfolding underlies dysmyelination in mutant mice. *J. Neurosci.* 16:7920–29

Kagitani-Shimono K, Mohri I, Oda H, Ozono K, Suzuki K, et al. 2006. Lipocalin-type prostaglandin D synthase (beta-trace) is upregulated in the alphaB-crystallin-positive oligodendrocytes and astrocytes in the chronic multiple sclerosis. *Neuropathol. Appl. Neurobiol.* 32:64–73

Kassmann CM, Lappe-Siefke C, Baes M, Brugger B, Mildner A, et al. 2007. Axonal loss and neuroinflammation caused by peroxisome-deficient oligodendrocytes. *Nat. Genet.* 39:969–76

Kidd GJ, Hauer PE, Trapp BD. 1990. Axons modulate myelin protein messenger RNA levels during central nervous system myelination in vivo. *J. Neurosci. Res.* 26:409–18

Kirkpatrick LL, Witt AS, Payne HR, Shine HD, Brady ST. 2001. Changes in microtubule stability and density in myelin-deficient shiverer mouse CNS axons. *J. Neurosci.* 21:2288–97

Klämbt C, Hummel T, Granderath S, Schimmelpfeng K. 2001. Glial cell development in *Drosophila. Int. J. Dev. Neurosci.* 19:373–78

Klugmann M, Schwab MH, Puhlhofer A, Schneider A, Zimmermann F, et al. 1997. Assembly of CNS myelin in the absence of proteolipid protein. *Neuron* 18:59–70

Kumar S, Kahn MA, Dinh L, de Vellis J. 1998. NT-3-mediated TrkC receptor activation promotes proliferation and cell survival of rodent progenitor oligodendrocyte cells in vitro and in vivo. *J. Neurosci. Res.* 54:754–65

Lai C, Brow MA, Nave KA, Noronha AB, Quarles RH, et al. 1987. Two forms of 1B236/myelin-associated glycoprotein, a cell adhesion molecule for postnatal neural development, are produced by alternative splicing. *Proc. Natl. Acad. Sci. USA* 84:4337–41

Lappe-Siefke C, Goebbels S, Gravel M, Nicksch E, Lee J, et al. 2003. Disruption of Cnp1 uncouples oligodendroglial functions in axonal support and myelination. *Nat. Genet.* 33:366–74

Lassmann H, Bartsch U, Montag D, Schachner M. 1997. Dying-back oligodendrogliopathy: a late sequel of myelin-associated glycoprotein deficiency. *Glia* 19:104–10

Lee J, Gravel M, Zhang R, Thibault P, Braun PE. 2005. Process outgrowth in oligodendrocytes is mediated by CNP, a novel microtubule assembly myelin protein. *J. Cell Biol.* 170:661–73

Lee J, O'Neill RC, Park MW, Gravel M, Braun PE. 2006. Mitochondrial localization of CNP2 is regulated by phosphorylation of the N-terminal targeting signal by PKC: implications of a mitochondrial function for CNP2 in glial and nonglial cells. *Mol. Cell Neurosci.* 31:446–62

Leimeroth R, Lobsiger C, Lussi A, Taylor V, Suter U, Sommer L. 2002. Membrane bound neuregulin1 type III actively promotes Schwann cell differentiation of multipotent progenitor cells. *Dev. Biol.* 246:245–58

Li C, Tropak MB, Gerlai R, Clapoff S, Abramow-Newerly W, et al. 1994. Myelination in the absence of myelin-associated glycoprotein. *Nature* 369:747–50

Li S, Jiang Q, Stys PK. 2000. Important role of reverse Na$^{(+)}$-Ca$^{(2+)}$ exchange in spinal cord white matter injury at physiological temperature. *J. Neurophysiol.* 84:1116–19

Linker RA, Maurer M, Gaupp S, Martini R, Holtmann B, et al. 2002. CNTF is a major protective factor in demyelinating CNS disease: a neurotrophic cytokine as modulator in neuroinflammation. *Nat. Med.* 8:620–24

Magistretti PJ. 2006. Neuron-glia metabolic coupling and plasticity. *J. Exp. Biol.* 209:2304–11

Marcus J, Honigbaum S, Shroff S, Honke K, Rosenbluth J, Dupree JL. 2006. Sulfatide is essential for the maintenance of CNS myelin and axon structure. *Glia* 53:372–81

Marrosu MG, Vaccargiu S, Marrosu G, Vannelli A, Cianchetti C, Muntoni F. 1998. Charcot-Marie-Tooth disease type 2 associated with mutation of the myelin protein zero gene. *Neurology* 50:1397–401

Martini R, Schachner M. 1986. Immunoelectron microscopic localization of neural cell adhesion molecules (L1, N-CAM, and MAG) and their shared carbohydrate epitope and myelin basic protein in developing sciatic nerve. *J. Cell Biol.* 103:2439–48

Maurel P, Einheber S, Galinska J, Thaker P, Lam I, et al. 2007. Nectin-like proteins mediate axon Schwann cell interactions along the internode and are essential for myelination. *J. Cell Biol.* 178:861–74

Maurel P, Salzer JL. 2000. Axonal regulation of Schwann cell proliferation and survival and the initial events of myelination requires PI 3-kinase activity. *J. Neurosci.* 20:4635–45

McFerran B, Burgoyne R. 1997. 2′,3′-Cyclic nucleotide 3′-phosphodiesterase is associated with mitochondria in diverse adrenal cell types. *J. Cell Sci.* 110:2979–85

McPhilemy K, Griffiths IR, Mitchell LS, Kennedy PG. 1991. Loss of axonal contact causes down-regulation of the PLP gene in oligodendrocytes: evidence from partial lesions of the optic nerve. *Neuropathol. Appl. Neurobiol.* 17:275–87

McTigue DM, Horner PJ, Stokes BT, Gage FH. 1998. Neurotrophin-3 and brain-derived neurotrophic factor induce oligodendrocyte proliferation and myelination of regenerating axons in the contused adult rat spinal cord. *J. Neurosci.* 18:5354–65

Meier C, Bischoff A. 1975. Oligodendroglial cell development in jimpy mice and controls. An electron-microscopic study in the optic nerve. *J. Neurol. Sci.* 26:517–28

Merzbacher L. 1909. Gesetzmaessigkeiten in der Vererbung und Verbreitung verschiedener hereditaer-familiaerer Erkrankungen. *Arch. Rass. Ges. Biol.* 6:172–98

Michailov GV, Sereda MW, Brinkmann BG, Fischer TM, Haug B, et al. 2004. Axonal neuregulin-1 regulates myelin sheath thickness. *Science* 304:700–3

Misgeld T, Kerschensteiner M, Bareyre FM, Burgess RW, Lichtman JW. 2007. Imaging axonal transport of mitochondria in vivo. *Nat. Methods* 4:559–61

Montag D, Giese KP, Bartsch U, Martini R, Lang Y, et al. 1994. Mice deficient for the myelin-associated glycoprotein show subtle abnormalities in myelin. *Neuron* 13:229–46

Moore MW, Klein RD, Farinas I, Sauer H, Armanini M, et al. 1996. Renal and neuronal abnormalities in mice lacking GDNF. *Nature* 382:76–79

Morrissey TK, Levi AD, Nuijens A, Sliwkowski MX, Bunge RP. 1995. Axon induced mitogenesis of human Schwann cells involves heregulin and p185erbB2. *Proc. Natl. Acad. Sci. USA* 92:1431–35

Murray MA. 1968. An electron microscopic study of the relationship between axon diameter and the initiation of myelin production in the peripheral nervous system. *Anat. Rec.* 161:337–52

Mutsaers SE, Carroll WM. 1998. Focal accumulation of intra-axonal mitochondria in demyelination of the cat optic nerve. *Acta Neuropathol.* 96:139–43

Nave KA, Boespflug-Tanguy O. 1996. Developmental defects of myelin formation: from X-linked mutations to human dysmyelinating diseases. *Neuroscientist* 2:33–43

Nave KA, Lai C, Bloom FE, Milner RJ. 1986. Jimpy mutant mouse: a 74-base deletion in the mRNA for myelin proteolipid protein and evidence for a primary defect in RNA splicing. *Proc. Natl. Acad. Sci. USA* 83:9264–68

Nave KA, Salzer JL. 2006. Axonal regulation of myelination by neuregulin 1. *Curr. Opin. Neurobiol.* 16:492–500

Nave KA, Sereda MW, Ehrenreich H. 2007. Mechanisms of disease: inherited demyelinating neuropathies—from basic to clinical research. *Nat. Clin. Pract. Neurol.* 3:453–64

Nixon RA. 1982. Increased axonal proteolysis in myelin-deficient mutant mice. *Science* 215:999–1001

Ogata T, Iijima S, Hoshikawa S, Miura T, Yamamoto S, et al. 2004. Opposing extracellular signal-regulated kinase and Akt pathways control Schwann cell myelination. *J. Neurosci.* 24:6724–32

Paratcha G, Ledda F, Ibanez CF. 2003. The neural cell adhesion molecule NCAM is an alternative signaling receptor for GDNF family ligands. *Cell* 113:867–79

Paratore C, Goerich DE, Suter U, Wegner M, Sommer L. 2001. Survival and glial fate acquisition of neural crest cells are regulated by an interplay between the transcription factor Sox10 and extrinsic combinatorial signaling. *Development* 128:3949–61

Pelizaeus F. 1885. Ueber eine eigentumliche Form spastischer Lahmung mit Cerebralerscheinungen auf hereditarer Grundlage (Multiple Sklerose). *Arch. Psychiat. Nervenkr.* 16:698–710

Peters A, Palay SL, Webster H deF. 1991. *The Fine Structure of the Nervous System.* New York: Oxford Univ. Press. 3rd ed.

Pichel JG, Shen L, Shang HZ, Granholm AC, Drago J, et al. 1996. Defects in enteric innervation and kidney development in mice lacking GDNF. *Nature* 382:73–76

Raine CS, Cross AH. 1989. Axonal dystrophy as a consequence of long-term demyelination. *Lab. Invest.* 60:714–25

Rasband MN, Tayler J, Kaga Y, Yang Y, Lappe-Siefke C, et al. 2005. CNP is required for maintenance of axon-glia interactions at nodes of Ranvier in the CNS. *Glia* 50:86–90

Raskind WH, Williams CA, Hudson LD, Bird TD. 1991. Complete deletion of the proteolipid protein gene (PLP) in a family with X-linked Pelizaeus-Merzbacher disease. *Am. J. Hum. Genet.* 49:1355–60

Readhead C, Schneider A, Griffiths I, Nave KA. 1994. Premature arrest of myelin formation in transgenic mice with increased proteolipid protein gene dosage. *Neuron* 12:583–95

Riethmacher D, Sonnenberg-Riethmacher E, Brinkmann V, Yamaai T, Lewin GR, Birchmeier C. 1997. Severe neuropathies in mice with targeted mutations in the ErbB3 receptor. *Nature* 389:725–30

Roach A, Takahashi N, Pravtcheva D, Ruddle F, Hood L. 1985. Chromosomal mapping of mouse myelin basic protein gene and structure and transcription of the partially deleted gene in shiverer mutant mice. *Cell* 42:149–55

Rosenberg SS, Ng BK, Chan JR. 2006. The quest for remyelination: a new role for neurotrophins and their receptors. *Brain Pathol.* 16:288–94

Rosenbluth J. 1980. Central myelin in the mouse mutant shiverer. *J. Comp. Neurol.* 194:639–48

Rosenbluth J, Nave KA, Mierzwa A, Schiff R. 2006. Subtle myelin defects in PLP-null mice. *Glia* 54:172–82

Rosenfeld J, Friedrich VL Jr. 1983. Axonal swellings in jimpy mice: Does lack of myelin cause neuronal abnormalities? *Neuroscience* 10:959–66

Roy K, Murtie JC, El-Khodor BF, Edgar N, Sardi SP, et al. 2007. Loss of erbB signaling in oligodendrocytes alters myelin and dopaminergic function, a potential mechanism for neuropsychiatric disorders. *Proc. Natl. Acad. Sci. USA* 104:8131–36

Sahenk Z, Nagaraja HN, McCracken BS, King WM, Freimer ML, et al. 2005. NT-3 promotes nerve regeneration and sensory improvement in CMT1A mouse models and in patients. *Neurology* 65:681–89

Saher G, Brugger B, Lappe-Siefke C, Mobius W, Tozawa R, et al. 2005. High cholesterol level is essential for myelin membrane growth. *Nat. Neurosci.* 8:468–75

Salzer JL, Williams AK, Glaser L, Bunge RP. 1980. Studies of Schwann cell proliferation. II. Characterization of the stimulation and specificity of the response to a neurite membrane fraction. *J. Cell Biol.* 84:753–66

Sanchez MP, Silos-Santiago I, Frisen J, He B, Lira SA, Barbacid M. 1996. Renal agenesis and the absence of enteric neurons in mice lacking GDNF. *Nature* 382:70–73

Sandell JH, Peters A. 2003. Disrupted myelin and axon loss in the anterior commissure of the aged rhesus monkey. *J. Comp. Neurol.* 466:14–30

Saugier-Veber P, Munnich A, Bonneau D, Rozet JM, Le Merrer M, et al. 1994. X-linked spastic paraplegia and Pelizaeus-Merzbacher disease are allelic disorders at the proteolipid protein locus. *Nat. Genet.* 6:257–62

Scherer SS, Vogelbacker HH, Kamholz J. 1992. Axons modulate the expression of proteolipid protein in the CNS. *J. Neurosci. Res.* 32:138–48

Schneider A, Montague P, Griffiths I, Fanarraga M, Kennedy P, et al. 1992. Uncoupling of hypomyelination and glial cell death by a mutation in the proteolipid protein gene. *Nature* 358:758–61

Schroering A, Carey DJ. 1998. Sensory and motor neuron-derived factor is a transmembrane heregulin that is expressed on the plasma membrane with the active domain exposed to the extracellular environment. *J. Biol. Chem.* 273:30643–50

Schweigreiter R, Roots BI, Bandtlow CE, Gould RM. 2006. Understanding myelination through studying its evolution. *Int. Rev. Neurobiol.* 73:219–73

Senderek J, Hermanns B, Lehmann U, Bergmann C, Senderek J, et al. 2000. Charcot–Marie–Tooth neuropathy type 2 and P0 point mutations: two novel amino acid substitutions (Asp61Gly; Tyr119Cys) and a possible "hotspot" on Thr124Met. *Brain Pathol.* 10:235–48

Sendtner M, Carroll P, Holtmann B, Hughes RA, Thoenen H. 1994. Ciliary neurotrophic factor. *J Neurobiol.* 25:1436–53

Shah NM, Marchionni MA, Isaacs I, Stroobant P, Anderson DJ. 1994. Glial growth factor restricts mammalian neural crest stem cells to a glial fate. *Cell* 77:349–60

Sherwood CC, Stimpson CD, Raghanti MA, Wildman DE, Uddin M, et al. 2006. Evolution of increased glia-neuron ratios in the human frontal cortex. *Proc. Natl. Acad. Sci. USA* 103:13606–11

Simons M, Krämer EM, Thiele C, Stoffel W, Trotter J. 2000. Assembly of myelin by association of proteolipid protein with cholesterol- and galactosylceramide-rich membrane domains. *J. Cell Biol.* 151:143–54

Sinha K, Karimi-Abdolrezaee S, Velumian AA, Fehlings MG. 2006. Functional changes in genetically dysmyelinated spinal cord axons of shiverer mice: role of juxtaparanodal Kv1 family K^+ channels. *J. Neurophysiol.* 95:1683–95

Smith KJ, Blakemore WF, Murray JA, Patterson RC. 1982. Internodal myelin volume and axon surface area. A relationship determining myelin thickness? *J. Neurol. Sci.* 55:231–46

Southwood CM, Garbern J, Jiang W, Gow A. 2002. The unfolded protein response modulates disease severity in Pelizaeus-Merzbacher disease. *Neuron* 36:585–96

Spencer PS, Sabri MI, Schaumburg HH, Moore CL. 1979. Does a defect of energy metabolism in the nerve fiber underlie axonal degeneration in polyneuropathies? *Ann. Neurol.* 5:501–7

Spiegel I, Adamsky K, Eshed Y, Milo R, Sabanay H, et al. 2007. A central role for Necl4 (SynCAM4) in Schwann cell-axon interaction and myelination. *Nat. Neurosci.* 10:861–69

Sporkel O, Uschkureit T, Bussow H, Stoffel W. 2002. Oligodendrocytes expressing exclusively the DM20 isoform of the proteolipid protein gene: myelination and development. *Glia* 37:19–30

Springer JE, Seeburger JL, He J, Gabrea A, Blankenhorn EP, Bergman LW. 1995. cDNA sequence and differential mRNA regulation of two forms of glial cell line-derived neurotrophic factor in Schwann cells and rat skeletal muscle. *Exp. Neurol.* 131:47–52

Stecca B, Southwood CM, Gragerov A, Kelley KA, Friedrich VL Jr, Gow A. 2000. The evolution of lipophilin genes from invertebrates to tetrapods: DM-20 cannot replace proteolipid protein in CNS myelin. *J. Neurosci.* 20:4002–10

Stevens B, Fields RD. 2000. Response of Schwann cells to action potentials in development. *Science* 287:2267–71

Stevens B, Porta S, Haak LL, Gallo V, Fields RD. 2002. Adenosine: a neuron-glial transmitter promoting myelination in the CNS in response to action potentials. *Neuron* 36:855–68

Stys PK, Waxman SG, Ransom BR. 1992. Ionic mechanisms of anoxic injury in mammalian CNS white matter: role of Na^+ channels and $Na^{(+)}$-Ca^{2+} exchanger. *J. Neurosci.* 12:430–39

Sussman CR, Vartanian T, Miller RH. 2005. The ErbB4 neuregulin receptor mediates suppression of oligodendrocyte maturation. *J. Neurosci.* 25:5757–62

Suter U, Scherer SS. 2003. Disease mechanisms in inherited neuropathies. *Nat. Rev. Neurosci.* 4:714–26

Suter-Crazzolara C, Unsicker K. 1994. GDNF is expressed in two forms in many tissues outside the CNS. *NeuroReport* 20:2486–88

Takahashi R, Yokoji H, Misawa H, Hayashi M, Hu J, Deguchi T. 1994. A null mutation in the human CNTF gene is not causally related to neurological diseases. *Nat. Genet.* 7:79–84

Taniike M, Mohri I, Eguchi N, Beuckmann CT, Suzuki K, Urade Y. 2002. Perineuronal oligodendrocytes protect against neuronal apoptosis through the production of lipocalin-type prostaglandin D synthase in a genetic demyelinating model. *J. Neurosci.* 22:4885–96

Taveggia C, Zanazzi G, Petrylak A, Yano H, Rosenbluth J, et al. 2005. Neuregulin-1 type III determines the ensheathment fate of axons. *Neuron* 47:681–94

Thippeswamy T, McKay JS, Morris R, Quinn J, Wong LF, Murphy D. 2005. Glial-mediated neuroprotection: evidence for the protective role of the NO-cGMP pathway via neuron-glial communication in the peripheral nervous system. *Glia* 49:197–210

Trapp BD, Andrews SB, Cootauco C, Quarles R. 1989. The myelin-associated glycoprotein is enriched in multivesicular bodies and periaxonal membranes of actively myelinating oligodendrocytes. *J. Cell Biol.* 109:2417–26

Trapp BD, Bernier L, Andrews SB, Colman DR. 1988. Cellular and subcellular distribution of 2′,3′-cyclic nucleotide 3′-phosphodiesterase and its mRNA in the rat central nervous system. *J. Neurochem.* 51:859–68

Trapp BD, Nave KA. 2008. Axonal degeneration in multiple sclerosis. *Annu. Rev. Neurosci.* 31:247–69

Trapp BD, Peterson J, Ransohoff RM, Rudick R, Mork S, Bo L. 1998. Axonal transection in the lesions of multiple sclerosis. *N. Engl. J. Med.* 338:278–85

Trupp M, Rydén M, Jörnvall H, Funakoshi H, Timmusk T, et al. 1995. Peripheral expression and biological activities of GDNF, a new neurotrophic factor for avian and mammalian peripheral neurons. *J. Cell Biol.* 130:137–48

Ueda H, Levine JM, Miller RH, Trapp BD. 1999. Rat optic nerve oligodendrocytes develop in the absence of viable retinal ganglion cell axons. *J. Cell Biol.* 146:1365–74

Van Vught PW, Van Wijk J, Bradley TE, Plasmans D, Jakobs ME, et al. 2007. Ciliary neurotrophic factor null alleles are not a risk factor for Charcot-Marie-Tooth disease, hereditary neuropathy with pressure palsies and amyotrophic lateral sclerosis. *Neuromuscul. Disord.* 17:964–67

Vartanian T, Fischbach G, Miller R. 1999. Failure of spinal cord oligodendrocyte development in mice lacking neuregulin. *Proc. Natl. Acad. Sci. USA* 96:731–35

Vartanian T, Goodearl A, Viehover A, Fischbach G. 1997. Axonal neuregulin signals cells of the oligodendrocyte lineage through activation of HER4 and Schwann cells through HER2 and HER3. *J. Cell Biol.* 137:211–20

Vega C, Martiel JL, Drouhault D, Burckhart MF, Coles JA. 2003. Uptake of locally applied deoxyglucose, glucose and lactate by axons and Schwann cells of rat vagus nerve. *J. Physiol.* 546:551–64

Vogel US, Thompson RJ. 1988. Molecular structure, localization, and possible functions of the myelin-associated enzyme 2′,3′-cyclic nucleotide 3′-phosphodiesterase. *J. Neurochem.* 50:1667–77

Voyvodic JT. 1989. Target size regulates calibre and myelination of sympathetic axons. *Nature* 342:430–33

Watanabe M, Sakurai Y, Ichinose T, Aikawa Y, Kotani M, Itoh K. 2006. Monoclonal antibody Rip specifically recognizes 2′,3′-cyclic nucleotide 3′-phosphodiesterase in oligodendrocytes. *J. Neurosci. Res.* 84:525–33

Weinberg HJ, Spencer PS. 1976. Studies on the control of myelinogenesis. Evidence for neuronal regulation of myelin production. *Brain Res.* 113:363–78

Weiss MD, Hammer J, Quarles RH. 2000. Oligodendrocytes in aging mice lacking myelin-associated glycoprotein are dystrophic but not apoptotic. *J. Neurosci. Res.* 62:772–80

Werner HB, Kuhlmann K, Shen S, Uecker M, Schardt A, et al. 2007. Proteolipid protein is required for transport of sirtuin 2 into CNS myelin. *J. Neurosci.* 27:7717–30

Wilkins A, Majed H, Layfield R, Compston A, Chandran S. 2003. Oligodendrocytes promote neuronal survival and axonal length by distinct intracellular mechanisms: a novel role for oligodendrocyte-derived glial cell line-derived neurotrophic factor. *J. Neurosci.* 23:4967–74

Willem M, Garratt AN, Novak B, Citron M, Kaufmann S, et al. 2006. Control of peripheral nerve myelination by the beta-secretase BACE1. *Science* 314:664–66

Windebank AJ, Wood P, Bunge RP, Dyck PJ. 1985. Myelination determines the caliber of dorsal root ganglion neurons in culture. *J. Neurosci.* 5:1563–69

Yin X, Baek RC, Kirschner DA, Peterson A, Fujii Y, et al. 2006. Evolution of a neuroprotective function of central nervous system myelin. *J. Cell Biol.* 172:469–78

Yin X, Crawford TO, Griffin JW, Tu P, Lee VM, et al. 1998. Myelin-associated glycoprotein is a myelin signal that modulates the caliber of myelinated axons. *J. Neurosci.* 18:1953–62

Yin X, Kidd GJ, Pioro EP, McDonough J, Dutta R, et al. 2004. Dysmyelinated lower motor neurons retract and regenerate dysfunctional synaptic terminals. *J. Neurosci.* 24:3890–98

Yin X, Peterson J, Gravel M, Braun PE, Trapp BD. 1997. CNP overexpression induces aberrant oligodendrocyte membranes and inhibits MBP accumulation and myelin compaction. *J. Neurosci. Res.* 50:238–47

Yool DA, Klugmann M, McLaughlin M, Vouyiouklis DA, Dimou L, et al. 2001. Myelin proteolipid proteins promote the interaction of oligodendrocytes and axons. *J. Neurosci. Res.* 63:151–64

Zanazzi G, Einheber S, Westreich R, Hannocks MJ, Bedell-Hogan D, et al. 2001. Glial growth factor/neuregulin inhibits Schwann cell myelination and induces demyelination. *J. Cell Biol.* 152:1289–99

Signaling Mechanisms Linking Neuronal Activity to Gene Expression and Plasticity of the Nervous System

Steven W. Flavell[1,2] and Michael E. Greenberg[1]

[1] F.M. Kirby Neurobiology Center, Children's Hospital Boston, and Departments of Neurology and Neurobiology, Harvard Medical School, Boston, Massachusetts 02115; email: Flavell@fas.harvard.edu, Michael.Greenberg@childrens.harvard.edu

[2] Program in Neuroscience, Harvard Medical School, Boston, Massachusetts 02115

Annu. Rev. Neurosci. 2008. 31:563–90

First published online as a Review in Advance on April 2, 2008

The *Annual Review of Neuroscience* is online at neuro.annualreviews.org

This article's doi:
10.1146/annurev.neuro.31.060407.125631

0147-006X/08/0721-0563$20.00

Key Words

activity-regulated transcription, synapse development, synaptic plasticity, CREB, MEF2, c-fos

Abstract

Sensory experience and the resulting synaptic activity within the brain are critical for the proper development of neural circuits. Experience-driven synaptic activity causes membrane depolarization and calcium influx into select neurons within a neural circuit, which in turn trigger a wide variety of cellular changes that alter the synaptic connectivity within the neural circuit. One way in which calcium influx leads to the remodeling of synapses made by neurons is through the activation of new gene transcription. Recent studies have identified many of the signaling pathways that link neuronal activity to transcription, revealing both the transcription factors that mediate this process and the neuronal activity–regulated genes. These studies indicate that neuronal activity regulates a complex program of gene expression involved in many aspects of neuronal development, including dendritic branching, synapse maturation, and synapse elimination. Genetic mutations in several key regulators of activity-dependent transcription give rise to neurological disorders in humans, suggesting that future studies of this gene expression program will likely provide insight into the mechanisms by which the disruption of proper synapse development can give rise to a variety of neurological disorders.

Contents

INTRODUCTION

While intrinsic genetic programs play a critical role in shaping both early and postnatal neural development, extrinsic sensory cues are essential for the proper development of neural circuitry during early postnatal life. In addition to their role during early postnatal development, sensory experiences can lead to the formation of long-lasting memories and other adaptations in adult organisms. Both in infancy and in old age, the memories and adaptations that humans make in response to sensory experiences are ultimately mediated by experience-driven changes in neural connectivity. These changes are central to our everyday lives and are tragically disrupted in numerous psychiatric, neurological, and neurodegenerative disorders.

Complex Behaviors Are Altered by Sensory Experience Both During Development and in Adulthood

The acquisition of language provides an excellent example of the importance of sensory experience during development. Many studies examining the correlation between age and the ability to acquire new language skills have suggested that this ability progressively declines after early adolescence (reviewed in Doupe & Kuhl 1999). This observation prompted researchers to propose that there is a developmental critical period during which exposure to language is necessary for proper language development. Evidence for a critical period in vocal learning is supported by studies that have directly tested this hypothesis in songbirds. In many species, male songbirds learn one song during development and continue to sing only this song throughout their lives. In these cases, there is also a critical period during which the bird must hear its song to reproduce it properly as an adult (reviewed in Brainard & Doupe 2002). For example, white-crowned sparrows raised in isolation and subsequently exposed to recordings of their native song (after 100 days of age) never learn to sing the correct song.

Although many additional instances can be cited in which experience alters behavior most effectively when it occurs during early postnatal development, other adaptations that organisms make in response to sensory experience may remain quite plastic and reversible through adulthood. One clear example is the entrainment of circadian rhythms by light exposure. Neurons in the suprachiasmatic nucleus (SCN) of the hypothalamus exhibit periodic oscillations in their firing rates (with cycles close to 24 h) that persist even when these cells are grown in dissociated cultures (reviewed in Reppert & Weaver 2002). In the intact brain, these cells coordinate their cycles with the solar cycle by receiving synaptic inputs from retinal ganglion neurons that detect light primarily with the photopigment melanopsin. In response to light, these retinal inputs alter the electrical activity of SCN neurons, thereby setting their pacemaking activity. Although we do not know yet how SCN neurons go on to orchestrate complex circadian behaviors, the importance of sensory input in setting the clock is well established.

Experience Can Alter Neural Circuitry Both During Early Postnatal Development and in Adulthood

These examples and many others illustrate the now well-accepted fact that humans and other organisms routinely alter their behaviors in response to experience. Ever since the pioneering work of Drs. David Hubel and Torsten Wiesel was published more than 40 years ago, neurobiologists have appreciated that experience also plays an essential role in shaping neural connectivity. Specifically, a large number of studies have provided evidence that neuronal activity

plays a critical role in (*a*) dendritic outgrowth, (*b*) synaptic maturation/potentiation, (*c*) synapse elimination, and (*d*) synaptic plasticity in adult organisms.

Dendritic outgrowth. Dendrite elaboration is one of the first key steps in defining the synaptic inputs that a neuron will receive. Many distinct neuronal cell types in the brain have unique dendritic arbors that are suited to their function. Although we have known for quite some time that environmental stimuli can alter dendritic outgrowth, recent advances in fluorescent imaging have made it possible to examine in greater detail the interplay between environmental stimuli and dendritic growth. The Cline laboratory utilized in vivo time-lapse imaging in the tadpole *Xenopus laevis* to examine dendritic development in tectal neurons, which receive synaptic inputs from retinal ganglion cells. They noted that the dendritic arbors of these neurons elaborate as synapses begin to form onto tectal cells (Rajan & Cline 1998). As has been observed in other mammalian cell types, these experiments show that dendritic outgrowth is a highly dynamic process: Dendrites commonly sprout new branches and extend or retract existing branches. Blockade of N-methyl-D-aspartate subtype glutamate receptors (NMDARs), the predominant glutamate receptor in developing tectal neurons, slows both new dendritic branch addition and branch extension (Rajan & Cline 1998), whereas blockade of either NMDARs or α-amino-3-hydroxy-5-methylisoxazole-4-propionic acid receptors (AMPARs) in more mature tectal cells (where both types of glutamate receptors can be found) destabilizes existing branches (Haas et al. 2006). Most strikingly, visual stimulation of young tadpoles, which increases synaptic activity onto tectal neurons, causes an increase in new branch addition and extension in an NMDA receptor–dependent manner (Sin et al. 2002). These experiments, which are supported by additional in vitro data (Redmond et al. 2002), suggest that synaptic activity promotes dendritic outgrowth.

SCN: suprachiasmatic nucleus

N-methyl-D-aspartate receptor (NMDAR): ionotropic glutamate receptors that regulate gene expression and synaptic plasticity by mediating glutamate- and depolarization-dependent calcium influx

α-amino-3-hydroxy-5-methylisoxazole-4-propionic acid receptor (AMPAR): ionotropic glutamate receptors that mediate fast synaptic transmission; their trafficking to the postsynaptic membrane is tightly controlled during development/plasticity

Synapse maturation. As the dendrites of a neuron begin to elaborate, synapses form on these branches and then undergo an extensive maturation process. Although neuronal activity is not absolutely required for synapse formation (Verhage et al. 2000), sensory experience and synaptic activity are important for the subsequent maturation of synapses. On the presynaptic side of the neuromuscular junction and other synapses, this maturation involves the arborization and elaboration of the axon (Antonini & Stryker 1993, Sanes & Lichtman 1999). On the postsynaptic side, synapse maturation can be detected electrophysiologically as an increase in synaptic current. At central nervous system (CNS) synapses, such as the retinogeniculate synapse (the synapse made by retinal ganglion cell axons onto thalamic neurons), the AMPAR-mediated current is the predominant component of the postsynaptic response that is strengthened (Chen & Regehr 2000). A recent study of this synapse demonstrated that both spontaneous retinal activity and visual activity after eye opening are critical for the strengthening of the AMPAR-mediated current (Hooks & Chen 2006).

Synapse elimination. Although neuronal activity is critical for synapse maturation, in most circuits an excess of synapses initially forms, and only a subset of those synapses are strengthened while others are eliminated. This elimination process depends on sensory experience and synaptic activity. Drs. Hubel and Wiesel and others showed that in adult mammals the terminal axonal arbors of thalamic neurons that receive innervation from the retinal ganglion cells of the two eyes segregate from one another in layer IV of the primary visual cortex, forming ocular dominance columns (ODCs) (reviewed in Wiesel 1982). However, early in development, these axonal arbors overlap with one another. During early postnatal development these inputs segregate in a neuronal activity–dependent manner owing to the loss of synapses in the overlapped areas. Monocular deprivation in one eye causes inputs from the other eye to maintain more territory in layer IV, whereas blockade of input activity altogether by tetrodotoxin infusion into both eyes prevents ODC refinement (LeVay et al. 1980, Stryker & Harris 1986). A similar observation has also been made for retinogeniculate projections (Penn et al. 1998).

The role of synaptic activity in synapse elimination has been especially well characterized at the neuromuscular junction (NMJ), the synapse made by motor neurons onto muscle cells. At the time of birth, individual muscle fibers are typically innervated by multiple motor neuron axons. However, during the first weeks of postnatal development, multiply innervated muscle fibers are converted to singly innervated fibers as a result of synapse elimination. This process occurs gradually, with many inputs decreasing in synaptic strength as a single input increases its strength (Colman et al. 1997). A similar skewing in synaptic inputs has been detected during the development of the retinogeniculate synapse and other CNS synapses (Chen & Regehr 2000). Preventing individual motor neuron axons from activating postsynaptic neurotransmitter receptors leads to the elimination of that input, suggesting that axons must be effective in depolarizing the muscle in order to be maintained (Buffelli et al. 2003). A number of studies have also shown that blockade of synaptic activity throughout an entire junction prevents synapse elimination, whereas increasing activity accelerates the process (reviewed in Sanes & Lichtman 1999). In addition, driving synchronous activity in all the fibers that innervate a single NMJ prevents synapse elimination, indicating that the unequal ability of axons to activate the muscle is essential for developmental input elimination (Busetto et al. 2000).

Synaptic plasticity. Although there are many examples where experience shapes neural connectivity specifically during a critical period in development, it is clear that experience can also alter neural connectivity in the adult brain. Much of this evidence comes from studies of associative learning, including Pavlovian fear conditioning, a paradigm in which an electrical shock to the foot is administered to an animal shortly after an auditory tone is presented. After several repetitions, animals learn that the tone and shock are paired and display a fear response (involving freezing, etc.) when presented with the tone even in the absence of the electrical shock. Recordings from the dorsolateral nucleus of the amygdala (dLA), which receives direct inputs from the auditory system and is required for this form of learning, have shown that neurons in this area have an increased firing rate in response to the tone after associative learning has occurred (reviewed in Maren & Quirk 2004). This increase in firing rate is likely due to alterations in synaptic connectivity because it shares many characteristics with long-term potentiation (LTP), an activity-dependent model for synaptic strengthening that can be induced by high-frequency stimulation of the afferents to the dLA in a slice preparation (reviewed in Maren 2005). The mechanisms underlying LTP at this synapse and others (most notably the Schaffer collateral-CA1 synapse in the hippocampus) have been studied in great detail and involve the activity-dependent insertion of AMPA receptors into the postsynaptic membrane as well as the growth of new dendritic spines, which serve as the sites of most excitatory CNS synapses (Engert & Bonhoeffer 1999, Malinow & Malenka 2002). In contrast, long-term depression (LTD), an activity-dependent model for synaptic weakening in mature neural circuits, involves removing AMPA receptors from the cell membrane and deconstructing existing dendritic spines (Nagerl et al. 2004).

Activity-Dependent Gene Expression: Importance in the Development of Neural Connectivity and Human Neurological Development

The cellular and molecular mechanisms that underlie these experience-driven changes in neural connectivity have been the subject of intense investigation in recent years. Sensory experience results in neurotransmitter release at synapses within a neural circuit, which in turn leads to membrane depolarization and calcium influx into individual neurons. This action then triggers a wide variety of cellular changes within these neurons, which are capable of altering the synaptic connectivity of the circuit. Some of these changes, such as the activation of calcium-sensitive signaling cascades, which lead to posttranslational modifications of proteins, or the regulation of mRNA translation, resulting in the production of new proteins, occur locally at the sites of calcium entry and play critical roles in altering synaptic function in a synapse-specific manner (reviewed in Malinow & Malenka 2002, Sutton & Schuman 2006).

In addition to these local effects, calcium influx into the postsynaptic neuron can alter cellular function by activating new gene transcription. Calcium influx activates a number of signaling pathways that converge on transcription factors within the nucleus, which in turn control the expression of a large number of neuronal activity-regulated genes. Recent work has revealed a number of the signaling pathways that mediate activity-dependent transcription and has identified their roles in experience-dependent neural development and plasticity. The remaining portion of this review discusses the signal transduction pathways by which neuronal activity regulates the activity-dependent gene expression program and summarizes recent findings that have clarified the function of this signaling network.

The work reviewed here, together with recent advances in human genetics, has revealed links between the signaling networks that control activity-dependent transcription and human neurological disorders, such as mental retardation and autism-spectrum disorders. The identification of these links highlights the importance of this subject and raises the possibility that future studies in this field will not only provide insights into the mechanisms that control activity-dependent brain development but may also enhance our understanding of how the disruption of proper brain development can lead to a wide variety of neurological disorders.

LTP: long-term potentiation

LTD: long-term depression

CHARACTERIZATION OF THE NEURONAL ACTIVITY-REGULATED GENE EXPRESSION PROGRAM

Early Insights: *c-fos* as the Prototypical Immediate Early Gene

The observation that extracellular stimuli can trigger rapid changes in gene transcription came from studies of quiescent fibroblasts stimulated by growth factors to reenter the cell cycle. The addition of platelet-derived growth factor (PDGF) or other growth factors to quiescent 3T3 fibroblasts led to the extremely rapid induction of the *c-fos* proto-oncogene (Greenberg & Ziff 1984). The increased transcription of *c-fos* and other genes, termed immediate early genes (IEGs), occurs rapidly (in the case of *c-fos*, within ~5 min) and transiently (for *c-fos*, undetectable ~30 min later) and does not require new protein synthesis. A separate group of extracellular stimuli-responsive genes (discussed below) is induced with slower kinetics, often showing peak transcriptional activity several hours after stimulation.

The importance of stimulus-induced transcription in neuronal cells was first suggested by the observation that *c-fos* transcription could be induced in response to a number of different stimuli in neuronal cell types. In the rat pheochromocytoma cell line PC12, activation of the nicotinic acetylcholine receptor, as well as an increase in the concentration of extracellular potassium chloride that leads to membrane depolarization and calcium influx via L-type voltage-gated calcium channels (L-VGCCs), triggers *c-fos* transcription (Greenberg et al. 1986). Upregulation of *c-fos* in neurons of the intact brain was subsequently observed in specific brain regions in response to seizures and a wide range of physiological stimuli (Morgan et al. 1987). For instance, neurons in the dorsal horn of the spinal cord show increased Fos immunoreactivity in response to various modes of tactile sensory stimulation (Hunt et al. 1987), whereas neurons in the SCN show increased Fos expression upon visual stimulation (Rusak et al. 1990). In fact, the expression of *c-fos* and other IEGs is now routinely used to mark neurons that have been recently activated. These studies indicate that IEGs are induced in individual neurons in response to physiological levels of neurotransmitter release.

The observation that the *c-fos* mRNA is robustly induced by synaptic activity resulting from sensory experience raised the possibility that the Fos protein, which together with Jun family members comprises the AP-1 transcriptional complex, is critical for the adaptive responses that organisms make in response to experience. Indeed, analyses of mice harboring a brain-specific deletion of the *c-fos* gene have shown that these animals display deficits in synaptic plasticity as well as defects in several forms of learning and memory (Fleischmann et al. 2003). Loss of Fos-dependent transcription can give rise to additional behavioral deficits because mice lacking the gene encoding the fos family member FosB display a defect in nurturing behaviors, as well as altered sensitivity to drugs of abuse, such as cocaine (Brown et al. 1996, Hiroi et al. 1997). Because the presence of a newborn pup induces FosB expression in the preoptic area of the mother's hypothalamus, while chronic cocaine administration induces FosB expression in the striatum, it is possible that the lack of an appropriate transcriptional response in these brain circuits in FosB null mice leads to altered behavioral responses in these animals.

The Activity-Regulated Gene Expression Program in Neurons: Genes that Directly Regulate Synaptic Function

Since the discovery of *c-fos* induction, many additional activity-regulated genes have been discovered (Bartel et al. 1989, Nedivi et al. 1993, Saffen et al. 1988, Yamagata et al. 1993). However, because of technical limitations, a genome-wide analysis of activity-dependent gene expression has only recently become possible. High-density oligonucleotide microarrays and differential analysis of library expression (DAzLE) have been used to identify genes

---

Immediate early gene (IEG): a gene whose expression is increased rapidly and transiently in a protein synthesis–independent manner in response to extracellular stimuli

in cortical and hippocampal neurons whose expression is acutely altered in response to membrane depolarization, NMDA application, LTP induction, and seizure induction (Altar et al. 2004, Hong et al. 2004, Li et al. 2004, Park et al. 2006). These experiments confirm the presence of several hundred neuronal activity-regulated genes in these cells. Although many neuronal activity-regulated genes, such as *c-fos*, encode transcription factors, a large number of these genes encode proteins that have important functions in dendrites and synapses and are likely to regulate circuit connectivity directly. Detailed studies investigating the functions of a few of these neuronal activity-regulated genes have yielded insights into how this activity-regulated transcriptional program controls neuronal development.

Candidate plasticity gene 15 (Cpg15). A number of neuronal activity-regulated genes encode cell surface proteins or secreted proteins that are transported to dendrites and synapses. Several of these proteins promote both dendritic outgrowth and synapse maturation. For instance, expression of *cpg15* (also called *neuritin*) enhances dendritic outgrowth, presynaptic axonal elaboration, and AMPA receptor insertion in *Xenopus* tectal neurons (Cantallops et al. 2000, Nedivi et al. 1998). In these experiments, dendritic outgrowth and synapse maturation were enhanced in neurons adjacent to those overexpressing CPG15. Because CPG15 is a glycosylphosphatidylinositol (GPI)-anchored membrane protein, its maturation-promoting effects are likely mediated by intercellular interactions.

Brain-derived neurotrophic factor (BDNF). The activity-regulated gene *bdnf* encodes a neurotrophin that is secreted at synapses in an activity-dependent manner and can bind to tyrosine kinase receptor B (trkB) and p75 neurotrophin receptors located in both pre- and postsynaptic membranes. Like CPG15, the induction of BDNF promotes both dendritic outgrowth and synapse maturation, but recent evidence suggests that the effects of

BDNF are more complex. Bath application of BDNF to organotypic cortical slices has cell type–specific effects on dendritic outgrowth: Whereas BDNF application enhances dendritic outgrowth in layer II/III and layer IV neurons, it can inhibit outgrowth in layer VI neurons (McAllister et al. 1995, 1997; Niblock et al. 2000). Similar seemingly contradictory results can be found within a single cell. Retinal ganglion cells in *Xenopus* display enhanced dendritic outgrowth when BDNF is applied to the distal tips of their axons and decreased dendritic outgrowth when BDNF is applied locally to the dendrites (Lom et al. 2002). These results suggest that BDNF secreted from different sources might signal differently (either via a different receptor or different downstream effectors) to elicit different effects. Given that the expression and secretion of BDNF are normally tightly controlled in neurons, bath application of BDNF might not recapitulate the effects of endogenous BDNF. Nevertheless, studies of a *bdnf* conditional knockout mouse confirm that BDNF regulates dendritic outgrowth in layer II/III cortical neurons (Gorski et al. 2003). Moreover, a decrease in dendritic complexity was also observed in a *bdnf* knock-in mouse designed to carry a valine-to-methionine amino acid substitution at valine-66 (val66met) in the BDNF protein (Chen et al. 2006). Because a single nucleotide polymorphism found in the human *bdnf* gene, resulting in the val66met substitution, confers susceptibility to memory deficits and other neurological and psychiatric disorders (Bath & Lee 2006), these data suggest that abnormalities in neural connectivity may underlie the neurological deficits found in patients carrying the val66met polymorphism.

The role of BDNF in synapse maturation is also complex. Numerous studies have reported that application of recombinant BDNF to neurons can lead to an increase in synaptic strength (Kang & Schuman 1995, Kovalchuk et al. 2002, Lohof et al. 1993). Investigators have reported both short- and long-term effects (minutes vs. hours) of BDNF application on synaptic strength and have suggested that both pre- and postsynaptic mechanisms

BDNF: brain-derived neurotrophic factor

underlie the strengthening process. In addition, in several instances, BDNF application strengthens only select synapses or synapses that have been recently activated (Kovalchuk et al. 2002, Schinder et al. 2000). The reasons for these synapse-specific differences are not yet clear, although a better understanding of the complexity of BDNF processing and signaling may eventually provide an answer. Analyses of *bdnf* knockout animals demonstrated a decrease in basal synaptic transmission in some genetic backgrounds and a deficit in tetanus-induced LTP at the Schaffer collateral-CA1 synapse in the hippocampus in several backgrounds (Korte et al. 1995, Patterson et al. 1996). However, synapse and LTP deficits in a total knockout mouse could arise from earlier developmental abnormalities, such as altered dendritic outgrowth or neuronal survival.

For this reason, a recent study of a *trkB* receptor conditional knockout mouse provides some additional insight into the importance of neurotrophins for synaptic development. Deletion of TrkB receptors after synaptic development has no effect on the number of synapses in the CA1 region of the hippocampus, but deletion of TrkB earlier leads to clear deficits in synaptic development that arise within the first weeks after birth (Luikart et al. 2005). Presynaptic loss of TrkB decreases the density of axon varicosities and presynaptic markers, whereas postsynaptic loss of TrkB decreases the density of dendritic spines and postsynaptic markers and leads to clear electrophysiological defects. These results suggest that BDNF has both pre- and postsynaptic modes of action, although in interpreting these results, one must consider that neurotrophin-4 (NT-4) also binds to the TrkB receptor and BDNF has additional receptors, such as p75.

In fact, recent experiments suggest that BDNF may promote synapse weakening by a TrkB-independent mechanism. The BDNF protein is initially synthesized as a precursor protein, termed proBDNF, which is proteolytically cleaved to form mature BDNF (mBDNF). Both forms of BDNF appear to be secreted from neurons, from either pre- or postsy-

naptic sites (Hartmann et al. 2001, Kohara et al. 2001), but they bind different receptors: proBDNF acts through the p75 receptor, whereas mBDNF signals via the TrkB receptor (Teng et al. 2005). The studies discussed above utilized mBDNF application and TrkB deletion and support a role for this signaling pathway in synaptic strengthening. However, application of proBDNF to neurons enhances LTD (Woo et al. 2005). In addition, hippocampal neurons in p75 receptor knockout mice display an increase in dendritic spine density, a phenotype opposite to the TrkB knockout animals (Zagrebelsky et al. 2005). These results support the idea that a proBDNF-p75 signaling pathway may negatively regulate excitatory synapse development, though this possibility needs to be further explored.

Class I major histocompatibility complex (MHC) molecules. In addition to the proBDNF-p75 signaling pathway, a separate group of neuronal activity-regulated signaling molecules have been suggested to restrict synapse number. Corriveau et al. (1998) identified a cDNA encoding class I MHC antigen in a screen for activity-regulated genes in the visual system, and subsequent experiments showed that mice with reduced levels of surface class I MHC display deficits in retinogeniculate eye-specific layer refinement, a process that involves synapse loss (Huh et al. 2000). Because MHC molecules have been studied primarily in the immune system, the molecular mechanisms by which the MHC proteins control synapse number are not yet clear. However, recent work in cultured hippocampal neurons suggests that class I MHC proteins are present in postsynaptic membranes and also negatively regulate excitatory synaptic function in these neurons (Goddard et al. 2007).

Homer1a and serum-inducible kinase (SNK). Further studies of the activity-regulated gene expression program in hippocampal neurons have revealed that, in addition to cell membrane and secreted proteins, neuronal activity induces the expression of genes whose products regulate intracellular signaling pathways in the postsynaptic compartment. Two recent reports highlight the ability of the activity-regulated gene expression program to destabilize synapses by regulating postsynaptic signaling. Although studies have not yet shown that these molecules mediate synapse elimination in vivo, they represent good candidates that may regulate this process. The first report describes studies of Homer1a, an activity-regulated gene product produced by the *homer1* gene locus. Studies of this gene revealed a special feature of activity-regulated transcription. In addition to promoting the transcription of the *homer1* gene, calcium influx causes a switch in *homer1* pre-mRNA processing. This switch promotes the use of a premature polyadenylation site in the fifth intron of the *homer1* gene, which results in the activity-dependent production of a truncated mRNA encoding only the N-terminus of Homer1 (Bottai et al. 2002). Whereas full-length Homer1 proteins that are present in the absence of neuronal activity form dimers via their C-terminal coiled-coil domains and act as scaffolding molecules, bridging synaptic molecules that bind to their N-terminal EVH domains, Homer1a encodes only the EVH domain. This truncated protein interferes with full-length Homer1's function and thereby disrupts synaptic protein complexes and reduces synapse number (Sala et al. 2003). A second report describes the characterization of another activity-regulated gene product, the serum-inducible kinase (SNK). SNK phosphorylates SPAR, a postsynaptic signaling molecule that promotes dendritic spine growth. Phosphorylation of SPAR leads to its ubiquitylation, which in turn promotes SPAR degradation by the ubiquitin-proteasome complex (Pak & Sheng 2003). The loss of SPAR results in the destabilization of synapses. Thus, SNK induction triggers SPAR phosphorylation and degradation, leading to a decrease in synapse number.

Activity regulated cytoskeletal-associated protein (Arc). Activity-regulated gene products can alter synaptic function not only by pro-

moting synapse loss but also by directly regulating the number of AMPA receptors inserted in the postsynaptic membrane. Studies of the activity-regulated gene product Arc have shown that this protein interacts with endophilins and promotes the internalization of AMPA receptors, thus inhibiting the number of functional glutamatergic synapses formed onto neurons (Chowdhury et al. 2006, Rial Verde et al. 2006, Shepherd et al. 2006). In contrast with many mRNAs, the subcellular localization of the *arc* mRNA is tightly controlled. After *arc* mRNA expression is induced by increased levels of synaptic activity, *arc* mRNAs are localized or stabilized specifically at synapses that were activated (Steward et al. 1998). Although this effect is quite striking, this property of the *arc* mRNA has not yet been related to the Arc protein's ability to promote AMPA receptor internalization. However, AMPA receptor internalization is likely enhanced only at synapses where *arc* mRNAs are localized and translated. Although these results suggest a mechanism by which the regulation of the subcellular localization of mRNAs produced in response to activity-regulated transcription can lead to the modification of specific synapses, additional experiments will be required to determine not only the effect of neuronal activity on *arc* mRNA localization but also whether the increased presence of *arc* mRNAs at synapses correlates with an increase in Arc protein and an increase in AMPAR internalization.

MicroRNA-134 (miR-134). Additional studies of the activity-regulated transcriptional program have recently uncovered a new mechanism by which calcium-dependent gene induction alters the function of specific synapses. This work stems from the observations that the translation of select mRNAs can occur at individual synapses (Sutton & Schuman 2006). One way in which this might work is through the actions of microRNAs (miRNAs), which are small RNAs known to inhibit the translation of mRNAs that have nucleotide sequences closely matching the miRNA. The level of miR-134 is increased by neuronal activity

MicroRNA (miRNA): small noncoding RNAs (~19–24 base pairs in length) that control the translation or degradation of target mRNAs

(Schratt et al. 2006). In neurons, miR-134 expression restricts excitatory synapse development, whereas its inhibition promotes dendritic spine growth. Because miR-134 localizes to dendrites and synapses, it may act locally to control mRNA translation and thereby control circuit development. Indeed, miR-134 represses the translation of the *limk1* mRNA at synapses until exposure of synapses to BDNF relieves miR-134-dependent repression. Therefore, although the activity-dependent transcription of miR-134 and other miRNAs likely results in miRNA expression throughout the cell, if the mRNA target of the miRNA is localized to synapses, then the miRNA could be a component of the local mRNA translation machinery that allows proteins to be translated in a synapse-specific manner.

MOLECULAR MECHANISMS UNDERLYING THE TRANSCRIPTIONAL CONTROL OF ACTIVITY-REGULATED GENES

Only a small number of the activity-regulated genes have been studied at the level of detail described above. Additional work will be necessary to reveal the neuronal functions of the large number of uncharacterized activity-regulated gene products. Nonetheless, the studies carried out so far demonstrate that this transcriptional program is critical in coordinating both dendritic and synaptic remodeling. Because this program is important for neuronal development and function, the transcriptional mechanisms that control the expression of these genes have been intensely investigated. These mechanistic studies have focused primarily on a limited number of activity-regulated genes, including *c-fos* and *bdnf*. We review the key discoveries that revealed how calcium influx leads to the transcription of these and other genes. In addition to uncovering the mechanisms underlying activity-dependent transcription, these studies have identified the transcription factors that coordinate the expression of numerous activity-regulated genes. We also discuss research aimed at understanding the functions of these transcriptional regulators in neuronal development.

Lessons from *c-fos*: The Cyclic Adenosine Monophosphate (cAMP) Response Element Binding Protein (CREB) in Activity-Regulated Transcription

To understand how calcium influx leads to an increase in the transcription of *c-fos*, *bdnf*, and other genes, investigators first identified *cis*-acting regulatory elements in the proximal promoters of these genes. The discovery of the key regulatory elements paved the way for the identification of the transcription factors that bind these elements, as well as the upstream signaling cascades that lead to the calcium-dependent modification of these factors. This approach, which essentially traces the calcium-regulated signaling pathway in reverse, has elucidated many of the signaling mechanisms that link calcium influx through the neuronal cell membrane to transcriptional activation within the nucleus. As many of these pathways have now been uncovered, more recent experiments aimed at characterizing these transcriptional networks have involved genomic and proteomic approaches that were previously unavailable.

The first *cis*-acting regulatory element that was shown to regulate calcium-dependent transcription is located within 100 base pairs of the *c-fos* transcriptional start site and was termed the calcium response element (CaRE) (**Figure 1**). The CaRE in the *c-fos* promoter is similar in sequence and function to the cAMP response element (CRE) initially identified in the *somatostatin* gene promoter (Montminy et al. 1986). DNA affinity chromatography was used to purify the nuclear protein that bound to the CRE of *somatostatin*, which subsequently allowed investigators to clone the cDNA encoding CREB and demonstrate that this transcription factor binds to the CRE and mediates transcription in response to elevated cAMP levels (Gonzalez et al. 1989, Montminy & Bilezikjian 1987).

c-fos promoter deletion analysis in PC12 cells identified the CaRE in the *c-fos* promoter as the first regulatory element capable of conferring calcium-dependent gene activation (Sheng et al. 1988). Purification of the transcription factor that binds to the *c-fos* CaRE revealed that this protein is identical to CREB and led to the conclusion that CREB mediates both calcium- and cAMP-dependent transcription within the nucleus (Sheng et al. 1990). A mechanism for cAMP/calcium-dependent transcription was suggested by the finding that elevated levels of cAMP lead to CREB phosphorylation at serine-133 and mutation of this site abolishes CREB-dependent reporter gene activation (Gonzalez & Montminy 1989). Membrane depolarization-induced calcium influx into PC12 cells also causes CREB serine-133 phosphorylation, and phosphorylation of this residue is required for calcium-dependent CREB activation; however, the kinase that phosphorylates CREB is likely different in this case (Sheng et al. 1991). The identity of the CREB serine-133 kinases has been the subject of many studies. Although an exhaustive review of this literature is not possible here (reviewed in Lonze & Ginty 2002), protein kinase A likely mediates cAMP-dependent phosphorylation, and calcium/calmodulin-dependent kinases (CamKs), ribosomal S6 kinases (RSKs), and mitogen- and stress-activated protein kinases (MSKs) likely mediate calcium-dependent phosphorylation. Importantly, the development of antisera that recognize the CREB protein only when phosphorylated at serine-133 allowed investigators to show that CREB is phosphorylated in neurons in the brain in response to physiological levels of neurotransmitter release (Ginty et al. 1993).

The discovery of CREB binding to the *c-fos* CaRE and to CaREs within the regulatory regions of other activity-regulated genes prompted researchers to investigate the function of the CREB protein in neuronal development. It is important to note that investigating the function of a transcription factor represents an approach that is distinct from examining the function of a single activity-

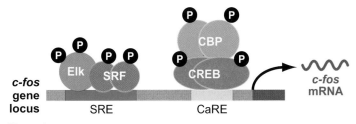

Figure 1

Regulatory mechanisms that control calcium-dependent *c-fos* transcription in neurons. At least two separate *cis*-acting regulatory elements are critical for calcium-dependent *c-fos* transcription: the CaRE and the SRE. These elements, as well as the protein complexes that are recruited to each of these elements, are shown. The transcribed region (*dark green*) and the *c-fos* mRNA produced by the *c-fos* gene (*dark green*) are also shown.

regulated gene. Alterations in CREB levels (either loss- or gain-of-function) will likely alter the expression of its numerous activity-regulated target genes and reveal the function(s) of this transcriptional program as a whole.

These experiments have revealed a role for CREB in both dendritic outgrowth and synaptic potentiation. Consistent with the *Xenopus* work described above, membrane depolarization-induced calcium influx or stimulation with other agents that enhance synaptic activity leads to an increase in dendritic outgrowth in cultured cortical or hippocampal neurons (Redmond et al. 2002, Wayman et al. 2006). In these cases, both CaMK activation and CREB activation are required for activity-dependent dendritic outgrowth. The CREB targets that mediate this effect likely include *cpg15*, *bdnf*, *miR-132*, and *wnt-2*, although many more likely remain to be discovered (Fujino et al. 2003, Tao et al. 1998, Vo et al. 2005, Wayman et al. 2006).

Evidence supporting a role for CREB in synaptic potentiation originally comes from work in the snail *Aplysia californica* (reviewed in Kandel 2001). In mature snails, a physical touch to the siphon activates sensory neurons that innervate motor neurons, which trigger a defensive response where the animal withdraws its gill. This gill withdrawal reflex is enhanced when the tail of the animal is shocked before the siphon is touched. This process, called sensitization, is mediated by serotonergic inputs to the sensory neurons that are activated by the shock to the tail. Sensitization presumably allows the organism to amplify its withdrawal response under stressful environmental conditions. When the tail is shocked a single time, sensitization lasts only for minutes or hours. However, when the tail is shocked four or five times, sensitization can last for days or weeks. Whereas short-term sensitization involves a protein synthesis–independent increase in neuronal excitability caused by serotonin release onto the sensory neurons, long-term sensitization (caused by repeated serotonin release onto the sensory neurons) requires new gene transcription and protein synthesis and involves increased arborization and synaptic growth in sensory neurons (Bailey & Chen 1983, Montarolo et al. 1986). Studies utilizing CRE oligonucleotide injections have suggested that *Aplysia* CREB-1, which is activated in sensory neurons when the tail is stimulated repeatedly, is required for synaptic growth and long-term sensitization (Dash et al. 1990, Kaang et al. 1993).

Because *Aplysia* sensitization can be studied in a cell culture model of this circuit, Martin et al. (1997) tested the axonal branch specificity of long-term sensitization. When a sensory neuron extends two branches and innervates two separate motor neurons, it is possible to apply serotonin repeatedly to one branch of the sensory neuron and mimic long-term sensitization. This protocol leads to long-term synaptic strengthening along the stimulated branch (also called long-term facilitation) but has no effect on synapses along the other branch. However, when a single pulse of serotonin on the second branch (which normally elicits only a short-term response) is paired with long-term facilitation of the first branch, both branches display a long-term response. Thus, the induction of long-term facilitation at one synapse converts a short-term response at a separate synapse to a long-term response. Because this process requires new gene transcription, activity-regulated gene products are likely distributed throughout the cell but utilized only at synapses that have been stimulated with serotonin. This process, termed synaptic capture, also occurs during synapse-specific LTP of synaptic inputs formed onto CA1 pyramidal neurons in the mammalian hippocampus (Frey & Morris 1997).

These experiments have been influential in establishing a role for CREB in plasticity, as well as in learning and memory. However, CREB's role in mammalian synaptic plasticity has been more controversial. The role of CREB and other factors in LTP at the Schaffer collateral-CA1 synapse in the hippocampus has been the subject of considerable interest because plasticity at this synapse as well as others in the hippocampus is critical for the acquisition of certain types of long-lasting memories, including forms of spatial memory and episodic memory. Although some studies have reported a role for CREB in LTP at the Schaffer collateral-CA1 synapse in the hippocampus (Bourtchuladze et al. 1994), subsequent studies have not reproduced this finding (Balschun et al. 2003, Gass et al. 1998). Moreover, any LTP deficit in a total knockout could be due to developmental defects caused by deletion of CREB family members, such as altered dendritic outgrowth or increased neuronal apoptosis. Nevertheless, one result that supports a role for CREB in mammalian plasticity comes from studies of a transgenic mouse expressing constitutively active CREB (VP16-CREB; Barco et al. 2002). In this animal, a stimulus that normally elicits a short-term form of LTP at the Schaffer collateral-CA1 synapse leads to a long-lasting form of LTP. These results are consistent with the *Aplysia* work, suggesting that CREB activation specifically strengthens recently activated synapses.

Given evidence that CREB plays a role in neural circuit development and a wide range of other processes during mammalian development, additional experiments were carried out to determine the mechanism by which serine-133 phosphorylation enhances CREB transcriptional activity. These experiments identified additional transcriptional regulators that control activity-dependent gene expression.

The CREB binding protein (CBP) was identified in a screen for proteins that bind specifically to phosphorylated CREB and was shown to be critical for stimulus-induced CREB activation (Chrivia et al. 1993). Although CBP and p300 (its close mammalian paralog) are ubiquitous transcriptional coactivators, the finding that mutations in the *cbp* gene in humans causes Rubenstein-Taybi syndrome, a severe form of mental retardation, suggests a unique role for CBP in the brain (Petrij et al. 1995). Like CREB, CBP itself undergoes calcium-dependent phosphorylation in neurons (Hu et al. 1999, Impey et al. 2002). However, it is unclear how phosphorylation affects CBP function.

CBP and p300 are large multidomain proteins that possess intrinsic histone acetyltransferase (HAT) activity. Their recruitment to the CaRE in the *c-fos* promoter could promote *c-fos* transcription either by (*a*) recruiting general transcription machinery and RNA polymerase II to the *c-fos* promoter via their multiple protein-protein interaction domains or (*b*) promoting acetylation of the histone N-terminal tails, which is likely required for gene activation. Although both mechanisms likely occur, recent studies have provided compelling support for the acetylation hypothesis. First, several studies have shown that stimuli known to induce activity-dependent gene transcription lead to an increase in histone acetylation at the *c-fos* promoter and at other activity-regulated gene promoters (Tsankova et al. 2004). Second, in a transgenic mouse that expresses a mutant version of CBP rendered HAT-inactive by two point mutations, *c-fos* induction is impaired (Korzus et al. 2004). Expression of the transgene, which is predicted to act dominantly to interfere with endogenous CBP, is limited to neurons in the CA1 region of the hippocampus in this animal. These mice are also deficient in specific aspects of long-term memory. A particularly nice feature of this experiment is that the phenotypes can be reversed both by suppression of transgene expression (using a tetracycline-inducible system) or by treatment with the histone deacetylase (HDAC) in-

hibitor Trichostatin A (TSA), suggesting that the phenotypes are due to the acute loss of activity-dependent histone acetylation rather than to developmental defects.

In contrast with CREB serine-133, which becomes phosphorylated upon exposure of neurons to a wide variety of extracellular stimuli, serines-142 and -143 of CREB are phosphorylated specifically in response to stimuli that induce calcium influx into neurons (Kornhauser et al. 2002). Mutations of CREB serine-142 and serine-143 to alanines significantly attenuate calcium-dependent CREB activation but not cAMP-dependent CREB activation. CREB phosphorylation at serine-142 inhibits CREB's interaction with CBP, suggesting that phosphorylation of these residues in response to calcium influx might activate a novel CREB-dependent signaling pathway (Kornhauser et al. 2002). Generation of a knock-in mouse harboring a serine-to-alanine mutation at serine-142 demonstrates that phosphorylation of this site is important in vivo, as light entrainment of the mouse circadian cycle and expression of the clock gene *mPer1* (processes that require CREB activation in SCN neurons) are disrupted in this mutant (Gau et al. 2002). Assuming that CREB phosphorylated at serine-133 and serine-142 does not interact with CBP, it will be worthwhile to identify the cofactors that are recruited to CREB in response to this novel mode of activation.

Recently, a screen for additional activity-regulated transcription factors revealed other CREB- and CBP-interacting neuronal transcription factors. By fusing a cDNA library to the Gal4 DNA binding domain and testing pools of cDNAs for calcium-dependent activation of a Gal4-dependent reporter gene in primary neurons, several putative activity-regulated transcription factors were identified (Aizawa et al. 2004). Although the specific signaling pathways upstream of these factors, which include CREST, NeuroD2, and Lmo4, are not yet fully characterized, CREST is known to interact with CBP and Lmo4 interacts with CREB (Aizawa et al. 2004; Kashani et al. 2006). Analyses of the knockouts for these

CBP: CREB binding protein

SRF: serum response factor

three factors indicate that they play important roles in neuronal circuit development (Aizawa et al. 2004, Ince-Dunn et al. 2006, Kashani et al. 2006).

Lessons from *c-fos*: Serum Response Factor (SRF) in Activity-Regulated Transcription

Work on the CaRE in the *c-fos* promoter has yielded a significant amount of information regarding the mechanisms underlying activity-dependent transcription and its importance in circuit development. In addition to the CaRE, however, promoter deletion analyses identified a second regulatory element in the *c-fos* promoter, termed the serum response element (SRE), which mediates the *c-fos* transcriptional response to serum application (Greenberg et al. 1987, Treisman 1985) (**Figure 1**). Subsequent experiments showed that the SRE is also required for calcium-dependent *c-fos* transcription (Misra et al. 1994). DNA affinity chromatography yielded the purification of the serum response factor (SRF) and allowed for the subsequent cloning of its cDNA (Norman et al. 1988, Treisman 1987). However, eventually studies found that more than one protein binds to the SRE. A separate DNA affinity purification resulted in the identification of a ternary complex containing the SRE, SRF, and a separate protein later named Elk-1 (Shaw et al. 1989). Closer analysis of the SRE showed that it consists of an inner DNA sequence required for SRF binding, as well as a 5' sequence dispensible for SRF binding but necessary for Elk-1 binding. Elk-1 interacts with SRF and requires the presence of both SRF and its own DNA binding site to form a stable ternary complex. In addition to Elk-1, several other related factors (known as ternary complex factors, all of which contain Ets-domain DNA binding domains) were identified that also form a stable ternary complex with the SRE and SRF (Posern & Treisman 2006). Both SRF and a ternary complex factor must bind to the SRE to induce a maximal transcriptional response upon calcium influx into neurons, though binding of SRF alone can confer some calcium-dependent activation (Xia et al. 1996).

Calcium influx into neurons leads to the phosphorylation of both SRF and Elk-1. Phosphorylation of SRF at serine-103 by RSKs and CamKs enhances binding of SRF to the *c-fos* promoter (Rivera et al. 1993). Elk-1 phosphorylation by extracellular signal-related kinases (ERKs) is critical for glutamate-mediated *c-fos* activation, although the mechanism(s) by which phosphorylation activates Elks remains unclear (Xia et al. 1996). In addition, serum stimulation activates SRF via a unique signaling pathway in which Rho GTPases promote actin polymerization, followed by the recruitment of a distinct set of ternary complex factors to SRF (Posern & Treisman 2006). Although the relevance of this signaling pathway to neuronal activity-dependent gene expression remains unclear, actin polymerization in dendritic spines could signal to the nucleus through this type of pathway. The finding that a large number of cofactors work together with SRF indicates that SRF may mediate neuronal responses to a variety of different stimuli.

Examination of the role of SRF in synapse development and function has recently been facilitated by the generation of mice carrying conditional deletions of the *srf* gene. Although these studies support a role for SRF in the activity-dependent transcription of *c-fos* and other activity-regulated genes, experiments aimed at dissecting the biological function of SRF have yielded conflicting results. Whereas a specific deficit in LTP at the Schaffer collateral-CA1 synapse in the hippocampus was reported in one strain, a separate study indicated that *srf* conditional knockouts have deficits in both LTP and LTD at this synapse (Etkin et al. 2006, Ramanan et al. 2005). The interpretation of these results is complicated by the fact that these animals may also have axonal targeting defects in the hippocampus (Knoll et al. 2006). These studies indicate an important role for SRF in hippocampal synaptic plasticity, but further studies will be necessary to clarify the function of SRF and its cofactors in neural circuit development and plasticity.

Lessons from *bdnf*: Multiple Promoters, Late Transcriptional Response, and Calcium Specificity

The transcription of *c-fos* and many other IEGs increases in many cells of the body in response to extracellular factors that induce proliferation or differentiation of the cells. These IEGs mediate cellular responses to changes in the cell's environment that vary depending on the cell's developmental history. In the nervous system, *c-fos* and other IEGs likely play specific roles in neuronal differentiation as well as in synaptic development and plasticity.

In contrast with *c-fos* and a variety of other IEGs that are induced in a wide range of cell types, recent studies have identified a subset of genes that is activated specifically in response to excitatory synaptic transmission that triggers calcium influx into the postsynaptic neuron. One gene whose expression is specifically induced by neuronal activity in neurons is *bdnf*, which encodes a neurotrophin that plays a critical role in many aspects of neural development (discussed above). The level of the *bdnf* mRNA increases in neurons in response to a variety of physiological stimuli, such as fear conditioning and seizure induction (Rattiner et al. 2004, Timmusk et al. 1993). As with *c-fos*, the induction of *bdnf* mRNA is likely due to an increase in transcription of the *bdnf* gene (Tao et al. 1998). However, unlike the *c-fos* gene, the genomic structure of *bdnf* is complex (**Figure 2**). To date, as many as nine distinct promoters in the mouse *bdnf* gene have been reported (Aid et al. 2007), six of which have been well characterized (Timmusk et al. 1993). Transcripts initiated at each of these promoters splice from their first exon to a common downstream exon, which contains the entire open reading frame encoding the BDNF protein. Because there are two polyadenylation sites in the 3′ untranslated region (UTR) of the common *bdnf* exon and additional alternatively spliced exons, a large number of distinct *bdnf* transcripts can be produced. The biological importance of this diversity remains a mystery, though the fact that each of the *bdnf* transcripts

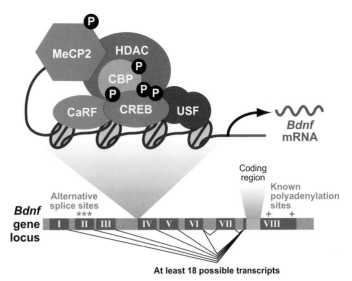

Figure 2

The genomic structure and transcriptional regulation of *bdnf*. The genomic structure of the mouse *bdnf* gene is shown here. Exons are depicted in *dark green* and their numbers are indicated by roman numerals; introns are shown in *gray*. At least six alternative promoters can be clearly detected at this gene (although more may exist—see text for a more detailed explanation). Each of these promoters controls the expression of a unique mRNA that consists of one unique exon (numbered I through VI) that is directly spliced to a common exon (exon VIII), which contains the entire *bdnf* coding region (*orange*). The diversity of *bdnf* transcripts is made even greater because the second exon can be spliced from three alternative splice sites. Moreover, transcripts initiated at exon VI can sometimes include an additional exon (exon VII). Finally, two alternative polyadenylation sites can be utilized within the 3′ UTR. Thus the *bdnf* gene can produce a large number of *bdnf* mRNAs that differ only in their 5′ and 3′ UTRs. The regulatory elements that are critical for calcium-dependent transcription from *bdnf* promoter IV are shown along with the protein complexes known to assemble at this specific promoter.

encodes an identical protein suggests that the distinct 5′ and 3′ UTRs of *bdnf* mRNAs likely confer additional regulation of BDNF production beyond its transcriptional regulation. For instance, the distinct *bdnf* mRNAs could be localized to different subcellular compartments or translated into proteins under different cellular conditions. It is tempting to speculate that this diversity could explain how BDNF can control such a large number of distinct processes during nervous system development.

CaREs 1–3. Studies of the mechanisms that regulate *bdnf* transcription have focused mostly on promoter IV (initially numbered promoter III), which is highly responsive to calcium influx both in vivo and in vitro. Promoter deletion studies identified a functional CRE in this promoter (also termed CaRE3) that binds CREB (Shieh et al. 1998, Tao et al. 1998). Expression of dominant-negative CREB mutants disrupts transcriptional activation driven by this promoter. However, *bdnf* mRNA induction occurs with a slower time course than that of several other CREB target genes, as the peak of *bdnf* transcriptional activation occurs at least one hour after stimulation (compared to a peak within minutes for both *c-fos* activation and CREB serine-133 phosphorylation; Tao et al. 1998). Moreover, although both calcium influx and elevated levels of cAMP induce CREB serine-133 phosphorylation and activation, *bdnf* promoter IV is activated only by calcium influx (Tao et al. 2002). Finally, whereas CREB activity can be induced in both primary neurons and PC12 cells, *bdnf* expression is induced only by calcium influx into neurons (Tao et al. 2002). These features of activity-dependent *bdnf* transcription suggested that the induction of this gene might involve a novel mechanism.

To clarify the mechanisms underlying calcium-dependent *bdnf* promoter IV transcription, two additional calcium-responsive elements (CaRE1 and CaRE2) were identified in *bdnf* promoter IV in close proximity to the CRE. Yeast one–hybrid experiments identified the upstream stimulatory factors (USFs) as the CaRE2 binding proteins (Chen et al. 2003b) and a novel calcium response factor (CaRF) as the CaRE1 binding protein (Tao et al. 2002). CaRF may confer at least part of the calcium-selective and neuronal-specific features of *bdnf* promoter IV transcription because reporter assays using a CaRE1 reporter gene or Gal4-CaRF together with a Gal4-dependent reporter gene show that CaRE1 responds specifically to calcium influx into neurons.

The methyl CpG binding protein 2 (MeCP2). Although promoter deletion analyses were very useful early on to identify transcriptional activators that control calcium-dependent gene transcription, additional components of the regulatory network that are recruited to promoters by binding to special features of the chromatin, such as modified histone tails or methylated CpGs, were not identified by deletion analysis. To begin to identify additional activity-regulated gene regulators, a chromatin immunoprecipitation (ChIP) assay was employed using antibodies that recognize specific proteins that were hypothesized to bind the *bdnf* genomic locus. This method demonstrated that MeCP2 binds to methylated CpGs in *bdnf* promoter IV (Chen et al. 2003a, Martinowich et al. 2003). Mutations in the *Mecp2* gene in humans lead to Rett syndrome, a severe neurological disorder that arises in girls within the first year of life (reviewed in Moretti & Zoghbi 2006). MeCP2 was initially identified as a repressor protein that binds to methylated CpGs. Once bound to DNA, MeCP2 recruits Sin3A, HDACs, and histone methyltransferases to DNA, leading to the formation of condensed chromatin. The prevailing view indicates that when MeCP2 is bound to DNA, the chromatin formed in the vicinity of MeCP2 is irreversibly silenced. However, in response to calcium influx into neurons, MeCP2 becomes phosphorylated at serine-421 with a time course that closely parallels that of *bdnf* induction (peaking at 30–60 min after membrane depolarization; Zhou et al. 2006). The loss of *Mecp2* by genetic deletion leads to an increase in *bdnf* promoter IV expression in unstimulated cells (Chen et al. 2003a), whereas mutation of serine-421 to alanine partially blocks calcium-induced *bdnf* promoter IV expression (Zhou et al. 2006). These findings suggest that MeCP2 represses *bdnf* transcription and that MeCP2 phosphorylation at serine-421 leads to derepression or some other switch in MeCP2 function. Consistent with the features of *bdnf* promoter IV activation, MeCP2 phosphorylation at serine-421 is highly enriched within the brain and is induced specifically by extracellular stimuli that lead to calcium influx.

Several lines of evidence suggest that MeCP2 serves an important function specifically within the brain. First, the conditional deletion of *Mecp2* in the brain alone recapitulates several aspects of Rett Syndrome (Chen et al. 2001). In addition, deficits in synaptic connectivity have recently been reported in *Mecp2* mutant mice, including impaired excitatory synaptic transmission and hippocampal LTP and LTD deficits (Dani et al. 2005, Moretti et al. 2006, Nelson et al. 2006). The mechanisms underlying these deficits are unclear, though a recent study showed that reintroduction of the *Mecp2* gene in immature or mature *Mecp2* null animals that already display Rett-like phenotypes can reverse these neurological symptoms as well as the hippocampal LTP deficit (Guy et al. 2007). A separate study showed that MeCP2 overexpression (which serves as a model of Rett syndrome in cases where the disorder is caused by *Mecp2* gene duplication) alters dendritic spine morphology in a manner that is dependent on its ability to be phosphorylated at serine-421 (Zhou et al. 2006). Therefore, activity-dependent modifications of MeCP2 may result in alterations in excitatory synaptic transmission, and disruption of this process may ultimately give rise to Rett syndrome.

A recent study addressed the issue of whether the Rett-like phenotypes found in *mecp2* knockout mice are due to the deregulation of *bdnf* in these animals (Chang et al. 2006). BDNF protein levels are actually reduced in the brains of *Mecp2* knockout mice. Although this seems to contradict the role of MeCP2 as a repressor of *bdnf* promoter IV transcription, a decrease in excitatory transmission in these animals may lead to an indirect decrease in BDNF production and secretion. Consistent with this possibility, transgenic expression of BDNF in *Mecp2* knockout animals ameliorates several of the Rett-like phenotypes that the *Mecp2* knockout mice display. However, the deregulation of yet-to-be-discovered activity-regulated target genes, in addition to *bdnf*, also likely contributes to Rett syndrome.

Myocyte enhancer factor 2 (MEF2). Another approach that has recently yielded an additional regulator of *bdnf* and other activity-regulated genes involves the genome-wide identification of transcription factor target genes. MEF2 family members (A–D) regulate the expression of muscle-specific genes, as well as IEGs such as *nur77* and *c-jun*, in diverse cell types (reviewed in McKinsey et al. 2002). However, a genome-wide screen conducted in our laboratory has also identified as MEF2 targets many neuronal activity-regulated genes, including *arc*, *homer1a*, and *bdnf* (Flavell et al. 2006; S.W. Flavell and M.E. Greenberg, unpublished observations). In the case of *bdnf*, MEF2 functions in an unusual manner, regulating the expression of each of the promoters so far tested at the *bdnf* genomic locus (i.e., promoter I, II, III, IV, etc.). These results are supported by the fact that changes in the levels of the mRNAs produced by each of these promoters can be detected as a result of MEF2 loss- and gain-of-function. Moreover, genome-wide mapping of MEF2D binding sites shows that there are several distinct sites of MEF2 binding at the *bdnf* gene locus.

MEF2 family transcription factors are activated by signaling pathways distinct from those that activate CREB. Class II HDACs, which act as corepressors by antagonizing HAT function, bind to MEF2 prior to membrane depolarization. Upon membrane depolarization and calcium influx, these HDACs are phosphorylated by CaMKs and exported from the nucleus, relieving MEF2 repression and allowing for the activation of MEF2-dependent transcription (Chawla et al. 2003, McKinsey et al. 2002). In addition to this mechanism of calcium-dependent MEF2 activation, calcium influx also leads to the calcium-dependent dephosphorylation (in contrast with CREB phosphorylation) of MEF2 proteins at several different serine residues, an effect that requires the calcium/calmodulin-regulated phosphatase calcineurin (Flavell et al. 2006, Shalizi et al. 2006). Although we do not know yet how MEF2 dephosphorylation leads to activation, inhibition of calcineurin activity prevents calcium-dependent MEF2 activation in neurons.

Reducing the expression of MEF2 in hippocampal neurons causes an increase in the

MEF2: myocyte enhancer factor 2

number of excitatory synapses formed onto neurons, suggesting that activation of this factor restricts synapse number (Flavell et al. 2006) Some MEF2 target genes may actively disassemble excitatory synapses formed onto MEF2-expressing neurons because acute transcriptional activation of MEF2 causes a decrease in synapse number. On the basis of their known functions, Arc and Homer1a are MEF2 targets that might mediate this effect. Arc induction by MEF2 could lead to AMPA receptor internalization, whereas Homer1a induction might lead to the deconstruction of scaffolding complexes at the synapse.

Although *bdnf* is also a target of MEF2 in neurons that are forming synapses, it is still unclear how BDNF mediates the MEF2-dependent restriction of synapse number. Because BDNF has been studied primarily in the context of synapse maturation, the finding that BDNF is a target of MEF2 might at first seem somewhat contradictory. However, as discussed above, more recent studies have shown that BDNF has diverse functions depending on its posttranscriptional processing and localization, as well as the receptor that it binds. Because MEF2 also regulates several activity-regulated genes whose products control BDNF cleavage from the proBDNF form to the mBDNF form (S.W. Flavell & M.E. Greenberg, unpublished observations), the MEF2 pathway might favor the production of proBDNF under certain circumstances, and under these conditions, proBDNF might contribute to the restriction of synapse number. Alternatively, BDNF and possibly some other MEF2 targets may be selectively deployed to a subset of synapses that will ultimately be retained rather than eliminated.

CONCLUSIONS AND FUTURE DIRECTIONS

Studies of individual genes, as well as genomic approaches, have characterized an activity-regulated transcriptional program in CNS neurons that consists of hundreds of genes (**Figure 3**). These genes encode transcriptional regulators as well as proteins that function in dendrites and at synapses. The detailed examination of the biological functions of select genes has revealed that this transcriptional program is critical for neural circuit development and activity-dependent changes in the neural connectivity of adult organisms. Each of these genes contributes to dendritic and synaptic remodeling in a unique way. Whereas some activity-regulated genes encode secreted factors that act as retrograde signals to the presynaptic cell (e.g., *bdnf*), others encode truncated, dominant-interfering forms of full-length proteins that function at the synapse (e.g., *homer1a*). Moreover, through the synaptic localization of activity-regulated mRNAs (e.g., *arc*) or through the activity-dependent transcription of miRNAs, which locally control mRNA translation (e.g., miR-134), this activity-regulated gene expression program may alter the function of specific synapses formed onto a neuron.

Because this gene expression program is important for brain development, the signaling mechanisms that link calcium influx to transcription have been extensively investigated. These studies have elucidated many of the mechanisms underlying activity-dependent transcription and have identified many of the transcription factors that coordinate the expression of activity-regulated genes. Investigation of each of these factors reveals that they too have quite specific functions. Whereas CREB is critical for dendritic outgrowth and synaptic potentiation, MEF2 plays a role in restricting excess synapses from forming onto a single neuron. Thus, each of these factors coordinates the expression of a group of target genes that likely act together to accomplish specific aspects of neural circuit development. In the future, the genome-wide definition of each of these subsets of the activity-regulated gene expression program will reveal important information regarding how these processes are controlled by activity.

Human genetic studies have revealed that the disruption of the activity-regulated gene expression program in humans gives rise to

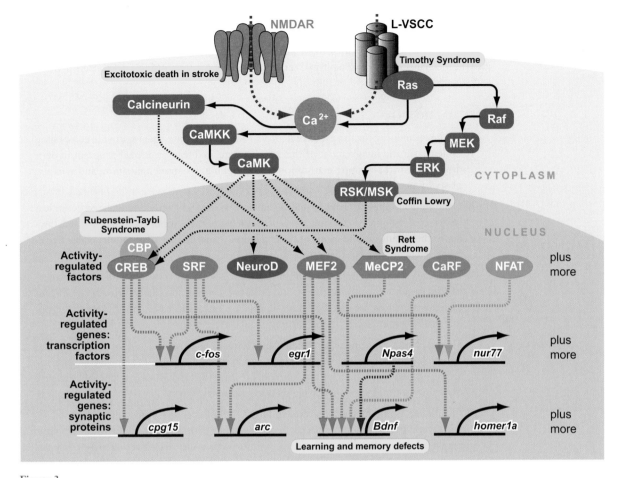

Figure 3

Signal transduction networks mediating neuronal activity-dependent gene expression. Calcium influx through either neurotransmitter receptors or voltage-gated calcium channels leads to the activation of many calcium-regulated signaling enzymes, which sets in motion several signal transduction cascades. These pathways converge on preexisting transcription factors in the nucleus and lead to their activation through direct posttranslational protein modifications. Several of the activity-regulated genes encode transcriptional regulators, which in turn promote the transcription of additional activity-regulated genes. Many other activity-regulated genes encode proteins that function in dendrites or at synapses and thereby coordinate activity-dependent dendritic and synaptic remodeling within the neuron. Genetic mutations in the genes that encode several of these signaling molecules give rise to neurological disorders in humans (*yellow boxes*). Only a subset of the signaling pathways that mediate activity-dependent transcription are shown here.

neurological disorders. Mutations in key regulators of this gene expression program have been identified as causing or conferring susceptibility to different types of mental retardation and autism-spectrum disorders. These mutations can be found in genes encoding the channels that mediate calcium influx in response to membrane depolarization [the L-VGCC subunit Ca(V)1.2 is disrupted in Timothy syndrome; Splawski et al. 2004], the calcium-regulated signaling enzymes that signal to the nucleus (the gene encoding the CREB serine-133 kinase Rsk2 is mutated in Coffin-Lowry syndrome; Hanauer & Young 2002), the transcriptional regulators that mediate gene induction (*cbp* mutations cause Rubenstein-Taybi syndrome and *mecp2* mutations lead to Rett syndrome; Moretti & Zoghbi 2006, Petrij et al. 1995), and the activity-regulated genes themselves (a polymorphism in the *bdnf* gene confers susceptibility to memory deficits; Bath & Lee 2006). Future studies of this gene expression program will therefore not only provide insight into the mechanisms by which neuronal activity controls neural circuit development and function, but might also reveal how the disruption of activity-dependent brain development gives rise to severe neurological disorders in humans.

SUMMARY POINTS

1. Studies examining the functions of neuronal activity-regulated genes provide compelling evidence for the involvement of this activity-regulated transcriptional program in dendritic outgrowth, synaptic maturation/strengthening, synapse elimination, and synaptic plasticity in adult organisms.

2. Genome-wide analyses have revealed that several hundred activity-regulated genes can be detected in neurons. The functions of only a few of these genes are well described. Owing to the technical limitations of these genomic experiments, many additional activity-regulated genes likely remain to be discovered.

3. Not all neuronal activity-regulated genes are alike. They are activated with distinct kinetics (fast response vs. slow response) and different stimulus specificity (growth factor– and calcium-responsive vs. calcium-specific) and are often expressed in different cell types (ubiquitous expression vs. brain specific).

4. Studies of *c-fos*, *bdnf*, and other model genes have revealed several molecular mechanisms by which calcium influx can lead to new gene transcription in neurons. Mechanisms for activity-dependent transcription commonly involve the calcium-dependent modification of a transcription factor, increasing its transcriptional activity.

5. Elucidation of the mechanisms controlling activity-dependent transcription has identified several transcription factors that coordinate the expression of the neuronal activity-regulated genes. Because each factor acts through a distinct *cis*-acting regulatory sequence, each of these proteins regulates a distinct (but overlapping) set of target genes. Functional experiments have uncovered important roles for many of these factors in dendritic and synaptic remodeling.

6. Identifying the signaling pathways that mediate calcium-dependent transcription in neurons has also revealed that several of these genes are commonly disrupted by mutations in the human population. These mutations cause or confer susceptibility to neurological disorders, such as mental retardation and autism-spectrum disorders.

FUTURE ISSUES TO BE RESOLVED

1. Genome-wide approaches have detected hundreds of activity-regulated genes in neurons. These numbers are likely to be underestimates because they do not include the analysis of noncoding RNAs (such as miRNAs) and are not optimized to detect activity-regulated changes in splicing, polyadenlyation, etc. The complete definition of the activity-regulated gene expression program is therefore incomplete, though this should now be possible through more recent advances in genomic technologies. It also remains unknown which of these genes are targets of which activity-regulated transcription factor.

2. It is now clear that several activity-regulated transcription factors bind to unique *cis*-acting regulatory elements. Each of these transcription factors is activated through a distinct signaling mechanism; in some cases, an individual factor can be activated through more than one mechanism. However, it remains unclear which patterns of endogenous neuronal activity favor the activation of which transcription factors (e.g., CREB vs. MEF2) and which mechanisms of activation are utilized in these different physiological contexts (e.g., serine-133 vs. serine-142 CREB phosphorylation).

3. How this gene expression program coordinates complicated developmental processes such as dendritic branching and the strengthening/elimination of specific synaptic inputs is still unknown. Each of these processes occurs concurrently in a given neuron while the activity-regulated gene expression program is being dynamically activated. Which target genes are utilized for which aspects of development and how these processes are coordinated within the cell are complex problems that remain to be understood.

DISCLOSURE STATEMENT

The authors are not aware of any biases that might be perceived as affecting the objectivity of this review.

ACKNOWLEDGMENTS

We thank members of the Greenberg lab for critical reading of the manuscript and J. Zieg for help with figures. Our apologies to authors whose work could not be cited or discussed here because of space limitations. Work on this subject in our laboratory is supported by the F.M. Kirby Foundation, as well as Mental Retardation Developmental Disabilities Research Center grant HD18655 and NIH grant NS048276.

LITERATURE CITED

Aid T, Kazantseva A, Piirsoo M, Palm K, Timmusk T. 2007. Mouse and rat BDNF gene structure and expression revisited. *J. Neurosci. Res.* 85:525–35

Aizawa H, Hu SC, Bobb K, Balakrishnan K, Ince G, et al. 2004. Dendrite development regulated by CREST, a calcium-regulated transcriptional activator. *Science* 303:197–202

Altar CA, Laeng P, Jurata LW, Brockman JA, Lemire A, et al. 2004. Electroconvulsive seizures regulate gene expression of distinct neurotrophic signaling pathways. *J. Neurosci.* 24:2667–77

Antonini A, Stryker MP. 1993. Development of individual geniculocortical arbors in cat striate cortex and effects of binocular impulse blockade. *J. Neurosci.* 13:3549–73

Bailey CH, Chen M. 1983. Morphological basis of long-term habituation and sensitization in Aplysia. *Science* 220:91–93

Balschun D, Wolfer DP, Gass P, Mantamadiotis T, Welzl H, et al. 2003. Does cAMP response element-binding protein have a pivotal role in hippocampal synaptic plasticity and hippocampus-dependent memory? *J. Neurosci.* 23:6304–14

Barco A, Alarcon JM, Kandel ER. 2002. Expression of constitutively active CREB protein facilitates the late phase of long-term potentiation by enhancing synaptic capture. *Cell* 108:689–703

Bartel DP, Sheng M, Lau LF, Greenberg ME. 1989. Growth factors and membrane depolarization activate distinct programs of early response gene expression: dissociation of fos and jun induction. *Genes Dev.* 3:304–13

Bath KG, Lee FS. 2006. Variant BDNF (Val66Met) impact on brain structure and function. *Cogn. Affect. Behav. Neurosci.* 6:79–85

Bottai D, Guzowski JF, Schwarz MK, Kang SH, Xiao B, et al. 2002. Synaptic activity-induced conversion of intronic to exonic sequence in Homer 1 immediate early gene expression. *J. Neurosci.* 22:167–75

Bourtchuladze R, Frenguelli B, Blendy J, Cioffi D, Schutz G, Silva AJ. 1994. Deficient long-term memory in mice with a targeted mutation of the cAMP-responsive element-binding protein. *Cell* 79:59–68

Brainard MS, Doupe AJ. 2002. What songbirds teach us about learning. *Nature* 417:351–58

Brown JR, Ye H, Bronson RT, Dikkes P, Greenberg ME. 1996. A defect in nurturing in mice lacking the immediate early gene fosB. *Cell* 86:297–309

Buffelli M, Burgess RW, Feng G, Lobe CG, Lichtman JW, Sanes JR. 2003. Genetic evidence that relative synaptic efficacy biases the outcome of synaptic competition. *Nature* 424:430–34

Busetto G, Buffelli M, Tognana E, Bellico F, Cangiano A. 2000. Hebbian mechanisms revealed by electrical stimulation at developing rat neuromuscular junctions. *J. Neurosci.* 20:685–95

Cantallops I, Haas K, Cline HT. 2000. Postsynaptic CPG15 promotes synaptic maturation and presynaptic axon arbor elaboration in vivo. *Nat. Neurosci.* 3:1004–11

Chang Q, Khare G, Dani V, Nelson S, Jaenisch R. 2006. The disease progression of Mecp2 mutant mice is affected by the level of BDNF expression. *Neuron* 49:341–48

Chawla S, Vanhoutte P, Arnold FJ, Huang CL, Bading H. 2003. Neuronal activity-dependent nucleocytoplasmic shuttling of HDAC4 and HDAC5. *J. Neurochem.* 85:151–59

Chen C, Regehr WG. 2000. Developmental remodeling of the retinogeniculate synapse. *Neuron* 28:955–66

Chen RZ, Akbarian S, Tudor M, Jaenisch R. 2001. Deficiency of methyl-CpG binding protein-2 in CNS neurons results in a Rett-like phenotype in mice. *Nat. Genet.* 27:327–31

Chen WG, Chang Q, Lin Y, Meissner A, West AE, et al. 2003a. Derepression of BDNF transcription involves calcium-dependent phosphorylation of MeCP2. *Science* 302:885–89

Chen WG, West AE, Tao X, Corfas G, Szentirmay MN, et al. 2003b. Upstream stimulatory factors are mediators of Ca^{2+}-responsive transcription in neurons. *J. Neurosci.* 23:2572–81

Chen ZY, Jing D, Bath KG, Ieraci A, Khan T, et al. 2006. Genetic variant BDNF (Val66Met) polymorphism alters anxiety-related behavior. *Science* 314:140–43

Chowdhury S, Shepherd JD, Okuno H, Lyford G, Petralia RS, et al. 2006. Arc/Arg3.1 interacts with the endocytic machinery to regulate AMPA receptor trafficking. *Neuron* 52:445–59

Chrivia JC, Kwok RP, Lamb N, Hagiwara M, Montminy MR, Goodman RH. 1993. Phosphorylated CREB binds specifically to the nuclear protein CBP. *Nature* 365:855–59

Colman H, Nabekura J, Lichtman JW. 1997. Alterations in synaptic strength preceding axon withdrawal. *Science* 275:356–61

Corriveau RA, Huh GS, Shatz CJ. 1998. Regulation of class I MHC gene expression in the developing and mature CNS by neural activity. *Neuron* 21:505–20

Dani VS, Chang Q, Maffei A, Turrigiano GG, Jaenisch R, Nelson SB. 2005. Reduced cortical activity due to a shift in the balance between excitation and inhibition in a mouse model of Rett syndrome. *Proc. Natl. Acad. Sci. USA* 102:12560–65

Dash PK, Hochner B, Kandel ER. 1990. Injection of the cAMP-responsive element into the nucleus of Aplysia sensory neurons blocks long-term facilitation. *Nature* 345:718–21

Doupe AJ, Kuhl PK. 1999. Birdsong and human speech: common themes and mechanisms. *Annu. Rev. Neurosci.* 22:567–631

Together with Martinowich et al. (2003), identified MeCP2 as a regulator of activity-dependent *bdnf* expression.

Provided the first evidence that CREB serine-133 phosphorylation leads to the recruitment of CBP and other coactivators.

Engert F, Bonhoeffer T. 1999. Dendritic spine changes associated with hippocampal long-term synaptic plasticity. *Nature* 399:66–70

Etkin A, Alarcon JM, Weisberg SP, Touzani K, Huang YY, et al. 2006. A role in learning for SRF: deletion in the adult forebrain disrupts LTD and the formation of an immediate memory of a novel context. *Neuron* 50:127–43

Flavell SW, Cowan CW, Kim TK, Greer PL, Lin Y, et al. 2006. Activity-dependent regulation of MEF2 transcription factors suppresses excitatory synapse number. *Science* 311:1008–12

Fleischmann A, Hvalby O, Jensen V, Strekalova T, Zacher C, et al. 2003. Impaired long-term memory and NR2A-type NMDA receptor-dependent synaptic plasticity in mice lacking c-Fos in the CNS. *J. Neurosci.* 23:9116–22

Frey U, Morris RG. 1997. Synaptic tagging and long-term potentiation. *Nature* 385:533–36

Fujino T, Lee WC, Nedivi E. 2003. Regulation of cpg15 by signaling pathways that mediate synaptic plasticity. *Mol. Cell Neurosci.* 24:538–54

Gass P, Wolfer DP, Balschun D, Rudolph D, Frey U, et al. 1998. Deficits in memory tasks of mice with CREB mutations depend on gene dosage. *Learn. Mem.* 5:274–88

Gau D, Lemberger T, von Gall C, Kretz O, Le Minh N, et al. 2002. Phosphorylation of CREB Ser142 regulates light-induced phase shifts of the circadian clock. *Neuron* 34:245–53

Ginty DD, Kornhauser JM, Thompson MA, Bading H, Mayo KE, et al. 1993. Regulation of CREB phosphorylation in the suprachiasmatic nucleus by light and a circadian clock. *Science* 260:238–41

Goddard CA, Butts DA, Shatz CJ. 2007. Regulation of CNS synapses by neuronal MHC class I. *Proc. Natl. Acad. Sci. USA* 104:6828–33

Gonzalez GA, Montminy MR. 1989. Cyclic AMP stimulates somatostatin gene transcription by phosphorylation of CREB at serine 133. *Cell* 59:675–80

Gonzalez GA, Yamamoto KK, Fischer WH, Karr D, Menzel P, et al. 1989. A cluster of phosphorylation sites on the cyclic AMP-regulated nuclear factor CREB predicted by its sequence. *Nature* 337:749–52

Gorski JA, Zeiler SR, Tamowski S, Jones KR. 2003. Brain-derived neurotrophic factor is required for the maintenance of cortical dendrites. *J. Neurosci.* 23:6856–65

Greenberg ME, Siegfried Z, Ziff EB. 1987. Mutation of the *c-fos* gene dyad symmetry element inhibits serum inducibility of transcription in vivo and the nuclear regulatory factor binding in vitro. *Mol. Cell Biol.* 7:1217–25

Greenberg ME, Ziff EB. 1984. Stimulation of 3T3 cells induces transcription of the c-fos proto-oncogene. *Nature* 311:433–38

Greenberg ME, Ziff EB, Greene LA. 1986. Stimulation of neuronal acetylcholine receptors induces rapid gene transcription. *Science* 234:80–83

Guy J, Gan J, Selfridge J, Cobb S, Bird A. 2007. Reversal of neurological defects in a mouse model of Rett syndrome. *Science* 315:1143–47

Haas K, Li J, Cline HT. 2006. AMPA receptors regulate experience-dependent dendritic arbor growth in vivo. *Proc. Natl. Acad. Sci. USA* 103:12127–31

Hanauer A, Young ID. 2002. Coffin-Lowry syndrome: clinical and molecular features. *J. Med. Genet.* 39:705–13

Hartmann M, Heumann R, Lessmann V. 2001. Synaptic secretion of BDNF after high-frequency stimulation of glutamatergic synapses. *EMBO J.* 20:5887–97

Hiroi N, Brown JR, Haile CN, Ye H, Greenberg ME, et al. 1997. FosB mutant mice: loss of chronic cocaine induction of Fos-related proteins and heightened sensitivity to cocaine's psychomotor and rewarding effects. *Proc. Natl. Acad. Sci. USA* 94:10397–402

Together with Kornhauser et al. (2002), demonstrated that the phosphorylation of CREB at serine-142 is important for brain function in vivo.

The first demonstration that *c-fos* is an IEG whose expression is induced by extracellular stimuli.

Hong SJ, Li H, Becker KG, Dawson VL, Dawson TM. 2004. Identification and analysis of plasticity-induced late-response genes. *Proc. Natl. Acad. Sci. USA* 101:2145–50

Hooks BM, Chen C. 2006. Distinct roles for spontaneous and visual activity in remodeling of the retinogeniculate synapse. *Neuron* 52:281–91

Hu SC, Chrivia J, Ghosh A. 1999. Regulation of CBP-mediated transcription by neuronal calcium signaling. *Neuron* 22:799–808

Huh GS, Boulanger LM, Du H, Riquelme PA, Brotz TM, Shatz CJ. 2000. Functional requirement for class I MHC in CNS development and plasticity. *Science* 290:2155–59

Hunt SP, Pini A, Evan G. 1987. Induction of c-fos-like protein in spinal cord neurons following sensory stimulation. *Nature* 328:632–34

Impey S, Fong AL, Wang Y, Cardinaux JR, Fass DM, et al. 2002. Phosphorylation of CBP mediates transcriptional activation by neural activity and CaM kinase IV. *Neuron* 34:235–44

Ince-Dunn G, Hall BJ, Hu SC, Ripley B, Huganir RL, et al. 2006. Regulation of thalamocortical patterning and synaptic maturation by NeuroD2. *Neuron* 49:683–95

Kaang BK, Kandel ER, Grant SG. 1993. Activation of cAMP-responsive genes by stimuli that produce long-term facilitation in Aplysia sensory neurons. *Neuron* 10:427–35

Kandel ER. 2001. The molecular biology of memory storage: a dialogue between genes and synapses. *Science* 294:1030–38

Kang H, Schuman EM. 1995. Long-lasting neurotrophin-induced enhancement of synaptic transmission in the adult hippocampus. *Science* 267:1658–62

Kashani AH, Qiu Z, Jurata L, Lee SK, Pfaff S, et al. 2006. Calcium activation of the LMO4 transcription complex and its role in the patterning of thalamocortical connections. *J. Neurosci.* 26:8398–408

Knoll B, Kretz O, Fiedler C, Alberti S, Schutz G, et al. 2006. Serum response factor controls neuronal circuit assembly in the hippocampus. *Nat. Neurosci.* 9:195–204

Kohara K, Kitamura A, Morishima M, Tsumoto T. 2001. Activity-dependent transfer of brain-derived neurotrophic factor to postsynaptic neurons. *Science* 291:2419–23

Kornhauser JM, Cowan CW, Shaywitz AJ, Dolmetsch RE, Griffith EC, et al. 2002. CREB transcriptional activity in neurons is regulated by multiple, calcium-specific phosphorylation events. *Neuron* 34:221–33

Korte M, Carroll P, Wolf E, Brem G, Thoenen H, Bonhoeffer T. 1995. Hippocampal long-term potentiation is impaired in mice lacking brain-derived neurotrophic factor. *Proc. Natl. Acad. Sci. USA* 92:8856–60

Korzus E, Rosenfeld MG, Mayford M. 2004. CBP histone acetyltransferase activity is a critical component of memory consolidation. *Neuron* 42:961–72

Kovalchuk Y, Hanse E, Kafitz KW, Konnerth A. 2002. Postsynaptic induction of BDNF-mediated long-term potentiation. *Science* 295:1729–34

LeVay S, Wiesel TN, Hubel DH. 1980. The development of ocular dominance columns in normal and visually deprived monkeys. *J. Comp. Neurol.* 191:1–51

Li H, Gu X, Dawson VL, Dawson TM. 2004. Identification of calcium- and nitric oxide-regulated genes by differential analysis of library expression (DAzLE). *Proc. Natl. Acad. Sci. USA* 101:647–52

Lohof AM, Ip NY, Poo MM. 1993. Potentiation of developing neuromuscular synapses by the neurotrophins NT-3 and BDNF. *Nature* 363:350–53

Lom B, Cogen J, Sanchez AL, Vu T, Cohen-Cory S. 2002. Local and target-derived brain-derived neurotrophic factor exert opposing effects on the dendritic arborization of retinal ganglion cells in vivo. *J. Neurosci.* 22:7639–49

Lonze BE, Ginty DD. 2002. Function and regulation of CREB family transcription factors in the nervous system. *Neuron* 35:605–23

Examined the roles of both spontaneous and patterned retinal activity in the development of the retinogeniculate synapse.

Luikart BW, Nef S, Virmani T, Lush ME, Liu Y, et al. 2005. TrkB has a cell-autonomous role in the establishment of hippocampal Schaffer collateral synapses. *J. Neurosci.* 25:3774–86

Malinow R, Malenka RC. 2002. AMPA receptor trafficking and synaptic plasticity. *Annu. Rev. Neurosci.* 25:103–26

Maren S. 2005. Synaptic mechanisms of associative memory in the amygdala. *Neuron* 47:783–86

Maren S, Quirk GJ. 2004. Neuronal signalling of fear memory. *Nat. Rev. Neurosci.* 5:844–52

Martin KC, Casadio A, Zhu H, Yaping E, Rose JC, et al. 1997. Synapse-specific, long-term facilitation of aplysia sensory to motor synapses: a function for local protein synthesis in memory storage. *Cell* 91:927–38

Martinowich K, Hattori D, Wu H, Fouse S, He F, et al. 2003. DNA methylation-related chromatin remodeling in activity-dependent BDNF gene regulation. *Science* 302:890–93

McAllister AK, Katz LC, Lo DC. 1997. Opposing roles for endogenous BDNF and NT-3 in regulating cortical dendritic growth. *Neuron* 18:767–78

McAllister AK, Lo DC, Katz LC. 1995. Neurotrophins regulate dendritic growth in developing visual cortex. *Neuron* 15:791–803

McKinsey TA, Zhang CL, Olson EN. 2002. MEF2: a calcium-dependent regulator of cell division, differentiation and death. *Trends Biochem. Sci.* 27:40–47

Misra RP, Bonni A, Miranti CK, Rivera VM, Sheng M, Greenberg ME. 1994. L-type voltage-sensitive calcium channel activation stimulates gene expression by a serum response factor-dependent pathway. *J. Biol. Chem.* 269:25483–93

Montarolo PG, Goelet P, Castellucci VF, Morgan J, Kandel ER, Schacher S. 1986. A critical period for macromolecular synthesis in long-term heterosynaptic facilitation in Aplysia. *Science* 234:1249–54

Montminy MR, Bilezikjian LM. 1987. Binding of a nuclear protein to the cyclic-AMP response element of the somatostatin gene. *Nature* 328:175–78

Montminy MR, Sevarino KA, Wagner JA, Mandel G, Goodman RH. 1986. Identification of a cyclic-AMP-responsive element within the rat somatostatin gene. *Proc. Natl. Acad. Sci. USA* 83:6682–86

Moretti P, Levenson JM, Battaglia F, Atkinson R, Teague R, et al. 2006. Learning and memory and synaptic plasticity are impaired in a mouse model of Rett syndrome. *J. Neurosci.* 26:319–27

Moretti P, Zoghbi HY. 2006. MeCP2 dysfunction in Rett syndrome and related disorders. *Curr. Opin. Genet. Dev.* 16:276–81

Morgan JI, Cohen DR, Hempstead JL, Curran T. 1987. Mapping patterns of c-fos expression in the central nervous system after seizure. *Science* 237:192–97

Nagerl UV, Eberhorn N, Cambridge SB, Bonhoeffer T. 2004. Bidirectional activity-dependent morphological plasticity in hippocampal neurons. *Neuron* 44:759–67

Nedivi E, Hevroni D, Naot D, Israeli D, Citri Y. 1993. Numerous candidate plasticity-related genes revealed by differential cDNA cloning. *Nature* 363:718–22

Nedivi E, Wu GY, Cline HT. 1998. Promotion of dendritic growth by CPG15, an activity-induced signaling molecule. *Science* 281:1863–66

Nelson ED, Kavalali ET, Monteggia LM. 2006. MeCP2-dependent transcriptional repression regulates excitatory neurotransmission. *Curr. Biol.* 16:710–16

Niblock MM, Brunso-Bechtold JK, Riddle DR. 2000. Insulin-like growth factor I stimulates dendritic growth in primary somatosensory cortex. *J. Neurosci.* 20:4165–76

Norman C, Runswick M, Pollock R, Treisman R. 1988. Isolation and properties of cDNA clones encoding SRF, a transcription factor that binds to the c-fos serum response element. *Cell* 55:989–1003

Examined the functions of pre- and postsynaptic TrkB receptors in hippocampal synapse development using conditional knockout mice.

Pak DT, Sheng M. 2003. Targeted protein degradation and synapse remodeling by an inducible protein kinase. *Science* 302:1368–73

Park CS, Gong R, Stuart J, Tang SJ. 2006. Molecular network and chromosomal clustering of genes involved in synaptic plasticity in the hippocampus. *J. Biol. Chem.* 281:30195–211

Patterson SL, Abel T, Deuel TA, Martin KC, Rose JC, Kandel ER. 1996. Recombinant BDNF rescues deficits in basal synaptic transmission and hippocampal LTP in BDNF knockout mice. *Neuron* 16:1137–45

Penn AA, Riquelme PA, Feller MB, Shatz CJ. 1998. Competition in retinogeniculate patterning driven by spontaneous activity. *Science* 279:2108–12

Petrij F, Giles RH, Dauwerse HG, Saris JJ, Hennekam RC, et al. 1995. Rubinstein-Taybi syndrome caused by mutations in the transcriptional coactivator CBP. *Nature* 376:348–51

Posern G, Treisman R. 2006. Actin' together: serum response factor, its cofactors and the link to signal transduction. *Trends Cell Biol.* 16:588–96

Rajan I, Cline HT. 1998. Glutamate receptor activity is required for normal development of tectal cell dendrites in vivo. *J. Neurosci.* 18:7836–46

Ramanan N, Shen Y, Sarsfield S, Lemberger T, Schutz G, et al. 2005. SRF mediates activity-induced gene expression and synaptic plasticity but not neuronal viability. *Nat. Neurosci.* 8:759–67

Rattiner LM, Davis M, French CT, Ressler KJ. 2004. Brain-derived neurotrophic factor and tyrosine kinase receptor B involvement in amygdala-dependent fear conditioning. *J. Neurosci.* 24:4796–806

Redmond L, Kashani AH, Ghosh A. 2002. Calcium regulation of dendritic growth via CaM kinase IV and CREB-mediated transcription. *Neuron* 34:999–1010

Reppert SM, Weaver DR. 2002. Coordination of circadian timing in mammals. *Nature* 418:935–41

Rial Verde EM, Lee-Osbourne J, Worley PF, Malinow R, Cline HT. 2006. Increased expression of the immediate-early gene arc/arg3.1 reduces AMPA receptor-mediated synaptic transmission. *Neuron* 52:461–74

Rivera VM, Miranti CK, Misra RP, Ginty DD, Chen RH, et al. 1993. A growth factor-induced kinase phosphorylates the serum response factor at a site that regulates its DNA-binding activity. *Mol. Cell Biol.* 13:6260–73

Rusak B, Robertson HA, Wisden W, Hunt SP. 1990. Light pulses that shift rhythms induce gene expression in the suprachiasmatic nucleus. *Science* 248:1237–40

Saffen DW, Cole AJ, Worley PF, Christy BA, Ryder K, Baraban JM. 1988. Convulsant-induced increase in transcription factor messenger RNAs in rat brain. *Proc. Natl. Acad. Sci. USA* 85:7795–99

Sala C, Futai K, Yamamoto K, Worley PF, Hayashi Y, Sheng M. 2003. Inhibition of dendritic spine morphogenesis and synaptic transmission by activity-inducible protein Homer1a. *J. Neurosci.* 23:6327–37

Sanes JR, Lichtman JW. 1999. Development of the vertebrate neuromuscular junction. *Annu. Rev. Neurosci.* 22:389–442

Schinder AF, Berninger B, Poo M. 2000. Postsynaptic target specificity of neurotrophin-induced presynaptic potentiation. *Neuron* 25:151–63

Schratt GM, Tuebing F, Nigh EA, Kane CG, Sabatini ME, et al. 2006. A brain-specific microRNA regulates dendritic spine development. *Nature* 439:283–89

Shalizi A, Gaudilliere B, Yuan Z, Stegmuller J, Shirogane T, et al. 2006. A calcium-regulated MEF2 sumoylation switch controls postsynaptic differentiation. *Science* 311:1012–17

Shaw PE, Schroter H, Nordheim A. 1989. The ability of a ternary complex to form over the serum response element correlates with serum inducibility of the human c-fos promoter. *Cell* 56:563–72

Sheng M, Dougan ST, McFadden G, Greenberg ME. 1988. Calcium and growth factor pathways of c-fos transcriptional activation require distinct upstream regulatory sequences. *Mol. Cell Biol.* 8:2787–96

Sheng M, McFadden G, Greenberg ME. 1990. Membrane depolarization and calcium induce c-fos transcription via phosphorylation of transcription factor CREB. *Neuron* **4:571–82**

Sheng M, Thompson MA, Greenberg ME. 1991. CREB: a $Ca^{(2+)}$-regulated transcription factor phosphorylated by calmodulin-dependent kinases. *Science* 252:1427–30

Shepherd JD, Rumbaugh G, Wu J, Chowdhury S, Plath N, et al. 2006. Arc/Arg3.1 mediates homeostatic synaptic scaling of AMPA receptors. *Neuron* 52:475–84

Shieh PB, Hu SC, Bobb K, Timmusk T, Ghosh A. 1998. Identification of a signaling pathway involved in calcium regulation of BDNF expression. *Neuron* 20:727–40

Sin WC, Haas K, Ruthazer ES, Cline HT. 2002. Dendrite growth increased by visual activity requires NMDA receptor and Rho GTPases. *Nature* **419:475–80**

Splawski I, Timothy KW, Sharpe LM, Decher N, Kumar P, et al. 2004. Ca(V)1.2 calcium channel dysfunction causes a multisystem disorder including arrhythmia and autism. *Cell* 119:19–31

Steward O, Wallace CS, Lyford GL, Worley PF. 1998. Synaptic activation causes the mRNA for the IEG Arc to localize selectively near activated postsynaptic sites on dendrites. *Neuron* 21:741–51

Stryker MP, Harris WA. 1986. Binocular impulse blockade prevents the formation of ocular dominance columns in cat visual cortex. *J. Neurosci.* 6:2117–33

Sutton MA, Schuman EM. 2006. Dendritic protein synthesis, synaptic plasticity, and memory. *Cell* 127:49–58

Tao X, Finkbeiner S, Arnold DB, Shaywitz AJ, Greenberg ME. 1998. Ca^{2+} influx regulates BDNF transcription by a CREB family transcription factor-dependent mechanism. *Neuron* 20:709–26

Tao X, West AE, Chen WG, Corfas G, Greenberg ME. 2002. A calcium-responsive transcription factor, CaRF, that regulates neuronal activity-dependent expression of BDNF. *Neuron* 33:383–95

Teng HK, Teng KK, Lee R, Wright S, Tevar S, et al. 2005. ProBDNF induces neuronal apoptosis via activation of a receptor complex of p75NTR and sortilin. *J. Neurosci.* 25:5455–63

Timmusk T, Palm K, Metsis M, Reintam T, Paalme V, et al. 1993. Multiple promoters direct tissue-specific expression of the rat BDNF gene. *Neuron* 10:475–89

Treisman R. 1985. Transient accumulation of c-fos RNA following serum stimulation requires a conserved 5′ element and c-fos 3′ sequences. *Cell* 42:889–902

Treisman R. 1987. Identification and purification of a polypeptide that binds to the c-fos serum response element. *EMBO J.* 6:2711–17

Tsankova NM, Kumar A, Nestler EJ. 2004. Histone modifications at gene promoter regions in rat hippocampus after acute and chronic electroconvulsive seizures. *J. Neurosci.* 24:5603–10

Verhage M, Maia AS, Plomp JJ, Brussaard AB, Heeroma JH, et al. 2000. Synaptic assembly of the brain in the absence of neurotransmitter secretion. *Science* 287:864–69

Vo N, Klein ME, Varlamova O, Keller DM, Yamamoto T, et al. 2005. A cAMP-response element binding protein-induced microRNA regulates neuronal morphogenesis. *Proc. Natl. Acad. Sci. USA* 102:16426–31

Wayman GA, Impey S, Marks D, Saneyoshi T, Grant WF, et al. 2006. Activity-dependent dendritic arborization mediated by CaM-kinase I activation and enhanced CREB-dependent transcription of Wnt-2. *Neuron* 50:897–909

Wiesel TN. 1982. Postnatal development of the visual cortex and the influence of environment. *Nature* 299:583–91

Demonstrated that CREB binding to the CaRE in the *c-fos* promoter is critical for calcium-dependent transcription in neuronal cell types.

Showed that visual stimulation enhances dendritic outgrowth in neurons of the visual system.

Woo NH, Teng HK, Siao CJ, Chiaruttini C, Pang PT, et al. 2005. Activation of p75NTR by proBDNF facilitates hippocampal long-term depression. *Nat. Neurosci.* 8:1069–77

Xia Z, Dudek H, Miranti CK, Greenberg ME. 1996. Calcium influx via the NMDA receptor induces immediate early gene transcription by a MAP kinase/ERK-dependent mechanism. *J. Neurosci.* 16:5425–36

Yamagata K, Andreasson KI, Kaufmann WE, Barnes CA, Worley PF. 1993. Expression of a mitogen-inducible cyclooxygenase in brain neurons: regulation by synaptic activity and glucocorticoids. *Neuron* 11:371–86

Zagrebelsky M, Holz A, Dechant G, Barde YA, Bonhoeffer T, Korte M. 2005. The p75 neurotrophin receptor negatively modulates dendrite complexity and spine density in hippocampal neurons. *J. Neurosci.* 25:9989–99

Zhou Z, Hong EJ, Cohen S, Zhao WN, Ho HY, et al. 2006. Brain-specific phosphorylation of MeCP2 regulates activity-dependent BDNF transcription, dendritic growth, and spine maturation. *Neuron* 52:255–69

Cumulative Indexes

Contributing Authors, Volumes 22–31

Dasen JS, 24:327–55
David S, 26:411–40
David SV, 29:477–505
Davis GW, 29:307–23
Davis RL, 28:275–302
Dawson TM, 28:57–84
Dawson VL, 28:57–84
Dayan P, 26:381–410
DeAngelis GC, 24:203–38
de Bono M, 28:451–501
De Camilli P, 26:701–28
deCharms RC, 23:613–47
de Leon RD, 27:145–67
Dellovade T, 29:539–63
De Vos KJ, 31:151–73
Dhaka A, 29:135–61
DiAntonio A, 27:223–46
Dickinson A, 23:473–500
Dijkhuizen P, 25:127–49
Dillon C, 28:25–55
DiMauro S, 31:91–123
Donoghue JP, 23:393–415
Douglas RJ, 27:419–51
Doupe AJ, 22:567–631
Dunwiddie TV, 24:31–55

E

Eatock RA, 23:285–314
Edgerton VR, 27:145–67
Ehlers MD, 29:325–62
Eichenbaum H, 30:123–52
Emery P, 24:1091–119
Emoto K, 30:399–423

F

Feldman JL, 26:239–66
Feller MB, 31:479–509
Fernald RD, 27:697–722
Ferster D, 23:441–71
Field GD, 30:1–30
Fields HL, 30:289–316
Fishell G, 25:471–90
Fisher SE, 26:57–80
Flavell SW, 31:563–90
Fortini ME, 26:627–56
Francis NJ, 22:541–66
Fregni F, 28:377–401
Friedrich RW, 24:263–97
Friston K, 25:221–50
Fritzsch B, 25:51–101

Fuchs AF, 24:981–1004
Fukuchi-Shimogori T, 26:355–80

G

Gage FH, 29:77–103
Gaiano N, 25:471–90
Gainetdinov RR, 27:107–44
Gallant JL, 29:477–505
Gao Q, 30:367–98
Garbers DL, 23:417–39
Garner CC, 28:251–74
Gegenfurtner KR, 26:181–206
Ghosh A, 25:127–49
Gibson AD, 23:417–39
Ginty DD, 26:509–63;
    28:191–222
Gitlin JD, 30:317–37
Glebova NO, 28:191–222
Glimcher PW, 26:133–79
Goda Y, 28:25–55
Goedert M, 24:1121–59
Gold JI, 30:535–74
Goldberg JL, 23:579–612
Goldberg ME, 22:319–50
Goldstein LSB, 23:39–72
González-Scarano F, 22:219–40
Goodwin AW, 27:53–77
Gordon JA, 27:193–222
Gottlieb DI, 25:381–407
Graybiel AM, 31:359–87
Graziano M, 29:105–34
Greenberg ME, 31:563–90
Greenspan RJ, 27:79–105
Grierson AJ, 31:151–73
Grill-Spector K, 27:649–77
Grimwood PD, 23:649–711
Grove EA, 26:355–80

H

Han V, 31:1–25
Harris KM, 31:47–67
Harter DH, 23:343–91
Hatten ME, 22:511–39;
    28:89–108
Häusser M, 28:503–32
He Z, 27:341–68
Heintz N, 28:89–108
Hen R, 27:193–222
Hendry SHC, 23:127–53
Hensch TK, 27:549–79

Hicke L, 27:223–46
Hickey WF, 25:537–62
Hjelmstad GO, 30:289–316
Hobert O, 26:207–38
Holtzman DM, 31:175–93
Honarpour N, 23:73–87
Horton JC, 28:303–26
Horvath TL, 30:367–98
Howe CL, 24:1217–81
Huang EJ, 24:677–736
Huber AB, 26:509–63
Huberman AD, 31:479–509
Hyman SE, 25:1–50; 29:565–98

I

Ikeda S, 26:657–700
Insel TR, 27:697–722

J

Jan L, 23:531–56
Jan Y-N, 23:531–56; 30:399–423
Jessell TM, 22:261–94
Jones EG, 22:49–103; 23:1–37
Joyner AL, 24:869–96
Julius D, 24:487–517

K

Kandel ER, 23:343–91
Karschin A, 23:89–125
Kastner S, 23:315–41
Katoh A, 27:581–609
Katz LC, 22:295–318
Kauer JS, 24:963–79
Kelley MW, 29:363–86
Kennedy MJ, 29:325–62
Kevrekidis IG, 28:533–63
Kiehn O, 29:279–306
Kim MD, 30:399–423
King DP, 23:713–42
King-Casas B, 29:417–48
Kiper DC, 26:181–206
Knowlton BJ, 25:563–93
Knudsen EI, 30:57–78
Kolodkin AL, 26:509–63
Konishi M, 26:31–55
Kopan R, 26:565–97
Kopnisky KL, 25:1–50
Koprivica V, 27:341–68
Koulakov AA, 27:369–92

Chapter Titles, Volumes 22–31

ANNUAL REVIEWS
Intelligent Synthesis of the Scientific Literature

Annual Reviews – Your Starting Point for Research Online
http://arjournals.annualreviews.org

- Over 1150 Annual Reviews volumes—more than 26,000 critical, authoritative review articles in 35 disciplines spanning the Biomedical, Physical, and Social sciences— available online, including all Annual Reviews back volumes, dating to 1932

- Current individual subscriptions include seamless online access to full-text articles, PDFs, Reviews in Advance (as much as 6 months ahead of print publication), bibliographies, and other supplementary material in the current volume and the prior 4 years' volumes

- All articles are fully supplemented, searchable, and downloadable — see http://neuro.annualreviews.org

- Access links to the reviewed references (when available online)

- Site features include customized alerting services, citation tracking, and saved searches

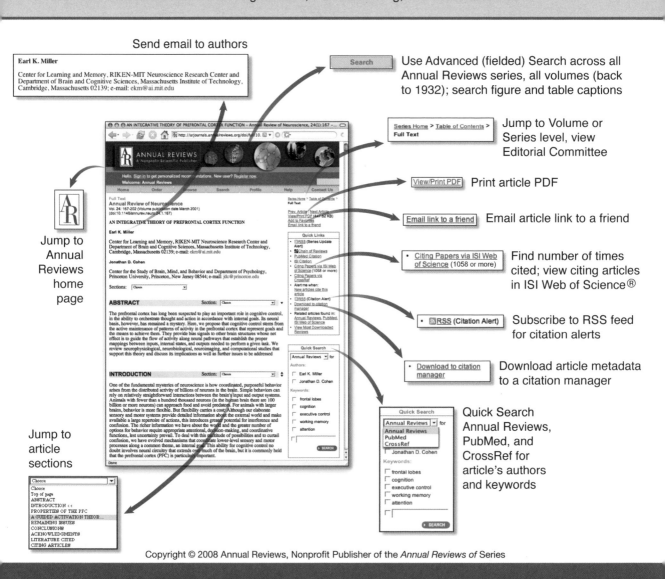

Send email to authors

Use Advanced (fielded) Search across all Annual Reviews series, all volumes (back to 1932); search figure and table captions

Jump to Volume or Series level, view Editorial Committee

Print article PDF

Email article link to a friend

Find number of times cited; view citing articles in ISI Web of Science®

Subscribe to RSS feed for citation alerts

Download article metadata to a citation manager

Quick Search Annual Reviews, PubMed, and CrossRef for article's authors and keywords

Jump to Annual Reviews home page

Jump to article sections